ORGANIC REACTION MECHANISMS · 1969

*An annual survey covering the literature
dated December 1968 through November 1969*

Edited by

B. CAPON University of Glasgow
C. W. REES University of Liverpool

INTERSCIENCE PUBLISHERS
a division of John Wiley & Sons Ltd.
London · New York · Sydney · Toronto

Library of Congress Catalog Card Number 66–23143
ISBN 0 471 13474 0

Printed in Great Britain by
William Clowes & Sons Limited
London, Colchester and Beccles

ORGANIC REACTION MECHANISMS · 1969

WITHDRAWN

Contributors

R. W. ALDER   School of Chemistry, University of Bristol

R. BAKER   Department of Chemistry, The University, Southampton

J. M. BROWN   School of Molecular Sciences, University of Warwick, Coventry

A. R. BUTLER   Department of Chemistry, St. Salvator's College, University of St. Andrews

B. CAPON   Department of Chemistry, The University, Glasgow

R. S. DAVIDSON   Department of Chemistry, The University, Leicester

T. L. GILCHRIST   Department of Chemistry, The University, Leicester

M. J. P. HARGER   Department of Chemistry, The University, Leicester

M. J. PERKINS   Department of Chemistry, King's College, University of London

C. W. REES   Department of Chemistry, The University, Liverpool

R. C. STORR   Department of Chemistry, The University, Liverpool

Preface

This fifth volume of the series is a survey of the work on organic reaction mechanisms published in 1969. For convenience, the literature dated from December 1968 to November 1969, inclusive, was actually covered. The principal aim has again been to scan all the chemical literature and to summarize the progress of work on organic reaction mechanism generally and fairly uniformly, and not just on selected topics. Therefore, certain of the sections are somewhat fragmentary and all are concise. Of the 4000 and more papers which have been reported, those which seemed at the time to be more significant are normally described and discussed, and the remainder are listed.

Our other major aim, second only to the comprehensive coverage, has been early publication since we felt that the immediate value of such a survey as this, that of current awareness, would diminish rapidly with time. In this way we have been fortunate to have the expert cooperation of the English office of John Wiley and Sons.

The organization of the earlier volumes has been retained, though there are more sub-headings and the subject index, which is cumulative, has been greatly enlarged. We welcome suggestions for improvement in future volumes.

June 1970   B.C.
                                                                                 C.W.R.

Contents

CHAPTER 1

Carbonium Ions[1]

B. Capon

Chemistry Department, Glasgow University

Bicyclic and Polycyclic Systems

Derivatives of Norbornane and Related Compounds

It has now proved possible to cool solutions of the norbornyl cation in SbF_5–SO_2ClF–SO_2F_2 to −154°, under which conditions the 3,2-hydride shift is slow on the NMR time-scale. The lowfield peak at $\delta = 5.2$ from the four protons at C-1, C-2, and C-6 [2a] is split into two signals at $\delta = 3.05$ and $\delta = 6.59$, and the high-field signal at $\delta = 2.1$ resulting from the six protons at C-3, C-5, and C-7 [2] develops a shoulder at $\delta = 1.70$ ppm. This result excludes the edge-protonated nortricyclene structure (1) proposed last year [2a] but is consistent

[1] (a) S. Winstein, "Nonclassical Ions and Homoaromaticity", *Quart. Rev.*, **23**, 141 (1969); (b) C. J. Collins, "Protonated Cyclopropanes", *Chem. Rev.*, **69**, 543 (1969); (c) W. Kraus, "Bicyclic Cations", *Mitt. Deut. Pharm. Ges.*, **37**, 245 (1967); *Chem. Abs.*, **69**, 18339 (1968); (d) D. N. Kirk and M. P. Hartshorn, *Steroid Reaction Mechanisms*, Elsevier, Amsterdam, 1968; (e) W. Kitching, "Carbonium-ion and Carbanion producing Heterolyses of Carbon-Metal Bonds", *Rev. Pure Appl. Chem.*, **19**, 1(1969).
[2a] See *Org. Reaction Mech.*, **1965**, 24; **1966**, 12; **1968**, 5.

with either non-classical structure (**2**), referred to by Olah and White as a "corner protonated nortricyclene", or with a rapidly equilibrating pair of classical ions. Only the former was considered to be consistent with the Raman spectrum[2a] and the ^{13}C NMR spectrum.[2b] The similarity of structure (**2**) to structure (**3**) proposed for the norbornadienyl cation[2b] and to that of CH_5^+ (see p. 153) should be noted. The activation barrier for the 6,2-hydride shift is

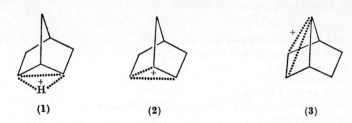

(**1**)   (**2**)   (**3**)

5.9 kcal mole^{-1} compared to 10.8 kcal mole^{-1} for the 2,3-shift[3] (see also p. 64 of this volume.)

It was concluded from a detailed analysis of its 100 MHz NMR spectrum that the 2-phenyl-2-norbornyl cation has a classical structure.[4] This is supported by the ^{13}C NMR spectrum, in which the shift of C-2 is similar to that of the positive carbon of the dimethylphenylmethyl cation.[3] The ^{13}C NMR spectra of the 2-methyl- and 2-ethyl-norbornyl cations suggest that these have bridged structures, as the signals of C-2 appear at higher fields than expected for classical structures. However, the Raman spectra are similar to those of norbornyl compounds, and it was suggested that the molecular geometry is similar to that of a norbornyl derivative but that there is a "degree of σ participation".[3]

The 1,2-dimethoxy-2-norbornyl cation has been shown to be classical and its degenerate rearrangement studied. The first-order rate constant for this reaction is 251 sec^{-1} at 7° ($\Delta G^{\ddagger} = 13.2$ kcal mole^{-1}).[5a]

Jensen and Smart have suggested that the strained carbon–carbon bonds of norbornane derivatives have an enhanced hyperconjugative effect as a result of their increased p-character. This hypothesis was used to explain the high rates of benzoylation of phenylnorbornanes (see p. 244), and it was suggested that it could also explain, at least in part, the high rates of solvolysis of *exo*-2-norbornyl derivatives. Thus, whereas the C-1 to C-6 bond is ideally situated to hyperconjugate with the developing carbonium ion in the transition

[2b] G. A. Olah and A. M. White, *J. Am. Chem. Soc.*, **91**, 6883 (1969).

[3] G. A. Olah and A. M. White, *J. Am. Chem. Soc.*, **91**, 3954, 3956, 5801 (1969); G. A. Olah, J. R. De Member, C. Y. Liu, and A. M. White, *ibid.*, p. 3958; G. A. Olah, C. L. Jeyell, and A. M. White, *ibid.*, p. 3961.

[4] D. G. Farnum and G. Mehta, *J. Am. Chem. Soc.*, **91**, 3256 (1969).

[5a] A. Nickon and Y. Lin, *J. Am. Chem. Soc.*, **91**, 6863 (1969).

state for solvolysis of *exo*-norbornyl derivatives, the C-1 to C-7 bond is much less favourably situated to do so in the reactions of their *endo*-isomers. In 2-bicyclo[2.2.2]octyl derivatives the C-1 to C-6 bond is also ideally situated but since these compounds are less strained there is less *p*-character, less hyperconjugation, and hence a decreased rate, as found.[5b]

Two new investigations of the acetolysis of the 7-chloro-2-norbornyl toluene-*p*-sulphonates have been reported.[6,7] The 7-chloro substituent reduces the rate of reaction of both the *exo*- and *endo*-isomers, and the *exo*:*endo* rate ratio remains high, viz. 246 for the *syn*-isomers and 80 for the *anti*-isomers (see Scheme 1). This suggests that there can be little charge delocalization to C-1 in the transition states for the reactions of the *exo*-isomers since if this were so their rates should be decreased much more than those of the *endo*-isomers, which is not found.[6] The products are mainly the *syn*- and *anti-exo*-acetates (see Scheme 1), and clearly 6,2-hydride shifts take place easily in the intermediate ions. The *syn-exo*-toluene-*p*-sulphonate is converted into the *anti*-isomer concurrently with acetolysis but no conversion of the *anti*- into

	Relative rate	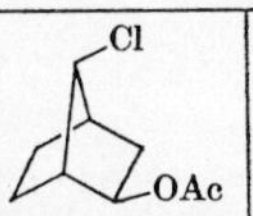	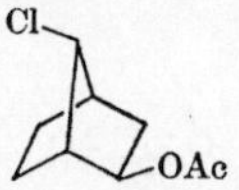	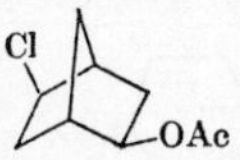	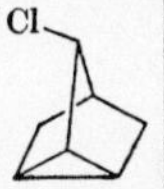	Other
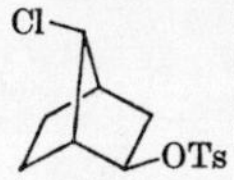	160	14	70	9	4	3
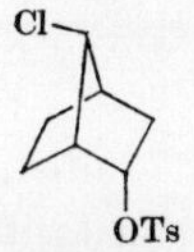	2	11	69	8	7	5
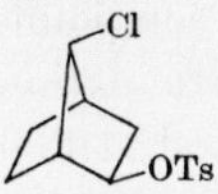	246	45	42	9	2	2
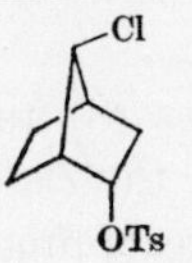	1	24	42	8	16	9

Scheme 1

Percentage composition of acetolysis mixture

[5b] F. R. Jensen and B. E. Smart, *J. Am. Chem. Soc.*, **91**, 5688 (1969).
[6] P. G. Gassman and J. M. Hornback, *J. Am. Chem. Soc.*, **91**, 4280 (1969).
[7] H. L. Goering and M. J. Degani, *J. Am. Chem. Soc.*, **91**, 4506 (1969).

the *syn*-isomer could be detected. The *syn–anti* conversion was shown to be mainly (> 90%) intramolecular by carrying out the reaction in the presence of sodium [^{14}C]toluene-*p*-sulphonate. It was written as involving conversion of ion-pair (4) into (5) through migration of the toluene-*p*-sulphonate anion from the *exo* face of C-2 to the *exo* face of C-6.

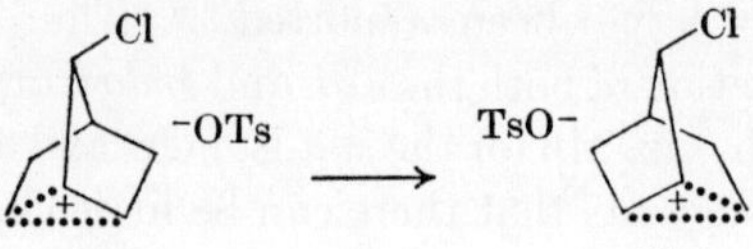

(4)   (5)

Gassman and Macmillan have investigated the acetolysis of the *exo*- and *endo*-toluene-*p*-sulphonates (6) and (7). The *exo*-compound reacts 1940 times more slowly than *exo*-2-norbornyl toluene-*p*-sulphonate and only 11 times faster than its *endo*-isomer. However, this result is difficult to interpret since the *endo*-isomer reacts with a substantial amount of fragmentation (equation

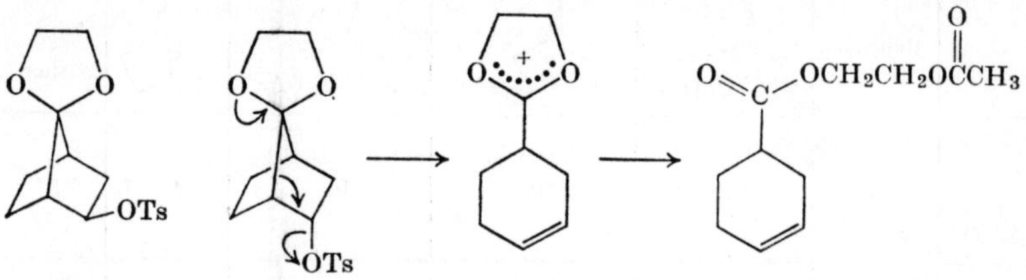

                                                                              ... (1)

(6)   (7)

1).[8] It was also shown that a plot of log k for the acetolysis of eight 7-substituted *endo*-2-norbornyl *p*-bromobenzenesulphonates against the σ^* constants was almost a random distribution, probably because some of the compounds react with fragmentation and others with neighbouring-group participation.[9] In contrast the corresponding plot for the *exo*-isomers was a good straight line with $\rho^* = -2.33$, but unfortunately this result is consistent with both bridged and non-bridged transition states.[10]

Acetolysis of the 5-oxo-2-norbornyl *p*-bromobenzenesulphonates (8) and (9) yields the mixtures shown. The *exo*-isomer reacts faster than the *endo*-isomer by a factor of about 2 at 100° and 40 (est.) at 25°, and both react much more slowly than the analogous norbornyl and 7-oxonorbornyl compounds. It was thought that reaction proceeds via classical ion-pairs and that the

[8] P. G. Gassman and J. G. Macmillan, *J. Am. Chem. Soc.*, **91**, 5527 (1969).
[9] Cf. *Org. Reaction Mech.*, **1966**, 10; **1968**, 71.
[10] P. G. Gassman, J. L. Marshall, J. G. Macmillan, and J. M. Hornback, *J. Am. Chem. Soc.*, **91** 4282 (1969).

deactivating effect of the carbonyl group depends on the orientation of its dipole with respect to the ionizing *p*-bromobenzenesulphonyloxy group.[11, 12]

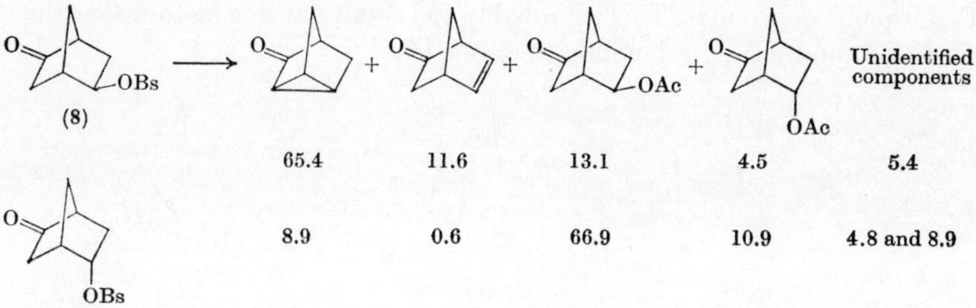

| | 65.4 | 11.6 | 13.1 | 4.5 | 5.4 |

| | 8.9 | 0.6 | 66.9 | 10.9 | 4.8 and 8.9 |

The solvolysis of 2,7,7-trimethyl-*exo*-2-norbornyl *p*-nitrobenzoate (**10a**) in 80% aqueous acetone is only 6.1 times faster than that of its *endo*-isomer (**10b**) whereas that of 2-methyl-*exo*-2-norbornyl *p*-nitrobenzoate (**11a**) is 885 times faster than that of its *endo*-isomer (**11b**), and that of 2,2,6-trimethyl-*exo*-2-norbornyl *p*-nitrobenzoate (**12a**) is 3,630,000 times faster than that of its *endo*-isomer (**12b**). These differences result mainly from the high rates of

		Relative rates *exo : endo*
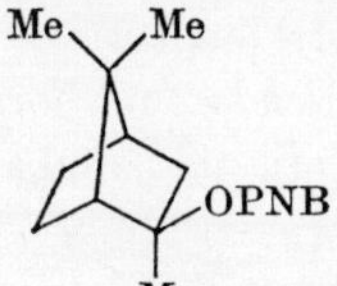 Me Me, OPNB, Me (10a)	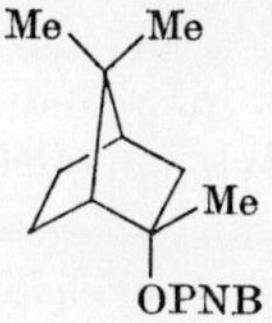 Me Me, Me, OPNB (10b)	6.1
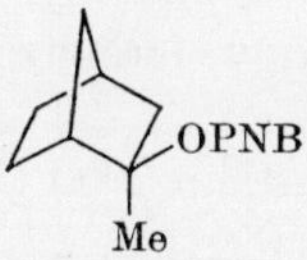 OPNB, Me (11a)	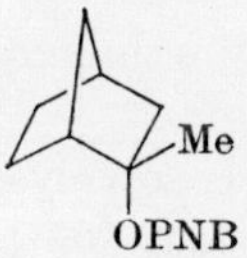 Me, OPNB (11b)	885
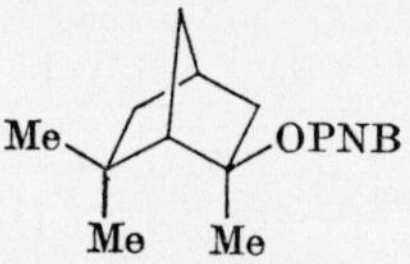 Me, OPNB, Me Me (12a)	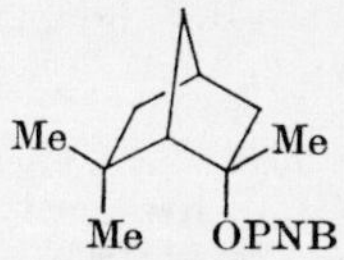 Me, Me, Me OPNB (12b)	3,630,000

[11] J. C. Greever and D. E. Gwynn, *Tetrahedron Letters*, **1969**, 813.
[12] Cf. *Org. Reaction Mech.*, **1967**, 14—15.

solvolysis of compounds (10b) and (12a), which presumably arise from release of steric strain in the transition state.[13a, 13b]

The *exo-p*-nitrobenzoates (13) are solvolysed several hundred times faster than their *endo*-isomers. This is probably the result of some factor other than participation by the 1,6-bonding electrons.[14a]

PNBO—⟨ R

(13)

(R = Me or Ph)

Ph OPNB

(14)

PNBO Ph

(15)

The *exo-p*-nitrobenzoate (14) is hydrolysed 11,600 times faster than its *endo*-isomer (15) in 80% aqueous acetone. This difference probably results partly from steric hindrance to ionization in the reaction of the *endo*-isomer and partly from release of steric strain in the reaction of the *exo*-isomer.[14b]

The deamination of *endo*-2-amino-1,3,3-trimethylnorbornane and *endo*-2-amino-1,7,7-trimethylnorbornane, and the buffered methanolysis of 1,3,3-trimethyl-*endo*-2-norbornyl toluene-*p*-sulphonate (*endo*-fenchyl) and 1,7,7-trimethyl-*endo*-2-norbornyl toluene-*p*-sulphonate (bornyl), yield small amounts of ring-contracted (i.e. bicyclo[3.1.1]heptyl) and ring-opened products.[15] The solvolysis of 3,3-dimethyl-*endo*-2-norbornyl toluene-*p*-sulphonate (*endo*-camphenylyl) in methanol, acetic acid, and aqueous acetone has been investigated.[16]

The isomerization of 2-methyl-*endo*-2-norbornanol into 1-methyl-*exo*-2-norbornanol in aqueous perchloric acid was shown to proceed mainly via 2-methyl-*exo*-norbornanol.[17] The conversion of 2-methyl-*exo*- and -*endo*-norbornyl formate into 1-methyl-*exo*-norbornyl formate has also been investigated.[18]

exo-2-Norbornyl chloride is solvolysed more than 7000 times faster than 1,*exo*-2-dichloronorbornane, and 200 times faster than 1,*exo*-3-dichloronorbornane in aqueous ethanol.[19]

[13a] H. C. Brown and S. Ikegami, *J. Am. Chem. Soc.*, **90**, 7122 (1968).
[13b] S. Ikegami, D. L. Vander Jagt, and H. C. Brown, *J. Am. Chem. Soc.*, **90**, 7124 (1968).
[14a] H. C. Brown, D. L. Vander Jagt, P. von R. Schleyer, R. C. Fort, and W. E. Watts, *J. Am. Chem. Soc.*, **91**, 6848 (1969).
[14b] H. C. Brown and D. L. Vander Jagt, *J. Am. Chem. Soc.*, **91**, 6850 (1969).
[15] W. Hückel, C. M. Jennewein, H. J. Kern, and O. Vogt, *Ann. Chem.*, **719**, 157 (1968); W. Hückel and H. J. Kern, *ibid.*, **728**, 49 (1969).
[16] W. Hückel and A. Majumdar, *Suomen Kemistilehti*, B **42**, 115 (1969); cf. *Org. Reaction Mech.*, **1965**, 18—19.
[17] J. Paasivirta, *Acta Chem. Scand.*, **22**, 2200 (1968).
[18] P. Hirsjärvi, H. L. Kauppinen, and S. Paavolainen, *Suomen Kemistilehti*, B **42**, 236 (1969).
[19] A. J. Fry and W. B. Farnham, *J. Org. Chem.*, **34**, 2314 (1969).

The rate constant for solvolysis of 1,2-dimethyl-*exo*-norbornyl *p*-nitrobenzoate in 90% aqueous acetone at 78.47° is only 1.02 times that for the corresponding 6,6-dideuterated compound. Since this reaction is known to proceed mainly without participation by the 1,6-electrons,[20] this result supports the view[21] that the isotope effects, $k_H/k_D = 1.09$ and 1.11, for the solvolyses of [6-*exo*-^{2}H]- and [6-*endo*-^{2}H]-norbornyl *p*-bromobenzenesulphonate do result from such participation.[22] The absolute configurations of optically active 1,2-dimethyl-*exo*- and -*endo*-norbornanols have been reported.[23]

The small amount of 3,2-hydride shift detected in the acetolysis of the 2-norbornyl *p*-bromobenzenesulphonates[24] probably arises from a rearrangement of the product acetate.[25]

Baker and Mason have reported further work on the solvolysis of compounds (16) and (17).[26] The *exo*:*endo* rate ratios are 1.7 for acetolysis and 11.4 for formolysis, and for the corresponding dimethoxy-compounds (18) and (19) they are 1.76 and 9.4 respectively. No *endo*-acetate or -formate could be detected in the products from any of these reactions.

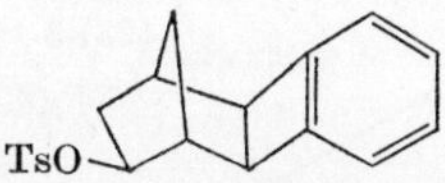

(16) 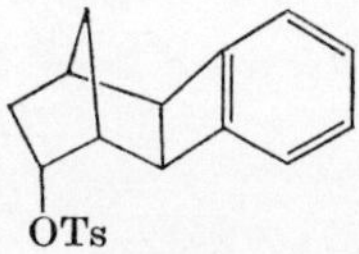   (17)

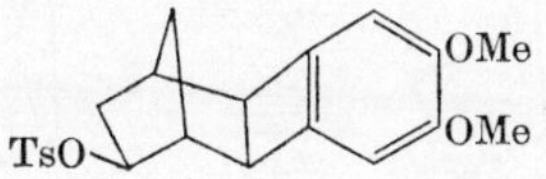

(18) 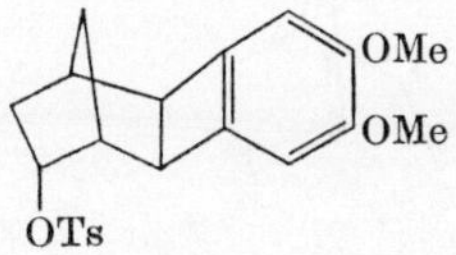   (19)

2-*p*-Anisylnorbornane-2,3-diol (21) reacts in 70% perchloric acid to yield 3-*endo-p*-anisyl-2-norbornanone (20),[27] but with 30% HCl in tetrahydrofuran it yields epoxide (22).[28]

[20] See *Org. Reaction Mech.*, **1968**, 1.
[21] See *Org. Reaction Mech.*, **1967**, 9.
[22] H. L. Goering and K. Humski, *J. Am. Chem. Soc.*, **91**, 4594 (1969).
[23] H. L. Goering, C. Brown, S. Chang, J. V. Clevenger, and K. Humski, *J. Org. Chem.*, **34**, 624 (1969); see *Org. Reaction Mech.*, **1968**, 1.
[24] See *Org. Reaction Mech.*, **1966**, 6.
[25] C. C. Lee, B. S. Hahn, and L. K. M. Lam, *Tetrahedron Letters*, **1969**, 3049; see also ref. 30.
[26] R. Baker and T. J. Mason, *Chem. Comm.*, **1969**, 120; see *Org. Reaction Mech.*, **1967**, 10, 11; see also M. Avram, I. Pogany, F. Badea, I. G. Dinulescu, and C. D. Nenitzescu, *Tetrahedron Letters*, **1969**, 3851.
[27] See *Org. Reaction Mech.*, **1966**, 8—9.
[28] D. C. Kleinfelter and J. H. Long, *Tetrahedron Letters*, **1969**, 347.

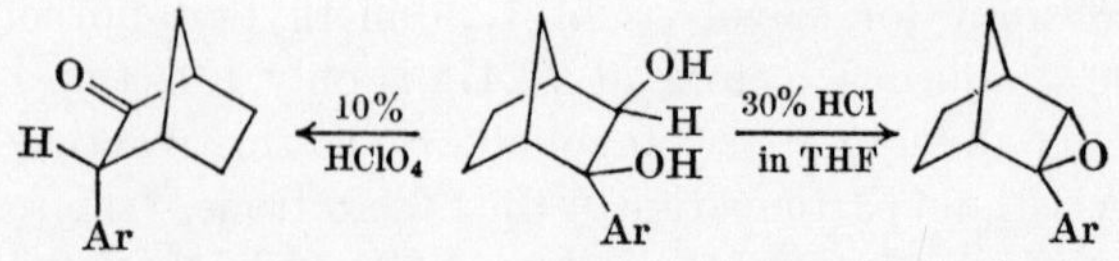

(20)   (21)   (22)

Lee and Hahn have reported a new investigation of the scrambling of ^{14}C label which occurs on solvolysis of 2-(cyclopent-3-enyl)-[2-^{14}C]ethyl p-nitro-benzenesulphonate.[29] The solvents used were aqueous acetone buffered with sodium hydrogen carbonate and acetic acid buffered with urea. More label was found at C-3 of the norbornyl products than at C-7, and this was inter-preted as indicating formation of products from an edge-protonated as well as a corner-protonated nortricyclene. The interconversion of these was formulated as occurring as shown in Scheme 2, and the ratios $k_{H'}/k_S$, k_H/k_S, and $k_{H'}/k_H$ calculated. The last of these is the equilibrium constant for the interconversion of corner- and edge-protonated nortricyclenes, and has values 116 and 4.6 respectively for the two solvents used.[30]

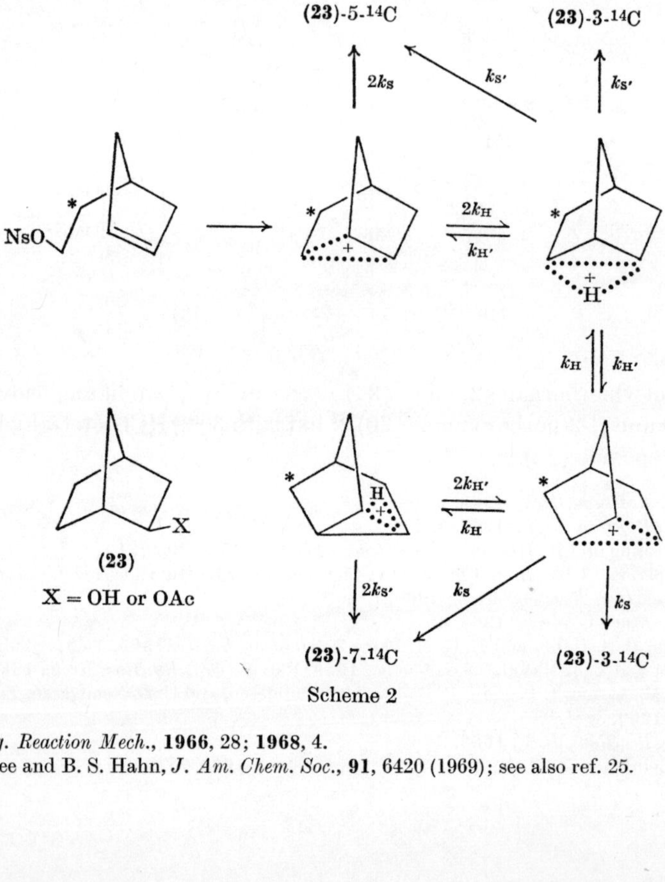

Scheme 2

[29] See *Org. Reaction Mech.*, **1966**, 28; **1968**, 4.
[30] C. C. Lee and B. S. Hahn, *J. Am. Chem. Soc.*, **91**, 6420 (1969); see also ref. 25.

When 2-(cyclopent-3-enyl)ethyl thiocyanate (**24**) is heated in a variety of solvents it is converted into a mixture of 2-(cyclopent-3-enyl)ethyl isothiocyanate (**26**) and *exo*-norbornyl thiocyanate (**28**) and isothiocyanate (**27**). 2-Cyclopentylethyl thiocyanate did not rearrange under the conditions used for this reaction, which suggests that conversion of (**24**) into (**26**) involves participation by the double bond. 2-Norbornyl thiocyanate (**28**) rearranges slowly into 2-norbornyl isothiocyanate (**27**) under the same conditions but yields no cyclopentenylethyl derivatives. It was suggested that these reactions involved two isomeric ion-pairs (**25a**) and (**25b**). Ion-pair (**25a**), formed from cyclopentenylethyl thiocyanate, collapses to (**24**) and (**26**) and rearranges to the isomeric ion-pair (**25b**) at comparable rates, but once (**25b**) is formed it does not revert to (**25a**).[31]

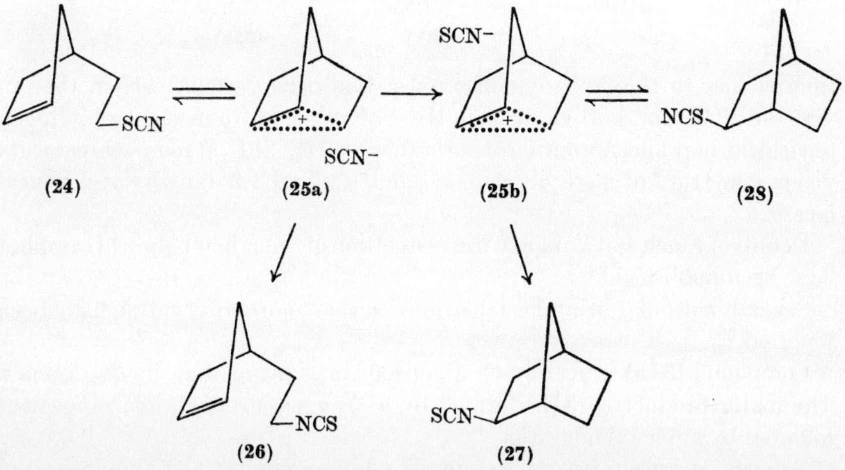

The addition of isothiocyanic acid to norbornene occurs *exo* and *cis* to yield *exo*-2-norbornyl isothiocyanate.[32]

Analysis of the 220 MHz NMR spectrum of the *exo*-2-norbornyl chloride formed on addition of DCl to norbornene in chloroform at $-78°$ shows the deuterium to be wholly in the 3-*exo*- and 7-*syn*-positions in the ratio 55:43 $\pm 3\%$, in disagreement with the earlier results of H. C. Brown and Liu.[33] Under identical conditions nortricyclene yields a product with deuterium 50% 6-*endo*, 43 $\pm$ 3% 6-*exo*, and 6% in the 7-*anti* position. It was suggested

[31] L. A. Spurlock and W. G. Cox, *J. Am. Chem. Soc.*, **91**, 2961 (1969); cf. *Org. Reaction Mech.*, **1966**, 28—29; **1968**, 4—5.
[32] W. R. Diverly, G. A. Buntin, and A. D. Lohr, *J. Org. Chem.*, **34**, 616 (1969).
[33] See *Org. Reaction Mech.*, **1967**, 6.

that these results indicate that the non-classical ion (corner-protonated nortricyclene) is more stable than an edge-protonated nortricyclene or a classical ion.[34, 35]

The 2-*p*-anisylbornyl cation (**29**) reacts with Ph_3SiH to yield 68% *p*-bornyl-anisole (**30a**) and 32% *p*-isobornylanisole (**30b**). Triphenyltin hydride gave 98% (**30a**) and 2% (**30b**) while $PhSiH_3$ gave only (**30a**).[36] Hydride transfer

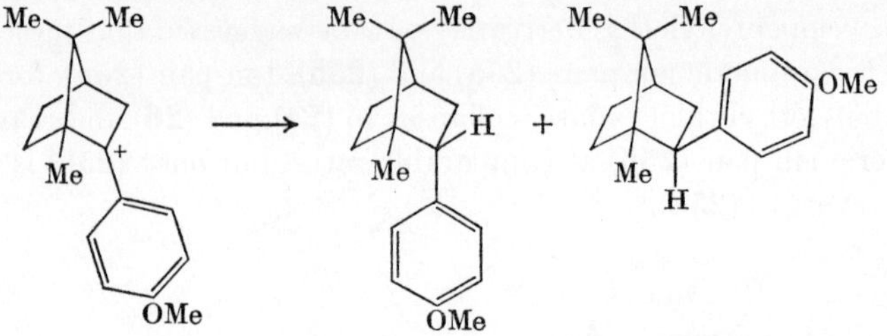

(**29**)   (**30a**)   (**30b**)

from silanes to the 2-phenylnorbornyl cation occurs exclusively in the *exo* direction.[37] The norbornyl cation in $HF-SbF_5$ reacts with molecular hydrogen to yield norbornane. A solution of norbornane in $HF-SbF_5$ at room temperature yields a mixture of methylcyclohexyl and 1,2- and 1,3-dimethylcyclopentyl ions.[38]

Details of Finch and Vaughan's investigation of the sulphonation of camphor have been published.[39a]

Several molecular-orbital calculations on the 2-norbornyl cation have been reported.[39b].

Compound (**31a**) is acetolysed about 60 times faster than its *anti*-isomer. The major product is (**31b**) formed by a Wagner–Meerwein rearrangement followed by a phenyl migration.[40a]

Wagner–Meerwein rearrangements in addition reactions of apopinene and *cis*-δ-pinene,[40b] and the rearrangement of bromonitrocamphane into anhydro-bromonitrocamphane,[41] have been investigated.

[34] J. M. Brown and M. C. McIvor, *Chem. Comm.*, **1968**, 238.
[35] For a discussion of electrophilic addition to norbornenes, see T. G. Traylor, *Accounts Chem. Res.*, **2**, 152 (1969).
[36] F. A. Carey and H. S. Tremper, *Tetrahedron Letters*, **1969**, 1645.
[37] F. A. Carey and H. S. Tremper, *J. Org. Chem.*, **34**, 4 (1969).
[38] H. Hogeveen and C. J. Gaasbeek, *Rec. Trav. Chim.*, **88**, 719 (1969).
[39a] A. M. T. Finch and W. R. Vaughan, *J. Am. Chem. Soc.*, **91**, 1416 (1969); see *Org. Reaction Mech.*, **1966**, 7.
[39b] H. O. Ohorodnyk and D. P. Santry, *J. Am. Chem. Soc.*, **91**, 4711 (1969); G. Klopman, *ibid.*, p. 89; N. S. Isaacs, *Tetrahedron*, **25**, 3555 (1969).
[40a] M. A. Battiste and J. W. Nebzydoski, *J. Am. Chem. Soc.*, **91**, 6887 (1969).
[40b] M. Barthélémy, J. P. Monthéard, and Y. Bessiére-Chrétien, *Bull. Soc. Chim. France*, **1968**, 4881; cf. *Org. Reaction Mech.*, **1967**, 2; **1968**, 3—4.
[41] S. Ranganathan and H. Raman, *Tetrahedron Letters*, **1969**, 3747.

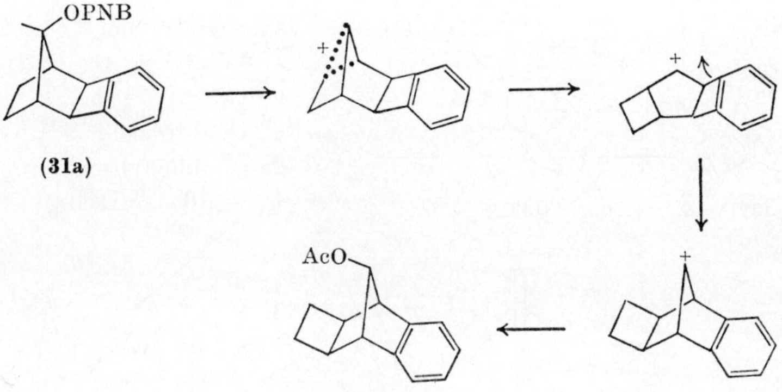

OPNB

(31a)

AcO

(31b)

Berson, Gajewski, and Donald have reported details of their investigation of the ring-expansion of norborn-2-en-7-ylmethyl derivatives.[42] Although the *syn*-isomer yields predominantly compounds of the G-series and the *anti*-isomer compounds of the L-series, there is substantial crossover (see Scheme 3). Thus the L/G ratio from acetolysis of the *anti-p*-nitrobenzenesulphonate is 2.7—3.9, and the G/L ratio from the *syn-p*-nitrobenzenesulphonate is 4.2—4.6.[43, 44] Since crossover is not observed on acetolysis of *exo-* and *endo-*bicyclo-[2.2.2]octyl toluene-*p*-sulphonates there must be an extra intermediate in the ring-expansions which permits crossover. Possible origins of the memory effect were discussed in detail. At present the simplest explanation still seems to be that ring-expansion of two isomers (**32**) and (**35**) leads respectively to conformationally isomeric cations (**33**) and (**36**) which are interconnected via a symmetrical ion (**34**).[45] (See, however, references 51 and 52.)

Ring-expansion of the β-methyl substituted compounds (**38**) and (**48**) was also investigated. Unlike the case of the unsubstituted compounds, crossover occurs at the stage of the twice rearranged ions (**40**) and (**50**) as well as the once rearranged ions (**39**) and (**49**). This was demonstrated by studying the solvolysis of compounds (**37**) and (**53**) directly related to the twice rearranged ions (**40**) and (**50**). Compound (**37**) yields only G′-products (**41**)—(**44**) but (**53**) yields these as well as the L′-products (**51**) and (**52**). Presumably cation (**50**) is converted into (**40**) by way of the tertiary ion (**45**). Despite this complication some preservation of memory occurs on deamination of the amines (**38**) and (**48**) which yield G′/L′ ratios of 18 and 10 respectively.[46]

[42] See *Org. Reaction Mech.*, **1966**, 15.
[43] These values are substantially different from those reported by Bly and Bly for acetolysis of the corresponding *p*-bromobenzenesulphonates.
[44] The G/L ratio obtained from deamination of the *syn*-amine is similar (2.7—4.9) but the L/G ratio obtained from deamination of the *anti*-amine is larger (32—68).
[45] J. A. Berson, J. J. Gajewski, and D. S. Donald, *J. Am. Chem. Soc.*, **91**, 5550 (1969).
[46] J. A. Berson, D. S. Donald, and W. J. Libbey, *J. Am. Chem. Soc.*, **91**, 5580 (1969).

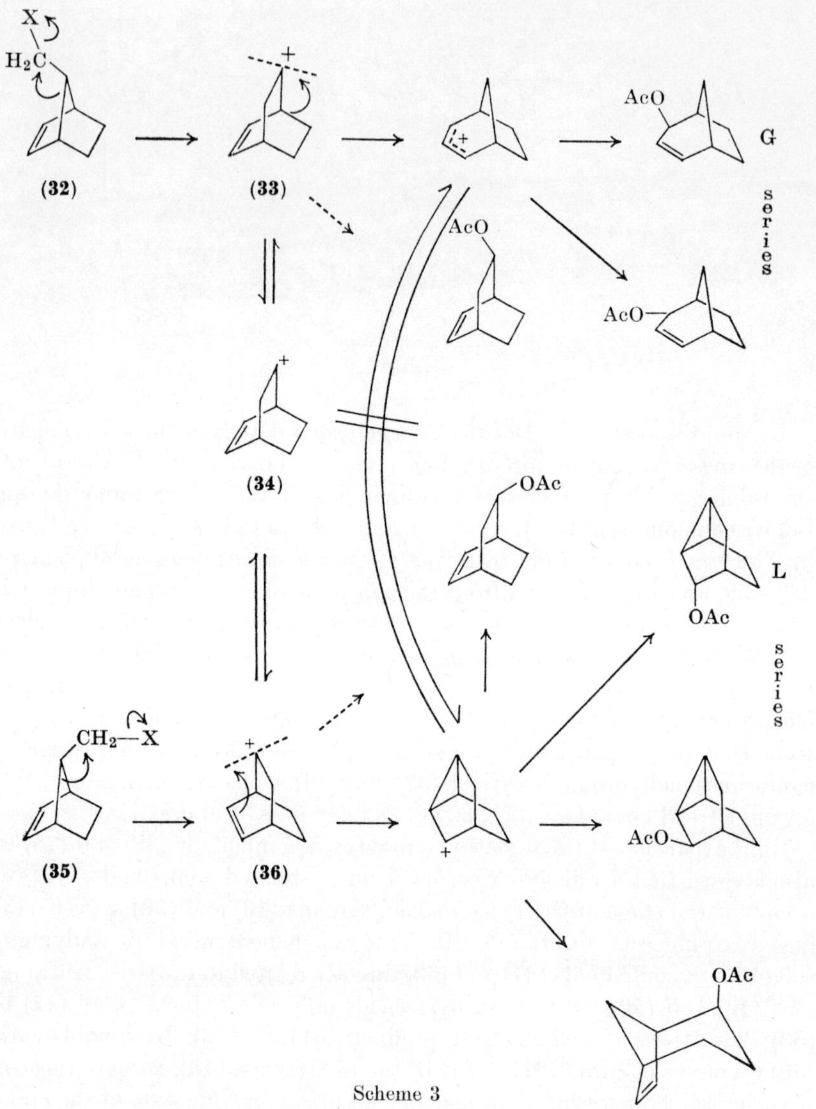

Scheme 3

Details have been published of investigation by Berson and his coworkers of memory effects in the ring-expansion of 7-norbornylmethyl derivatives.[47] The reduction of several norbornyl epoxides has been investigated.[48]

[47] J. A. Berson, M. S. Poonian, and W. J. Libbey, *J. Am. Chem. Soc.*, **91**, 5567 (1969); see *Org. Reaction Mech.*, **1966**, 15.

[48] R. S. Bly and G. B. Konizer, *J. Org. Chem.*, **34**, 2346 (1969).

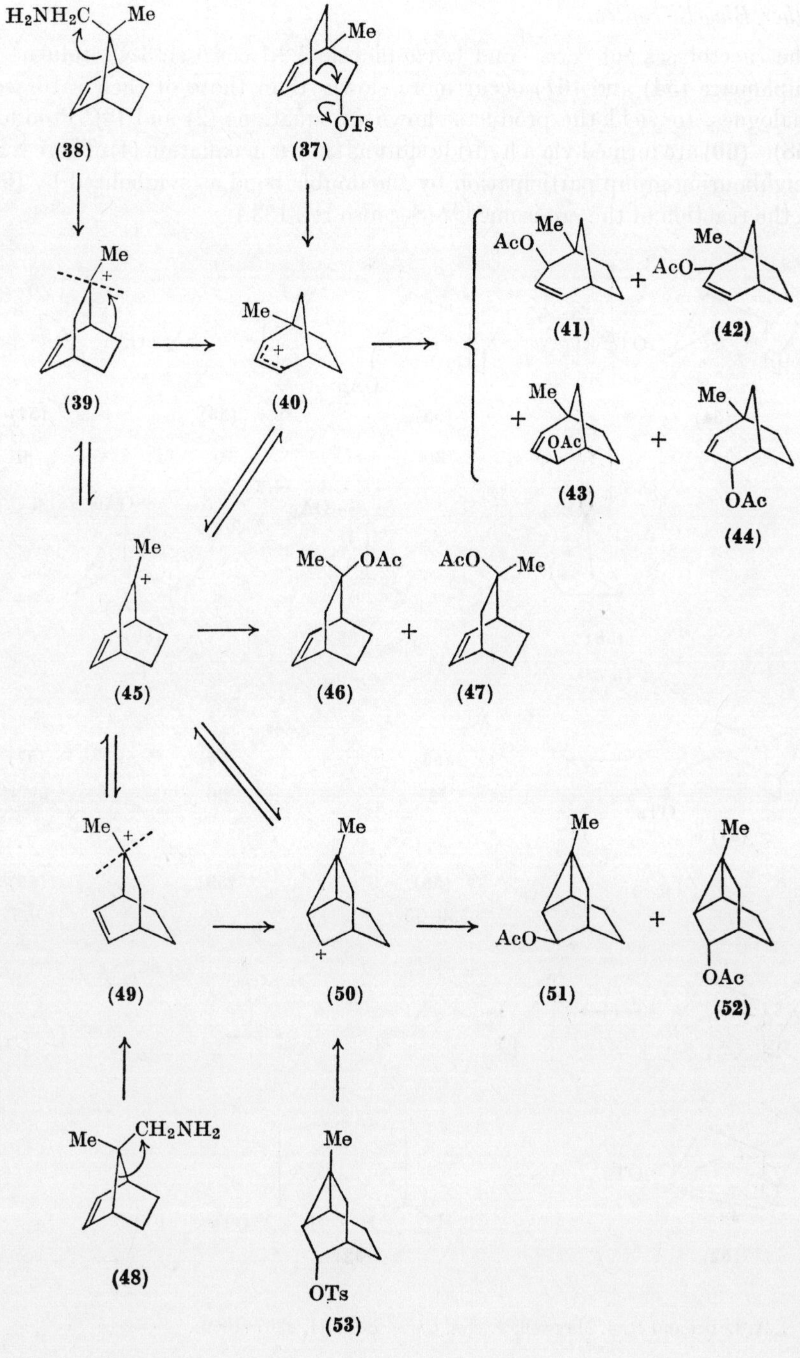

H₂NH₂C   Me

(38)

Me

OTs

(37)

Me

(39)

Me

(40)

AcO   Me

(41)

Me

AcO

(42)

Me

OAc

(43)

Me

OAc

(44)

Me

(45)

Me   OAc

(46)

AcO   Me

(47)

Me

(49)

Me

(50)

Me

AcO

(51)

Me

OAc

(52)

Me   CH₂NH₂

(48)

Me

OTs

(53)

Other Bicyclic Systems

The acetolyses of *exo-* and *endo-*bicyclo[3.2.1]oct-6-en-3-yl toluene-*p*-sulphonate (54) and (61) occur more slowly than those of their saturated analogues, to yield the products shown in equations (2) and (3). Products (58)—(60) are formed via a hydride shift as shown in equation (4). There is no neighbouring-group participation by the double bond as symbolized by (62) in the reaction of the *exo-*isomer.[49] (See also ref. 133.)

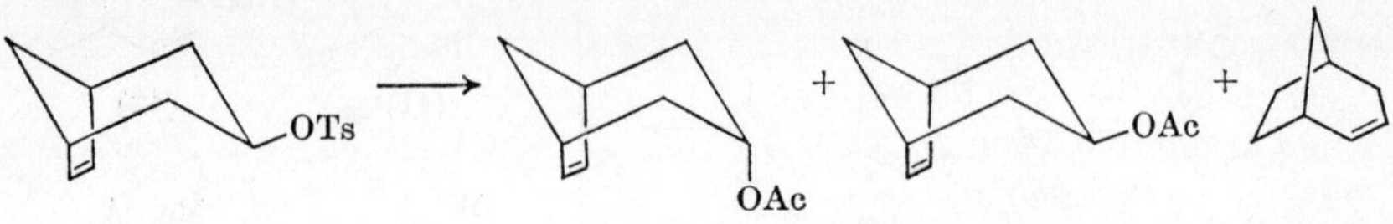

	(54)		(55)		(56)		(57)
			30		0		50

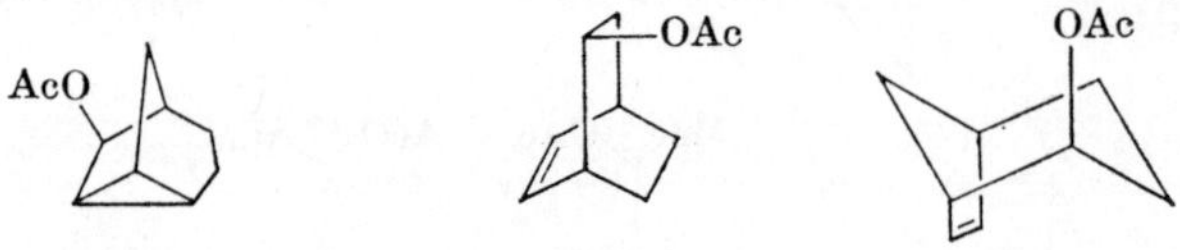

... (2)

(58)	(59)	(60)
15–20	3	1

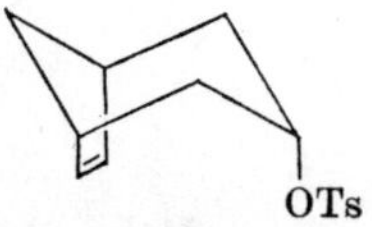

(61) →	(55)	+	(56)	+	(57)
	3		30		30–35

... (3)

(58)	+	(59)	+	(60)
30–35		4		0.5

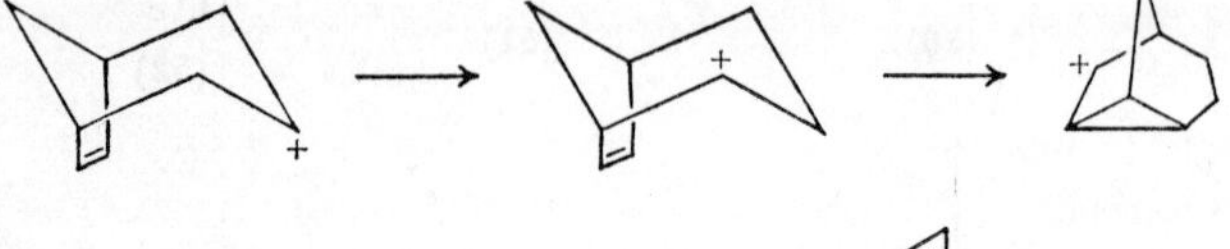

... (4)

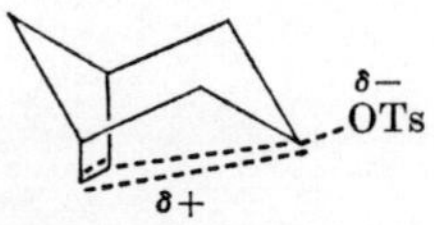

(62) 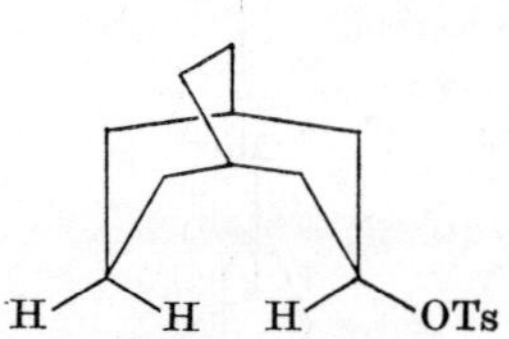   (63)

[49] N. A. LeBel and R. J. Maxwell, *J. Am. Chem. Soc.*, **91**, 2307 (1969).

Acetolysis of bicyclo[3.3.2]dec-3-*exo*-yl toluene-*p*-sulphonate (63) occurs 7.5 times faster than that of bicyclo[3.3.1]non-3-*exo*-yl toluene-*p*-sulphonate and yields mainly olefin. The enhanced rate was ascribed to release of the transannular interaction between the hydrogens at C-3 and C-7 in the transition state, and the formation of only small amounts of acetates to the fact that capture of the intermediate ion by acetic acid would re-establish this interaction.[50]

The reaction of bicyclo[2.2.2]octylsulphoxonium chloride in acetic acid has been studied (see p. 75). An X-ray crystallographic determination of the structure of 2-*p*-bromobenzoyl-1,5,5-trimethylbicyclo[2.2.2]octane-6,8-dione suggests that the bicyclic system is slightly twisted.[51] Details of Ermer and Dunitz's determination of the structure of bicyclo[2.2.2]octane-1,4-dicarboxylic acid and calculations of the conformation of the bicyclo[2.2.2]octane skeleton have been reported.[52]

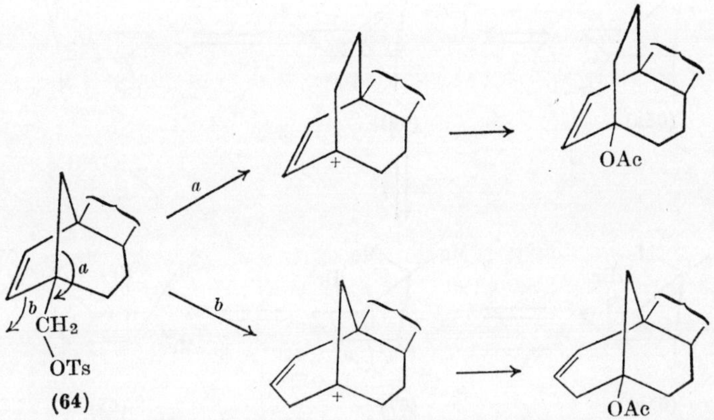

Acetolysis of the toluene-*p*-sulphonate of erythroxylol B [partial formula (64)] is about 50 times slower than that of its saturated analogue. As with norbornenyl-1-methyl toluene-*p*-sulphonate[53a] the double bond is not correctly oriented to participate. The products are derived from migration of the methylene bridge (pathway *a*, 55%) and the unsaturated bridge (pathway *b*, 45%).[53b]

[50] M. P. Doyle and W. Parker, *Chem. Comm.*, 1969, 319.
[51] A. F. Cameron and G. Ferguson, *J. Chem. Soc.* (B), 1969, 1009.
[52] O. Ermer and J. D. Dunitz, *Helv. Chim. Acta*, 52, 1861 (1969); see *Org. Reaction Mech.*, 1968, 16.
[53a] See *Org. Reaction Mech.*, 1968, 19—20.
[53b] J. C. Fairlie, R. McCrindle, and R. D. H. Murray, *J. Chem. Soc.* (C), 1969, 2115.

The rearrangement of bicyclo[2.2.0]hex-1-ylmethyl *p*-nitrobenzoate into 1-norbornyl *p*-nitrobenzoate which occurs concurrently with its solvolysis in aqueous acetone involves no scrambling of the oxygens of the carboxy group (see equation 5).[53c]

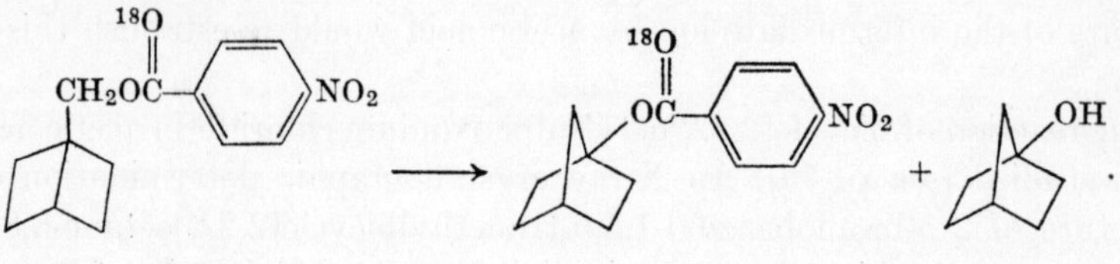

$$\cdots (5)$$

It has been suggested that leakage occurs between the bicyclo[2.1.1]hexyl ions (66) and (67), since compounds (65a) and (65b) are both converted into (68) during gas chromatography.[54]

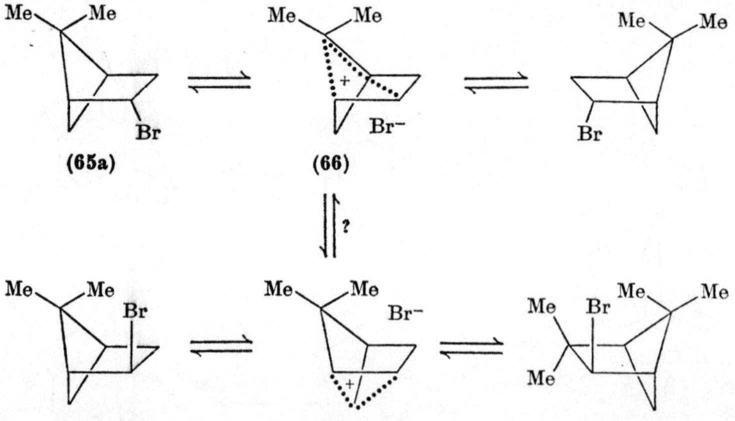

(65a)   (66)

(65b)   (67)   (68)

Acetolysis of the spiro-toluene-*p*-sulphonates (69)—(72) yields the products shown. Those from compounds (69), (71), and (72) result from migration of the group which is antiparallel to the leaving toluene-*p*-sulphonate but compound (70) must presumably be converted into another conformation before migration occurs. This would yield a carbonium ion (74) which is isomeric with (73) derived from (69), and hence the differences in the proportions of olefins obtained from (69) and (70) could be explained. The greater rate of reaction of (69) compared to (71) presumably results from release of ring-strain on going to the transition state for expansion of the five-membered ring. The compounds without t-butyl groups were also studied.[55]

[53c] W. G. Dauben and J. L. Chitwood, *J. Org. Chem.*, **34**, 726 (1969); see *Org. Reaction Mech.*, **1968**, 22—23.

[54] E. A. Hill, R. J. Theissen, and K. Taucher, *J. Org. Chem.*, **34**, 3061 (1969).

[55] H. Christol, A. P. Krapcho, R. C. H. Peters, and C. Arnel, *Tetrahedron Letters*, **1969**, 2799.

(69) 95 4 1

(70) 74 25 1

(71) 58 9 33

(72) 35 55 10

(73)

(74)

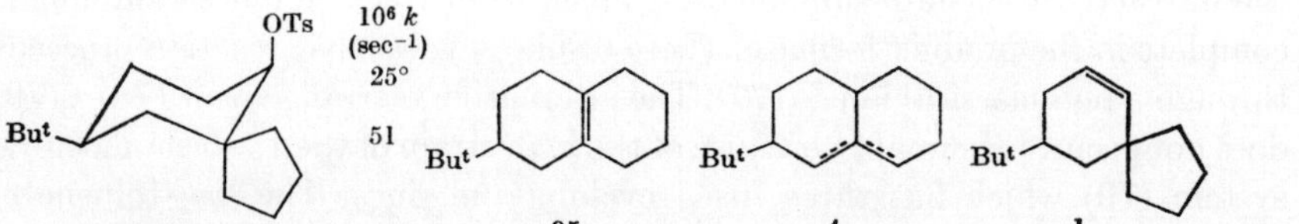

Solvolysis of *cis*- and *trans*-9-decalyl *p*-nitrobenzoates and chlorides yields different mixtures of products, which suggests that the initially formed ions are conformational isomers.[56]

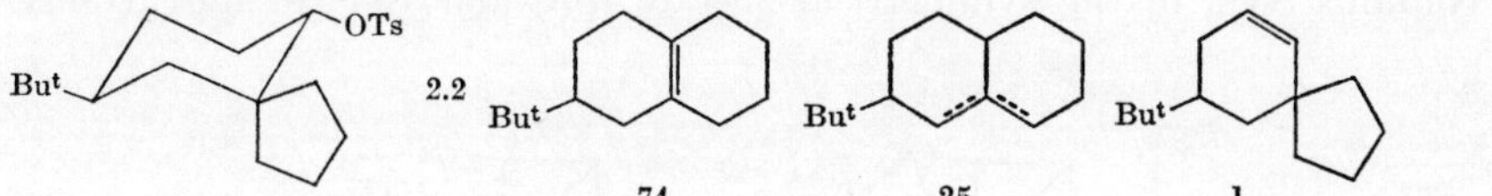

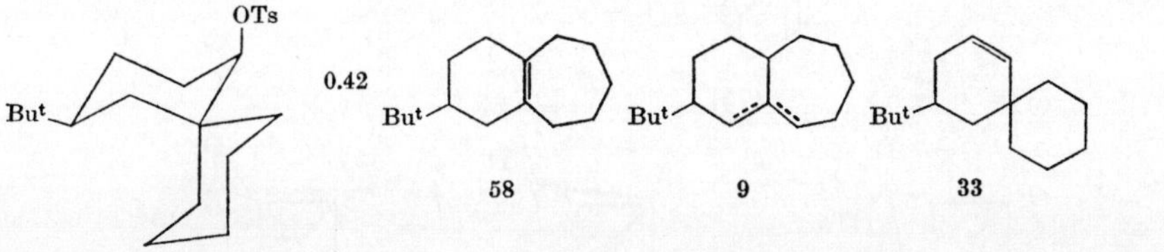

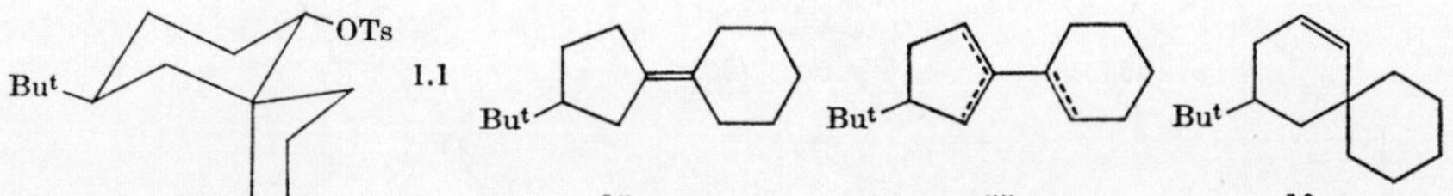

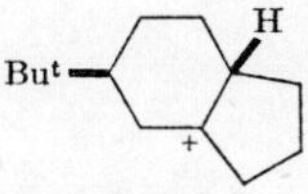

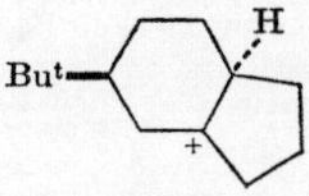

Polycyclic Systems

Acetolysis of *syn*-1,3-bishomocubyl toluene-*p*-sulphonate (**75**) proceeds with retention of configuration to yield the *syn*-acetate. The α-deuterated compound (**76**) yields acetate in which the deuterium is scrambled between the α- and β-positions, and similar scrambling was found with the unreacted toluene-*p*-sulphonate isolated after partial acetolysis. On formolysis, rearrangement of

[56] R. C. Fort and R. E. Hornish, *Chem. Comm.*, **1969**, 11; A. F. Boschung, M. Geisel, and C. A. Grob, *Tetrahedron Letters*, **1968**, 5169.

the unreacted toluene-*p*-sulphonate is much less extensive but scrambling is complete in the product formate. These results suggest that reaction proceeds through a non-classical ion as (**77**). The alternative rearrangement (**78**)→(**79**) does not occur, presumably because of the high strain of the 1,2-bishomocubyl system (**79**) which has three fused cyclobutane rings. The *anti*-toluene-*p*-sulphonate (**80**) also reacts with rearrangement to yield a mixture which contains 85% of the symmetrical acetate (**83**) and 15% of unrearranged

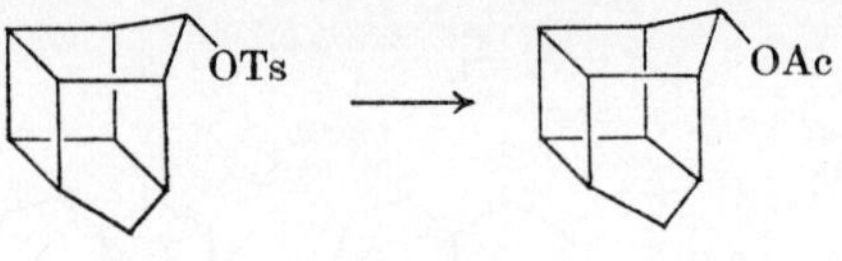

(**75**)

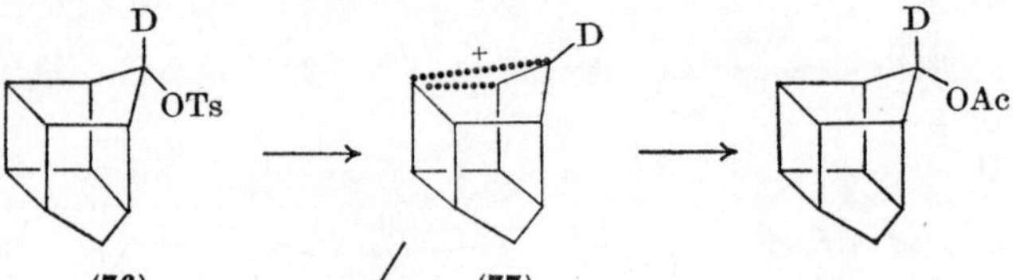

(**76**)   (**77**)

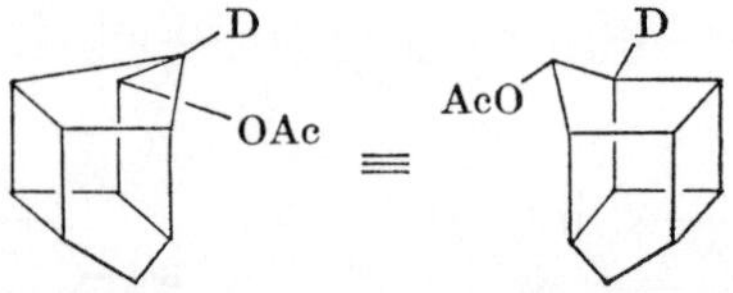

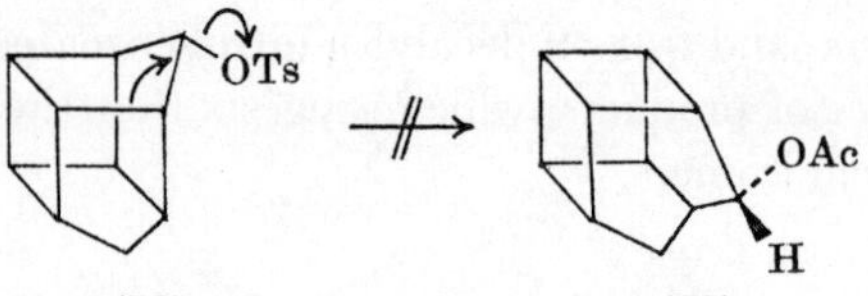

(**78**)   (**79**)

anti-acetate. This reaction is also accompanied by extensive rearrangement to the symmetrical toluene-*p*-sulphonate (**82**). These reactions can be rationalized as proceeding through the non-classical ion (**81**). The alternative rearrangement (**84**)→(**85**) does not occur again presumably because (**85**) is highly strained owing to the presence of three four-membered rings. Application of the Foote–Schleyer correlation led to the conclusion that the acetolyses of the *syn*- and *anti*-toluene-*p*-sulphonates (**75**) and (**80**) are accelerated by factors of 1.3×10^4 and 5×10^3 respectively.[57]

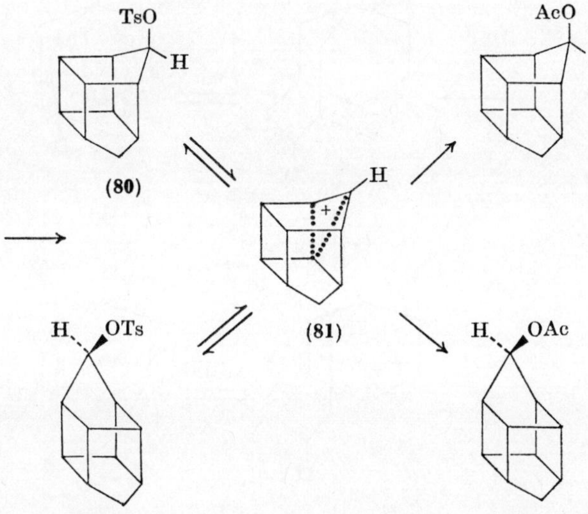

Acetolysis of [4-²H]homoadamant-4-yl toluene-*p*-sulphonate (**86**) yields 75% of the corresponding acetate which retains 38% of the deuterium label at position 4 and the remainder at other positions in the molecule. This result can be explained in terms of consecutive Wagner–Meerwein and 1,2-hydride shifts with the former very rapid and the latter occurring at a rate (k_h) competitive with that for reaction of the ion with solvent (k_p) ($k_h/k_p = 0.37$). The rate of solvolysis of non-deuterated (**86**) is less than that calculated from the

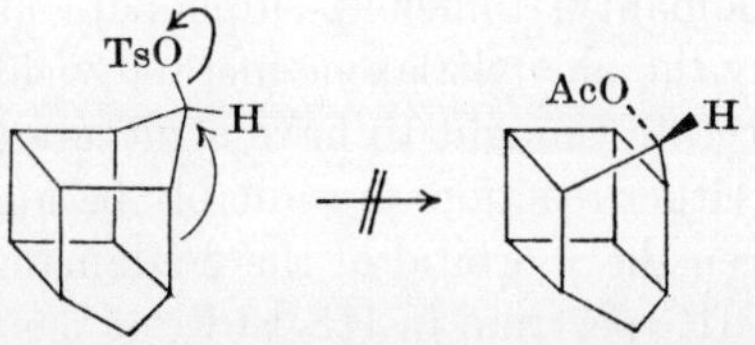

57 W. L. Dilling, C. E. Reineke, and R. E. Plepys, *J. Org. Chem.*, **34**, 2605 (1969); W. L. Dilling, R. A. Plepys, and R. D. Kroenig, *J. Am. Chem. Soc.*, **91**, 3404 (1969); see also *Org. Reaction Mech.*, **1967**, 20—21, and L. A. Paquette, G. V. Meehan, and L. D. Wise, *J. Am. Chem. Soc.*, **91**, 3231 (1969).

Foote–Schleyer correlation, and it therefore seems likely that there is little anchimeric assistance.[58, 59]

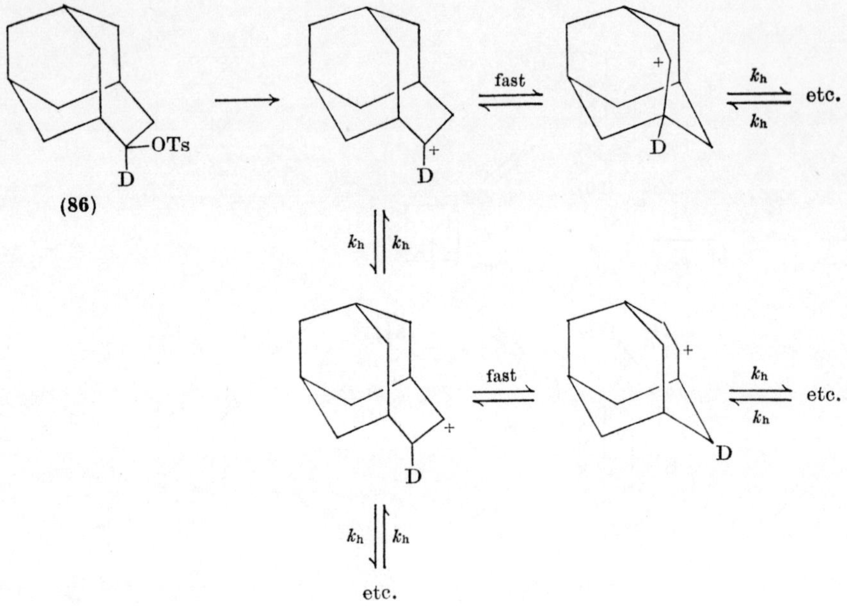

(86)

etc.

Acetolysis of 2-adamantyl toluene-p-sulphonate and deamination of 2-aminoadamantane by the phenyltriazene method yield 0.4—0.5% and 7.5% respectively of an acetate thought to have structure **(87)**. No hydride shift could be detected in either reaction, presumably because of the unfavourable dihedral angle between the p-orbital of the 2-adamantyl cation and the C-1 to H bond.[60] The NMR spectrum in HF–SbF$_5$ of the adamantyl cation also indicates that hydride shifts are slow.[61]

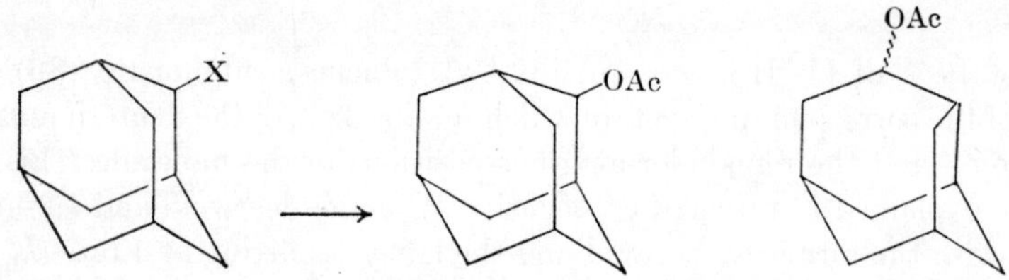

(87)

[58] J. E. Nordlander, F. Y. H. Wu, S. P. Jindal, and J. B. Hamilton, *J. Am. Chem. Soc.*, **91**, 3962 (1969).
[59] P. von R. Schleyer, E. Funke, and S. H. Liggero, *J. Am. Chem. Soc.*, **91**, 3965 (1969).
[60] M. L. Sinnott, H. J. Storesund, and M. C. Whiting, *Chem. Comm.*, **1969**, 1000.
[61] H. Hogeveen and C. J. Gaasbeek, *Rec. Trav. Chim.*, **88**, 719 (1969).

Solvolysis of 1-bromoadamantane in hydroxyalkylamines yields predominantly the product of O-alkylation.[62] The reaction of 1-fluoro- and 1-chloroadamantane with bromine to yield 1-bromoadamantane has been described.[63] Dehydrogenations by the adamantyl cation have been investigated.[64]

Participation by Aryl Groups

1-*Phenyl*-2-*propyl*, 3-*Phenyl*-2-*butyl*, *and Related Compounds*

Lancelot, Harper, and Schleyer have shown that the products and the rates of acetolysis and formolysis of 1-aryl-2-propyl toluene-p-sulphonates (88) can be explained in terms of competing aryl-assisted and solvent-assisted ioniza-

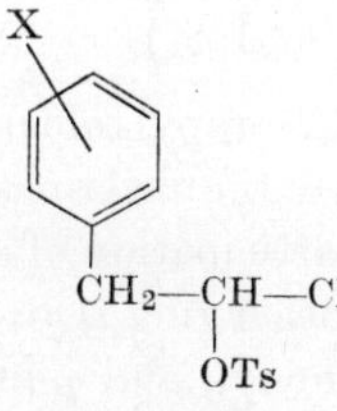

CH₂—CH—CH₃   CH₂—CH—CH₂

(88) 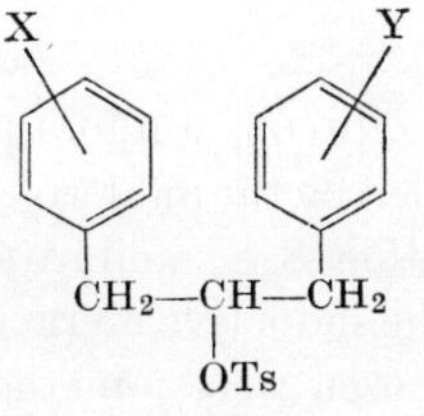   (89)

tions. Thus, the titrimetrically determined rate constants, k_t, may be written

$$k_t = Fk_\Delta + k_s$$

where k_s and k_Δ are the rate constants for the solvent-assisted and aryl-assisted ionization and F is the fraction of the aryl-assisted ionization which yields product (the aryl-assisted ionization may also be followed by ion-pair return to yield starting material). Three approaches were used. In the first, $\log k_t$ for acetolysis and formolysis of compounds (88) was plotted against σ. Both plots were straight lines with $\rho = -0.7$ and -0.84 respectively, when X was an electron-withdrawing substituent. These values are of the size expected if the aryl ring was influencing the rate solely through an inductive effect, and here Fk_Δ makes a negligible contribution to k_t. However, the points for the compounds with X = H, p-Me, and p-MeO fell above these lines, which indicates that here there is a significant contribution by k_Δ. It was estimated, *inter alia*, that the value of k_t/k_s for the acetolysis of 1-phenyl-2-propyl toluene-p-sulphonate was 1.6 and that 36% of this reaction proceeded by way of a phenyl-assisted pathway.[65] The second approach was to plot $\log k_t$ for the acetolysis and formolysis of a series of secondary alkyl toluene-p-sulphonates

[62] J. K. Chakrabarti, M. J. Foulis, and S. S. Szinai, *Tetrahedron Letters*, **1968**, 6249.
[63] M. R. Peterson and G. H. Wahl, *Chem. Comm.*, **1968**, 1552.
[64] W. H. Lunn and E. Farkas, *Tetrahedron*, **24**, 6773 (1968).
[65] C. J. Lancelot and P. von R. Schleyer, *J. Am. Chem. Soc.*, **91**, 4291 (1969).

against σ^*. The points for compounds (**88**) when X is an electron-withdrawing substituent (e.g. p-NO_2, p-Cl) fell on the same straight line as those for the other secondary alkyl compounds with $\rho^* = -2.32$ and -2.85 for acetolysis at $100°$ and formolysis at $75°$ respectively, but the points for the compounds with X = H, p-Me, and p-MeO fell above the lines so defined. Again the deviations give a measure of the assistance, and a value of $k_t/k_s = 1.6$ was again estimated for the acetolysis of 1-phenyl-2-propyl toluene-p-sulphonate.[66] The third approach involved an investigation of the solvolysis of compounds (**89**). Clearly the effects of the substituents X and Y should be additive if the aryl rings influenced the rates solely through inductive effects, but since only one ring can participate at any one time they should not be when there is participation. The titrimetric rate constant can then be written

$$k_t = k_0(S_x S_y + \varDelta_x S_y + \varDelta_y S_x) \tag{6}$$

where k_0 is the rate constant for the solvolysis of 2-propyl toluene-p-sulphonate which is modified by the inductive/steric effects of the aryl rings on the solvent-assisted ionization, $S_x S_y$, and by the effect of participation of either ring itself modified by the inductive/steric effect of the other ring $\varDelta_x S_y$ and $\varDelta_y S_x$. The aryl groups chosen were p-methoxyphenyl, phenyl, and p-nitrophenyl, and the solvolysis of the six possible symmetrical and unsymmetrical 1,3-diaryl-2-propyl toluene-p-sulphonates studied. Since each aryl group introduces two unknowns (an S and a $\varDelta$) there were six unknowns to be extracted from six equations like equation (6). The percentage reaction calculated as proceeding via the aryl-assisted pathway was in excellent agreement with values calculated by the method utilizing the Hammett and Taft equations (see Table 1).[67]

Table 1. Percentage aryl-assisted reaction in the solvolysis of $XC_6H_4CH_2CH(OTs)CH_3$ as determined by different methods

			$(Fk_\varDelta/k_t) \times 100$		
X	Solvent	Temp. (°C)	Hammett	Taft	Mult. subst.
p-H	CH_3CO_2H	100	35	38	38
	HCO_2H	75	78	79	72
p-MeO	CH_3CO_2H	100	91	93	92
	HCO_2H	75	99	99	99

Brown and Kim have made a very detailed investigation of the acetolysis of *threo*-3-aryl-2-butyl p-bromobenzenesulphonates and *trans*-2-arylcyclopentyl toluene-p-sulphonates.[68] It seems to us that their results (see Tables 2

[66] C. J. Lancelot, J. J. Harper, and P. von R. Schleyer, *J. Am. Chem. Soc.*, **91**, 4294 (1969).
[67] C. J. Lancelot and P. von R. Schleyer, *J. Am. Chem. Soc.*, **91**, 4297 (1969).
[68] C. J. Kim and H. C. Brown, *J. Am. Chem. Soc.*, **91**, 4286, 4287, 4289 (1969); see also P. Villa, *Compt. Rend.*, *C* **267**, 1365 (1968).

and 3) could be explained in terms of competing aryl-assisted and solvent-assisted processes as proposed by Schleyer for the 3-aryl-2-propyl compounds.

Table 2. Rates and products of acetolysis of *threo*-3-aryl-2-butyl
p-bromobenzenesulphonates

Subst.	Rel. rate at 25°	Temp. (°C)	Product Olefin	*s*-Acetate	Acetate config. *threo* (%)	*erythro*
p-MeO	80	50	0.3	—	100	0
p-Me	6.0	—	—	—	—	—
m-Me	1.6	—	—	—	—	—
H	1.0	50	52.6	47.3	96	4
p-Cl	0.19	75	48	51	86	14
m-Cl	0.11	100	72	26	57	43
m-CF$_3$	0.069	100	76	21	30	70
p-NO$_2$	0.015	100	68^a	13^a	7	93

^{a} Isolated yields reported in reference 69.

Table 3. Rates and products of acetolysis of *trans*-2-arylcyclopentyl toluene-*p*-sulphonates

Subst.	Rel. rate at 25°	Temp. (°C)	Product Olefin	2-Acetate	Acetate config. *trans* (%)	*cis*
p-MeO	6.02	50	27	73	98	2
p-Me	1.61	50	69	31	65	35
m-Me	1.24	—	—	—	—	—
H	1.00	50	82	18	18	82
p-Cl	0.366	—	—	—	—	—
m-Cl	0.298	75	81	19	<1	>99
m-CF$_3$	0.264	—	—	—	—	—
p-NO$_2$	0.119	75	71	29	0	100

The aryl-assisted process competes less successfully with the solvent-assisted process in the reactions of the *trans*-2-arylcyclopentyl compounds than in those of the *threo*-3-aryl-2-butyl compounds, and the rates of reaction of the former are all less than those of the corresponding *cis*-isomers. *cis*-2-Phenyl-cyclopentyl toluene-*p*-sulphonate yields 4.1% of the *trans*-acetate, 89% of 1-phenylcyclopentene, and 6.9% of 3-phenylcyclopentene. The phenyl-cyclopentenes obtained from the deuterated compound (**90**) (see equation 7) show that the major pathway for their formation involves the 1-phenyl-cyclopentyl cation. A Hammett ρ–σ plot for the acetolysis of seven *cis*-2-

69 J. A. Thompson and D. J. Cram, *J. Am. Chem. Soc.*, **91**, 1778 (1969); see *Org. Reaction Mech.*, **1968**, 25.

arylcyclopentyl toluene-*p*-sulphonates yielded a ρ-value of −1.66. It was considered that this arose from the inductive effects of the aryl rings on a rate-determining ionization to a 2-arylcyclopentyl cation which rearranged rapidly to a 1-arylcyclopentyl cation. It was thought that if ionization occurred directly to the 1-arylcyclopentyl cation the transition state should resemble this and hence a much larger ρ-value should be obtained.[68]

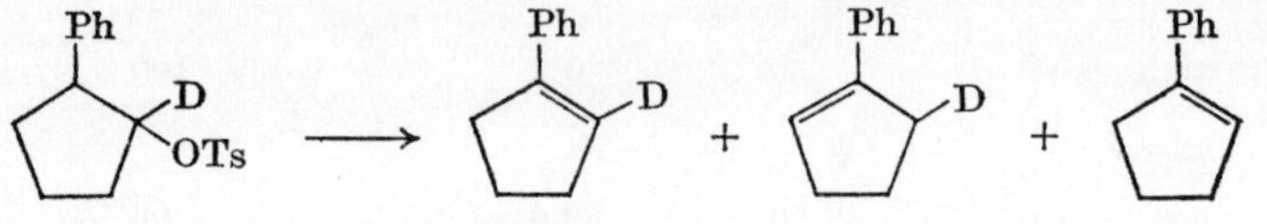

(90)   31%   59%   10%

Details of Cram and Thompson's investigation of the solvolysis of the 3-*p*-nitrophenyl-2-butyl toluene-*p*-sulphonates have been published.[69]

Diaz and Winstein have estimated the proportions of the solvolyses of 1-phenyl-2-propyl toluene-*p*-sulphonate proceeding through solvent-assisted and phenyl-assisted pathways from the steric course of the reaction. The results were in good agreement with those of Lancelot, Harper, and Schleyer.[70] Nordlander and Kelly have published details of their investigation of the trifluoroacetolysis of 1-phenyl-2-propyl toluene-*p*-sulphonate. This reaction proceeds 20 times faster than trifluoroacetolysis of 2-propyl toluene-*p*-sulphonate, and the product of 1-phenyl-2-propyl trifluoroacetate is formed with retention of configuration.[71]

Isotope effects on the polarimetric rate constants for acetolysis and trifluoroacetolysis of optically active $CH_3CH(Ph)CH(CD_3)OTs$ and $CD_3CH(Ph)-CH(CH_3)OTs$ have been measured.[72]

A linear correlation has been noted between the logarithm of the rate constant for solvolysis of some β-arylalkyl toluene-*p*-sulphonates with the ionization potential of the corresponding aromatic compounds C_6H_5R and the charge-transfer transition energy of the compounds $(p\text{-}RC_6H_4CH_2)_3B$.[73]

The decomposition of ^{14}C-labelled and optically active *erythro*- and *threo*-3-phenyl-2-butylsulphoxonium ions is discussed on p. 74.

Acetolysis of *p*-nitroneophyl *p*-bromobenzenesulphonate occurs about 1000 times more slowly than that of neophyl *p*-bromobenzensulphonate, and it yields 72.2% of products formed with *p*-nitrophenyl migration and 25.8% of products formed with methyl migration.[74]

[70] A. F. Diaz and S. Winstein, *J. Am. Chem. Soc.*, **91**, 4300 (1969).
[71] J. E. Nordlander and W. J. Kelly, *J. Am. Chem. Soc.*, **91**, 996 (1969).
[72] S. L. Loukas, M. R. Velkou, and G. A. Gregoriou, *Chem. Comm.*, **1969**, 1199.
[73] B. G. Ramsey and N. K. Das, *J. Am. Chem. Soc.*, **91**, 6191 (1969).
[74] H. Tanida, T. Tsuji, H. Ishitobi, and T. Irie, *J. Org. Chem.*, **34**, 1086, (1969).

Aryl participation in the reduction of α-bromo-4-hydroxy-3,5-di-t-butyl-acetophenones has been investigated further.[75] An example of aryl participation in the concerted cyclization of a cyclopropyl ketone has been reported.[76]

Phenethyl Compounds

Coke, McFarlane, Mourning, and Jones have measured the titrimetric rate constant (k_t) for the acetolysis of phenethyl toluene-p-sulphonate under a variety of conditions and, by ^{14}C-labelling studies, the extent of phenyl migration in the starting toluene-p-sulphonate and the product acetate as a function of time. It was thus possible to dissect k_t into $k_{\varDelta}$ and k_s. Typically at 75° in acetic acid without added salts, $k_s = 2.07 \times 10^{-7}$ and $k_{\varDelta} = 2.39 \times 10^{-7}$ sec^{-1}; k_t, presumed to be equal to k_s, for the acetolysis of ethyl toluene-p-sulphonate at 75° is 7.72×10^{-7}. Introduction of a phenyl group at position 2 therefore causes a decrease in reaction rate, and although part of the reaction of phenethyl toluene-p-sulphonate involves phenyl participation there is no overall rate enhancement. The entropy of activation for the phenyl-assisted reaction (-14.6 e.u.) is substantially more positive than that for the solvent-assisted reaction (-21.5 e.u.). The extent of phenyl migration is decreased in the presence of sodium acetate owing to an increase in the rate of the direct displacement reaction.[77]

When $k_{\varDelta}$ for the acetolysis of 2-phenyl-, 2-p-chlorophenyl-, 2-p-methyl-phenyl-, and 2-p-methoxyphenyl-ethyl toluene-p-sulphonate is plotted against $k_{\varDelta}$ for the acetolysis of the corresponding 2-aryl-2-methyl-1-propyl toluene-p-sulphonate (a reaction which is well established to involve aryl participation), a good straight line is obtained. This supports the view that the $k_{\varDelta}$ value for the 2-arylethyl compound obtained in this way is a reliable measure of aryl participation.[78]

Jablonski and Snyder have shown that the extent of retention of configuration in the solvolysis of *erythro*-[1,2-^{2}H$_2$]phenethyl p-bromobenzenesulphonate is always twice that of label rearrangement on solvolysis of [1-^{14}C]phenethyl toluene-p-sulphonate under conditions where rearrangement accounts for 5, 15, 44, and 50% of the product.[79] Similar results were obtained on deamination of phenethylamine, which suggests that the phenonium ion which is an intermediate in this reaction is similar to that which is an intermediate in the solvolyses.[80]

[75] L. H. Schwartz and R. V. Flor, *J. Org. Chem.*, **34**, 1499 (1969); see *Org. Reaction Mech.*, **1968**, 29.
[76] G. Stork and M. Gregson, *J. Am. Chem. Soc.*, **91**, 2373 (1969).
[77] J. L. Coke, F. E. McFarlane, M. C. Mourning, and M. G. Jones, *J. Am. Chem. Soc.*, **91**, 1154 (1969).
[78] M. G. Jones and J. L. Coke, *J. Am. Chem. Soc.*, **91**, 4284 (1969).
[79] R. J. Jablonski and E. I. Snyder, *J. Am. Chem. Soc.*, **91**, 4445 (1969).
[80] E. I. Snyder, *J. Am. Chem. Soc.*, **91**, 5118 (1969).

Acetolysis of 2-[1-^{14}C]phenylethyl *p*-nitrobenzenesulphonate and 2-(*p*-methoxy[1-^{14}C]phenylethyl *p*-nitrobenzenesulphonate yields isotope effects k^{12}/k^{14} of 1.002 and 1.028 respectively. This is consistent with the *p*-methoxyphenyl compound reacting with participation and the phenyl compound without.[81a]

The rates of acetolysis of some polycyclic arylethyl mercuric perchlorates have been correlated with the rates for the corresponding arylmethyl perchlorates and interpreted in terms of aryl participation.[81b]

Participation by a β-Ferrocenyl Group

Participation by a β-ferrocenyl group has been investigated further.[82] β-Ferrocenylethyl toluene-*p*-sulphonate (91) is acetolysed 3120 times faster than phenethyl toluene-*p*-sulphonate. [1,1-^{2}H$_2$]-β-Ferrocenylethyl toluene-*p*-sulphonate yields unrearranged acetate with the label unscrambled, and (−)-1-ferrocenyl-2-propyl toluene-*p*-sulphonate yields the corresponding acetate with retention of configuration. These results (which are very similar to those found on solvolysis of paracyclophanylethyl toluene-*p*-sulphonate[83]) suggest that there is participation by the ferrocenyl group but that the resulting ion only undergoes attack by nucleophiles in one direction. Evidence on the structure of the ion was provided by an investigation of the acetolysis of compounds (92) and (94). The *endo*-isomer (92) reacted 2780 times faster than the *exo*-isomer (94) and yielded unrearranged acetate (93) whereas the *exo*-isomer yielded the rearranged acetate (95). It was suggested that partici-

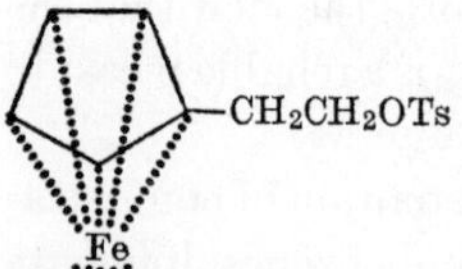

(91)

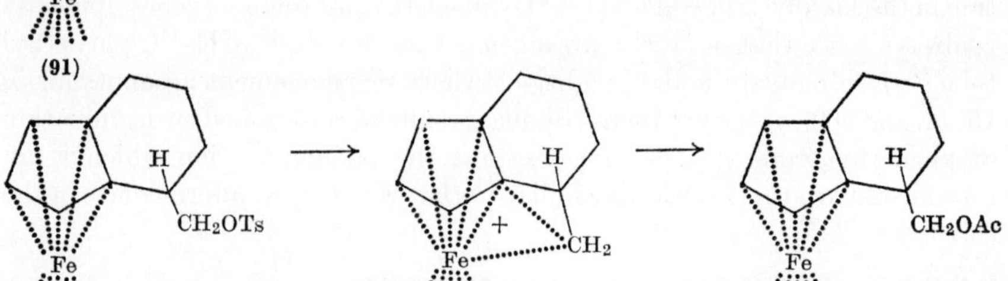

(92)   (93)

[81a] Y. Yukawa, T. Ando, K. Token, M. Kawada, and S. G. Kim, *Tetrahedron Letters*, **1969**, 2367.
[81b] B. G. van Leuwen and R. J. Ouellette, *J. Am. Chem. Soc.*, **90**, 7061 (1968).
[82] Cf. *Org. Reaction Mech.*, **1967**, 53.
[83] See *Org. Reaction Mech.*, **1967** 23—24, and the following section of this volume.

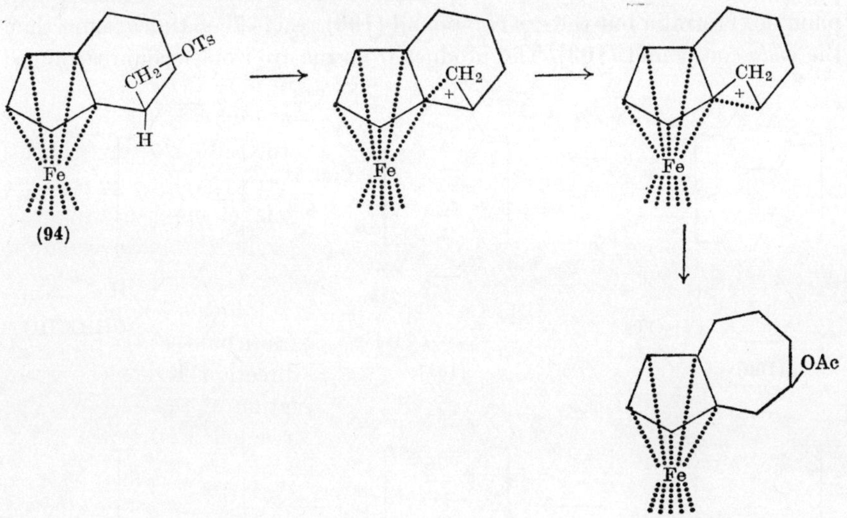

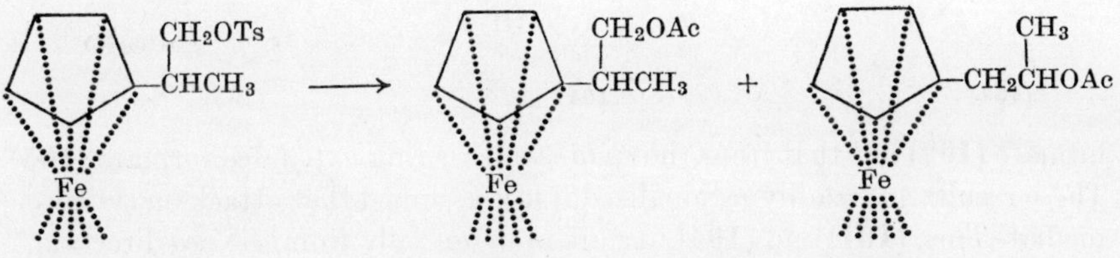

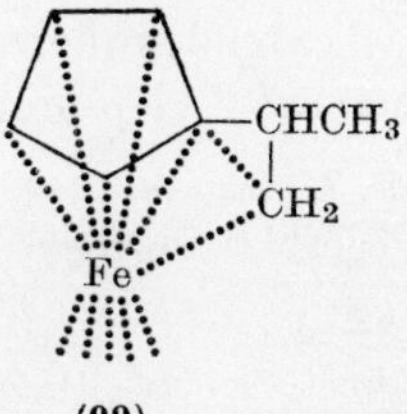

pation by the inter-annular electrons was important in the reaction of the *endo*-isomer but that the *exo*-compound reacted with participation by the extra-annular electrons. Acetolysis of 2-ferrocenyl-1-propyl toluene-*p*-sulphonate (**96**) yields a 1:1 mixture of the acetates (**97**) and (**98**). It is not clear if this reaction proceeds with concurrent participation by the extra- and inter-annular electrons or whether the ion formed by participation of the interannular electrons (99) reacts with attack at C-1 and C-2 when free of the constraint of the ring.[84]

The reactions of α-ferrocenylalkyl derivatives is discussed on p. 63.

Participation by the [2.2]Paracyclophanyl Group

The preference for participation by the outside electrons of the [2.2]para-cyclophanyl ring[85] has been investigated by studying the formolysis of compounds (100) and (103). Both compounds react faster than 1-toluene-*p*-sulphonyloxytetralin but the *exo*-compound (100) reacts 7—8 times faster than the *endo*-compound (103). The product from the *exo*-isomer is unrearranged

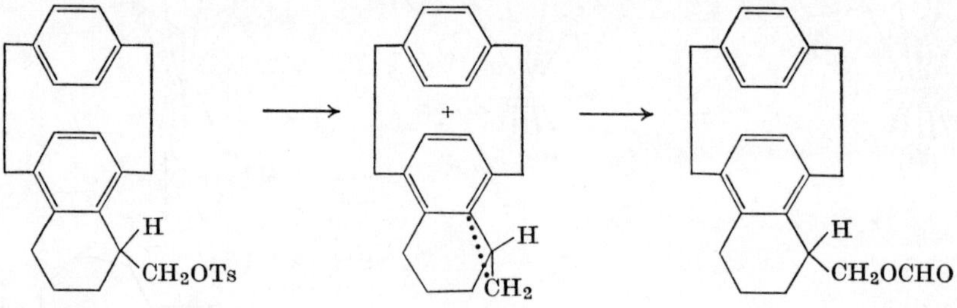

(100)   (101)   (102)

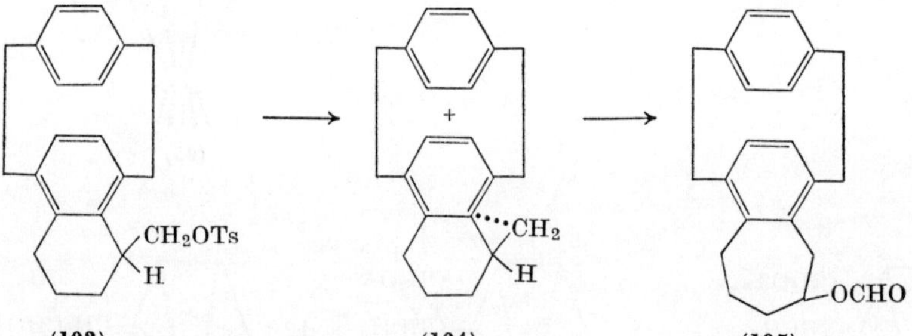

(103)   (104)   (105)

formate (102) but that from the *endo*-isomer is ring-expanded formate (105). These results are readily rationalized if it is assumed that attack on the intermediate ions, (101) and (104), occurs preferentially from the *exo* direction.[86]

Benzonorbornen-2-yl Compounds

Tanida and his coworkers have extended their investigation of the acetolysis of substituted benzonorbornen-*exo*-2-yl *p*-bromobenzenesulphonates.[87] The

[84] M. J. Nugent and J. H. Richards, *J. Am. Chem. Soc.*, **91**, 6138 (1969); M. J. Nugent, R. Kummer, and J. H. Richards, *ibid.*, p. 6141; M. J. Nugent, R. E. Carter, and J. H. Richards, *ibid.*, p. 6145
[85] See *Org. Reaction Mech.*, **1967**, 23—24.
[86] M. J. Nugent and T. L. Vigo, *J. Am. Chem. Soc.*, **91**, 5483 (1969).
[87] See *Org. Reaction Mech.*, **1968**, 27—28.

7-methoxy-6-nitro- and 6,7-dinitro-compounds (106) and (107) react approximately 1000 and 100,000 times more slowly than the unsubstituted compounds respectively. The unsubstituted compound yields exclusively *exo*-acetate, the 7-methoxy-6-nitro-compound yields 97% *exo*-acetate and 1% of the benzonorbornadiene, and the 6,7-dinitro-compound yields 41% *exo*-acetate, 35% *endo*-acetate, and 21% of the benzonorbornadiene. The rate of racemization of the optically active 6,7-dinitro-compound was 4.1 times the rate of acetolysis,[88] which indicates that it is being racemized by way of internal return concurrent with acetolysis. The *endo*-acetate in the product retained 25.7% of the original optical activity, and it was calculated that this was consistent with its being formed by an S_N2 process, partly from active and partly from racemized *p*-bromobenzenesulphonate. The *exo*-acetate in the product retains 4.51% of the optical activity. Since this is formed mainly from racemized *p*-bromobenzenesulphonate it seems that the activity of the *exo*-acetate formed directly from active *p*-bromobenzenesulphonate is higher than this. This suggests that *exo*-acetate is formed, at least in part, from an asymmetric (classical?) ion. It was also reported that the 6,7-dinitro-*p*-bromobenzenesulphonate (107) reacts only four times faster than its *endo*-isomer. Hence, when the benzene ring of a benzonorbornenyl compound is as strongly deactivated as this ($\Sigma\sigma^+ = 1.464$) the *exo*:*endo* rate ratio becomes very small. This suggests that the *exo*:*endo* rate ratio of 15,000 (at 25°) reported for the unsubstituted compounds[87] is largely the result of aryl participation in the reaction of the *exo*-isomer.[89]

(106)   (107)

The effect of substituents on the rates of solvolysis of 2-methyl- and 2-phenyl-benzonorbornen-2-yl derivatives has also been studied. The effect is quite large in the solvolysis of the 2-methyl-*exo*-2-*p*-nitrobenzoates (108) in 50% aqueous acetone with relative rates 1:16:0.74:0.025 when X = H, 6-MeO, 7-MeO, and 6- and 7-nitro, respectively. Clearly, the reactions of the methoxy-compounds and the unsubstituted compound are anchimerically assisted. The amount of assistance in the reactions of the 2-phenyl-*exo*-2-*p*-nitrobenzoates (109) is more difficult to decide. Here the 6-methoxy-compound reacts 4.7 times faster than the unsubstituted compound in 80% aqueous

[88] The value for benzonorbornen-*exo*-2-yl *p*-bromobenzenesulphonate is 4.0 (ref. 89) or 4.16 (J. P. Dirlam, A. Diaz, S. Winstein, W. P. Giddings, and G. C. Hanson, *Tetrahedron Letters*, 1969, 3133.)

[89] H. Tanida, H. Ishitobi, T. Irie, and T. Tsushima, *J. Am. Chem. Soc.*, **91**, 4512 (1969).

acetone at 75°, and it was estimated from a consideration of the ρ-values for the reactions shown in Scheme 4 that k_Δ contributes between 50 and 100% of the total rate constant. It is difficult to judge the validity of this argument since the solvent, the leaving group, and the temperature vary so widely.[90]

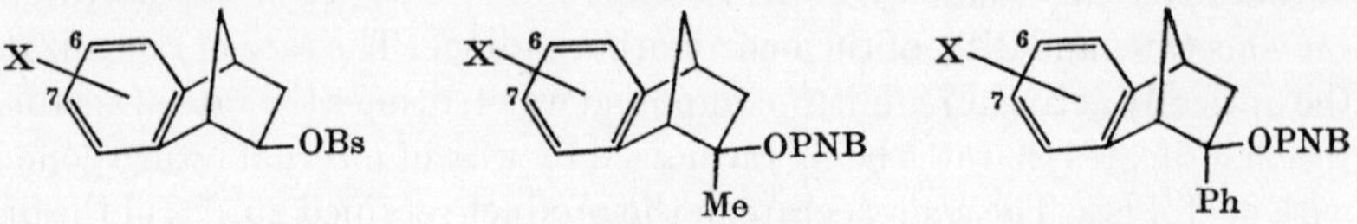

(108)	**(109)**	

Solvent	AcOH	50% Aqueous acetone	80% Aqueous acetone
Temperature	75°	125°	75°
ρ	−3.3	−1.7	−0.9

Scheme 4

A detailed investigation of the solvolysis in 80% aqueous acetone of 2-aryl-6-methoxybenzonorbornen-2-yl *p*-nitrobenzoates (**110**) and (**111**), and 2-aryl-benzonorbornen-2-yl *p*-nitrobenzoates (**112**) and (**113**), has also been described. The rate of solvolysis of the *exo*-2-*p*-methoxy-compound (**112**; $Ar = p\text{-MeOC}_6\text{H}_4$) is increased by a factor of only 2 on going to the 6-methoxy-

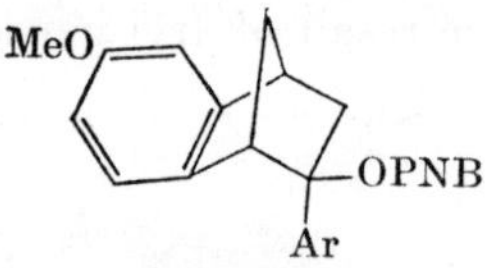

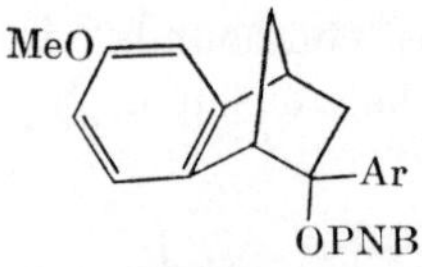

(110)	**(111)**	
ρ −3.72	−4.05	

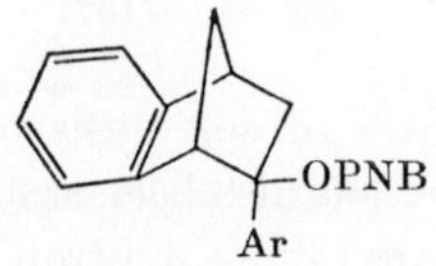

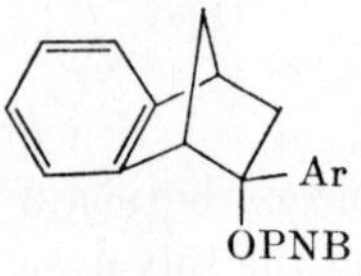

(112)	**(113)**
−4.50	−4.52

compound (**110**; $Ar = p\text{-MeOC}_6\text{H}_4$). Nevertheless the *exo*:*endo* rate ratio for compounds (**112**; $Ar = p\text{-MeOC}_6\text{H}_4$) and (**113**; $Ar = p\text{-MeOC}_6\text{H}_4$) is 3300, which suggests that other factors besides aryl participation are playing an important part in determining this ratio.[91]

[90] J. P. Dirlam and S. Winstein, *J. Am. Chem. Soc.*, **91**, 5905, 5907 (1969).
[91] H. C. Brown and K. T. Liu, *J. Am. Chem. Soc.*, **91**, 5909 (1969); H. C. Brown, S. Ikegami, and K. T. Liu, *ibid.*, p. 5911.

Addition of HBr to benzonorbornadiene in methylene chloride yields 93.6% of *exo-* and 6.4% of *endo*-benzonorbornen-2-yl bromide. The *exo*-bromide from the addition of DBr is 37% deuterated at position 7 and the remainder presumably at position 3. All the deuterium in the *endo*-form is at position 3. These results were explained by the initial formation of a classical ion which collapsed to a 4:1 mixture of *exo-* and *endo*-3-deuterated bromides and rearranged to a non-classical ion at similar rates.[92]

Stable Phenonium Ions

When a 3-aryl-2,3-dimethyl-2-butyl chloride (**114**) or an aryl-t-butyl-methyl-carbinyl chloride (**115**) is dissolved in SbF_5–SO_2 at $-78°$ the same carbonium ions are formed but the nature of these depends on the aryl group (p-XC_6H_4).

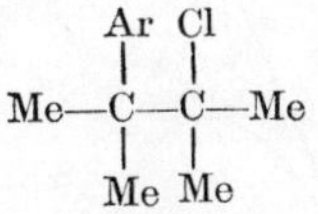

(**114**) 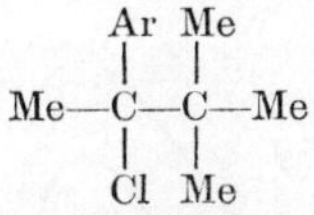   (**115**)

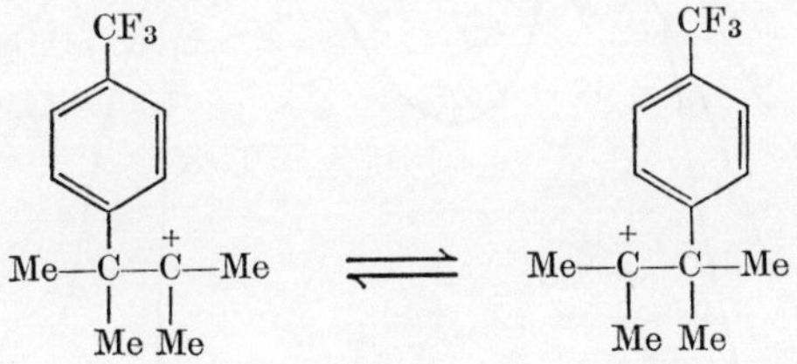

(**116**)   (**117**)

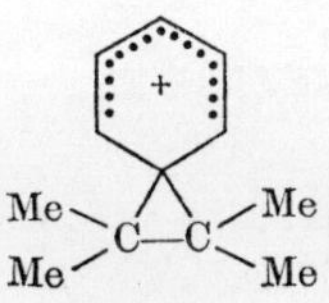

(**118**) 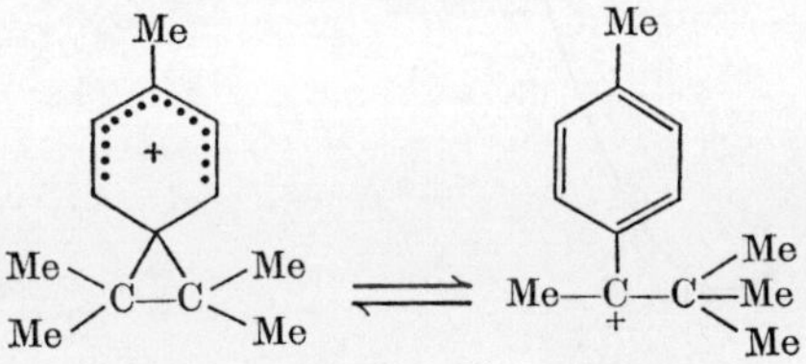   (**119**)   (**120**)

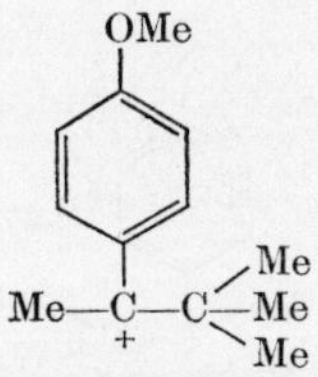

(**121**)

[92] L. E. Barstow and G. A. Wiley, *Tetrahedron Letters*, **1968**, 6309.

From an analysis of the NMR spectrum and the products of quenching the solution in methanol, it was concluded that when X = CF₃ a pair of equilibrating ions (**116**) and (**117**) are formed, when X = H the predominant ion is the phenonium ion (**118**), when X = Me a mixture of phenonium ion (**119**) and benzylic ion (**120**) is formed, and when X = MeO the only ion is the benzylic ion (**121**). This is readily explained on the assumption that the energy of the phenethyl ion (**123**) is unaffected by the group X, but that the energy of the

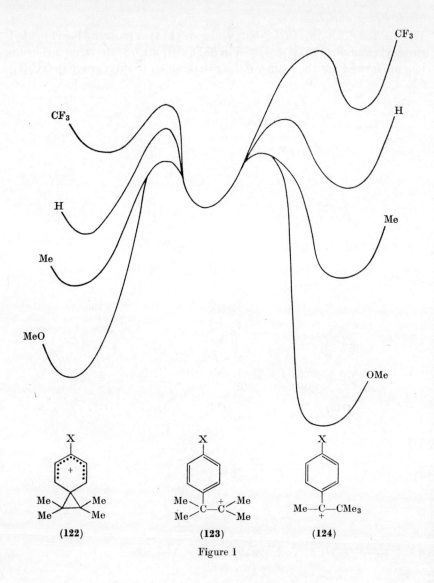

Figure 1

benzylic ion (**124**) is affected strongly and that of the phenonium ion (**122**) moderately, as shown diagrammatically in Figure 1.[93, 94]

Isomeric ions (**125**) and (**126**) have been generated by protonation of the corresponding dienones in fluorosulphonic acid at −65° and their NMR spectra recorded. When the temperature was raised to −40° they were not interconverted but converted into the benzylic ion (**127**).[95]

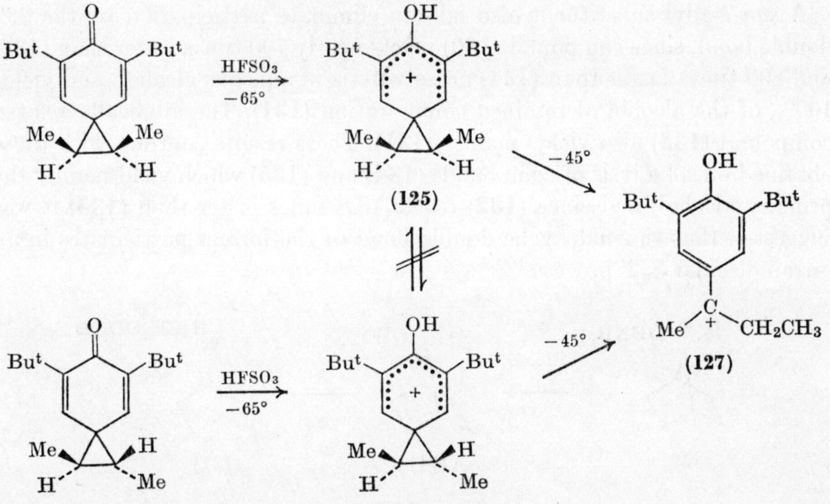

(125)

(126)

(127)

Participation by Double and Triple Bonds[96, 97]

Gassman has continued his work on the effect of *syn*-7-aryl substituents on the reactivity of norbornen-*anti*-7-yl derivatives. As reported last year,[98] the

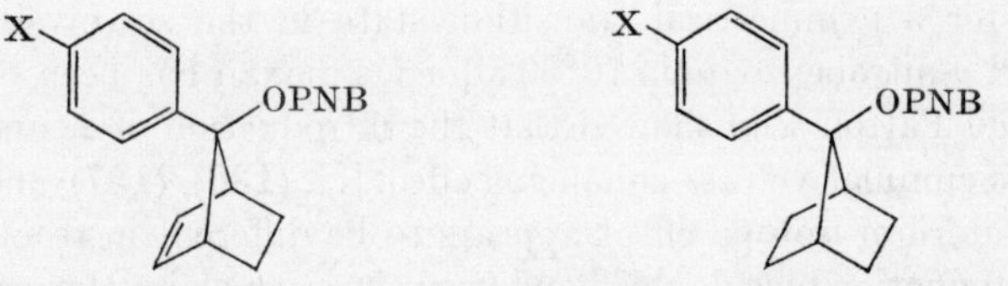

(128)　　　　(129)

(a: X = MeO; b: X = H; c: X = CF₃; PNB = *p*-NO₂·C₆H₄·CO—)

[93] G. A. Olah, M. B. Comisarow, and C. J. Kim, *J. Am. Chem. Soc.*, **91**, 1458 (1969).
[94] B. G. Ramsey and J. Cook, *Tetrahedron Letters*, **1969**, 535.
[95] D. Chamot and W. H. Pirkle, *J. Am. Chem. Soc.*, **91**, 1569 (1969).
[96] M. Hanack, *Trans. N.Y. Acad. Sci.*, **31**, 139 (1969).
[97] M. Banciu, *Stud. Cercet. Chim.*, **16**, 509 (1968); *Chem. Abs.*, **70**, 10720 (1969).
[98] See *Org. Reaction Mech.*, **1968**, 30.

syn-p-methoxyphenyl compound (**128a**) is solvolysed only about 3 times faster than the compound without the double bond (**129a**), and it has now been shown that the *syn*-phenyl and *syn-p*-trifluoromethylphenyl compounds (**128b**) and (**128c**) react 40 and 3×10^4 times faster in aqueous dioxan than (**129b**) and (**129c**) respectively. Therefore, with these compounds, anchimeric assistance is a function of electron demand at the incipient carbonium ion centre.[99]

A *syn*-7-allyl substituent also fails to eliminate participation by the 2,3-double bond, since compound (**130**) reacts nearly 700 times faster than (**133**) and 800 times faster than (**134**) on solvolysis in aqueous alcohol, and yields 100% of the alcohol of retained configuration (**131**). The allylically related compound (**132**) also yields mainly (**131**). These results contrast with those obtained on solvolysis of compounds (**134**) and (**135**) which yield mainly the primary alcohol, and since (**132**) reacts 18.8 times faster than (**135**) it was suggested that the endocyclic double bond of the former participates in an intramolecular S_N2' process.[100]

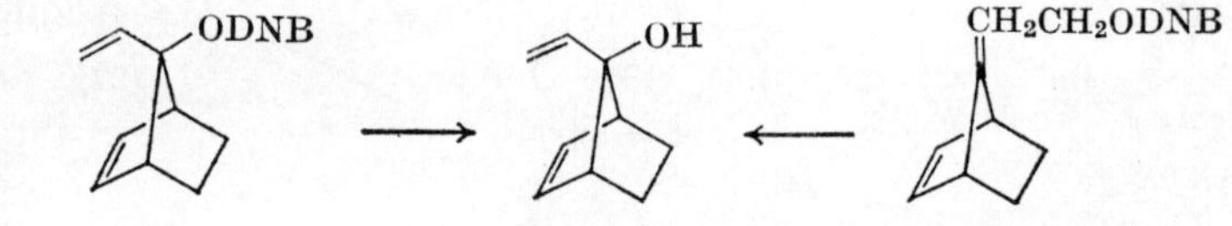

ODNB   OH   CH₂CH₂ODNB

(**130**)   →   (**131**)   ←   (**132**)

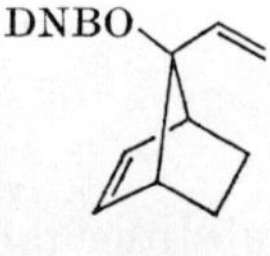

DNBO   ODNB   CH₂CH₂ODNB

(**133**) 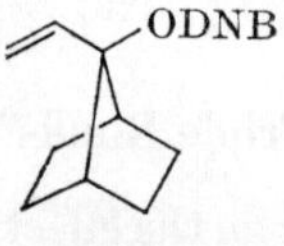   (**134**) 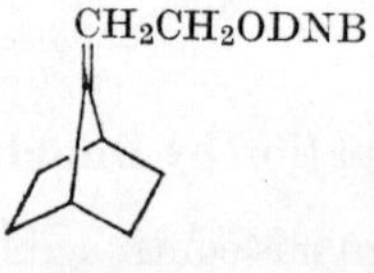   (**135**)

Support for a symmetrical transition state in the solvolysis of *anti*-7-norbornenyl *p*-nitrobenzoate in 70:30 aqueous dioxan has been obtained by Gassman and Patton who showed that the introduction of 2- and 3-methyl groups has a cumulative rate-enhancing effect [cf. (**136**), (**137**), and (**138**)].[101]

The α-deuterium isotope effect appears to be different in reactions which involve participation by a double bond from those which involve participation by other neighbouring groups. Thus for the acetolysis of [7-^{2}H]-7-*anti*-norbornenyl toluene-*p*-sulphonate $k_H/k_D = 1.13$ whereas for the ethanolysis of [1,1-^{2}H$_2$]-4-methoxy-1-pentyl *p*-bromobenzenesulphonate $k_H/k_D = 1.01$.[102]

[99] P. G. Gassman and A. F. Fentiman, *J. Am. Chem. Soc.*, **91**, 1545 (1968).
[100] G. D. Sargent, J. A. Hall, M. J. Harrison, W. H. Demisch, and M. A. Schwartz, *J. Am. Chem. Soc.*, **91**, 2379 (1969).
[101] P. G. Gassman and D. S. Patton, *J. Am. Chem. Soc.*, **91**, 2160 (1969).
[102] R. Eliason, M. Tomic, S. Borcic, and D. E. Sunko, *Chem. Comm.*, **1968**, 1490.

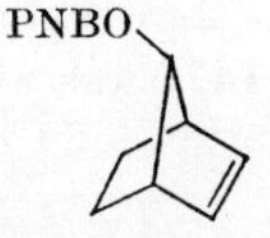

(136)	(137)	(138)

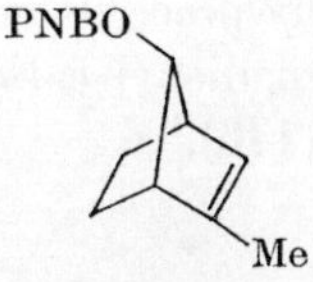

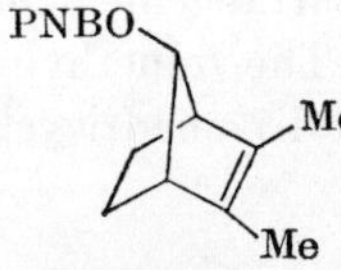

Relative rate of
solvolysis at 140°   1   13.3   148

Solvolysis of the tricyclic p-methoxybenzoate[103] and p-nitrobenzoate[104] (139) in aqueous acetone is accompanied by extensive ion-pair return to form *anti*-norbornenyl esters (140). The product from the solvolysis was the *anti*-norbornenol (141) and no tricyclic alcohol (142) could be detected. The p-methoxybenzoate was estimated to react 10^{10}—10^{11} times faster and the p-nitrobenzoate ca. 10^{12} times faster than the corresponding *anti*-norbornenyl ester. This reactivity difference is much larger than that found in the acid-catalysed cleavage of the corresponding methyl ethers which is 7×10^6.[105] It was estimated that the partitioning of the intermediate ion on reaction with the p-nitrobenzoate anion (and other anions) strongly favours formation of tricyclic products but that on reaction with methanol and water it strongly favours formation of bicyclic norbornenyl products.

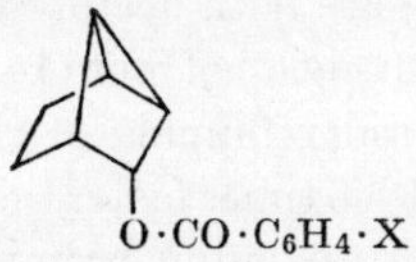

(139)	(140)	(141)	(142)

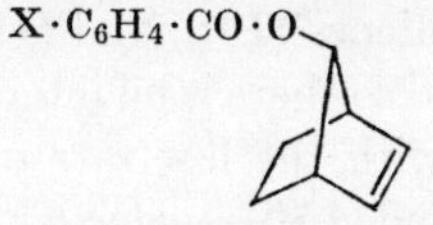

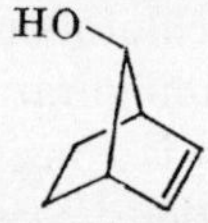

(X = p-MeO or p-NO₂)

Lustgarten, Brookhart, and Winstein have extended their investigation of the rearrangement of the bicyclo[3.2.0]heptadienyl cation into the norbornadienyl cation[106] to the methyl-substituted ions (145) and (147). Unlike the case of the unsubstituted ion, the NMR spectrum of the methyl-substituted ion (144) generated from the alcohol (143) can be observed in FSO_3H–SO_2ClF at −130° prior to its rearrangement. At −105° it is converted into a 55:45 mixture of the 1-methyl- and 5-methyl-norbornadienyl ions (145) and (147). At −75° the 1-methyl ion rearranges "fairly completely" into the 5-methyl ion.

[103] J. J. Tufariello and R. J. Lorence, *J. Am. Chem. Soc.*, **91**, 1546 (1969).
[104] J. Lhomme, A. Diaz, and S. Winstein, *J. Am. Chem. Soc.*, **91**, 1548 (1969).
[105] See *Org. Reaction Mech.*, **1966**, 25.
[106] See *Org. Reaction Mech.*, **1967**, 32—36.

This rearrangement probably proceeds via the 2-methylbicyclo[3.2.0]-cation (**144**). The 5-methylnorbornadienyl cation (**147**) was also generated from 7-methyl-7-quadricyclanol (**146**).[107]

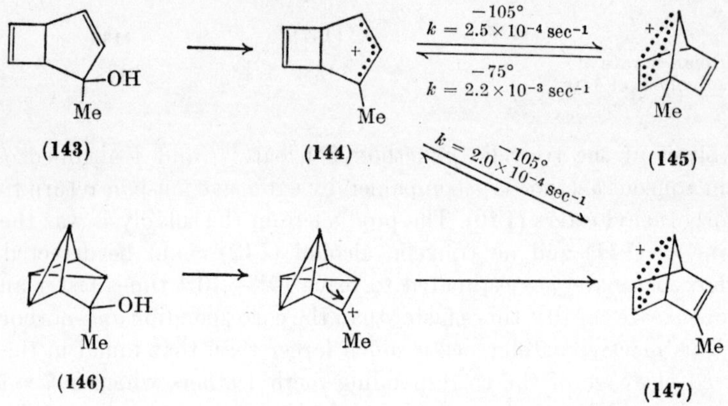

Molecular-orbital calculations on the norbornenyl[108] and 9-benzonorbornenyl[109] cation have been reported. There has been an X-ray crystallographic structure determination on *anti*-9-benzonorbornenyl *p*-bromobenzenesulphonate.[110]

The epimeric 7-chloronorbornadienes (**148**) and (**150**) are both acetolysed with retention of configuration, and so there is no interconversion of ions (**149**) and (**151**).[111] This is consistent with the low rate of bridge-flipping of the norbornadienyl cation.[112] Compound (**150**) reacts about 1000 times faster than (**148**). Similar results were obtained on solvolysis of the corresponding bromide in aqueous dioxan.[113] Compounds (**152**) and (**153**) also underwent silver-ion assisted acetolysis with retention of configuration. (**153**) reacts about six times faster than (**152**).[111]

The methanolysis of *syn*-7-chloro-7-azabenzonorbornadiene (**154**) occurs much more readily than that of its *anti*-isomer (**156**) and yields compound (**155**), probably as shown. The *anti*-isomer also yields (**155**) probably via initial isomerization to the *syn*-isomer. When the *anti*-isomer is treated with silver nitrate in methanol it yields a compound thought to be (**157**).[114]

[107] R. K. Lustgarten, M. Brookhart, and S. Winstein, *J. Am. Chem. Soc.*, **90**, 7364 (1968).
[108] T. Tsushima and H. Tanida, *J. Chem. Soc. Japan*, **90**, 650 (1969).
[109] H. Tanida and T. Tsushima, *J. Chem. Soc. Japan*, **89**, 1418 (1968).
[110] H. Koyama and K. Okada, *J. Chem. Soc.* (B), **1969**, 940.
[111] S. J. Cristol and G. W. Nachtigall, *J. Am. Chem. Soc.*, **90**, 7132, 7133 (1968).
[112] See *Org. Reaction Mech.*, *1967*, 34.
[113] J. W. Wilt and P. J. Chenier, *J. Am. Chem. Soc.*, **90**, 7366 (1968).
[114] V. Rautenstrauch, *Chem. Comm.*, **1969**, 1122.

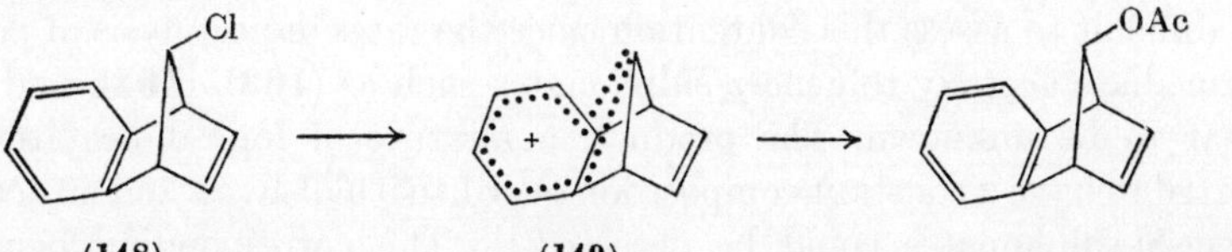

(148)   (149)

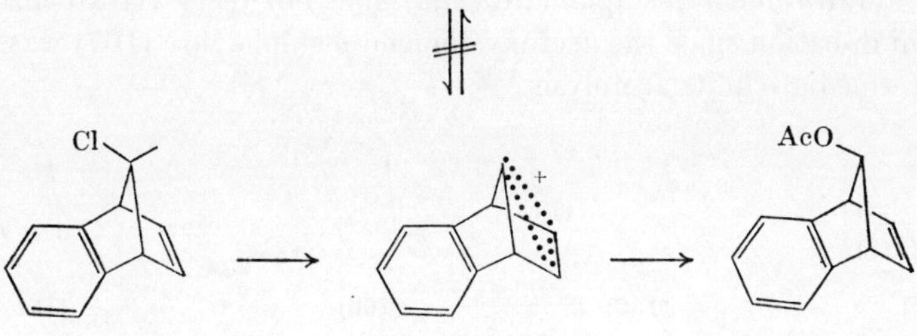

(150)   (151)

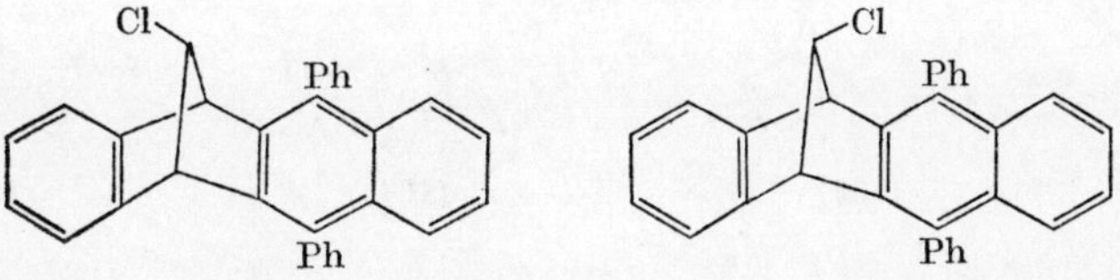

(152)   (153)

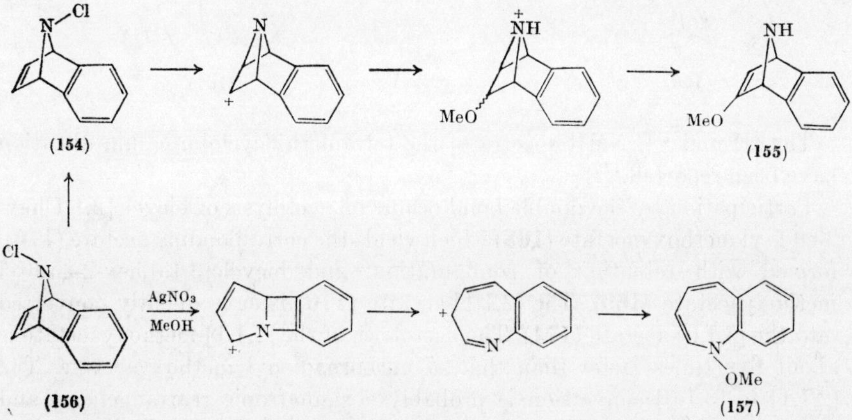

(154)   (155)

(156)   (157)

Acetolysis of compound (158) with loss of both toluene-p-sulphonyloxy groups is about 500 times faster than that of compound (159) whereas acetolysis of (160) is 3—4 times slower than that of (161). It was suggested that (158) reacts with "nearly concerted loss of both tosylate groups to form a cyclobutenium-stabilized dicarbonium ion" as (162), and the constancy of the first-order rate constant with time was quoted as supporting evidence.[115]

[115] J. B. Lambert and A. G. Holcomb, *J. Am. Chem. Soc.*, **91**, 1572 (1969).

It is difficult to assess this contention since the rates of acetolysis of possible intermediate acetoxy-toluene-p-sulphonates such as **(163)**, **(164)**, and **(165)** appear to be unknown. The product, a mixture of four di-acetates, was reported to have a constant composition over 1 to 5 half-lives, and no acetoxy-toluene-p-sulphonates could be detected.[110] The corresponding benzonor-bornadienyl di-toluene-p-sulphonate **(166)** does not react via a benzocyclo-butenium di-cation since the acetoxy-toluene-p-sulphonate **(167)** was shown to be intermediate in its acetolysis.[116]

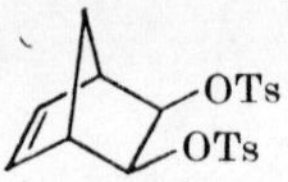

(158) 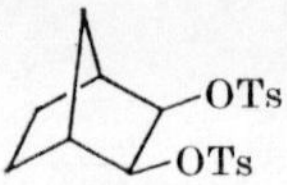   (159) 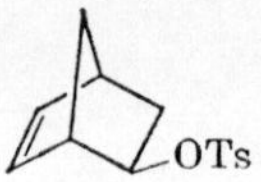   (160) 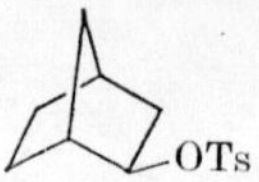   (161)

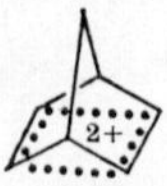

(162) 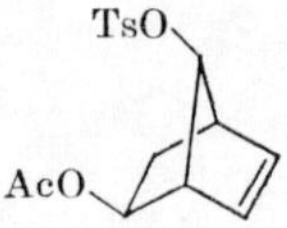   (163) 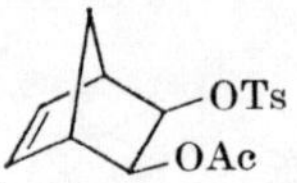   (164) 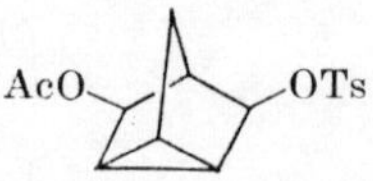   (165)

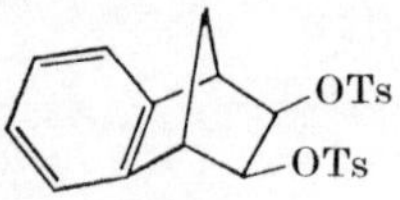

(166) 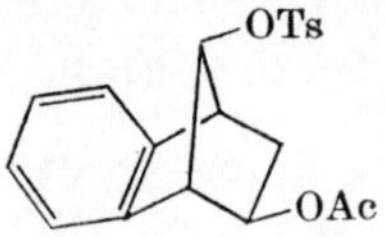   (167)

The ^{1}H and ^{13}C NMR spectra of the tetramethylcyclobutenium di-cation have been reported.[117]

Participation by the double bond occurs on acetolysis of bicyclo[2.1.1]hex-2-en-5-yl methoxyacetate **(168)** which yields the corresponding acetate **(170)**, formed with retention of configuration, and bicyclo[3.1.0]hex-2-en-6-yl methoxyacetate **(169)**. The [2.1.1]-acetate **(170)** is subsequently converted into the [3.1.0]-acetate **(171)**. The acetolysis of the [2.1.1]-methoxyacetate is about five times faster than that of norbornadienyl methoxyacetate. The [2.1.1] to [3.1.0] conversion is probably a sigmatropic rearrangement and does not involve ion-pairs since the [2.1.1]-acetate reacts at similar rates in dodecane and in acetic acid.[118]

The bicyclo[2.1.1]hexenyl cations **(172)** and **(173)** have been generated by dissolving hexamethyl-Dewar benzene and hexamethylprismane in FSO$_3$H–

116 H. Tanida and T. Tsushima, *Tetrahedron Letters*, **1969**, 3647.
117 G. A. Olah, J. M. Bollinger, and A. M. White, *J. Am. Chem. Soc.*, **91**, 3667 (1969).
118 S. Masamune, S. Takada, N. Nakatsuki, R. Vukov, and E. N. Cain, *J. Am. Chem. Soc.*, **91**, 4322 (1969).

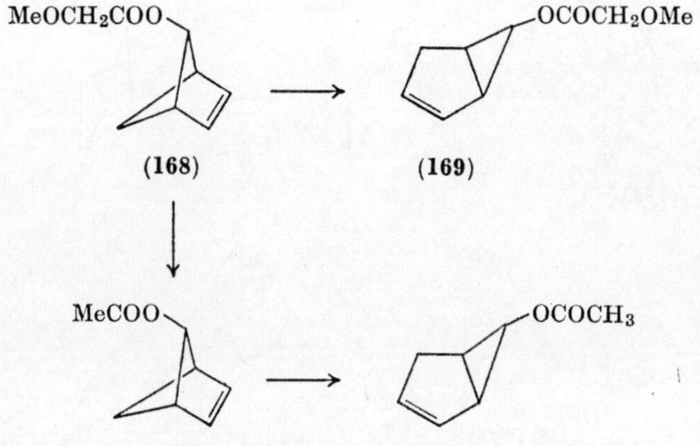

(168) → **(169)**

↓

(170) → **(171)**

SbF_5–SO_2 and HF–BF_3, and hexamethylprismane in FSO_3H–SbF_5–SO_2 at $-78°$.[119,120] The kinetics of the interconversion of these ions were also studied.[120]

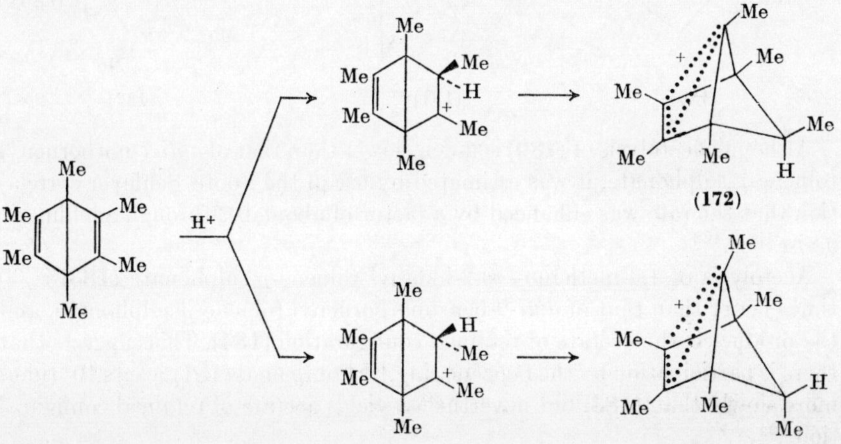

(172)

(173)

Acetolysis of *endo*-bicyclo[3.2.1]oct-6-en-8-yl toluene-*p*-sulphonate **(174)** occurs 1.9×10^5 times faster than that of its saturated analogue, and it yields unrearranged acetate of retained configuration. Presumably the non-classical ion **(175)** is an intermediate. The relative rates of acetolysis of compounds **(176)**—**(179)** show how sensitive participation by the double bond is to the geometry of the compound in which it occurs.[121]

[119] L. A. Paquette, G. R. Krow, J. M. Bollinger, and G. A. Olah, *J. Am. Chem. Soc.*, **90**, 7147 (1968).

[120] H. Hogeveen and H. C. Volger, *Rec. Trav. Chim.*, **88**, 353 (1969).

[121] B. A. Hess, *J. Am. Chem. Soc.*, **91**, 5657 (1969).

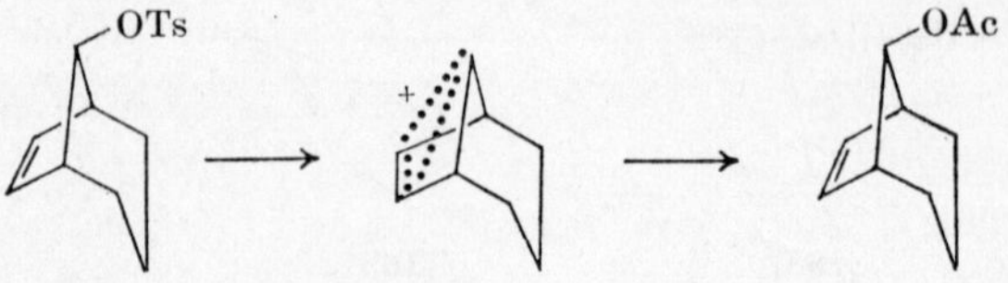

(174)   (175)

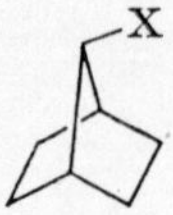

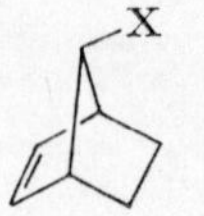

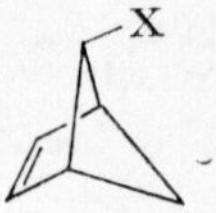

(176)   (177)   (178)   (179)

Relative   1 2×10^8 10^{11} 5×10^{14}
rate

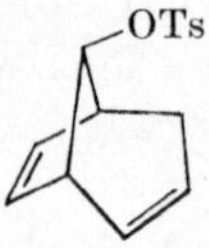

(180) 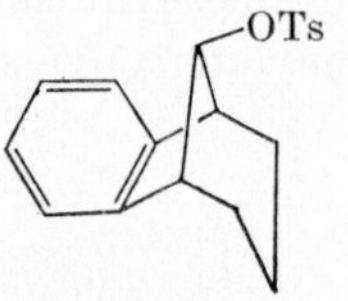   (181) 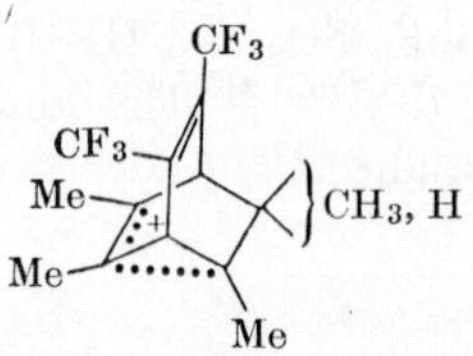   (182)

Although acetolysis of (180) is much slower than that of *anti*-7-norbornenyl toluene-*p*-sulphonate, it was estimated by use of the Foote–Schleyer correlation that the rate was enhanced by a factor of about 10^5 through anchimeric assistance.[122]

Acetolysis of 1,3-methano-*exo*-2-indanyl toluene-*p*-sulphonate (183) is 30 times faster than that of *anti*-9-benzonorbornenyl toluene-*p*-sulphonate, and the product is the acetate of retained configuration (184). This suggests that there is participation by the benzene ring.[123] Compound (181) reacts 10^4 times more slowly than (183) but nevertheless yields acetate of retained configuration.[124]

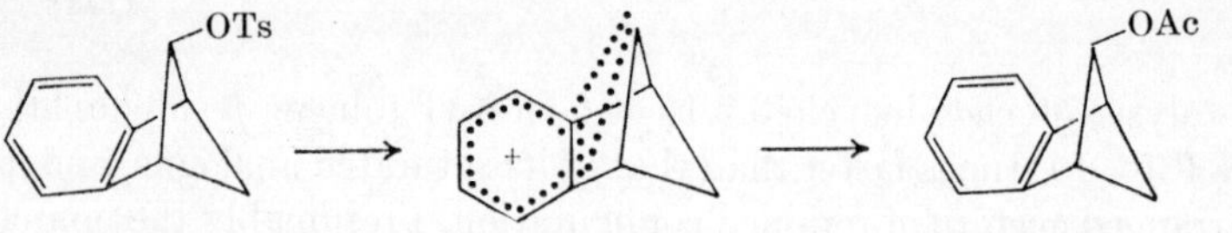

(183)   (184)

Ion (182) has been prepared by dissolving the corresponding triene in SO_2ClF–HF–SbF_5, and its NMR spectrum was measured.[125]

[122] G. W. Klumpp, G. Ellen, and F. Bickelhaupt, *Rec. Trav. Chim.*, **88**, 474 (1969).
[123] Y. Hata and H. Tanida, *J. Am. Chem. Soc.*, **91**, 1170 (1969).
[124] Result communicated by Y. Takano to the authors of ref. 123.
[125] H. Hogeveen and C. J. Gaasbeek, *Rec. Trav. Chim.*, **88**, 367 (1969).

Acetolysis of compound (**185**) occurs 15 times faster than that of the compound without the olefinic double bond, and yields the compounds shown, which are presumably formed via consecutive participation by the double bond and the phenyl group, as shown in equation (8). Similar products are formed from (**189**) except that there is no (**186**), which suggests that ion (**188**) does not revert to ion (**187**).[126–128]

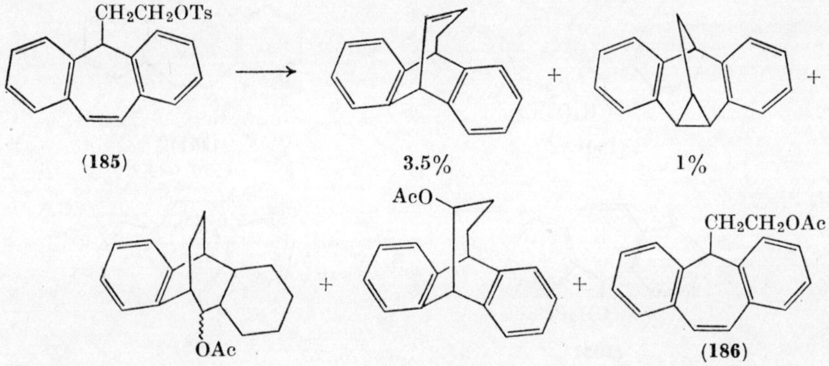

(**185**) 3.5% 1%

81.5% 3.5% (**186**) 10.5%

(**187**) (**189**) (**188**) . . . (8)

[126] I. Ciorănescu, M. Banciu, R. Jelescu, M. Rentzea, M. Elian, and C. D. Nenitzescu, *Rev. Roum. Chim.*, **14**, 911 (1969).
[127] M. Voicu and F. Badea, *Rev. Roum. Chim.*, **14**, 929 (1969).
[128] E. Ciorănescu, M. Banciu, R. Jelescu, M. Rentzea, M. Elian, and C. D. Nenitzescu, *Tetrahedron Letters*, **1969**, 1871.

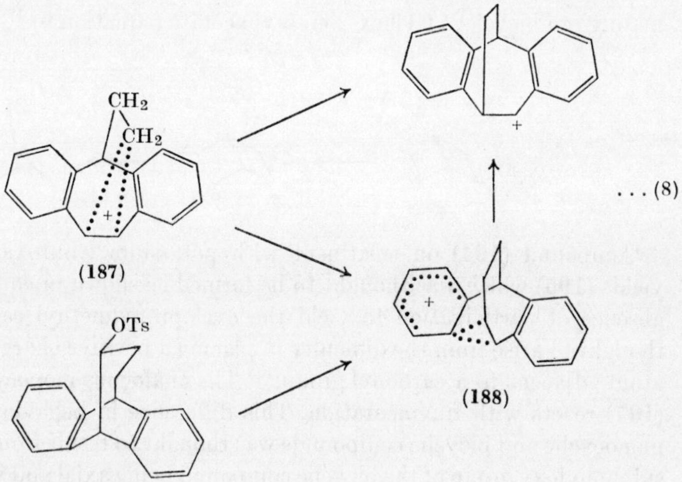

Double-bond participation competes ineffectively with phenyl participation in the acetolysis of **(190)**. An attempt to acetolyse the bromide **(191)** led to isomerization into **(192)** but solvolysis in aqueous acetone occurred 9×10^4 times faster than that of the compound without the olefinic double bond. It was suggested that this reaction proceeds via the dibenzohomotropylium ion **(193)**.[129]

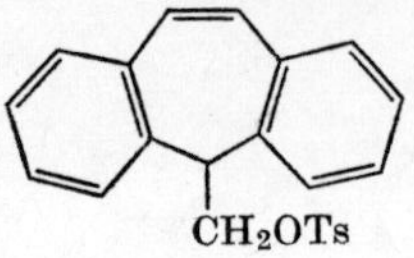

CH₂OTs

(190)

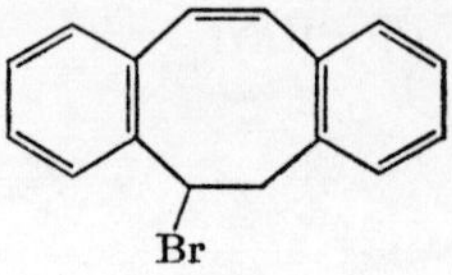

(191)

Br

CH₂Br

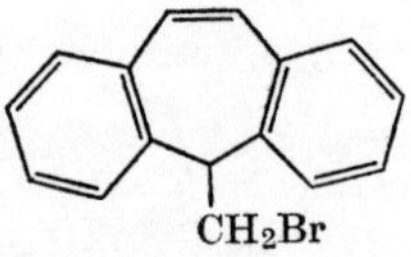

(192)

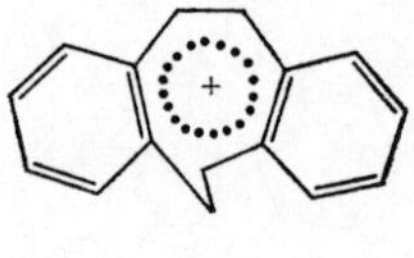

(193)

Acetolysis of bicyclo[2.2.0]hex-2-en-5-yl *p*-nitrobenzenesulphonate yields mainly *exo*-bicyclo[2.1.1]hex-2-en-5-yl acetate (equation 9).[130]

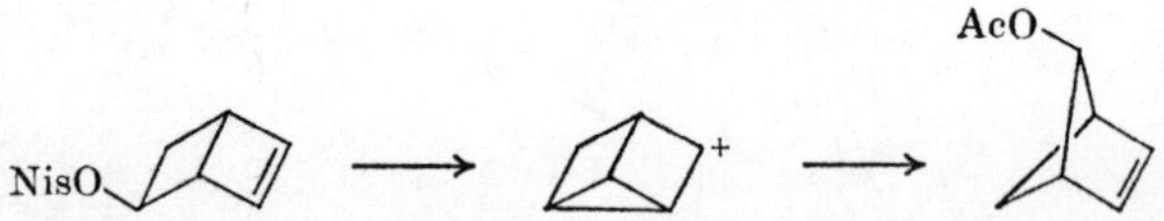

NisO   →   +   →   AcO   ...(9)

Compound **(194)** on treatment with potassium t-butoxide in t-butanol yields **(195)** which was thought to be formed as shown in equation (10). The absence of participation to yield the cyclopropylmethyl cation **(196)** was thought to arise from the difficulty in placing a positive charge on the carbon atom adjacent to a carbonyl group.[131] The analogous monocyclic compound **(197)** reacts with fragmentation. This difference in behaviour between the monocyclic and bicyclic compounds was thought to result from the toluene-*p*-sulphonyloxy group of the bicyclic compound being axial and well situated for participation by the double bond.[132]

129 E. Ciorănescu, A. Bucur, F. Badea, M. Rentzea, and C. D. Nenitzescu, *Tetrahedron Letters*, **1969**, 1867.

130 S. Masamune, E. N. Cain, R. Vukov, S. Takada, and N. Nakatsuka, *Chem. Comm.*, **1969**, 243.

131 P. C. Mukarji and A. N. Ganguly, *Tetrahedron*, **25**, 5281 (1969).

132 P. C. Mukharji, P. K. Sen Gupta, and G. S. Sambamurti, *Tetrahedron* **25**, 5287 (1969).

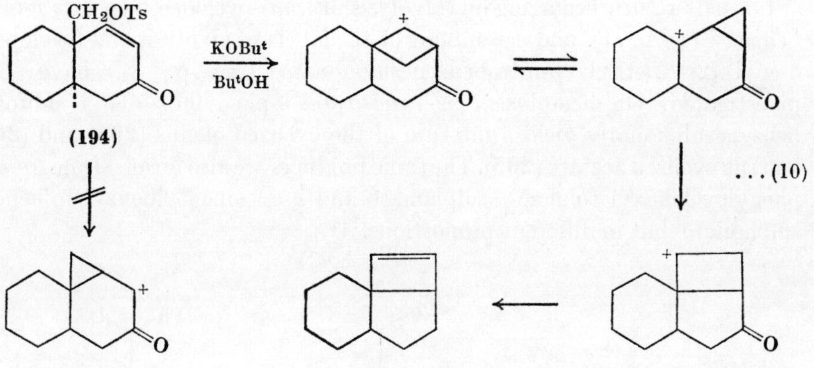

KOBu^t
Bu^{t}OH

... (10)

(194)

(196)   (195)

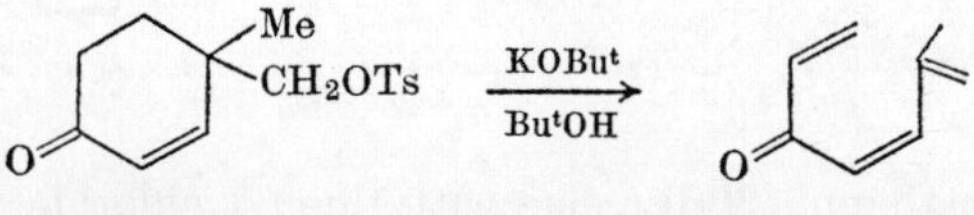

Me
CH$_2$OTs
KOBu^t
Bu^{t}OH

(197)

An example of participation by a double bond to form a four-membered ring has been reported by Berson, Donald, and Libbey[133] who showed that compound (198) yielded five unknown products in a total yield of 45% as well as those which result from ring-expansion (see p. 11). One of these was identified as the acetate of alcohol (199), and it was shown that the toluene-*p*-sulphonate of this yielded the same five products as (198). The anchimeric assistance appears to be small since compound (198) reacts only about twice as fast as (200). There is no participation by the double bond on acetolysis of compound (201)[134] or compound (202) (see p. 14).

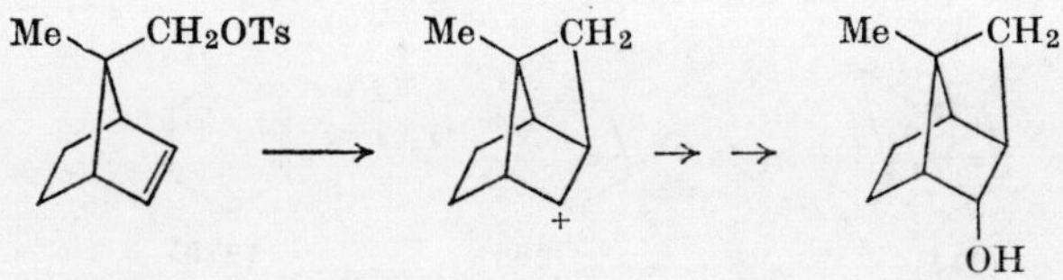

Me   CH$_2$OTs   Me   CH$_2$ Me   CH$_2$

OH

(198)   (199)

Me   CH$_2$OTs   H   CH$_2$OBs

OTs

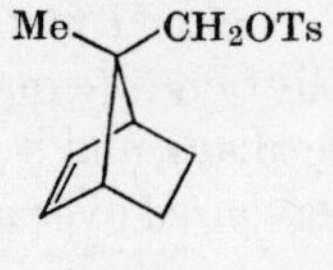

(200) 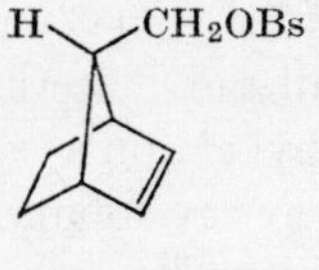   (201) 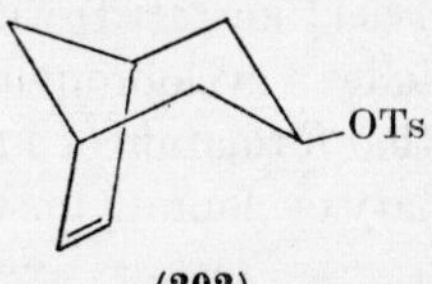   (202)

[133] J. A. Berson, D. S. Donald, and W. J. Libbey, *J. Am. Chem. Soc.*, **91**, 5580 (1969).
[134] See *Org. Reaction Mech.*, **1966**, 30.

Ion-pair return occurring on solvolysis of *trans*-cyclodec-5-en-1-yl *p*-nitro-benzoate (see p. 71), and scrambling of the label on solvolysis of 2-(cyclopent-3-enyl)-[2-^{14}C]ethyl *p*-nitrobenzenesulphonate (see p. 8), have been investigated. On acetolysis, *cis*- and *trans*-6-phenylhex-5-enyl *p*-bromo-benzenesulphonates yield a mixture of the cyclized olefins (203) and (204), and the cyclized acetate (205). The same products are also formed from *trans*-2-phenylcyclohexyl toluene-*p*-sulphonate and α-cyclopentylbenzyl toluene-*p*-sulphonate, but in different proportions.[135]

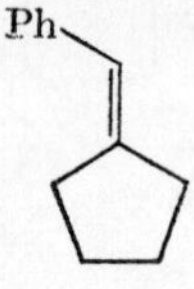

(203) 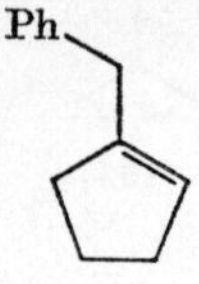   (204) 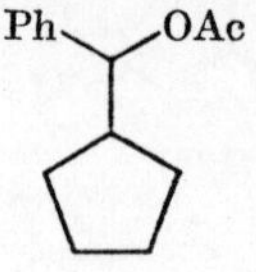   (205)

Cyclopent-2-enylmethanol (206) reacts with thionyl chloride to yield the products shown. The formation of twice as much *cis*-isomer (207) as of *trans*-isomer (208) was attributed to the intervention of an ion-pair. *cis*-Bicyclo-[3.1.0]hexanol (211) also yields (207) and (208) in the ratio 89:11, but no (209) and (210).[136]

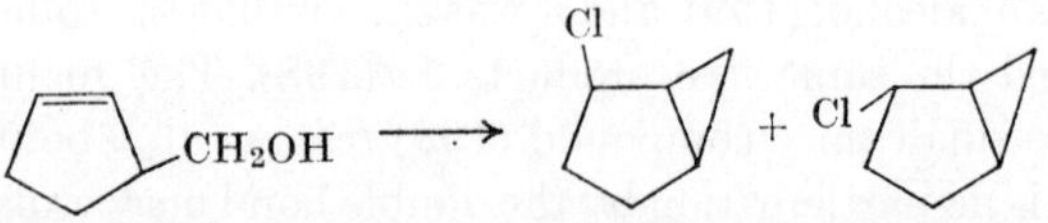

(206)   (207)   (208)

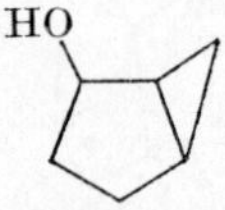

(211) 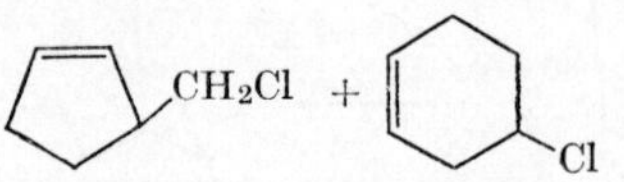   (209)   (210)

Deamination of 3-methyl[1,1-^{2}H$_2$]but-3-enylamine yields 1-methylcyclo-butanol (yield unstated) with all the deuterium at position 3 (equation 11). This excludes a cyclopropylmethyl cation as an intermediate, and it therefore appears that formation of a tertiary cyclobutyl ion is favoured over formation of a primary cyclopropylmethyl ion.[137]

[135] S. A. Roman and W. D. Closson, *J. Am. Chem. Soc.*, **91**, 1701 (1969).
[136] P. K. Freeman, F. A. Raymond, and J. N. Blazevich, *J. Org. Chem.*, **34**, 1175 (1969).
[137] W. B. Kover and J. D. Roberts, *J. Am. Chem. Soc.*, **91**, 3687 (1969).

$$H_2C = \overset{\overset{\displaystyle Me}{|}}{C} \overset{\displaystyle }{\underset{\displaystyle H_2}{C}} - CD_2NH_2 \longrightarrow \overset{\overset{\displaystyle Me}{|}}{\underset{\displaystyle CD_2-CH_2}{CH_2 - C^+}} \longrightarrow \overset{\overset{\displaystyle Me}{|}}{\underset{\displaystyle CD_2-CH_2}{CH_2 - C - OH}}$$

$$\dots (11)$$

$$\overset{\displaystyle CH_2}{\underset{\displaystyle CD_2}{\diagdown}} \overset{\displaystyle Me}{\underset{\displaystyle CH_2^+}{C}} \longrightarrow \overset{\overset{\displaystyle Me}{|}}{\underset{\displaystyle CH_2-CH_2}{CD_2 - C - OH}}$$

Several examples of participation by double bonds in biosynthetic-like cyclizations have been reported.[138-143]

Details have been published of Jacobs and Macomber's[144] and Bly and Koock's[145] extensive investigations of homoallenylic participation. Penta-3,4-dienyl toluene-*p*-sulphonate is acetolysed about twice as fast as pentyl toluene-*p*-sulphonate, and the effect of alkyl substitution is to increase $k_{\varDelta}$ by the factors shown with (**212**).[146] The relatively small effects of substituents at positions 3 and 5 suggest that most of the positive charge is delocalized to position 4 in the transition state, and that the initial intermediate resembles a cyclopropylvinyl cation. On the other hand Jacobs and Macomber prefer to formulate it as a bicyclobutonium ion, and some support for this has been obtained by Santelli and Bertrand (see below).

The reaction products are complex mixtures of cyclized and uncyclized materials, and a typical example from a compound without *gem*-dimethyl substituents at position 2 is given in equation (12). When there are *gem*-dimethyl substituents at positions 1 and 3, rearranged acyclic compounds are formed (e.g. equation 13). If there is also a substituent at C-1 a mixture of rearranged and unrearranged acyclic compounds are formed, but only if there is also a substituent at C-3 are cyclobutyl compounds formed. No cyclopropyl derivatives are formed from arenesulphonates with *gem*-dimethyl substituents. Presumably the *gem*-dimethyl cyclopropylvinyl cation or its non-classical analogue undergoes ring-opening too easily, possibly via a tertiary cation.[147]

138 G. Stork and M. Marx, *J. Am. Chem. Soc.*, **91**, 2371 (1969); G. Stork and P. A. Grieco, *ibid.*, p. 2407.
139 E. E. van Tamelen and J. P. McCormick, *J. Am. Chem. Soc.*, **91**, 1847, (1969); K. B. Sharpless and E. E. van Tamelen, *ibid.*, p. 1848.
140 J. L. Fourrey, J. Polonsky, and E. Wenkert, *Chem. Comm.*, **1969**, 714.
141 R. C. Haley, J. A. Miller, and H. C. S. Wood, *J. Chem. Soc.* (C), **1969**, 264.
142 D. J. Goldsmith and C. F. Phillips, *J. Am. Chem. Soc.*, **91**, 5862 (1969).
143 G. P. Moss and S. A. Nicolaidis, *Chem. Comm.*, **1969**, 1072.
144 T. L. Jacobs and R. S. Macomber, *J. Am. Chem. Soc.*, **91**, 4824 (1969); see *Org. Reaction Mech.*, **1967**, 37.
145 R. S. Bly and S. U. Koock, *J. Am. Chem. Soc.*, **91**, 3292, 3299 (1969).
146 See also M. Santelli and M. Bertrand, *Tetrahedron Letters*, **1969**, 3699.
147 Cf. *Org. Reaction Mech.*, **1968**, 37—38.

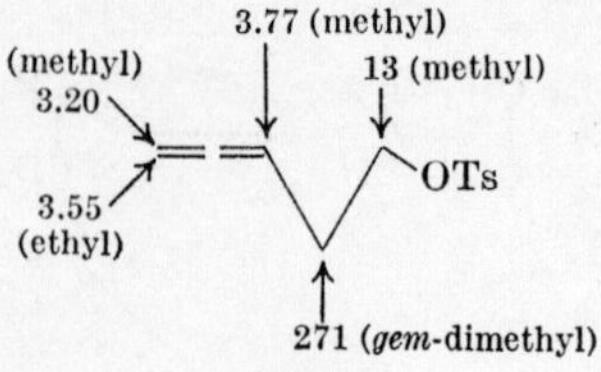

3.77 (methyl)

(methyl)   13 (methyl)
3.20

3.55
(ethyl)

OTs

271 (*gem*-dimethyl)

(212)

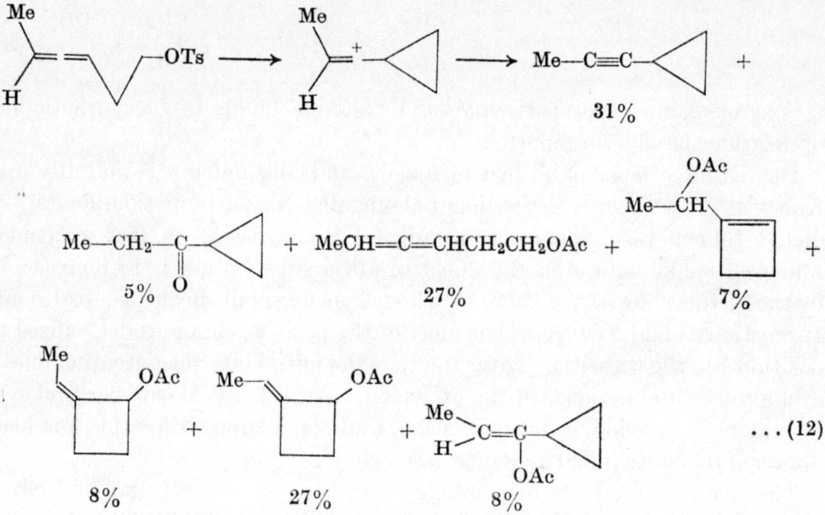

$$
\text{(212)} \longrightarrow \text{Me—C}\!\equiv\!\text{C}\!-\!\triangle \quad + \quad 31\%
$$

Me—CH$_2$—C($\triangle$)=O   5%   +   MeCH=C=CHCH$_2$CH$_2$OAc   27%   +   (OAc on cyclobutene)   7%   +

Me (OAc cyclobutane) 8%   +   Me— (OAc cyclobutane) 27%   + $\underset{H}{\overset{Me}{>}}$C=C$\overset{\triangle}{\underset{OAc}{}}$ 8%   ...(12)

$$
\underset{\underset{\text{Me Me}}{}}{\overset{Me}{\underset{H}{>}}=\!\!=\!-\!\text{OBs}} \longrightarrow \overset{Me}{\underset{H}{>}}\overset{+}{=}\!\!\triangle\!-\!\underset{Me}{\text{Me}} \longrightarrow
$$

Me $\overset{H}{>}$ /\=/\= Me   21%   +   Me /\=/\ (Me, Me)   6.1%   +   Me /\=/\ (Me, Me, OAc)   73%   ...(13)

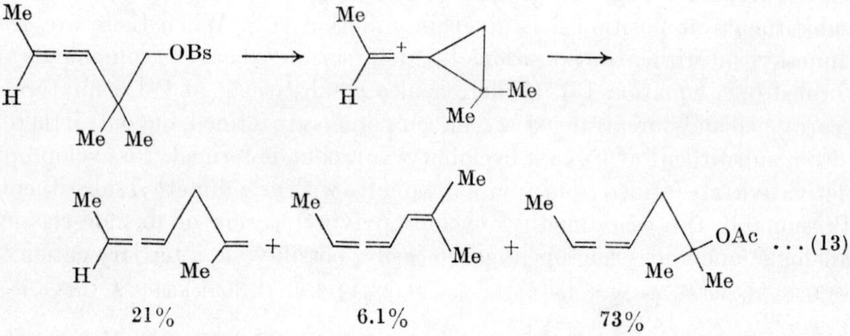

An argument in favour of formulating the initial intermediate as non-classical has been advanced by Santelli and Bertrand who showed that compounds of type **(213)** and **(216)** did not yield identical mixtures of the two possible butane derivatives **(214)** and **(215)** as would be expected if they were reached via a cyclopropylvinyl cation which should be the same. It was therefore suggested that reaction proceeded via two non-classical ions **(217)**

and (218) which were interconverted and converted into the allylic cations (219) and (220) at comparable rates.[148]

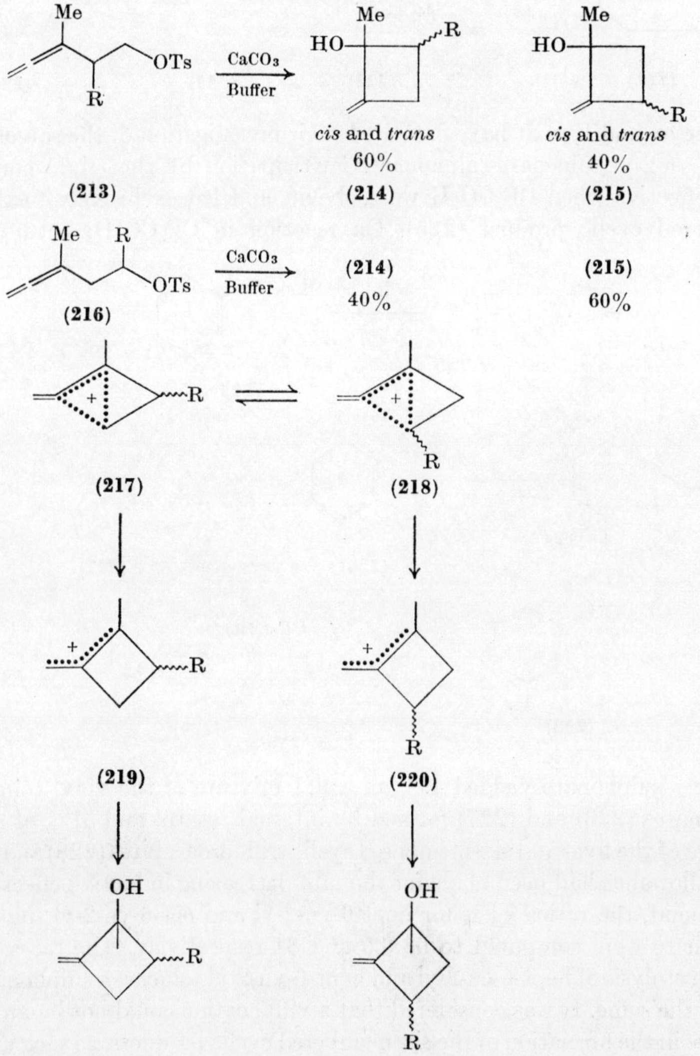

Me

═⟨⟩—OTs $\xrightarrow{\text{CaCO}_3}$ HO—⟨Me⟩—R   HO—⟨Me⟩
     | Buffer
     R

(213)

cis and trans   cis and trans
60%   40%

(214)   (215)

Me R

═⟨⟩—OTs $\xrightarrow{\text{CaCO}_3}$ (214)   (215)
           Buffer
(216)   40%   60%

(217) ⇌ (218)

(219)   (220)

OH   OH

(219)   (220)

Cyclonona-1,2-dien-5-yl toluene-*p*-sulphonate (221) is hydrolysed with participation by the allenic double bonds to yield the ketones (222) and (223). Compound (224) reacts without participation by the double bonds.[149]

[148] M. Santelli and M. Bertrand, *Tetrahedron Letters*, **1969**, 2511, 2515.
[149] C. Santelli-Rouvier, P. Archier, and M. Bertrand, *Compt. Rend., C*, **269**, 252 (1969).

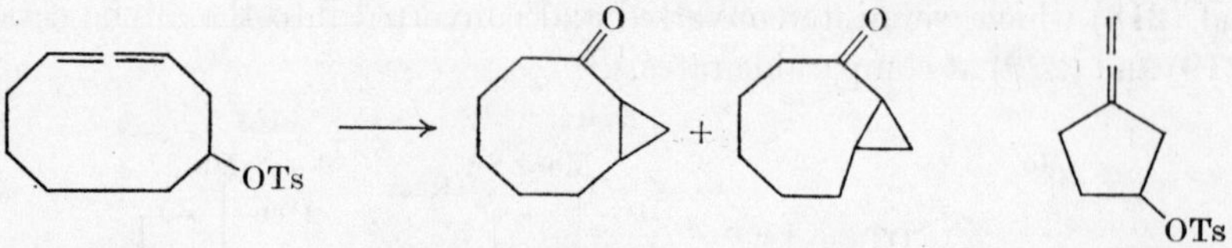

(221)   (222)   (223)   (224)

Peterson and Kamat have extended their investigation of the solvolysis of hept-6-yn-2-yl toluene-*p*-sulphonate. Participation by the triple bond was most effective when CF_3CO_2H was solvent, and led exclusively to the six-membered cyclic product (225). On reaction in CF_3CO_2H, oct-6-yn-2-yl

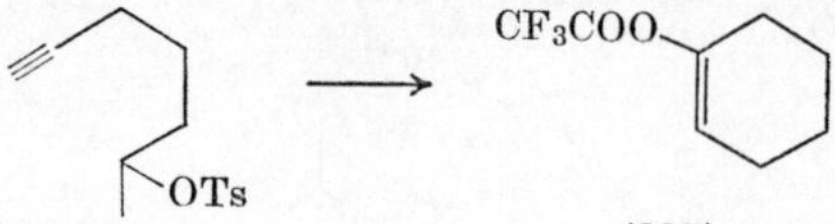

(225)

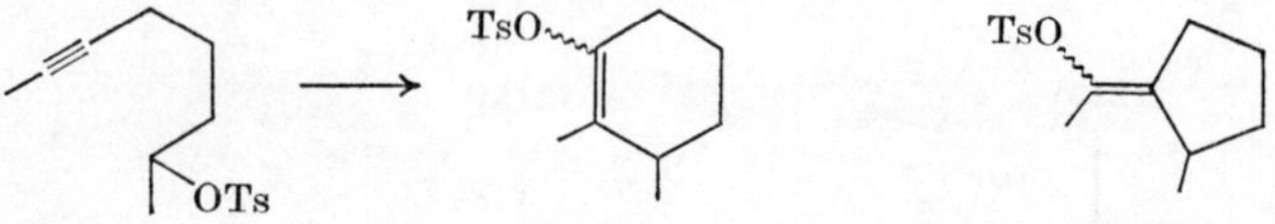

(226)   (227)

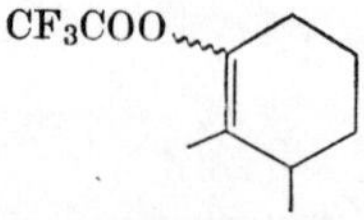

(228) 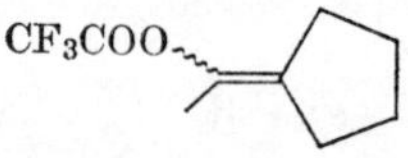   (229)

toluene-*p*-sulphonate yielded 49% of a 9:1 mixture of the vinyl toluene-*p*-sulphonates (226) and (227) formed by internal return, and 51% of an 8:2 mixture of the five- and six-membered cyclic trifluoroacetates (228) and (229). After allowance had been made for the rate-decreasing inductive effect of the triple bond, the ratios k_Δ/k_s for hept-6-yn-2-yl and oct-6-yn-2-yl toluene-*p*-sulphonate were computed to be 6.5 and 84 respectively. The rates of trifluoroacetolysis of hept-6-en-2-yl and hept-6-yn-2-yl toluene-*p*-sulphonate are almost the same. It was considered that a vinyl cation could not be an intermediate in the formation of the six-membered cyclic products, as such a "bent structure" would have too high an energy.[150]

Details of Wilson's investigation of the formolysis of 2,2-dimethylpent-3-yn-1-yl toluene-*p*-sulphonate have been published.[151]

[150] P. E. Peterson and R. J. Kamat, *J. Am. Chem. Soc.*, **91**, 4521 (1969); see *Org. Reaction Mech.*, **1966**, 31.

[151] J. W. Wilson, *J. Am. Chem. Soc.*, **91**, 3238 (1969); see *Org. Reaction Mech.*, **1968**, 37—38.

Molecular-orbital calculations have been reported for the three- and four-membered cyclic cations which result from participation in the reactions of homopropargylic compounds.[152]

Reactions of Small-ring Compounds[153]

Cyclopropylmethyl Derivatives

The solvolyses of compounds (230a) and (230b) are 600 and 150 times slower than those of compounds (232a) and (232b) respectively. Although they react via the cyclopropylmethyl cation (231) this has the perpendicular conformation (233) rather than the favoured[154] bisected conformation (234). There is therefore no conjugative interaction between the cyclopropyl ring and the developing p-orbital of the incipient carbonium ion in the transition state. The slow reaction presumably results from the electron-withdrawing inductive effect of the cyclopropyl ring.[155, 156] The allylic compound (235), which has an unfavourable geometry for overlap between the double bond and the developing cationic centre in the transition state, reacts more than 10^4 times more slowly than (232b), which indicates the strong electron-withdrawing inductive effect of a double bond.[156]

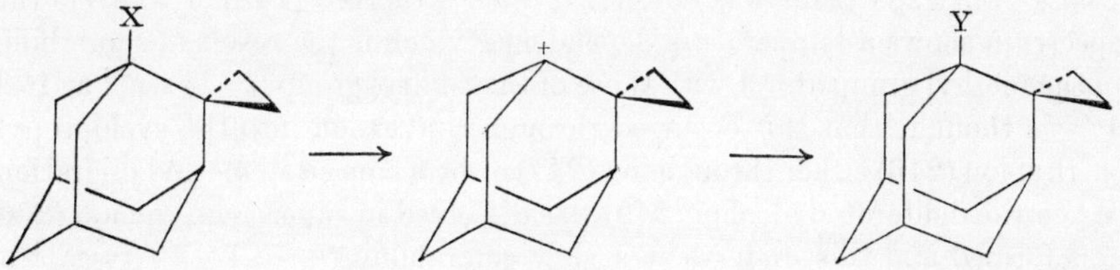

(230a) X = Cl
(230b) X = OTs

(231)

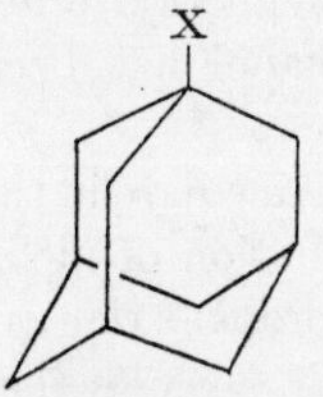

(232a) X = Cl
(232b) X = OTs

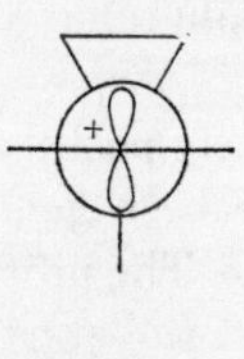

(233)

(234)

[152] H. Fischer, K. Hummel, and M. Hanack, *Tetrahedron Letters*, **1969**, 2169.
[153] Ring-expansion reactions via carbonium ions have been reviewed: C. D. Gutsche and D. Redmore, *Carbocyclic Ring Expansion Reactions*, Academic Press, New York, 1969.
[154] See *Org. Reaction Mech.*, **1965**, 43.
[155] P. von R. Schleyer and V. Buss, *J. Am. Chem. Soc.*, **91**, 5880 (1969).
[156] J. C. Martin and B. R. Ree, *J. Am. Chem. Soc.*, **91**, 5882 (1969).

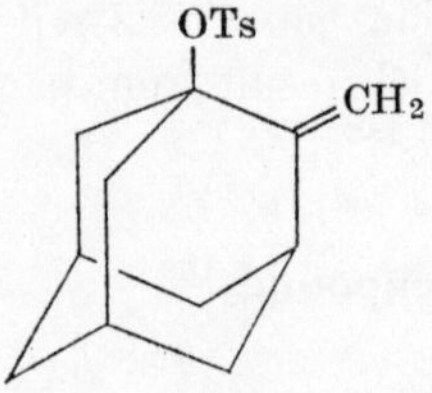

(235)

The estimated rate of solvolysis of hexamethylcyclopropylmethyl *p*-nitro-benzoate **(236a)** in 80% aqueous acetone is 300 times greater than that for cyclopropyldimethylmethyl *p*-nitrobenzoate.[157] The products of methanolysis of **(236b)** are 41% of the cyclopropylmethyl ether **(239)**, 55% of the homo-allylic methyl ether **(240)**, and 4% of diene **(241)**. The analogous homoallylic benzoate **(238)** reacts 10^5 times more slowly than **(236b)** and yields a similar mixture of products which suggests that both compounds react through the cation **(237)**. Solvolysis of the deuterated compound **(242)** led to no scrambling of the label, which indicates that no degenerate rearrangement of the type **(243)**→**(244)** had occurred.[158] Attempts to observe the cation **(246)**≡**(237)** in super-acid media were unsuccessful since the alcohol **(245)** yielded solutions whose NMR spectrum was consistent with structure **(248)**. However, this spectrum shows a temperature dependence which is the result of scrambling of the methyl group at C-1 with those of the t-butyl group ($k = 4 \sec^{-1}$ at 16°). It was thought that this occurred through conversion into the cyclopropyl-methyl ion **(246)** either through ion **(247)** or by a concerted methyl migration. When the deuterated alcohol **(249)** was dissolved in super-acid, the ion **(250)** was formed and this underwent a slow scrambling ($k = 3.1 \times 10^{-3} \sec^{-1}$ at 16°) of the label. This presumably involves the degenerate rearrangement of the type **(243)**→**(244)**.[159]

The distribution of the deuterium label in the cyclopropylmethyl ether obtained on treatment of bicyclobutane with acidic methanol has been reported.[160, 158]

The ion **(252)** which was previously postulated to be an intermediate in the solvolysis of compound **(251)** has been generated by protonation of olefin **(253)** in FSO_3H at low temperature. The protons of the cyclopropane ring are deshielded 2—3 ppm compared to the starting olefin, which suggests that much of the charge is delocalized. At −33.5° the ion **(252)** is slowly converted into the pentamethylbenzenonium ion **(254)**, and irradiation of this regenerates **(252)**. Orbital symmetry considerations indicate that the interconversion of

157 Cf. *Org. Reaction Mech.*, **1966**, 31.
158 Cf. *Org. Reaction Mech.*, **1968**, 39—40.
159 C. D. Poulter and S. Winstein, *J. Am. Chem. Soc.*, **91**, 3649, 3650 (1969).
160 W. G. Dauben, J. H. Smith, and J. Saltiel, *J. Org. Chem.*, **34**, 261 (1969).

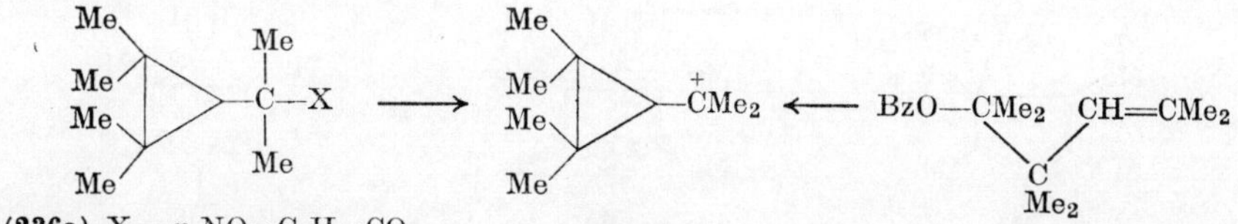

(236a) X = p-NO$_2$·C$_6$H$_4$·CO$_2$
(236b) X = C$_6$H$_5$CO$_2$

(237)

(238)

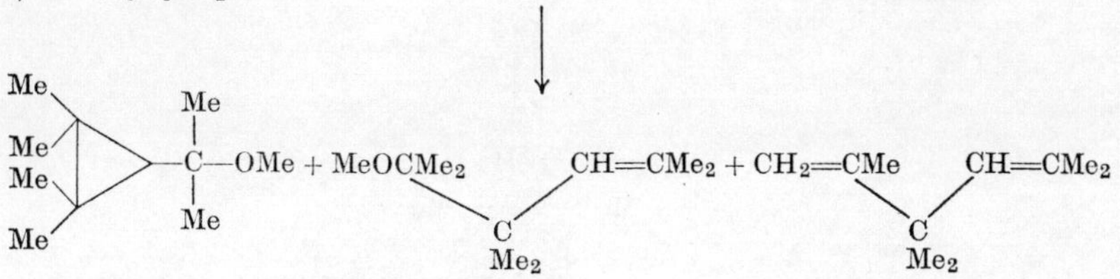

(239)

(240)

(241)

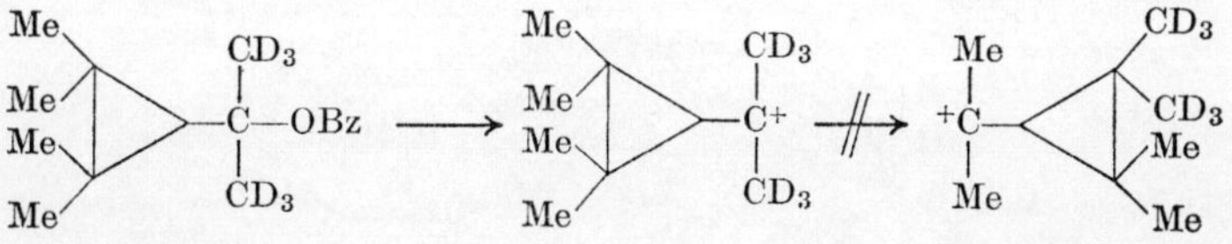

(242)

(243)

(244)

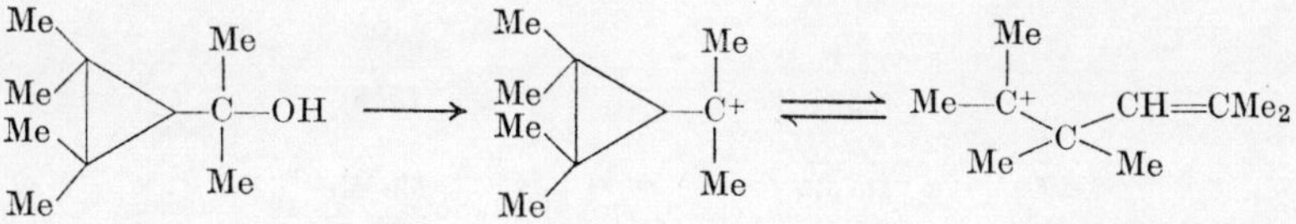

(245)

(246)

(247)

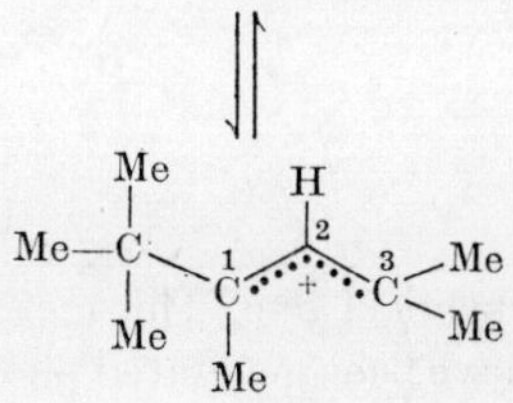

(248)

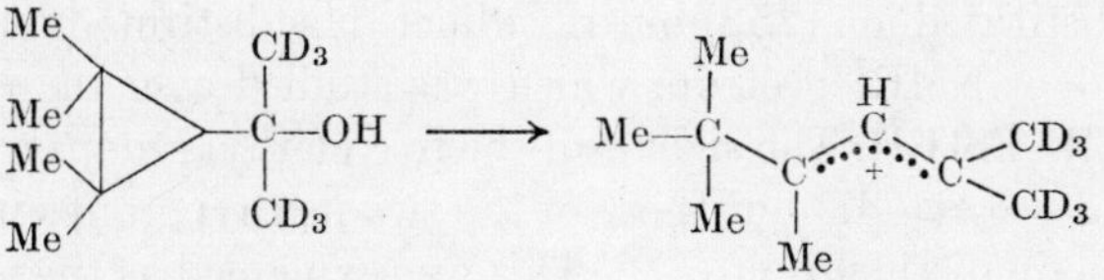

(249)

(250)

(252) and (254) should occur by a conrotatory process when thermal, and disrotatory when photochemical. Since a conrotatory conversion of (252) into (254) is difficult, this explains the slow thermal reaction and the facile photo-

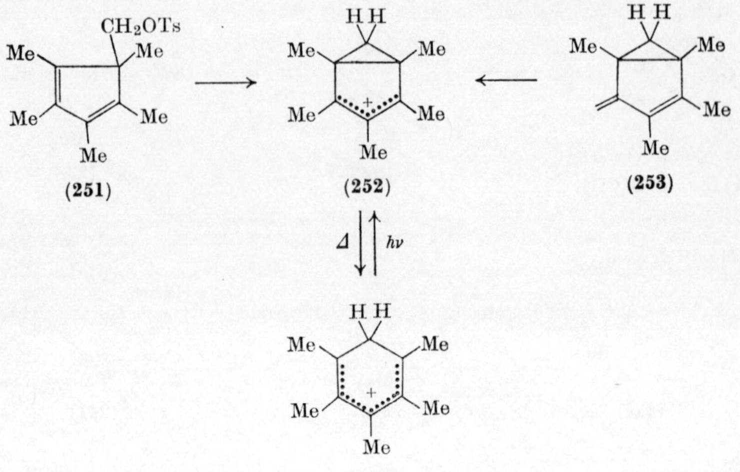

(251)   (252)   (253)

$\Delta \Big\updownarrow h\nu$

(254)

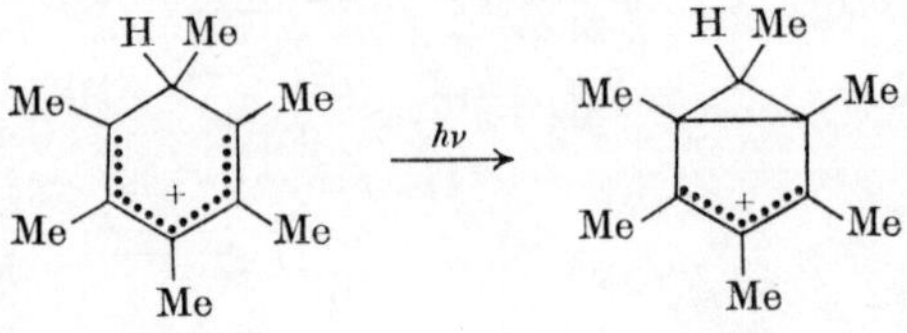

(255)   (256)

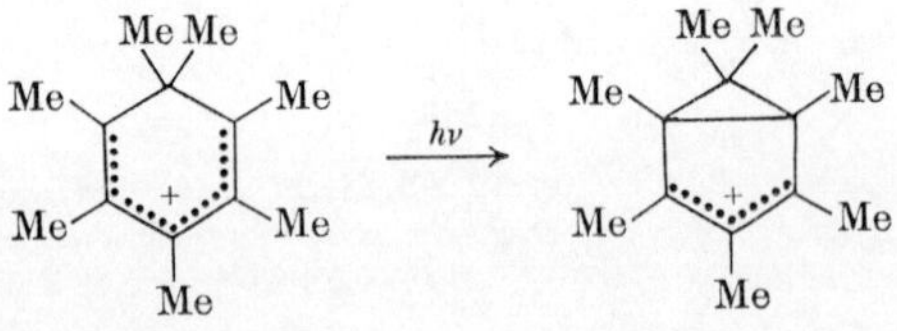

(257)   (258)

chemical one. The methyl- and dimethyl-substituted ions (256) and (258) have been generated photochemically from the hexamethyl- and heptamethyl-benzenonium ions (255) and (257). The monomethyl ion (256) behaves similarly to the unsubstituted ion (252) but the dimethyl-substituted ion has a temperature-dependent NMR spectrum which was studied over the range $-110°$ to $-9°$ in SO_2ClF–FSO_3H. The signals of the five methyl groups on the cyclopentene ring coalesce at $-87°$ but those of the two methyl groups in the cyclopropane ring remain distinct up to $-9°$. This was explained as the result of migration of the cyclopropane carbon C-6 around the cyclopentene ring as shown in Scheme 4. This is a suprafacial 1,4-sigmatropic shift, and if the transition state is regarded as having a radical or cationic centre at C-6 interacting with a butadiene system, the orbital symmetry is as indicated in

(259). Hence C-6 employs both lobes of its antisymmetric p-orbital and undergoes inversion at each shift so that the "inside" and "outside" methyl groups remain "inside" and "outside". It was estimated that exchange of the inside and outside methyl groups must be at least 50,000 times slower than exchange of the cyclopentene methyl groups. The fact that the unsubstituted and monomethyl cations do not undergo this rearrangement suggests that C-6 bears considerable positive charge in the transition state.[161, 162]

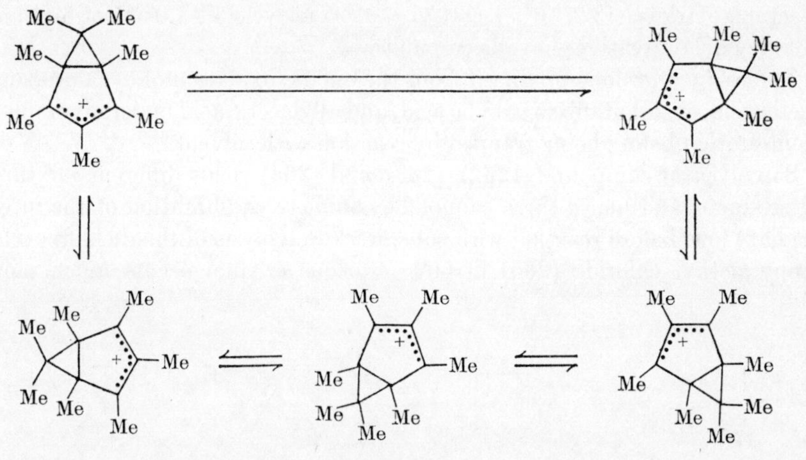

Scheme 4

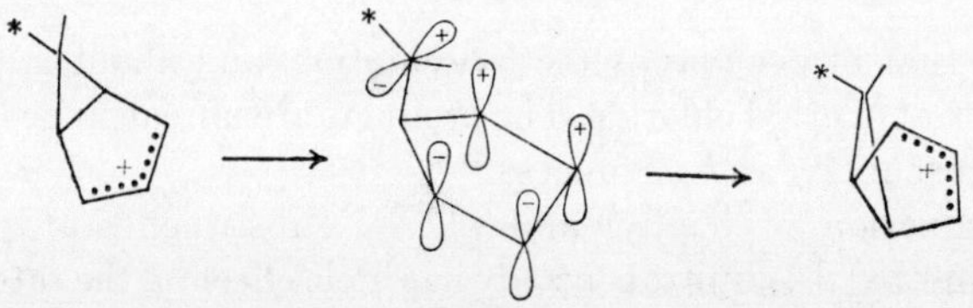

(259)

The analogous rearrangement of the protonated bicyclo[3.1.0]hexenones[163] also occurs without interchange of the inside and outside alkyl groups. This was demonstrated by showing that compounds **(260)** and **(261)** were not interconverted under conditions where deuterium labelling studies showed that migration of the three-membered ring around the perimeter of the five-membered ring was occurring.[164]

[161] R. F. Childs, M. Sakai, S. Winstein, *J. Am. Chem. Soc.*, **91**, 7144 (1968).
[162] R. F. Childs and S. Winstein, *J. Am. Chem. Soc.*, **90**, 7146 (1968).
[163] See *Org. Reaction Mech.*, **1967**, 43—44.
[164] H. Hart, T. R. Rodgers, and J. Griffiths, *J. Am. Chem. Soc.*, **91**, 754 (1969).

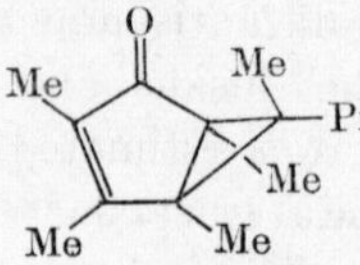

(260) 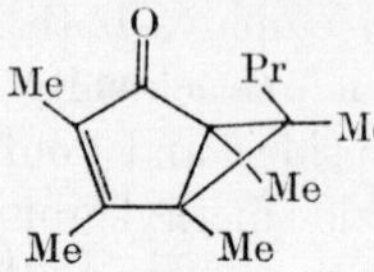   (261)

Details of Berson, Wege, Clarke, and Bergman's investigation of the reactions of tricyclo[3.2.1.0^{2,7}]oct-4-yl, *exo-exo*-tricyclo[3.2.1.0^{2,4}]oct-6-yl, and nortricyclyl derivatives have been published.[165]

The carbonium ions produced from the anodic oxidation of cyclopropane-acetic acid, cyclobutanecarboxylic acid, and allylacetic acid undergo complete equilibration before being trapped by reaction with solvent.[166]

Solvolysis of compounds (262), (263), and (264) yields different mixtures of products, and hence there cannot be complete equilibration of the inter-mediate ions before reaction with solvent.[167] Solvolysis of the dichlorocyclo-propylmethyl chloride (265) in 50% aqueous ethanol occurs much more

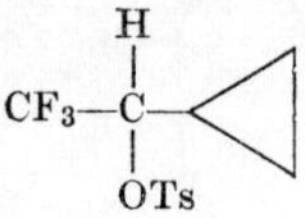

(262) 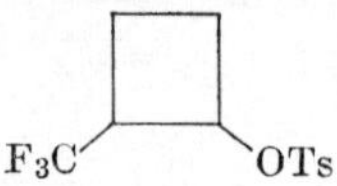   (263) 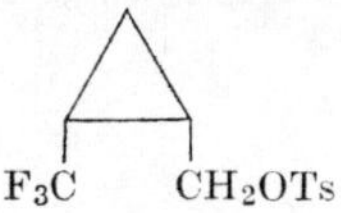   (264) 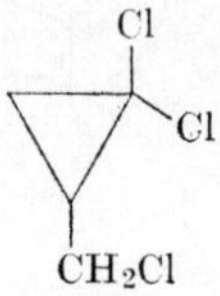   (265)

slowly than that of cyclopropylmethyl chloride, and about five times more slowly than that of ethyl chloride. The products are unrearranged and reaction presumably occurs by an S_N2 process.[168]

The introduction of methyl and phenyl substituents at position 1 of cyclopropylmethyl 2,4-dinitrobenzoate has little effect on the rate of solvolysis but methyl and phenyl substituents at position 1 of cyclobutyl 2,4-dinitro-benzoate have a large rate-enhancing effect.[169]

Solvolysis of the methylenecyclopropylmethyl toluene-*p*-sulphonate (266) yields the products (268), (269), (271), and (272) which are presumably formed via the allylic and homoallylic cations (267) and (270).[170]

[165] J. A. Berson, D. Wege, G. M. Clarke, and R. G. Bergman, *J. Am. Chem. Soc.*, **91**, 5594, 5601 (1969); see *Org. Reaction Mech.*, **1968**, 46—48; G. N. Fickes, *J. Org. Chem.*, **34**, 1513 (1969), and ref. 192.
[166] J. T. Keating and P. S. Skell, *J. Am. Chem. Soc.*, **91**, 695 (1969).
[167] M. Hanack and H. Meyer, *Ann. Chem.*, **720**, 81 (1968).
[168] G. C. Robinson, *J. Org. Chem.*, **34**, 2517 (1969).
[169] D. D. Roberts, *J. Org. Chem.*, **34**, 285 (1969).
[170] H. Monti and M. Bertrand, *Compt. Rend.*, *C*, **269**, 612 (1969).

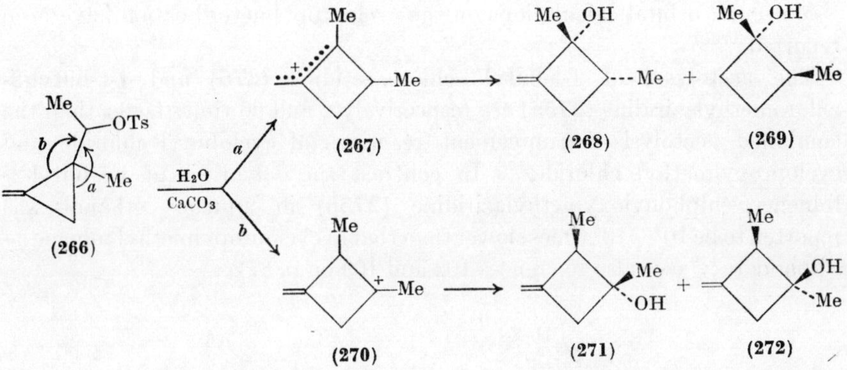

(266) (267) (268) (269) (270) (271) (272)

The silver-ion assisted ethanolysis of the methylenedimethylcyclopropyl chloride (273) yields (274) as the primary product. A mechanism (equation 14) analogous to oxymercuration was proposed.[171]

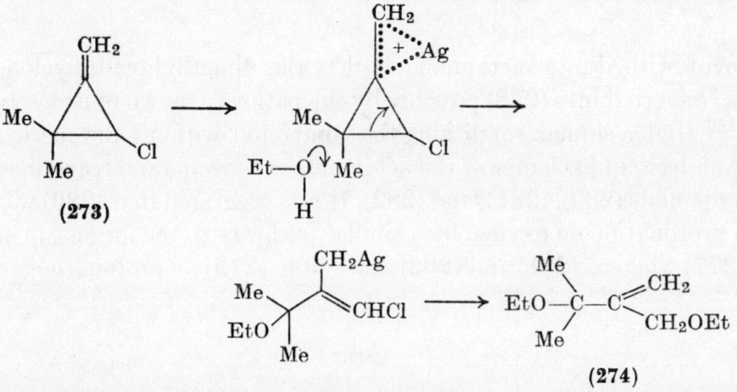

(273)

$$... (14)$$

(274)

The solvolyses of cyclopropylmethyl and cyclobutyl methanesulphonate in aqueous diglyme in the presence of sodium borohydride,[172] and the effect of pressure on the hydrolysis of cyclopropylmethyl and cyclobutyl chloride,[173] have been investigated.

The bisected conformation of the cyclopropyl boranes does not appear to be especially stable, unlike that of the isoelectronic cyclopropylmethyl cation.[174, 154]

Hydride transfers by silanes to cyclopropylmethyl cations have been investigated.[175]

[171] R. M. Babb and P. D. Gardner, *Tetrahedron Letters*, **1968**, 6197.
[172] Z. Majerski, S. Borčić, and D. E. Sunko, *Tetrahedron*, **25**, 301 (1969).
[173] S. Hariya and S. Terasawa, *J. Chem. Soc. Japan*, **90**, 765 (1969).
[174] A. H. Cowley and T. A. Furtsch, *J. Am. Chem. Soc.*, **91**, 39, (1969).
[175] F. A. Carey and H. S. Tremper, *J. Am. Chem. Soc.*, **91**, 2967 (1969).

Molecular-orbital calculations on the cyclopropylmethyl cation have been reported.[176, 177]

The acetolyses of 1-t-butyl-3-chloroazetidine **(276)** and 1-t-butyl-2-chloromethylaziridine **(275a)** are respectively 2 and 80 times faster than the combined acetolysis–rearrangement reactions of cyclobutyl chloride and cyclopropylmethyl chloride.[178] In contrast the solvolysis of 1-t-butyl-2-toluene-*p*-sulphonyloxymethylaziridine **(275b)** in aqueous ethanol was reported to be 10^4—10^5 times slower than that of cyclopropylmethyl toluene-*p*-sulphonate[179] (see also references 102 and 103 on p. 87).

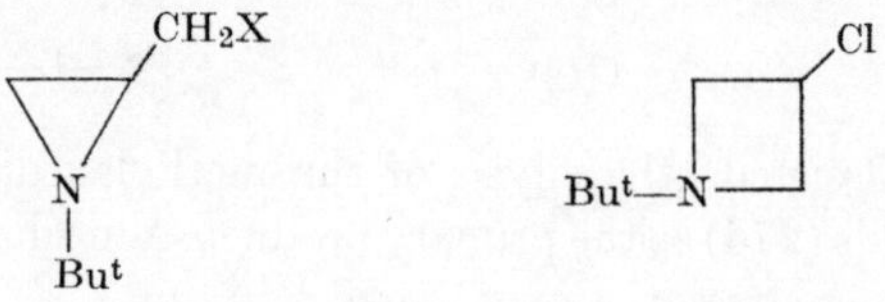

(275a) X = Cl   (276)
(275b) X = OTs

On treatment with dilute methanolic H_2SO_4 the dimethylquadricyclone ketal **(277)** is converted into **(278)** possibly by the pathway shown or a closely related one.[180] Under similar conditions the compound without the methyl groups **(279)** underwent exchange of the ketal methoxy groups and rearrangement into compounds **(280)**, **(281)**, and **(282)**. It was suggested that **(280)** was formed from protonation on oxygen by a similar pathway to the formation of **(278)** from **(277)** whereas **(281)** and **(282)** arose from **(279)** by protonation on carbon.[181]

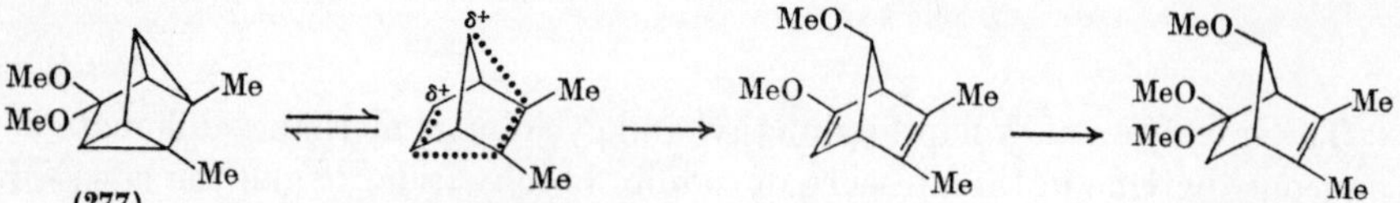

(277)   (278)

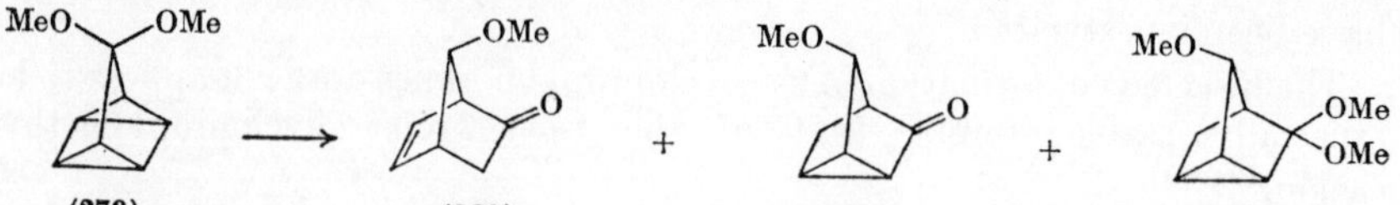

(279)   (280)   +   (281)   +   (282)

176 H. S. Tremper and D. D. Shillady, *J. Am. Chem. Soc.*, **91**, 6341 (1969).
177 C. Trindle and O. Sinanoglu, *J. Am. Chem. Soc.*, **91**, 4054 (1969).
178 V. R. Gaertner, *Tetrahedron Letters*, **1968**, 5919.
179 J. A. Deyrup and C. L. Moyer, *Tetrahedron Letters*, **1968**, 6179.
180 P. G. Gassman, D. H. Aue, and D. S. Patton, *J. Am. Chem. Soc.*, **90**, 7271 (1968).
181 P. G. Gassman and D. S. Patton, *J. Am. Chem. Soc.*, **90**, 7276 (1968).

Other studies on cyclopropylmethyl cations are described in references 182—184.

There have been several discussions of conjugation by a cyclopropyl ring.[185-189]

Participation by More Remote Cyclopropyl Rings

Details of Haywood-Farmer and Pincock's investigation of the solvolysis of compound (283), which occurs 10^{12} times faster than that estimated for 7-norbornyl *p*-nitrobenzoate, have been reported. The acetolysis of the *exo-syn-p*-bromobenzenesulphonate (284) also appears to be anchimerically assisted as this occurs 10^4 times faster than acetolysis of 7-norbornyl toluene-*p*-sulphonate. Possibly there is concerted formation of the cyclopropylmethyl cation (285) which would have the correct conformation for overlap between the developing *p*-orbital and the cyclopropyl ring. The *endo-syn*-isomer (286), which reacts only 10 times faster than 7-norbornyl *p*-bromobenzenesulphonate, would yield the same ion but in a conformation (287) unfavourable for overlap.[190]

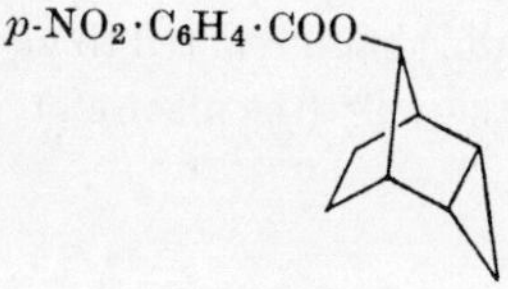

(283) 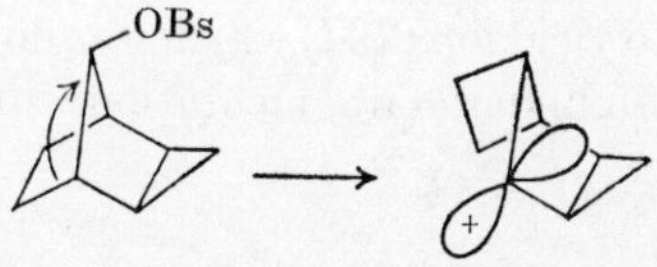　　(284)　　(285)

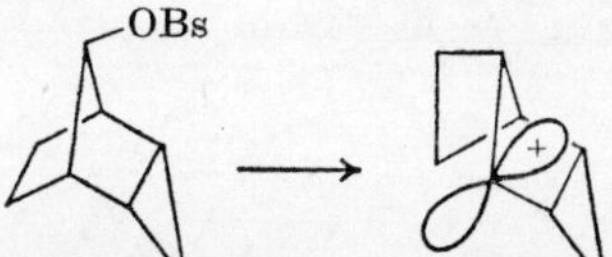

(286)　　(287)

[182] B. Fraser-Reid and B. Radatus, *Can. J. Chem.*, **47**, 4095 (1969).
[183] G. Just, C. Simonovitch, F. H. Lincoln, W. P. Schneider, U. Axen, G. B. Spero, and J. E. Pike, *J. Am. Chem. Soc.*, **91**, 5364 (1969).
[184] W. Cocker, D. P. Hanna, and P. V. R. Shannon, *J. Chem. Soc.* (C), **1969**, 1302.
[185] R. G. Pews and N. D. Ojha, *J. Am. Chem. Soc.*, **91**, 5769 (1969).
[186] C. Agami and J. L. Pierre, *Bull. Soc. Chim. France*, **1969**, 1963.
[187] S. A. Monti, D. J. Bucheck, and J. C. Shepard, *J. Org. Chem.*, **34**, 3080 (1969).
[188] J. M. Stewart and D. R. Olsen, *J. Org. Chem.*, **33**, 4534 (1968); J. M. Stewart and G. K. Pagenkopf, *ibid.*, **34**, 7 (1969).
[189] R. C. Hahn, P. H. Howard, S. M. Kong, G. A. Lorenzo, and N. L. Miller, *J. Am. Chem. Soc.*, **91**, 3558 (1968).
[190] J. S. Haywood-Farmer and R. E. Pincock, *J. Am. Chem. Soc.*, **91**, 3020 (1969); see *Org. Reaction Mech.*, **1967**, 39; cf. refs. 155 and 156.

The relative rates of solvolysis of compounds (288) and (289) are 1.25 in acetic acid at 99.8°, 1.18 in 70% aqueous dioxan at 70°, and 3.01 in trifluoroethanol at 75°. It was suggested that the *syn*-compound (288) reacted with participation in trifluoroethanol (Δ_5 participation). The product was reported to be a complex mixture.[191]

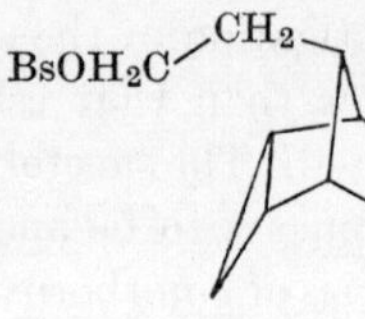

(288) 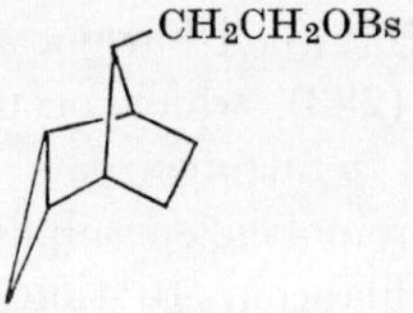   (289)

On treatment with BF_3 the *exo-exo*-epoxide (290) is converted cleanly into aldehyde (292) but under the same conditions the *exo-endo*-epoxide (293) reacts more slowly and yields much polymeric material as well as a 3:1 mixture of fluorohydrin (295) and aldehyde (292). It was suggested that the *exo-exo*-epoxide reacts via the delocalized ion (291) which has the correct geometry to fragment to form a *cis*-double bond. However, the *exo-endo*-epoxide was thought to yield ion (294) which was not delocalized and reacted to yield (295) rather than undergo fragmentation which was slow[192] (see also reference 165, p. 54).

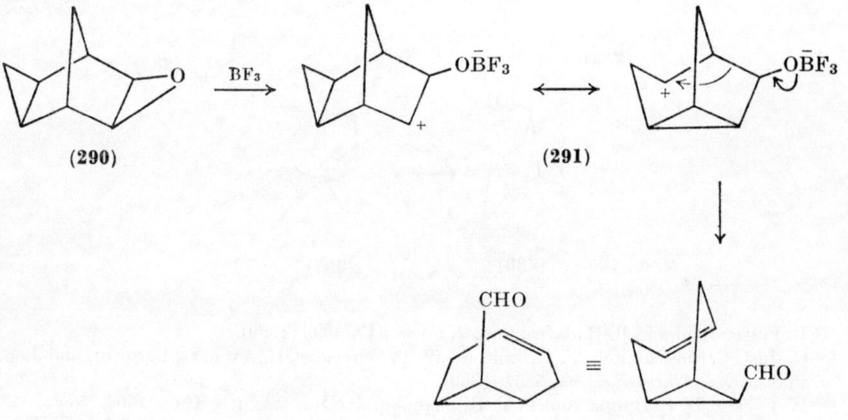

(290)   (291)

(292)

An electron diffraction investigation of the structure of 4-chloronortricyclene has been reported.[193]

[191] R. Muneyuki, T. Yano, and H. Tanida, *J. Am. Chem. Soc.*, **91**, 2408 (1969).
[192] G. D. Sargent, M. J. Harrison, and G. Khoury, *J. Am. Chem. Soc.*, **91**, 4937 (1969); B. C. Henshaw, D. W. Rome, and B. L. Johnson, *Tetrahedron Letters*, **1968**, 6049.
[193] J. F. Chiang, C. F. Wilcox, and S. H. Bauer, *Tetrahedron*, **25**, 369 (1969).

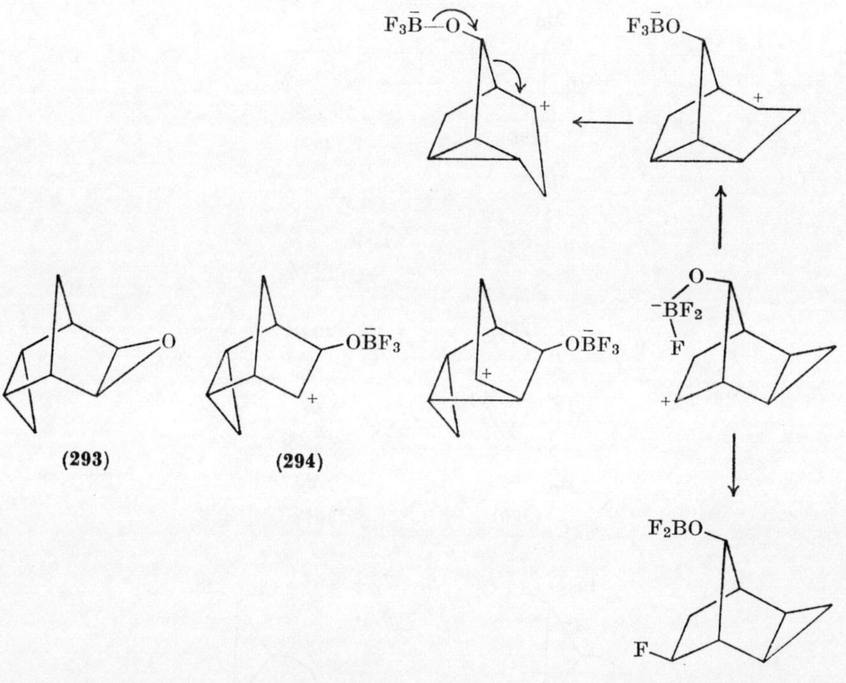

(293)   (294)   (295)

Ring-opening Reactions of Cyclopropyl Derivatives[194] and of Chloro-aziridines

The cyclopropyl chlorides (296), (298), and (300) yield the expected allylic ions (297), (299), and (301) when they are dissolved in SbF_5–SO_2ClF at $-100°$. At higher temperatures these are interconverted.[195]

It was previously suggested that the ring-opening of *exo*-bicyclo[*n*.1.0]alkyl toluene-*p*-sulphonates proceeded by a disrotatory process to form a partially ring-opened cation when $n = 5$ or 6.[196] Support for this view has come from a study of the effect of α-aryl substituents on the acetolysis of bicyclo[*n*.1.0] chlorides. Thus an α-phenyl substituent increases the rate of reaction of the *exo*-compounds (302) by a factor 10^6—10^8 when $n = 3$ or 4 but only 143-fold when $n = 5$. The effect of a phenyl substituent on the rate of reaction of the *endo*-compounds (303) ($n = 4$ or 5) is also small ($<$ca. 100).[197]

[194] R. Bartlet and Y. Vo-Quang, *Bull. Soc. Chim. France*, **1969**, 3729.
[195] P. von R. Schleyer, T. M. Sun, M. Saunders, and J. C. Rosenfeld, *J. Am. Chem. Soc.*, **91**, 5174 (1969).
[196] See *Org. Reaction Mech.*, **1967**, 51.
[197] D. T. Clark and G. Smale, *Chem. Comm.*, **1969**, 868, 1050; D. B. Ledlie and E. A. Nelson, *Tetrahedron Letters*, **1969**, 1175; D. T. Clark and D. R. Armstrong, *Chem. Comm.*, **1969**, 850, *Theor. Chim. Acta*, **13**, 365 (1969); D. T. Clark and G. Smale, *Tetrahedron*, **25**, 13 (1969).

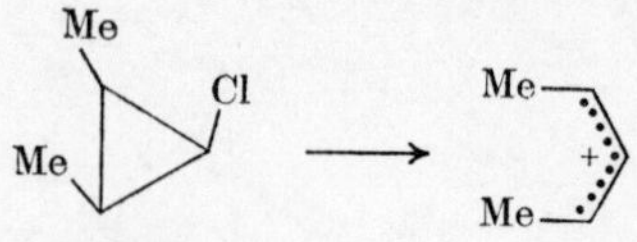

Me
Cl
Me
(296)

Me
Me
+
(297)

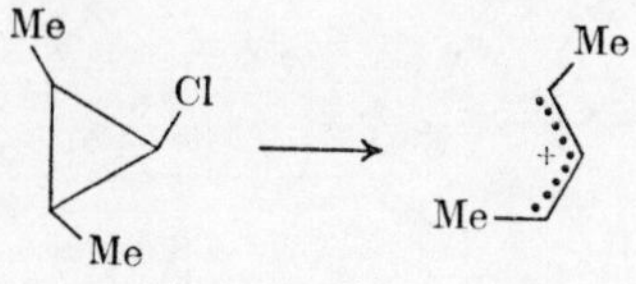

Me
Cl
Me
(298)

Me
Me
+
(299)

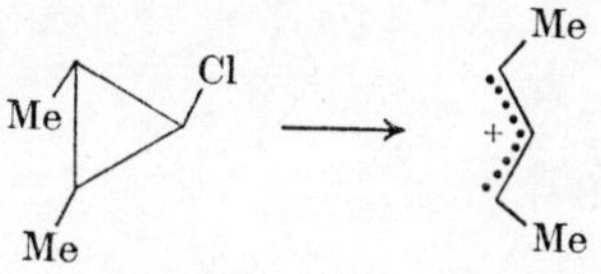

Me
Cl
Me
(300)

Me
+
Me
(301)

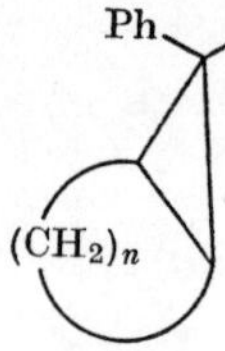

Ph　　Cl
$(CH_2)_n$
(302)

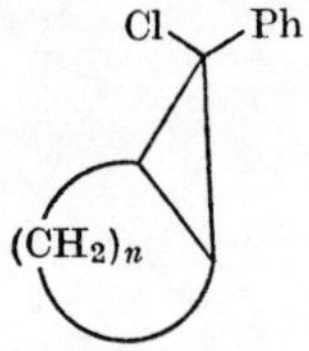

Cl　　Ph
$(CH_2)_n$
(303)

Use has been made of the good leaving-group ability of the trifluoromethyl-sulphonyloxy group to study the reactivity of some inert cyclopropyl systems. It was found that *exo*-bicyclo[3.1.0]hex-6-yl trifluoromethanesulphonate reacted 26,000 times slower than cyclopropyl trifluoromethanesulphonate, and that 1-nortricyclyl trifluoromethanesulphonate reacted 840,000 times slower than 7-methyl-3-noradamantyl trifluoromethanesulphonate.[198]

Other examples of ring-opening reactions of halogenocyclopropanes are discussed in references 199—204.

N-Chloroaziridines are hydrolysed according to equation (15), and the relative reactivities of compounds (**304**)—(**306**) follow the pattern expected

[198] T. M. Su, W. F. Sliwinski, and P. von R. Schleyer, *J. Am. Chem. Soc.*, **91**, 5386 (1969).
[199] M. S. Baird and C. B. Reese, *Tetrahedron Letters*, **1969**, 2117; *J. Chem. Soc.* (C), **1969**, 1803.
[200] D. B. Ledlie and W. H. Hearne, *Tetrahedron Letters*, **1969**, 4837.
[201] S. R. Sandler, *J. Org. Chem.*, **33**, 4537 (1968).
[202] T. Ando, H. Hosaka, H. Yamanaka, and W. Funasaka, *Bull. Chem. Soc. Japan*, **42**, 2013 (1969).
[203] B. Graffe, M. C. Sacquet, G. Fontaine, and P. Maitte, *Compt. Rend.*, *C*, **269**, 992 (1969).
[204] W. M. Horspool, R. G. Sutherland, and B. J. Thomson, *Tetrahedron Letters*, **1968**, 6045.

if the first step was disrotatory ring-opening concerted with N–Cl bond breaking.[205] A similar reaction of compound (307) yields isoquinoline which is presumably formed via the ion (308).[206]

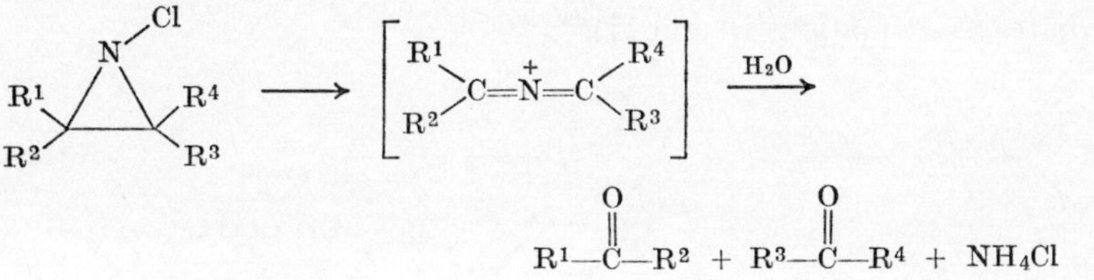

$$R^1\text{–}\overset{O}{\overset{\|}{C}}\text{–}R^2 + R^3\text{–}\overset{O}{\overset{\|}{C}}\text{–}R^4 + NH_4Cl \quad \ldots (15)$$

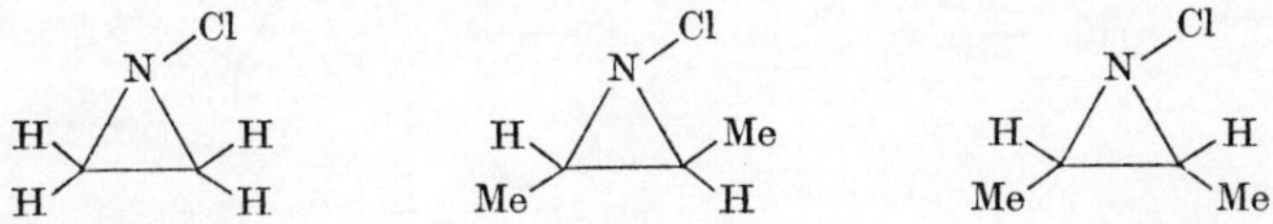

Relative rate 1	1490	155,000
(304)	(305)	(306)

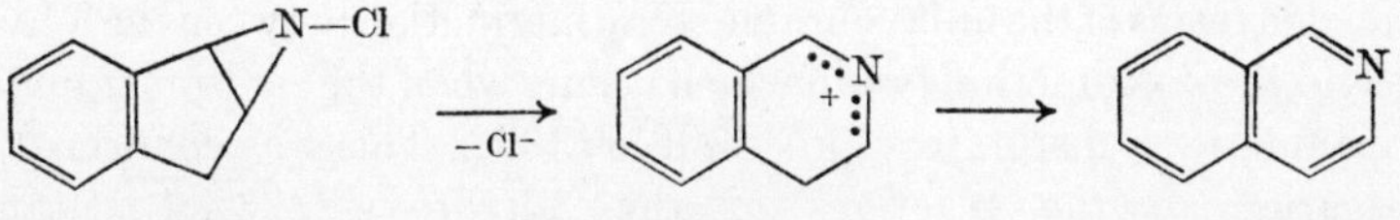

(307)   (308)

Ring-opening Reactions of Cyclobutyl Derivatives

The acetolyses of *cis*- and *trans*-3-arylcyclobutyl toluene-*p*-sulphonates are slower than that of cyclobutyl toluene-*p*-sulphonate, and both yield a ρ-value of −1.55 (at 110° and 75° respectively). The *trans*-compounds react more slowly than the *cis*-compounds, and the products are mainly but-3-enyl derivatives. *cis*- and *trans*-3-Ethoxycyclobutyl toluene-*p*-sulphonate react 2300 and 190 times more slowly than cyclobutyl toluene-*p*-sulphonate, and *cis*- and *trans*-3-chlorocyclobutyl toluene-*p*-sulphonate react 50,000 and 15,000 times more slowly. The 3-ethoxy-compounds yield mixtures of *cis*- and *trans*-3-ethoxycyclobutyl acetate, and the 3-chloro-compounds yield the acetate of inverted configuration. These results suggest that the transition state does not resemble the but-3-enyl cation.[207] The greater rate of reaction of the *cis*-compounds is probably due to a steric effect which arises from the *cis*-3-substituent coming close to the axial hydrogens on C-2 and C-4 in the transition state. This effect is particularly large in the reaction of *cis*-t-butylcyclobutyl toluene-*p*-sulphonate whose solvolysis in aqueous acetone is about 100 times

[205] P. G. Gassman and D. K. Dygos, *J. Am. Chem. Soc.*, **91**, 1543 (1969).
[206] D. C. Horwell and C. W. Rees, *Chem. Comm.*, **1969**, 1428.
[207] K. B. Wiberg and G. L. Nelson, *Tetrahedron Letters*, **1969**, 4385.

slower than that of cyclobutyl toluene-*p*-sulphonate. *cis*- and *trans*-3-t-Butyl-cyclobutyl toluene-*p*-sulphonate yield similar products to *cis*- and *trans*-2-t-butylcyclopropylmethyl 2,4-dinitrobenzoate respectively, and it was suggested that both series of compounds involve t-butylcyclopropylmethyl cations as intermediates (see equations 16 and 17).[207, 208]

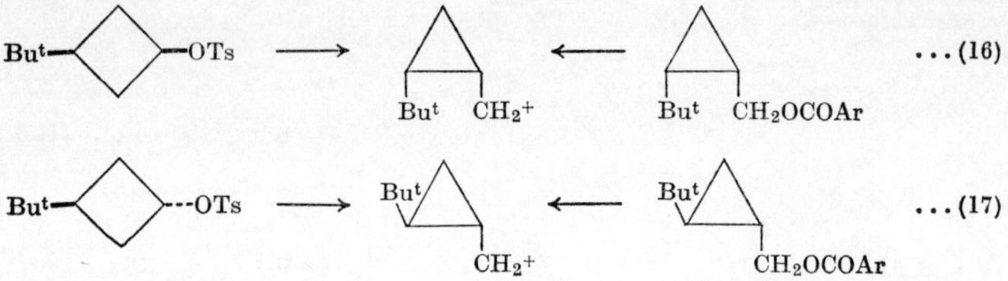

Dolby and Wilkins have published details of their investigation of the solvolysis of 3-hydroxy-2,2,4,4-tetramethylcyclobutyl toluene-*p*-sulphonates. The 1500-fold greater rate with the *trans*- compared to the *cis*-isomer is best explained in terms of the unfavourable steric interaction between the 3-hydroxy group and the *cis*-2-methyl group which occurs when the *cis*-isomer undergoes ring-opening by a disrotatory process [see (**309**)]. This was confirmed by an investigation of the reactions of the 3-hydroxy-2,2,3,4,4-pentamethyl compounds. Here disrotatory ring-opening of the *trans*-isomer would lead to an unfavourable interaction between the methyl groups at positions 2 and 3 [see (**310**)] and the rate is reduced. The rate for the *cis*-isomer is increased presumably as a result of release of steric interactions between the 2- and 3-methyl groups. The ratio of the rates is now only 5.[209]

Deamination of *cis*- and *trans*-3-methylcyclobutylamine[210] and of a mixture of *cis*- and *trans*-3-isopropyl[1-²H]cyclobutylamine[211] have been investigated.[212]

208 P. von R. Schleyer, P. Le Perchech, and D. J. Raber, *Tetrahedron Letters*, **1969**, 4389.
209 L. J. Dolby and C. L. Wilkins, *Tetrahedron*, **25**, 2381 (1969).
210 I. Lillien and L. Handloser, *J. Org. Chem.*, **34**, 3058 (1969).
211 I. Lillien and L. Handloser, *Tetrahedron Letters*, **1969**, 1035.
212 See *Org. Reaction Mech.*, **1967**, 84; **1968**, 52.

Metallocenylmethyl Cations[213]

It has been noted that although α-ferrocenyl-α-phenylmethyl acetate is solvolysed only 67 times more rapidly than triphenylmethyl acetate, the α-ferrocenyl-α-phenylmethyl cation has a pK_{R+} nearly 7 units more positive than the triphenylmethyl cation. This and similar results lead to the suggestion that kinetic studies do not provide reliable information about the relative stabilities of ferrocenylmethyl cations.[214] However, it seems to us that it may be the triphenylmethyl cation which is showing anomalous behaviour (cf. p. 64).

In contrast to these results there appears to be a good correlation between the stability of the tricarbonylchromiumbenzyl cation and the reactivity of tricarbonylchromiumbenzyl chloride. Thus, the pK_{R+} of the tricarbonyl-chromiumbenzyl cation is 5.5 units more positive than that of the benzyl cation, and the corresponding chloride reacts 10^5 times faster than benzyl chloride in aqueous acetone.[215, 216] Complexing with tricarbonylchromium has a much smaller effect on the stability of hydroxy-carbonium ions, e.g. PhC^+HOH, presumably because much of the positive charge is now carried by the oxygen.[215]

The 1-ferrocenyl-3-phenylallyl cation reacts with methoxide ion to yield 88% of the product of attack at C-1.[217a]

The NMR spectra of a series of ferrocenylmethyl cations have been interpreted as indicating that they are *not* stabilized by metal participation.[217b]

5-*exo*-Halogenomethyl derivatives of cyclopentadienylcobalt (**311**) ring-expand to cyclopentadienyl cyclohexadienyl salts (**312**).[218]

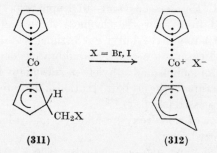

$$X = Br, I$$

(**311**)   (**312**)

[213] See *Org. Reaction Mech.*, **1965**, 47—48; **1966**, 41—42; **1967**, 53—54; **1968**, 56. The reactions of 2-ferrocenylethyl compounds are discussed on p. 26.

[214] E. A. Hill and R. Wiesner, *J. Am. Chem. Soc.*, **91**, 509 (1969).

[215] W. S. Trahanovsky and D. K. Wells, *J. Am. Chem. Soc.*, **91**, 5870 (1969); D. K. Wells and W. S. Trahanovsky, *ibid.*, p. 5871.

[216] J. D. Holmes, D. A. K. Jones, and R. Pettit, *J. Organometal. Chem.*, **4**, 324 (1965).

[217a] M. J. A. Habib and W. E. Watts, *J. Organometal. Chem.*, **18**, 361 (1969).

[217b] J. Feinberg and M. Rosenblum, *J. Am. Chem. Soc.*, **91**, 4324 (1969); cf. *Org. Reaction Mech.*, **1966**, 42.

[218] G. F. Herberich and J. Scharzer, *Angew. Chem. Internat. Ed. Engl.*, **8**, 143 (1969).

Deprotonation of ferrocenylmethyl cations to yield ferrocenylcarbenes has been investigated.[219]

Other Stable Carbonium Ions and their Reactions

Olah and White have published an important paper on the ^{13}C NMR spectra of carbonium ions. The ^{13}C chemical shift of the positively charged carbon of the t-butyl cation in SO_2ClF–SbF_5 is 135.4 ppm downfield from $^{13}CS_2$, and that of the isopropyl cation is 125.0 ppm downfield (−125.0). On the assumption that the ^{13}C chemical shift is a measure of electron density, the central carbon of the t-butyl cation is therefore slightly more positive than that of the isopropyl cation. The isopropyl cation generated from 2-chloro[2-^{13}C]propane underwent equilibration of the label at −60° with a half-life of 1 hour probably via a protonated cyclopropane intermediate.

The ^{13}C chemical shift of the hydroxycarbonium ions Me_2C^+OH, $MeC^+(OH)_2$, and $C^+(OH)_3$ are −55.7, −42.6, and +28.0 ppm respectively, which indicates that successive replacement of a methyl group by a hydroxy group causes a decrease in electron density at the central carbon. However, successive replacement of a methyl by a phenyl group does not have this effect, as the chemical shifts of PhC^+Me_2, Ph_2C^+Me, and Ph_3C^+ are −61.1, −5.6, and −18.1 respectively, and it was suggested that the electron density on the central carbon of the triphenylmethyl cation was higher than on that of the diphenylmethyl cation because the three phenyl rings of the former cannot lie in the same plane. The cyclopropyl group appears to delocalize positive charge less effectively than the phenyl group, as the chemical shifts of cyclo-$C_3H_5C^+Me_2$ and PhC^+Me_2 are −86.8 and −61.1 ppm respectively.

It was concluded from their ^{13}C chemical shifts and ^{13}C–^1H coupling constants that the 2,3-dimethyl-2-butyl, 2,3,3-trimethyl-2-butyl, 2-butyl, and 1-methylcyclopentyl cations are rapidly equilibrating degenerate ions, and that the ethylenebromonium and ethylene-*p*-anisonium ions have bridged structures. The structures of the norbornyl cation and of substituted norbornyl cations were also discussed (see p. 1).[220]

The NMR spectra of α-hydroxy-, α-alkoxy-, and α-halogeno-carbonium ions have been reported.[221, 222]

Protonation of *syn*- and *anti*-octamethyltricyclo[4.2.0.0^{2,5}]octa-3,7-diene (313) and (314), octamethylsemibullvalene (316), and octamethylcyclo-octatetraene (317) yield the octamethylbicyclo[3.3.0]octadienyl di-cation (315), identified by its NMR spectrum.[223]

[219] P. Ashkenazi, S. Lupan, A. Schwarz, and M. Cais, *Tetrahedron Letters*, **1969**, 817.
[220] G. A. Olah and A. M. White, *J. Am. Chem. Soc.*, **91**, 5801 (1969).
[221] A. M. White and G. A. Olah, *J. Am. Chem. Soc.*, **91**, 2943 (1969).
[222] G. A. Olah and M. B. Comisarow, *J. Am. Chem. Soc.*, **91**, 2955 (1969).
[223] J. M. Bollinger and G. Olah, *J. Am. Chem. Soc.*, **91**, 3380 (1969).

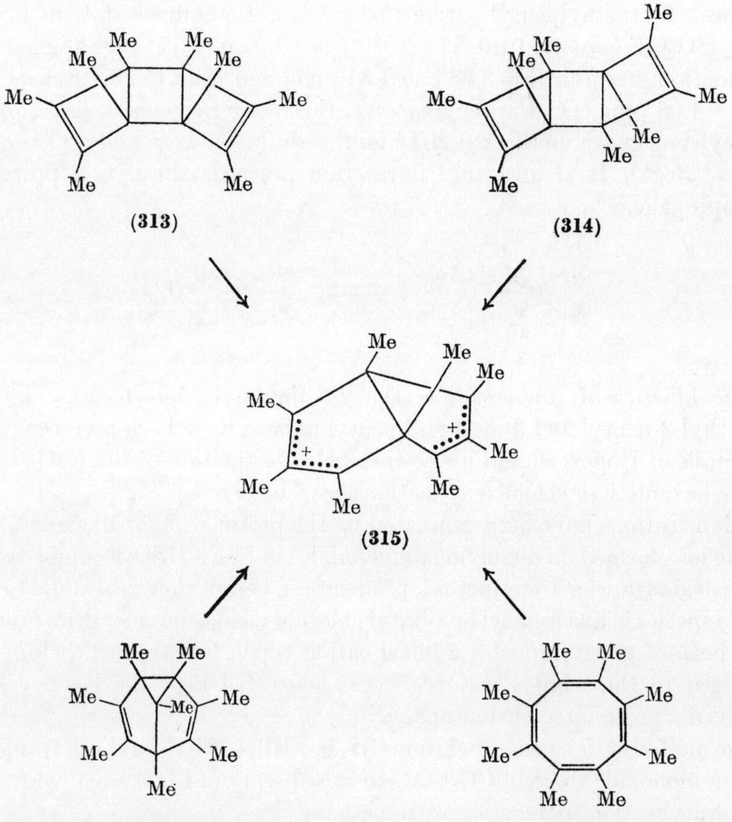

(313) (314)

(315)

(316) (317)

Cyclohexenone undergoes rearrangement in SbF$_5$–HF to 3-methylcyclo-pentenone, probably via the di-cation ion (318) as shown in equation (18).[224]

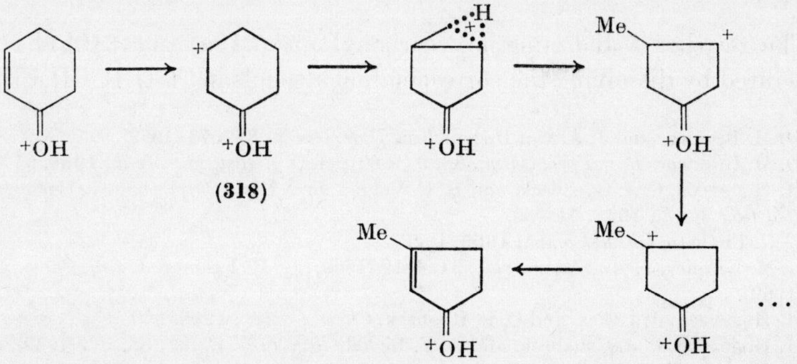

(318)

... (18)

[224] H. Hogeveen, *Rec. Trav. Chim.*, 87, 1295 (1968).

3

The 2,4-dimethylpentyl cation undergoes a 1,3-hydride shift in FSO_3H–SbF_5–SO_2ClF (equation 19). The NMR spectrum at $-117°$ shows signals from the methyl groups at $\delta = 3.88$ and 1.34 ppm, and from the methylene group at $\delta = 4.08$ ppm. On warming to $-80°$ the methyl signals coalesce but the methylene signal is unaffected. $\Delta H^{\neq}$ for this shift is 6.5 kcal, and $\Delta S^{\neq}$ is -11 cal mole^{-1} deg^{-1}. It is uncertain if reaction proceeds through a protonated cyclopropane.[225]

$$\ldots (19)$$

The kinetics of conversion of the 2,3-dimethyl-2-butyl cation into the 2-methyl-2-pentyl and 3-methyl-3-pentyl cations have been investigated.[226]

Details of Hogeveen and his coworkers' investigation of the NMR spectra of the hexamethylcyclohexenyl cations have been reported.[227]

Allylic cations have been generated by the protonation of allenes.[228]

The ions formed on dissolving alkyl halides in SbF_5–HSO_3F can be trapped by hydride transfer from methylcyclopentane before they rearrange to more stable species. Thus n-butyl or 2-butyl chloride yield n-butane. It was thought that both of these formed a 2-butyl cation which was trapped before it rearranged to the t-butyl cation. It was suggested that the latter reaction involved a protonated cyclopropane.[229]

The methyl cation generated from CH_4 in $FHSO_3$–SbF_5 has been trapped by carbon monoxide to yield CH_3CO^+. In the absence of CO it is converted into the t-butyl cation by reaction with methane.[230]

The kinetics of the reduction of carbonium ions by hydrogen to alkanes have been studied.[231]

When 1,1-di-(p-methoxyphenyl)ethylene is mixed with $SbCl_5$ the chloromethyldi-(p-methoxyphenyl)methyl cation is formed through transfer of Cl^+.[232]

The α-carboxy- and α-methoxycarbonyl-ions (**319a**) and (**319b**) have been generated by dissolving the corresponding alcohols in $ClSO_3H$–CH_2Cl_2.[233]

[225] D. M. Brouwer and J. A. van Doorn, *Rec. Trav. Chim.*, **88**, 573 (1969).

[226] D. M. Brouwer, *Rec. Trav. Chim.*, **88**, 9 (1969); cf. *Org. Reaction Mech.*, **1968**, 57—58.

[227] H. Hogeveen, C. J. Gaasbeek, and H. C. Volger, *Rec. Trav. Chim.*, **88**, 379 (1969); see *Org. Reaction Mech.*, **1968**, 54—55.

[228] C. U. Pittman, *Chem. Comm.*, **1969**, 122.

[229] G. M. Kramer, *J. Am. Chem. Soc.*, **91**, 4819 (1969); G. M. Kramer, *J. Org. Chem.*, **34**, 2919 (1969).

[230] H. Hogeveen, J. Lukas, and C. F. Roobeek, *Chem. Comm.*, **1969**, 920.

[231] H. Hogeveen, C. J. Gaasbeek, and A. F. Bickel, *Rec. Trav. Chim.*, **88**, 703 (1969).

[232] W. Bracke, W. J. Cheng, J. M. Pearson, and M. Szwarc, *J. Am. Chem. Soc.*, **91**, 203 (1969).

[233] G. P. Nilles and R. D. Schuetz, *Tetrahedron Letters*, **1969**, 4313.

The kinetics of the rearrangement of cyclohexenyl cations into the more stable cyclopentenyl cations,[234] and the steric course of the rearrangement of pentadienyl cations into cyclopentenyl cations,[235] have been investigated.

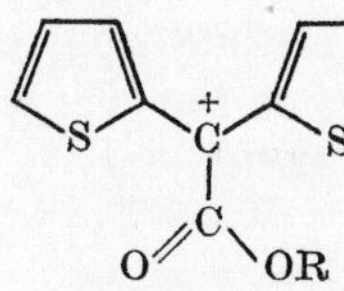

(319a) R = H

(319b) R = Me

Large differences in the enthalpies of formation of benzenonium ions from methyl-, ethyl-, isopropyl-, and t-butyl-benzene in SbF_5–HSO_3F have been reported. They vary from -5.00 kcal mole^{-1} for toluene to -1.24 kcal mole^{-1} for t-butylbenzene.[236]

It was concluded by double-irradiation NMR experiments that the degenerate rearrangement of the heptamethylbenzenonium ion occurs via successive 1,2-shifts rather than by a mechanism involving an unlocalized π-complex.[237]

The degenerate rearrangement of the 9,9,10-trimethylphenanthrenium ion has been investigated.[238]

Stable perhalogenated arenonium ions have been prepared.[239]

The reduction of mono-, di-, and tri-carbonium ions to the corresponding radicals,[240, 241] and the polarographic reduction of tri-arylmethyl cations have been investigated.[242]

The generation of carbonium ions through the anodic oxidation of alkanecarboxylic acids, alkaneboronates, and alkyl halides has been investigated.[243] Transient primary carbonium ions have also been postulated to be formed on oxidation of carboxylic acids with peroxymonosulphuric acid.[244] Alkoxy-

[234] T. S. Sorensen, *J. Am. Chem. Soc.*, **91**, 6398 (1969).

[235] P. H. Campbell, N. W. K. Chiu, K. Deugau, I. J. Miller, and T. S. Sorensen, *J. Am. Chem. Soc.*, **91**, 6404 (1969).

[236] E. M. Arnett and J. W. Larsen, *J. Am. Chem. Soc.*, **91**, 1438 (1969).

[237] B. G. Derendyaev, V. I. Mamatyuk, and V. A. Koptyug, *Tetrahedron Letters*, **1969**, 5.

[238] V. G. Shubin, D. V. Korchaginci, A. I. Resvukhin, and V. A. Koptyug, *Dokl. Akad. Nauk SSSR*, **179**, 119 (1968); *Chem. Abs.*, **69**, 58715 (1969).

[239] V. D. Shteingarts, Y. V. Pozdnyakovich, and G. G. Yakobson, *Chem. Comm.*, **1969**, 1264.

[240] G. Kothe, W. Sümmermann, H. Baumgärtel, and H. Zimmermann, *Tetrahedron Letters*, **1969**, 2185; F. Strohbusch and H. Zimmermann, *ibid.*, 1705; W. Sümmermann, G. Kothe, H. Baumgärtel, and H. Zimmermann, *ibid.*, p. 3807.

[241] H. Volz and W. Lotsch, *Tetrahedron Letters*, **1969**, 2275.

[242] M. Feldman and W. C. Flythe, *J. Am. Chem. Soc.*, **91**, 4577 (1969).

[243] J. T. Keating and P. S. Skell, *J. Org. Chem.* **34**, 1479 (1969).

[244] N. C. Deno, W. E. Billups, J. S. Bingman, R. R. Lastomirsky, and R. G. Whalen, *J. Org. Chem.*, **34**, 3207 (1969).

carbonium ions have been generated by the reactions of acetals and orthoesters with bromine in SO_2.[245]

The rate of interconversion of the two conformers of ion (**320**) through rotation of the C–N bond is increased when X is electron-releasing (e.g. MeO) and decreased when it is electron-withdrawing (e.g. CF_3), which indicates that substituents have a greater effect on the energy of the transition state (**321**) than on that of the ground state (**320**).[246]

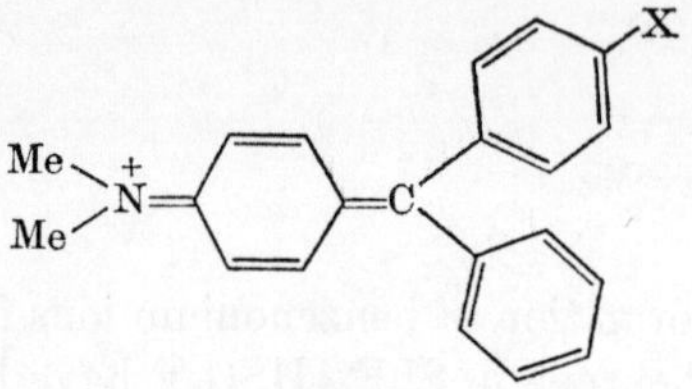

(**320**) 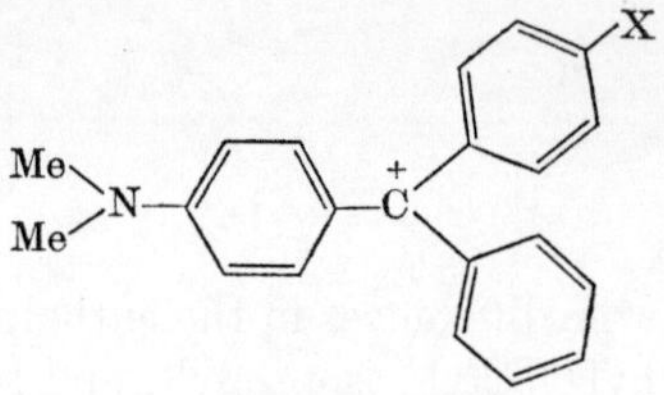   (**321**)

Charge-transfer complexes of carbonium ions have been studied.[247]

Calculations have been reported which support the view that carbonium ions tend to a planar structure (cf. footnote 36 of ref. 220).[248] Calculations of the dependence of potential energy of phenyl-substituted carbonium ions on the angle of twist of the phenyl ring,[249] on ketones as models for carbonium ions,[250] and other calculations on carbonium ions[251] have been reported.

The action of bases on the tropylium cation[252] and the intervention of the homotropylium cation in addition reactions of cyclooctatetraene[253] have been studied.

There have been several investigations of the UV spectra of carbonium ions.[254-257]

The isomerization of carbonium ions in the gas phase has been studied.[258]

[245] C. H. V. Dusseau, S. E. Schaafsma, H. Steinberg, and T. J. de Boer, *Tetrahedron Letters*, **1969**, 467.
[246] J. W. Rakshys, S. V. McKinley, and H. H. Freedman, *Chem. Comm.*, **1969**, 1180.
[247] H. J. Dauben and J. D. Wilson, *Chem. Comm.*, **1968**, 1629.
[248] J. E. Williams, R. Sustmann, L. C. Allen, and P. von R. Schleyer, *J. Am. Chem. Soc.*, **91**, 1037 (1969).
[249] R. Hoffmann, R. Bissell, and D. G. Farnum, *J. Phys. Chem.*, **73**, 1789 (1969).
[250] R. F. Davis, C. R. C. Pfaffenberger, D. J. Grosse, and J. V. Morris, *Tetrahedron*, **25**, 1175 (1969).
[251] R. Sustmann, J. E. Williams, M. J. S. Dewar, L. C. Allen, and P. von R. Schleyer, *J. Am. Chem. Soc.*, **91**, 5350 (1969); D. T. Clark, *Chem. Comm.*, **1969**, 637.
[252] S. G. McGeachin, *Can. J. Chem.*, **47**, 151 (1969).
[253] L. A. Paquette, J. R. Malpass, and T. J. Barton, *J. Am. Chem. Soc.*, **91**, 4714 (1969).
[254] R. Zahradnik, A. Kröhn, J. Pančiv and J. Snobl, *Coll. Czech. Chem. Comm.*, **34**, 2553 (1969); R. Zahradnik, J. Pančiv, and A. Kröhn, *ibid.*, p. 2831.
[255] B. Föhlisch, *Ann. Chem.*, **721**, 48 (1969).
[256] R. N. Young, *J. Chem. Soc. (B)*, **1969**, 896.
[257] V. Bertoli and P. H. Plesch, *J. Chem. Soc. (B)*, **1968**, 1500.
[258] W. G. Cole and D. H. Williams, *Chem. Comm.*, **1969**, 784.

The protonation of amides[259] and of thiocarboxylic acids and S-alkyl esters[260] in super-acid media has been studied.

The following carbonium ions have also been investigated: aryloxocarbonium ions,[261] triarylmethyl ions[262] (as their pentahalogenostannates), cyclohexadienyl cations[263] (generated by protonation of cyclohexadienones), triphenylcyclopropenyl cation,[264] and the di-t-butylcyclopropenyl cation.[265]

[259] J. L. Sudmeier and K. E. Schwartz, *Chem. Comm.*, **1968**, 1646.
[260] G. A. Olah, A. T. Ku, and A. M. White, *J. Org. Chem.*, **34**, 1827 (1969).
[261] D. A. Tomalia, *J. Org. Chem.*, **34**, 2583 (1969).
[262] K. M. Harmon, L. L. Hesse, L. P. Klemann, C. W. Kocher, S. V. McKinley, and A. E. Young, *Inorg. Chem.*, **8**, 1054 (1969).
[263] V. P. Vitullo, *J. Org. Chem.*, **34**, 224 (1969).
[264] I. A. Dyakonov, R. R. Kostikov, and A. P. Holchanov, *Zh. Org. Khim.*, **5**, 175 (1969).
[265] J. Ciabattoni and E. C. Nathan, *J. Am. Chem. Soc.*, **91**, 4766 (1969).

CHAPTER 2

Nucleophilic Aliphatic Substitution

B. Capon

Chemistry Department, Glasgow University

Ion-pair Phenomena and Borderline Mechanisms[1–3]

Goering and Myers have studied the steric course of the rearrangement of *trans*-cyclodec-5-enyl *p*-nitrobenzoate (**1**) into *trans-cis*-1-decalyl *p*-nitrobenzoate (**2**) in 90% aqueous acetone at 100°, and also measured the extent of ^{18}O scrambling. Working with (**1**) of known absolute configuration it was demonstrated that both the solvolysis product (**3**) and the rearrangement product (**2**) were formed with inversion of configuration [e.g. as shown in (**1**)]. However, despite the long distance that the *p*-nitrobenzoate anion has to migrate, there is only 70% equilibration of the oxygens as demonstrated by working with ether and carbonyl ^{18}O-labelled (**1**). The oxygen atoms are not

[1] M. Szwarc, *Accounts Chem. Res.*, **2**, 87 (1969).
[2] M. Szwarc, *Svensk Kem. Tidskr.*, **81**, No. 3, 14 (1969).
[3] J. E. Prue, *J. Chem. Educ.*, **46**, 12 (1969).

equivalent in the ion-pair, and it is interesting to speculate that this may be a general property of ion-pairs from secondary alkyl esters.[4]

Rearrangement of bicyclo[2.2.0]hexylmethyl p-nitro[*carbonyl*-^{18}O]benzoate into 1-norbornyl p-nitrobenzoate occurs without any detectable scrambling of the label (see p. 15).

The isomerization of *syn*-7-chloro-*exo*-2-norbornyl toluene-p-sulphonate into the *anti*-7-chloro-isomer in acetic acid proceeds with little incorporation of label in the presence of sodium [^{14}C]toluene-p-sulphonate (see p. 3).[5]

Sneen and Larsen have reported details of their demonstration that the solvolysis and reaction with sodium azide of 2-octyl methanesulphonate in aqueous dioxan proceed via reversibly formed ion-pairs.[6] Similar mechanisms were also demonstrated to be consistent with the kinetics of the concurrent elimination and substitution reactions of 1-phenylethyl bromide,[7] of the reaction of 4-methoxybenzyl chloride in 70% aqueous acetone in the presence of sodium azide, and of the reaction of benzoyl chloride with water and o-nitroaniline in 80% aqueous acetone.[8]

Albery and Robinson have measured $\Delta C_p^{\ddagger}$ and the variation of $\Delta C_p^{\ddagger}$ with temperature for the hydrolysis of t-butyl chloride, and suggested that their results could be explained by a mechanism which involved capture of a reversibly formed ion-pair by water.[9] The variation of the rate constant with solvent composition in H_2O–D_2O mixtures is also consistent with this mechanism.[10a]

Volumes of activation for the solvolysis of benzyl halides (p-XC$_6$H$_4$CH$_2$Cl; X = Me, H, Cl, NO$_2$) in aqueous acetone fall in the range −18 to −24 cc mole^{-1}. These were interpreted as indicating that the mechanism is S_N2.[10b]

Smith and Goon have demonstrated that the ratios of the overall rate constants (k_{obs}) for ethanolysis of cumyl chloride, p-nitrobenzoate, and

[4] H. L. Goering and R. F. Myers, *J. Am. Chem. Soc.*, **91**, 3386 (1969).

[5] H. L. Goering and M. J. Degani, *J. Am. Chem. Soc.*, **91**, 4506 (1969).

[6] R. A. Sneen and J. W. Larsen, *J. Am. Chem. Soc.*, **91**, 362 (1969); see *Org. Reaction Mech.*, **1966**, 44.

[7] R. A. Sneen and H. M. Robbins, *J. Am. Chem. Soc.*, **91**, 3100 (1969).

[8] R. A. Sneen and J. W. Larsen, *J. Am. Chem. Soc.*, **91**, 6031 (1969).

[9] W. J. Albery and B. H. Robinson, *Trans. Faraday Soc.*, **65**, 980 (1969); see also G. J. Hills and C. A. Viana in ref. 27, p. 261, and ref. 61a.

[10a] W. J. Albery and B. H. Robinson, *Trans. Faraday Soc.*, **65**, 1623 (1969); see also L. Treindl, R. E. Robertson, and S. E. Sugamori, *Can. J. Chem.*, **47**, 3397 (1969).

[10b] K. J. Laidler and R. Martin, *Int. J. Chem. Kinetics*, **1**, 113 (1969).

thionbenzoate to those for the corresponding benzhydryl compounds are 5.5, 23, and 4700. The percentages of olefin from the cumyl derivatives are 11.8, 49.6, and 90.6 respectively. This indicates that the products are formed from ion-pairs, and it was suggested that they are formed reversibly as shown in equation (1). This scheme leads to the expression given in equation (2) for

$$RX \underset{k_{-1}}{\overset{k_1}{\rightleftarrows}} R^+X^- \xrightarrow{k_S} \text{Substitution} \qquad \qquad \dots(1)$$

$$\downarrow k_E$$

$$\text{Elimination}$$

$$k_{obs} = k_1(k_E + k_S)/(k_{-1} + k_E + k_S) \qquad \qquad \dots(2)$$

k_{obs}, and it was suggested that the variation of the ratio of k_{obs} for the cumyl and benzhydryl compounds with leaving group resulted from variations in k_{-1}. The proportions of olefins formed from the deuterated cumyl derivatives $PhCX(CD_3)CH_3$ and $PhCX(CD_3)_2$ are less than those from the undeuterated compounds. The value of k_{obs} is also decreased on deuteration, and it was shown that the isotope effect on proton loss from the carbonium ion makes a substantial contribution to the measured overall kinetic isotope effect.[11a] Thus the isotope effect of k_{obs} is not a good measure of the isotope effect on k_1 (see equation 2). It would be interesting to find if this conclusion is a general one, for if so it has important consequences for the theory of kinetic isotope effects.[11a, 11b]

The rate of decomposition of t-butyl nitrate relative to that of t-butyl chloride is 2890 in acetonitrile and 28–51 in aqueous ethanol and aqueous dioxan. It was suggested that this resulted from an increase in the amount of ion-pair return in acetonitrile.[12]

Fava and his coworkers have continued their investigation of ion-pair return in the reactions of 4,4'-dimethylbenzhydryl thiocyanate (RSCN).[13] In 95% aqueous acetone this compound is isomerized into the isothiocyanate (RNCS) four times faster than it is hydrolysed. The rate constant for hydrolysis (k_s) is increased more than that for isomerization (k_i) by adding lithium perchlorate. In solutions in which the ionic strength is maintained at 0.2M by the addition of sodium perchlorate, k_s is decreased more with increasing sodium thiocyanate concentration than k_i. In the presence of labelled sodium thiocyanate exchange takes place and the rate constant of this, k_{ex}, increases with [NaCNS] in a non-linear fashion and approaches a steady value, equal to the rate constant for solvolysis at zero sodium thiocyanate concentration ($k_s{}^0$). The sum of the rate constants for exchange and hydrolysis at each

11a S. G. Smith and D. J. W. Goon, *J. Org. Chem.*, **34**, 3127 (1969).
11b Cf. *Org. Reaction Mech.*, **1965**, 62.
12 D. N. Kevill and R. F. Sutthoff, *J. Chem. Soc. (B)*, **1969**, 366.
13 See *Org. Reaction Mech.*, **1966**, 45—46.

concentration is equal to the rate constant for hydrolysis when $[NaCNS] = 0$, i.e. $k_{ex} + k_s = k_s^0$. This indicates that hydrolysis and exchange proceed via a common intermediate, which must be a free carbonium ion. With $[RSCN] = 5 \times 10^{-2}M$ and $[NaSCN] = 2 \times 10^{-2}M$ it was calculated from the specific activities of RSCN and RNCS at 2—3% isomerization that the relative rates of capture of the 4,4'-dimethylbenzhydryl cation by water and by the N and S atoms of the thiocyanate ion were 1:33:275 respectively. The fraction of ion-pairs which dissociate to free ions is 0.085, and so most of the isomerization must occur by way of the former. The decrease in the rate of isomerization with increasing $[NaCNS]$ could not be explained but it was suggested that this was associated with the dependence on $[LiClO_4]$ of the S:N ratio for attack by the thiocyanate ion which was also not understood.[14] The exchange reaction between 4,4'-dimethylbenzhydryl thiocyanate and several alkali-metal thiocyanates in acetone was also investigated.[15]

The rearrangement of ethyl cyanate into ethyl isocyanate is intermolecular, since when the ^{15}N-labelled compound is allowed to rearrange in the presence of butyl cyanate the label is distributed almost equally between both iso-cyanates. It was suggested that reaction proceeds through solvent-separated ion-pairs.[16]

The decomposition of ^{14}C-labelled and optically active *erythro-* and *threo-*3-phenyl-2-butylsulphoxonium ions, generated by the reaction of the corresponding dinitrobenzenesulphonates with chlorine, has been investigated. Phenyl,

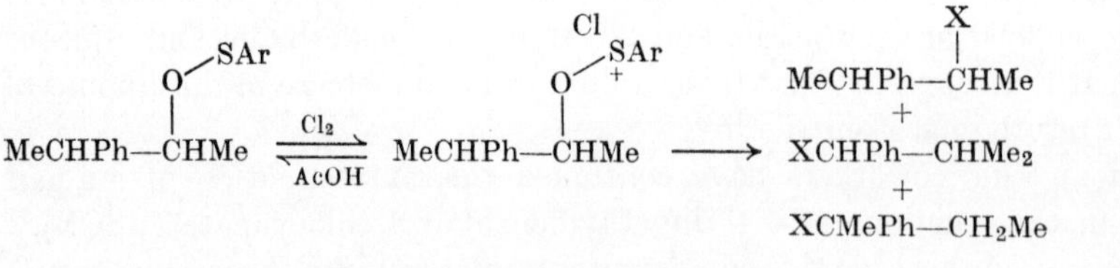

$$(X = Cl \text{ or } OAc; Ar = 2,4\text{-dinitrophenyl})$$

methyl, and hydrogen migration occur (equation 3), and the results show that phenyl migration occurs in the formation of *threo*-products from *threo*-reactant and of *erythro*-product from *erythro*-reactant. In the absence of lithium perchlorate leakage between the *erythro* and *threo* series is 7—17% but this is halved in the presence of 0.08M-lithium perchlorate. Lithium perchlorate reduces the amount of methyl and hydrogen migrations but increases the amount of phenyl migration. The *erythro*:*threo* leakage occurs with predominant inversion at C-α, which indicates that it occurs mainly via classical

14 A. Ceccon, A. Fava, and I. Papa, *J. Am. Chem. Soc.*, **91**, 5547 (1969); *Chim. Ind.* (Milan), **51**, 53 (1969).
15 A. Ceccon and I. Papa, *J. Chem. Soc.* (B), **1969**, 703.
16 D. Martin, H. J. Niclas, and D. Habisch, *Ann. Chem.*, **727**, 10 (1969).

ions and not via phenonium ions. The methyl migration occurs with net retention of configuration, which indicates that the acetate attacks from the same side as the methyl group migrates. This probably involves a classical isopropylphenylmethyl cation (equation 4). The ratio of chloride to acetate in

$$
\begin{array}{c}
\overset{Y}{\underset{Ph}{\overset{|}{C}}}\!\!-\!\!\overset{|}{\underset{Me}{C}}\!\cdots\!H
\quad\longrightarrow\quad
H\cdots\overset{Me}{\underset{Ph}{\overset{|}{C}}}\!\!-\!\!\overset{Y^-}{\underset{Me}{\overset{|}{\overset{+}{C}}}}\!\cdots\!H + \text{other conformers}
\end{array}
$$

$$
\begin{array}{c}
H\cdots\overset{-OAc}{\underset{Ph}{\overset{+}{C}}}\!\!-\!\!CHMe_2
\quad\longrightarrow\quad
H\cdots\overset{OAc}{\underset{Ph}{\overset{|}{C}}}\!\!-\!\!CHMe_2
\quad\cdots(4)
\end{array}
$$

the product is changed from 2.65—2.85 to 0.81 by the addition of 0.08M-lithium perchlorate. These results were interpreted in terms of the previously suggested scheme[17] involving transfer of solvent structure from the sulphoxonium ion to the carbonium ion. In the absence of lithium perchlorate the predominant species is the intimate sulphoxonium-ion–chloride-ion pair which yields the intimate carbonium-ion–chloride-ion pair, and this yields predominantly chlorides and undergoes a high proportion of hydrogen and methyl migration. In the presence of lithium perchlorate the predominant species is a solvent-separated sulphoxonium ion-pair which yields a solvent-separated carbonium ion-pair, and this yields mainly acetates and undergoes less methyl and hydrogen migration but more phenyl migration. On this interpretation the *erythro–threo* leakage must be mainly a reaction of the intimate ion-pairs.[18]

The reaction of bicyclo[2.2.2]octylsulphoxonium chloride (4) in acetic acid yields 43.1% of a mixture of chlorides and 56.9% of acetates. The ratio of bicyclo[2.2.2]oct-2-yl to bicyclo[3.2.1]oct-*exo*-2-yl product in both is similar to the ratio of acetates obtained from acetolysis of bicyclo[2.2.2]oct-2-yl *p*-bromo-benzenesulphonate (i.e. approximately 1:1). The addition of lithium perchlorate (0.06M) strongly diminishes the amount of chloride in the product (from 43 to 2%) but the ratio of [2.2.2]- to [3.2.1]-chloride and acetate is not appreciably changed. In the presence of lithium perchlorate (0.06M) and lithium chloride (0.12M) the percentage of chloride is again increased to 34%. It was suggested that the chloride is formed from intimate sulphoxonium ion-pairs via intimate carbonium ion-pairs, and that the concentration of these is reduced by lithium perchlorate and increased by lithium chloride.[19]

[17] See *Org. Reaction Mech.*, **1967**, 57—58.
[18] H. Kwart, E. N. Givens, and C. J. Collins, *J. Am. Chem. Soc.*, **90**, 7162 (1968); **91**, 5532 (1969).
[19] H. Kwart and J. L. Irvine, *J. Am. Chem. Soc.*, **91**, 5541 (1969).

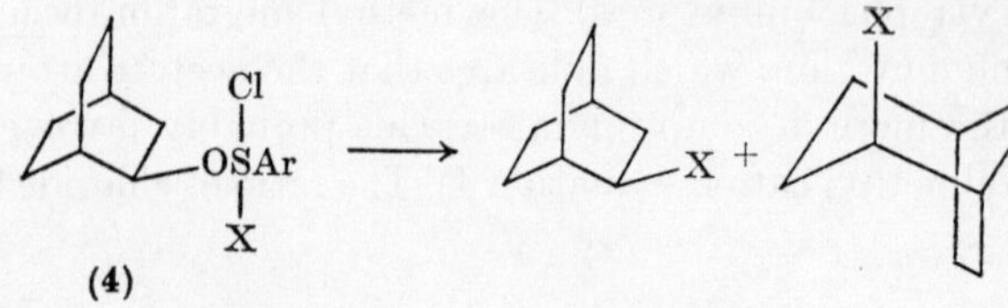

(4)

(X = Cl or OAc)

The role of ion-pairs in the reaction of allylic alcohols with thionyl chloride has been discussed.[20]

There have been several spectroscopic investigations of ion-pairs.[21]

The thermodynamics of ion-pair formation have been discussed.[22]

Details of Gassman and Cryberg's investigation of the silver-ion-assisted rearrangement of 4,7,7-trimethyl-2-chloro-2-azabicyclo[2.2.1]heptane have been published.[23] The analogous reactions shown in equations (5)—(7) have also been reported.[24]

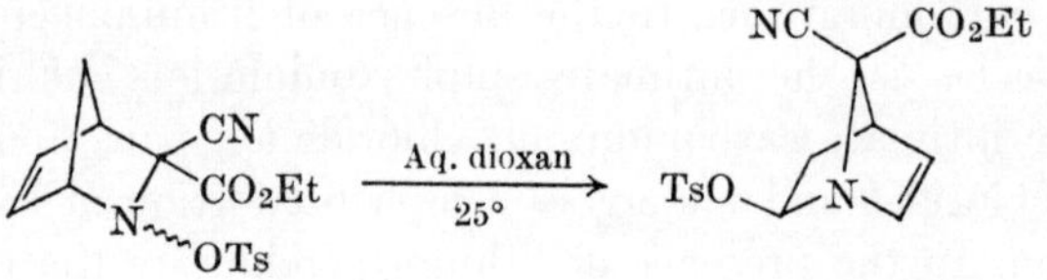

... (5)

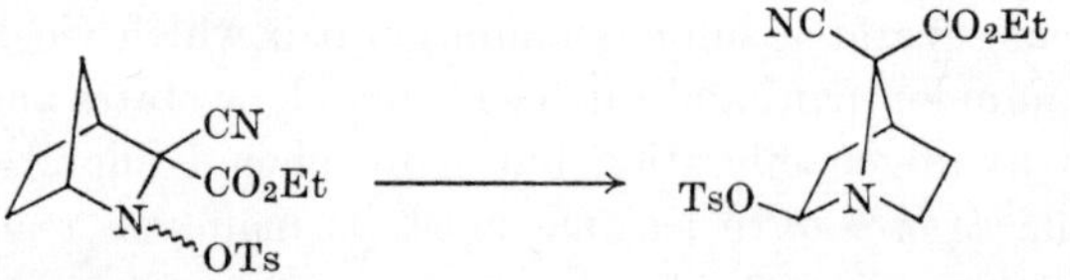

... (6)

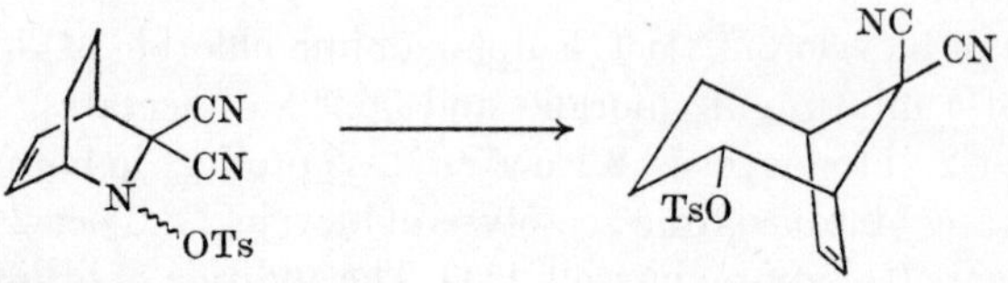

... (7)

[20] S. Czernecki, C. Georgoulis, J. Labertrande, G. Fusey, and C. Prevost, *Bull. Soc. Chim. France*, **1969**, 3568.

[21] D. W. Larsen, *J. Am. Chem. Soc.*, **91**, 2920 (1969); K. Höfelmann, J. Jagur-Grodzinski, and M. Szwarc, *ibid.*, p. 4645; T. E. H. Esch and J. Smid, *ibid.*, p. 4580; R. P. Taylor and I. D. Kuntz, *ibid.*, p. 4006; A. Fratiello, R. E. Lee, and R. E. Schuster, *Chem. Comm.*, **1969**, 37; W. F. Edgell and N. Pauuwe, *ibid.*, p. 284; E. Warhurst and A. M. Wilde, *Trans. Faraday Soc.*, **65**, 1413 (1969); R. F. Adams and N. M. Atherton, *ibid.*, p. 649; P. K. Ludwig, *J. Chem. Phys.*, **50**, 1787 (1969); D. Casson and B. J. Tabner, *J. Chem. Soc. (B)*, **1969**, 572; L. L. Bohm and G. V. Schulz, *Ber. Bunsenges. Phys. Chem.*, **73**, 260 (1969).

[22] G. Anderegg, *Chimia*, **22**, 477 (1968).

[23] P. G. Gassman and R. L. Cryberg, *J. Am. Chem. Soc.*, **91**, 2047 (1969); see *Org. Reaction Mech.*, **1968**, 65; cf. P. G. Gassman and R. L. Cryberg, *J. Am. Chem. Soc.*, **91**, 5176 (1969).

[24] J. P. Fleury, J. M. Biehler, and M. Desbois, *Tetrahedron Letters*, **1969**, 4091.

Solvent and Medium Effects[25]—[30b]

The decreased energies of activation of the reaction of SCN⁻, 2,4-dinitro-phenoxide, and 4-nitrophenoxide with methyl iodide, and of some other nucleophilic substitution reactions on going from MeOH to DMF, have been dissected into the enthalpies of transfer of the reactants and transition state. The enthalpy of transfer of the nucleophile is not always the dominant factor in the changes in the enthalpy of activation.[31] Other measurements of the enthalpies of transfer of nucleophiles from water and methanol to dipolar aprotic solvents have been reported.[32, 33]

The chlorine NMR shifts of lithium and tetramethylammonium chloride in DMSO–H_2O mixtures show a fairly regular change with solvent composition, which suggests that the chloride ion shows no strong preference for solvation by water.[34]

The rate of solvolysis of t-butyl chloride in mixtures of water with DMSO, dioxan, acetone, and DMF have been correlated with the chemical shifts of the protons.[35]

The volumes of activation for the hydrolysis of 1-chloro-2-nitroethane, benzhydryl chloride, and benzoyl chloride in dioxan–water mixtures have been determined.[36a]

The variation with solvent of the free energies of activation for the reaction of 4-nitrobenzyl chloride and methyl iodide with trimethylamine have been dissected into the variations in the free energies of each of the reactants and of the transition state.[36b]

[25] G. Reichardt, *Losungsmittel-Effekte in der Organischen Chemie*, Verlag Chemie, Weinheim, 1969.

[26] A. J. Parker, "Protic-Dipolar Aprotic Solvent Effects on Rates of Bimolecular Reactions", *Chem. Rev.*, **69**, 1 (1969).

[27] A. K. Covington and P. Jones (Ed.), *Hydrogen-Bonded Solvent Systems*, Taylor and Francis, London, 1968.

[28] R. A. Horne, "The Structure of Water and Aqueous Solutions", *Survey Progr. Chem.*, **4**, 2 (1968).

[29] E. Tommila, "The Influence of Solvent on Substituent Effects in Chemical Reactions", *Ann. Acad. Sci. Fennicae*, Ser. A II, No. 139 (1967).

[30a] H. Normant, "Hexamethylphosphortriamide in Organic Chemistry", *Bull. Soc. Chim. France*, **1968**, 791.

[30b] C. Agami, "Spectroscopic Studies of Aprotic Solvents", *Bull. Soc. Chim. France*, **1969**, 2183.

[31] P. Haberfield, L. Clayman, and J. S. Cooper, *J. Am. Chem. Soc.*, **91**, 787 (1969); see *Org. Reaction Mech.*, **1968**, 787.

[32] R. Fuchs, J. L. Bear, and R. F. Rodewald, *J. Am. Chem. Soc.*, **91**, 5797 (1969); see *Org. Reaction Mech.*, **1968**, 67.

[33] G. Choux and R. L. Benoit, *J. Am. Chem. Soc.*, **91**, 6221 (1969).

[34] C. H. Langford and T. R. Stengle, *J. Am. Chem. Soc.*, **91**, 4014 (1969).

[35] W. Köhler and R. Radeglia, *Z. Chem.*, **9**, 193, 392 (1969); L. Neuheiser, W. Köhler, and P. Reiner, *ibid.*, **8**, 275 (1968).

[36a] D. Büttner and H. Heydtmann, *Ber. Bunsenges. Phys. Chem.*, **73**, 640 (1969).

[36b] M. H. Abraham, *Chem. Comm.*, **1969**, 1307.

The rate of reaction of methyl iodide with sodium phenoxide in acetonitrile is first increased and then decreased on addition of phenol.[37]

The nucleophilic reactivity of alkali-metal phenoxides in mixtures of dioxan with dipolar aprotic solvents has been studied.[38]

The effect of solvent composition on the rates of the following reactions has been studied: benzyl chloride with thiosulphate in methanol–water and ethanol–water mixtures;[39] methyl iodide and CN^- in $DMSO-H_2O$ and Me_2CO-H_2O mixtures;[40] the hydrolysis of n-hexyl, n-heptyl, and n-octyl bromide in $EtOH-H_2O$ mixtures,[41a] and of allyl halides and sulphonates in dioxan–water mixtures.[41b] The effect of solvent on the reaction of alcohols with HCl[42] and on the Menschutkin reaction[43] have been studied.

Koppel and Palm have given a theoretical discussion of the effect of solvent composition on reaction rate.[44a]

It has been demonstrated that the decrease in the second-order constant for the reaction of methyl chloride with lithium radiochloride in 95% aqueous acetone from 7.5×10^{-4} l mole^{-1} sec^{-1} with $[LiCl] = 5.0 \times 10^{-6}$M to 2.34×10^{-4} l mole^{-1} sec^{-1} with $[LiCl] = 1.24 \times 10^{-2}$M is not the result of conversion of nucleophilic chloride ions into less nucleophilic chloride ion-pairs with increasing concentration. A number of other salts with non-nucleophilic anions cause similar decreases, and it was suggested that the phenomenon "is most reasonably assumed to reflect the operation of a simple medium effect, consistent with Debye–Hückel theory".[44b] In contrast, the decrease in the second-order constant for the reaction of *p*-nitrobenzyl chloride with lithium radiochloride in anhydrous acetone has been analysed in terms of the Acree equation, and the dissociation constant obtained for lithium chloride ion-pairs, 2.83×10^{-6} mole l^{-1}, is in reasonable agreement with that obtained conductometrically, 3.39×10^{-6} mole l^{-1}.[45]

Investigations which may be relevant to the effect of solvents on reaction mechanisms have been published on the following topics: proton exchange

[37] H. Ginsburg, G. Le Ny, O. Parguez, and B. Tchoubar, *Bull. Soc. Chim. France*, **1969**, 301.
[38] V. S. Karavan, L. M. Volkova, S. S. Istovik, and T. I. Temnikova, *Zh. Org. Khim.*, **5**, 694 (1969).
[39] E. Tommila and M. L. Savolainen, *Suomen Kemistilehti*, B **42**, 111 (1969).
[40] E. Tommila and E. Airo, *Suomen Kemistilehti*, B **42**, 104 (1969).
[41a] B. Singh, K. Behari, and B. Krishna, *Austral. J. Chem.*, **22**, 137 (1969).
[41b] R. W. Visghert and R. W. Sendeha, *Organic Reactivity* (Tartu), **6**, 212 (1969).
[42] M. Savolainen, E. Tommila, and E. Lindquist, *Ann. Acad. Sci. Fennicae*, Ser. A, No. 148 (1969).
[43] H. Hartmann and A. P. Schmidt, *Z. Phys. Chem.* (Frankfurt), **62**, 312 (1968); **66**, 183 (1969); I. S. Lyubovskii, A. L. Shapiro, V. I. Romanova, and S. Z. Levin, *Zh. Obshch. Khim.*, **39**, 478 (1969); *Chem. Abs.*, **71**, 21419 (1969); Y. Drougard and D. Decroocq, *Bull. Soc. Chim. France*, **1969**, 2972.
[44a] I. A. Koppel and V. A. Palm, *Organic Reactivity* (Tartu), **6**, 526 (1969).
[44b] R. A. Sneen and F. R. Rolle, *J. Am. Chem. Soc.*, **91**, 2140 (1969).
[45] P. Beronius, *Acta Chem. Scand.*, **23**, 1175 (1969).

reactions in Bu^tOH-H_2O mixtures;[46] NMR spectra of Bu^tOH-H_2O mixtures, and of solutions of electrolytes in methanol;[47] ultrasonic relaxation of water–alcohol[48] and water–dioxan[49] mixtures; physico-chemical studies of solutions in *N*-methyl-2-pyrrolidone,[50] of mixtures of sulpholan with water, methanol, and ethanol,[51] and of DMSO with methanol, ethanol, and benzene;[52] solvation of alkali-metal ions in DMSO by infrared spectroscopy;[53] spectroscopic studies of Bu^tOH-H_2O mixtures;[54] hydration numbers in solutions of $NaClO_4$, HCl, and $HClO_4$ by NMR spectroscopy.[55]

There has been a theoretical discussion of the structure of water.[56] (See also refs. 27 and 28.)

Isotope Effects

It has been estimated that the α-deuterium isotope effects for the limiting (unimolecular) solvolysis of alkyl chlorides, bromides, iodides, fluorides, and arenesulphonates are 1.14, 1.11, ca. 1.09, ca. 1.22, and ca. 1.20 respectively.[57] In agreement with the last value Streitwieser and Dafforn report a value of 1.22 at 25° for trifluoroacetolysis of isopropyl toluene-*p*-sulphonate.[58]

The formolysis of optically active [α-^{2}H]benzyl fluoride yields racemic formate. The α-deuterium iotope effect is $k_H/k_D = 1.15$, and the solvent isotope effect $k_{HCOOH}/k_{HCOOD} = 1.43$. The solvent isotope effect for the reaction of several substituted benzyl fluorides with thiophenoxide ion was also determined.[59]

The α-deuterium isotope effects (per 3 deuteriums) for the reactions of CD_3Br, CD_3I, CD_3OMs, and CD_3OTs with sodium thiosulphate in 50% v/v ethanol–water at 25° are 1.03, 1.06, 1.11, and 1.12 respectively.[60]

[46] E. K. Ralph and E. Grunwald, *J. Am. Chem. Soc.*, **91**, 2426, 2429 (1969).
[47] R. G. Anderson and M. C. R. Symons, *Trans. Faraday Soc.*, **65**, 2550 (1969); R. N. Butler and M. C. R. Symons, *ibid.*, p. 2559.
[48] M. J. Blandamer, N. J. Hidden, M. C. R. Symons, and N. C. Treloar, *Trans. Faraday Soc.*, **65**, 1805 (1969); *ibid.*, **64**, 3242 (1968); M. J. Blandamer, M. J. Foster, N. J. Hidden, and M. C. R. Symons, *ibid.*, **64**, 3247 (1968).
[49] K. Arakawa and N. Takenaka, *Bull. Chem. Soc. Japan*, **42**, 5 (1969).
[50] P. O. I. Virtanen and P. Ilvesluoto, *Suomen Kemistilehti*, B **41**, 354 (1969); P. O. I. Virtanen and R. Kerkela, *ibid.*, B **42**, 29 (1969).
[51] E. Tommila, E. Lindell, M. L. Virtalaine, and R. Laasko, *Suomen Kemistilehti*, B **42**, 95 (1969); E. Tommila and E. Lindell, *ibid.*, p. 93; H. Schott, *J. Pharm. Sci.*, **58**, 946 (1969).
[52] E. Tommila and T. Autio, *Suomen Kemistilehti*, B **42**, 107 (1969).
[53] B. W. Maxey and A. I. Popov, *J. Am. Chem. Soc.*, **91**, 20 (1969).
[54] M. J. Blandamer, M. F. Fox, M. C. R. Symons, K. W. Wood, and M. J. Wootten, *Trans. Faraday Soc.*, **64**, 3210 (1968).
[55] P. S. Knapp, R. O. Waite, and E. R. Malinowski, *J. Chem. Phys.*, **49**, 5459 (1968).
[56] K. Arakawa and K. Sasaki, *Bull. Chem. Soc. Japan*, **42**, 303 (1969).
[57] V. J. Shiner, M. W. Rapp, E. A. Malevi, and M. Wolfsberg, *J. Am. Chem. Soc.*, **90**, 7171 (1968).
[58] A. Streitwieser and G. A. Dafforn, *Tetrahedron Letters*, **1969**, 1263.
[59] C. Béguin, *Bull. Soc. Chim. France*, **1969**, 372; C. Béguin and J. J. Delpuech, *ibid.*, p. 378.
[60] G. E. Jackson and K. T. Leffek, *Can. J. Chem.*, **47**, 1537 (1969).

The isotope effect for the solvolysis of $[^2H_9]$-t-butyl chloride is larger in trifluoroethanol–water mixtures than in ethanol–water mixtures of the same Y-value. Since the percentage of olefin is also larger in the former, this difference in isotope effect may be the result of the rate being partly controlled by the rate of elimination from a reversibly formed ion-pair (see also ref. 9). The β-deuterium isotope effect for solvolysis of 1-phenylethyl chloride, a reaction which yields only a small amount of olefin, is almost the same in ethanol–water and trifluoroethanol–water mixtures. The α-deuterium isotope effect for the solvolysis of isopropyl p-bromobenzenesulphonate in trifluoroethanol–water mixtures is less than expected for a limiting mechanism, and is considered to be consistent with "nucleophilic attack on the tight ion pair" (cf. ref. 9).[61a]

The heat of solution of t-butyl chloride in absolute ethanol is 332 ± 8 cal mole^{-1}, and that for $[^2H_6]$-t-butyl chloride is 310 ± 5 cal mole^{-1}.[61b]

The ^{13}C isotope effect, $(k_{12}/k_{13} - 1) \times 100$, for the methanolysis of p-XC$_6$H$_4$CHBrCH$_3$ is $-0.05, 0.50$, and 1.27 respectively when $X = $ Me, H, and Br respectively.[62]

The deuterium isotope effect for the reaction of $[^2H_7]$benzyl chloride with radioactive chloride ion falls in the range 1.08—1.11, and that for the reaction of $[1,1,1,3,3,3-^2H_6]$-2-propyl chloride in the range 1.11—1.13.[63] Willi has presented a theoretical discussion of isotope effects in S_N2 reactions.[64]

The difference in $\Delta G^{\ddagger}$, $\Delta H^{\ddagger}$, and $\Delta S^{\ddagger}$ for the solvolysis of alkyl halides and benzenesulphonates in H_2O and D_2O has been discussed.[65]

The ^{14}C isotope effect for the hydrolysis of N,N-dimethyl-o-toluidine $[^{14}C]$methiodide has been determined.[66]

There have been investigations of steric isotope effects in the racemization of biphenyls,[67] and of the isotope effect on the ionization of $[^2H_9]$trimethylamine.[68]

Neighbouring Group Participation[69]

Participation by Ether and Hydroxyl Groups

Acetolysis of 5-methoxypentyl p-bromobenzenesulphonate is accompanied by extensive conversion into methyl p-bromobenzenesulphonate. Since this is

61a V. J. Shiner, W. Dowd, R. D. Fisher, S. R. Hartshorn, M. A. Kessick, L. Milakofsky, and M. W. Rapp, *J. Am. Chem. Soc.*, **91**, 4838 (1969).
61b J. O. Stoffer, W. C. Duer, and G. L. Bertrand, *Chem. Comm.*, **1969**, 1188.
62 J. Bron and J. B. Stothers, *Can. J. Chem.*, **47**, 2506 (1969).
63 H. Strecker and H. Elias, *Chem. Ber.*, **102**, 1270 (1969).
64 A. V. Willi, *Z. Phys. Chem.* (Frankfurt), **66**, 317 (1969).
65 L. Treindl, R. E. Robertson, and S. E. Sugamori, *Can. J. Chem.*, **47**, 3397 (1969).
66 J. McKenna, J. M. McKenna, and J. M. Stuart, *J. Chem. Soc.* (B), **1969**, 698.
67 R. E. Carter and L. Dahlgren, *Acta Chem. Scand.*, **23**, 504 (1969).
68 D. Northcott and R. E. Robertson, *J. Phys. Chem.*, **73**, 1559 (1969).
69 Neighbouring group participation in reactions of carbohydrates has been discussed: B. Capon, *Chem. Rev.*, **69**, 407 (1969).

prevented by the addition of lithium perchlorate it was considered to proceed via solvent-separated ion-pairs.[70] The trifluoroethanolysis of 5-methoxypentyl toluene-*p*-sulphonate is accompanied by 28.5% conversion into methyl toluene-*p*-sulphonate but this is only reduced to 19% by the addition of lithium perchlorate. This suggests that two-thirds of the methyl toluene-*p*-sulphonate is formed from intimate ion-pairs, and that intimate ion-pair (5) is converted into intimate ion-pair (6) without passing through a solvent-separated ion-pair.[71]

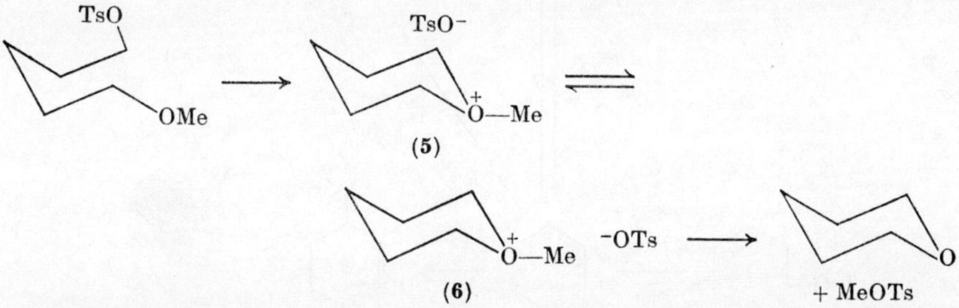

The NMR spectra of several cyclic oxonium ions have been published.[72]

The acetolysis of *cis*-5-methoxypent-3-en-1-yl *p*-bromobenzenesulphonate (7) is anchimerically assisted and yields compounds (8)—(11). It was shown that the corresponding toluene-*p*-sulphonate yields methyl toluene-*p*-sulphonate but that formation of this is prevented by the addition of lithium perchlorate. It was thought that lithium perchlorate would also prevent capture to yield the allylic toluene-*p*-sulphonate (13), and that products (10) and (11) must therefore be formed from allylic cation (12).[73]

The entropies of activation for the cyclization in aqueous acetone of 4-chlorobutanol and 5-chloropentanol are 5—6 cal deg^{-1} mole^{-1} less negative than that for the S_N2 solvolysis of butyl chloride.[74]

It has been shown that hydrolysis of 4-methoxy[1-^{14}C]butyl bromide in water yields 4-methoxybutanol with the label scrambled equally between C-1 and C-4. This result is readily explained if the cyclic oxonium ion (14) is an intermediate.[70] The absence of products of methyl–oxygen fission does not exclude the intervention of this ion as claimed.[75] Five-membered rings are generally opened as well as formed more rapidly than six-membered ones,[76]

---

[70] See *Org. Reaction Mech.*, **1967**, 66.
[71] J. R. Hazen, *Tetrahedron Letters*, **1969**, 1897.
[72] R. J. Gargiulo and D. S. Tarbell, *Proc. Nat. Acad. Sci.*, **62**, 52 (1969).
[73] J. R. Hazen and D. S. Tarbell, *Tetrahedron Letters*, **1968**, 5927.
[74] G. Kohnstam and M. Penty, in *Hydrogen-Bonded Solvent Systems* (Ed. A. K. Covington and P. Jones), Taylor and Francis, London, 1968, p. 275.
[75] M. J. Blandamer, H. S. Golinkin, and R. E. Robertson, *J. Am. Chem. Soc.*, **91**, 2678 (1969).
[76] Cf. *Org. Reaction Mech.*, **1968**, 74; B. Capon and D. Thacker, *J. Chem. Soc.*(B), **1967**, 1322.

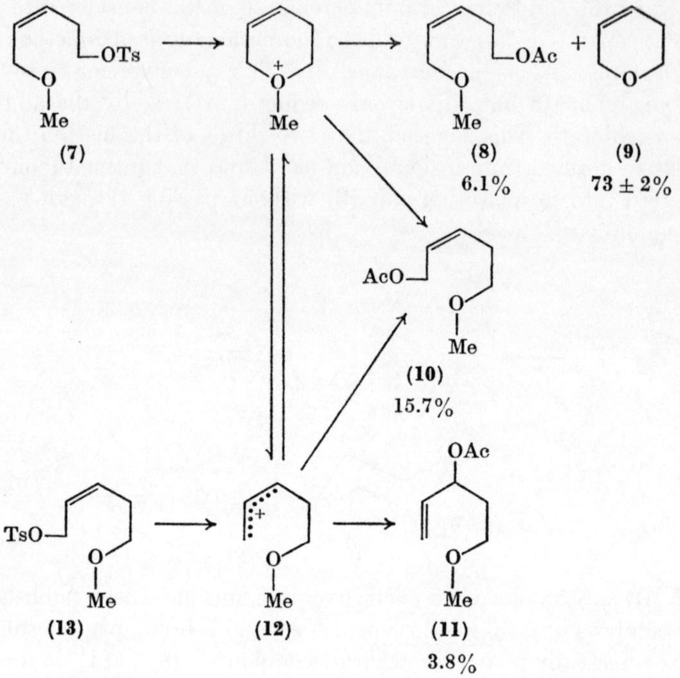

and the failure of the five-membered cyclic oxonium ion (14) to undergo methyl–oxygen fission, whereas the analogous six-membered ion does, can easily be explained as the result of this difference. The fact that 4-methoxy-butyl bromide reacts about nine times faster than 4-chlorobutanol suggests that there is little O–H bond breaking in the transition state for the latter.[75]

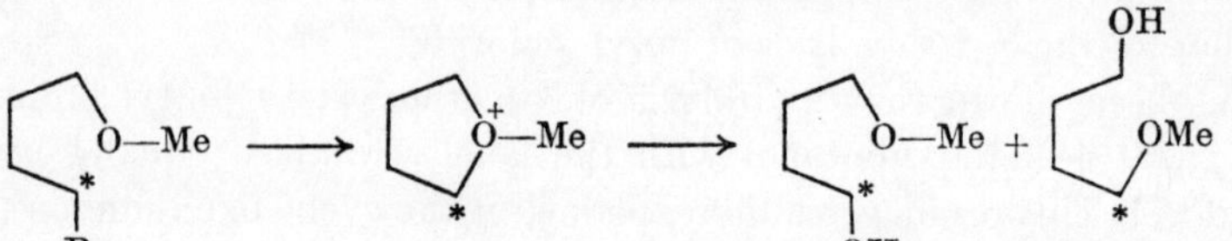

The deuterium isotope effect for the ethanolysis of 4-methoxy-1-[1,1-^{2}H$_2$]-pentyl p-bromobenzenesulphonate is $k_H/k_D = 1.01$ (see p. 34).[77]

Solvolysis of methyl 4-O-p-nitrobenzenesulphonyl-β-D-xylopyranoside (15) in aqueous acetate buffer yields methyl 3,4-anhydro-α-L-arabinoside (16) formed by HO-2 participation, and 4-O-methyl-L-arabinose (17) formed by MeO-5 participation. Methyl 4-O-p-nitrobenzenesulphonyl-β-D-glucoside

77 R. Eliason, M. Tomič, S. Borčić, and D. E. Sunko, *Chem. Comm.*, 1968, 1490.

reacts similarly to yield methyl 3,4-anhydro-β-D-galactopyranoside and 4-*O*-methylgalactose but also yields some methyl α-L-altrofuranoside formed through participation by the ring oxygen.[78]

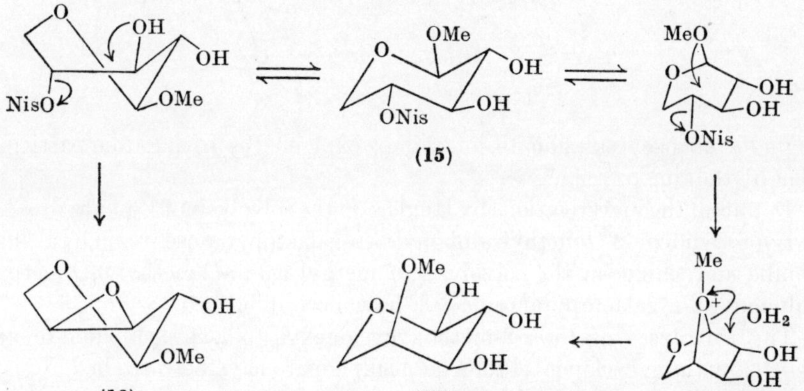

(15)

(16)   (17)

The relative rates of epoxide formation from (18), (19), and (20) are 1:180:23.3.[79] Other examples of O-participation in reactions of carbohydrates are described in refs. 80 and 81.

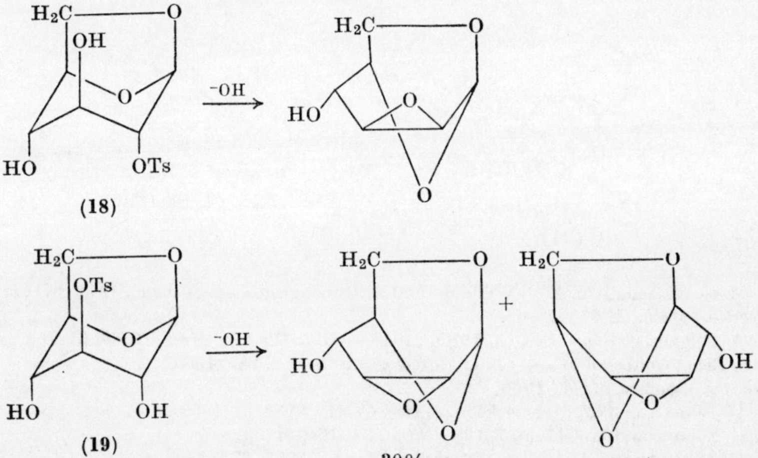

(18)

(19)   30%   70%

The products from the deamination of methyl 4-amino-4-deoxy-α-D-glucopyranoside are similar to those obtained from solvolysis of the correspond-

[78] J. G. Buchanan, A. R. Edgar, and D. G. Large, *Chem. Comm.*, **1969**, 558.
[79] M. Černý, J. Staněk, and J. Pacák, *Coll. Czech. Chem. Comm.*, **34**, 849 (1969).
[80] J. G. Buchanan and A. R. Edgar, *Carbohydrate Res.*, **10**, 295 (1969).
[81] M. Jarman and W. C. J. Ross, *Carbohydrate Res.*, **9**, 139 (1969).

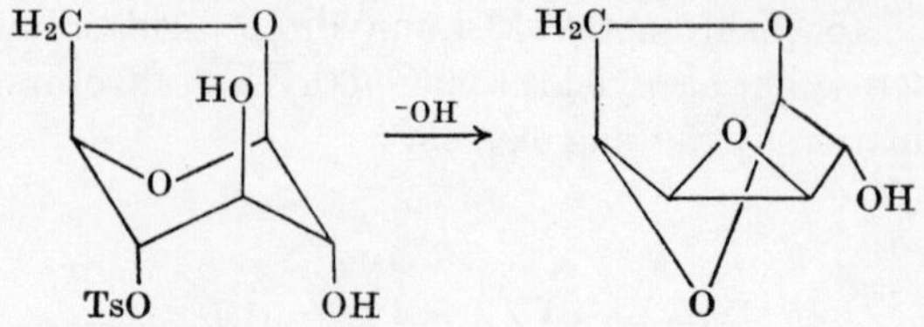

(20)

ing p-nitrobenzenesulphonate, and can be explained as arising from participation by the ring oxygen.[82]

Details of the investigations by Hughes on the solvolysis of 1,6-anhydro-3,4-O-isopropylidene-2-O-methylsulphonyl-β-D-galactopyranose,[83] and by Brimacombe and Ching on the solvolysis of methyl 2,3-di-O-methyl-6-O-methyl-sulphonyl-β-D-galactopyranoside[84, 85] have been published.

The solvolyses of *trans*-2-methoxycyclohexyl toluene-p-sulphonate and *trans*-2-methoxycyclopentyl toluene-p-sulphonate have been studied.[86]

Other examples of participation by hydroxyl and methoxyl groups are given in refs. 87—93.

The solvolyses of compounds (**21**) and (**22**) in 95% aqueous ethanol and 80% aqueous acetone respectively are not anchimerically assisted.[94, 95]

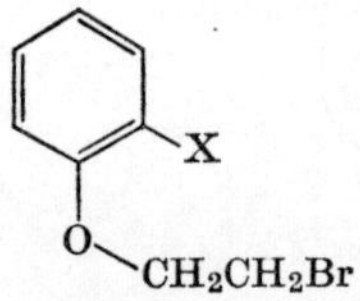

(X = NO₂, MeO, Cl)

(21)

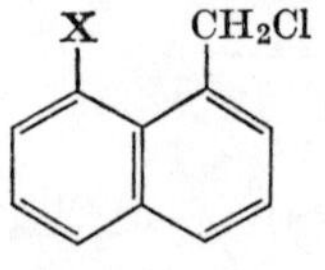

(X = Cl, Br, OMe)

(22)

[82] N. M. K. Ng Ying Kin, J. M. Williams, and A. Horsington, *Chem. Comm.*, **1969**, 971; cf. *Org. Reaction Mech.*, **1967**, 67—68.

[83] N. A. Hughes, *J. Chem. Soc.*(C), **1969**, 2263; see *Org. Reaction Mech.*, **1967**, 67—68.

[84] J. S. Brimacombe and O. A. Ching, *Carbohydrate Res.*, **9**, 287 (1969).

[85] See *Org. Reaction Mech.*, **1968**, 72.

[86] D. D. Roberts and W. Hendrickson, *J. Org. Chem.*, **34**, 2415 (1969).

[87] L. H. Zalkow and M. Ghosal, *J. Org. Chem.*, **34**, 1646 (1969).

[88] R. M. Moriarty and T. Adams, *Tetrahedron Letters*, **1969**, 3715.

[89] G. Bakassian, F. Chizat, D. Sinou, and G. Descotes, *Bull. Soc. Chim. France*, **1969**, 621.

[90] P. F. Vogt and D. F. Tavares, *Can. J. Chem.*, **47**, 2875 (1969).

[91] J. W. Apsimon and H. Krehm, *Can. J. Chem.*, **47**, 2859 (1969).

[92] D. C. Kleinfelter and J. H. Long, *Tetrahedron Letters*, **1969**, 347.

[93] G. E. McCasland, M. O. Naumann, and L. J. Durham, *J. Org. Chem.*, **34**, 1382 (1969).

[94] V. Dauksas and I. Dembinskiene, *Zh. Vses. Khim. Obshch.*, **14**, 119 (1969); *Chem. Abs.*, **70**, 114328 (1969).

[95] D. C. Kleinfelter and P. H. Chen, *J. Org. Chem.*, **34**, 1741 (1969).

Participation by Thioether and Thiol Groups[96]

An interesting example of S-4 participation has been reported by Paquette, Meehan, and Wise who have shown that compound (**23**) is acetolysed about 5000 times faster than compound (**28**) and yields the corresponding acetate (**24**) with retention of configuration. The closely related compound (**25**) probably does not react with participation as it is acetolysed only four times faster than its epimer (**29**). The product does however contain 79% of the acetate of retained configuration, but this was explained as arising from ion (**26**). The other product from this reaction is compound (**27**) formed by fragmentation. Compounds (**28**) and (**29**), in which the sulphur is not stereochemically oriented for participation, react very similarly to the bishomocubyl compounds discussed on p. 17.[97]

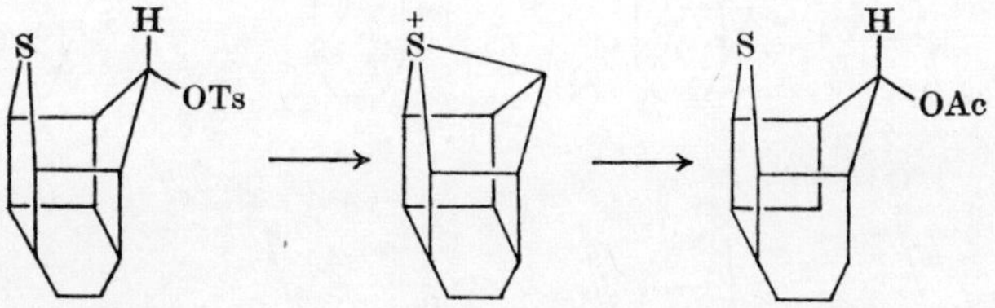

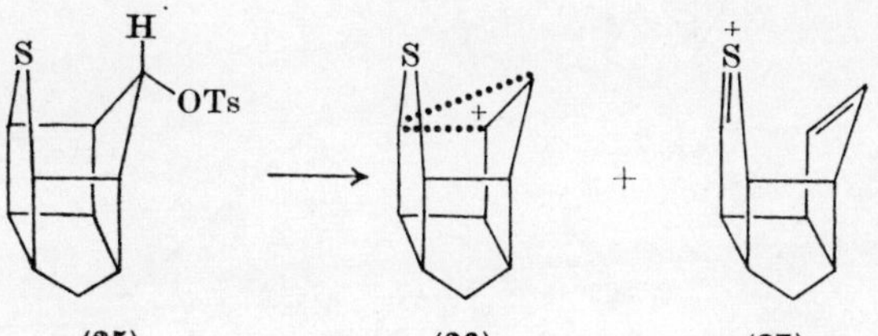

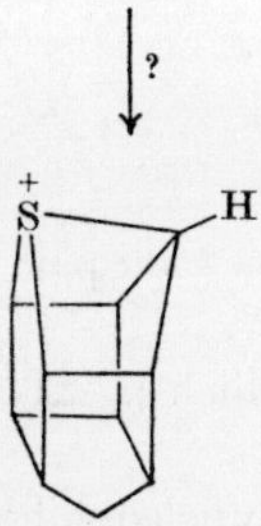

Compound (**30**) is racemized on heating in ethyl methyl ketone. Under the same conditions the deuterated compound (**31**) yields a product in which the

[96] W. H. Mueller, "Thiuronium Ions as Reaction Intermediates", *Angew. Chem. Internat. Ed. Engl.*, **8**, 482 (1969).

[97] L. A. Paquette, G. V. Meehan, and L. D. Wise, *J. Am. Chem. Soc.*, **91**, 3231 (1969).

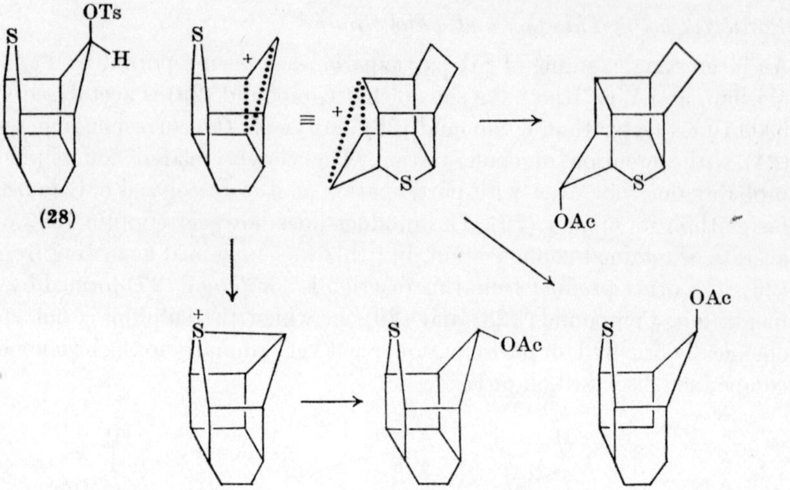

deuterium is scrambled. The reaction possibly involves ion-pair return from
(**32**) or a four-centred transition state (**33**).[98]

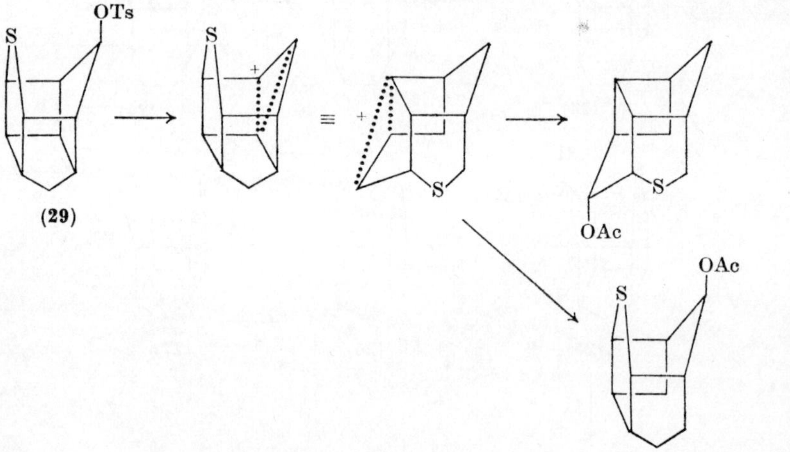

S-participation occurs in the acetolysis of (**34**), which is 220 times faster
than that of its *para*-isomer.[99]

The reaction of the episulphonium salt (**35**) with nucleophiles in a 2:1:1
mixture of acetonitrile, cyclohexene, and methylene chloride has been studied.
Most of these (e.g. I⁻, Br⁻, Cl⁻, F⁻, and MeS⁻) reacted with nucleophilic attack
at sulphur and generation of cyclooctene (**36**). (**35**) was recovered unchanged

[98] A. R. Dunn and R. J. Stoodley, *Chem. Comm.*, **1969**, 1169.
[99] M. Hojo, T. Ichi, Y. Tamaru, and Z. Yoshida, *J. Am. Chem. Soc.*, **91**, 5170 (1969).

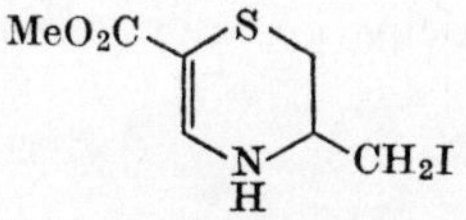

(30)

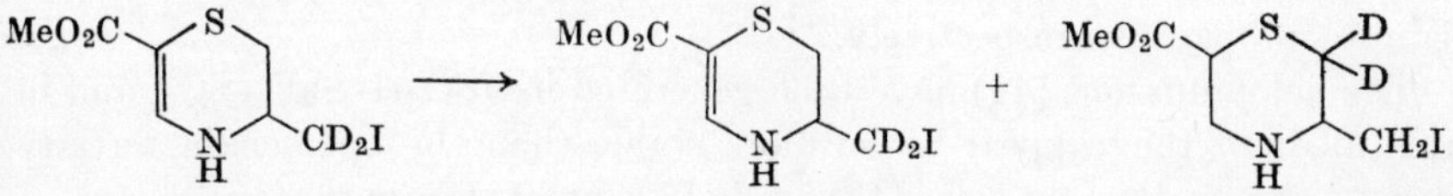

(31)

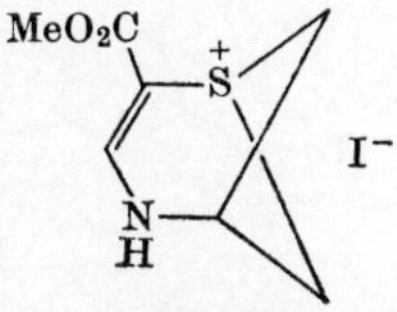

(32) 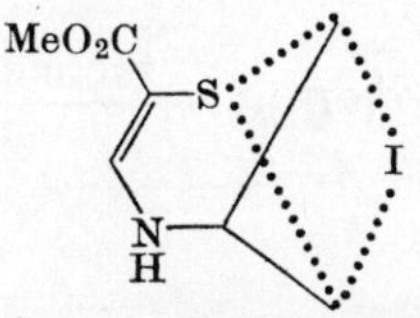   (33) 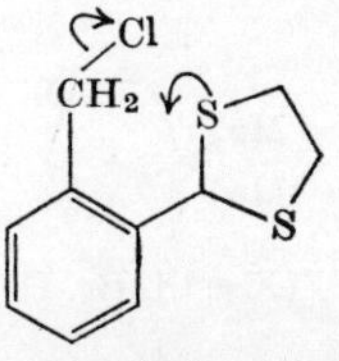   (34)

after 30 minutes in a mixture of acetonitrile, acetic acid, and methylene chloride (2:1:1).[100] It was shown by NMR spectroscopy that (35) reacts with Cl⁻ in CD_3NO_2 to form (37).[101]

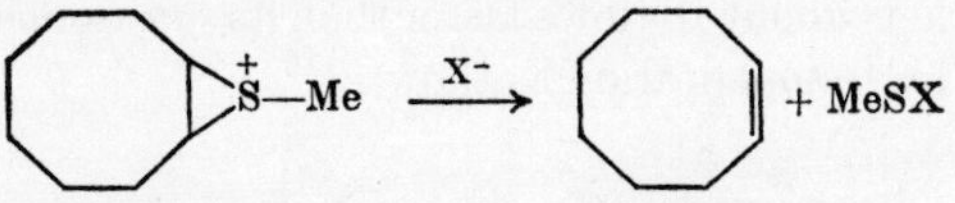

(35)   (36) 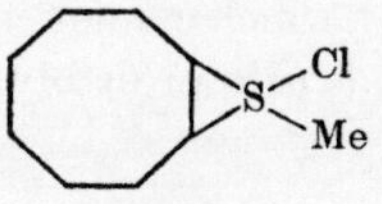   (37)

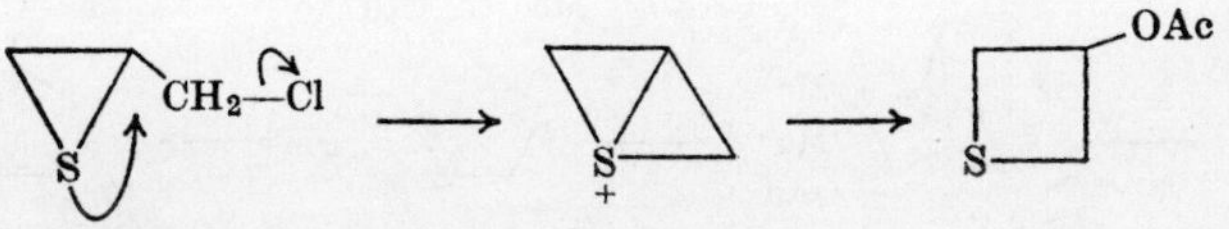

(38)   (39)   (40)

3-Chloropropene sulphide (38) is acetolysed 1000 times faster than cyclopropylmethyl chloride, and one of the products is 3-acetoxythietane (40) which is presumably formed via ion (39).[102] 3-Chloropropene oxide may react similarly, since its rate of acetolysis is about 3 times that of cyclopropylmethyl chloride[102, 103] (see also Chapter 1, refs. 178 and 179).

[100] D. C. Owsley, G. K. Helmkamp, and S. N. Spurlock, *J. Am. Chem. Soc.*, **91**, 3606 (1969).
[101] D. C. Owsley, G. K. Helmkamp, and M. F. Rettig, *J. Am. Chem. Soc.*, **91**, 5239 (1969).
[102] H. Morita and S. Oae, *Tetrahedron Letters*, **1969**, 1347.
[103] H. G. Richey and D. V. Kinsman, *Tetrahedron Letters*, **1969**, 2505.

Other examples of S-participation are reported in refs. 104—106.

Participation by Halogens

Deamination of 2-chloropropylamine and 2-bromopropylamine proceeds with migration of the halogen to yield 53 and 76% of 1-chloropropan-2-ol and 1-bromopropan-2-ol respectively.[107]

The halonium ions (41) have been generated in $FSO_3H–SbF_5–SO_2$, and in this medium they appear to be more stable than the open-chain tertiary cations. An ion thought to be (43) has been generated from the iodo-acetylene (42).[108]

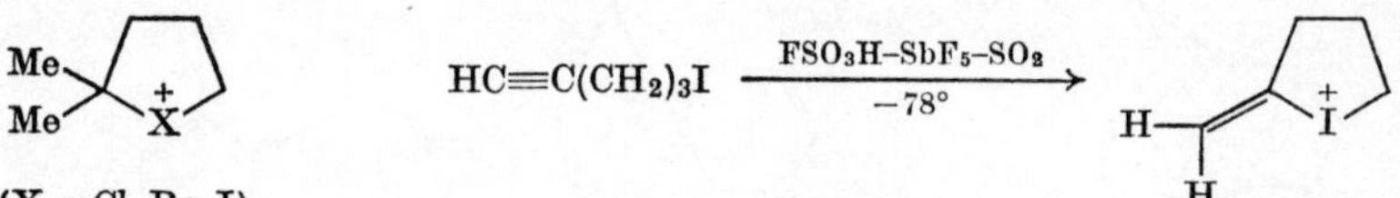

(41)   (42)   (43)

Participation by Carbonyl Groups

Details of Gassman and his coworkers' investigation of participation by the double bond of the enol in the acetolysis of *anti*-7-toluene-*p*-sulphonyloxynor-bornanone have been published.[109] The corresponding 3,3-dimethyl compound (44), which cannot enolize, reacts about 18 times faster than its *syn*-isomer and yields (45) probably formed by fragmentation as shown.[110]

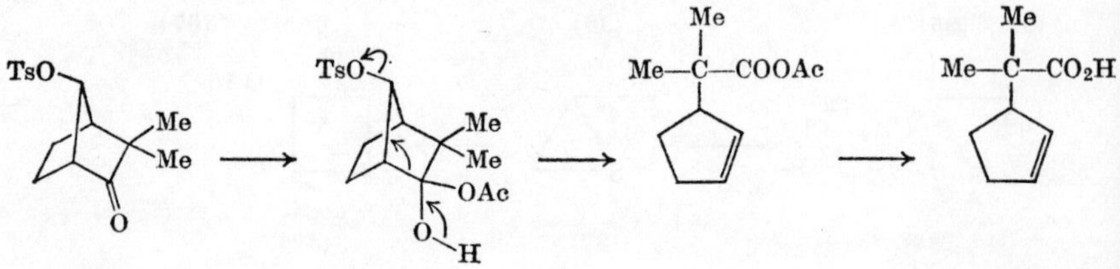

(44)   (45)

104 W. E. Parham and D. G. Weetman, *J. Org. Chem.*, **34**, 56 (1969).
105 H. Nimz, *Chem. Ber.*, **102**, 3803 (1969).
106 A. de Groot, J. A. Boerma, and H. Wynberg, *Rec. Trav. Chim.*, **88**, 994 (1969).
107 O. A. Reutov, A. S. Gudkova, G. M. Ostapchuk, and V. G. Okimova, *Izv. Akad. Nauk SSSR, Ser. Khim.*, **1968**, 1922; *Chem. Abs.*, **69**, 105627 (1968).
108 G. A. Olah, J. M. Bollinger, and J. Brinich, *J. Am. Chem. Soc.*, **90**, 6988 (1968).
109 P. G. Gassman, J. L. Marshall, and J. M. Hornback, *J. Am. Chem. Soc.*, **91**, 5811 (1969); see *Org. Reaction Mech.*, **1966**, 53.
110 P. G. Gassman and J. M. Hornback, *Tetrahedron Letters*, **1969**, 1325; *J. Am. Chem. Soc.*, **91**, 5817 (1969).

Two successive participations by an enolate anion occur when compound (**46**) is treated with potassium t-butoxide in dimethylformamide containing a little t-butyl alcohol, to yield (**47**).[111]

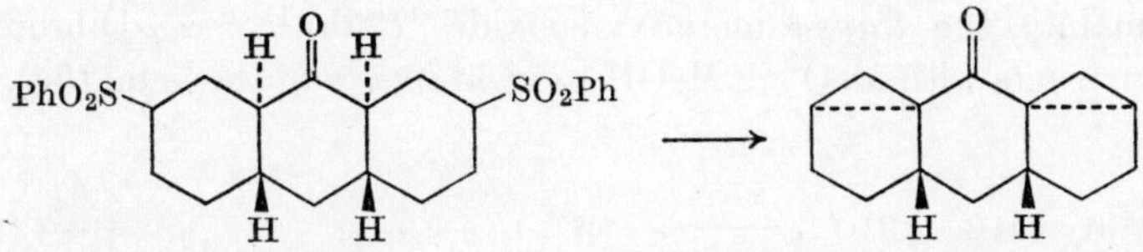

(**46**)　　　　　　　　　　(**47**)

Two examples of neighbouring enolate participation to yield four-membered rings have been reported (see equations 8 and 9).[112]

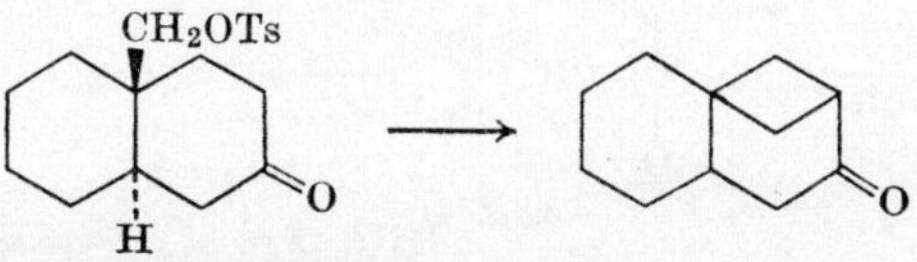

$$\ldots (8)$$

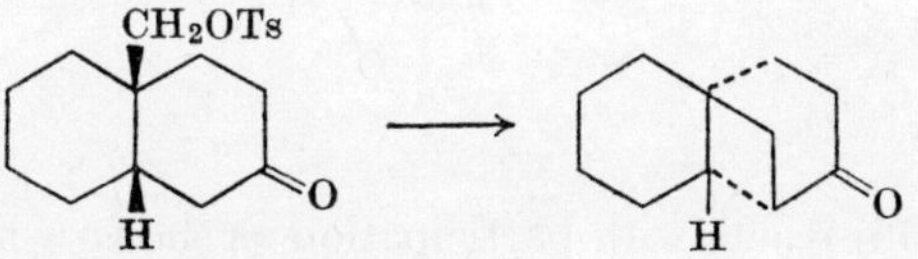

$$\ldots (9)$$

Ion (**49**) has been generated by dissolving compound (**48**) in sulphuric acid.[113]

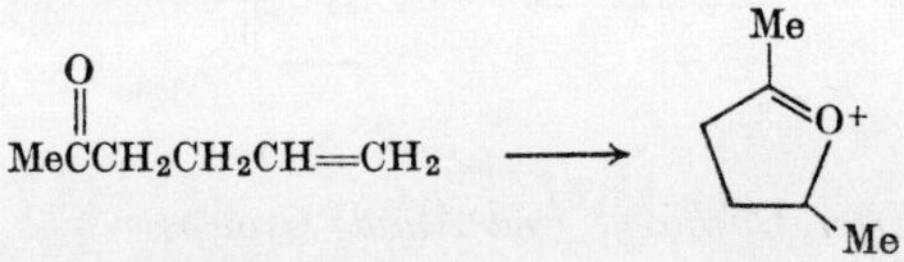

$$MeCCH_2CH_2CH{=}CH_2 \longrightarrow$$

(**48**)　　　　　　　　　　(**49**)

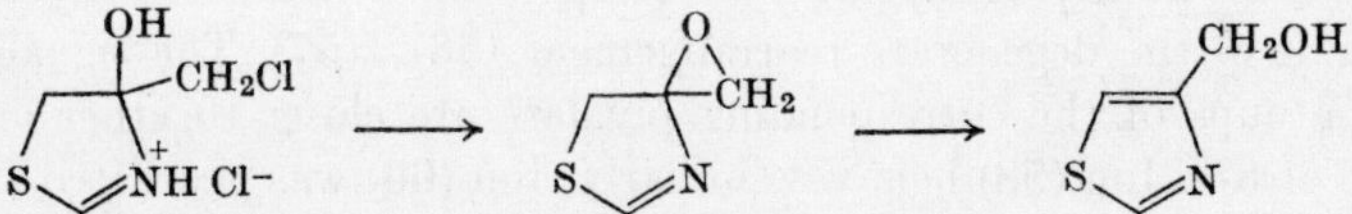

(**50**)　　　　　(**51**)　　　　　(**52**)

[111] W. L. Parker and R. B. Woodward, *J. Org. Chem.* **34**, 3085 (1969).
[112] P. C. Mukharji and A. N. Ganguly, *Tetrahedron*, **25**, 5267 (1969).
[113] C. U. Pittman and S. P. McManus, *Chem. Comm.*, **1968**, 1479; see also D. M. Brouwer, *Rec. Trav. Chim.*, **88**, 530 (1969).

When *sym*-dichloroacetone is treated with thiobenzamide it yields compound (50). This compound reacts with sodium hydroxide in methanol to yield (52) which is thought to be formed via epoxide (51).[114]

The α-chloro-aldehydes (53a) react with sodium methoxide in ether to yield mainly the *trans*-α-methoxy-epoxides (53b).[115a] α,α-Dibromoaceto-phenone reacts with MeO⁻ in MeOH to yield *inter alia* the ketal (54).[115b]

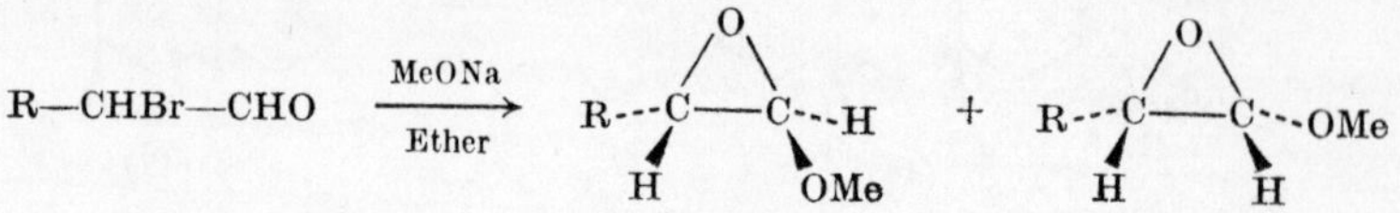

$$R—CHBr—CHO \xrightarrow[\text{Ether}]{\text{MeONa}} R \cdots \overset{\text{O}}{\underset{\text{H}}{C}}—\underset{\text{OMe}}{C} \cdots H \quad + \quad R \cdots \overset{\text{O}}{\underset{\text{H}}{C}}—\underset{\text{H}}{C} \cdots OMe$$

(53a) (53b)

(R = C₂H₅, C₄H₉, C₅H₁₁, C₆H₅) Major product

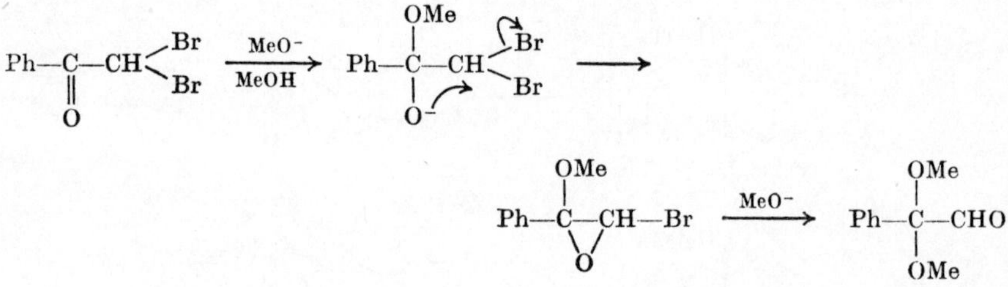

$$Ph—\overset{\text{O}}{\underset{\|}{C}}—CH \overset{Br}{\underset{Br}{<}} \xrightarrow[\text{MeOH}]{\text{MeO}^-} Ph—\overset{\text{OMe}}{\underset{\text{O}^-}{C}}—CH \overset{Br}{\underset{Br}{<}} \longrightarrow$$

$$Ph—\overset{\text{OMe}}{\underset{\text{O}}{C}}—CH—Br \xrightarrow{\text{MeO}^-} Ph—\overset{\text{OMe}}{\underset{\text{OMe}}{C}}—CHO$$

(54)

The Schiff base (55) reacts with participation as shown when it is treated with sodium methoxide in dioxan.[116]

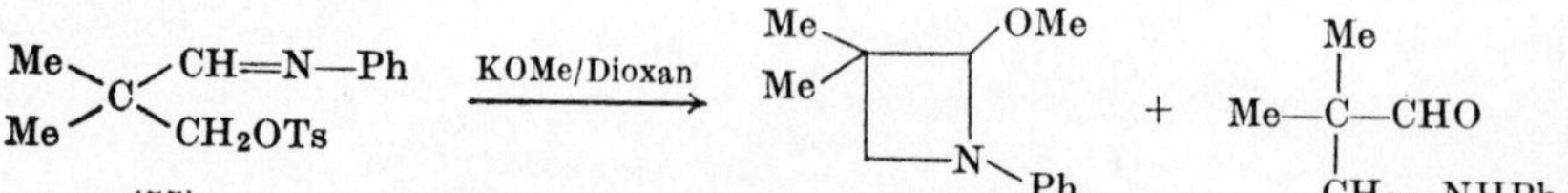

$$\underset{Me}{\overset{Me}{>}}C\overset{CH=N—Ph}{\underset{CH_2OTs}{<}} \xrightarrow{\text{KOMe/Dioxan}} \begin{array}{c} Me \\ Me \end{array} \rlap{\text{OMe}} \underset{N—Ph}{\square} \quad + \quad Me—\overset{Me}{\underset{CH_2—NHPh}{C}}—CHO$$

(55)

Participation by Ester, Carboxyl, and Amide Groups

The NMR spectrum of the dioxolenium cation (56) in CD₃NO₂ shows separate signals for the methyl groups at room temperature but these coalesce at 105° as a result of the degenerate rearrangement (56)⇌(57). The signals of the t-butyl groups of the corresponding pivalate are closer together and they coalesce at 87°. Ion (58) behaves similarly. Ion (60) was generated from the dioxolan (59) by hydride transfer to Ph₃C⁺BF₄⁻. At low temperatures the NMR spectrum in CH₃CN showed the presence of three distinct methyl groups.

[114] K. Brown and R. A. Newberry, *Tetrahedron Letters*, **1969**, 2797.
[115a] J. J. Riehl and L. Thil, *Tetrahedron Letters*, **1969**, 1913.
[115b] C. Raulet and E. Levas, *Compt. Rend., C*, **269**, 996 (1969).
[116] F. Nerdel, P. Weyerstahl, and K. Zabel, *Chem. Ber.*, **102**, 1606 (1969); F. Nerdel, H. Kaminski, and P. Weyerstahl, *ibid.*, p. 3679.

At 80° the signals broadened, indicative of a rapid interchange, but decomposition set in before the temperature could be raised sufficiently for coalescence to be observed.[117] The low-temperature NMR spectrum of ion (61) showed signals from two t-butyl groups which coalesced at 110°. Ion (62) derived from pivaloyl pentaerythritol behaved similarly and all three t-butyl groups became equivalent at 110°.[118]

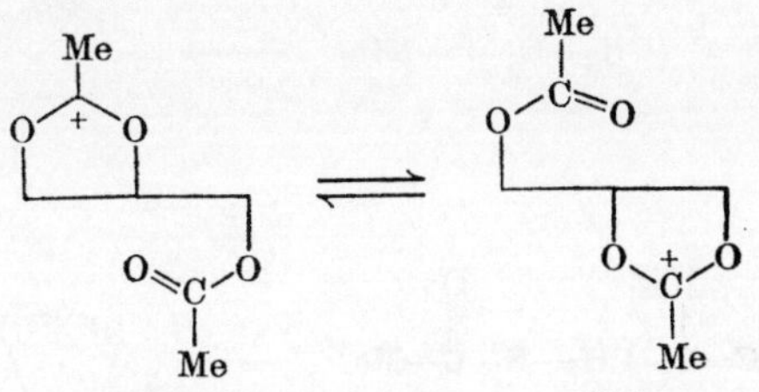

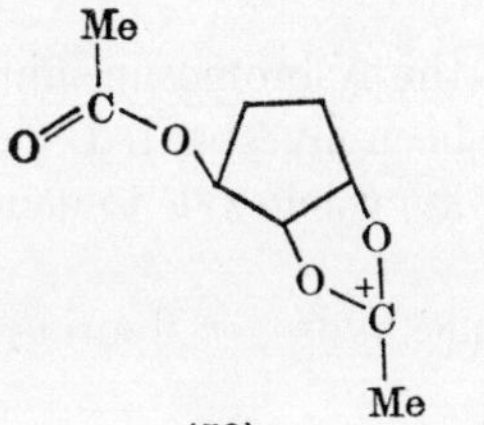

Details of King and Allbutt's investigation of the stereochemistry of the opening of dioxolenium ions have been published.[119]

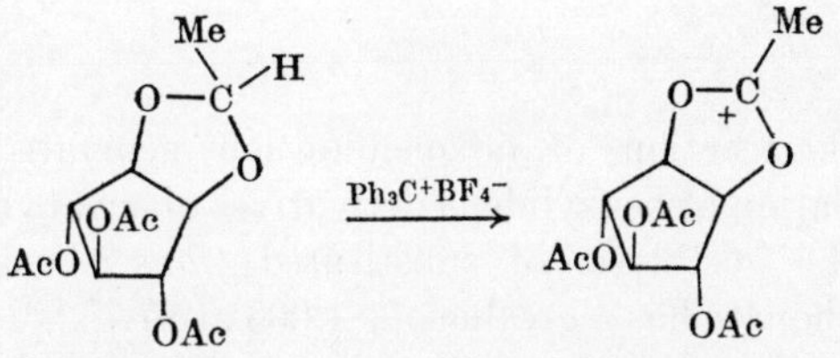

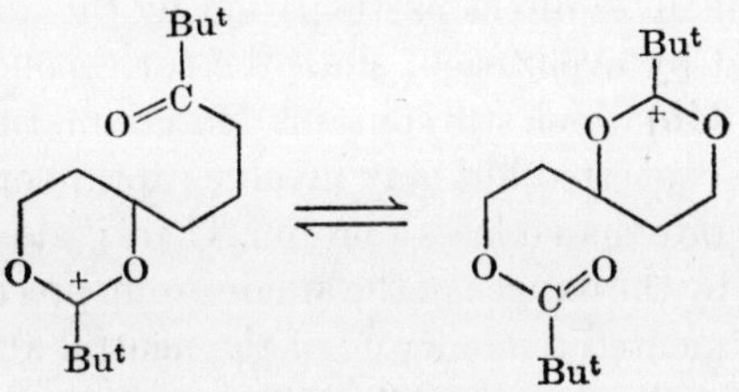

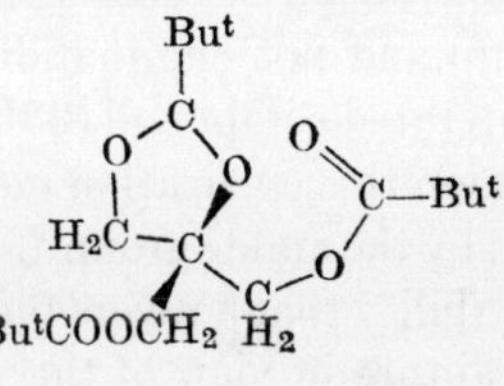

Dioxolenium ions and oxathiolenium ions have been generated from the corresponding allyl esters in H_2SO_4 and FSO_3H (see equations 10 to 12).[120]

[117] H. Paulsen and H. Behre, *Angew. Chem. Internat. Ed. Engl.*, 8, 886, 887 (1969).
[118] H. Paulsen, M. Meyborg, and H. Behre, *Angew. Chem. Internat. Ed. Engl.*, 8, 888 (1969).
[119] J. F. King and A. D. Allbutt, *Can. J. Chem.*, 47, 1445 (1969); see *Org. Reaction Mech.*, 1966, 55—60.
[120] C. U. Pittman and S. P. McManus, *Tetrahedron Letters*, 1969, 339.

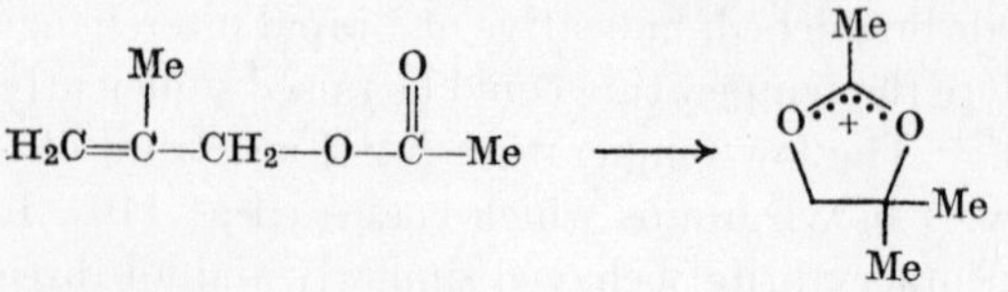

$$\underset{\underset{H_2C=\overset{|}{C}-CH_2-O-\overset{||}{C}-Me}{\text{Me}\qquad\qquad\text{O}}}{} \longrightarrow \qquad \text{(see structure)} \qquad \ldots (10)$$

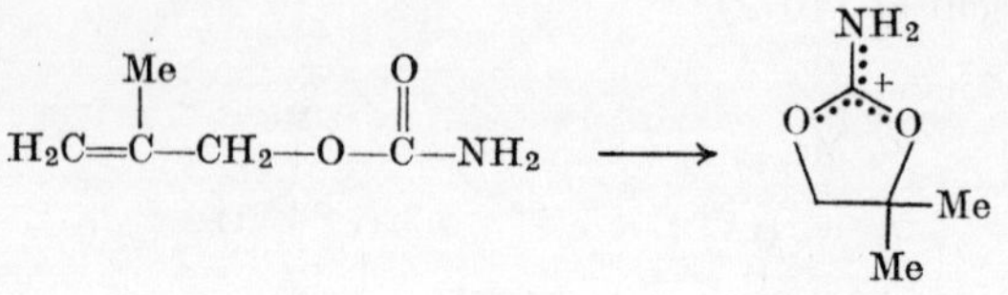

$$\underset{\underset{H_2C=\overset{|}{C}-CH_2-O-\overset{||}{C}-NH_2}{\text{Me}\qquad\qquad\text{O}}}{} \longrightarrow \qquad \text{(see structure)} \qquad \ldots (11)$$

$$\underset{\underset{H_2C=\overset{|}{C}-CH_2-S-\overset{||}{C}-Me}{\text{Me}\qquad\qquad\text{O}}}{} \longrightarrow \qquad \text{(see structure)} \qquad \ldots (12)$$

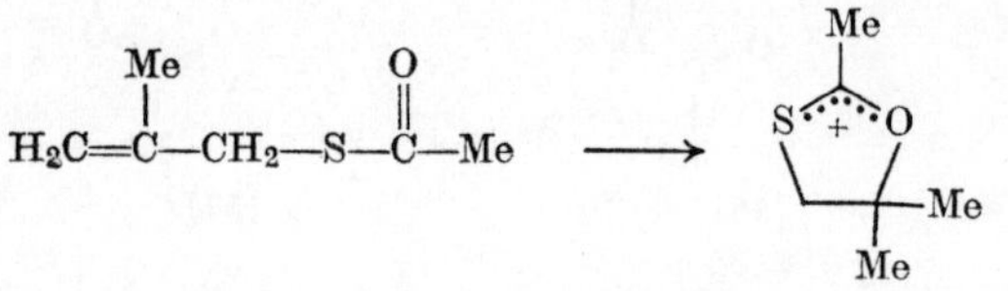

The reactions of dioxolenium ions generated by the N-bromosuccinimide oxidation of benzylidene derivatives of sugars have been investigated.[121]

The acetolysis of substituted *trans*-2-benzoyloxycyclohexyl toluene-*p*-sulphonate has a ρ-value of -1.003 at $97.7°$.[122]

Participation by acyl groups in reactions in liquid hydrogen fluoride has been reviewed.[123]

Other examples of participation by acyl groups are described in ref. 124.

Conversion of (63a) into (64a) involves direct participation by the amide group and not elimination followed by cyclization, since the corresponding deuterated compound (63b) yields (64b) which still contains deuterium. In our opinion this cyclization may not be concerted but may involve capture of ion (65) by the amide group at a greater rate than it loses a proton. There is also the possibility that a reversible capture by the oxygen of the amide group precedes N-capture in view of the well documented tendency[125] of the neutral amide group to react with O-attack. The phenylhydrazone (66) is also cyclized without loss of deuterium.[126]

Neighbouring sulphonamide participation occurs in the conversion of (67) and (69) into (68) and (70).[127]

[121] S. Hanessian and N. R. Plessas, *J. Org. Chem.*, **34**, 1053 (1969).
[122] K. B. Gash and G. U. Yuen, *J. Org. Chem.*, **34**, 720 (1969).
[123] J. Lenard, *Chem. Rev.*, **69**, 625 (1969).
[124] J. A. Montgomery and K. Hewson, *Chem. Comm.*, **1969**, 15.
[125] See *Org. Reaction Mech.*, **1966**, 55.
[126] C. L. Mao, F. E. Henoch, and C. R. Hauser, *Chem. Comm.*, **1968**, 1595.
[127] W. W. Paudler, A. G. Zeiler, and G. R. Gapski, *J. Org. Chem.*, **34**, 1001 (1969).

(63) a: A = H; b: A = D

(64) a: A = H; b: A = D

(65)

(66)

(67)

(68)

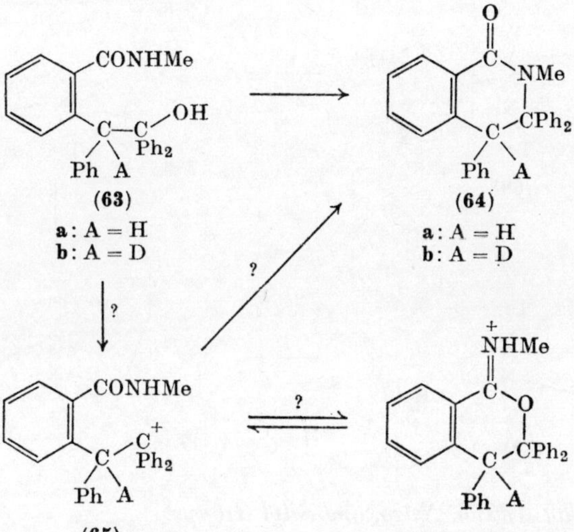

Other examples of participation by amide[128, 129] and carbamate[130] groups have been reported.

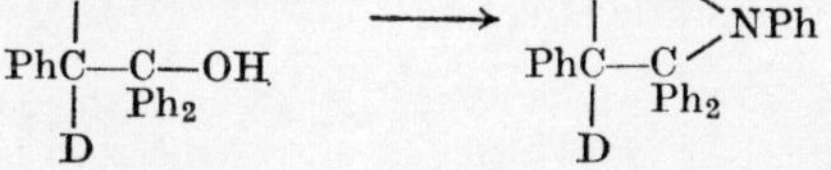

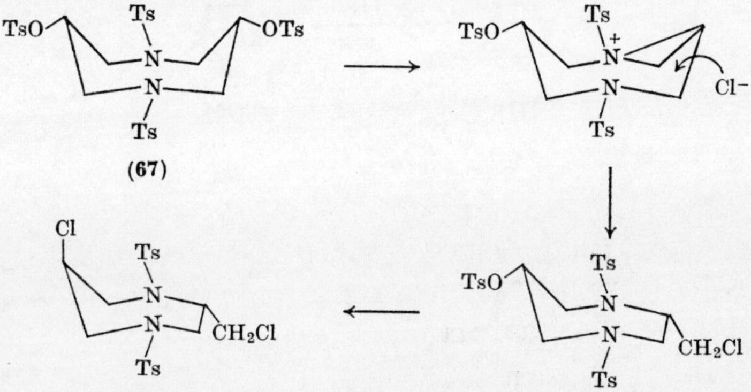

[128] S. H. Pines, M. A. Kozlowski, and S. Karady, *J. Org. Chem.*, **34**, 1621 (1969).

[129] V. N. Bochenkov and V. F. Borodkin, *Izv. Vyssh. Ucheb. Zaved., Khim. Khim. Tekhnol.* **12**, 429 (1969); *Chem. Abs.*, **71**, 49821 (1969); R. B. Morin, B. G. Jackson, E. H. Flynn, R. W. Roeske, and S. L. Andrews, *J. Am. Chem. Soc.*, **9**, 1396 (1969).

[130] T. A. Foglia and D. Swern, *J. Org. Chem.*, **34**, 1680 (1969).

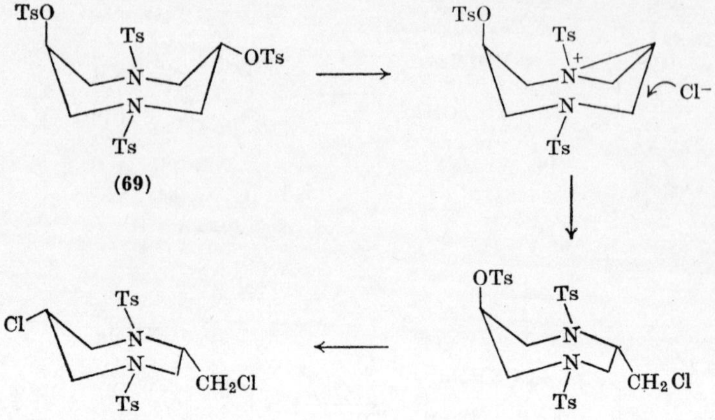

(69)

(70)

Participation by Amino, Nitro, and Silyl Groups

Cyclization of **(71)** with ethylamine in acetonitrile occurs readily to yield **(72)**, but the analogous reaction of **(73)** does not take place. It was suggested that this was the result of a steric effect.[131]

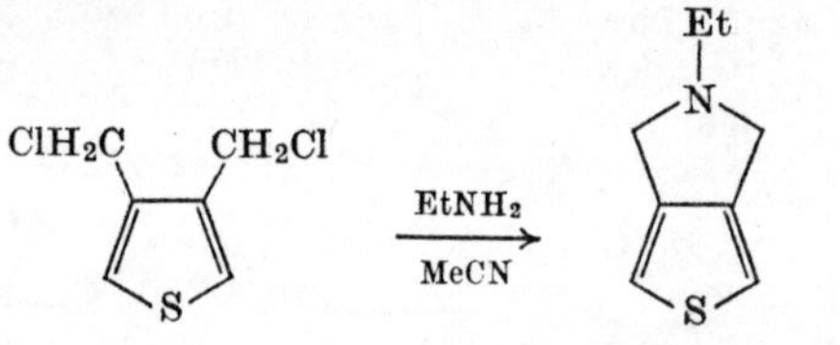

(71)   **(72)**

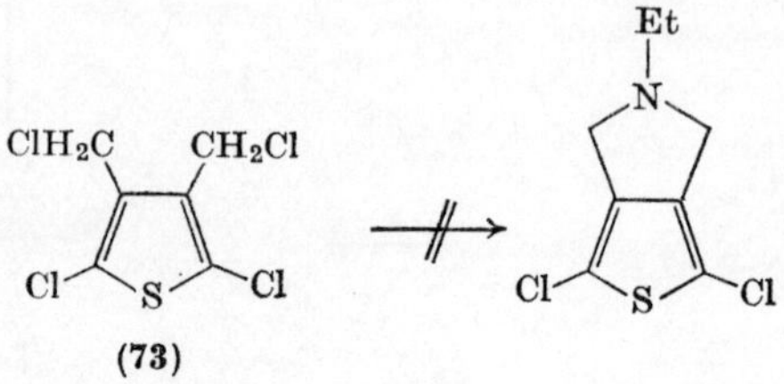

(73)

Other examples of participation by amino groups are reported in refs. 132—136.

[131] D. J. Zwanenburg and H. Wynberg, *J. Org. Chem.*, **34**, 333, 341 (1969).
[132] A. M. Farid, J. McKenna, J. M. McKenna, and E. N. Wall, *Chem. Comm.*, **1969**, 1222.
[133] P. Bravo, G. Guadiano, and A. Umani-Ronchi, *Tetrahedron Letters*, **1969**, 679.
[134] A. Hassner, G. J. Matthews, and F. W. Fowler, *J. Am. Chem. Soc.*, **91**, 5046 (1969).
[135] J. R. Sowa and C. C. Price, *J. Org. Chem.*, **34**, 474 (1969).
[136] D. R. Brown, J. McKenna, and J. M. McKenna, *J. Chem. Soc.*(B), **1969**, 567.

The reaction of *erythro*-1,2-dibromopropyl-trimethylsilane in aqueous ethanol yields *cis*-1-bromopropene by an *anti*-elimination. A mechanism involving participation by the silyl group was proposed (equation 13).[137]

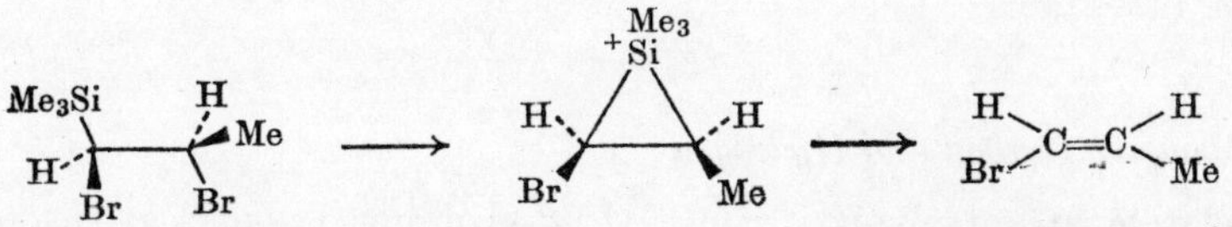

$$\text{Me}_3\text{Si} \quad \text{H} \qquad \overset{+\text{Si}\text{Me}_3}{\triangle} \qquad \overset{\text{H}}{\underset{\text{Br}}{}}\text{C}{=}\text{C}\overset{\text{H}}{\underset{\text{Me}}{}}$$

$$\cdots (13)$$

A possible example of participation by a nitro group is described in ref. 138.

Participation by Neighbouring Carbanion

When compound (74) is treated with aqueous potassium hydroxide at 100° it is converted into (76), presumably through participation by the carbanion (75). This involves inversion of configuration at both reaction centres, and it is another example of a 1,3-elimination in which the transition state has the W-geometry.[139]

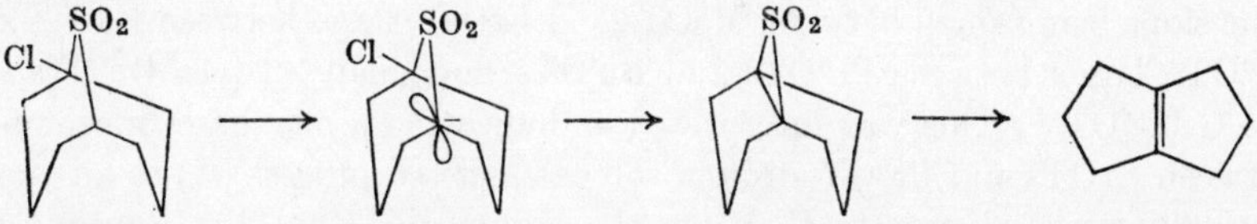

(74) (75) (76)

Baker and Spillett have published details of their investigation of the cyclization of compounds (77) through carbanion participation. The yield of cyclopropane increases from 0% when R = H to 97% when R = Bu$^\text{t}$. It was suggested that this resulted from a decrease in the initial-state rotational entropy with increasing size of the group R. Compound (78) yielded no cyclopropane.[140] The compounds (77) were demonstrated to undergo hydrogen–deuterium exchange in deuterated dimethyl sulphoxide, and the kinetics of this process were studied.[141]

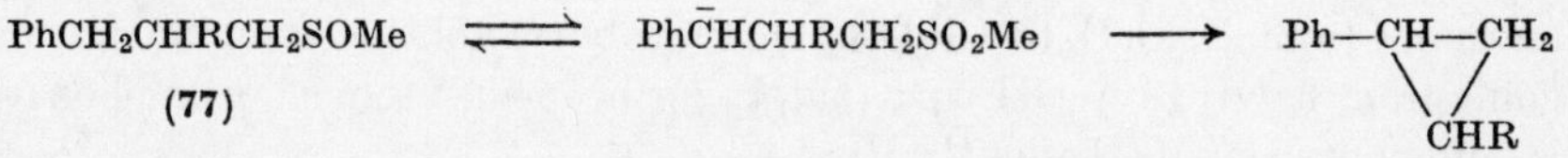

$$\text{PhCH}_2\text{CHRCH}_2\text{SOMe} \rightleftharpoons \text{Ph}\bar{\text{C}}\text{HCHRCH}_2\text{SO}_2\text{Me} \longrightarrow \text{Ph}{-}\text{CH}{-}\text{CH}_2$$
$$\overset{\text{CHR}}{\diagdown\diagup}$$

(77)

$$\text{PhCH}_2\text{CMe}_2\text{CH}_2\text{SO}_2\text{Me}$$

(78)

[137] A. W. P. Jarvie, A. Holt, and J. Thompson, *J. Chem. Soc.*(B), **1969**, 852.
[138] T. Fujisawa, T. Kobori, and G. Tsuchihashi, *Tetrahedron Letters*, **1969**, 4291.
[139] L. A. Paquette and R. W. Houser, *J. Am. Chem. Soc.*, **91**, 3870 (1969); cf. *Org. Reaction Mech.*, **1968**, 83; **1967**, 74.
[140] R. Baker and M. J. Spillett, *J. Chem. Soc.*(B), **1969**, 581; cf. *Org. Reaction Mech.*, **1968**, 67.
[141] R. Baker and M. J. Spillett, *J. Chem. Soc.*(B), **1969**, 880.

Electrolytic reduction of 1,3-dibromopropane and 1,4-dibromobutane yields cyclopropane and cyclobutane.[142]

Other examples of participation by neighbouring carbanion are described in refs. 143—145.

Neighbouring Carbon and Hydrogen

The relative rates of solvolysis of methyl, ethyl, propyl, isobutyl, and neopentyl toluene-p-sulphonate in EtOH, AcOH, CF_3CO_2H, H_2SO_4, and FSO_3H are shown in Table 1. The increase in rate with increasing β-methyl substitution in CF_3CO_2H, H_2SO_4, and FSO_3H clearly shows the importance of hydrogen or carbon participation in the propyl, isobutyl, and neopentyl compounds. Trifluoroacetolysis of propyl toluene-p-sulphonate yields 11.4% of propyl trifluoroacetate and 28.6% of isopropyl trifluoroacetate at 100°. Isobutyl toluene-p-sulphonate yields t-butyl trifluoroacetate and 20% 2-butyl trifluoroacetate, and neopentyl toluene-p-sulphonate yields t-pentyl trifluoroacetate. The plot of log k_{obs} against σ^* for the trifluoroacetolyses is S-shaped, "the slope being small between MeOTs and EtOTs, large between EtOTs and Pr^nOTs, larger between Pr^nOTs and Bu^iOTs, and small between Bu^iOTs and neo-$C_5H_{11}OTs$". This was interpreted as indicating a change in mechanism between EtOTs and Pr^nOTs from a solvent-assisted process (k_s) to an anchimerically assisted process (k_Δ). On the assumption that the n-propyl trifluoroacetate arises from a k_s process and the isopropyl trifluoroacetate from a k_Δ process, k_{obs} for the trifluoroacetolysis of n-propyl toluene-p-sulphonate was considered to be made up of 87.4% k_Δ and 12.6% k_s at 75°, and the β-deuterium isotope effects were calculated as $k_{\Delta(H)}/k_{\Delta(D)} = 1.85$ and $k_{s(H)}/k_{s(D)} = 1.15$. The plot of log k_Δ estimated in this way for the various solvents against log k for the solvolysis of neophyl toluene-p-sulphonate was a straight line. Since the latter reaction is well established to be anchimerically assisted, this supports the view that the k_Δ value for n-propyl toluene-p-sulphonate has been estimated correctly. Analysis of all the reactions in this way led to the conclusion that the mechanistic shift from k_s to k_Δ for propyl toluene-p-sulphonate occurs on going from HCO_2H to CF_3CO_2H, for isobutyl toluene-p-sulphonate on going from EtOH to AcOH, and that k_Δ for neopentyl toluene-p-sulphonate is dominant even in ethanol.[146, 147]

The solvolysis of ethyl toluene-p-sulphonate in FSO_3H was shown by

142 M. R. Rifi, *Tetrahedron Letters*, **1969**, 1043.
143 H. Watanabe, F. N. Jones, and C. R. Hauser, *J. Org. Chem.*, **34**, 2393 (1969).
144 J. Gosselck and G. Schmidt, *Tetrahedron Letters*, **1969**, 2615.
145 S. Z. Taits, E. A. Krasnyanskaya, Y. L. Gol'dfarb, *Izvest. Akad. Nauk SSSR, Ser. Khim.*, **1968**, 762.
146 I. L. Reich, A. Diaz, and S. Winstein, *J. Am. Chem. Soc.*, **91**, 5635, 5637 (1969).
147 P. C. Myhre and K. S. Brown, *J. Am. Chem. Soc.*, **91**, 5639 (1969).

deuterium-labelling to proceed with some intramolecular migration. The β-isotope effect $k_H/k_{(\beta\text{-}D_3)}$ was 1.58 at 0°, and it was suggested that concurrent k_s and k_Δ processes were occurring.[148]

Table 1. The relative rates of solvolysis of alkyl toluene-*p*-sulphonates in various solvents[146]

Solvent	Temp. (°C)	Me	Et	Pr^n	Bu^i	neo-C_5H_{11}
EtOH	75	4040	1750	1140	78	1.0
AcOH	75	10	9.2	7.3	2.8	1.0
HCO_2H	75	1.0	1.8	1.2	2.2	1.8
CF_3CO_2H	75	1.0	12.5	93	3060	6000
H_2SO_4	30	1.0	26	530	7500	50,000
FSO_3H	−44	1.0	118	3.3×10^4	5.4×10^5	1.14×10^6

Acetolysis of optically active 1-adamantyl[^{2}H]methyl toluene-*p*-sulphonate yields 1-adamantyl[^{2}H]methyl acetate with retention of configuration, which suggests that non-classical ion **(79)** is an intermediate. The major product homoadamantyl acetate is presumably also formed from this.[149]

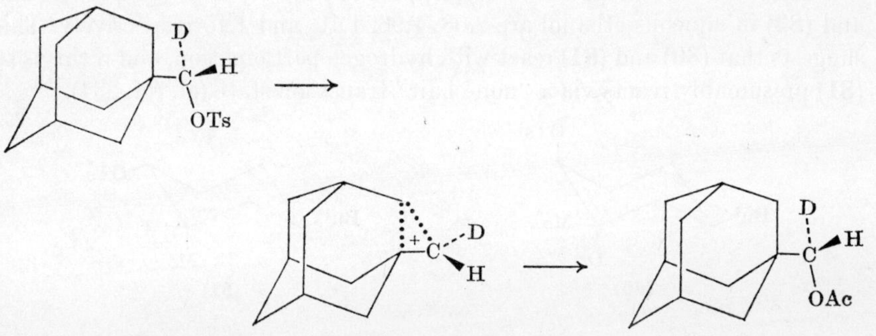

(79)

The silver-ion-assisted reaction of neopentyl iodide in nitromethane has been studied.[150]

When methyl 3-acetamido-3,6-dideoxy-2-*O*-methylsulphonyl-α-L-gluco-pyranoside is heated with methanolic sodium methoxide it reacts with carbon-participation and ring-contraction (see equation 14).[151]

[148] P. C. Myhre and E. Evans, *J. Am. Chem. Soc.*, **91**, 5641 (1969).
[149] S. H. Liggero, R. Sustmann, P. von R. Schleyer, *J. Am. Chem. Soc.*, **91**, 4571 (1969); cf. *Org. Reaction Mech.*, **1966**, 67.
[150] D. N. Kevill, G. H. Johnson, and V. V. Likhite, *Chem. Ind.* (London), **1969**, 1555.
[151] K. Čapek, J. Jarý, and Z. Samek, *Chem. Comm.*, **1969**, 1162.

4

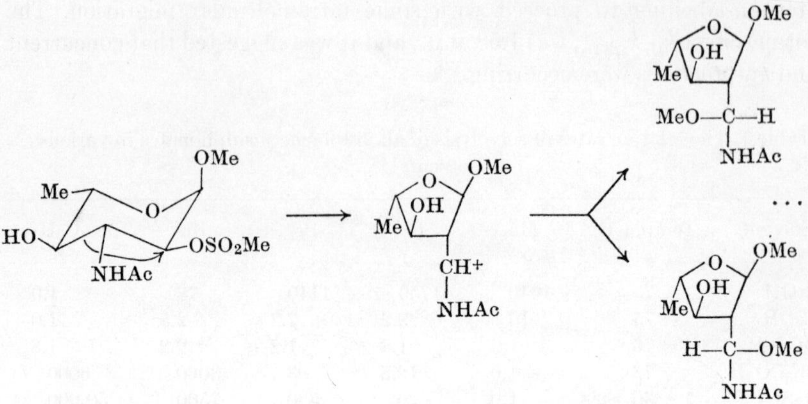

... (14)

Acetolysis of [1-^{2}H]cyclooctyl toluene-p-sulphonate is accompanied by ion-pair return to form a cyclooctyl toluene-p-sulphonate with a proton at C-1; 20% of the unreacted toluene-p-sulphonate consists of this material after one half-life. It is probably formed by a transannular 1,5-hydride shift. The rate of acetolysis of [3,3,4,4,5,5,6,6,7,7-^{2}H$_{10}$]cyclooctyl toluene-p-sulphonate is about 0.8 times that of the undeuterated compound. This suggests that hydride participation provides little anchimeric assistance.[152]

The deuterium isotope effects (k_H/k_D) for the solvolysis of (80), (81), (82), and (83) in aqueous ethanol are 2.08, 1.96, 1.19, and 1.15 respectively. This suggests that (80) and (81) react with hydrogen participation, and if this is so (81) presumably reacts via a "non-chair" transition state (cf. ref. 294).[153]

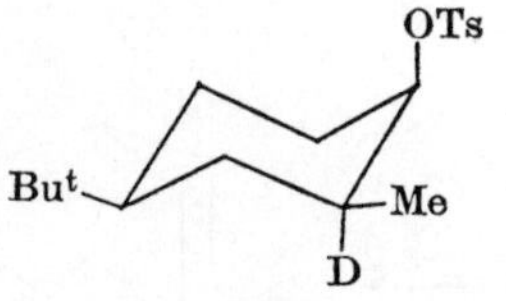

(80)

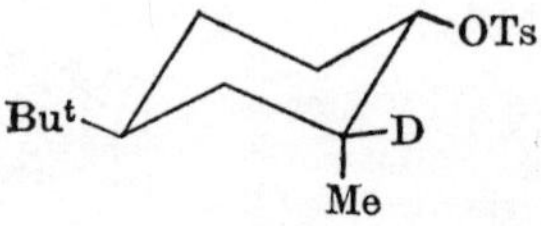

(81)

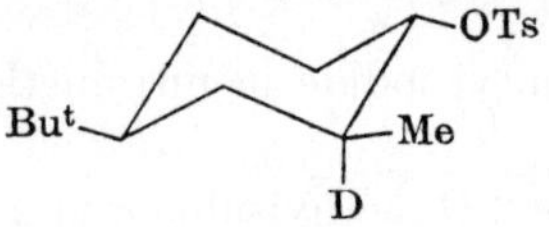

(82)

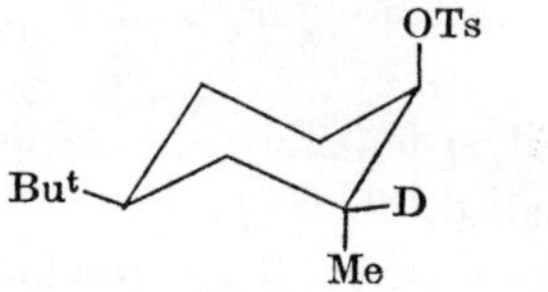

(83)

A 1,5-deuteride migration occurs when (84) is dissolved in AcOH–H$_2$SO$_4$ to yield a 3:1 mixture of (85) and (86).[154]

[152] A. A. Roberts and C. B. Anderson, *Tetrahedron Letters*, **1969**, 3883.

[153] M. Tichý, J. Hapala, and J. Sicher, *Tetrahedron Letters*, **1969**, 3739.

[154] P. T. Lansbury, J. B. Bieber, F. D. Saeva, and K. R. Fountain, *J. Am. Chem. Soc.*, **91**, 399 (1969).

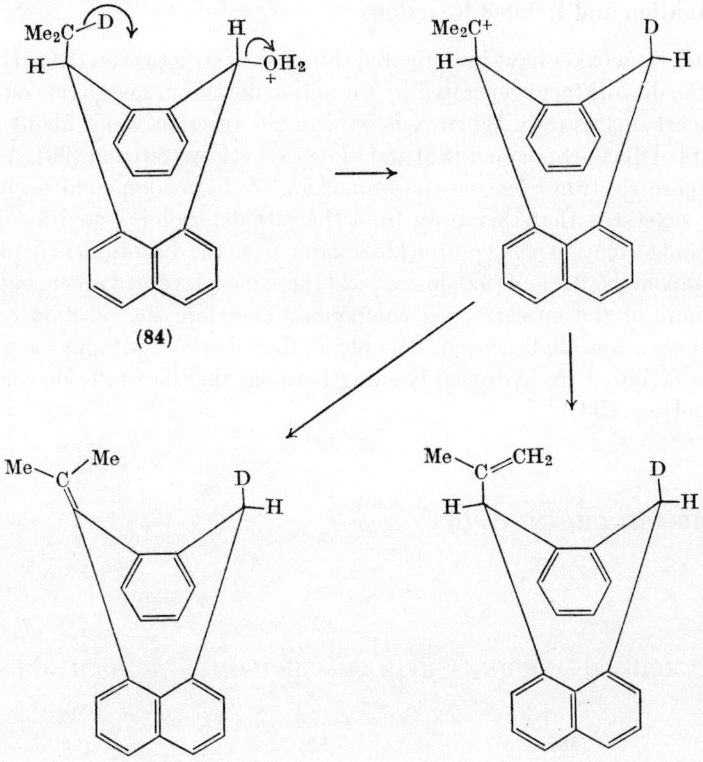

Me₂C–D

H–––OH₂
⁺

Me₂C⁺   D

H–––H

(84)

Me–C⟨Me   D

–H

Me–C=CH₂   D

H–––   –H

(85)   (86)

More examples of hydride participation in the reactions of *exo*-6-hydroxy-norborn-*endo*-2-ylmethyl halides have been reported.[155]

There have been further studies on the rearrangement of the cyclohexyl cation.[156–158] Other examples of 1,2-hydride shifts are reported in refs. 159 and 160. Acetolysis and formolysis of [1-¹⁴C]propyl toluene-*p*-sulphonate yields propyl acetate and formate with all the label at C-1.[161]

[155] D. W. Rome and B. L. Johnson, *Tetrahedron Letters*, **1968**, 6053; see *Org. Reaction Mech.*, **1968**, 85–86.

[156] See *Org. Reaction Mech.*, **1968**, 59.

[157] T. N. Shatkina and O. A. Reutov, *Doklady Akad. Nauk SSSR*, **183**, 612 (1968).

[158] T. N. Shatkina, E. V. Leont'eva, and O. A. Reutov, *Izv. Akad. Nauk SSSR, Ser. Khim.*, **1968**, 2838.

[159] I. G. Bundel, I. I. Levina, I. R. Prokhorenko, and O. A. Reutov, *Doklady Akad. Nauk SSSR*, **188**, 348 (1969).

[160] I. G. Bundel, I. I. Levina, S. M. Funtova, and O. A. Reutov, *Zh. Org. Khim.*, **5**, 15 (1969); *Chem. Abs.*, **70**, 86750 (1969).

[161] T. N. Shatkina, A. N. Lovtsova, and O. A. Reutov, *Izv. Akad. Nauk SSSR, Ser. Khim.*, **1967**, 2748; *Chem. Abs.*, **69**, 26411 (1968).

Deamination and Related Reactions

Hart and Brewbaker have investigated the effect of structure on the partitioning of the diazotic acid generated by the action of sodium methoxide on alkyl nitrosocarbamates (**87**). When X is *m*-nitro the reaction yields about equal amounts of diazo-compound (**88**) and of methyl ethers (**89**) and (**90**). As X is made more electron-releasing the proportion of diazo-compound decreases. It was suggested that this arose from the rate-determining step for diazo-compound formation being proton abstraction from the diazotic acid (equation 15). Compounds (**91**) and (**92**) do not yield the same mixture of ethers but each gives more of the unrearranged compound. Therefore the reaction cannot proceed via a free allylic cation. Possibly methanol attack without rearrangement is favoured by hydrogen-bonding between the diazotate ion and the methanol [e.g. (**93**)].[162]

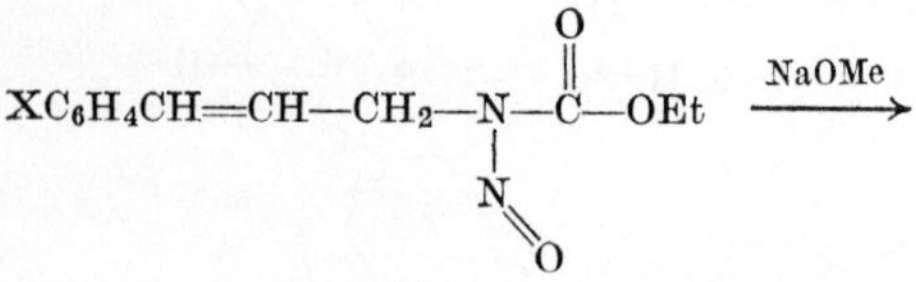

$$XC_6H_4CH{=}CH{-}CH_2{-}\underset{\underset{O}{\overset{|}{N}}}{\overset{\overset{O}{\|}}{N}}{-}\overset{\overset{O}{\|}}{C}{-}OEt \xrightarrow{\text{NaOMe}}$$

(**87**)

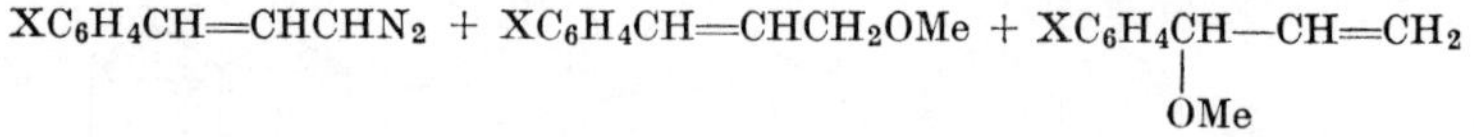

$$XC_6H_4CH{=}CHCHN_2 \ + \ XC_6H_4CH{=}CHCH_2OMe \ + \ XC_6H_4\underset{\underset{OMe}{|}}{CH}{-}CH{=}CH_2$$

(**88**)   (**89**)   (**90**)

$$RCH_2N{=}NOH + {}^-OMe \xrightarrow{-MeOH} RCH{-}N{=}N{-}OH \longrightarrow RCHN_2 + O\bar{H}$$

$$\dots (15)$$

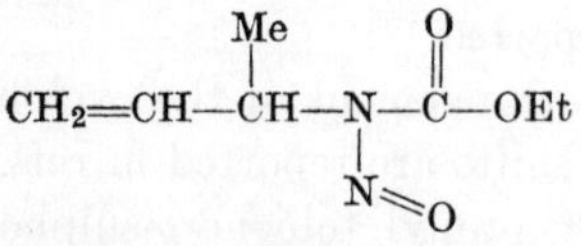

$$CH_2{=}CH{-}\underset{\underset{N{\overset{\|}{O}}}{\overset{Me}{|}}}{CH}{-}\overset{\overset{O}{\|}}{C}{-}OEt \qquad\qquad MeCH{=}CH{-}CH_2{-}\underset{\underset{N{\overset{\|}{O}}}{|}}{N}{-}\overset{\overset{O}{\|}}{C}{-}OEt$$

(**91**) 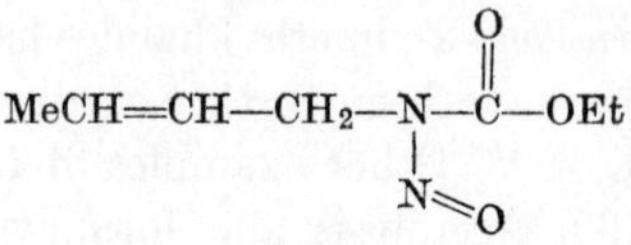   (**92**)

$$R{-}\underset{N{=}N}{\overset{\overset{Me}{|}}{\underset{}{O{-}H{\cdots}O^-}}}$$

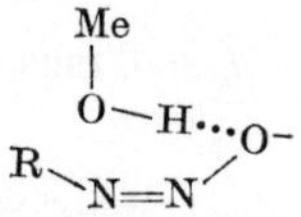

(**93**)

Reaction of the epoxy-*N*-nitrosocarbamate (**94**) with sodium methoxide led to compounds (**95**), (**96**), and (**97**). It was suggested that the carbonyl compounds (**95**) and (**96**), which are interconverted under the reaction conditions,

[162] H. Hart and J. L. Brewbaker, *J. Am. Chem. Soc.*, **91**, 716 (1969).

are formed by a mechanism in which opening of the epoxide ring is synchronous with attack of methoxide on the carbonyl carbon (see equation 16). Compound (**98**) reacts similarly except that it yields two epoxides (**99**) and (**100**). (**100**) is possibly formed by neighbouring group participation by the epoxide [see (**101**)→(**102**)] or possibly from (**99**) by an epoxide migration[163]

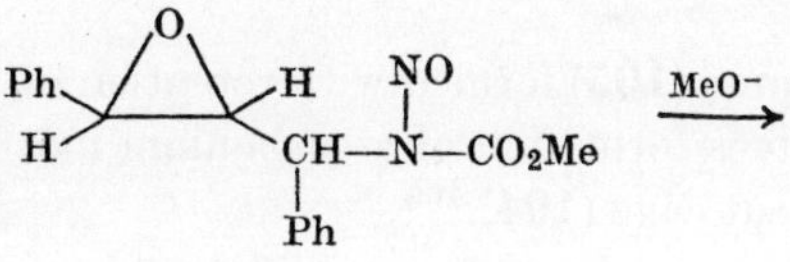

(94)

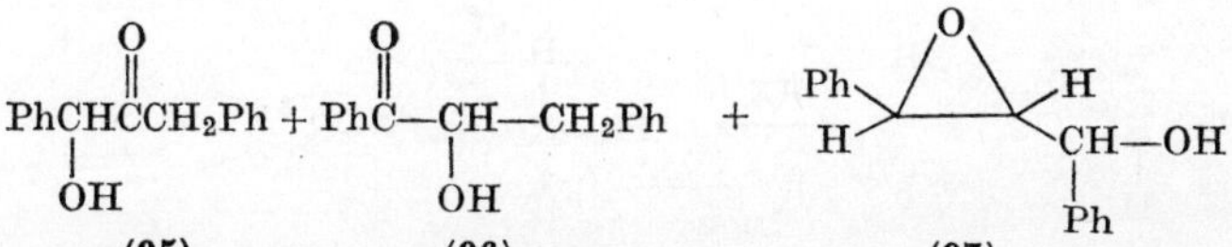

PhCHCCH$_2$Ph + PhC—CH—CH$_2$Ph +
        | |
       OH   OH
       (95)   (96)

(97)

(96) ⇌ (95)

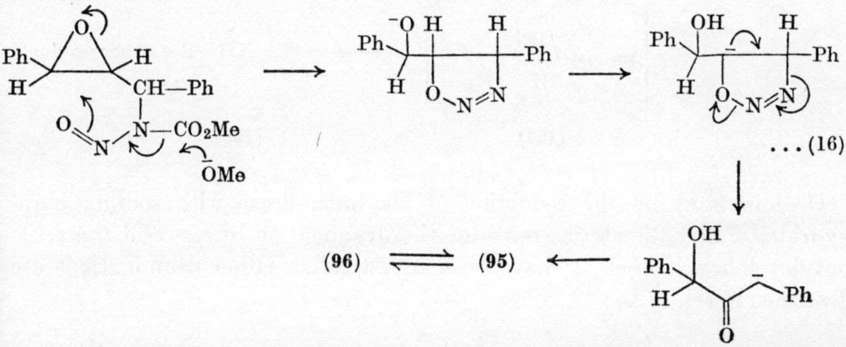

... (16)

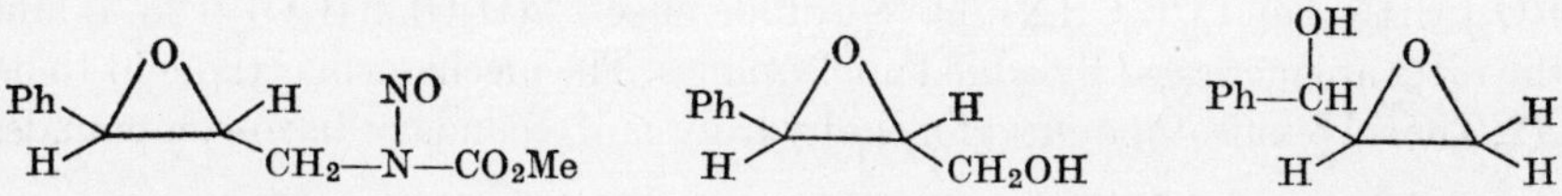

(98)   (99)   (100)

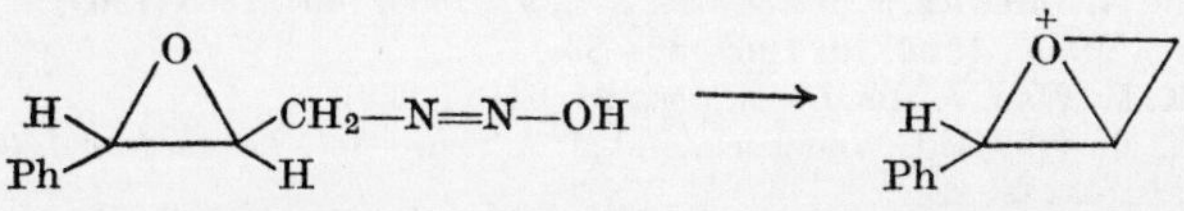

(101) → (102)

[163] A. Padwa, N. C. Das, and D. Eastman, *J. Am. Chem. Soc.*, **91**, 5178 (1969).

Details of Friedman and his coworkers' investigation of aprotic diazotization have been published.[164–167]

Lee and Wan have confirmed that deamination of 1-[1-^{14}C]propylamine gives propan-1-ol with 2% of the label at positions 2 and 3. This indicates that some of the propan-1-ol is formed from a protonated cyclopropane. The propane that is formed in this reaction and in the deoxidation of potassium 1-[1-^{14}C]propoxide has no label at C-2, and so it is not formed from a protonated cyclopropane.[168]

Deamination of triazene (**103**) yields amine (**105**) formed with retention of configuration. The reaction possibly involves formation of a carbonium ion, counter-ion, and aromatic amine in a solvent cage (**104**).[169]

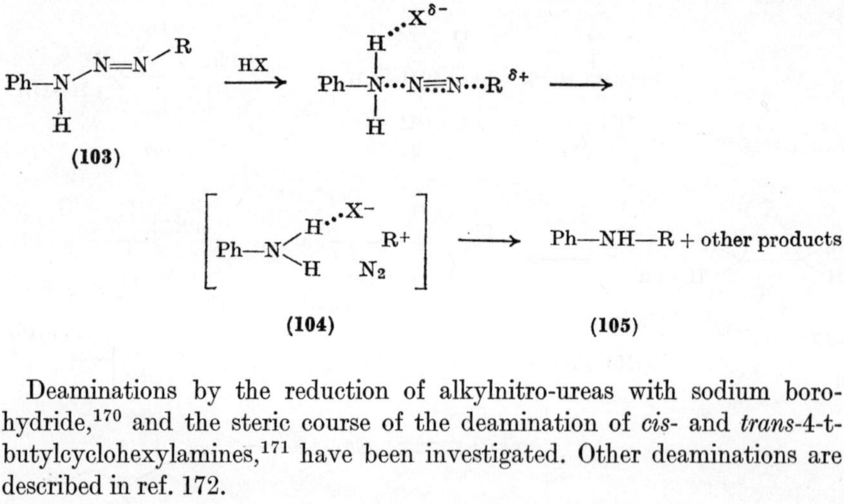

Deaminations by the reduction of alkylnitro-ureas with sodium borohydride,[170] and the steric course of the deamination of *cis-* and *trans-*4-t-butylcyclohexylamines,[171] have been investigated. Other deaminations are described in ref. 172.

Reactions of Aliphatic Diazo-compounds

The acid-catalysed hydrolysis of primary diazo-compounds, e.g. $RCOCHN_2$, RO_2CCHN_2, and CF_3CHN_2, have isotope effects $k(D_2O)/k(H_2O) = 2$—4, and the rates are increased by added nucleophiles. The mechanism is thought to be $A2$. The solvent isotope effect for the hydrolysis of secondary diazo-compounds,

[164] L. Friedman and J. H. Bayless, *J. Am. Chem. Soc.*, **91**, 1790 (1969).
[165] L. Friedman, A. T. Jurewicz, and J. H. Bayless, *J. Am. Chem. Soc.*, **91**, 1795 (1969).
[166] L. Friedman and A. Jurewicz, *J. Am. Chem. Soc.*, **91**, 1800, 1803, 1808 (1969).
[167] See *Org. Reaction Mech.*, **1966**, 70; **1967**, 82—83.
[168] C. C. Lee and K. M. Wan, *J. Am. Chem. Soc.*, **91**, 6416 (1969).
[169] E. H. White, D. J. Maskill, D. J. Woodcock, and M. A. Schroeder, *Tetrahedron Letters*, **1969**, 1713.
[170] W. Kirmse and H. Schutte, *Ann. Chem.*, **718**, 86 (1969).
[171] C. W. Shoppee, C. Culshaw, and R. E. Lack, *J. Chem. Soc.*(C), **1969**, 506.
[172] F. Lingens, J. Rau, and R. Sussmuth, *Z. Naturforsch.*, **23b**, 1565 (1968).

e.g. $RCOCN_2R'$, RO_2CCN_2R', and F_3CCN_2Me, $k(D_2O)/k(H_2O) = 0.4$—0.6 and the mechanism is thought to be $A\text{-}S_E2$.[173]

The rate of hydrolysis of $MeO_2CCH(Ph)COCHN_2$ in 1.7M-$HClO_4$ in 60% aqueous dioxan is enhanced by Br^- and the mechanism is presumably $A2$.[174]

The rates of hydrolysis of *p*-methoxy-*ω*-diazoacetophenone and of *ω*-diazoacetophenone in dioxan–water mixtures containing $HClO_4$ vary similarly with dioxan and $HClO_4$ concentration, which suggests that both react by the same mechanism, probably $A2$.[175]

The variation of the rate of acid-catalysed hydrolysis of ethyl diazoacetate in H_2O–D_2O mixtures with solvent composition has been shown to be consistent with an $A2$ mechanism.[176]

The *ρ*-value for the hydrolysis of a series of arylsulphonyldiazomethanes thought to proceed by an $A2$ mechanism is -1.0.[177]

The acid-catalysed reaction of 2-diazo-4,4-dimethylcholestan-3-one[178] and of salicylaldehyde with methyl isocyanate and [^{2}H]diphenyldiazomethane[179] have been investigated.

Fragmentation Reactions[180, 181]

The volume of activation for the fragmentation of *β*-bromoangelic acid (equation 17) is $+17.7$ cm^3 mole^{-1}, which is much larger than the previously reported values for decarboxylation reactions. This is good supporting evidence that the fragmentation is concerted and does not involve a rate-limiting decarboxylation.[182]

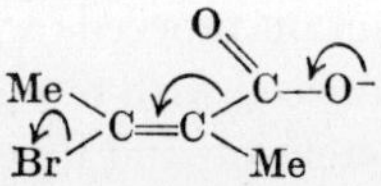

$$Me\text{—}C\equiv C\text{—}Me + Br^- + CO_2 \quad \ldots (17)$$

[173] H. Dahn, H. Gold, M. Ballenegger, J. Lenoir, G. Diderich, and R. Malherbe, *Helv. Chim. Acta*, **51**, 2065 (1968); see *Org. Reaction Mech.*, **1968**, 87.
[174] H. Dahn and J. P. Leresche, *Bull. Soc. Vaudoise Sci. Nat.*, **70**, 31 (1968); *Chem. Abs.* **69**, 76094 (1968).
[175] L. L. Leveson and C. W. Thomas, *J. Chem. Soc.*(B), **1969**, 1051.
[176] W. J. Albery and M. H. Davies, *Trans Faraday Soc.*, **65**, 1066 (1969).
[177] J. B. F. N. Engberts, G. Zuidema, B. Zwanenburg, and J. Strating, *Rec. Trav. Chim.*, **88**, 641 (1969); cf. *Org. Reaction Mech.*, **1968**, 87.
[178] M. Avaro and J. Levisalles, *Bull. Soc. Chim. France*, **1969**, 3166, 3173, 3180.
[179] L. Capuano, M. Dürr, and R. Zander, *Ann. Chem.*, **721**, 75 (1969).
[180] C. A. Grob, "Mechanism and Stereochemistry of Heterolytic Fragmentations", *Angew. Chem. Internat. Ed. Engl.*, **8**, 535 (1969).
[181] J. A. Marshall, "Fragmentation of Decalyl Boranes", *Record Chem. Progr.*, **30**, 3 (1969).
[182] W. J. le Noble, R. Goitien, and A. Shurpik, *Tetrahedron Letters*, **1969**, 895.

Fragmentation reactions of the following classes of compounds have also been studied: β-hydroxy-ketones,[183, 184] γ-aldehydo-toluene-p-sulphonates,[185] δ-keto-toluene-p-sulphonates.[186] toluene-p-sulphonyloxy-epoxides,[187a] and peroxides.[187b]

Displacement Reactions at Elements other than Carbon

Silicon, Germanium, Tin, and Lead[188]

Triphenylchlorosilane undergoes exchange with Bu_4NCl in benzene about 10^6 times faster than triphenylchloromethane. The reaction of the chlorosilane is of order 1.6 with respect to salt, which was interpreted as indicating that the reaction involved ion quadrupoles as well as ion pairs. It was predicted that the reaction with ion pairs and ion quadrupoles should proceed with retention of configuration as the transition state must be symmetrical, and this was confirmed by showing that the rate of exchange of 1-naphthylphenyl-methylchlorosilane was ca. 100 times greater than the rate of racemization. The order of racemization with respect to salt appears to be 0.5, which suggests that it involves a reaction of the undissociated chloride ion.[189]

Exchange reactions of R_3SiCl with lithium chloride in acetone and dioxan have been studied. When R is aliphatic the order with respect to lithium chloride is 0.5, which suggests that the reaction is between the chlorosilane and a free chloride ion. When R is Ph, $PhCH_2$, or $PhCH_2CH_2$ this order is 1.0, and it was suggested that the reactive species was the LiCl ion-pair, possibly with the lithium solvated by the benzene ring.[190]

The reactions of a large number of optically active silicon compounds with BF_3 and BCl_3 have been studied. Generally reaction proceeds with inversion of configuration but the reactions of R_3SiOMe proceed with retention. Transition states (**106**) and (**107**) were proposed for retention and inversion respectively.[191]

[183] R. C. Cambie, D. R. Crump, and R. N. Dave, *Chem. Comm.*, **1969**, 523.

[184] B. L. Yates and J. Quijano, *J. Org. Chem.*, **34**, 2506 (1969).

[185] F. Nerdel, D. Frank, and G. Barth, *Chem. Ber.*, **102**, 395 (1969); F. Nerdel, G. Barth, D. Frank, and P. Weyerstahl, *ibid.*, p. 407.

[186] P. C. Mukharji and T. K. Das Gupta, *Tetrahedron*, **25**, 5275 (1969).

[187a] J. M. Coxon, R. P. Garland, M. P. Hartshorn, and G. A. Lane, *Chem. Comm.*, **1968**, 1506.

[187b] W. H. Richardson and R. S. Smith, *J. Am. Chem. Soc.*, **91**, 3610 (1969); see *Org. Reaction Mech.*, **1967**, 123.

[188] R. Belloli, "Resolution and Stereochemistry of Asymmetric Silicon, Germanium, Tin, and Lead Compounds", *J. Chem. Educ.*, **46**, 640 (1969).

[189] M. W. Grant and R. H. Prince, *Nature*, **222**, 1163 (1969).

[190] M. W. Grant and R. H. Prince, *J. Chem. Soc.*(A), **1969**, 1138.

[191] L. H. Sommer, J. D. Citron, and G. A. Parker, *J. Am. Chem. Soc.*, **91**, 4729 (1969).

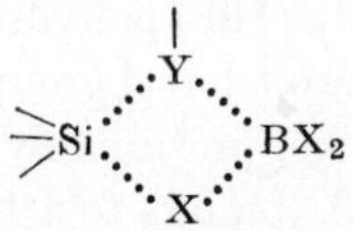

(+)-Methyl-1-naphthyl-phenylsilane reacts with lithium aluminium deuteride in $Et_2O-Bu^n_2O$ with retention of configuration but in tetrahydrofuran with inversion. The corresponding germanium compound did not react.[192]

Corriu and Masse have reported details of some of their work on the steric course of the reactions of 2-naphthyl-2-methoxy-1,3,4-trihydro-2-silanaphthalene with nucleophiles.[193]

The rate constants for the hydrolysis of triethylphenoxysilanes are correlated better by the Yukawa–Tsuno equation than by the Hammett equation to yield a ρ-value of 3.52.[194]

Alkoxysilanes are protonated on oxygen in $FSO_3H-SbF_5-SO_2$, and undergo a slow cleavage, probably through attack by F^-. There was no evidence for the formation of silyl cations.[195]

The following reactions have also been studied: propanolysis of Ph_3SiCl in CCl_4;[196] hydrolysis of 1-alkyl-1-alkoxysilanthranes;[197] the reaction between Me_3SiCl and sodium acetate in dioxan;[198] base-catalysed condensation of diorganosilanediols in methanol;[199] the reaction of alkoxyphosphazenes with chlorosilanes;[200] acid-catalysed cleavage of methylphenylsiloxanes;[201] and the alkaline hydrolysis of methyl ethoxysilanes and tetraethoxysilanes.[202]

Phosphorus, Arsenic, and Antimony[203]

Like the hydrolysis of phosphinate esters,[204] the rate of hydrolysis of phosphonium salts is strongly retarded when two bulky substituents are directly

[192] G. J. Peddle, J. M. Shafir, and S. G. McGeechin, *J. Organometal. Chem.*, **15**, 505 (1968).
[193] R. Corriu and J. Masse, *Bull. Soc. Chim. France*, **1969**, 3491; see *Org. Reaction Mech.*, **1968**, 90—91.
[194] A. A. Humffray and J. J. Ryan, *J. Chem. Soc.*(B), **1969**, 1138.
[195] G. A. Olah, D. H. O'Brien, and C. Y. Lui, *J. Am. Chem. Soc.*, **91**, 701 (1969)
[196] A D. Allen and S. J. Lavery, *Can. J. Chem.*, **47**, 1263 (1969).
[197] M. G. Voronkov and G. I. Zelchan, *Khim. Geterotsikl. Soedin.*, **1969**, 450.
[198] H. Bentkowsta and A. Nagorska, *Rocz. Chem.*, **43**, 1150 (1969).
[199] B. Dejak, Z. Lasocki, and W. Mogilnicki, *Bull. Acad. Pol. Sci.*, *Ser. Sci. Chim.*, **17**, 7 (1969).
[200] S. I. Belykh, S. M. Zhivukhin, V. V. Kineer, and G. G. Kolesnikov, *Zh. Obshch. Khim.*, **39**, 799 (1969); *Chem. Abs.*, **71**, 50062 (1969).
[201] A. G. Kuznetsova, S. A. Golubtsov, and V. I. Ivanov, *Plast. Massy*, **1969**, 26; *Chem. Abs.*, **71**, 21369 (1969).
[202] M. G. Voronkov and L. Zagata, *Zh. Obshch. Khim.*, **37**, 2551 (1967); *Chem. Abs.*, **69**, 35087 (1968).
[203] M. J. Gallagher and I. D. Jenkins, "Stereochemical Aspects of Phosphorus Chemistry", *Topics Stereochem.*, **3**, 1 (1968).
[204] See *Org. Reaction Mech.*, **1968**, 405.

attached to the phosphorus. Thus although salt (**108**) is hydrolysed only 50 times more slowly than (**109**), salt (**110**) with two t-butyl groups reacts much slower and yields benzyl-t-butylphenylphosphine, presumably formed by an elimination reaction. It was suggested that HO^- prefers to attack opposite a t-butyl group to give the penta-covalent intermediate (**111a**), and that this becomes very difficult when there are two t-butyl groups present since one of these must be equatorial. If it is correct that HO^- prefers to attack opposite a t-butyl group, a mono-t-butyl compound should react with retention of configuration. This was confirmed by showing that the Wittig reaction and alkaline hydrolysis of (**111b**) yielded t-butylmethylphenylphosphine oxide with rotation of the same sign. On the assumption that the Wittig reaction proceeds with 100% retention, the alkaline hydrolysis proceeds with at least 79% retention.[205]

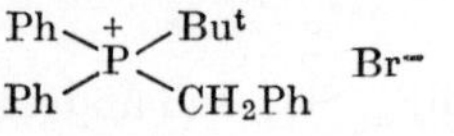

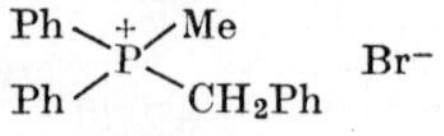

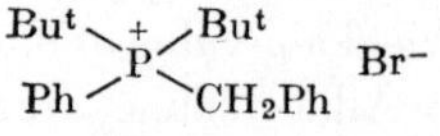

In contrast, the reaction of equation (18) proceeds with inversion of configuration despite the presence of a t-butyl group. The steric course of these reactions thus depends on the leaving group, and it was suggested that when both the nucleophile (X) and the leaving group (Y) are electronegative the lowest-energy pathway leads to an intermediate in which X and Y occupy apical positions. When the rate of expulsion of Y ($= OEt$) is faster than pseudorotation this leads to inversion of configuration, but when it is slower and when Y is not electronegative (e.g. benzyl) "pseudorotation may, *when steric factors favour the process*, lead to racemization and/or retention."[206]

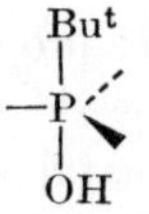

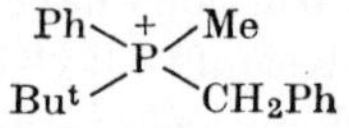

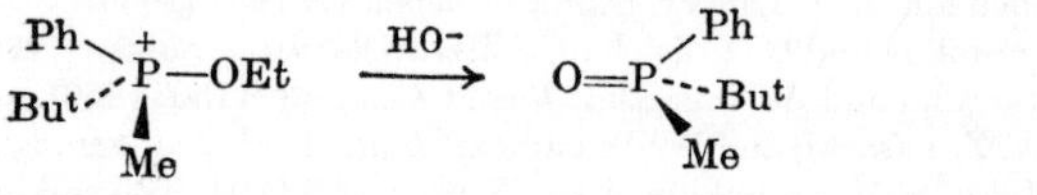

Details of Hawes and Trippett's demonstration that the hydrolysis of (**112**) proceeds with retention of configuration have been published.[207] Similarly the hydrolysis of optically active (**113**) and of (**115**) also proceeded with

[205] N. J. De'ath and S. Trippett, *Chem. Comm.*, **1969**, 172.
[206] R. A. Lewis, K. Naumann, K. E. De Bruin, and K. Mislow, *Chem. Comm.*, **1969**, 1010.
[207] W. Hawes and S. Trippett, *J. Chem. Soc.*(C), **1969**, 1465; see *Org. Reaction Mech.*, **1968**, 94.

retention of configuration, but hydrolysis and Wittig olefin synthesis of (117) [isomeric with (115)] yielded a mixture of the two phosphine oxides (116) and (118). This was shown to arise through the conversion of (117) into isomer (115) prior to reaction, possibly via an ylid or more likely [208] via a penta-covalent intermediate. A similar result was obtained with (114) isomeric with (112).[209]

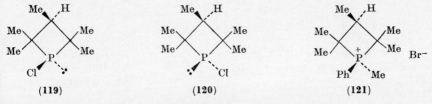

Compounds (119) and (120) react with nucleophiles with predominant inversion of configuration. Since the four-membered ring must occupy an

[208] S. E. Cremer, R. J. Chorvat, and B. C. Trivedi, *Chem. Comm.*, **1969**, 769.
[209] J. R. Corfield, J. R. Shutt, and S. Trippett, *Chem. Comm.*, **1969**, 789.

apical-equatorial position, these reactions may involve equatorial attack by
the nucleophile and equatorial loss of the leaving group.[210]

An X-ray structure determination has been reported for compound (**121**).[211]

The alkaline hydrolyses of salts (**122**) and (**124**) follow the third-order rate
law and yield ring-opened products (**123**) and (**125**). The energies of activation
for the compounds with R = Me are 18.9 and 21.8 kcal mole^{-1} respectively,
which are considerably lower than that for the hydrolysis of $Ph_2P^+Me_2Br^-$
(32 kcal mole^{-1}) and those for the hydrolysis of the saturated compounds (**126**)
and (**127**) which react with expulsion of the exocyclic phenyl group. It was
suggested that the low energies of activation for hydrolysis of (**122**) and (**124**)
resulted from the ring occupying an apical-equatorial position such that
apical attack would lead to a penta-covalent intermediate from which apical
expulsion of a phenyl anion was possible. In contrast, compound (**126**) was
thought to react with apical attack and equatorial expulsion of the phenyl

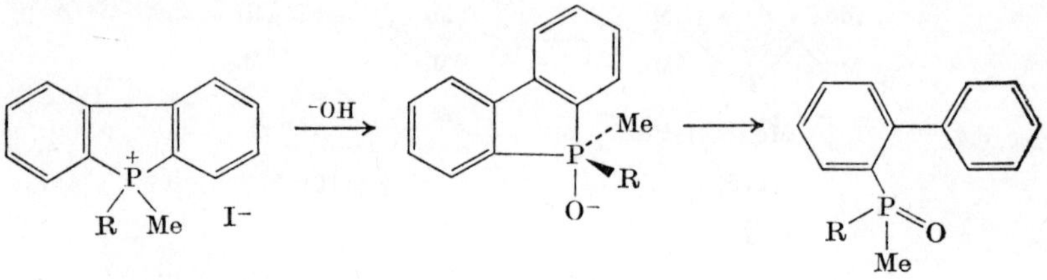

(**122**)   (**123**)

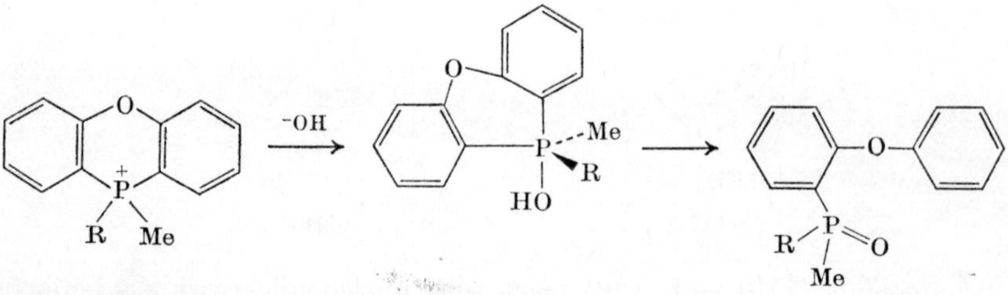

(**124**)
(R = Ph, Me)   (**125**)

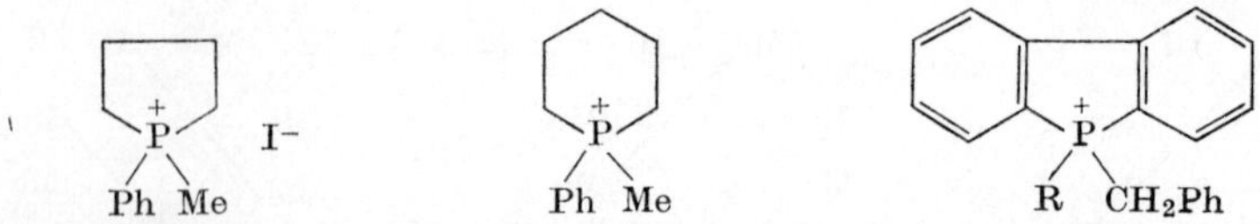

(**126**)   (**127**)   (**128**)
                                                          (R = Me, Ph)

[210] D. J. H. Smith and S. Trippett, *Chem. Comm.*, **1969**, 855.
[211] C. Moret and L. M. Trefonas, *J. Am. Chem. Soc.*, **91**, 2255 (1969).

anion which is a much higher energy process.[212, 213] The benzyl compounds (128), analogous to (122), react with expulsion of the benzyl group.[214]

Details of Marsi's demonstration that the alkaline hydrolyses of compounds (129) and (130) with loss of the benzyl groups proceed with retention of configuration have been reported.[215]

(129)　　　　　　　　(130)

Optically active phosphine oxides are racemized by $LiAlH_4$ prior to reduction. It was suggested that this involved a reversible addition to form a penta-covalent intermediate.[216]

Several examples of nucleophilic substitution at a thiophosphoryl centre with inversion of configuration have been reported.[217-219]

The alkaline hydrolysis of bisphosphonium salts has been investigated.[220]

The reaction of alkyldiarylarsine sulphides with alkyl halides,[221] and the hydrolyses of the $SbCl_6^-$ and SbF_5OH ions,[222a, 222b] have been investigated.

Sulphur[223]

Nucleophilic substitution at sulphinyl and sulphonyl sulphur have been compared by studying the solvolyses of aryl α-disulphones (equation 19) with those of arylsulphinyl sulphones (equation 20) in aqueous dioxan. The solvent isotope effect, entropies of activation, and the effect of structure and solvent composition on reaction rate are very similar for both reactions but the sulphinyl sulphones react about 10,000 times faster. It was suggested that the

[212] D. W. Allen and I. T. Millar, *J. Chem. Soc.*(B), **1969**, 263.

[213] D. W. Allen and I. T. Millar, *J. Chem. Soc.*(C), **1969**, 252.

[214] B. R. Ezzell and L. D. Freedman, *J. Org. Chem.*, **34**, 1777 (1969).

[215] K. L. Marsi, *J. Am. Chem. Soc.*, **91**, 4724 (1969).

[216] P. D. Henson, K. Naumann, and K. Mislow, *J. Am. Chem. Soc.*, **91**, 5645 (1969).

[217] J. Michalski, M. Mikolajczyk, B. Mlotkowska, and J. Omelánczuk, *Tetrahedron*, **25**, 1743 (1969).

[218] M. Mikolajczyk, J. Omelánczuk, and J. Michalski, *Bull. Acad. Pol. Sci., Ser. Sci. Chim.*, **16**, 615 (1968).

[219] B. Pliszka-Krawiecka, M. Mikolajczyk, and J. Michalski, *Bull. Acad. Pol. Sci., Ser. Sci. Chim.*, **17**, 75 (1969).

[220] J. J. Brophy and M. J. Gallagher, *Austral. J. Chem.*, **22**, 1385, 1399 (1969).

[221] Y. F. Gatilov, *Zh. Obshch. Khim.*, **38**, 372 (1968); *Chem. Abs.*, **69**, 76078 (1968).

[222a] S. B. Willis and H. M. Neumann, *J. Am. Chem. Soc.*, **91**, 2924 (1969); see also L. Kolditz, R. Dlaske, and G. Heller, *Z. Chem.*, **99**, 348 (1969).

[222b] A. L. Gehala, W. L. Johnson, and M. M. Jones, *J. Inorg. Nucl. Chem.*, **31**, 3495 (1965).

[223] E. Ciuffarin and A. Fava, *Progr. Phys. Org. Chem.*, **6**, 81 (1968).

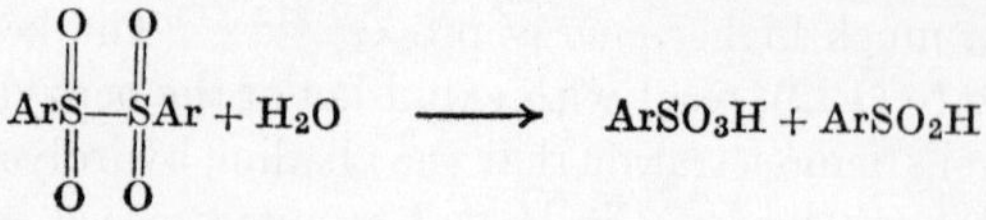

$$\underset{\underset{O}{\overset{O}{||}}\quad\underset{O}{\overset{O}{||}}}{ArS-SAr} + H_2O \longrightarrow ArSO_3H + ArSO_2H \qquad \dots (19)$$

$$\underset{\underset{O}{\overset{O}{||}}\quad\underset{O}{\overset{}{}}}{ArS-SAr} + H_2O \longrightarrow 2ArSO_2H \qquad \dots (20)$$

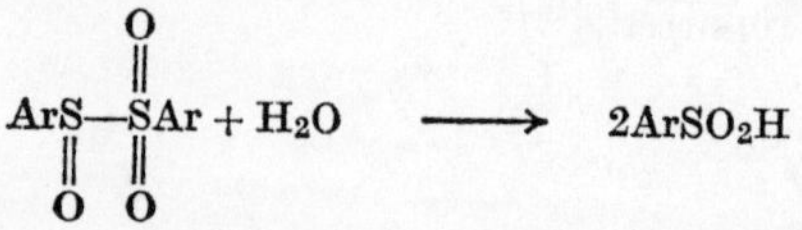

reactions proceeded by similar mechanisms involving nucleophilic attack "concerted with scission of the S–S bond and transfer of a proton from the incoming water molecule to the departing $ArSO_2$ group".[224]

Nucleophilic catalysis of the hydrolysis of the α-disulphones was also studied, and the following order of nucleophilicities determined: Bu^nNH_2 $(5.9 \times 10^3) > Bu^iNH_2 (2.9 \times 10^3) > Et_2NH (4.4 \times 10^2) > N_3^- (3.3 \times 10^2) > F^-$ $(59) > NO_2^- (10) > AcO^- (1.0) > Cl^- (0.0016) > Br^- (0.0009)$. This is approximately the reverse of the order reported last year [225] for attack at sulphinyl sulphur, and can be explained in terms of sulphonyl sulphur being a harder electrophile than sulphinyl sulphur, and being more reactive towards hard nucleophiles.[226] The results suggest that sulphonyl sulphur is about as hard an electrophilic centre as is carbonyl carbon.[226]

Nucleophilicities towards sulphur and oxygen have been discussed.[227]

Sabol and Andersen have carried out the series of reactions (133)→(131)→ (132) and obtained sulphoxide (132) with the same sign of rotation as that previously found by Stirling who obtained it from optically active (134). If it

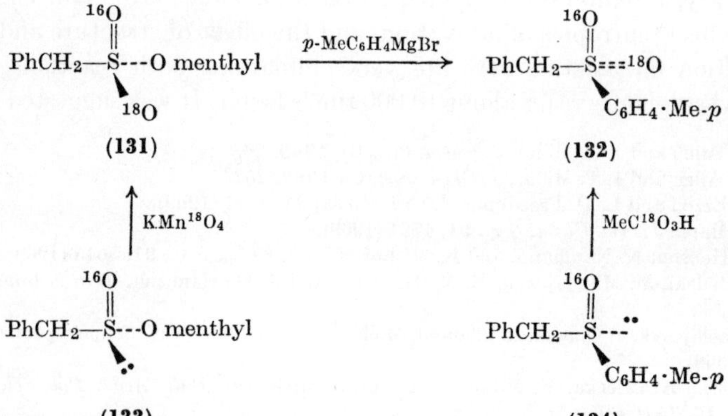

(131)   (132)

(133)   (134)

[224] J. L. Kice and G. J. Kasperek, *J. Am. Chem. Soc.*, **91**, 5510 (1969); see also P. Allen and P. J. Conway, *Can. J. Chem.*, **47**, 873 (1969).
[225] See *Org. Reaction Mech.*, **1968**, 97.
[226] J. L. Kice, G. J. Kasperek, and D. Patterson, *J. Am. Chem. Soc.*, **91**, 5516 (1969).
[227] R. E. Davis, S. P. Molmar, and R. Nehring, *J. Am. Chem. Soc.*, **91**, 97 (1969).

is assumed that the oxidations (**133**)→(**131**) and (**134**)→(**132**) proceed by the same stereochemical course, it then follows that the conversion (**131**)→(**132**), a substitution at sulphonyl sulphur, proceeds with inversion.[228]

The base-catalysed hydrolysis of *cis*- and *trans*-1-ethoxy-3-methylthietanium hexachloroantimonates (**135**) and (**136**) occurs with inversion of configuration. This result contrasts with those obtained with analogous phosphorus compounds (cf. refs. 207—209) and suggests that these reactions proceed through an intermediate or transition state in which either the four-membered ring or the entering and leaving groups occupy two equatorial positions.[229]

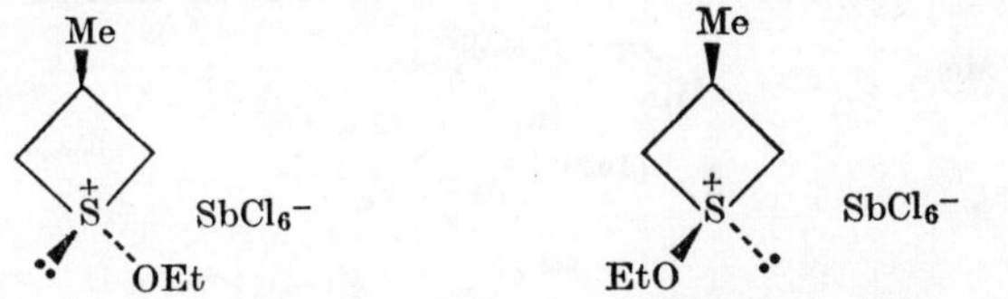

(135)   (136)

Christensen and Kjaer have carried out the nucleophilic substitution on sulphur (**137**)→(**138**). (**138**) was then oxidized to give (**140**) which after removal of the protecting groups yielded 85% of the 2(*S*),S(*R*)-compound (**139**)

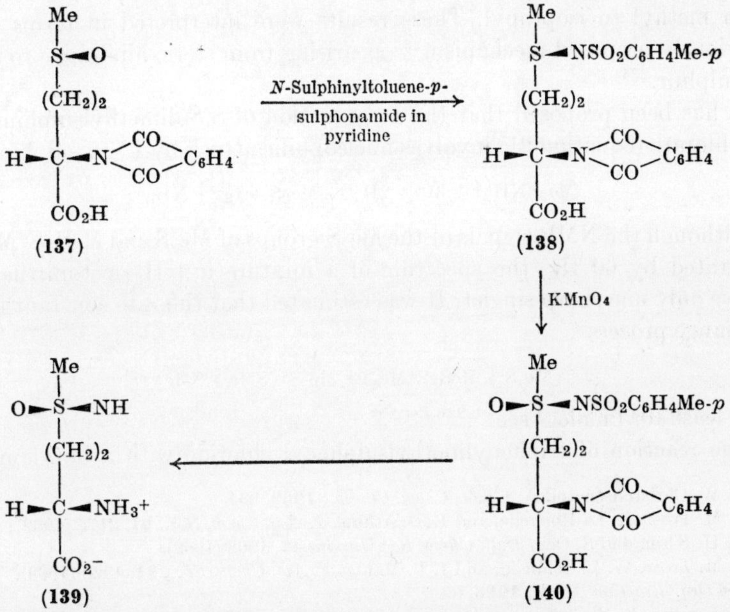

(137)   (138)

(139)   (140)

[228] M. A. Sabol and K. K. Andersen, *J. Am. Chem. Soc.*, **91**, 3603 (1969).
[229] R. Tang and K. Mislow, *J. Am. Chem. Soc.*, **91**, 5644 (1969).

and 15% of its 2(*S*),S(*S*)-isomer. Thus, on the assumption that the oxidation step **(138)**→**(140)** proceeds with retention of configuration, it follows that the substitution step **(137)**→**(138)** also proceeds with predominant retention.[230]

See ref. 231 for a discussion of penta-covalent sulphur intermediates.

Conversion of **(141)** into **(143)** is thought to involve formation of **(142)** which then reacts either by a direct fragmentation or cleavage to the tri-methylene **(144)**.[232]

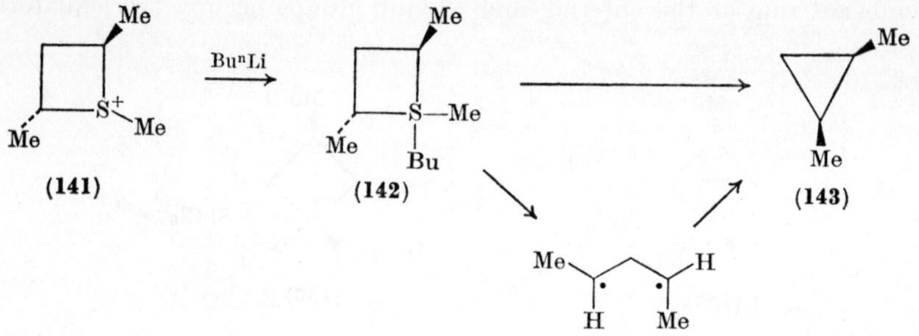

The rate of reduction of sulphoxide PhSOR by I^-, and the rate of racemiza-tion promoted by Cl^- and Br^-, are reduced 60—100-fold when R is changed from methyl to isopropyl. These results were interpreted in terms of the previously proposed mechanism[233] as arising from steric hindrance to attack on sulphur.[234]

It has been proposed that the I^- reduction of *S,S*-dimethylsulphiminium perchlorate (equation 21) involves nucleophilic attack by I^- on sulphur.[235]

$$Me_2SNH_2^+ + 3I^- + 2H^+ \rightarrow Me_2S + I_3^- + NH_4^+ \quad \quad \dots (21)$$

Although the NMR signals of the Me_2S groups of Me_2S and $MeS-S^+Me_2$ are separated by 60 Hz, the spectrum of a mixture in [$^2H_3$]acetonitrile at 0° shows only one sharp singlet. It was estimated that the rate constant for the exchange process

$$Me_2S + MeS-S^+Me_2 \rightleftharpoons Me_2S^+-SMe + SMe_2$$

is at least 10^5 l mole^{-1} sec^{-1}.[236]

The reaction of triphenylmethylsulphenyl chloride with n-butylamine in

[230] B. W. Christensen and A. Kjaer, *Chem. Comm.*, **1969**, 934.
[231] B. M. Trost, R. La Rochelle, and R. C. Atkins, *J. Am. Chem. Soc.*, **91**, 2175 (1969); see also Y. H. Khim and S. Oae, *Bull. Chem. Soc. Japan*, **42**, 1968 (1969).
[232] B. M. Trost, W. L. Schinski, and I. B. Mantz, *J. Am. Chem. Soc.*, **91**, 4320 (1969).
[233] See *Org. Reaction Mech.*, **1968**, 99.
[234] D. Landini, F. Montanari, G. Modena, and G. Scorrano, *Chem. Comm.*, **1969**, 1.
[235] J. H. Krueger, *J. Am. Chem. Soc.*, **91**, 4974 (1969).
[236] J. L. Kice, and N. A. Favstritsky, *J. Am. Chem. Soc.*, **91**, 1751 (1969).

benzene at 25° is second-order in amine. The rate is strongly increased by salts, and the reaction then shows mixed second- and third-order kinetics:

$$\text{Rate} = k_2[\text{RSCl}][\text{BuNH}_2] + k_3[\text{RSCl}][\text{BuNH}_2]^2$$

The second-order dependence on amine was not thought to arise from a monomer–dimer equilibrium since it extended to very low concentrations (5×10^{-5}M). It was suggested that reaction might proceed via a trivalent sulphur intermediate.[237]

It has been concluded from cryoscopic and conductometric measurements on solutions of 2,4-dinitrobenzenesulphenyl chloride in 100% H_2SO_4 that the species present is the protonated chloride and not a sulphenium ion.[238]

The hydrolyses of disulphides[239, 240] and the reactions of tetra-O-acetyl-β-D-glucopyranosylsulphenyl bromide[241] have been studied.

The conversion of (145) into (146) proceeds without appreciable incorporation of ^{18}O when the reaction is carried out in ^{18}O-enriched water, and it was suggested that both oxygens of the sulphinate group of (146) came from the nitro group of (145)[242] (see p. 262).

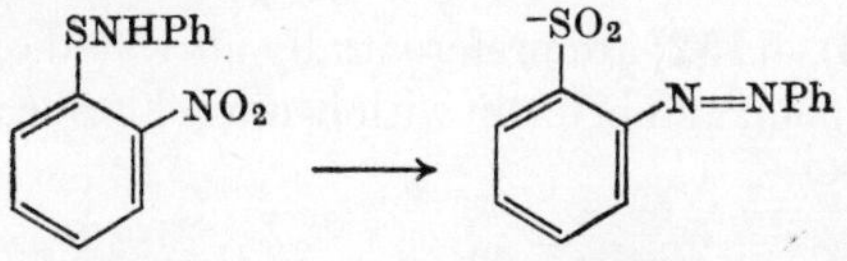

(145)   (146)

Other Elements

The rate of N-chlorination of a large number of amides follows the order $\text{MeCO}_2\text{Cl}, \text{Cl}_2 > ^-\text{OCl} > \text{HOCl}$. It was suggested that the reaction with $^-$OCl initially involves formation of an H-bonded complex with the amide, but that the other reagents react via a direct displacement on chlorine.[243]

The N-chloro-N-methylcyclopropylamine (147) reacts with methanol to yield (148) and (149) which are probably formed as shown.[244]

Details of Blackborow and Lockhart's investigation of the reaction of dinitroaniline and dinitronaphthylamine with the BCl_3 and BBr_3 addition complexes of acetonitrile have been reported.[245] The rates of hydrolysis of

[237] E. Ciuffarin and G. Guaraldi, *J. Am. Chem. Soc.*, **91**, 1745 (1969).
[238] E. A. Robinson and S. A. A. Zaidi, *Can. J. Chem.*, **46**, 3927 (1968).
[239] B. C. Pal, M. Uziel, D. G. Doherty, and W. E. Cohn, *J. Am. Chem. Soc.*, **91**, 3634 (1969).
[240] T. Sugimura and Y. Tanaka, *J. Chem. Soc. Japan*, **88**, 145 (1968); *Chem. Abs.*, **69**, 66567 (1968).
[241] R. H. Bell and D. Horton, *Carbohydrate Res.*, **9**, 187 (1969).
[242] C. Brown, *Chem. Comm.*, **1969**, 100.
[243] M. Wayman and E. W. C. W. Thomm, *Can. J. Chem.*, **47**, 2561, 3289 (1969).
[244] P. G. Gassman and A. Carrasquillo, *Chem. Comm.*, **1969**, 495.
[245] J. R. Blackborow and J. C. Lockhart, *J. Chem. Soc.*(A), **1968**, 3015.

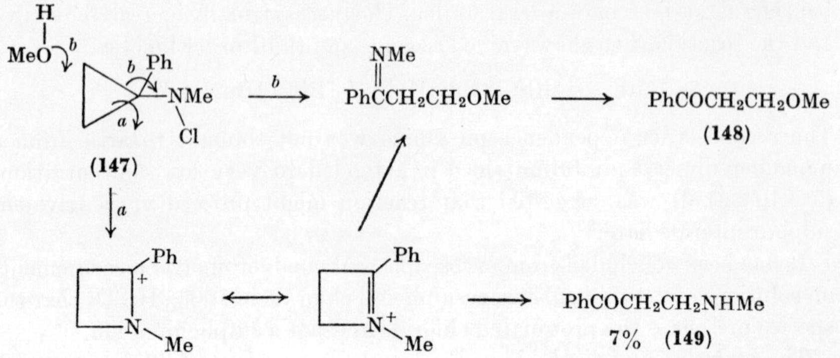

H
MeO
b
Ph
NMe
Cl
a
b
NMe
PhCCH2CH2OMe
PhCOCH2CH2OMe
a
Ph
N
Me
Ph
N+
Me
PhCOCH2CH2NHMe
7%

(147) (148) (149)

triethylamine–borane and quinuclidine–borane complexes are very similar.[246]

Other investigations of displacement reactions on chlorine[247] and on oxygen[248, 249] have also been described.

Ambident Nucleophiles[250]

The di-anions (**150**)—(**152**) are preferentially alkylated on carbon.[251] Alkylation of acetoin, adipoin, and benzoin with benzyl chloride and methyl iodide in

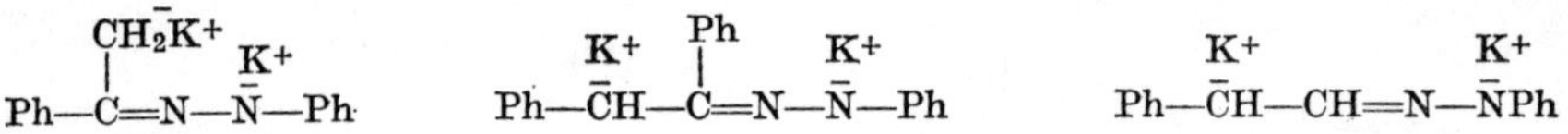

(150) (151) (152)

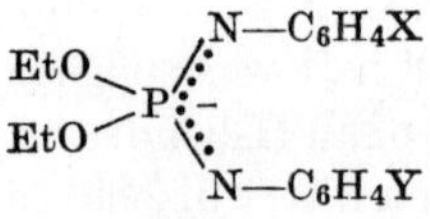

N—C6H4X
EtO
P
EtO
N—C6H4Y

(153)

1,2-dimethoxyethane in the presence of sodium hydride occurs on carbon.[252] The silver salts of diazoacetic ester and of diazo-ketones are alkylated on carbon by allyl halides and by benzyl bromide.[253] Methylation of the anion of

246 H. C. Kelly and J. A. Underwood, *Inorg. Chem.*, **8**, 1202 (1969).
247 K. Tsujihara, N. Furukawa, K. Oae, and S. Oae, *Bull. Chem. Soc. Japan*, **42**, 2631 (1969).
248 W. Adam, R. J. Ramirez, and S. C. Tsai, *J. Am. Chem. Soc.*, **91**, 1254 (1969).
249 N. M. Beileryan, O. A. Chaltykyan, S. A. Grigoryan, and Z. Z. Meliksetyan, *Armyan. Khim. Zh.*, **21**, 7 (1968).
250 Hard and soft acids and bases have been reviewed: R. G. Pearson, *Survey Progr. Chem.*, **5**, 1 (1969).
251 F. F. Henoch, K. G. Hampton, and C. R. Hauser, *J. Am. Chem. Soc.*, **91**, 676 (1969).
252 J. H. van de Sande and K. R. Kopecky, *Can. J. Chem.*, **47**, 163 (1969).
253 U. Schöllkopf and N. Rieber, *Chem. Ber.*, **102**, 488 (1969).

4,4-dimethylcyclopent-2-enone in dioxan occurs on carbon.[254] Alkylation of the anion (153) by methyl iodide in tetrahydrofuran has been studied.[255]

Alkylation of the anion of *anti*-benzaldoxime with triphenylmethyl chloride occurs on oxygen and not on nitrogen as previously reported.[256]

The products of O-attack on trimethylsilyl chloride by the enolate anions derived from unsymmetrical ketones have been determined.[257]

Other reactions of ambident ions which have been investigated include the alkylation of the di-anion of benzophenone,[258] 8-pyridylpurines,[259] triphenylarsine oxide,[260] and of the enolates of acetoacetic ester and acetylacetone.[261–263]

Substitution at Vinylic Carbon[264]

Most of the work this year has been concerned with the detection of vinyl cations as reaction intermediates. Compounds (154)—(157) all react faster in 80% aqueous ethanol than α-bromo-4-methoxystyrene, and the rates are insensitive to the addition of triethylamine. The *cis*- and *trans*-forms of (158) react much more slowly, which indicates that (159) must be an important canonical form of the intermediate cation. Compound (154) yields 55% of the ethoxy-allene derived from this.[265]

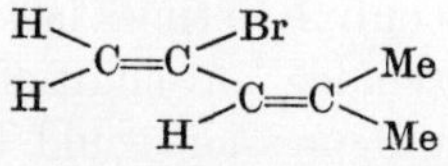

(154) 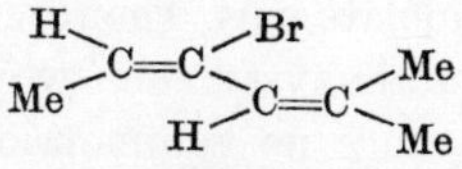   (155) 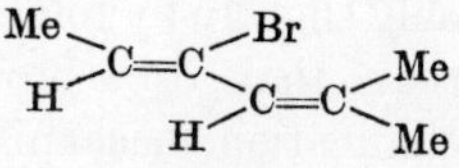   (156)

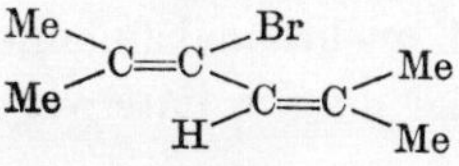

(157) 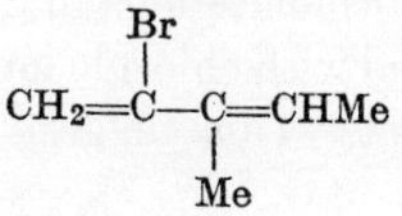   (158) 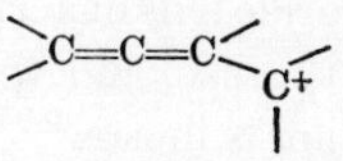   (159)

[254] A. J. Bellamy, *J. Chem. Soc.*(B), **1969**, 449.

[255] M. I. Kabachnik and V. A. Gilyarov, *Izv. Akad. Nauk SSSR, Ser. Khim.*, **1968**, 2036; *Chem. Abs.*, **70**, 19409 (1969).

[256] E. J. Grubbs and J. A. Villarreal, *Tetrahedron Letters*, **1969**, 1841; see *Org. Reaction Mech.*, **1967**, 95.

[257] H. O. House, L. J. Czuba, M. Gall, and H. D. Olmstead, *J. Org. Chem.*, **34**, 2324 (1969).

[258] W. S. Murphy and D. J. Buckley, *Tetrahedron Letters*, **1969**, 2975.

[259] F. Bergman and M. Rashi, *J. Chem. Soc.*(C), **1969**, 1831.

[260] B. D. Chernokal'skii, R. M. Bairamov, R. R. Shagidullin, I. A. Lamanova, and G. Kamai, *Zh. Obshch. Khim.*, **39**, 618 (1969); *Chem. Abs.*, **71**, 48951 (1969).

[261] A. L. Kurts, I. P. Beletskaya, A. Masias, S. S. Yufit, and O. A. Reutov, *Izv. Akad. Nauk SSSR, Ser Khim.*, **1968**, 1473; *Chem. Abs.*, **69**, 95588 (1968).

[262] A. L. Kurts, N. K. Genkina, I. P. Beletskaya, and O. A. Reutov, *Doklady Akad. Nauk SSSR*, **188**, 597 (1969).

[263] A. L. Kurts, A. Masias, N. K. Genkina, I. P. Beletskaya, and O. A. Reutov, *Doklady Akad. Nauk SSSR*, **187**, 807 (1969).

[264] Z. Rappoport, *Adv. Phys. Org. Chem.*, **7**, 1 (1969).

[265] C. A. Grob and R. Spaar, *Tetrahedron Letters*, **1969**, 1439.

Solvolysis of 1-anisyl-2,2-diphenyliodoethylene in 70.4% aqueous DMF is much faster than that of 1,2,2-triphenyliodoethylene, and there is a common-ion rate depression. This was interpreted as indicating the intervention of a free vinyl cation as shown in equation (21).[266] A similar mechanism was proposed for the solvolysis of trianisylvinyl bromide in 80% ethanol whose rate is

$$
\begin{array}{c}
\underset{Ph}{\overset{Ph}{\diagdown}}\!\!C\!\!=\!\!C\!\!\overset{C_6H_4OMe}{\diagup}\quad\underset{\underset{I^-}{\overset{-I^-}{\rightleftharpoons}}}{}\quad \underset{Ph}{\overset{Ph}{\diagdown}}\!\!C\!\!=\!\!\overset{+}{C}\!\!-\!\!C_6H_4OMe\quad\longrightarrow\quad \underset{Ph}{\overset{Ph}{\diagdown}}\!\!H\!\!-\!\!C\!\!-\!\!\overset{\overset{O}{\|}}{C}\!\!-\!\!C_6H_4OMe
\end{array}
$$

$$\cdots (21)$$

$$
\underset{Ph}{\overset{MeO\cdot C_6H_4}{\diagdown}}\!\!C\!\!=\!\!C\!\!\underset{C_6H_4\cdot OMe}{\overset{Br}{\diagup}}
$$

(160)

$$
\underset{Ph}{\overset{MeO\cdot C_6H_4}{\diagdown}}\!\!C\!\!=\!\!C\!\!\underset{Br}{\overset{C_6H_4\cdot OMe}{\diagup}}
$$

(161)

$$
\underset{Ph}{\overset{Ph}{\diagdown}}\!\!C\!\!=\!\!C\!\!\underset{Ph}{\overset{OSO_2R}{\diagup}}
$$

(162)

independent of the concentration of added sodium hydroxide and increased only slightly by toluene-p-thiolate ions. The reaction is only 1.75 times faster than that of α-bromo-4-methoxystyrene, which excludes an addition–elimination mechanism.[267] Since no common-ion rate depression could be detected there is no external return.[267] The reactions of **(160)** and **(161)** with benzylthiolate and toluene-p-thiolate ions in 80% ethanol, with acetate and chloride ions in acetic acid, and with chloride ion in DMF, yield equal amounts of the *cis-* and *trans*-products. This suggests that the intermediate vinyl cation is linear.[268]

The relative rates of acetolysis of the vinyl sulphonates **(162)** are 13,500, 41,700, and 1 respectively when R = F, CF$_3$, and p-tolyl. This sensitivity to leaving group contrasts with the results reported last year for the solvolysis of cyclohex-1-enyl and *cis*-but-2-enyl arenesulphonates, and supports a mechanism involving ionization to a vinyl cation.[269]

The reaction of **(163)** to yield the benzothiophen **(165)** shows a common-ion rate depression in acetone. When the reaction was carried out in the presence of ^{35}S-labelled trinitrobenzenesulphonic acid there was incorporation of label into unreacted **(163)**. The results support the view that a vinyl cation **(164)** is an intermediate.[270]

[266] L. L. Miller and D. A. Kaufman, *J. Am. Chem. Soc.*, **90**, 7282 (1968); see also D. Kaufman and L. L. Miller, *J. Org. Chem.*, **34**, 1495 (1965).
[267] Z. Rappoport and A. Gal, *J. Am. Chem. Soc.*, **91**, 5246 (1969).
[268] Z. Rappoport and Y. Apeloig, *J. Am. Chem. Soc.*, **91**, 6734 (1969).
[269] W. M. Jones and D. D. Maness, *J. Am. Chem. Soc.*, **91**, 4314 (1969).
[270] G. Modena and U. Tonellato, *Chem. Comm.*, **1968**, 1676; see *Org. Reaction Mech.*, **1968**, 108.

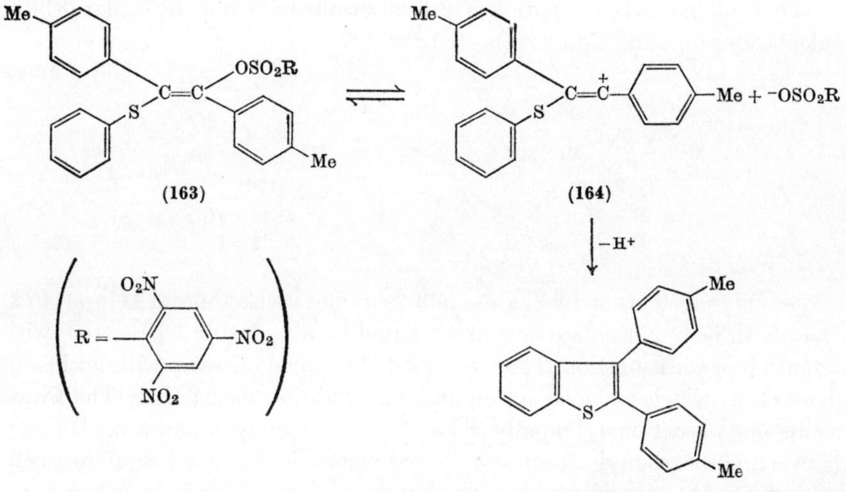

(163) (164)

$$R = \left(\begin{array}{c} O_2N \\ \\ NO_2 \end{array} \right.\!\!-\!\!NO_2 \left.\right)$$

(165)

Hanack and Bässler have shown that 1-cyclopropylvinyl chloride (166) reacts with silver perchlorate in acetic acid in the presence of sodium acetate to yield predominantly 1-cyclopropylvinyl acetate. 2-Chloro-3,3-dimethylbutene (167) does not react under these conditions. On reaction with silver perchlorate in unbuffered acetic acid the identified products from (166) were cyclopropyl methyl ketone (80%) and cyclopropylacetylene (15%). These reactions were thought to involve formation of a vinyl cation (168) which is strongly stabilized by the adjacent cyclopropyl ring.[271] Similar results were obtained by Sherrod and Bergman for the reaction of 1-cyclopropylvinyl iodide with silver acetate.[272]

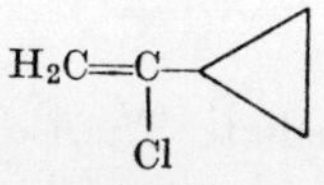

(166) 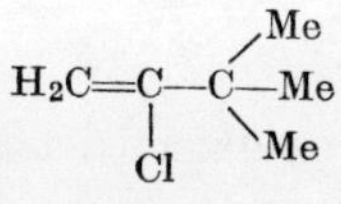 (167) 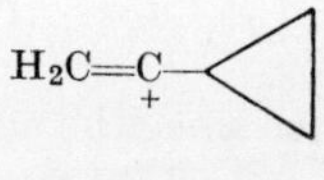 (168)

The products of solvolysis of (169a) in aqueous ethanol are 58% dimethylacetylene, 33% butan-2-one, and 9% methylallene. It was suggested that formation of the butan-2-one proceeds via a vinyl cation, and the isotope effect, $k_H/k_D = 1.20$, observed with (169b) supports this. In contrast, the *trans*-isomer (170a) reacts 40 times faster than (169a) and yields $98 \pm 3\%$ dimethylacetylene. The isotope effect, $k_H/k_D = 2.09$, observed with (170b) suggests that reaction proceeds via a concerted elimination.[273]

[271] M. Hanack and T. Bässler, *J. Am. Chem. Soc.*, **91**, 2117 (1969).
[272] S. A. Sherrod and R. G. Bergman, *J. Am. Chem. Soc.*, **91**, 2115 (1969).
[273] P. J. Stang and R. Summerville, *J. Am. Chem. Soc.*, **91**, 4600 (1969).

The 1-adamantylvinyl cation has been studied,[274] and molecular-orbital calculations on vinyl cations reported.[275]

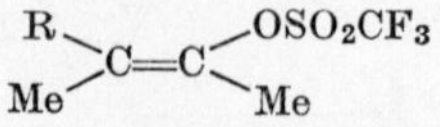

$$\underset{Me}{\overset{R}{>}}C=C\underset{Me}{\overset{OSO_2CF_3}{<}}$$

(169)

a: R = H
b: R = D

$$\underset{R}{\overset{Me}{>}}C=C\underset{Me}{\overset{OSO_2CF_3}{<}}$$

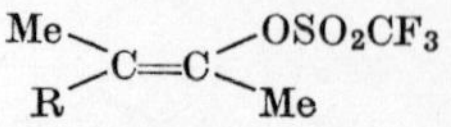

(170)

a: R = H
b: R = D

cis- and *trans-p-*Anisyl-2-chloro- and -2-bromovinyl ketones **(171)** and **(172)** react with benzenethiolate ions in methanol by a second-order reaction with retention of configuration. The 1-deuterated compounds react without loss of deuterium, which excludes an elimination–addition mechanism. The *trans-*compounds react more rapidly than the *cis*, possibly because coplanarity between the carbonyl group and the ethylenic double bond is more easily achieved in the transition state. The bromo-compound reacts less than twice as fast as the corresponding chloro-compound. The mechanism therefore involves addition–elimination or is a direct displacement. In contrast, the *cis-*isomers **(171)** react with methoxide ions by an elimination–addition

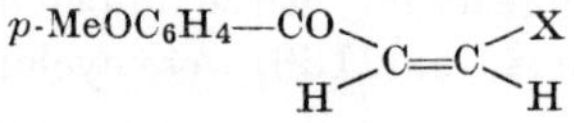

$$\underset{H}{\overset{p\text{-}MeOC_6H_4-CO}{>}}C=C\underset{H}{\overset{X}{<}}$$

(171)

(X = Cl, Br)

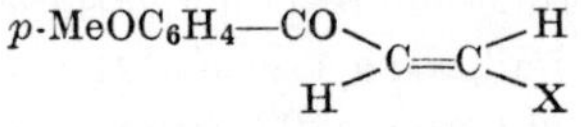

$$\underset{H}{\overset{p\text{-}MeOC_6H_4-CO}{>}}C=C\underset{X}{\overset{H}{<}}$$

(172)

(X = Cl, Br)

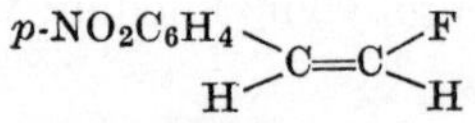

$$\underset{H}{\overset{p\text{-}NO_2C_6H_4}{>}}C=C\underset{H}{\overset{F}{<}}$$

(173)

$$\underset{H}{\overset{p\text{-}NO_2C_6H_4}{>}}C=C\underset{F}{\overset{H}{<}}$$

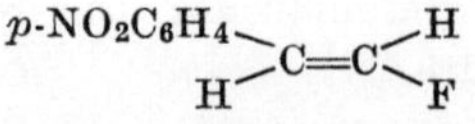

(174)

sequence since the acetylene was detected as an intermediate. The 1-deuterocompounds show appreciable isotope effects, and some exchange of deuterium with the solvent could be detected with the chloro-compounds. The bromo-compound reacts about 50 times faster than the chloro-compound. No acetylene could be detected in the reactions of the *trans-*isomers with methoxide, which presumably proceed via addition–elimination or a direct displacement.[276]

cis- and *trans-β-*Fluoro-4-nitrostyrenes **(173)** and **(174)** both undergo substitution rather than elimination on reaction with methoxide ions. In contrast, the corresponding *cis-*bromo- and -chloro-compounds react mainly by

274 K. Bott, *Chem. Comm.*, **1969**, 1349.
275 H. Fischer, K. Hummel, and M. Hanack, *Tetrahedron Letters*, **1969**, 2169.
276 D. Landini, F. Montanari, G. Modena, and F. Naso, *J. Chem. Soc.*(B), **1969**, 243.

elimination. The fluoro-compounds react 5—50 times faster with the benzene-thiolate ion than the corresponding chloro- and bromo-compounds, and yield a mixture of the *cis-* and *trans-*thioethers. This suggests that the anionic intermediate in these reactions has a longer lifetime than those derived from the chloro- and bromo-compounds which react with retention of configuration.[277]

1,1-Diaryl-2-fluoroethylenes react faster than the corresponding chloro- and bromo-compounds with toluene-*p*-thiolate ions in DMF. The ρ-value is 4.3.[278]

cis- and *trans-*2-Chloro- and -2-bromo-1-*p*-nitrophenyl-1-phenylethylene (175) and (176) react with sodium toluene-*p*-thiolate in DMF with retention of configuration. The *trans-*isomer (176) reacts 2.9 and 1.4 times faster than the *cis-*isomer (175) when X = Cl and Br respectively.[279]

$$\underset{p\text{-}NO_2C_6H_4}{\overset{Ph}{\diagdown}}C{=}C\overset{H}{\underset{X}{\diagup}}$$

(175)

(X = Cl, Br)

$$\underset{p\text{-}NO_2C_6H_4}{\overset{Ph}{\diagdown}}C{=}C\overset{X}{\underset{H}{\diagup}}$$

(176)

(X = Cl, Br)

The reactions of 2-halogenovinyl ketones with PhS^- and $PhSO_2^-$ ions proceed with retention of configuration. An elimination–addition mechanism was excluded by deuterium labelling.[280a] The reaction of [*trans-*^{2}H]vinyl chloride with sodium methoxide in methanol yields vinyl methyl ether which contains no deuterium. Reaction therefore proceeds by elimination–addition.[280b] The NMR spectra of the carbanions produced by attack of nucleophiles on α-cyano-4-nitrostilbenes have been determined. These ions have structures similar to those proposed as intermediates in nucleophilic substitution reactions at vinylic carbon.[281a] The reactions of chloronorbornadienes and chloronor-bornenes with sodium thiomethoxide have been studied.[281b]

Nucleophilic substitution reactions of 1-chloro-4-methylcyclohexene with amines and the benzenethiolate ion yield a mixture of 4-methylcyclohex-1-enyl and 3-methylcyclohex-1-enyl products which are possibly formed via a cyclohexyne (equation 22).[282]

[277] G. Machese, F. Naso, and G. Modena, *J. Chem. Soc.*(B), **1969**, 290.
[278] P. Beltrame, P. L. Beltrame, M. L. Cereda, and G. Lazzerini, *J. Chem. Soc.*(B), **1969**, 1100.
[279] P. Beltrame, P. L. Beltrame, and L. Bellotti, *J. Chem. Soc.*(B), **1969**, 932.
[280a] B. Cavalchi, D. Landini, and F. Montanari, *J. Chem. Soc.* (C), **1969**, 1204.
[280b] H.-J. Anacker, D. Beyer, C. Duscheck, M. Hampel, and W. Pritzkow, *Z. Chem.*, **9**, 420 (1969).
[281a] C. A. Fyfe, *Can. J. Chem.*, **47**, 2331 (1969).
[281b] D. I. Davies and P. J. Rowley, *J. Chem. Soc.*(C), **1969**, 288.
[282] P. Caubere and J. J. Brunet, *Tetrahedron Letters*, **1969**, 3323.

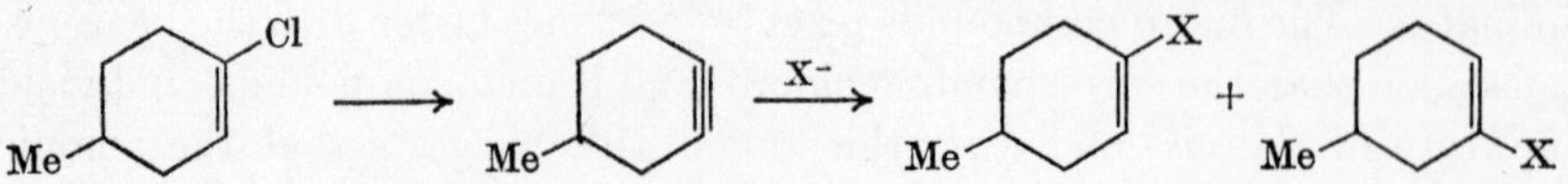

$$\dots (22)$$

Reactions of α-Halogenocarbonyl Compounds[283]

The reactions of phenacyl bromide with substituted anilines yield a Taft ρ-value of −2.66.[284] The reactions of α-bromobenzyl phenyl ketones with amines,[285–288] of α-bromo- and α-chloro-benzyl phenyl ketone with triphenylphosphine,[289] and of 1-phenyl-2,3-dimethyl-4-iodoacetylpyrazolone with iodide ions[290] have also been investigated.

The α-chloro-ketone (**177**) reacts with sodium methoxide to yield (**179**) which is probably formed through rearrangement of the enolic form of the chloride (**178**).[291] 1-Chloro-3-phenylthiopropan-2-one undergoes a similar rearrangement.[292]

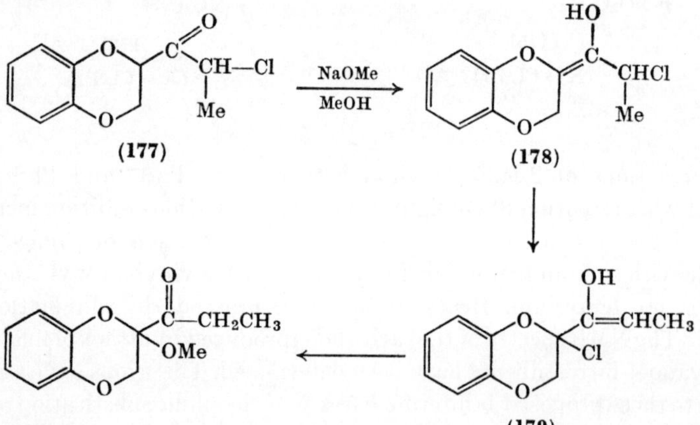

[283] See also p. 88.

[284] L. M. Litvinenko and L. A. Perelman, *Zh. Org. Khim.*, **4**, 2132 (1969); *Chem. Abs.*, **70**, 67302 (1969); see also S. Mishra, J. N. Kar, G. B. Behera, and M. K. Rout, *Indian J. Chem.*, **7**, 890 (1969).

[285] V. S. Karavan, T. E. Zhesko, L. P. F. Ionenko, and T. I. Temnikova, *Zh. Org. Khim.*, **5**, 1085 (1969).

[286] V. S. Karavan, L. M. Spitsyna, and T. I. Temnikova, *Zh. Org. Khim.*, **5**, 468 (1969).

[287] V. S. Karavan, T. E. Zhesko, and T. I. Temnikova, *Zh. Org. Khim.*, **4**, 1969 (1968); *Chem. Abs.*, **70**, 28134 (1969).

[288] V. S. Karavan, T. E. Zhesko, and T. I. Temnikova *Zh. Org. Khim.*, **4**, 1000 (1968); *Chem. Abs.*, **69**, 43577 (1968).

[289] I. J. Borowitz, P. E. Rusek, and R. Virkhaus, *J. Org. Chem*, **34**, 1595 (1969); cf. *Org. Reaction Mech.*, **1965**, 77; **1966**, 80; **1967**, 93; **1968**, 109.

[290] E. Kórös, M. Orbán, and Z. Bedó, *Magyar Kém. Foly.*, **75**, 93 (1969).

[291] V. Rosnati, F. Annicolo, and G. Pagani, *Gazzetta*, **99**, 152 (1969); see also *Org. Reaction Mech.*, **1967**, 73.

[292] V. Rosnati, F. Sannicolo, and G. Pagani, *Gazzetta*, **99**, 152 (1969); see also V. Rosnati, F. Sannicolo, and G. Zecchi, *ibid.*, p. 651.

Other Reactions

Claims[293] that the solvolysis of cyclohexyl toluene-*p*-sulphonate proceeds via a boat transition state have been rendered unlikely by the observation that 4,4-dimethylcyclohexyl toluene-*p*-sulphonate reacts slightly more rapidly than cyclohexyl toluene-*p*-sulphonate. If the 4,4-dimethyl compound reacted via a boat transition state as (180) or (181) this should be of higher energy than that derived from cyclohexyl toluene-*p*-sulphonate, and hence a reduced rate would be expected.[294]

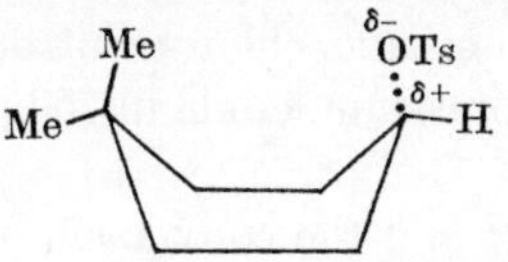

(180) 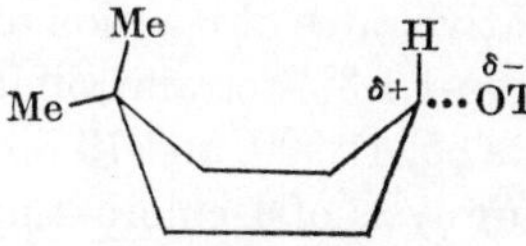   (181)

Solvolyses of *cis*- and *trans*-4-t-butyl-2,2-dimethylcyclohexyl toluene-*p*-sulphonate have been studied.[295]

The reactions of methyl, isopropyl, allyl, and benzyl iodides with hydrazine, morpholine, imidazole, and piperidine may be correlated with the basicities of the amines. Hydrazine did not show a positive deviation from the Brønsted plots and so the α-effect is not operative.[296]

The rate constants for the reactions of methyl perchlorate with a large number of nucleophiles in water and in methanol have been correlated by the Swain–Scott equation.[297a] The rate of reaction of methyl benzenesulphonate is less sensitive to the nature of the nucleophile than is that of methyl perchlorate.[297b]

The plot of volume of activation for the hydrolysis of methyl bromide against pressure shows striking discontinuities at high pressures, and it was suggested that under these conditions a second mechanism is followed.[298] The variation with pressure of the volume of activation for the hydrolysis of allyl chloride has been measured.[299]

The heat capacities of solution ($\Delta C_p{}^0$) of twenty-one low molecular weight alcohols in water have been reported. The values are large and positive, and

[293] See *Org. Reaction Mech.*, **1965**, 62.

[294] J. E. Nordlander, J. M. Blank, and S. P. Jindal, *Tetrahedron Letters*, **1969**, 3477; cf. ref. 153.

[295] J.-C. Richer and P. Belanger, *Can. J. Chem.*, **47**, 3281 (1969).

[296] S. Oae, Y. Kadona, and Y. Yano, *Bull. Chem. Soc., Japan*, **42**, 1110 (1969).

[297a] J. Koskikallio, *Acta Chem. Scand.*, **23**, 1477, 1490 (1969).

[297b] E. Yrjänheikki and J. Koskikallio, *Suomen Kemistilehti, B* **42**, 195 (1969).

[298] B. T. Baliga and E. Whalley, *J. Phys. Chem.*, **74**, 654 (1969).

[299] A. B. Lakef and J. B. Hyne, *Can. J. Chem.*, **47**, 1369 (1969).

the possibility was emphasized that rate and equilibrium differences in aqueous solution may have large contributions from heat capacity differences.[300]

The rates of the substitution reactions of methylene halides with alkoxide ions are changed only slightly on going from MeO^- to Pr^iO^- and Bu^tO^-. In contrast, the rate of alcoholysis of chloroform by an α-elimination mechanism, and the rates of deuterium exchange of the methylene halides, are increased markedly. It was therefore concluded that the substitution reactions of the methylene halides follow an S_N2 mechanism.[301]

The hydrolyses of chloroform and carbon tetrachloride on anion-exchange resins have been studied.[302]

The relative rates of reaction of tetranitromethane, chlorotrinitromethane, and 1-chloro-1,1,3,3-tetranitrobutane with potassium iodide in 75% aqueous ethanol are 85, 41,900, and 79 respectively.[303]

The hydrolysis of 1-chloro-2-nitroethane[304] and the conversion of methyl β-benzoyl-β-bromopropionate into methyl β-benzoyl-α-methoxypropionate in methanol containing potassium acetate[305] proceed by elimination–addition mechanisms.

The ^{13}C and ^{18}O isotope effects for production of carbon dioxide in the decomposition reaction of benzhydryl triphenylphosphoniumacetate in acetic acid are $k_{12}/k_{13} = 1.045$ and $k_{16}/k_{18} = 1.015$. A mechanism (equation 23) involving concerted carbon–carbon and carbon–oxygen bond breaking was proposed, and calculated values of the isotope effects in good agreement with the experimental values were reported. The corresponding n-octyl and s-octyl compounds yielded isotope effects $k_{12}/k_{14} = 1.020$ and $k_{16}/k_{18} = 1.000$, and it was suggested that they react via triphenylphosphoniumacetic acetic anhydride.[306]

$$\overset{+}{Ph_3}PCH_2\overset{O}{\overset{\|}{C}}OCHPh_2 \xrightarrow{HOAc} \left[\overset{+}{Ph_3}PCH_2 \cdots \overset{\delta^-}{C} \overset{O}{\overset{\|}{\cdots}} O \cdots \overset{\delta^+}{C}HPh_2 \atop \begin{matrix} \vdots & \vdots \\ HOAc & O \\ & \| \\ H{\diagdown}O{-}C{-}Me \end{matrix} \right] \longrightarrow$$

$$\overset{+}{Ph_3}PMe + CO_2 + AcOCHPh_2 \quad \ldots (23)$$

[300] E. M. Arnett, W. B. Kover, and J. V. Carter, *J. Am. Chem. Soc.*, **91**, 4028 (1969).

[301] J. Hine, R. B. Duke, and E. F. Glod, *J. Am. Chem. Soc.*, **91**, 2316 (1969).

[302] H. F. Ryan, *Austral. J. Chem.*, **21**, 2933 (1968).

[303] V. I. Slovetskii, M. S. L'vova, A. A. Fainzil'berg, and T. I. Chaeva, *Izv. Akad. Nauk SSSR, Ser. Khim.*, **1968**, 80; *Chem. Abs.*, **69**, 35191 (1968); see also M. S. L'vova, V. I. Slovetskii, and A. A. Fainzil'berg, *ibid.*, p. 323; *Chem. Abs.*, **69**, 51325 (1968).

[304] A. Talvik and I. Talvik, *Organic Reactivity* (Tartu), **6**, 484 (1969).

[305] N. Sugiyama, T. Gasha, H. Kataoka, and C. Kashima, *J. Chem. Soc.*(B), **1969**, 1060.

[306] S. Seltzer, A. Tsolis, and D. B. Denney, *J. Am. Chem. Soc.*, **91**, 4236 (1969).

Examples of the S_N2' mechanism are given in refs. 307 and 308.

The equilibrium constants for the exchange reactions of N-benzyl-N-methylpiperidinium halides with pyridine to form N-benzylpyridinium salts and N-methylpiperidine fall in the range 0.3—0.4. The difference in carbon basicity between pyridine and N-methylpiperidine is therefore much smaller than the difference in hydrogen basicity.[309]

The reactions of alkyl bromides with pyridine in DMF,[310] of methyl picrate with substituted dimethylanilines to yield trimethylanilinium ions,[311] and of methyl iodide, ethyl iodide, benzyl chloride, and 4-nitrobenzyl chloride with aromatic amines[312] have also been studied. The effect of pressure on the reaction of methyl iodide with pyridine in benzene–ethanol mixtures has been determined.[313] The quaternization of substituted isoquinolines[314, 315] and of substituted naphthyridines[316] has been investigated. There have been more investigations of the steric course of quaternization reactions,[317–320] and of sulphonium salt formation.[321] Other quaternization reactions are described in refs. 322 and 323.

The ρ-value for the conversion of N-benzyl-N,N-dimethylanilinium iodides into benzyl iodide and N,N-dimethylaniline in chloroform is +1.4.[324]

The nucleophilic reactivity of fluorenyl anions,[325, 326] and of the anions of aliphatic nitro-compounds,[327] has been measured.

Cleavage of the central ether group of 1,2,3-trimethoxybenzenes by HBr in acetic acid is faster than that of the flanking groups, possibly as a result of

[307] H. Goldwhite and C. M. Valdez, *Chem. Comm.*, **1969**, 7.
[308] G. Maury and N. H. Cromwell, *Tetrahedron Letters*, **1969**, 1716.
[309] R. E. J. Hutchison and D. S. Tarbell, *J. Org. Chem.*, **34**, 66 (1969).
[310] K. Murai and C. Kimura, *J. Chem. Soc. Japan*, **90**, 503 (1969); *Chem. Abs.*, **71**, 48947 (1969).
[311] Y. Kondo, T. Matsui, and N. Tokura, *Bull. Chem. Soc. Japan*, **42**, 1037 (1969).
[312] P. S. Radhakrishnamurti and G. P. Panigrahi, *J. Indian Chem. Soc.*, **46**, 318 (1969).
[313] H. Heydtmann and D. Büttner, *Z. Phys. Chem.* (Frankfurt), **63**, 39 (1969).
[314] J. Kóbor, G. Bernáth, L. Radics, and M. Kajtár, *Acta Chim. Acad. Sci. Hung.*, **60**, 255 (1969).
[315] L. Radics, M. Kajtár, J. Kóbor, and G. Bernáth, *Acta Chim. Acad. Sci. Hung.*, **60**, 381 (1969).
[316] R. A. Y. Jones and N. Wagstaff, *Chem. Comm.*, **1969**, 56.
[317] D. R. Brown, J. McKenna, and J. M. McKenna, *Chem. Comm.*, **1969**, 186.
[318] D. R. Brown and J. McKenna, *J. Chem. Soc.*(B), **1969**, 570.
[319] J. McKenna and J. M. McKenna, *J. Chem. Soc.*(B), **1969**, 644.
[320] R. Brettle, D. R. Brown, J. McKenna, and R. Mason, *Chem. Comm.*, **1969**, 339.
[321] M. J. Cook, H. Dorn, and A. R. Katritzky, *J. Chem. Soc.*(B), **1968**, 1467.
[322] L. Pentimalli and L. Greci, *Gazzetta.*, **98**, 1360 (1968).
[323] M. Shamma, C. D. Jones, and J. A. Weiss, *Tetrahedron*, **25**, 4347 (1969).
[324] J. T. Burns and K. T. Leffek, *Can. J. Chem.*, **47**, 3725 (1969).
[325] A. F. Cockerill and R. T. Hargreaves, *Chem. Comm.*, **1969**, 915.
[326] K. Bowden and R. S. Cook, *J. Chem. Soc.*(B) **1968**, 1529.
[327] V. M. Belikov, Y. N. Belokon, N. G. Faleev, and T. B. Korchemnaya, *Izv. Akad. Nauk SSSR, Ser. Khim.*, **1968**, 1477; *Chem. Abs.*, **69**, 95575 (1968).

steric acceleration.[328] Cleavage of ethers by sulphite,[329] and the acetolysis of ethers,[330] have also been investigated.

The rates of the reactions between cyclic alcohols and HBr have been measured.[331]

Catalysis of the hydrolysis of benzyl fluoride,[332] t-butyl fluoride,[333] and t-butyl chloride[333] by metal ions has been investigated. Heterogeneous catalysis of the reaction between ethyl iodide and silver nitrate has been studied.[334, 335]

The ρ-value of the solvolysis of 5-substituted 2-furylmethyl p-nitrobenzoates in aqueous ethanol is -7.5.[336] 2- and 4-Methyl substituents have a large rate-enhancing effect on the methanolysis 3-chloromethylfuran.[337]

The solvolysis of α-phenoxyalkyl chlorides[338] and the methanolysis of glucopyranosyl bromides having a benzyloxy group at C-2[339a] have been studied. Nucleophilic displacement reactions of carbohydrate sulphonates have been discussed.[339b]

The solvolyses of substituted cyclohexyl and decalyl methanesulphonates have been investigated.[340–342]

The nucleophilic reactivity of vitamin B_{12s}, cobaloximes, and other Co(I) chelates towards a large number of alkyl halides has been measured.[343]

Substitution reactions of propargyl halides have been studied.[344–347]

[328] C. F. Wilcox and M. A. Seager, *J. Org. Chem.*, **34**, 2319 (1969).
[329] J. Gierer and B. Koutek, *Acta Chem. Scand.*, **23**, 1343 (1969).
[330] S. Coffi-Nketsia, A. Kergomard, and H. Tanton, *Compt. Rend.*, *C*, **267**, 1495 (1968).
[331] P. S. Radhakrishnamurti and T. P. Visvanathan, *Proc. Nat. Inst. Sci. India*, *A*, **35**, 146 (1969); *Chem. Abs.*, **71**, 112083 (1969).
[332] H. R. Clark and M. M. Jones, *J. Am. Chem. Soc.*, **91**, 4302 (1969).
[333] V. P. Tretyakov, E. S. Rudakov, and V. B. Bistrenko, *Organic Reactivity* (Tartu), **6**, 558 (1969); E. S. Rudakov and I. V. Kozchevnikov, *ibid.*, p. 572; E. S. Rudakov, I. V. Kozchevnikov, and V. V. Zamashchikov, *ibid.*, p. 579.
[334] P. S. Walton and M. Spiro, *J. Chem. Soc.*(B) **1969**, 42.
[335] J. M. Austin, O. D. E. S. Ibrahim, and M. Spiro, *J. Chem. Soc.*(B), **1969**, 669.
[336] D. S. Noyce and G. V. Kaiser, *J. Org. Chem.*, **34**, 1008 (1969).
[337] M. A. Gal'bershtam, G. T. Khachaturova, N. E. Bairamova, K. Y. Novitskii, and Y. K Uyr'ev, *Izv. Vyssh. Ucheb. Zaved., Khim. Khim. Tekhnol.*, **11**, 1395 (1968); *Chem. Abs.*, **71**, 21370 (1969).
[338] K. K. Zikherman, *Khim. Atset.*, **1968**, 307; *Chem. Abs.*, **71**, 2687 (1969).
[339a] T. Ishikawa and M. G. Fletcher, *J. Org. Chem.*, **34**, 563 (1969).
[339b] A. C. Richardson, *Carbohydrate Res.*, **10**, 395 (1969); Y. Ali and A. C. Richardson, *J. Chem. Soc.*(C), **1969**, 320; B. Capon, *Chem. Rev.*, **69**, 469 (1969).
[340] D. S. Noyce and B. E. Johnston, *J. Org. Chem.*, **34**, 1253 (1969).
[341] D. S. Noyce, B. E. Johnston, and B. Weinstein, *J. Org. Chem.*, **34**, 463 (1969).
[342] D. S. Noyce, B. Bastian, P. T. S. Lau, R. S. Monson, and B. Weinstein, *J. Org. Chem.*, **34**, 1247 (1969).
[343] G. N. Schrauzer and E. Deutsch, *J. Am. Chem. Soc.*, **91**, 3341 (1969).
[344] G. F. Hennion and J. F. Motier, *J. Org. Chem.*, **34**, 1319 (1969).
[345] R. V. Vizgert and R. V. Sendeka, *Organic Reactivity* (Tartu), **5**, 126 (1968).
[346] R. V. Vizgert and R. V. Sendeka, *Organic Reactivity* (Tartu), **5**, 362 (1968); *Chem. Abs.*, **70**, 19321 (1969).
[347] T. A. Azizyan, P. S. Chobanyan, Z. A. Abramyan, and A. T. Babayan, *Uch. Zap. Erevan. Gos. Univ.*, **1968**, No. 2, 58; *Chem. Abs.*, **71**, 38096 (1969).

Propene, styrene, and isobutene episulphoxide undergo acid-catalysed methanolysis with predominant attack at the secondary or tertiary carbon. This was interpreted as indicating that the mechanism is $A1$.[348]

The acid-catalysed methanolysis[349] and the reaction with lithium aluminium hydride[350] of 2-aryloxetans have been studied. There have been many investigations of the ring-opening reactions of epoxides.[351]

Solvolyses of the following compounds have also been studied: arylmethyl-mercuric ions (cf. ref. 81*b*, p. 26),[352] hydrazonyl bromides,[353] aminohydrazidic bromides,[354] steroidal toluene-*p*-sulphonates,[355] s-pentyl bromide,[356] and n-propyl chlorosulphate.[357]

Other nucleophilic substitution reactions which have been studied include those between benzyl chloride and alkoxide ions,[358] butyl chloride and sodium isopentylate,[359] propyl and butyl iodide and zinc iodide,[360] ketonic Mannich bases and triethyl phosphite,[361] trialkyloxonium salts and esters,[362] thiamine and sulphite ions,[363] allyl chloride and cuprous acetylide,[364] β-propionolactone[365] and 5-acetoxymethyluracil[366] and nucleophiles, xanthates, and

[348] K. Kondo, A. Negishi, and G. Tsuchihashi, *Tetrahedron Letters*, **1969**, 3173; cf. *Org. Reaction Mech.*, **1968**, 113—114.

[349] P. O. I. Virtanen and H. Ruotsalainen, *Suomen Kemistilehti*, B**42**, 69 (1969).

[350] J. Seyden-Penne and C. Schaal, *Bull. Soc. Chim. France*, **1969**, 3653.

[351] P. O. I. Virtanen, *Suomen Kemistilehti*, B**42**, 199 (1969); A. Chatterjee and D. Banerjee, *Tetrahedron Letters*, **1969**, 4559; Z. A. Musharov, *Kinet. Katal.*, **9**, 845 (1968); *Chem. Abs.*, **70**, 76991 (1969); M. Vanderwalle, K. L. Seghal, and V. Sipodo, *Bull. Soc. Chim. Belges*, **77**, 611 (1968); T. Nakajima, S. Suga, T. Sugita, and K. Ichikawa, *Tetrahedron*, **25**, 1807 (1969); R. R. Benerito, H. M. Ziifle, and R. J. Berni, *J. Phys. Chem.*, **73**, 1216 (1969); F. Mareš, J. Hetflejš, and V. Bažant, *Coll. Czech. Chem. Comm.*, **34**, 3086, 3098 (1969); E. J. Langstaff, R. Y. Moir, R. A. B. Bannard, and A. A. Casselman, *Can. J. Chem.*, **46**; 3649 (1968); A. E. Audier, J. F. Dupin, and J. Julien, *Bull. Soc. Chim. France*, **1968**, 3850; R. M. Laird and R. E. Parker, *J. Chem. Soc.*(B), **1969**, 1062; H. C. Brown and N. M. Yoon, *Chem. Comm.*, **1968**, 1549; Z. A. Musharov, I. G. Kaufman, and S. G. Entelis, *Zh. Fiz. Khim.*, **43**, 95 (1969); *Chem. Abs.*, **70**, 95 (1969).

[352] B. G. van Leuwen and R. J. Ouellette, *J. Am. Chem. Soc.*, **90**, 7056, 7061 (1968).

[353] J. B. Aylward and F. L. Scott, *J. Chem. Soc.*(B), **1969**, 1080.

[354] F. L. Scott, J. A. Cronin, and J. Donovan, *Tetrahedron Letters*, **1969**, 4615.

[355] R. Baker, J. Hudec, and K. L. Rabone, *Chem. Comm.*, **1969**, 197.

[356] K. Behari and B. Krishan, *Chim. Anal.* (Paris), **51**, 87 (1969); *Chem. Abs.*, **71**, 2683 (1969).

[357] E. Buncel and J. P. Millington, *Can. J. Chem.*, **47**, 2145 (1969).

[358] I. G. Murgulescu, D. Oancea, and G. Minou, *Rev. Roum. Chim.*, **14**, 3 (1969).

[359] F. Barbulescu and R. Sternberg, *An. Univ. Bucuresti Ser. Stiint. Natur. Chim.*, **16**, 19 (1967); *Chem. Abs.*, **70**, 114324 (1969).

[360] B. F. Howell and D. E. Troutner, *J. Org. Chem.*, **33**, 4146 (1968).

[361] B. E. Ivanov and V. F. Zheltukhin, *Izv. Akad. Nauk SSSR, Ser. Khim.*, **1969**, 1016, 1022.

[362] L. M. Bogdanova, A. I. Efremova, B. A. Rozenberg, and N. S. Enikolopyan, *Izv. Akad. Nauk SSSR, Ser. Khim.*, **1969**, 801; A. I. Efremova, V. P. Roshchupkin, B. A. Rozenberg, and N. S. Enikolopyan, *ibid.*, p. 807; *Chem. Abs.*, **71**, 21422, 21423 (1969).

[363] J. Leichter and M. A. Joslyn, *Biochem. J.*, **113**, 611 (1969).

[364] K. Y. Odintsov, V. I. Trofimova, and R. M. Flid, *Zh. Fiz. Khim.*, **42**, 1525 (1968); *Chem. Abs.*, **69**, 95593 (1968).

[365] R. E. Davis, L. Suba, P. Klimishin, and J. Carter, *J. Am. Chem. Soc.*, **91**, 104 (1969).

[366] D. V. Santi and A. L. Pogolotti, *Tetrahedron Letters*, **1968**, 6159.

trimethylamine,[367] phosphonates and alkyl iodides,[368] bromoacetanilides and piperidine,[369] and the reaction of bischloromethylphosphinic acid and its phenyl ester with benzylamine and aniline.[370]

Molecular-orbital calculations on the transition state for S_N2 reactions have been reported.[371]

[367] H. Yoshida, *Bull. Chem. Soc. Japan*, **42**, 1948 (1969).

[368] H. M. Bell, *J. Org. Chem.*, **34**, 681 (1969).

[369] J. P. Chupp, J. F. Olin, and H K Landwehr, *J. Org. Chem.*, **34**, 1192 (1969).

[370] M. K. Il'ina and I. M. Shermergorn, *Izv. Akad. Nauk SSSR, Ser. Khim.*, **1969**, 175; *Chem. Abs.*, **70**, 105608 (1969).

[371] O. Martensson, *Acta Chem. Scand.*, **23**, 335 (1969).

CHAPTER 3

Carbanions and Electrophilic Aliphatic Substitution

J. M. Brown

School of Molecular Sciences, University of Warwick

Carbanion Structure and Stability

Information on the structure of a variety of carbanionic species in solution has been accrued by nuclear magnetic resonance. Unlike the parent hydrocarbon, the di-anion of [16]annulene is conformationally rigid on an NMR time-scale and has chemical shifts consistent with an eighteen-electron diamagnetic ring-current.[1] In contrast, a paramagnetic $4n$-electron ring-current is suggested by chemical shift data on di-anions of some polycyclic aromatic hydrocarbons in tetrahydrofuran,[2] and is required by the 25 ppm *downfield* shift of internal methyl group signals on conversion of a dimethyldihydropyrene into its di-anion (1).[3] Other studies include oxapentalenyl[4] and azapentalenyl[5] anions, the di-anion of dicyclopenta[*a,f*]naphthalene (where most of the charge resides on the five-membered rings, in accordance with HMO calculations),[6] and the stable intermediates formed by addition of

[1] J. F. M. Oth, G. Anthoine, and J. M. Gilles, *Tetrahedron Letters*, **1968**, 6265.
[2] R. G. Lawler and C. V. Ristagno, *J. Am. Chem. Soc.*, **91**, 1534 (1969).
[3] R. H. Mitchell, C. E. Klopfenstein, and V. Boekelheide, *J. Am. Chem. Soc.*, **91**, 4931 (1969).
[4] T. S. Cantrell and B. L. Harrison, *Tetrahedron Letters*, **1969**, 1299.
[5] H. Volz and B. Messner, *Tetrahedron Letters*, **1969**, 4111.
[6] R. S. Schneider and E. F. Ullman, *Tetrahedron Letters*, **1969**, 3249.

methoxide or acetonide anion to α-cyano-4-nitrostilbenes.[7] Deprotonation
of a variety of 1,4-enynes by butyl-lithium has been shown to occur in two
steps; a rapid deprotonation to a mono-anion, and with an excess of butyl-
lithium in [²H₈]THF, the dilithiated derivative (3) is formed more slowly.[8]
The chemical shift of the vinyl protons in (2) and (3) suggests that the latter
has *less* negative charge delocalized on to the vinyl group. Electronic spectra
of the (carbanionic) species formed by interaction of sodium hydride and a
variety of aromatic substrates in dimethylformamide are recorded.[9] Solvent
effects on the NMR and UV spectra of 1-nitroindenylide and 9-nitrofluorenylide
anions have been examined and are consistent with solvation at nitronate
oxygen in protic solvents, and consequently more delocalization of negative

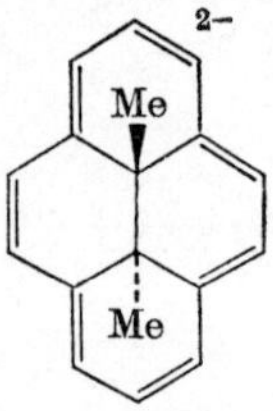

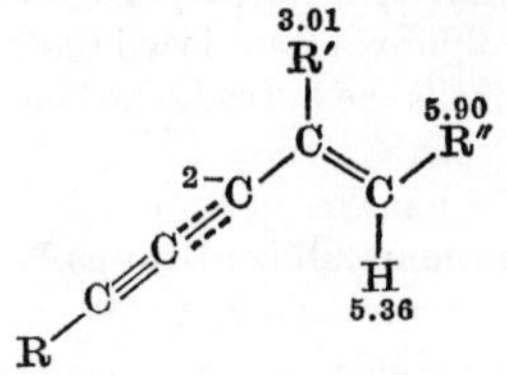

(τ-values given for R = Ph; R′ = R″ = H)

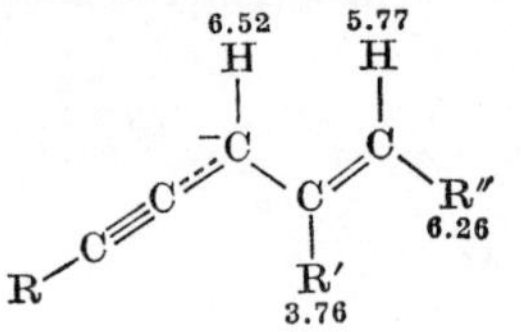

(1)   (2)   (3)

charge to ring positions in aprotic media.[10] Further work on the nature of
ion-pairing in salts of delocalized carbanions is recorded; 4,5-benzofluorenylide
anions[11] and 1,3-diphenyl-1-methylallyl[12] anions have been shown to follow
trends in the distribution between contact and solvent-separated species
expected from Smid's results on fluorenylide anions. In the presence of added
tetraglyme, the polymerization of styrene in tetrahydropyran by sodium may
be propagated by free "living polymer" anion, by its contact ion-pair, or by
the tetraglyme separated ion-pair. By variation of tetraglyme concentration,
rate constants for propagation processes were estimated.[13] A most interesting
study has been made on the UV spectrum of barium difluorenylide.[14] This
exists purely as a contact ion-pair in tetrahydrofuran solution. Addition of
the chelating cyclic ether (4) changes the spectrum to that expected for
equimolar amounts of contact and solvent-separated ion-pairs, and with
additional evidence from NMR shielding of the ether methylene protons, this

[7] C. A. Fyfe, *Can. J. Chem.* **47**, 2331 (1969).
[8] J. Klein and S. Brenner, *J. Am. Chem. Soc.*, **91**, 3094 (1969).
[9] L. N. Guseva, Y. M. Mikheev, and M. G. Kuzmin, *Zh. Org. Khim.*, **5**, 306 (1969); *Chem. Abs.*, **70**, 105740 (1969).
[10] R. C. Kerber and A. Porter, *J. Am. Chem. Soc.*, **91**, 366 (1969).
[11] D. Casson and B. J. Tabner, *J. Chem. Soc.*(B), **1969**, 572.
[12] J. W. Burley and R. N. Young, *Chem. Comm.*, **1969**, 1127
[13] M. Shinohara, J. Smid, and M. Szwarc, *Chem. Comm.*, **1969**, 1232.
[14] T. E. H. Esch and J. Smid, *J. Am. Chem. Soc.*, **91**, 4580 (1969).

was interpreted in terms of an asymmetric ion-pair with one fluorene residue exposed to the barium ion at any given time, and one shielded by the oxygen atoms of the ether (5). Proton abstraction from 3,4-benzofluorene by the unsolvated ion-pair is 30,000 times slower than by the sodium fluorenylide ion-pair.

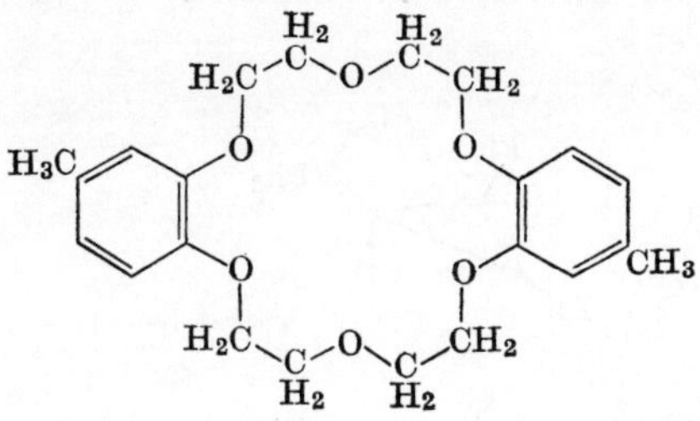

(4) 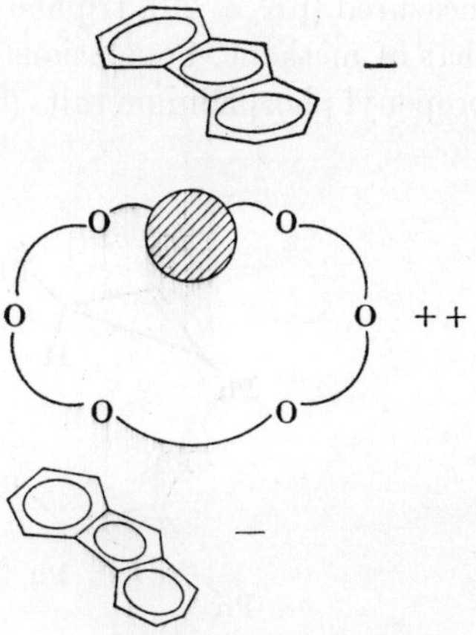   (5)

Theoretical contributions to carbanion chemistry include extended Hückel calculations on the equilibrium geometries of a variety of conjugated anions [15] (and radicals and cations). The energy well for twist about the C–phenyl bonds in e.g. benzhydryl anion is predicted to be much deeper than in benzhydryl cation. New CNDO/2 calculations on alkyl anions [16] predict an inversion barrier in pyramidal CH_3^- of 17.8 kcal mole^{-1}, and a stabilization of anionic carbon (in the gas phase) by alkylation at the carbanion site. An equation for hydrocarbon pK_a [17] [$= 83.12 - 1.33$ (% s-character)] was derived by a scaled maximum overlap method. An *a priori* calculation on cyclopropenyl anion [18] predicts a breakdown in degeneracy by movement of one C–H bond out of the molecular plane, with an inversion barrier of 52 kcal mole^{-1}; the cyclopropene→cyclopropenyl anion ionization process is calculated to be more than 10 kcal mole^{-1} less favourable than the cyclopropane→ cyclopropyl anion process. A direct approach [19] to the thermodynamic acidity of triphenylcyclopropene has been made by rapid-scan polarimetric techniques. In hexamethylphosphorus triamide solution, triphenylmethyl cation undergoes two successive one-wave reductions with $E = -0.91$ v, and triphenyl-

[15] R. Hoffmann, R. Bissell, and D. G. Farnum, *J. Phys. Chem.*, **73**, 1789 (1969).
[16] T. P. Lewis, *Tetrahedron*, **25**, 4117 (1969).
[17] Z. B. Maksic and M. Eckert-Macsic, *Tetrahedron*, **25**, 5113 (1969).
[18] D. T. Clark, *Chem. Comm.*, **1969**, 637; D. T. Clark and D. R. Armstrong, *ibid.*, p. 850.
[19] R. Breslow and K. Balasubramanian, *J. Am. Chem. Soc.*, **91**, 5182 (1969).

cyclopropenyl cation under the same conditions undergoes two successive one-wave reductions with $E = -2.67$v. Knowing the pK_R values of Ph_3C^+ and $Ph_3C_3^+$, and assuming that the thermodynamics of the process R–OH→R–H are comparable in the two cases, triphenylcyclopropene is thereby shown to be 18 pK_a units less acidic than triphenylmethane. Since the latter has been measured ($pK_a = 29$), triphenylcyclopropene has a pK_a of 47, comparable to that of methane. Conclusions [20] based on the direction of cleavage of cyclopropenyl phosphonium salts (Scheme 1) are therefore questionable.

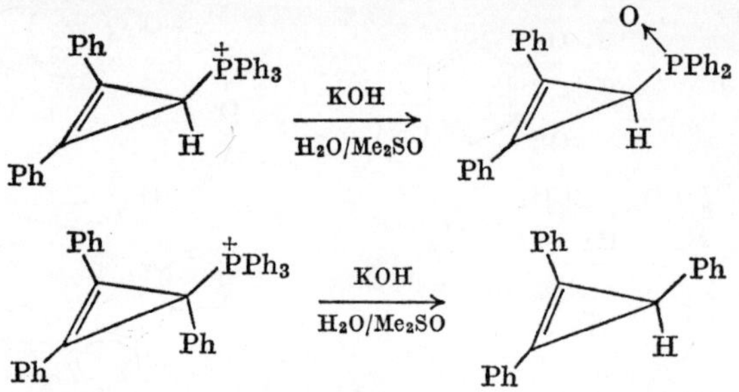

Scheme 1

Treatment of the tetrathia-adamantane (6) with butyl-lithium–tetramethylethylenediamine allows quantitative conversion into a di-anion; the possibility of one sulphur atom stabilizing two adjacent carbanionic centres was discussed. [21] Selenium-stabilized carbanions $(PhSe)_nCH^-_{(3-n)}Li^+$ have been prepared. [22] The NMR spectrum of α-^{13}C-enriched benzyl-lithium shows $J(^{13}C\text{–}H) = 116$ Hz in benzene and 132 in tetrahydrofuran; [23] this coupled with chemical shift data suggests rehybridization towards sp^3 and increased covalency in benzene. The carbanion (7), which mediates in the reaction of the

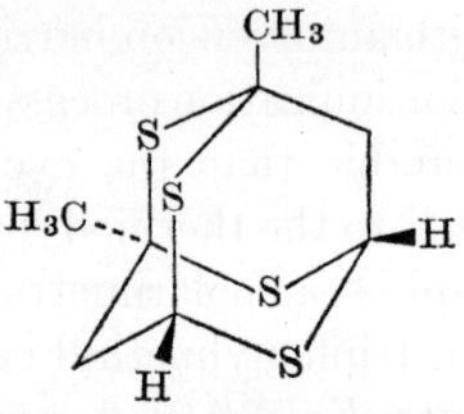

(6) 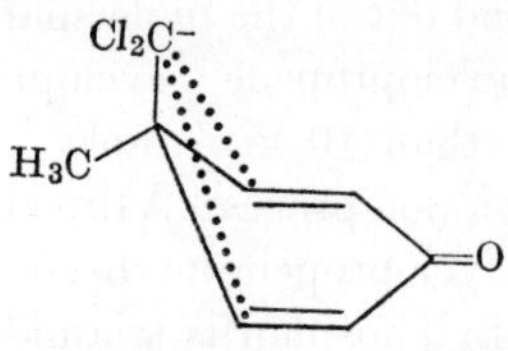   (7)

[20] M. A. Battiste and C. T. Spouse, *Tetrahedron Letters*, **1969**, 3165.
[21] K. C. Bank and D. L. Coffen, *Chem. Comm.*, **1969**, 8.
[22] D. Seebach and N. Peleties, *Angew. Chem. Intern. Ed. Engl.*, 8, 450 (1969).
[23] R. Waack, L. D. McKeever, and M. A. Doran, *Chem. Comm.*, **1969**, 117.

corresponding trichloride with trisdibutylaminophosphine, is considerably stabilized by homoconjugation, and reaction is $>10^6$ faster than with 1,1,1-trichloroethane. Presumably a symmetrical intermediate is involved, although this was apparently not considered.[24]

pK_a's of 1,1-dinitroalkanes have been measured,[25] and those of p-methoxy-(28.6) and p-methyl- (27.1) diphenylmethane measured in liquid ammonia,[26] assuming $K_s = 3.3 \times 10^{-33}$. The relationship between the pK_a of RH and the rate of electrochemical reduction of R_2Hg is discussed.[27]

Reactions of Carbanions

Orbital symmetry criteria permit the prediction of a low-energy disrotatory path for cyclization of pentadienyl anions, and this process has been observed in cyclooctadienyl anion (8) which cyclizes to *cis*-bicyclo[3.3.0]oct-2-en-4-yl anion (9) with a half-life of 80 minutes at 35°.[28] Hunter and Sim have described a very informative study of the base-catalysed conversion of hydrobenzamide into amarine (Scheme 2). Since the thermodynamically unstable isomer is produced, the authors consider that a disrotatory cyclization via the preferred form of the hydrobenzamide anion is occurring.[29] The intermediate anion may be prepared *in situ* at −70° and has a half-life for rearrangement of about 7 hours. The amount of isoamarine in the product may be increased by photo-irradiation, but no distinction was made between possible mechanisms. The reverse reaction (cyclopentenyl→pentadienyl) has been observed in two instances. Treatment of 2,5-dihydrofuran with potassamide in liquid ammonia leads to the oxapentadienyl anion (10),[30] and reaction of 6,6-diphenyl-bicyclo[3.1.0]hexene with potassium t-pentoxide in refluxing t-pentanol leads to isomers of diphenylcyclohexadiene, presumably via the anion (11).[31]

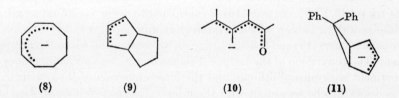

    (8)   **(9)**   **(10)**   **(11)**

[24] B. Miller, *J. Am. Chem. Soc.*, **91**, 751 (1969).
[25] V. I. Slovetskii and I. S. Ivanova, *Izv. Akad. Nauk SSSR, Ser. Khim.*, **1968**, 1735; *Chem. Abs.*, **70**, 28282 (1969).
[26] J. H. Takemoto and J. J. Lagowski, *J. Am. Chem. Soc.*, **91**, 3785 (1969).
[27] K. P. Butin, A. N. Kashin, I. P. Beletskaya, and O. A. Reutov, *J. Organometal. Chem.*, **16**, 27 (1969).
[28] R. B. Bates and D. A. McCombs, *Tetrahedron Letters*, **1969**, 977.
[29] D. H. Hunter and S. K. Sim, *J. Am. Chem. Soc.*, **91**, 6202 (1969).
[30] H. Kloosterziel, J. A. A. van Drunen, and P. Galama, *Chem. Comm.*, **1969**, 885.
[31] D. J. Atkinson, M. J. Perkins, and P. Ward, *Chem. Comm.*, **1969**, 1390.

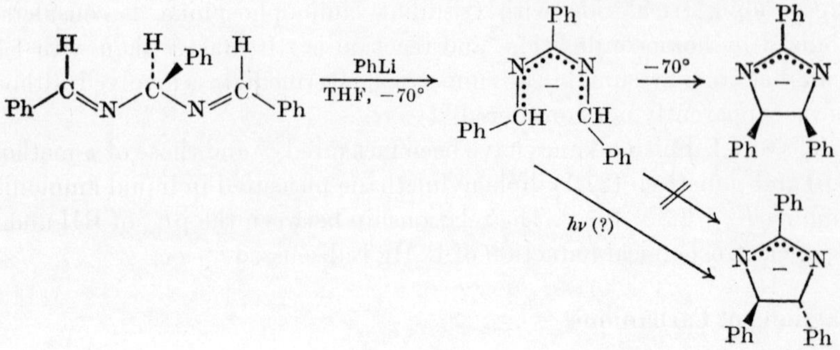

Scheme 2

The analogous heptatrienyl–cycloheptadienyl rearrangement has been characterized. Treatment of hepta-1,3,6-triene with butyl-lithium in [^{2}H$_8$]THF at $-50°$ gave the all-*trans*-heptatrienyl anion (**12**) which cyclized quantitatively at $-30°$ ($t_{\frac{1}{2}} = 13$ minutes) to cycloheptadienyl-lithium [see (**13**)].[32] Methylation at C-2, C-4, or C-6 in the olefin reduced the rate of cyclization but failed to change the position of equilibrium. 2,3-Dihydro-oxepin on treatment with potassamide in liquid ammonia underwent (conrotatory?) ring-opening to an oxaheptadienyl anion, presumably via (**14**).[30]

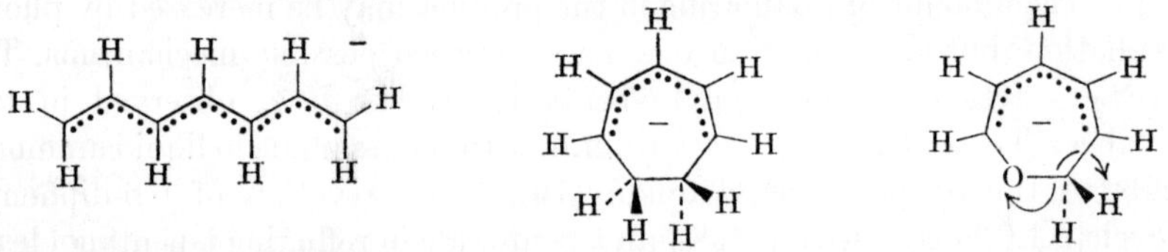

(**12**)   (**13**)   (**14**)

The rearrangement of (**15**) in Bu^{t}OK–Bu^{t}OH to a mixture of C$_6$-epimers of the tricyclo[3.2.1.0^{2,7}]octene (**16**) is considered to occur via an intermediate non-classical carbanion which invariably reprotonates at C-6.[33] The cyclopropane isomers (**17**) and (**18**) exchange with retention and without rearrangement. The conversion of the lactone (Scheme 3) into *two* acyclic diketones on treatment with phenyl-lithium, and the interconversion of these with base, was deemed to be consistent with the intervention of two homoenolate ions on the rearrangement pathway.[34] A mechanism for the Smiles rearrangement[35] (diaryl sulphone→arylsulphinic acid) involving a deprotonation–cyclization sequence has been suggested, and an intermediate trapped and identified.

[32] R. B. Bates, W. H. Deines, D. A. McCombs, and D. E. Potter, *J. Am. Chem. Soc.*, **91**, 4608 (1969); H. Kloosterziel and J. A. A. van Drunen, *Rec. Trav. Chim.*, **88**, 1084 (1969).

[33] W. Eberbach and H. Prinzbach, *Chem. Ber.*, **102**, 4164 (1969).

[34] P. Yates, G. D. Abrams, and S. Goldstein, *J. Am. Chem. Soc.*, **91**, 6869 (1969).

[35] V. N. Drodz and L. I. Zefirova, *Zh. Org. Khim.*, **4**, 1682 (1968); V. N. Drodz, L. A. Nikonova, L. I. Zefirova, and K. A. Pak, *Izv. Timiryasev. Sel'skok. Akad.*, **1969**, 179; *Chem. Abs.*, **70**, 38735 (1969).

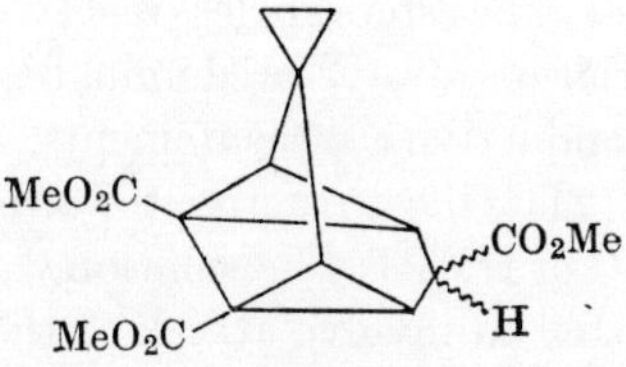

(15)

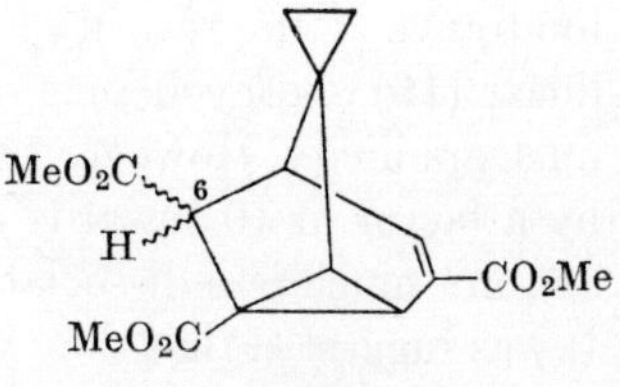

(16)

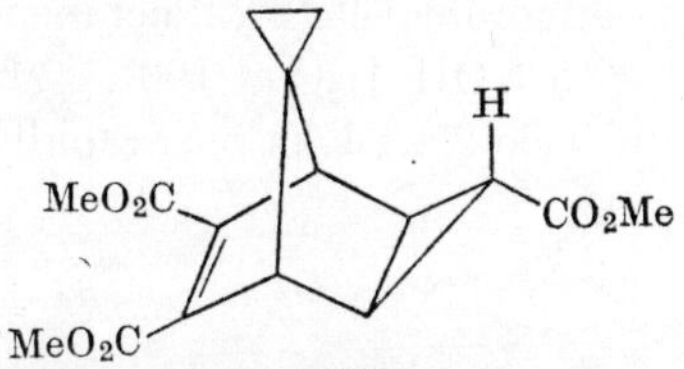

(17)

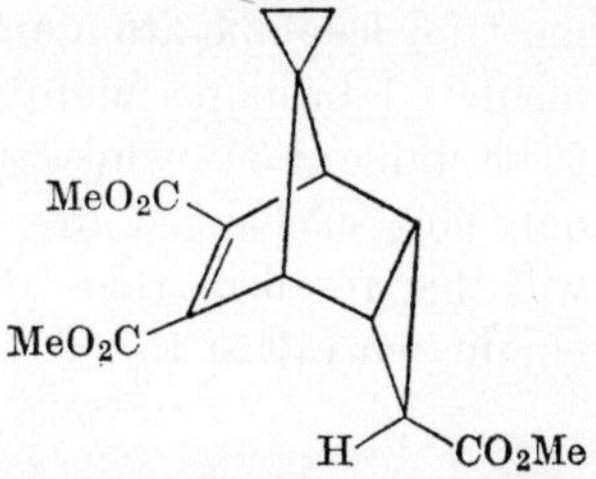

(18)

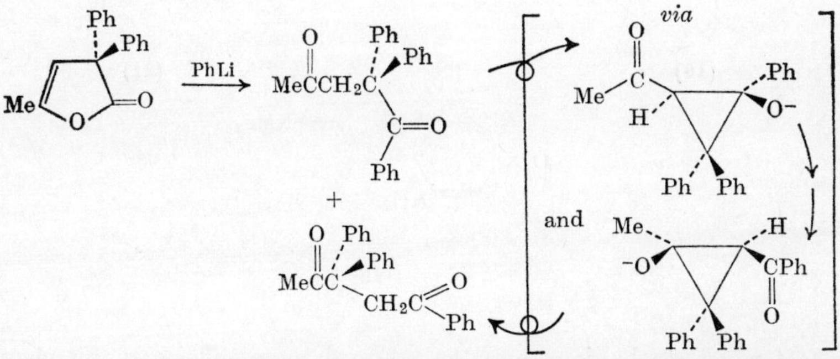

Scheme 3

A study has been made of the (S_N2) nucleophilicity of 9-cyanofluorenylide,[36] and the observation made that nucleophilic reactions of fluorenyl carbanions are kinetically affected in differing ways by addition of DMSO to a t-butanol reaction medium.[37] Anthracene di-anion reacts with alkyl halides to give predominantly (~5:1) cis-9,10-dialkyldihydroanthracenes, and a mechanism is proposed.[38] A variety of protonation reactions of 4-cyano- and 4-ethoxy-carbonyl-t-butylcyclohexyl anions have been studied, and the direction of approach of the proton has been reconciled with minimization of steric

[36] K. Bowden and R. S. Cook, *J. Chem. Soc.*(B), **1968**, 1529.
[37] A. F. Cockerill and R. T. Hargreaves, *Chem. Comm.*, **1969**, 915.
[38] R. G. Harvey and L. Arzadon, *Tetrahedron*, **25**, 4887 (1969).

hindrance at the transition state.[39] In ethanol the sodium enolate anions of linear (19) and cyclic, e.g. (20), β-keto-esters have comparable basicity and nucleophilicity. However, in DMSO or HMPT the cyclic esters are less reactive by a factor of 10 towards alkyl halides (C-alkylation always predominates) and are much less dissociated as judged from conductivity measurements.[40] It was suggested that the "W" form of the anion (21) is most reactive towards alkyl halides and this is available to (19) in DMSO or HMPT. Fluorination via the reaction of carbanions with $FClO_3$ is suggested to involve attack of the most nucleophilic atom of the carbanion at chlorine and internal rearrangement.[41] 1-Lithioperfluorobicyclo[2.2.2]octane has been prepared and found to be stable at 35°, whilst the parent (1-H) tridecafluorobicyclo[2.2.2]octane is only 60% exchanged after 6 hours' treatment with KOH–D_2O at 100°.[42a] It will be recalled that the 1-lithioperfluorobicyclo[2.2.1]heptane rapidly eliminated LiF at 15°.[42b]

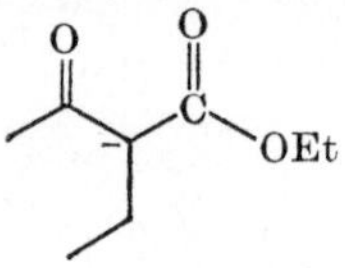

(19) 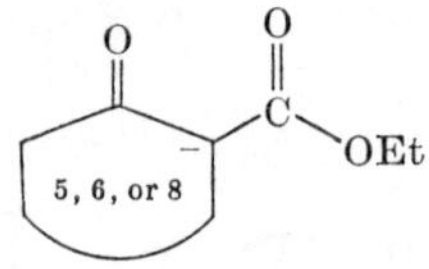   (20) 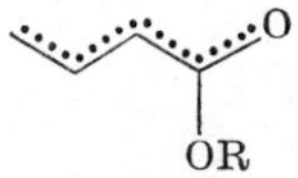   (21)

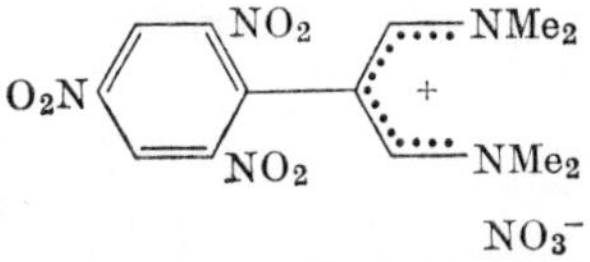

(22)

Once formed, carbanions may be trapped more efficiently by electron transfer to nitro-arenes than by proton transfer from alcohols,[43] and the technique may be used to measure the rate of ionization of fluorene in DMSO–Bu^{t}OH. A mechanistic study has been made of the system containing styrene dimer di-anion and polycyclic aromatic hydrocarbon;[44] in the case of pyrene, rapid formation of adduct was faster than direct electron transfer.

[39] R. A. Abramovitch, M. M. Rogic, S. S. Singer, and N. Venkateswaran, *J. Am. Chem. Soc.*, **91**, 1571 (1969).
[40] S. J. Rhoads and W. R. Holder, *Tetrahedron*, **25**, 5443 (1969).
[41] W. A. Sheppard, *Tetrahedron Letters*, **1969**, 83.
[42a] W. B. Hollyhead, R. Stephens, J. C. Tatlow, and W. T. Westwood, *Tetrahedron*, **25**, 1777 (1969).
[42b] See *Org. Reaction Mech.*, **1965**, 99.
[43] R. D. Guthrie, *J. Am. Chem. Soc.*, **91**, 6201 (1969); G. A. Russell and S. A. Weiner, *Chem. Ind.* (London), **1969**, 659.
[44] S. C. Chadha, J. Jagur-Grodzinski, and M. Szwarc, *Trans. Faraday Soc.*, **65**, 1074 (1969).

Cyclononatetraenide anion is made more basic by photoexcitation, when it may abstract a proton from hex-1-yne.[45] 2,4,6-Trinitrotoluene reacts under Vilsmeier conditions ($POCl_3$–DMF) to form the salt (22), presumably via 2,4,6-trinitrobenzyl anion since 2,4-dinitrotoluene is unaffected.[46] Three reviews[47, 48] and a book[49] on carbanion chemistry have been published.

Ylids

Further activity has been evident in the development of allylsulphur ylid rearrangements (see Chapter 8), including a synthesis of squalene in three stages from difarnesyl sulphide.[50] The rearrangement is thought to be completely concerted,[51] and the only product from (23) is (24), where a five-centre transition-state and 2,3-migration is required. Two diastereomers of product are produced in 40:60 ratio, and this presumably reflects the relative ease of removal of the diastereotopic protons in (23). Completely different behaviour is noted in allylammonium ylids, which undergo 1,4-migration to give enamine-derived products,[51] e.g. (25)→(26). Radical pairs may very well be involved in the latter reaction, and the Stevens rearrangement [(27)→(28)] under appropriate conditions gives a CIDNP effect through emission from the

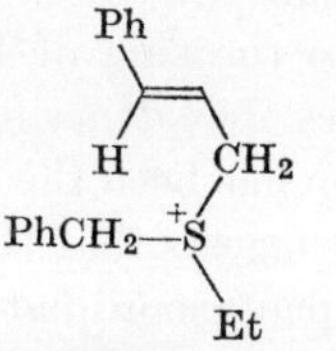

(23) 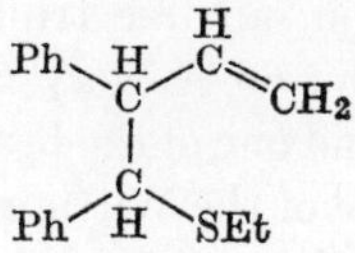 (24)

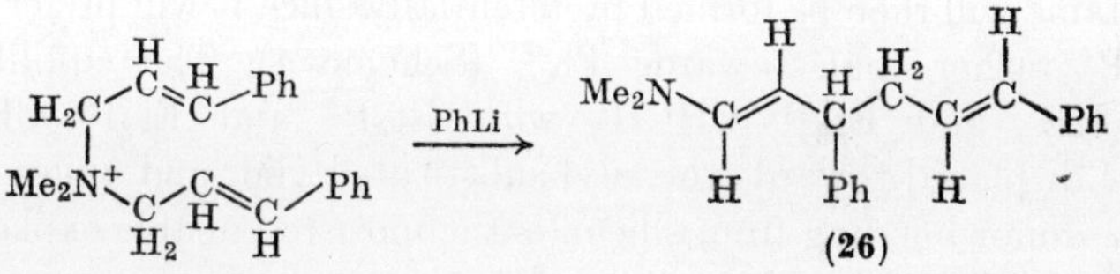

(25) PhLi→ (26)

[45] J. Schwartz, *Chem. Comm.*, **1969**, 833.
[46] V. L. Zbarshii, G. M. Shutov, V. F. Zhilin, and E. Y. Orlova, *Zh. Org. Khim.*, **4**, 1879 (1968).
[47] H. F. Ebel, *Fortschr. Chem. Forsch.*, **12**, 387 (1969). (Structure and reactivity of carbanions.)
[48] Influence of the solvent on the electrophilic substitution reactions of carbanions at the saturated carbon atom: N. Radulescu, *Rev. Chim.* (Roumania), **20**, 273 (1969); general review, D. J. Cram, *Survey Progr. Chem.*, **4**, 45 (1968).
[49] M. Szwarc, *Carbanions, Living Polymers, and Electron-Transfer Processes*, Interscience, New York, 1968.
[50] G. M. Blackburn, W. D. Ollis, C. Smith, and I. O. Sutherland, *Chem. Comm.*, **1969**, 99.
[51] R. W. C. Cose, A. M. Davies, W. D. Ollis, C. Smith, and I. O. Sutherland, *Chem. Comm.*, **1969**, 293; R. W. Jemison and W. D. Ollis, *ibid.*, p. 294.

PhCH$_2$ hydrogens.[52] In the mechanistically related Wittig rearrangement, 1,4-shifts have been observed for the first time [53] in e.g. the conversion of butyl

(27)   (28)

allyl ether into n-heptanal in 30% yield by reaction with the tetramethyl-ethylenediamine–n-propyl-lithium complex. Reaction is rather insensitive to the nature of the allyl substitution pattern, and this is suggestive of a radical rather than an ionic mechanism. Stabilized sulphur ylids have been isolated and characterized,[54] and a 1,2-benzyl shift has been observed in the ylid PhCH$_2$S$^+$(Me)C$^-$(CO$_2$Me)$_2$ at 200°.[55] The reaction of sulphur ylids with Grignard reagents has been studied.[56] Barriers to rotation about the ylid "double bond" have been examined in phenacyl dimethyl sulphurane[57] and in *para*-substituted α-methoxycarbonylbenzyltriphenylphosphoranes.[58]

An X-ray study on salt-free triphenylphosphinemethylene shows an ylid P–C bond length of 1.66 Å, very close to the sum of Pauling–Schomaker double-bond radii, and one of the best pieces of evidence for $p\pi$–$d\pi$ bonding.[59] An enigmatic feature of the Wittig reaction has been the frequent preference for formation of *cis*-alkenes. Schneider[60] suggests that the Wittig betaine is properly considered as a derivative of pentacoordinate phosphorus with oxygen at an axial site. If the oxygen–phosphorus bond is formed first, the *erythro*-betaine will then be formed preferentially since R will prefer to rotate towards R' rather than towards Ph[60] (Scheme 4). The equilibrium of Et$_3$P$^+$CH$_2$SiH$_3$ and Et$_3$P^{+-}CHCH$_3$ with Et$_4$P$^+$ and Et$_3$P^{+-}CHSiH$_3$ is completely displaced towards the silyl-substituted ylid, and anion-stabiliza-tion by $d\pi$–donor bonding from silicon is the most reasonable explanation.[61] 1,3-Silyl transfer in phosphorus ylids[62a] appears to be an intermolecular

[52] U. Schoollkopf, U. Ludwig, G. Ostermann, and M. Patsch, *Tetrahedron Letters*, **1969**, 3415.
[53] H. Felkin and A. Tambute, *Tetrahedron Letters*, **1969**, 821.
[54] A. W. Johnson and R. T. Amel, *J. Org. Chem.*, **34**, 1240 (1969).
[55] W. Ando, T. Yagihara, and T. Migita, *Tetrahedron Letters*, **1969**, 1983.
[56] H. Nozaki, K. Nakamura, and M. Takaku, *Tetrahedron*, **25**, 3675 (1969).
[57] S. H. Smallcombe, R. J. Holland, R. H. Fishard, and M. C. Caserio, *Tetrahedron Letters*, **1968**, 5987.
[58] H. I. Zeliger, J. P. Snyder, and H. J. Bestmann, *Tetrahedron Letters*, **1969**, 2199.
[59] J. C. J. Bart, *J. Chem. Soc.*(B), **1969**, 350.
[60] W. P. Schneider, *Chem. Comm.*, **1969**, 785.
[61] H. Schmidbaur and W. Malisch, *Angew. Chem. Intern. Ed. Engl.*, **8**, 372 (1969).
[62a] See *Org. Reaction Mech.*, **1968**, 120.

reaction[62b] since mono- and tris-trimethylsilylated ylids equilibrate to the disubstituted product under transfer conditions (150°).

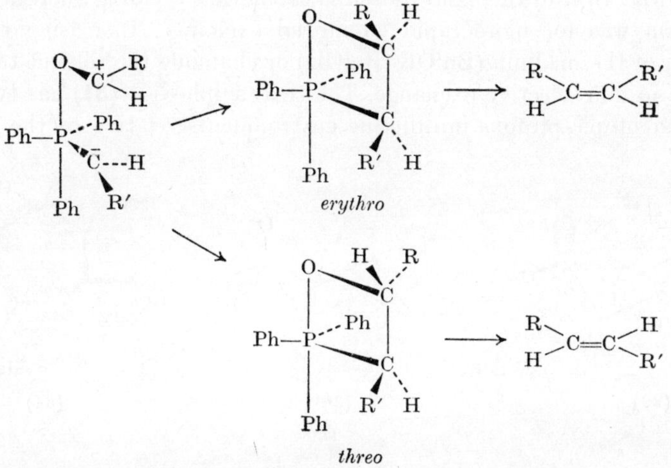

Scheme 4

Proton Transfer, Hydrogen Isotope Exchange, and Related Reactions

Sulphoxide-stabilized Carbanions

A stimulating debate on the nature, and particularly the stereochemistry, of sulphoxide–α-carbanion interactions has been generated by the non-empirical MO studies on the model system $HSOCH_2^-$ by Rauk, Wolfe, and Csizmadia.[63] These authors examined the energy profile for rotation about the C–S bond and for changes in H–C–H bond angle. They concluded that the carbon hybridization differed significantly from sp^2 (with a C–S bond length held at 1.80 Å!), and that the most stable torsional angle was with carbanion lone-pair *gauche* to both S–O and S lone-pair, with a considerable barrier to rotation.[63] These calculations were used to justify conclusions reached earlier on the stereochemical course of base-catalysed isotope exchange in benzyl methyl sulphoxide,[64a] and similar conclusions[64b] reached in the case of the sulphone-stabilized anion $HSO_2CH_2^-$. However, evidence has accumulated from a number of sources which may demand modification of these conclusions. In a rigid system, the course of exchange may be defined without the need for arguments about preferred conformations, and Katritzky's group has examined isotope exchange in (29) and (30)[65] with values of $k_1 \times 10^7$ for

[62b] H. Schmidbaur and W. Malisch, *Chem. Ber.*, **102**, 83 (1969).
[63] A. Rauk, S. Wolfe, and I. G. Csizmadia, *Can. J. Chem.*, **47**, 113 (1969).
[64a] See *Org. Reaction Mech.*, **1966**, 92.
[64b] S. Wolfe, A. Rauk, and I. G. Csizmadia, *J. Am. Chem. Soc.*, **91**, 1567 (1969).
[65] B. J. Hutchinson, K. K. Andersen, and A. R. Katritzky, *J. Am. Chem. Soc.*, **91**, 3839 (1969).

exchange in basic deuteromethanol as indicated. The equatorial proton in
(29) is exchanged much faster, and the proton *trans* to the lone-pair only
very slowly. In (30) the axial proton is exchanged rapidly, whereas the MO
prediction was for more rapid equatorial exchange. Increasing the base
strength of the medium (Bu^tOK–Bu^tOD) or changing the solvent to DMSO
resulted in non-selective exchange. The rigid sulphoxide (31) has two pairs
of diastereotopic protons in differing environments relative to the sulphur

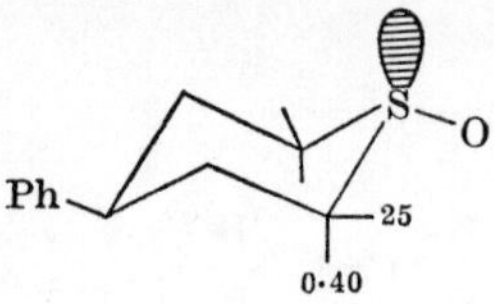

(29) 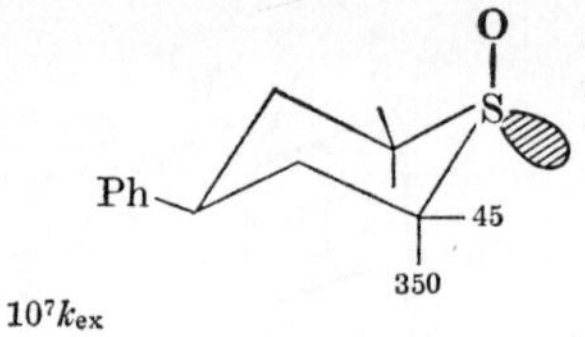   (30) 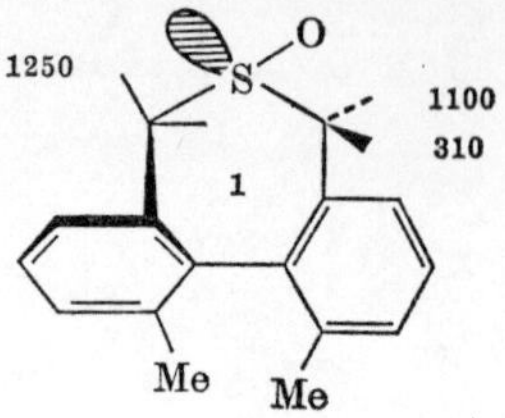   (31)

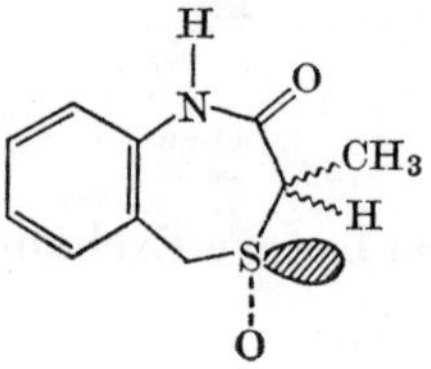

$10^7 k_{ex}$

(32)

which were assigned on the basis of NMR analogies, and which exchanged
at the relative rates indicated.[66a] Again, and in contrast to the MO prediction,
the protons *anti* to oxygen (H_2) exchanged faster than the proton on the
O–S lone-pair bisector. The sulphone analogue of (31)[66b] exhibits selective
exchange of α-hydrogens in sodium phenoxide–[$^2H_0$]-t-butanol with the
protons on the oxygen–oxygen bisector exchanging three times faster. Only
the *cis*-diastereomer of (32)[67] exchanges rapidly in D_2O–NaOH; in the *trans*-
diastereomer the α-proton is at 180° to the sulphur lone-pair. It is recorded[68]
that both diastereomers of 2,4-dithiapentane 2,4-dioxide exchange rapidly
in D_2O.

A serious setback to Wolfe's interpretation of the benzyl methyl sulphoxide
exchange came on redetermination of the absolute configuration of benzyl
methyl sulphoxide[69] by a rigorous method, and demonstration of an earlier

[66a] R. R. Fraser and F. J. Schuber, *Chem. Comm.*, 1969, 397.
[66b] R. R. Fraser and F. J. Schuber, *Chem. Comm.*, 1969, 1474.
[67] P. C. Thomas, I. C. Paul, T. Williams, G. Grelle, and M. Uskokovic, *J. Org. Chem.* 34, 365 (1969).
[68] R. Louw and H. Nieuwenhuyse, *Chem. Comm.*, 1968, 1561.
[69] J. E. Baldwin, R. E. Hackler, and R. M. Scott, *Chem. Comm.*, 1969, 1415.

error. Assuming the validity of the earlier conformational assignments (by NMR and dipole moment studies), stereoselective exchange must have given (**33**) and not its diastereomer. Chlorination of benzyl methyl sulphone by PhICl$_2$[70] involves replacement of the opposite diastereotopic proton, H$_a$, by chlorine. Deprotonation of PhSOCH$_2$Cl by butyl-lithium[71] gave an anion with one highly preferred conformation, judging by the stereospecific reaction with acetone.

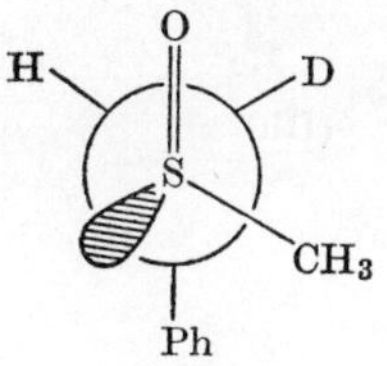

(**33**) 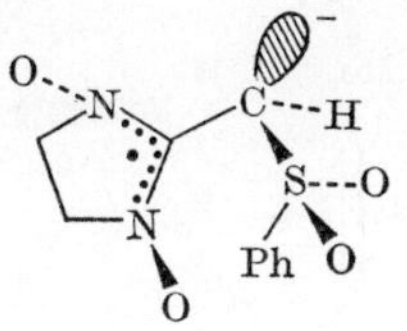 (**34**) 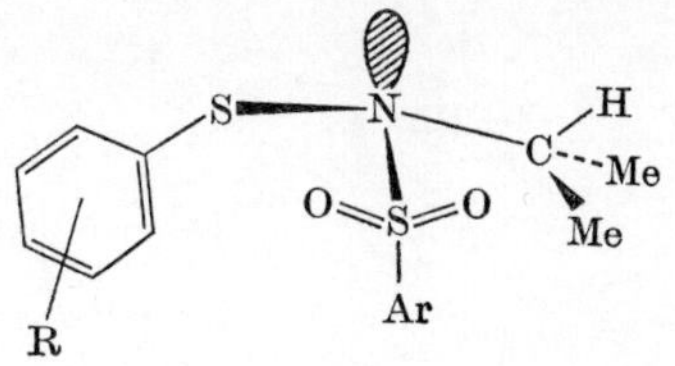 (**35**)

The most reasonable explanation[72] for ESR non-equivalent nitrogens in (**34**) is that the carbanion centre is pyramidal, with a rotational barrier of about 6 kcal mole^{-1} in 50% H$_2$O–EtOH (coalescence temperature 58°). The rate-determining stage in NMR equivalencing of the non-equivalent methyl groups in (**35**) is considered to be pyramidal inversion at nitrogen, and the effect of varying R was studied.[73]

Exchange in Hydrocarbons

The rates of exchange of ring protons in cyclic hydrocarbons in caesium cyclohexylamide–cyclohexylamine correlate well with $J(^{13}\text{C–H})$ values. There are no special effects in medium-ring hydrocarbons, and Streitwieser suggests that this means that the carbanion is sp^3-hybridized.[74] Further work has appeared on exchange and isomerization in optically active indenes;[75] it is possible to assess the relative extent of these two processes and obtain the relative rates of rotation and collapse in the intermediate (Scheme 5). There is a strong preference for proton over deuteron transfer in the ion-pair collapse; thus (**Id**) isomerizes with 95% exchange in PrNH$_2$–THF, whereas (**Ih**) isomerizes with only 36% exchange in PrND$_2$–THF. The bridgehead proton of triptycene[76]

[70] M. Cinquini, S. Colonna, and F. Montanari, *Chem. Comm.*, **1969**, 607.
[71] T. Durst, *J. Am. Chem. Soc.*, **91**, 1034 (1969).
[72] R. Darcy and E. F. Ullman, *J. Am. Chem. Soc.*, **91**, 1024 (1969).
[73] M. Raban and F. B. Jones, *J. Am. Chem. Soc.*, **91**, 2180 (1969).
[74] A. Streitwieser, W. R. Young, and R. A. Caldwell, *J. Am. Chem. Soc.*, **91**, 529 (1969).
[75] G. Bergson and L. Ohlsson, *Acta Chem. Scand.*, **23**, 2175 (1969); J. Almy and D. J. Cram, *J. Am. Chem. Soc.*, **91**, 4459, 4468 (1969).
[76] A. Streitwieser and G. R. Ziegler, *J. Am. Chem. Soc.*, **91**, 5081 (1969).

exchanges 10^7 times faster than a proton of cyclohexane in CsCHA–cyclohexylamine, and 0.24 times as fast as a proton in benzene. The factor of 10^7
is broken down as $10^{4.5}$ due to inductive effects and $10^{2.5}$ due to increased
s-character.

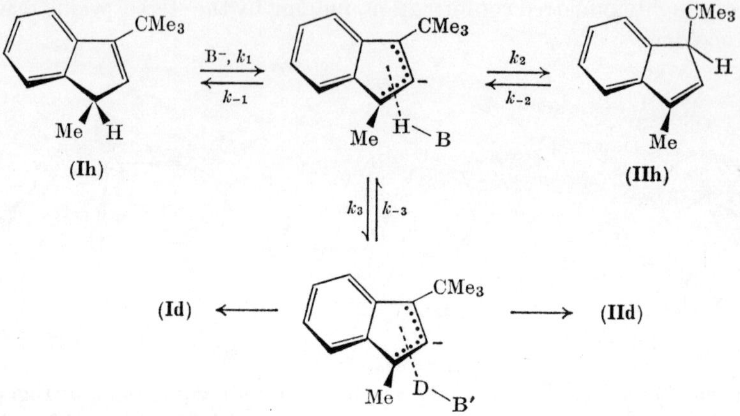

Scheme 5

Compound (**36**) exchanges 750 times faster than (**37**) in KOBu^t–DMSO,[77]
and its conjugate base was thought to be "bicycloaromatic" on this basis.
Other factors may be important, when it is remembered that bicyclo[2.2.2]-
octatriene is enormously more strained than bicyclo[2.2.2]octadiene. (**36**) is
isomerized to 3-methylbarbaralane on warming with KNH$_2$–NH$_3$ to 50°.
Little anchimeric assistance is evident in the exchange of (**38**), which is
complicated by a facile competing intramolecular ene-reaction.[78] Isomerization of benzylacetylene and/or phenylallene to phenyl methyl acetylene is
almost completely intramolecular when achieved by Na$^+$–CH$_2$SOMe in
DMSO.[79] The rates of isomerization of cycloalkenes have been followed by
internal ^{14}C-labelling.[80] With KOBu^t–DMSO at 90° the rates are: cyclopentene
$k_1 = 3 \times 10^{-6}$; cyclohexene $k_1 = 59 \times 10^{-6}$; cycloheptane $k_1 = 296 \times 10^{-6}$.
This was explained on stereoelectronic grounds but the present writer considers
the central internal angle of the allylic anion to be more important. If this is
large, destabilizing interaction between the allylic termini is minimized.

Equilibration of 2,4,4-trimethylpent-1-ene (70%) with 2,4,4-trimethylpent-
2-ene (30%) was studied in Bu^{t}OK–Bu^{t}OD at 215°.[81] Protonation occurs

[77] S. W. Staley and D. W. Reichard, *J. Am. Chem. Soc.*, **91**, 3998 (1969).
[78] J. M. Brown, *J. Chem. Soc.*(B), **1969**, 868.
[79] J. Klein and S. Brenner, *Chem. Comm.*, **1969**, 1020.
[80] S. B. Tjan, H. Steinberg, and T. J. de Boer, *Rec. Trav. Chim.*, **88**, 680 (1969).
[81] D. H. Hunter and R. W. Mair, *Can. J. Chem.*, **47**, 2361 (1969).

(36)　　　　　　(37)　　　　　　(38)

preferentially at the least-substituted end of the allylic anion intermediate, and in the pent-2-ene exchange in the methyl group *cis* to t-butyl was faster than that in that *trans* to t-butyl by a factor of 11. Isomerization of 1,8-diarylocta-3,5-diynes[82] has been studied; in the diphenyl compound the product is the all-*trans*-octatetraene, but the steric effect of the methyl groups in the di-(2,3,5,6-tetramethylphenyl) compound directs the isomerization to the 1,3-diyne. The "new" structure for triphenylmethyl dimer has been proved by isomerization to diphenyl-(*p*-tritylphenyl)methane[83] in base. The highly strained hydrocarbon bicyclo[3.2.0]hepta-1,3,6-triene may be formed transiently by base-elimination routes,[84] and it dimerizes[85] rapidly. If elimination is carried out with $KOBu^t$-$[^2H_6]DMSO$, then 2.0 atoms of deuterium are incorporated into the dimer;[84] the monomer is therefore rather acidic in accordance with HMO predictions. The benzo-analogue (6,7-benzo-bicyclo[3.2.0]hepta-1,3,6-triene) has been investigated similarly, and in this case the conjugate base is stable in DMSO solution (λ_{max} 500, 590 nm).[86]

The high isotope effect ($k_H/k_D = 6.5$) for cyclohexane exchange promoted by caesium cyclohexylamide in cyclohexylamine shows that proton transfer is the rate-determining stage,[87] and suggests that the reverse reaction is not diffusion-controlled. In contrast, isotope effects on exchange reactions in DMSO are uniformly low with $LiOBu^t$ or $KOBu^t$ as base; for diphenylmethane, or the 2-position of thiophen or furan, $k_D/k_T \sim 1.4$.[88] [α-^2H]Toluene exchanges 100 times slower in $K^{+-}CH_2SOCH_3$-DMSO than at a comparable concentration of $KOBu^t$-DMSO[89a] (cf. ref. 89b). The proton basicity of the conjugate base of Carbowax-350, $K^{+-}OCH_2CH_2(OCH_2CH_2)_nOCH_3$, has a proton basicity in Et_2O many hundreds of times higher than that of $K^{+-}OBu^t$, observed by

[82] A. J. Hubert and A. J. Anciaux, *Bull. Soc. Chim. Belges*, **77**, 513 (1968).

[83] R. D. Guthrie and G. R. Weisman, *Chem. Comm.*, **1969**, 1316.

[84] R. Breslow, W. Washburn, and R. G. Bergman, *J. Am. Chem. Soc.*, **91**, 196 (1969).

[85] N. L. Bauld, C. E. Dahl, and Y. S. Rim, *J. Am. Chem. Soc.*, **91**, 2787 (1969).

[86] M. P. Cava, K. Narasimhan, W. Zeiger, L. J. Radenovitch, and M. D. Glick, *J. Am. Chem. Soc.*, **91**, 2379 (1969).

[87] A. Streitwieser, W. R. Young, and R. A. Caldwell, *J. Am. Chem. Soc.*, **91**, 527, 529 (1969).

[88] I. O. Shapiro, F. S. Yakushin, I. A. Romanskii, and A. I. Shatenshtein, *Kinet. Katal.*, **9**, 1011 (1968); *Chem. Abs.*, **70**, 36740 (1969).

[89a] I. A. Romanskii, I. O. Shapiro, and A. I. Shatenshtein, *Org. Reactivity* (Tartu), **5**, 452 (1968); *Chem. Abs.*, **70**, 19287 (1969).

[89b] See *Org. Reaction Mech.*, **1968**, 122.

comparison of the relative rates at which these bases isomerize allylbenzene to propenylbenzene.[90] Presumably the polyether chain creates a region of higher polarity in the region of the reaction centre. Studies have been made on the effect of solvent on base-catalysed isomerization of hept-1-ene[91] and on deuterium exchange promoted by bases in HMPA.[92] Equilibrium mixtures in the isomerization of octalins by lithium ethylenediamide in ethylenediamine are recorded.[93]

Other Systems

Abundant data have been presented for exchange at aryl and arylmethyl sites, particularly in heterocyclic systems. In basic media thioanisole undergoes exchange at the methyl group 5×10^7 times faster than anisole, and dimethylphenylphosphine 3×10^4 times faster than dimethylaniline.[94] The partial rate factors shown for rates of exchange in ammonia (Scheme 6) are interpreted in favour of $d\pi$–$p\pi$ bonding at phosphorus.[95] The ring proton in N-substituted tetrazoles exchanges very rapidly, and an important role for $d\pi$–$p\pi$ bonding in the 2-thiazolium carbanion is claimed by comparison of exchange rates and J(C–H) values for the exchanging proton with corresponding oxazolium and N-methylimidazolium systems.[96] Rates of exchange of ring protons in oxazole and related systems are reported.[97] Base-catalysed

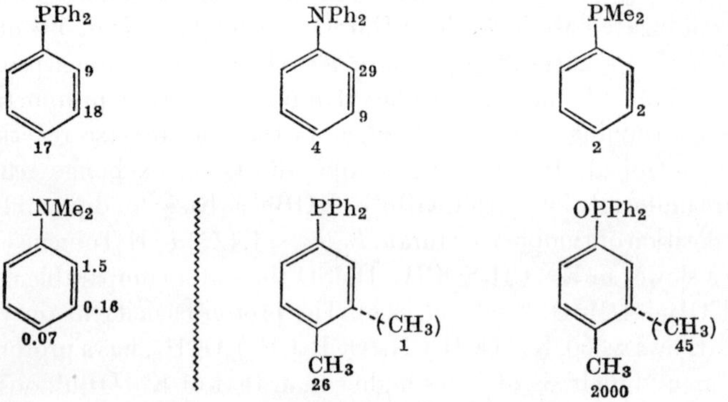

Scheme 6

[90] O. A. Rokstad, J. Ugelstad, and H. Lid, *Acta Chem. Scand.*, **23**, 782 (1969).
[91] C. Cerceau, M Laroche, A. Pazdzerski, and B. Blouri, *Bull. Soc. Chim. France*, **1969**, 921.
[92] H. Normant, T. Cuvigny, and G. J. Martin, *Bull. Soc. Chim. France*, **1969**, 1605.
[93] P. Oberhansli and M. C. Whiting, *J. Chem. Soc.*(B), **1969**, 467.
[94] A. I. Shatenshtein and H. Gvozdeva, *Tetrahedron*, **25**, 2749 (1969).
[95] E. A. Yakovleva, E. N. Tsvetkov, D. I. Lobanov, A. I. Shatenshtein, and M. I. Kabachnik, *Tetrahedron*, **25**, 1173 (1969).
[96] A. C. Rochat and R. A. Olofson, *Tetrahedron Letters*, **1969**, 3377; P. Haake, L. P. Busher, and W. B. Miller, *J. Am. Chem. Soc.*, **91**, 1113 (1969).
[97] D. J. Brown and P. B. Ghosh, *J. Chem. Soc.*(B), **1969**, 271.

exchange of 2-methylpyrylium salts occurs rapidly in D_2O, and mechanisms are discussed.[98] Parallel studies have been carried out by Russian[99] and American[100] authors on exchange at ring protons in pyridines and pyridine *N*-oxides. In the latter, an attempt was made to assign partial rate factors to $-N=$ and N^+-O^-, treated as simple substituents. The latter has an extreme acidifying effect on *ortho*-protons (log $p_f = 9.58$) and a considerable effect on *para*-protons. In contrast, ring-nitrogen has a more strongly acidifying effect at *meta*- and *para*-positions [log $p_f = 1.31$ (*ortho*), 2.43 (*meta*), and 2.46 (*para*)] presumably due to competition at the *ortho*-positions between a stabilizing inductive effect and destabilizing lone-pair interaction. Rates are recorded for a variety of six-membered ring heterocycle exchanges.[101] There is a good correlation between HMO predictions of relative acidity, and rates of exchange of methylquinolines and picolines in NaOMe–MeOH.[102] In [$^2H_6$]DMSO–D_2O, methyl exchange in methylisothiazoles occurs rapidly with 3- or 5-methyl substitution but more than 10^4 more slowly for the 4-methyl compound.[103] Lumazine (39) undergoes facile exchange at the underlined methyl group in D_2O,[104] and the anion is presumably delocalized into the pyrimidone ring.

Rates of exchange of protons in di- and tri-halogenomethylenes at 33° are recorded for OMe$^-$–MeOD solutions.[105] The rate patterns suggest that α-fluoro- is less acidifying than α-chloro-, but that β-fluoro- is more acidifying than β-chloro-substitution. The course of reaction of 4-nitrobenzyl bromide with hydroxide ion is completely different from that for the chloride, where formation of α-chloro-4-nitrobenzyl anion is irreversible, and leads to carbene production. In the α-bromo case, the carbanion is more stable and may be trapped with *p*-nitrobenzaldehyde.[106]

(39)   (40)

[98] E. Gard, I. Stanoiu, and A. T. Balaban, *Rev. Roum. Chim.*, **14**, 247 (1969).

[99] I. F. Tupitsyn, N. N. Zatsepina, A. Y. Kirova and Y. M. Kapustin, *Org. Reactivity* (Tartu), **5**, 601 (1968); *Chem. Abs.*, **70**, 76940 (1969); *ibid.*, p. 613, 626, 636, 806 (1968); *Chem. Abs.*, **70**, 76941—76944 (1969).

[100] J. A. Zoltewicz, G. Grahe, and C. L. Smith, *J. Am. Chem. Soc.*, **91**, 5501 (1969); J. A. Zolte-wicz and G. M. Kauffmann, *J. Org. Chem.*, **34**, 1405 (1969).

[101] P. Beak and E. M. Monroe, *J. Org. Chem.*, **34**, 589 (1969).

[102] W. N. White and D. Lazdins, *J. Org. Chem.*, **34**, 2756 (1969).

[103] J. A. White and R. C. Anderson, *J. Heterocyclic Chem.*, **6**, 199 (1969).

[104] T. Paterson and H. C. S. Wood, *Chem. Comm.*, **1969**, 290.

[105] D. Daloze, H. G. Viehe, and G. Chiurdoglu, *Tetrahedron Letters*, **1969**, 3925.

[106] A. A. Abdallah, Y. Iskander, and Y. Riad, *J. Chem. Soc.*(B), **1969**, 1188.

Rates of exchange in (40) in basic D_2O are drastically affected by the size of R; thus relative rates are: R = H, 42,000; R = Me, 150; R = cyclo-Pr, 1.[107] A careful study of prototropic equilibria between pent-2-ynoic acid anion, penta-2,3-dienoic acid anion, and pent-3-ynoic acid anion shows that the last-named is most stable and also deprotonates most rapidly.[108] Isomerization of linoleic acid is recorded,[109] and exchange in other carboxylates reported.[110] Rappe has now proved that there is no significant difference between methyl and methylene α-ketone deuteration rates when the base is changed from deuteroxide to acetate,[111a] despite Warkentin's counter-claim.[111b] Adamantanone could not be induced to homoenolize.[112]

Other isotopic exchange studies deal with nickelocene[113] (rapid in $LiNMe_2$–Me_2ND at 25°), coordinated dimethyl sulphoxide,[114] glycylglycine coordinated to Pt(II),[115] dimethylsulphimine and related compounds,[116] and sulphonic acids.[117]

Mechanisms of Proton Transfer

Detailed studies on the general-base catalysed detritiation of malononitrile and t-butylmalononitrile have shown that a component is due to solvent water acting as a base.[118] In mixed solvents there is an initial increase in the proton basicity of water, reaching a maximum at about 80 mole % H_2O and interpreted as an increase in the amount of "unbound" water. The 2,6-dinitrobenzyl anion may be generated by flash photolysis of the hydrocarbon, since the very acidic *aci*-form is produced and reionizes.[119] Proton transfer from this to a variety of bases was studied, and even with $\Delta pK > 16$ the limiting rate is still 10^5 l $mole^{-1}$ sec^{-1}, nowhere near diffusion control, and this is interpreted as being a consequence of delocalization in the anion and a need for reorganization at the transition state. A plot of Brønsted α versus ΔpK was a smooth curve, in accord with Eigen's theories. Methylene-tritiated *trans*-1,4-dicyanobut-2-ene (41) exchanges with general-base catalysis and a similar

107 P. Belanger, J. G. Atkinson, and R. S. Stuart, *Chem. Comm.*, **1969**, 1067.
108 R. J. Bushby and G. H. Whitham, *J. Chem. Soc.*(B), **1969**, 67.
109 O. Korver and C. Boelhouwer, *Rec. Trav. Chim.*, **88**, 696 (1969).
110 C. van der Brink, L. D. C. Bok, and N. J. J. van Rensburg, *Z. Phys. Chem.*, **62**, 145 (1968); *Chem. Abs.*, **70**, 77072 (1969).
111a C. Rappe, *Acta Chem. Scand.*, **23**, 2305 (1969).
111b See *Org. Reaction Mech.*, **1968**, 365.
112 J. E. Norlander, S. P. Jindal, and D. J. Kitko, *Chem. Comm.*, **1969**, 1136.
113 D. N. Kursanov, E. V. Bykova, and V. N. Setkina, *Dokl. Akad. Nauk SSSR*, **184**, 100 (1969); *Chem. Abs.*, **70**, 77165 (1969).
114 D. A. Johnson, *Inorg. Nucl. Chem. Letters*, **5**, 225 (1969).
115 L. E. Erickson, H. L. Fritz, R. J. May, and D. A. Wright, *J. Am. Chem. Soc.*, **91**, 2510 (1969).
116 F. Knoll, J. Gronebaum, and R. Appel, *Chem. Ber.*, **102**, 848 (1969).
117 D. M. Brouwer and J. A. van Doorn, *Rec. Trav. Chim.*, **88**, 1041 (1969).
118 F. Hibbert, F. A. Long, and E. A. Walters, *J. Am. Chem. Soc.*, **91**, 2381 (1969).
119 M. E. Langmuir, L. Dogliotti, E. D. Black, and G. Wettermark, *J. Am. Chem. Soc.*, **91**, 2204 (1969).

Brønsted α for phenolates and secondary amines; however, hydroxide ion is 10^3 slower than expected on this correlation.[120] A stringent requirement for desolvation of OH^- was therefore proposed. Phenylacetylene, which is a weaker acid than (41), is nevertheless deprotonated by hydroxide ion 10^3 faster than (41), and therefore the reverse reaction between phenylacetylide and water is much closer to diffusion control than the reaction between (41)-anion and water. This is further evidence for the slowness of reprotonation of conjugated anions, and Ritchie[121] has shown that proton transfer between carbanion and hydrocarbon in DMSO is slower than would be inferred on a Brønsted correlation of reaction between hydrocarbon and phenoxides, amines, and benzoates. This was again attributed to a need for reorganization at the transition state. Ritchie[121] also showed that the reaction of methoxide ion with methyl fluorene-9-carboxylate was much slower in methanol than similar reactions in DMSO, and that highly delocalized carbanions were 6 pK_a units more acidic in DMSO than in methanol. This latter fact was thought to be due largely to dispersion interactions between carbanion and DMSO; increased dispersion interactions are probably responsible for the fact that (42)[122] is more completely converted into ylid in pyridine than in piperidine despite the relative basicities of these media. Counter-ions are important in determining the position of equilibrium;[123] in the reversible reaction between diphenylmethylide and ammonia in ether, potassium favoured the carbanion, and lithium the amide.

Ionization of nitro-compounds deviates very markedly from a normal Brønsted relationship,[124] and for (43a) and (43b) the rate of proton-abstraction

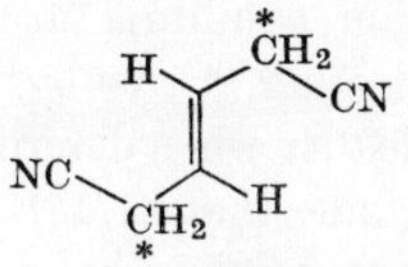

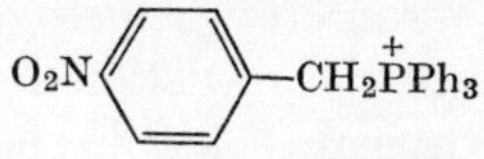

(or *o*-nitro)

(41)   (42)

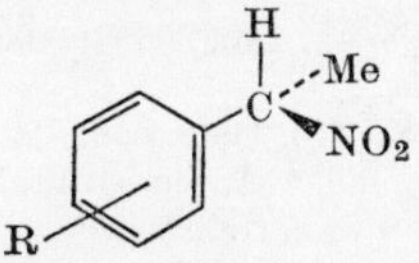

(43a) 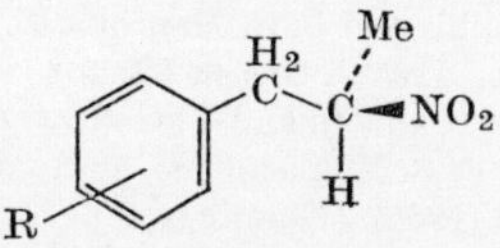   (43b)

[120] E. A. Walters and F. A. Long, *J. Am. Chem. Soc.*, **91**, 3733 (1969).
[121] C. D. Ritchie, *J. Am. Chem. Soc.*, **91**, 6749 (1969).
[122] T. A. Modro, *Bull. Acad. Polon. Sci., Ser Sci. Chim.*, **16**, 475 (1968).
[123] W. S. Murphy and C. R. Hauser, *Chem. Ind.* (London), **1969**, 832.
[124] F. G. Bordwell, W. J. Boyle, J. A. Haulala, and K. C. Yee, *J. Am. Chem. Soc.*, **91**, 4002 (1969).

by hydroxide ion has a Brønsted α (*vis-à-vis* variation of R) of >1.3, and there-fore ionization rate is *more* sensitive to substitution than is acidity. This is not really surprising when one considers that deprotonation at *carbon* is producing an anion in which nearly all the charge resides at *oxygen*. Rates of ionization of nitrocycloalkanes bear no systematic relationship to thermodynamic acidi-ties;[125] for example, nitrocyclobutane is weakly acidic but deprotonates rapidly. Ionization of nitroethane[126] in aqueous and alcoholic media has been studied.

The relationship between isotope effects and activation energies has been discussed.[127] Isotope effects for deprotonation of diethyl ketone by 2,6-lutidine $(k_H/k_D = 6.8)$ are higher than for deprotonation by pyridine $(k_H/k_D = 4.1)$.[128]

Reviews have appeared on the following topics: kinetic isotope effects on proton transfer;[129a] relative (kinetic) strengths of C–H acids;[129b] isotope exchange in hydrocarbons;[129c] and fast proton transfers.[129d]

Organometallics: Groups Ia, IIa, III

Structure and Spectra

t-Butyl-lithium is adjudged to be a tetrameric cube from the ^7Li–^{13}C NMR couplings observed in cyclohexane.[130] NMR spectra are recorded for mixtures of MeLi and Me_2Cd in ether and THF, where various complexes are formed.[131] In ether, methyl exchange between MeLi and Li_2CdMe_4 is faster than lithium exchange, for which the rate-determining step is dissociation of MeLi tetramer. In THF, MeLi forms a $2:1$ complex with Me_2Mg, $2:1$ and $1:1$ complexes with Me_2Zn, and a $1:1$ complex with Me_2Cd. In all these, lithium and methyl exchange occur at similar rates. Association equilibria are recorded for EtMgBr and Et_2Mg in ethyl 2-methylbutyl ether.[132] Ashby[133] stresses the importance of considering deviations from ideality when determining associa-tion equilibria of Grignard solutions; he concludes that EtMgBr is monomeric in THF and dimeric in Et_2O, and that arylmagnesium halides have a high

125 P. W. K. Flanagan, H. W. Amburn, H. W. Stone, J. G. Traynham, and H. Schechter, *J. Am. Chem. Soc.*, **91**, 2797 (1969).

126 D. Jannalhoudalhis and P. G. Mauridis, *Z. Naturforsch*, **24b**, 206 (1969).

127 D. B. Matthews, *Aust. J. Chem.*, **22**, 463 (1969).

128 J. P. Calmon, M. Calmon, and V. Gold, *J. Chem. Soc.*(B), **1969**, 659.

129a I. O. Shapiro, F. S. Yakushin, I. A. Romanskii, and A. I. Shatenshtein, *Org. Reactivity* (Tartu), **5**, 178 (1968). (Kinetic isotope effects on proton transfer.)

129b A. I. Shatenshtein and I. O. Shapiro, *Uspekhi Khim.*, **37**, 1946 (1968). (Relative strengths of C-H acids.)

129c A. Schriesheim, *Trans. N.Y. Acad. Sci.*, **31**, 97 (1969). (Review of the author's work.)

129d E. F. Caldin, *Chem. Rev.*, **69**, 135 (1969).

130 L. D. McKeever and R. Waack, *Chem. Comm.*, **1969**, 750.

131 L. M. Seitz and B. F. Little, *J. Organometal. Chem.*, **18**, 227 (1969).

132 P. Vink, C. Blomberg, A. D. Vreugdenhil, and F. Bickelhaupt, *J. Organometal. Chem.*, **15**, 273 (1968).

133 F. W. Walker and E. C. Ashby, *J. Am. Chem. Soc.*, **91**, 3845 (1969).

degree of solvent association. Alkyl exchange equilibria are recorded for cyclopentadienyldiethylaluminium in THF;[134] the cyclopentadienyl protons of cyclopentadienyl(di-isobutyl)aluminium show a sharp singlet in the NMR down to $-91°$,[135] so that 1,2-shifts are rapid in this system. A re-examination[136] of site-exchange reactions in trimethylaluminium shows that equilibration of bridging and terminal methyl groups occurs at the same rate as exchange with trimethylgallium; the latter reaction is therefore collision-controlled, and postulation of a cage effect is unnecessary. Both reactions occur faster in toluene than in cyclopentane. Exchange between trimethylgallium and its trimethylphosphine complex[137] occurs by a dissociative mechanism.

Reactions

Chemically induced dynamic nuclear polarization (CIDNP) is proving to be a generally useful probe for radical intermediates, and this year has seen applications to halogen–metal exchange between alkyl-lithiums and alkyl iodides, where polarization in both products suggests at least a portion of exchanges occur by a free-radical path,[138] and to the reactions of ethyl-lithium with 1,1-dimethyl-2,2-dichlorocyclopropane.[139] A mechanism is proposed for the reaction of N,N-dialkylarylamines with lithium alkyls,[140] whence α-alkylation of the amine is induced. Observation of a negative activation energy in ether cleavage by butyl-lithium[141] suggested a change in structure and association of BuLi with temperature. Rapid cyclization of various 6-lithiohex-1-enes, slow cyclization of corresponding Grignard reagents, and ring-opening of 2-bicyclo[2.2.1]hexylmagnesium chlorides are recorded.[142] The reaction of alkyl-lithium compounds with peroxides results in O–O cleavage.[143]

Holm has continued studies on the mechanism of addition of Grignard reagents to carbonyl compounds.[144] From kinetic results on the rate–concentration dependence of RMgBr in its reaction with acetone, he infers that coordination of the reactants destroys the reactivity of acetone. Acetone and benzophenone may well react by different mechanisms, given that the relative

[134] W. R. Kroll, *Chem. Comm.*, **1969**, 844.
[135] W. R. Kroll and W. Naegele, *Chem. Comm.*, **1969**, 246.
[136] E. A. Jeffrey and T. Mole, *Aust. J. Chem.*, **22**, 1129 (1966).
[137] K. L. Herold, J. B. De Roos, and J. P. Oliver, *Inorg. Chem.*, **8**, 2035 (1969).
[138] H. R. Ward, R. G. Lawler, and R. A. Cooper, *J. Am. Chem. Soc.*, **91**, 746 (1969); A. R. Lepley and R. L. Landau, *ibid.*, p. 748; A. R. Lepley, *Chem. Comm.*, **1969**, 64.
[139] H. R. Ward, R. G. Lawler, and H. Y. Loken, *J. Am. Chem. Soc.*, **90**, 7359 (1968).
[140] A. R. Lepley, *J. Org. Chem.*, **33**, 4362 (1968).
[141] Y. N. Baryshnikov, G. I. Vesnovskaya, and V. A. Shushunov, *Dokl. Akad. Nauk SSSR*, **185**, 580 (1969); *Chem. Abs.*, **71**, 21505.
[142] E. A. Hill, R. J. Thiessen, and K. Taucher, *J. Org. Chem.*, **34**, 3061 (1969); V. N. Drodz, Y. A. Ustynyuk, M. A. Tsel'eva, and L. B. Dmitriev, *Zh. Obshch. Khim.*, **38**, 2114 (1969).
[143] Y. N. Baryshnikov and G. I. Vesnovskaya, *Zh. Obshch. Khim.*, **39**, 529 (1969); *Chem. Abs.*, **71**, 38043 (1969).
[144] T. Holm, *Acta Chem. Scand.*, **23**, 579 (1969).

rates of addition of MeMgBr and Bu^tMgCl to acetone are 3.8:0.15, and to benzophenone under the same conditions 0.30:100.[144] In the addition of excess of PrMgBr to pinacolone in THF, the pseudo-first-order rate dependence on PhMgBr concentration supports[145a] the Swain mechanism.[145b] Addition, reduction, and enolization rate ratios for a variety of Grignard reagents with pinacolone have been treated quantitatively in terms of steric (E_s^0) and inductive (σ^*) parameters.[146] The reaction between neopentylmagnesium chloride and benzophenone in THF is slow enough to be followed by NMR, and shows the formation of an intermediate (RPh_2C–OMgCl or RPh_2C–OMgR)? which breaks down by reaction with RMgCl in a slow stage.[147] The reaction of ethyl diphenylphosphonate, $Ph_2P(O)OEt$, with PhMgBr followed by NMR reveals rapid formation of a 1:1 complex which slowly rearranges to products by a concentration-independent pathway.[148]

Work has been reported on asymmetric induction in the reaction of Grignard reagents with ketones,[149] and on the stereochemical course of additions to cyclic ketones by organomagnesium and related compounds.[150] Crotyl-magnesium bromide normally gives α-methylallyl products in reactions with unhindered ketones, but the amount of crotyl product, and also the *cis*:*trans* ratio therein, increases with increasing bulk of the ketonic substituents.[151a] The initial α-methylallyl product formed in these reactions is unstable and gradually rearranges to crotyl product, probably intramolecularly, since an intermediate cannot be trapped by external MeMgBr (Scheme 7).[151b] The cyclic orthoformate (**44**) reacts with Grignard reagents with stereospecific

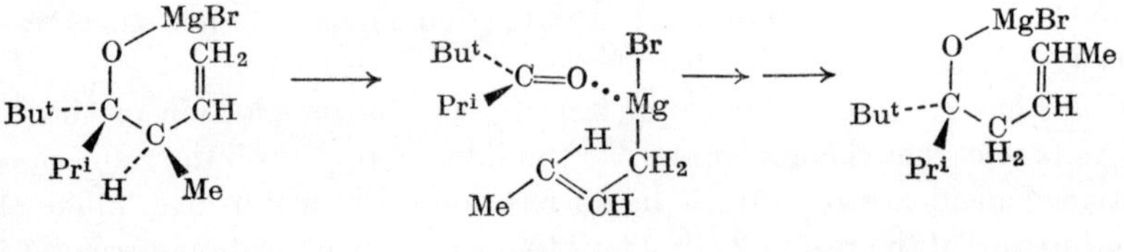

Scheme 7

145a I. Koppel, L. Margna, and A. Tuulmets, *Org. Reactivity* (Tartu), **5**, 1041 (1968); *Chem. Abs.*, **71**, 21405.

145b C. G. Swain and H. B. Boyles, *J. Am. Chem. Soc.*, **73**, 870 (1951).

146 A. Tuulmets, *Org. Reactivity* (Tartu), **4**, 17 (1968); *Chem. Abs.*, **69**, 43202.

147 C. Blomberg, R. M. Salinger, and H. S. Mosher, *J. Org. Chem.*, **34**, 2385 (1969).

148 H. R. Mays, *J. Am. Chem. Soc.*, **91**, 2736 (1969).

149 T. D. Inch, G. J. Lewis, G. I. Sainsbury, and D. I. Sellers, *Tetrahedron Letters*, **1969**, 3657; J. D. Morrison, A. Tomash, and R. W. Ridgway, *ibid.*, **1969**, 565, 569, 573.

150 P. R. Jones, E. J. Goller, and W. J. Kauffmann, *J. Org. Chem.*, **34**, 3566 (1969); W. H Glaze, C. M. Selman, A. L. Ball, and L. E. Bray, *ibid.*, p. 641; K. Suga, S. Watanabe, and Y. Yamaguchi, *Aust. J. Chem.*, **22**, 669 (1969); J. P. Battoni, M. L. Capman, and W. Chodkiewicz, *Bull. Soc. Chim. France*, **1969**, 976.

151a R. A. Benkeser, W. G. Young, W. E. Broxtermann, D. A. Jones, and S. J. Piaseczynski, *J. Am. Chem. Soc.*, **91**, 132 (1969).

151b R. A. Benkeser and W. E. Broxtermann, *J. Am. Chem. Soc.*, **91**, 5162 (1969).

replacement of OMe by R; the epimeric (at C-2) orthoformate is unreactive, and a carbonium ion intermediate formed with participation from *trans* lone-pairs on oxygen is implicated.[152] Formation of cyclobutane derivatives in the reaction of 6-ethoxy-4,5-dihydropyran with Grignard reagents,[153] competing 1,2- and 1,4-addition in the reaction of Grignard reagents with 3-isobutoxy-cyclohex-2-enone,[154] and reactivity of Grignard reagents[155] are described. The products from reaction between phenylmagnesium bromide and trityl [^{18}O]acetate include ditrityl peroxide lacking oxygen-18, and a mechanism involving electron transfer from PhMgBr to Ph$_3$C$^+$ is suggested.[156] A full paper[157a] on Tochtermann's work on the hydrolysis of acenaphthylene ate-complexes[157b] has appeared. A concerted front-side displacement mechanism (45) is suggested for the S_E2 reaction of benzylboronic acid with mercuric

R — OCH$_3$

R — O — ^{2}H

CH$_3$

(R = H or CH$_3$)

(44)

$$\left[\begin{array}{c} HO \quad O \\ B \\ ArCH_2 \quad O \quad CH_2OH \\ Cl \\ Hg \\ Cl \end{array} \right]^{\pm -}$$

(45)

chloride, which is catalysed by glycerol and shows a small positive ρ-value (0.93) on variation of ring substitutent.[158] Reduction can be a serious side-reaction in the alkylation of ketones by organoaluminium compounds.[159] Reaction of trimethylthallium with carbon acids such as cyclopentadiene results in replacement of one methyl group.[160]

Reviews in the field deal with NMR studies of organometallic exchange reactions,[161a] and alkyl derivatives of Group IIa metals.[161b]

Organometallics: Other Elements

Several papers concerned with the mechanisms of reactions of dialkylcadmium reagents have appeared.[162] Reaction with acid chlorides, particularly benzoyl

[152] E. L. Eliel and F. Nader, *J. Am. Chem. Soc.*, **91**, 536 (1969).
[153] J. d'Angelo, *Bull. Soc. Chim. France*, **1969**, 181.
[154] J. C. Richer and D. Eugene, *Can. J. Chem.*, **47**, 2387 (1969).
[155] A. Spassky-Pasteur and H. Riviere, *Bull. Soc. Chim. France*, **1969**, 811.
[156] K. D. Berlin, R. D. Shupe, and R. D. Grigsby, *J. Org. Chem.*, **34**, 2500 (1969).
[157a] B. Knickel and W. Tochtermann, *Chem. Ber.*, **102**, 3508 (1969).
[157b] See *Org. Reaction Mech.*, **1968**, 133.
[158] D. S. Matteson and E. Kramer, *J. Am. Chem. Soc.*, **90**, 7261 (1968).
[159] J. L. Namy, E. Henry-Basch, and P. Freon, *Compt. Rend., C*, **268**, 287 (1969).
[160] A. G. Lee and G. M. Sheldrick, *J. Organometal. Chem.*, **17**, 481 (1969).
[161a] N. S. Ham and T. Mole, *Progr. N.M.R. Spectroscopy*, **4**, 91 (1969).
[161b] B. J. Wakefield, *Adv. Inorg. Chem. Radiochem.*, **11**, 341 (1968).
[162] J. Thomas and P. Freon, *Compt. Rend., C*, **267**, 1850 (1968); J. Thomas, E. Henry-Basch, and P. Freon, *Bull. Soc. Chim. France*, **1969**, 109; J. Michel, E. Henry-Basch, and P. Freon, *ibid.*, **1968**, 4898, 4902.

chloride, in cyclohexane is much more effective in the presence of Lewis acids. Bispentamethylcyclopentadienylmercury has a static NMR at room temperature,[163] and complete collapse of the A_2B_2X proton NMR pattern of diallylmercury to AX_4 type could not be induced, even at 160°.[164a] Nevertheless, evidence for both inter- and intra-molecular shifts was presented. Site exchange between dimethylcadmium and Group II and Group III metals has been studied by NMR,[164b] and the conclusion made that exchange rates vary with Lewis acidity of the organometallic.

S_E2 and Related Reactions

A number of detailed mechanistic papers from the University College London group have appeared. The course of acidolysis of 4-pyridylmethylmercuric chloride in the presence of added chloride ion and $HgCl_2$ suggested protonation of a reversibly formed ylid (Scheme 8).[165] $HgCl_2$ inhibits the reaction, presumably through competition for this ylid. The S_E2 mercury-exchange reaction between this mercurial and $^{203}HgCl_2$ was examined. This does not compete with the concurrent S_E1 acidolysis except in the presence of added Cl⁻, and the rate showed a non-linear dependence on [Cl⁻]. This suggested one of the reactants to be (II) (Scheme 8) and the other ($HgCl_2 + nCl^-$).[166] Reaction of methylgold(triphenylphosphine) with Hg(II) salts and alkylmercurials is

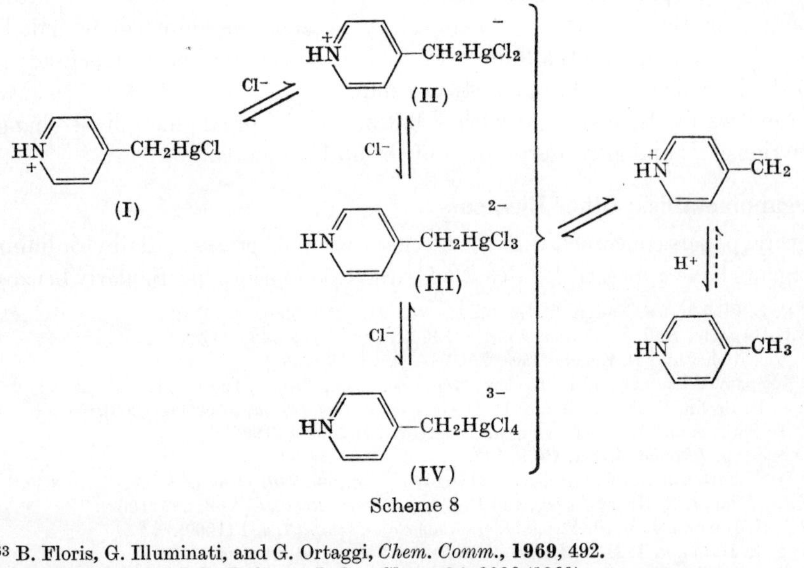

Scheme 8

[163] B. Floris, G. Illuminati, and G. Ortaggi, *Chem. Comm.*, **1969**, 492.
[164a] H. E. Zieger and J. D. Roberts, *J. Org. Chem.*, **34**, 2826 (1969).
[164b] K. Henold, J. Soulati, and J. P. Oliver, *J. Am. Chem. Soc.*, **91**, 3171 (1969).
[165] D. Dodd and M. D. Johnson, *J. Chem. Soc.*(B), **1969**, 1071.
[166] D. Dodd, C. K. Ingold, and M. D. Johnson, *J. Chem. Soc.*(B), **1969**, 1076.

deduced to have an S_E2 mechanism; however, (46) may react by an S_E1 pathway with mercurials in solvent mixtures containing DMSO, but by S_E2 in dioxan.[167] Protolysis of α-ethoxycarbonylbenzylmercuric chloride in 70%

(46)

aqueous dioxan requires a nucleophilic acid anion, since $HClO_4$ is not an effective catalyst.[168] A mechanism involving reaction between $RHgCl_3^-$ and H_3O^+ was suggested for the chloride ion catalysed reaction. Acetyldemercuration of benzylmercuric chloride in CH_2Cl_2 with $AlBr_3$ catalysis gives mainly *p*-acetylphenylacetone.[169] Discussion of the method of "even and odd cycles" in determining the course of S_E2 reactions at vinylic centres,[170] further comments on the reaction of *trans*-chlorovinylmercuric chloride with HCl,[171] and work on vinyl transfer from divinylmercury[172] to trityl bromide (which may involve radical intermediates) have appeared.

In the S_E2 reaction between tetramethyltin and mercuric chloride, solvent effects on the reactants are as important as solvent effects on the transition state in determining activation parameters.[173, 174] In 96% MeOH–4% H_2O, an S_E2 (open) mechanism is proposed for reaction between mercuric iodide and tetra-alkyltins, since a "steric" order of reactivity Me ≫ Et > Pr^n ≫ Bu^i ensues.[175] A positive kinetic salt effect was ascribed to destabilization of the reactants by $LiClO_4$. Insertion of SO_2 into propargyltriphenyltin[176] or allenyltrimethyltin is clearly an S_E2' reaction (Scheme 9). Similarly, allylmercuric acetate inserts SO_2 much faster than benzylmercuric acetate, and

[167] B. J. Gregory and C. K. Ingold, *J. Chem. Soc.*(B), **1969**, 276.

[168] J. R. Coad and C. K. Ingold, *J. Chem. Soc.*(B), **1968**, 1455.

[169] Y. G. Bundel, V. I. Rosenberg, A. L. Kurts, N. D. Antonova, and O. A. Reutov, *J. Organometal. Chem.*, **18**, 209 (1969).

[170] A. N. Nesmeyanov, A. E. Borisov, N. V. Novikova, and E. I. Fedin, *J. Organometal. Chem.*, **15**, 279 (1968).

[171] I. P. Beletskaya, O. A. Reutov, V. S. Petrosyan, and L. V. Savinykh, *Tetrahedron Letters*, **1969,** 485. (See *Org. Reaction Mech.*, **1968**, 133.)

[172] I. P. Beletskaya, V. B. Vol'eva, V. B. Golubev, and O. A. Reutov, *Izv. Akad. Nauk SSSR, Ser. Khim.* **1969**, 1197; *Chem. Abs.*, **71**, 49014.

[173] M. H. Abraham, F. Benbahany, M. J. Hogarth, R. J. Irving, and G. F. Johnston, *Chem. Comm.*, **1969**, 117.

[174] M. H. Abraham, G. F. Johnston, J. F. C. Oliver, and J. A. Richards, *Chem. Comm.*, **1969**, 930.

[175] M. H. Abraham and T. R. Spalding, *J. Chem. Soc.*(A), **1969**, 399, 784.

[176] W. Kitching, C. W. Fong, and A. J. Smith, *J. Am. Chem. Soc.*, **91**, 767 (1969).

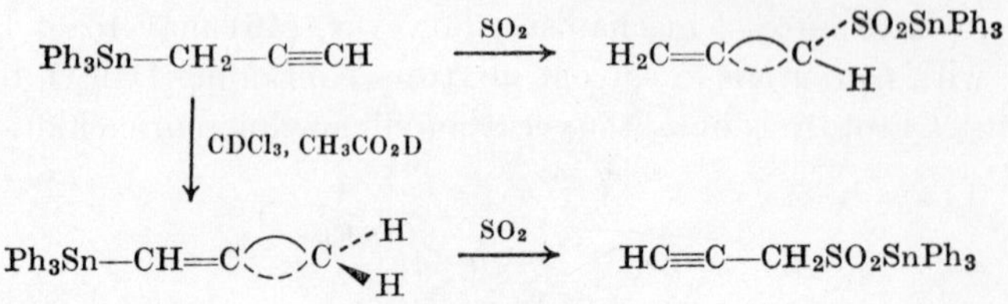

$$Ph_3Sn-CH_2-C\equiv CH \xrightarrow{SO_2} H_2C=C\overbrace{}^{}C\underset{H}{\overset{SO_2SnPh_3}{\diagup}}$$

$$\downarrow \text{CDCl}_3,\ \text{CH}_3\text{CO}_2\text{D}$$

$$Ph_3Sn-CH=C\overbrace{}^{}C\underset{H}{\overset{H}{\diagup}} \xrightarrow{SO_2} HC\equiv C-CH_2SO_2SnPh_3$$

Scheme 9

an S_E2' mechanism is inferred.[177] The rearrangement of triphenylpropargyltin (under conditions where the silicon and germanium analogues are unreactive) is rapid at 130° in decalin, concentration-dependent, and catalysed by Lewis acids.[178] An ion-pair mechanism was suggested. The reaction between allyl-trialkyltins and mercuric salts is, however, S_E2 rather than S_E2', and the allyl compound was 100 times as reactive as the corresponding tin crotyl. With $HgCl_2$ in acetone, $Et_3Sn(CH_2)_4Ph$ was cleaved immeasurably faster than Et_4Sn; presumably the phenyl group acts as a Lewis base in complexing with the incoming electrophile.[179]

An isotope effect $k_H/k_D = 2.5$ is observed in the protolysis of diethylcadmium by benzyl alcohol.[180]

Reviews in the field include: mechanisms of electrophilic displacement of metals at saturated carbon;[181a] mercuration;[181b,c] and electrophilic substitution reactions.[181d]

Electrophilic Reactions of Hydrocarbons

Further work has appeared on the reaction of superacids with saturated hydrocarbons,[182a,b] and differences in behaviour between the different acid systems are becoming clear. In HF–SbF_5,[182a] there is rapid exchange of hydrogen isotope with methane, and hexamethylethane is cleanly cleaved to isobutane and t-butyl carbonium ion. In SbF_5–HSO_3F,[182b] there is appreciable (oxidative) condensation of methane accompanying exchange at higher temperatures, and neopentane is rapidly cleaved to methane and t-butyl

177 W. Kitching, B. Hegarty, S. Winstein, and W. G. Young, *J. Organometal. Chem.*, **20**, 253 (1969).

178 M. Lequan and G. Guillerm, *Compt. Rend.*, *C*, **268**, 858 (1969); N. Lequan, G. Guillerm, and A. Jean, *ibid.*, p. 1542.

179 R. M. G. Roberts, *J. Organometal. Chem.*, **18**, 307 (1969).

180 A. Jubier, G. Emptoz, E. Henry-Basch, and P. Freon, *Bull. Soc. Chim. France*, **1969**, 2032.

181a D. S. Matteson, *Organometal. Chem. Rev.*, **4**, A263 (1969).

181b W. Kitching, *Organometal. Chem. Rev.*, **3**, A35, A61 (1968).

181c W. Kitching, *Rev. Pure. Appl. Chem.*, **19**, 1 (1969).

181d O. A. Reutov, *Pure. Appl. Chem.*, **17**, 79 (1968).

182a H. Hogeveen and A. F. Bickel, *Rev. Trav. Chim.*, **88**, 371 (1969).

182b G. A. Olah, G. Klopman, and R. H. Schlosberg, *J. Am. Chem. Soc.*, **91**, 3261 (1969).

carbonium ion. When the medium is diluted by SO_2ClF, neopentane liberates hydrogen and is converted into t-pentyl cation. Klopman[182b] provides quantum mechanical calculations on protonated alkanes, and concludes that for $CH_5{}^+$ the C_{2v} form (47) is more stable than the C_{3v} structure (48) or hydrogen-protonated species (49).

(47) (48) (49)

Many facile additions to highly strained cyclopropanes are reported,[183] and intramolecular capture of protonated cyclopropanes by neighbouring π-systems promises to be a useful synthetic cyclization procedure.[184] The stereochemistry of ring-opening of cyclopropanes by acids (i.e. whether protonation initially gives a corner-protonated or edge-protonated cyclopropane) seems to vary with the system under consideration. 1-Methylnortricyclene gives a mixture of products in the acid-catalysed addition of acetic [²H]acid but degradation to norbornanone established that deuteron attack was initially 62% *endo* and 38% *exo* at C-6.[185] Hendrickson finds an anti-Markownikow cleavage of the cyclopropane (50) by DBr in acetic acid (presumably due to the field-destabilizing effect of the anhydride) with *cis* stereochemistry explained by edge-protonation.[186] Acid-catalysed acetolysis of the *cis*- and *trans*-bicyclo[5.1.0]octanes gives very different product distributions, the *cis*-compound (51) giving rise mainly to methylcycloheptane derivatives, by cleavage of an exocyclic bond, and the *trans*-isomer (52) forming mainly cyclooctane derivatives through cleavage of the highly strained ring-junction.[187] No 1,3-dibromocyclopropane is formed in the addition of HBr to bromocyclopropane, ruling out a cyclopropanebromonium ion intermediate.[188]

Oxidation of phenylcyclopropanes by $Ti(OAc)_3$ or $Pb(OAc)_4$[189a] involves

[183] K. B. Wiberg and G. J. Burgmaier, *Tetrahedron Letters*, **1969**, 317; R. E. Pincock and E. J. Torupka, *J. Am. Chem. Soc.*, **91**, 4593 (1969); P. G. Gassman, A. Topp, and J. W. Keller, *Tetrahedron Letters*, **1969**, 1093; W. G. Dauben, J. H. Smith, and J. Saltiel, *J. Org. Chem.*, **34**, 261 (1969).

[184] G. Stork and M. Marx, *J. Am. Chem. Soc.*, **91**, 2371 (1969); G. Stork and M. Gregson, *ibid.*, p. 2373; G. Stork and P. A. Grieco, *ibid.*, p. 2407.

[185] J. H. Hammons, E. K. Probasco, L. A. Sanders, and E. J. Whalen, *J. Org. Chem.*, **33**, 4493 (1968).

[186] J. B. Hendrickson and R. K. Boeckman, *J. Am. Chem. Soc.*, **91**, 3269 (1969).

[187] K. B. Wiberg and A. de Meijere, *Tetrahedron Letters*, **1969**, 519.

[188] C. C. Lee, B. S. Hahn, K. M. Wan, and D. J. Woodcock, *J. Org. Chem.*, **34**, 3210 (1969).

[189a] R. J. Ouellette, D. Miller, A. South, and R. D. Robins, *J. Am. Chem. Soc.*, **91**, 971 (1969); A. South and R. J. Ouellette, *ibid.*, **90**, 7064.

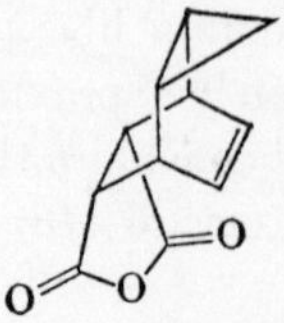

(50) 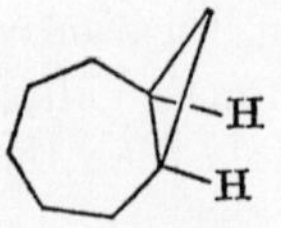　　　　　　　**(51)** 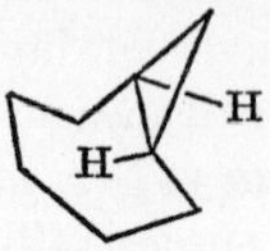　　　　　　　**(52)** 　　　　　　　**(53)**

an electrophilic attack of metal in the rate-determining stage; unlike the mercuric acetate reaction,[189b] no stable intermediate is formed and cinnamyl and phenylpropane diol acetates are produced. Addition of mercuric acetate to **(53)** gives the expected addition product and also the product resulting from 1,2-norbornyl shift.[190a] Protonated cyclopropanes have been reviewed.[190b]

Miscellaneous Reactions

Cram and Hoffman have studied the stereochemical course of cleavage of **(54)** and **(55)** in base.[191] The Haller–Bauer reaction of **(54)** occurs with some retention of configuration even in DMSO. The diastereomers of **(55)** equilibrate rapidly with dimsylsodium in DMSO at 25°; **(56)** retains considerable optical integrity, being 75% optically pure at 8% conversion in DMSO. Thus carbonyl capture of an intermediate carbanion competes successfully with external proton-donors, and this is apparent even in t-butanol.

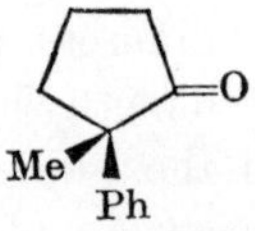

(54)

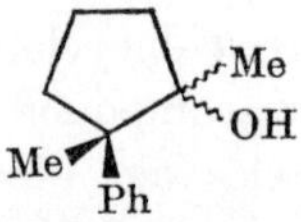

(55) *trans*-diastereomer
(56) *cis*-diastereomer

　　The carbanion intermediate in the mechanism of action of aldolases may be estimated by reaction with $C(NO_2)_4$, and spectrophotometric determination of $^-C(NO_2)_3$;[192] CCl_3^- is produced in the reaction of $P(NMe_2)_3$ with CCl_4, and may be trapped by ketones;[193] α-chlorination of bridgehead sulphides cannot be carried out for stereoelectronic reasons, and the reaction takes a different course.[194]

[189b] See *Org. Reaction Mech.*, **1968**, 137.
[190a] V. I. Sokolov, N. B. Rodina, and O. A. Reutov, *J. Organometal. Chem.*, **17**, 477 (1969).
[190b] C. J. Collins, *Chem. Rev*, **69**, 543 (1969).
[191] T. D. Hoffman and D. J. Cram, *J. Am. Chem. Soc.*, **91**, 1000, 1009 (1969).
[192] J. E. Riordan and P. Christen, *Biochemistry*, **8**, 2381 (1969).
[193] B. Castro, R. Burgada, G. Lavielle, and J. Villieras, *Compt. Rend., C*, **268**, 1067 (1969).
[194] L. A. Paquette and R. W. Houser, *J. Am. Chem. Soc.*, **91**, 3870 (1969).

CHAPTER 4

Elimination Reactions[1]

R. BAKER

Chemistry Department, The University, Southampton

Steric Course of *E2* Reactions

Further studies have been reported on the steric course in the elimination of cycloalkyltrimethylammonium salts confirming that *trans*-olefin is produced by a *syn*-mechanism.[2] The reaction of the *cis*- and *trans*-8-D derivatives of 1,1,4,4-tetramethylcyclodecyl-7-trimethylammonium chloride was studied in KOMe–MeOH, KOBu^t–BuOH, and Bu^{t}OK–DMSO (equation 1). For the *trans*-8-D isomer, k_H/k_D for formation of 1,1,4,4-tetramethyl-*trans*-cyclodec-7-ene was between 2.7 and 3.9 but was 1 for the *cis*-7-D isomer. Also the *trans*-

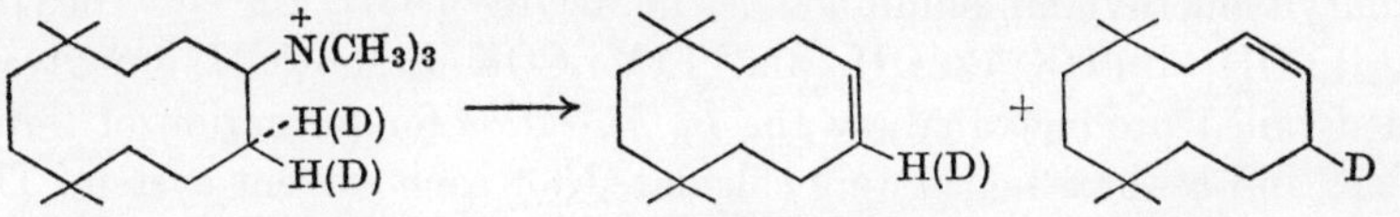

$$\cdots (1)$$

cis- and *trans*-8-D   *cis*-7   *cis*-6
derivatives   *trans*-7   *trans*-6

cyclodec-7-ene formed from *trans*-8-D was deuterium-free whereas the same isomer was formed from the *cis*-8-D isomer with retention of the deuterium label. It was apparent that the *trans*-cyclodecene arises from a *syn*-elimination. Similarly, the small amount of *cis*-cyclodec-7-ene produced in MeOH from the *trans*-8-D and *cis*-8-D isomers was associated with k_H/k_D of 3.6 and 1.0, confirming that the *cis*-cyclodecene was produced by an *anti*-elimination. Since the *syn*-elimination leading to *trans*-7-olefin has to overcome a deuterium isotope effect of 3—4, the energy difference between *syn*-elimination leading to *trans*-cyclodecene and *anti*-elimination to the same olefin (*syn*→*trans* and *anti*→*trans*) must be substantial (3—4 kcal mole^{-1}). The possibility of these

---

[1] A review has appeared "Olefin-forming elimination reactions", J. F. Bunnett, *Survey Progr. Chem.*, **5**, 53 (1969).

[2] J. Závada, M. Svoboda, and J. Sicher, *Coll. Czech. Chem. Comm.*, **33**, 4027 (1968); see *Org. Reaction Mech.*, **1968**, 145.

reactions taking place by an α',β-mechanism was positively discounted by the observation that the trimethylamine obtained from the *trans*-8-D labelled 'onium compound was essentially deuterium-free.

Evidence has been presented that a *syn*-elimination occurs in the reaction of 2-hexyl chloride, bromide, and iodide in Bu^tOK–DMSO.[3] The ratio of *trans*-hex-2-ene to *cis*-hex-2-ene produced for these substrates in DMSO was about 5.0 which is substantially greater than the *trans/cis* ratios in NaOMe– MeOH and Bu^tOK–BuOH. This ratio in DMSO was compared with the *trans/ cis* ratio obtained in the pyrolysis of 2-hexyl acetate (7.5 at 50°) which is known to be a *syn*-elimination.[4] High *trans/cis* ratios are expected in *syn*- eliminations owing to the large steric interactions produced by the eclipsing of alkyl groups in the transition state. The change from *anti*-elimination of 2-hexyl halides in MeONa–MeOH and Bu^tOK–BuOH to *syn*-elimination in Bu^tOK–DMSO parallels studies of Závada, Krupicka, and Sicher.[5] In the reactions of cycloalkyl bromides in Bu^tOK–Bu^tOH *trans*-cycloalkenes were considered to be formed by *syn*-elimination and *cis*-cycloalkenes by *anti*-elimination, but both *cis*- and *trans*-cycloalkenes appear to be formed by predominantly *anti*-elimination in EtOK–EtOH.

The effect of solvent upon the stereochemistry of eliminations from quater- nary ammonium salts has also been examined by Bailey and Saunders.[6] The olefin proportions and k_H/k_D values in eliminations from the reactions of *erythro*- and *threo*-1-ethyl[2-²H_1]butyltrimethylammonium [(1) and (2)] and 1-ethylbutyltrimethylammonium iodides in MeOK–MeOH, Bu^nOK–Bu^nOH, Bu^sOK–Bu^sOH, Bu^tOK–Bu^tOH, and $EtMe_2COK$–$EtMe_2COH$ are given. From a detailed product analysis the k_H/k_D ratios for formation of *trans*- hex-3-enes and *cis*-hex-3-enes were calculated for each solvent system. The formation of olefin by three possible modes was considered: (1) all *anti*- elimination; (2) all *syn*-elimination; and (3) *syn-anti*-elimination (*trans*- olefin via *syn*-elimination and *cis*-olefin via *anti*-elimination). These possibilities are summarized for (1) and (2). The isotope effects were used to distinguish between these possibilities since loss of deuterium would be associated with a substantial isotope effect but only small secondary effects would be expected when deuterium is retained. The results were considered to support a pre- dominant *anti*-mechanism in MeOK–MeOH and Bu^nOK–Bu^nOH but a *syn-anti*-mechanism as the major path in Bu^tOK–Bu^tOH and $EtMe_2COK$– $EtMe_2COH$. Further support for these conclusions was obtained by the measurement of the deuterium content of the olefin mixtures obtained from reaction in Bu^nOK–Bu^nOH and $EtMe_2COK$–$EtMe_2COH$ (Table 1).

[3] R. A. Bartsch and J. F. Bunnett, *J. Am. Chem. Soc.*, **91**, 1382 (1969).
[4] C. H. DePuy, C. A. Bishop, and C. N. Goeders, *J. Am. Chem. Soc.*, **83**, 2151 (1961).
[5] See *Org. Reaction Mech.*, **1967**, 116.
[6] D. S. Bailey and W. H. Saunders, Jr., *Chem. Comm.*, **1968**, 1598.

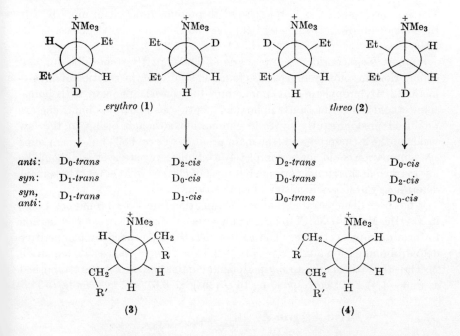

erythro (**1**)　　　　　　　　　threo (**2**)

anti:	D_0-*trans*	D_2-*cis*	D_2-*trans*	D_0-*cis*
syn:	D_1-*trans*	D_0-*cis*	D_0-*trans*	D_2-*cis*
syn, *anti*:	D_1-*trans*	D_1-*cis*	D_0-*trans*	D_0-*cis*

(**3**)　　　　　　　　　(**4**)

Table 1. Deuterium loss in eliminations from *erythro*- and *threo*-1-ethyl[2-^{2}H$_1$]butyltrimethylammonium iodides at 85°.

Reaction conditions	Reactant	% [^{2}H$_1$] loss calc. for		[% ^{2}H$_1$] loss observed
		anti	*syn-anti*	
Bu^{n}OH–Bu^{n}OK	[4-^{2}H$_1$]*erythro*	10.8	5.2	8.2
	[4-^{2}H$_1$]*threo*	20.3	38.1	20.1
EtMe$_2$COK–EtMe$_2$COH	[4-^{2}H$_1$]*erythro*	31.9	5.2	6.8
	[4-^{2}H$_1$]*threo*	8.9	26.9	20.3

Saunders and Bailey advanced a plausible, if not entirely convincing, hypothesis for the elimination results in acyclic systems. A change to stronger base (n-alkoxide to t-alkoxide) was assumed to be associated with a more reactant-like transition state and a decrease in the preference for *anti*-eliminations. Owing to the large bulk of the trimethylammonium group the conformation with the α- and β-alkyl group *trans* to each other was expected to be favoured (**3**). In this configuration the *anti*-β-H is shielded on both sides, particularly seriously when these alkyl groups are large. For the conformation leading to *cis*-olefin (**4**), the *anti*-β-hydrogen is shielded on only one side so that *trans*-olefin would be formed less readily than *cis*-olefin by *anti*-elimination.

The *syn*-mechanism was also expected to produce *trans*-olefin more readily than *cis*-olefin owing to greater eclipsing interaction of alkyl groups in the latter case.

Support for a consideration of steric effects as partly responsible for *syn*-eliminations in some systems is also found in studies of the reaction of *threo*-1-methyl[2-^{2}H$_1$]propyltrimethylammonium toluene-*p*-sulphonate with potassium ethoxide in dimethyl sulphoxide.[7] Both *cis*- and *trans*-but-2-ene are produced predominantly by *trans*-elimination although owing to the low yields of the 2-ene produced (the main product was *ca.* 98% but-1-ene) some *syn*-elimination could not be completely excluded. Any steric effects should be less marked in this simpler system but it is not clear to what extent the solvent determines the stereochemistry of elimination.

It has been demonstrated that *anti*-elimination from *cis*-4-bromo-oct-4-ene, in NaOMe–MeOH–DMSO, to form an allene occurs faster than *syn*-elimination to form dipropylacetylene.[8] The values of k_1/k_2 determined by competitive dehydrobromination at 25° were 43.8 ± 2.5 and $k_2/k_3 = 1500 \pm 300$ (equation 2). Thus the ratio of *anti*- to *syn*-periplanar elimination (k_1/k_3) was established as $(6.6 \pm 1.4) \times 10^4$ corresponding to $G^{\ddagger}(25°) = -6.6 \pm 0.2$ kcal mole^{-1}. This

$$
\begin{array}{l}
\mathrm{Pr^n}\!\!\diagdown\!\!\mathrm{C}\!\!=\!\!\mathrm{C}\!\!\diagup\!\!\mathrm{Br} \quad \xrightarrow{k_1} \quad \mathrm{Pr^n C}\!\!\equiv\!\!\mathrm{CPr^n} \\
\mathrm{H}\!\!\diagup \qquad\quad \diagdown\!\!\mathrm{Pr^n} \\[2mm]
\hspace{6cm}\xrightarrow{k_3} \qquad\qquad\qquad\qquad \cdots (2)\\[2mm]
\mathrm{H}\!\!\diagdown\!\!\mathrm{C}\!\!=\!\!\mathrm{C}\!\!\diagup\!\!\mathrm{Br} \quad \xrightarrow{k_2} \quad \mathrm{EtCH}\!\!=\!\!\mathrm{C}\!\!=\!\!\mathrm{CHPr^n} \\
\mathrm{Pr^n}\!\!\diagup \qquad\quad \diagdown\!\!\mathrm{Pr^n}
\end{array}
$$

is therefore in agreement with results from earlier work, that *anti*-elimination of hydrogen halide from *trans*-vinyl halides to give acetylenes is much faster than *syn*-elimination from the corresponding *cis*-vinyl halide.[9] Similar studies on the dehydrochlorinations of the three isomeric chlorodiphenylethylenes have been reported[10] (equation 3). This work was undertaken since it was

$$
\begin{array}{l}
\mathrm{Ph}\!\!\diagdown\!\!\mathrm{C}\!\!=\!\!\mathrm{C}\!\!\diagup\!\!\mathrm{Cl} \\
\mathrm{H}\!\!\diagup \qquad\quad \diagdown\!\!\mathrm{Ph} \\[2mm]
\hspace{4cm}\searrow^{k_{antt}}\\[2mm]
\mathrm{Ph}\!\!\diagdown\!\!\mathrm{C}\!\!=\!\!\mathrm{C}\!\!\diagup\!\!\mathrm{Ph} \quad \xrightarrow{k_{syn}} \quad \mathrm{PhC}\!\!\equiv\!\!\mathrm{CPh} \hspace{1cm}(3)\\
\mathrm{H}\!\!\diagup \qquad\quad \diagdown\!\!\mathrm{Cl} \\[2mm]
\hspace{4cm}\nearrow^{k_a}\\[2mm]
\mathrm{Ph}\!\!\diagdown\!\!\mathrm{C}\!\!=\!\!\mathrm{C}\!\!\diagup\!\!\mathrm{H} \\
\mathrm{Ph}\!\!\diagup \qquad\quad \diagdown\!\!\mathrm{Cl}
\end{array}
$$

[7] D. H. Froemsdorf, H. R. Pinnick, Jr., and S. Meyerson, *Chem. Comm.*, **1968**, 1600.

[8] S. W. Staley and R. F. Doherty, *Chem. Comm.*, **1969**, 288.

[9] S. I. Miller, *J. Org. Chem.*, **26**, 2619 (1961).

[10] S. J. Cristol and C. A. Whittemore, *J. Org. Chem.*, **34**, 705 (1969).

apparent from earlier studies that $k_{anti} > k_{syn} > k_\alpha$ in ethoxide–ethanol, whereas in the PhLi-induced reactions this order was reversed. Using a constant base, potassium t-butoxide, the solvent was varied and the relative rates for elimination were determined in each solvent: in ethanol (97°), $k_{anti}:k_{syn}:k_\alpha = 2 \times 10^4:100:1$; in Bu^tOH (95°), 3000:600:1; in 5M DMSO–Bu^tOH (25°), $10^8:6 \times 10^5:1$; and in DMSO (19°), 250:40:1. Taking into account the suggestion that *syn*-elimination would involve a more carbanion-like transition state than *anti*-elimination, it was hoped that the solvent changes would reduce the energy requirements of the *anti*:*syn* processes. That this is not the case can be seen from the fact that the relative rates for the three processes do not vary in a constant fashion. The comparable *syn*- to *anti*-elimination in Bu^tOH (*anti/syn* = 6) could well be a result of the increasing importance of the *syn*-process owing to ion-pairing[11] but no clear result emerges from the mixed-solvent studies.

The general pattern of high olefin yields from the deamination of axial amines but low yield from equatorial amines has been demonstrated for the decomposition of the *N*-nitrosocarbamates (5) and (6).[12] Decomposition of (5) in boiling cyclohexane yielded a product mixture composed of 80% 1- and 2-octalins and 20% of a mixture of the epimeric *trans*-2-decalyl carbonates. Under the same conditions the product mixture from the equatorial isomer contained only 41% olefin. In acetic acid, 78% olefin was obtained from (5) and 14% from (6). Although in the past this behaviour in nitrous acid deamination has been attributed to either an *E2 trans*-diaxial elimination or loss of the *trans*-axial proton from a carbonium ion, it was shown by deuterium labelling that the 2-octalin formed in the decomposition of (5) and (6) in cyclohexane is

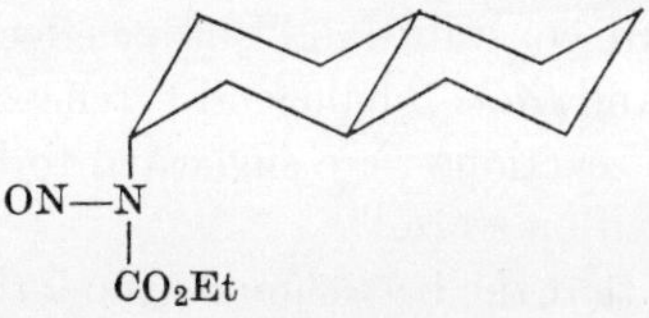

(5) 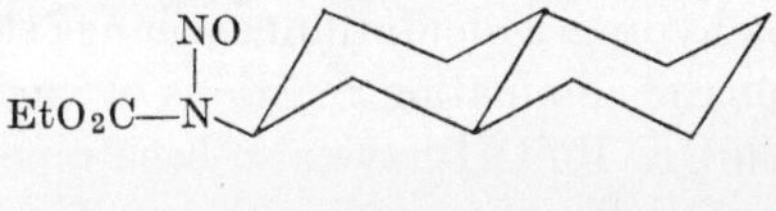   (6)

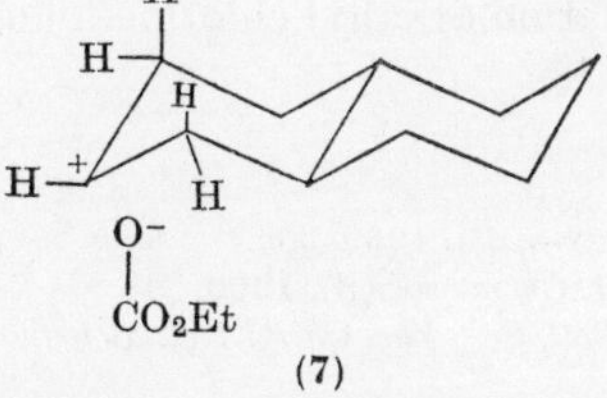

(7) 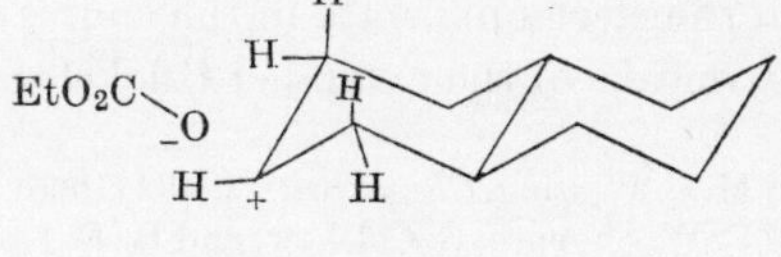   (8)

[11] S. J. Cristol and R. S. Bly, Jr., *J. Am. Chem. Soc.*, **83**, 4027 (1961); see *Org. Reaction Mech.*, **1968**, 146.
[12] T. Cohen and A. R. Daniewski, *J. Am. Chem. Soc.*, **91**, 533 (1969).

formed by *syn*-elimination. It was shown that the nitrous acid deamination of the epimeric *trans*-2-decalylamines also results in predominantly *syn*-elimination, although substantial *anti*-elimination was also found with the axial epimer. The incursion of ion-pairs in the *N*-nitrosocarbamate decomposition was suggested to account for the stereochemical results. The ion-pairs (7) and (8) are probably formed in the decomposition of (5) and (6).[13] In (7) the counterion can only remove a β-equatorial hydrogen because the *trans*-axial hydrogens are on the other side of the molecule. In (8), however, the anion is between the neighbouring axial and equatorial protons, and either can be removed; the preference for removal of the axial proton was suggested to be due to a possible stereoelectronic effect. This argument would account for the observed greater selectivity of the axial compared to the equatorial *N*-nitrosocarbamate decomposition. The elimination for the nitrous acid deamination of the axial decalylamine was suggested to involve a substantial proportion of *E*2 elimination.

More olefin was also found in the deamination of *cis*-4-t-butylcyclohexylamine in AcOH than from *trans*-4-t-butylcyclohexylamine.[14] Although this was suggested to be due to the occurrence of a base-induced *anti*-diaxial elimination of the diazonium ion, the formation of ion-pairs must also be considered in these reactions.

Studies on the reaction of *cis*- and *trans*-p-anisyl 2-chloro- and 2-bromo-vinyl ketones with methoxide ions in methanol have been reported.[15] With the *cis*-chloro- and *cis*-bromo-derivatives, reaction occurs by an elimination–addition process, whereas the corresponding *trans*-isomer reacts by direct substitution. The transition state for the elimination of *cis*-chlorovinyl ketone was shown to have more carbanion character than the *cis*-bromo-compound by hydrogen–deuterium exchange studies. Rate constants and isotope effects for the elimination reactions of various *cis*- and *trans*-β-halogenostyrenes in Bu^{t}OK–Bu^{t}OH have also been reported; the reactions were suggested to be essentially concerted, with a carbanionic transition state.[16]

Following earlier MO calculations, showing that the base should attack the *trans*-β-hydrogen most easily in *E*2 reactions,[17] a quantitative method to interpret the reactivity in *E*2 reactions has been proposed,[18] and the change in the electronic state in the course of *E*2 reaction of ethyl chloride followed by means of the extended Hückel calculation.[19]

[13] M. C. Whiting, *Chem. Brit.*, **2**, 482 (1966).
[14] C. W. Shoppee, C. Culshaw, and R. E. Lack, *J. Chem. Soc.*(C), **1969**, 506.
[15] D. Landini, F. Montanari, G. Modena, and F. Naso, *J. Chem. Soc.*(B), **1969**, 243.
[16] G. Marchese, G. Modena, F. Naso, and N. Tangari, *Boll. Sci. Fac. Chim. Ind. Bologna*, **26**, 209 (1969).
[17] K. Fukui and H. Fujimoto, *Tetrahedron Letters*, **1965**, 4303.
[18] K. Fukui, H. Hao, and H. Fujimoto, *Bull. Chem. Soc. Japan*, **48**, 348 (1969).
[19] H. Hao, H. Fujimoto, and K. Fukui, *Bull. Chem. Soc. Japan*, **42**, 1256 (1969).

Orientation in *E*2 Reactions

Bartsch and Bunnett have extended their studies on 2-substituted hexanes[20] to a whole series of leaving groups (halides, phenylseleno, phenylsulphonyl, and a series of arenesulphonyloxy groups) in the base–solvent systems Bu^{t}OK–Bu^{t}OH and MeONa–MeOH. In MeONa–MeOH the decrease in both the hex-2-ene/hex-1-ene ratio and the *trans/cis* hex-2-ene ratio with the poorer leaving group conforms to the pattern expected from the theory of the variable *E*2 transition state.[21] A striking observation in Bu^{t}OK–Bu^{t}OH is the strong predominance of Hofmann elimination from the halides (90% hex-1-ene for the bromide and chloride) and the formation of larger amounts of *cis*-hex-2-ene compared to *trans*-hex-2-ene with some substrates (arenesulphonates, 2,4,6-trimethylbenzoate, phenyl sulphone, diphenyl phosphate). More Hofmann elimination is obtained with the 2-hexyl halides in Bu^{t}OK–Bu^{t}OH than from 2-butyl halides; 54% but-1-ene is obtained from 2-butyl bromide but 87% from 2-hexyl bromide.[22] This difference was considered to be steric in origin owing to steric interactions between the n-propyl group on C$_\beta$ in the hexene-forming transition state and the t-butoxide ion; these types of interactions would be much less serious when the β-alkyl or base is smaller (methoxide in place of butoxide) or when the base attacks an alternative β-H to form hex-1-ene. A decision on whether these steric interactions were due to compressions or restricted rotation in the transition states could not be made on the basis of the reported results. Bartsch and Bunnett plotted log(hex-2-ene/hex-1-ene) against a logarithmic measure of reactivity in forming hex-1-ene (termed the reactivity index). In MeOH, the amount of hex-1-ene increased roughly with decreasing reactivity of the substrate, and although the correlation was not good the results were broadly in line with the predictions of the theory of the variable *E*2 transition state. In Bu^{t}OH a sharp minimum in the graph was found which cannot be explained in terms of the theory. The authors speculated that a new mechanism may operate when poor leaving groups are present. It seems that the question of *syn*-elimination should be considered in this context, particularly since in an accompanying paper Bartsch and Bunnett present evidence for *syn*-elimination in the reactions of the 2-substituted hexanes in Bu^{t}OK–DMSO.

trans-[2-^{2}H$_1$]Cyclohexyltrimethylammonium, *trans*-[2-^{2}H$_1$]cyclopentyltrimethylammonium, and 3-[2,2-^{2}H$_2$]pentyltrimethylammonium *p*-sulphonates have been prepared, and comparison with the reactions of the undeuterated substrates in NaOH–H$_2$O gave k_H/k_D for elimination: cyclohexyl, 4.33; cyclopentyl, 3.99; and 3-pentyl, 3.22.[23] Elimination competed very poorly

[20] R. A. Bartsch and J. F. Bunnett, *J. Am. Chem. Soc.*, **91**, 1376 (1969); see *Org. Reaction Mech.*, **1968**, 140.
[21] See reference 1.
[22] H. C. Brown and R. L. Klimisch, *J. Am. Chem. Soc.*, **88**, 1425 (1966).
[23] W. H. Saunders, Jr., and T. A. Ashe, *J. Am. Chem. Soc.*, **91**, 1376 (1969).

6

with substitution for the cyclohexyl system (15% olefin), and the rate of elimination was much slower compared to the other systems, probably owing to the difficulty of forcing the trimethylammonium group into an axial position for the elimination. In general, the results support the earlier view, advanced to explain the predominance of Saytzeff-product from a number of alicyclic quaternary ammonium salts, that steric compression in the transition state lengthens the carbon–nitrogen bond and results in considerable double-bond character instead of the carbanion character normally found in quaternary ammonium salt eliminations.[24] Isotope effects were also measured for the cyclopentyl and 3-pentyl systems in EtONa–EtOH, Pr^iOK–Pr^iOH, and Bu^tOK–Bu^tOH; in the event that an *anti*- or *syn*-mechanism might well operate in Pr^iOH and Bu^tOH the measured values could be unreliable. In the other solvents, however, the higher k_H/k_D for the cyclopentyl than for the 3-pentyl system (and in H_2O the higher value for the cyclohexyl) is in line with the hypothesis that the extent of carbon–nitrogen cleavage decreases in the order cyclohexyl < cyclopentyl < 3-pentyl. In an accompanying paper, Katz and Saunders have calculated the deuterium and sulphur isotope effects for $E2$ reactions of hydroxide ion with ethyldimethylsulphonium ion, and have examined the dependence of the isotope effects on transition-state force constants.[25]

The reaction of (−)-menthyl benzenesulphonate in Bu^tOK–DMSO has been reinvestigated and found to yield exclusively menth-2-ene with no trace of menth-3-ene;[26] this is contrary to earlier reports where equal amounts of menth-2- and -3-ene were reported which required a *syn*-elimination.[27] With other arenesulphonates (*p*-Me, *p*-Br, and *p*-Cl) exclusive formation of menth-2-ene was reported, showing that only Hofmann elimination occurs, probably by *anti*-elimination.

An $E1cB$-like mechanism has been suggested for the elimination reactions of a series of phenylethylmethyl sulphoxides in Bu^tOK–Bu^tOH on the basis of a ρ-value of +4.4, the absence of hydrogen-exchange, and $k_H/k_D = 2.7$.[28] Similar results were found for the elimination of HCN from diphenylmethyl thiocyanates in Pr^iONa–Pr^iOH.[29] $k_H/k_D = 3.81$ has been found for the bimolecular dehydrochlorination of t-butyl chloride in acetonitrile;[30] this value is significantly higher than that found for the $E1$ reaction.

Rates of dehalogenation reactions of several *meso*- and *erythro*-1,2-dibromo-1-(4-R^1-phenyl)-2-(4-R^2-phenyl)ethanes with I^-, Br^-, and Cl^- in methanol

24 T. H. Brownlee and W. H. Saunders, Jr., *Proc. Chem. Soc.*, **1961**, 314.
25 A. M. Katz and W. H. Saunders, Jr., *J. Am. Chem. Soc.*, **91**, 4469 (1969).
26 C. L. Snyder and H. H. Chang, *Chem. Comm.*, **1969**, 114.
27 C. L. Snyder and A. R. Soto, *J. Org. Chem.*, **29**, 742 (1964).
28 R. Baker and M. J. Spillett, *J. Chem. Soc.*(B), **1969**, 481.
29 A. Ceccon, U. Miotti, U. Tonellato, and M. Padovan, *J. Chem. Soc.*(B), **1969**, 1084.
30 D. N. Kevill and J. E. Dorsey, *J. Org. Chem.*, **34**, 1985 (1969).

and dimethylformamide have been reported.[31] Leaving-group and substituent effects indicate that the reactions occur by a concerted mechanism with substantial double-bond character in the transition state.

The elimination of the p-tolylsulphonyl group has been shown to occur to a greater extent than the elimination of the alkoxysulphonyl group in the reactions of alkyl esters of 2-p-tolylsulphonylethanesulphonic acid with alkoxides; this difference is suggested to be due to the greater activation by the alkoxysulphonyl group.[32] Elimination of hydrogen halides from stereoisomeric 1,2,3,4-tetrahalogenocyclobutanes has also been investigated.[33]

Kinetic evidence has been presented that S_N2 and $E2$ reactions of 1-phenylethyl bromide with EtONa–EtOH share a common ion-pair intermediate.[34] A rate decrease of greater than a factor of 2 for the reaction of 1-phenylethyl bromide was found on increasing the base strength from 0.114M to 0.533M compared to a 17% decrease with ethyl bromide. The experimental results were considered to be consistent with equations (4) but it remains to be seen how generally these observations will apply.

$$\begin{array}{ccc}
\underset{H}{\overset{Me}{Ph-C-Br}} \rightleftharpoons R^+ X^- & \begin{array}{l} \xrightarrow{k_{1s}} \quad PhCH(Me)OEt \\[2mm] \xrightarrow[k_{1e}]{} \quad PhCH=CH_2 \\[2mm] \xrightarrow[k_{2s}[OEt]]{} \quad PhCH(Me)OEt \\[2mm] \xrightarrow[k_{2e}[OEt]]{} \quad PhCH=CH_2 \end{array} & \cdots (4)
\end{array}$$

Strong evidence has been presented against the suggested *E2C* mechanism for elimination reactions.[35] On the assumption that the *E2C* mechanism should be sensitive to steric hindrance at C_α **(9)** the kinetics of chloride-induced elimination reactions of 2-bromo-2,3,3-trimethylbutane **(10)** compared to t-butyl bromide were studied in acetone and 1,4-dioxan in the presence of 2,6-lutidine.[36] Compound **(10)** was more reactive than t-butyl bromide by a factor of 2, whereas, on comparison of the S_N2 rates of two similarly related systems, ethyl bromide is 38,000 times more reactive than neopentyl bromide with lithium chloride in acetone.[37] The greater rates of chloride-ion induced

[31] E. Baciocchi and A. Schirolli, *J. Chem. Soc.*(B), **1969**, 554.

[32] E. J. Miller and C. J. M. Stirling, *J. Chem. Soc.*(C), **1968**, 2895.

[33] M. Avram, G. D. Mateescu, I. G. Dinulescu, I. Pogany, and C. D. Nenitzescu, *Rev. Roum. Chim.*, **13**, 1085 (1968).

[34] R. A. Sneen and H. M. Robbins, *J. Am. Chem. Soc.*, **91**, 3100 (1969).

[35] See *Org. Reaction Mech.*, **1968**, 149.

[36] D. Eck and J. F. Bunnett, *J. Am. Chem. Soc.*, **91**, 3099 (1969).

[37] E. D. Hughes, C. K. Ingold, and J. D. H. Mackie, *J. Chem. Soc.*, **1955**, 3173.

elimination were ascribed to steric acceleration but it was considered that, if the *E2C* mechanism applied, steric retardation would have depressed the rate of elimination of (**10**) by several powers of ten if covalent interaction of chloride ion with C_α occurred.

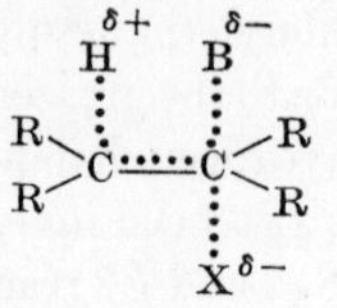

$$\text{(9)} \qquad\qquad\qquad\qquad \text{(10)}$$

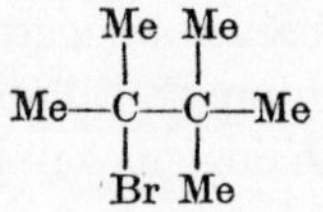

Protic and dipolar solvent effects on rates of bimolecular reactions have been reviewed.[38]

The *E1cB* Mechanism

A novel carbanion mechanism, (*E1cB*)ip or pre-equilibrium ion-pair mechanism, has been proposed for the reaction between *cis*-dibromoethylene and triethylamine in DMF.[39] A substitution–elimination process was discounted on the basis of the high rate of reaction of $BrCH{=}CHNMe_3{}^+Br^-$ towards triethylamine in DMF which indicates that it does not lie on the reaction path for formation of monobromoacetylene from *cis*-dibromoethylene. An *E2* process was ruled out on the basis of the $k_H/k_D = 1.00$, and the pre-equilibrium *E1cB* by the observation that, in the reaction of *cis*-dibromoethylene with triethylamine in the presence of triethylammonium deuterobromide, the unreacted dibromoethylene did not become deuterated. However, hydrogen exchange was observed in dehydrobromination in methanol.[40] On the basis of these observations the (*E1cB*)ip mechanism via an intimate ion-pair was proposed (equation 5). The difference from the established *E1cB* process is that,

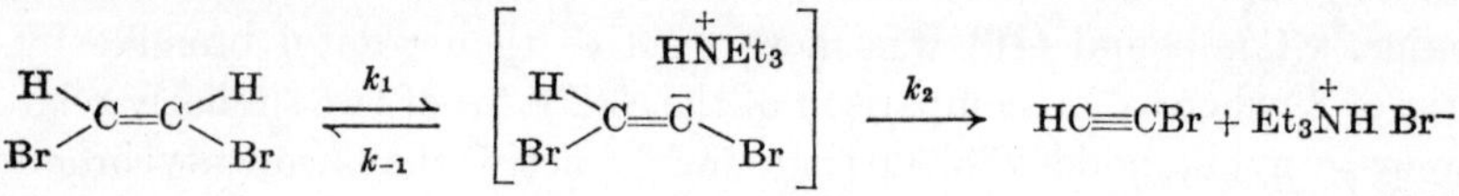

$$\cdots (5)$$

instead of a separated ion, an intimate ion-pair is initially formed which either reverts to reactants or collapses to product without exchanging trimethylammonium ions with those in the solvent. The precedent for tight or contact

[38] A. J. Parker, *Chem. Rev.*, **69**, 1 (1969).
[39] W. K. Kwok, W. G. Lee, and S. I. Miller, *J. Am. Chem. Soc.*, **91**, 468 (1969).
[40] S. I. Miller and W. G. Lee, *J. Am. Chem. Soc.*, **81**, 6313 (1959).

ion-pairs is found commonly in the studies by Cram and coworkers[41] of electrophilic substitution. It was suggested that Ingold's spectrum of transition states should be elaborated to include the ion-pair mechanism:

$$E1 - (E1)\text{ip} - E2 - (E1cB)\text{ip} - E1cB$$

Miller and coworkers suggest that other base-catalysed eliminations which display small isotope effects, but no hydrogen exchange, might well react by an $(E1cB)$ip mechanism, rather than a carbanion-like $E2$ elimination, e.g. 1,1,1-trifluoro-2-methyl-3-phenylpropane in Bu^tOK–Bu^tOH ($k_H/k_D = 1.2$),[42] and *erythro*-2,3-dimethyl-3-(trimethylammonio)nonane in Bu^tOK–Bu^tOH ($k_H/k_D = 0.9$—1.1).[43] *syn*-Eliminations of various cyclic trimethylammonium hydroxides were also suggested to be candidates for the $(E1cB)$ip mechanism.[44]

A carbanion mechanism has also been suggested for the formation of dibenzofulvene from the base-catalysed β-elimination of 9-fluorenylmethanol in $NaOH$–H_2O (equation 6).[45] Evidence for formation of a carbanion was the slightly greater rate of tritium exchange than the rate of elimination and kinetic induction periods when the reaction was conducted in D_2O, which could be interpreted quantitatively in terms of isotopic exchange. The carbanion was identified as an intermediate by a kinetic analysis of the induction periods.

$$^-OH + FlCHCH_2OH \xrightarrow{k_2} FlC{=}CH_2 + {}^-OH + H_2O \qquad \text{...(6)}$$

$$\overset{k_1}{\diagdown} \quad \overset{k_{-1}}{\diagdown} \qquad \diagup\, k$$

$$(FlCH_2 = \text{Fluorene}) \qquad Fl\bar{C}CH_2OH$$

Fedor has demonstrated that the β-elimination of methoxide ion from 4-methoxybutan-2-one and 4-methoxy-4-methylpentan-2-one in aqueous solution is specific-base catalysed.[46] When the reactions are carried out in D_2O, $k_{\text{exch}}/k_{\text{elim}} = 226$, which supports an $E1cB$ mechanism for the decomposition; this was expected since both compounds possess acidic hydrogen atoms, groups capable of stabilizing negative charge, and a poor leaving group. Deuterium solvent kinetic isotope effects and Arrhenius activation parameters provided additional evidence for the mechanism of rapid and general-base catalysed enolate anion formation, followed by rate-determining loss of methoxide ion from the enolate anion. Fedor has also demonstrated that the β-elimination of acetate from t-butyl 3-acetoxythiolpropionate, t-butyl

[41] D. J. Cram, *Fundamentals of Carbanion Chemistry*, Academic Press, New York, 1965, Ch. 4.
[42] D. J. Cram and A. S. Wingrove, *J. Am. Chem. Soc.*, **86**, 5490 (1964).
[43] M. Panková, J. Sicher, and J. Závada, *Chem. Comm.*, **1967**, 394.
[44] J. L. Coke, M. P. Cooke, and M. C. Mourning, *Tetrahedron Letters*, **1968**, 2247.
[45] R. A. More O'Ferrall and S. Slae, *Chem. Comm.*, **1969**, 486.
[46] L. R. Fedor, *J. Am. Chem. Soc.*, **91**, 908 (1969).

3-acetoxythiolbutyrate, and t-butyl 3-acetoxy-3-methylthiolbutyrate to give α,β-unsaturated thiol esters is general-base catalysed by tertiary amines and hydroxyl ion.[47]

Although in most cases only indirect evidence can be obtained for the presence of carbanions in elimination reactions, in one case the conjugate base of (11) has been isolated and the kinetics of its decomposition in solution have been investigated.[48] Compound (12) was obtained by the reaction of (11) with

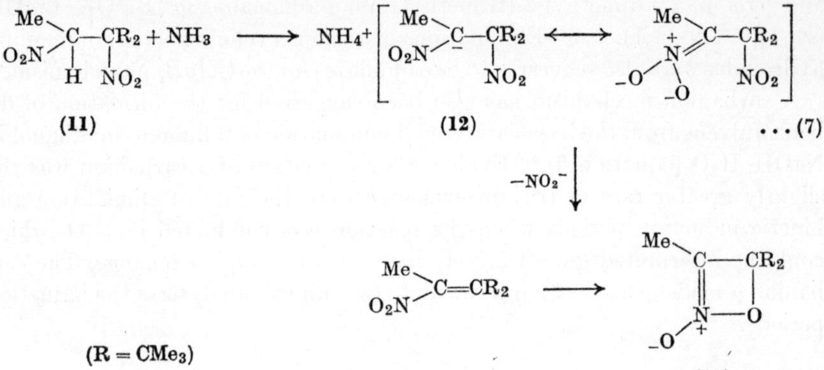

(11)　　　　　　　　　　　　　　　　(12)　　　　　　　　　　　... (7)

(R = CMe₃)

(13)

NH₃ in ether. The ammonium salt decomposed within a few minutes in methanol to the olefin which cyclized to (13); the salt also decomposes slowly in the solid state. The kinetics of decomposition of (12) were studied by NMR in CH_3OD and shown to be first-order with half-lives of ca. 10 minutes at 20° and ca. 2 minutes at 30° ($\Delta G^{\ddagger} = -22 \pm 5$ kcal mole⁻¹). The stability of (12) was considered to be due to steric hindrance since no intermediate could be isolated with 2-t-butyl-3,3-dimethyl-1,2-dinitrobutane.

Further investigations have been reported on the kinetics and mechanism of the reversible deamination of aspartic acid.[49] A carbanion intermediate was suggested by the observation that exchange of the methylene protons of aspartic acid took place in a buffered solution in D_2O at 116.3° at pH 6.1. Equal amounts of *threo-* and *erythro-*3-deutero-(±)-aspartic acid were obtained by the addition of ND_3 to fumaric acid in D_2O, indicating equal amounts of *cis-* and *trans-*addition. Both the exchange and addition experiments argue for a planar carbanion intermediate and an *E1cB* mechanism for the elimination reaction.

[47] L. R. Fedor, *J. Am. Chem. Soc.*, **91**, 913 (1969).
[48] A. Berndt, *Angew. Chem. Intern. Ed. Engl.*, **8**, 613 (1969).
[49] J. L. Bada and S. L. Miller, *J. Am. Chem. Soc.*, **91**, 3946 (1969).

Other Topics

Phenyldimethylcarbinyl chloride, *p*-nitrobenzoate, and thionbenzoate have been solvolysed in ethanol, and the rates and products determined.[50] The fraction of elimination increased in the order chloride (12%) < *p*-nitrobenzoate (50%) < thionbenzoate (91%), which suggests the intervention of an ion-pair in the reactions:

$$RX \underset{k_{-1}}{\overset{k_1}{\rightleftharpoons}} R^+X^- \quad \begin{array}{l} \xrightarrow{\ k_s\ } \text{Substitution} \\ \xrightarrow{\ k_e\ } \text{Elimination} \end{array}$$

substitution could occur from the ion-pair or from further stages of dissociation. Successive deuteration of the methyl groups decreased the fraction of elimination and increased substitution. For the [$^2H_3$]-derivatives the rates of the olefin product formed with loss of D or H provided a measure of the isotope effect for the elimination (at 50°: chloride, 2.9; *p*-nitrobenzoate, 2.4; and thionbenzoate, 1.68), and calculations based on these results predicted the observed products with fair accuracy. Rates of reaction were greater than those for the corresponding benzhydryl compounds; ratios were: chloride, 5.5; *p*-nitrobenzoate, 23; and thionbenzoate, 4700. The high reactivity of the phenyldimethylcarbinyl thionbenzoate compared to benzhydryl thionbenzoate was attributed to less ion-pair return in the former system.

Ion-pairs have also been suggested to be important in the elimination reaction of t-butyl chloride in 2,2,2-trifluoroethanol–water mixtures.[51] For [$^2H_9$]-t-butyl chloride, isotope effects and the olefin fraction are larger in the fluorinated solvent than in ethanol–water mixtures. Since more olefin was produced from the less nucleophilic solvent, a significant *E*2 contribution was excluded, but the results were suggested to be consistent with elimination from an ion-pair intermediate in agreement with the earlier conclusions of Winstein and Cocivera.[52] The larger *β*-deuterium isotope effect accompanying the larger olefin fraction in trifluoroethanol is strong evidence that proton loss from an ion-pair may be rate-determining. It is interesting that $k_H/k_D = 2.4$ is larger than the factor of 2 found from product studies for the elimination of deuterium from a carbonium ion intermediate compared to elimination of a proton. This was suggested to be due to the competition between rate-determining elimination and another process leading to substitution from the reversibly formed ion-pair.

Further evidence has been reported that the acid-catalysed dehydration of 1-aryl-2-phenylethanols proceeds by rate-determining proton transfer from

[50] S. G. Smith and D. J. W. Goon, *J. Org. Chem.*, **34**, 3127 (1969).
[51] V. J. Shiner, Jr., W. Dowd, R. D. Fischer, S. R. Hartshorn, M. A. Kessick, L. Milakofsky, and M. W. Rapp, *J. Am. Chem. Soc.*, **91**, 4838 (1969).
[52] S. Winstein and M. Cocivera, *J. Am. Chem. Soc.*, **85**, 1702 (1963).

carbonium ion to solvent.[53] The dehydration of 1-(2,4,6-trimethoxyphenyl)-2-phenylethanol was found to be specific-acid general-base catalysed. Rates for compounds *ortho*-substituted in the 2-phenyl ring were correlated by σ^+ constants with $\rho = -1.11$, showing that the reaction of *ortho*-substituted compounds proceeds by the same mechanism as that of the *meta*- and *para*-substituted compounds. It was concluded that these reactions did not proceed via π-complexes.

Kinetics of decarboxylative dehydration of *erythro*- and *threo*-β-anisyl-β-hydroxy-α-phenylpropionic acid have been studied.[54] *trans*-4-Methoxystilbene was formed from both diastereoisomers, and the mechanism was suggested to be the generation of a dipolar ion with subsequent loss of CO_2 more rapid than reaction with water.

Investigations have been made of the β- and γ-elimination reactions of a series of 3-phenyl-2-alkylpropyl sulphoxides in basic media[55] (equation 8).

$$
\begin{array}{c}
\underset{\text{PhCH}_2\text{CHRCH}_2\text{SOCH}_3}{\overset{\gamma\quad\beta\quad\alpha}{}}
\end{array}
\qquad
\begin{array}{c}
\overset{\beta}{\nearrow}\quad \text{PhCH}_2\text{C(R)}{=}\text{CH}_2 \xrightarrow{\text{isom.}} \text{PhCH}{=}\text{C(R)CH}_3 \\
\\
\underset{\gamma}{\searrow}\quad \text{Ph}\quad\text{R}
\end{array}
\qquad \ldots (8)
$$

It was found that the percentage of 1,3-elimination increased with size of alkyl group on the β-carbon, and that 1,3-elimination is of greater importance in DMSO than in Et_3N. The difference in relative proportions of cyclopropane and olefin in the two media was suggested to be due to the difference in importance of an $E1cB$ 1,3-elimination and a concerted 1,2-elimination in the two solvent systems. The high dielectric constant of dimethyl sulphoxide and its ability for specific solvation of the cation favour the carbanion mechanism. Deuterium exchange studies in $[^2H_6]$dimethyl sulphoxide confirmed the assigned pre-equilibrium step of carbanion formation in the 1,3-elimination.[56] The increase in rate of 1,3-elimination with increasing size of alkyl group was rationalized in terms of entropy loss on cyclization due to restricted rotation about the C-3–C-2 and C-2–C-1 bonds which would increase with size of alkyl group.

A carbanion intermediate has also been identified in the reaction of 2-methyl-3-phenylprop-2-en-1-ol and similar systems with lithium dimethoxyaluminium hydride.[57] Phenylcyclopropanes are formed by a 1,3-elimination from an

[53] G. M. Loudon and D. S. Noyce, *J. Am. Chem. Soc.*, **91**, 1433 (1969); see *Org. Reaction Mech.*, 1968, 151.
[54] D. S. Noyce and E. McGovan, *J. Org. Chem.*, **34**, 2558 (1969).
[55] R. Baker and M. J. Spillett, *J. Chem. Soc.*(B), **1969**, 581.
[56] R. Baker and M. J. Spillett, *J. Chem. Soc.*(B), **1969**, 880.
[57] M. J. Jorgenson and A. F. Thacker, *Chem. Comm.*, **1969**, 1290.

organoaluminium intermediate. The carbanionic character of the transition state was recognized from the ρ +3.8 for the reaction obtained by changing the substituents in the phenyl ring, rate, and stereochemical studies.

From deuterium exchange and rate data it has been shown that the Favorskii rearrangement of $ArCH_2COCH_2Cl$ to $ArCH_2CH_2CO_2Me$ in MeONa–MeOH involves reversible carbanion formation and that $\rho = -4.97$ for the first-order release of chloride ion from this carbanion.[58] The effect of salts on the reaction was examined, and dilution of the methanol with water gave a substantial rate increase. A mechanism was proposed involving ionization of the C–Cl bond aided by π-participation on the part of the neighbouring enolate ion (see Chapter 8). A 1,3-elimination of sulphinate ions from δ-keto-sulphones has been used to form a dicyclopropyl ketone of known fixed stereochemistry.[59]

The solvolysis of *erythro*-1,2-dibromopropyltrimethylsilane in aqueous ethanol results in stereospecific *anti*-elimination of trimethylbromosilane. The mechanism was suggested to involve initial ionization of the carbon–halogen bond with participation of the silicon with the developing positive charge on the carbon atom.[60]

The mechanism of the liquid-phase thermal decomposition of some dialkyl oxalates of tertiary alcohols has been further examined,[61] and more support obtained for the ion-pair mechanism obtained earlier.

$$R^1R^2CHC(R^3R^4)Ox \xrightarrow{140°} R^1R^2CHC^+R^3R^4 Ox^- \to R^1R^2C{=}CR^3R^4 + (CO_2H)_2$$

Small k_H/k_D values (1.6—1.8) argue strongly against any extensive C_β–H bond breakage at the transition state. Information was obtained indicating that collapse of the ion-pair to give olefin product occurred 5—10 times faster than any internal reorientation.

Studies on the mechanism of decomposition of alkyl diphenylphosphinates have been reported.[62] A concerted cyclic mechanism was shown to be unable to account for the product distribution with some systems in boiling DMSO or diphenyl ether, and an ionic process involving ion-pairs was suggested. The thermal decomposition of β-hydroxyphosphonamides has been shown to involve preferential *syn*-elimination, and the synthetic route extended to conjugated dienes.[63]

$$R^1R^2C(OH)CR^3R^4P(O)(NMe_2)_2 \to (Me_2N)_2PO_2H + R^1R^2C{=}CR^3R^4$$

β-Elimination reactions of methyl (methyl hexopyranoside)uranates[64] and

[58] F. G. Bordwell, R. G. Scamehorn, and W. R. Springer, *J. Am. Chem. Soc.*, **91**, 2087 (1969).
[59] W. L. Parker and R. B. Woodward, *J. Org. Chem.*, **34**, 3085 (1969).
[60] A. W. P. Jarvis, A. Holt, and J. Thompson, *J. Chem. Soc.*(B), **1969**, 852.
[1] G. J. Karabatsos and K. L. Krunel, *J. Am. Chem. Soc.*, **91**, 3324 (1969).
[2] K. D. Berlin, J. G. Morgan, M. E. Peterson, and W. C. Pivonka, *J. Org. Chem.*, **34**, 1266 (1969).
[3] E. J. Corey and D. E. Cane, *J. Org. Chem.*, **34**, 3053 (1969).
[4] H. W. H. Schmidt and H. Neukom, *Tetrahedron Letters*, **1969**, 2011.

the preparation of olefins by pyrolysis of O-alkyl dimethylthiocarbamates[65] have been studied.

Details of preliminary studies of reactions of benzaldehyde acetals with butyl-lithium have been reported; depending upon the acetal the reaction mechanisms involve $E2$ eliminations, modified Wittig rearrangements, and cyclo-eliminations.[66]

The first examples of the elimination of hydrogen cyanide and nitriles in retro-cycloaddition reactions have been described.[67] The reaction of oxazoles (14; R = H or Me) with dimethyl acetylenedicarboxylate produced the furan (16) and the appropriate nitrile without isolation of the adduct (15). Pyrone (17) was isolated from a similar reaction of the oxazoles with diphenylcyclopropenone in boiling toluene, again without isolation of the initially formed adduct.

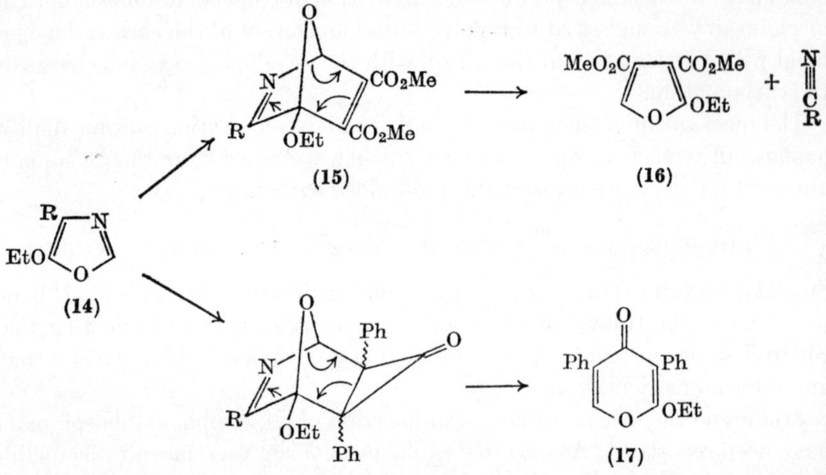

The fragmentation of (18) and (19) into SO_2 and octatriene occurs by an *anti*-concerted elimination (antarafacial process).[68] The pyrolysis of the episulphone of *cis*-dibenzoylstilbene has been reported.[69]

The following reactions have also been investigated: pyrolytic eliminations of steroidal sulphoxides,[70] optically active amine oxides to assign nitrogen chirality,[71] steroidal sulphonate esters to assign sulphur chirality,[72] N-oxides

[65] M. S. Newman and F. W. Hetzel, *J. Org. Chem.*, **34**, 3604 (1969).
[66] J. N. Hines, M. J. Peagram, G. H. Whitham, and M. Wright, *Chem. Comm.*, **1968**, 1593.
[67] R. Grigg, R. Hayes, and J. L. Jackson, *Chem. Comm.*, **1969**, 1167.
[68] W. L. Mock, *J. Am. Chem. Soc.*, **91**, 5682 (1969).
[69] D. C. Dittner, G. C. Levy, and G. E. Kuhlman, *J. Am. Chem. Soc.*, **91**, 2097 (1969).
[70] D. N. Jones and W. Higgins, *J. Chem. Soc.*(C), **1969**, 2159.
[71] S. I. Goldberg and Fak-Luen, *J. Am. Chem. Soc.*, **91**, 5113 (1969).
[72] D. N. Jones and W. Higgins, *Chem. Comm.*, **1968**, 1685.

(18)

(19)

of *trans*-pyrrolidinecyclohexanols,[73] organosilanes,[74] Δ^1-1,2-diazetines,[75] episulphides,[76] *cis*- and *trans*-2-methylcyclohexyl toluene-*p*-sulphonates,[77] and methyl xanthates;[78] dehydrohalogenation of 2-chloromethyl-4-methylene-1,3-dioxolans,[79] β-chloropropionitrile,[80] alkanesulphenyl chlorides,[81] 1-acyl-4-chlorooctane,[82] and *exo*-2-bromo-*exo*-3-deuterobenzonorbornene;[83] dehalogenation of 1,3-dihalogenoisoindenones;[84] acid-catalysed elimination of α,β-ditertiary ketones[85] and 2-t-butyl-4-dimethylamino-1-phenylbutan-2-ol;[86] dehydration of cyclic pinacols;[87] deamination of spiro[5,5]-1-undecylamine and spiro[4,5]-1-decylamine;[88] base-catalysed elimination of 2-chloroethyldipropylarsine,[89] β-methoxy-sulphides,[90] 17aα- and 17aβ-toluene-*p*-sulphonates of 17α-methyl-5α-D-homoandrostane,[91] and the tosylhydrazones of benzoin, benzoin acetate, and benzoin benzoate;[92] reaction of steroidal alcohols

[73] J. Gore, J. P. Drouet, and J. J. Barceux, *Tetrahedron Letters*, **1969**, 9.
[74] W. K. Musker and G. L. Larsen, *J. Am. Chem. Soc.*, **91**, 514 (1969).
[75] N. Rieber, J. Alberts, J. A. Lipsky, and D. M. Lemal, *J. Am. Chem. Soc.*, **91**, 5668 (1969).
[76] E. M. Lown, H. S. Sandhu, H. E. Gunning, and O. P. Strausz, *J. Am. Chem. Soc.*, **90**, 7164 (1968).
[77] C. Tatsumi and R. Kotani, *Bull. Univ. Osaka Prefect.*, *B*, **20**, 25 (1968); *Chem. Abs.*, **70**, 37253 (1969).
[78] K. Yoshira and T. Taguchi, *J. Chem. Soc. Japan*, **89**, 1225 (1969).
[79] H. J. Dietrich, R. J. Raynor, and J. N. Karabinos, *J. Org. Chem.*, **34**, 2975 (1969).
[80] G. Dienys, S. Jonaitis, and P. Buekus, *Org. Reactivity* (Tartu), **5**, 415 (1968); *Chem. Abs.*, **70**, 19410 (1969).
[81] J. F. King and T. W. S. Lee, *J. Am. Chem. Soc.*, **91**, 6524 (1969).
[82] J. K. Groves and N. Jones, *J. Chem. Soc.*(C) **1969**, 2350.
[83] L. E. Barstow and G. A. Wiley, *Tetrahedron Letters*, **1968**, 6309.
[84] K. Blatt and R. W. Hoffmann, *Angew. Chem. Intern. Ed. Engl.*, **8**, 606 (1969).
[85] P. Bauer and J. E. Dubois, *Chem. Comm.*, **1969**, 229.
[86] A. F. Casy and R. R. Ison, *Tetrahedron*, **25**, 641 (1969).
[87] H. Christol, A. P. Krapcho, and F. Pietrasanta, *Bull. Soc. Chim. France*, **1969**, 4059.
[88] J. M. Bessiere and H. Christol, *Bull. Soc. Chim. France*, **1969**, 4063.
[89] L. D. Pettit and D. Turner, *J. Chem. Soc.*(C), **1969**, 294.
[90] G. A. Russell and E. T. Sabourn, *J. Org. Chem.*, **34**, 2336 (1969).
[91] M. Leboeuf, A. Cave, and R. Goutarel, *Bull. Soc. Chim. France*, **1969**, 2100.
[92] T. Iwadare, I. Adachi, and M. Hayashi, *Tetrahedron Letters*, **1969**, 4447.

with DMSO–acetic anhydride[93] and formation of 2,6-benzodiazonine and dibenzo[c,l][1,6]diazocine by Hofmann elimination;[94] elimination of steroidal tosylates by PCl_5;[95] elimination of 2-tosyloxymethylcycloalkanone acetals,[96] lithium bromide from 1-lithio-2-bromocyclopentene,[97] β-acetoxy-organo-boranes,[98] and cyclooctyl sulphide;[99] thermal decomposition of t-butyl-lithium in hydrocarbon solution;[100] deamination of aziridines;[101] the stereochemistry of elimination of norcamphor and apocamphor tosylhydrazone with methyllithium.[102] The dehydration of atropine to apoatropine,[103] chlorismate synthetase reaction,[104] the dehydration by fumarate hydratase of malic acid,[105] and the aromatization reaction of androstenedione have been investigated.[106]

Formation of ketones by pyrolysis of ethyl and benzyl dichlorotetrafluoro-hemiketal acetates[107] and base-catalysed reaction of the bromolactone in the bicyclo[2.2.1]heptyl system have been reported.[108]

Detailed product studies have been made on the $KHSO_4$ and $ZnCl_2$ catalysed dehydration of *trans*-bicyclo[4.2.0]octan-3-ol[109] and *trans*-bicyclo[4.3.0]nonan-3-ol,[110] the pyrolysis of the corresponding xanthates, and base-catalysed elimination of the corresponding toluene-p-sulphonates in MeONa–MeOH and Bu^tOK–Bu^tOH.

The synthesis of alkynes by elimination of enol esters has been reviewed.[111]

The stereochemistry of fragmentation reactions of small-ring heterocycles has been investigated.[112] *cis*- and *trans*-But-2-ene episulphides decomposed with n-butyl-lithium stereospecifically to *cis*- and *trans*-but-2-ene; *threo*- and *erythro*-2-bromo-3-ethylthiobutane both yielded a mixture of *cis*- and *trans*-but-2-ene, indicating that the episulphide decomposed by a concerted mechanism and not through an intermediate carbanion.

[93] S. M. Ifzal and D. A. Wilson, *J. Chem. Soc.*(C), **1969**, 2168.
[94] P. Aeberli and W. J. Houlihan, *J. Org. Chem.*, **34**, 2715 (1969).
[95] A. Abad, M. Allard, and J. Levissalles, *Bull. Soc. Chim. France*, **1969**, 1236.
[96] W. Kirmse, *Chem. Ber.*, **102**, 2440 (1969).
[97] G. Wittig and J. Heyn, *Ann. Chem.*, **726**, 57 (1969).
[98] A. Suzuki, K. Ohmori, and M. Itoh, *Tetrahedron*, **25**, 3707 (1969).
[99] A. C. Cope and J. E. Engelhart, *J. Org. Chem.*, **34**, 3199 (1969).
[100] R. L. Eppley and J. A. Dixon, *J. Organometal Chem.*, **11**, 174 (1968).
[101] R. M. Carlson and S. Y. Lee, *Tetrahedron Letters*, **1969**, 4001.
[102] K. T. Lui and R. H. Shapiro, *J. Chinese Chem. Soc.* (Taipei), **16**, 30 (1969).
[103] W. Lund and T. Waaler, *Acta Chem. Scand.*, **22**, 3085 (1968).
[104] R. K. Hill and G. R. Newkome, *J. Am. Chem. Soc.*, **91**, 5893 (1969).
[105] D. E. Schmidt, Jr., W. G. Nigh, C. Tanzer, and J. H. Richards, *J. Am. Chem. Soc.*, **91**, 5849 (1969).
[106] M. J. Brodie, K. J. Kripalani, and G. Possanza, *J. Am. Chem. Soc.*, **91**, 1241 (1969).
[107] P. E. Newallis, P. Lombardo, and E. R. McCarthy, *J. Org. Chem.*, **33**, 4168 (1968).
[108] R. M. Moriarty and T. Adams, *Tetrahedron Letters*, **1969**, 3715.
[109] C. Largeau, A. Casadevall, and E. Casadevall, *Bull. Soc. Chim. France*, **1969**, 2734.
[110] J. C. Jallageas, A. Casadevall, and E. Casadevall, *Bull. Soc. Chim. France*, **1969**, 4047.
[111] J. C. Craig, M. D. Bergenthal, I. Fleming, and J. Harley-Mason, *Angew. Chem. Intern. Ed. Engl.*, **8**, 429 (1969).
[112] B. M. Trost and S. Ziman, *Chem. Comm.*, **1969**, 181.

The mechanisms of gas-phase elimination reactions of alkyl halides have been reviewed.[113]

Gas-phase pyrolysis of the following compounds has been investigated: 3,4-dihydro-2*H*-pyran,[114] triisobutylaluminium,[115] tetrahydropyran ethers,[116] 1-arylbut-3-enyl acetates,[117] neophyl (2-methyl-2-phenylpropyl) esters,[118] allylic formates,[119] α,α- and γ,γ-dimethylallyl chlorides,[120] ethyl, isopropyl, and t-butyl isothiocyanates,[121] ethyl fluoride,[122, 123] t-alkyl chlorides,[124] 2,2-difluoromethyldifluorosilane,[125] 3-chlorobutan-2-one, 3-chloropentan-2-one, and 4-chloropentan-2-one,[126] aliphatic alcohols,[127] and cyclopentyl and cyclohexyl acetates.[128]

Operational criteria for concerted bond-breaking in gas-phase molecular elimination reactions have been discussed.[129]

Rate constants and Arrhenius parameters for the pyrolysis of 1-arylethyl benzoates, 1-arylethyl acetates, and 1-arylethyl methyl carbonates were reported, and the rates correlated with σ^+ constants. Good correlations were also obtained for *ortho*-substituents, indicating the non-importance of proximity effects in these reactions.[130]

A number of investigations of heterogeneous dehydrations (refs. 131—139) and dehalogenations (refs. 140—142) have been reported.

[113] A. Maccoll, *Chem. Rev.*, **69**, 33 (1969).
[114] C. A. Wellington, *J. Chem. Soc.*(A), **1969**, 2584.
[115] K. W. Egger, *J. Am. Chem. Soc.*, **91**, 2867 (1969).
[116] J. Korrola and P. J. Malkonen, *Suomen Kemistilehti, B*, **42**, 282 (1969).
[117] M. Yoshida, H. Sugihara, S. Tsushima, and T. Miki, *Chem. Comm.*, **1969**, 1223.
[118] H. Kwart and H. G. Ling, *Chem. Comm.*, **1969**, 302.
[119] J. M. Vernon and D. J. Waddington, *Chem. Comm.*, **1969**, 623.
[120] C. J. Harding, A. Maccoll, and R. A. Ross, *J. Chem. Soc.*(B), **1969**, 634.
[121] N. Barroeta, A. Maccoll, and A. Fava, *J. Chem. Soc.*(B), **1969**, 347.
[122] M. Day and A. F. Trotman-Dickenson, *J. Chem. Soc.*(A), **1969**, 233.
[123] A. W. Kirk, A. F. Trotman-Dickenson, and B. L. Trus, *J. Chem. Soc.*(A), **1968**, 3058.
[124] A. Maccoll and S. C. Wong, *J. Chem. Soc.*(B), **1968**, 1492.
[125] D. Graham, R. N. Haszeldine, and P. J. Robinson, *J. Chem. Soc.*(B), **1969**, 652.
[126] M. Dakubu and A. Maccoll, *J. Chem. Soc.*(B), **1969**, 1248.
[127] J. L. Beauchamp, *J. Am. Chem. Soc.*, **91**, 5925 (1969).
[128] T. O. Bamkole and E. V. Emovon, *J. Chem. Soc.*(B), **1969**, 187.
[129] S. H. Bauer, *J. Am. Chem. Soc.*, **91**, 3688 (1969).
[130] G. G. Smith, K. K. Lum, J. A. Kirby, and J. Posposil, *J. Org. Chem.*, **34**, 2090 (1969).
[131] C. W. Spangler and N. Johnson, *J. Org. Chem.*, **34**, 1444 (1969).
[132] H. Knoezinger and H. Buchl, *Z. Phys. Chem.* (Frankfurt), **63**, 199 (1969).
[133] H. Knoezinger and H. Buchl, *Z. Naturforsch.*, **24b**, 290 (1969).
[134] K. V. Narayanan and C. N. Pillai, *Indian J. Chem.*, **7**, 409 (1969).
[135] H. Knoezinger and A. Scheglila, *Z. Phys. Chem.* (Frankfurt), **63**, 197 (1969).
[136] D. N. Misra, *J. Catal.*, **14**, 34 (1969).
[137] E. Essam, A. A. Balandin, and A. P. Rudenko, *Kinet. Katal,.* **9**, 1101 (1968).
[138] De Huy-Gao, A. Verdier, and A. Latles, *Bull. Soc. Chim. France*, **1969**, 2337.
[139] G. Descotes, J. C. Martin, and G. Labrit, *Bull. Soc. Chim. France*, **1969**, 4151.
[140] P. Andreu, J. Bellovin, G. Cunto, and H. Noller, *Z. Phys. Chem.* (Frankfurt), **64**, 71 (1969).
[141] A. Paulino, S. Zerain, and H. Noller, *Anales Quim.*, **65**, 141 (1969).
[142] J. Sitte, M. Hunger, E. Schmitz, and H. Noller, *Z. Naturforsch.*, **23b**, 1384 (1968).

CHAPTER 5

Addition Reactions

R. C. STORR

Department of Organic Chemistry, University of Liverpool

Electrophilic Additions

Halogen and Related Additions

A valuable and comprehensive account of the stereochemistry of electrophilic addition to all types of olefins and acetylenes has been presented.[1] Reviews of the mechanism of halogen addition to olefins,[2] and of electrophilic additions to strained olefins,[3] have also appeared, and the addition of electrophiles to olefins in HF as a route to a variety of substituted fluoro-compounds[4] has been discussed.

Further evidence has been obtained that fluoroxy-compounds react with olefinic bonds through electrophilic attack of the fluorine atom on the more nucleophilic terminus of the double bond to give an intimate ion-pair. This collapses as expected for a reactive carbonium ion, to exclusive *cis*-addition products.[5]

The mechanism of the anomalous reaction of chlorine with vinylidene chloride[6] and the vapour-phase addition of hypochlorous acid ($Cl_2 + H_2O$)

[1] R. C. Fahey, *Topics Stereochem.*, **3**, 237 (1968).

[2] H. Durand, *Bull. Union Physiciens*, **63**, 373 (1968); G. Heublein, *Z. Chem.*, **9**, 281 (1969).

[3] T. G. Traylor, *Accounts Chem. Res.*, **2**, 152 (1969).

[4] L. S. German and I. L. Knunyants, *Angew. Chem. Intern. Ed. Engl.*, 8, 349 (1969).

[5] D. H. R. Barton, L. J. Danks, A. K. Ganguly, R. H. Hesse, G. Tarzia, and M. M. Pechet, *Chem. Comm.*, **1969**, 227.

[6] A. I. Subbotin, G. A. Korchagina, and I. V. Bodrikov, *Zh. Org. Khim.*, **4**, 2078 (1968); *Chem. Abs.*, **70**, 67314 (1969).

to propene[7] have been studied. The reaction of hypochlorous acid with 2,3-dichloropropene in the presence of Ag^+ probably involves nucleophilic addition together with normal electrophilic addition.[8] The pseudohalogen 1-chlorobenzotriazole adds to olefins by initial electrophilic attack of the chlorine atom on the double bond.[9]

The addition of bromine to ring and side-chain substituted styrenes is very dependent on the presence of added salts, solvent polarity, and temperature. The addition is at best only *trans*-stereoselective. The observed behaviour indicates that solvent-separated as well as intimate ion-pairs are involved, and is rationalized in terms of initial formation of intimate pairs in which the cationic part is a weakly bridged, essentially benzilic, carbonium ion. However, some degree of bromine-bridging exists since the additions to *cis*- and *trans*-styrenes in no case give a common distribution of products. In contrast, addition to *cis*- and *trans*-butene is virtually completely *trans* even in very polar solvents, and unlike the addition to styrenes it involves no solvent incorporation. It is concluded that strongly bridged bromonium ions are involved in this case.[10, 11] Detailed kinetic studies with *m*- and *p*- and *α*- and *β*-substituted styrenes support the formulation of a weakly bridged cation. Separated rate constants for bromine and tribromide ion attack on the olefin were obtained. The data are consistent with competing attack of Br_2 and Br_3^- (rather than Br^- catalysed Br_2 attack) on the olefin leading to two distinct cationic type intermediates for each of which both solvent and bromide ion compete.[12]

Heublein has probed the effect of solvation of the cationic intermediates in bromine addition to olefinic compounds.[13, 14] Solvation of cationic intermediates from alkylolefins is stronger than for delocalized intermediates from arylolefins. This is reflected in a greater proportion of solvent incorporated in the former case.[14] These results appear to conflict with those above where solvent incorporation was greater for styrenes than for butenes.

Investigation of substituent effects on the rate of Br_2 addition to suitably chosen 1,1-diarylolefins indicates that aryl carbonium ion stability and the *σ*-value for a given aryl substituent are dependent on the amount of *p–π* overlap between the carbonium ion and the aryl ring.[15] No simple additivity

[7] M. Baerns and G. Sticken, *Tetrahedron Letters*, **1969**, 1479.
[8] D. L. H. Williams, *J. Chem. Soc.*(B), **1969**, 421.
[9] C. W. Rees and R. C. Storr, *J. Chem. Soc.*(C), **1969**, 1478.
[10] J. H. Rolston and K. Yates, *J. Am. Chem. Soc.*, **91**, 1469 (1969).
[11] J. H. Rolston and K. Yates, *J. Am. Chem. Soc.*, **91**, 1477 (1969); see also *Org. Reaction Mech.*, **1968**, 158, ref. 23.
[12] J. H. Rolston and K. Yates, *J. Am. Chem. Soc.*, **91**, 1483 (1969).
[13] G. Heublein and B. Rauscher, *Tetrahedron*, **25**, 3999 (1969).
[14] G. Heublein and I. Koch, *Z. Chem.*, **9**, 28 (1969); *Chem. Abs.*, **70**, 86772 (1969).
[15] J. E. Dubois and A. F. Hegarty, *J. Chem. Soc.*(B), **1969**, 638; M. Loizos, A. F. Hegarty, and J. E. Dubois, *Bull. Soc. Chim. France*, **1969**, 2747; see also *Org. Reaction Mech.*, **1968**, 158.

of substituent effects is observed in the addition of bromine to 1,1-diarylethyl-
enes. However, the effects of the two aryl rings are interdependent, and
the nature of this interdependence has been discussed.[16]

exo-3,4-Benzotricyclo[4.2.1.0^{2,5}]nona-3,7-diene undergoes normal *trans*-
addition of bromine in CH_2Cl_2, no products from *exo*-attack on a non-classical
ion being observed. However, the *endo*-isomer (1) gives mainly (2) and (3) by
a skeletal rearrangement possibly involving the non-classical ion (4). This
readily accounts for (3), and to explain the formation of dibromide (2) from
this non-classical ion a 4-membered cyclic bromonium ion (5), which can only
be attacked from the *endo*-side, was suggested.[17]

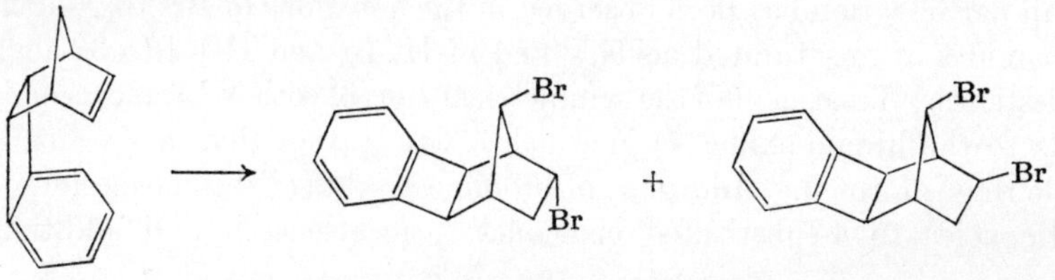

Addition of bromine to the highly sterically hindered adamantylidene-
adamantane in CCl_4 gives what is believed to be the first bromonium-ion
bromide which is sufficiently stable for isolation.[18] Studies of Br_2 addition to
several unsaturated acids, esters, and amides have been interpreted as
involving rate-determining attack of a Br_2 molecule on a Br_2–olefin π-complex. The structure of the π-complex was discussed in some detail.[19] Re-
investigation of the addition of bromine to butadiene indicates that free-radical
addition takes over from ionic addition at high butadiene concentration;[20] cf.

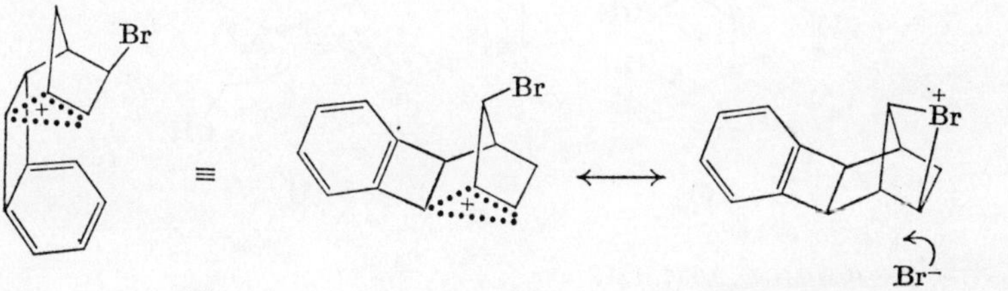

[16] E. D. Bergman, J. E. Dubois, and A. F. Hegarty, *Chem. Comm.*, **1968**, 1616.
[17] M. Avram, I. Pogany, F. Badea, I. G. Dinulescu, and C. D. Nenitzescu, *Tetrahedron Letters*, **1969**, 3851.
[18] J. Strating, J. H. Wieringa, and H. Wynberg, *Chem. Comm.*, **1969**, 907.
[19] R. Ganesan and S. Viswanathan, *Z. Phys. Chem.* (Frankfurt), **66**, 243 (1969); see also S. Viswanathan and R. Ganesan, *Current Sci.* (India), **37**, 102 (1968); *Chem. Abs.*, **69**, 2247 (1968).
[20] V. L. Heasley and S. K. Taylor, *J. Org. Chem.*, **34**, 2779 (1969).

chlorine addition.[21] Bromine addition to disubstituted alkyl terminal olefins in CH_2Cl_2 at $0°$ is complicated by the HBr produced in allylic substitution. This rapidly isomerizes the olefins to internal olefins.[22]

The kinetics of bromine additions to methyl acrylate, crotonate, and methacrylate,[23] enynic silanes,[24] *p*-substituted stilbenes,[25] and unsaturated sulphones,[26] and the influence of solvent on substituent effects in Br_2 additions to olefins[27] have been investigated. Additions of Br_2 to cyclohexene,[28] sodium and silver *cis*- and *trans*-stilbenecarboxylates,[29] substituted ethylenes,[30] and *cis*- and *trans*-butene and cyclohexene on activated carbon[31] have received attention. The use of ring-disc electrodes for study of fast irreversible reactions has been applied to the addition of bromine to allyl alcohol.[32] Neighbouring group participation has been observed in the additions of Br_2 to N-acylthio-carbamides of unsaturated acids,[33] and of HOBr and HOCl to 3β-acyloxy-cholestenes.[34] Treatment of the symmetrical diol (6) with N-bromosuccinimide leads to the bromo-oxide (7) in which a five- rather than a six-membered oxide ring is formed.[35] Additions of bromine acetate (N-bromosuccinimide in acetic acid) to 4-substituted chalcones,[36] acetylenes,[37] and additions to

(6)   (7)

[21] See *Org. Reaction Mech.*, **1967**, 131.
[22] J. Wolinsky, R. W. Novak, and K. L. Erickson, *J. Org. Chem.*, **34**, 490 (1969).
[23] D. Acharya and M. N. Das, *J. Org. Chem.*, **34**, 2828 (1969).
[24] N. N. Belyaev, M. D. Stadnichuk, and A. A. Petrov, *Zh. Obshch. Khim.*, **38**, 886 (1968); *Chem. Abs.*, **69**, 76166 (1969); M. D. Stadnichuk and N. N. Belyaev, *Zh. Obshch. Khim.*, **38**, 1658 (1968); *Chem. Abs.*, **69**, 95601 (1968).
[25] G. Heublein and P. Hallpap, *Z. Chem.*, **9**, 149 (1969); G. Heublein and E. Schuetz, *ibid.*, p. 147.
[26] A. Kaslová, M. Paleček, and M. Procházka, *Coll. Czech. Chem. Comm.*, **34**, 1826 (1969).
[27] E. Bienvenüe-Goetz and J. E. Dubois, *Tetrahedron*, **24**, 6777 (1968).
[28] G. B. Sergeev, T. V. Pokholok, and T. H. Ch'eng, *Kinet. Katal.*, **10**, 47 (1969); *Chem. Abs.*, **70**, 114342 (1969).
[29] C. C. Price and H. W. Blunt, *J. Org. Chem.*, **34**, 2484 (1969).
[30] J. Chrétien, M. Durand, and G. Mouvier, *Bull. Soc. Chim. France*, **1969**, 1966.
[31] S. H. Stoldt and A. Turk, *J. Org. Chem.*, **34**, 2370 (1969).
[32] W. J. Albery, M. L. Hitchman, and J. Ulstrup, *Trans. Faraday Soc.*, **65**, 1101 (1969).
[33] Y. V. Migalina, V. I. Staninets, and I. V. Smolanka, *Ukr. Khim. Zh.*, **35**, 526 (1969); *Chem. Abs.*, **71**, 49916 (1969).
[34] S. Julia and R. Lorne, *Compt. Rend., C*, **268**, 1617 (1969).
[35] N. L. Wendler, D. Taub, and C. H. Kuo, *J. Org. Chem.*, **34**, 1510 (1969).
[36] A. Jovcheff, S. L. Spassov, J. N. Stefanovsky, L. Stoilov, and V. Gocheva, *Monatsh.*, **100**, 51 (1969).
[37] A. Jovcheff and S. L. Spassov, *Monatsh.*, **100**, 328 (1969).

(−)-umbellulose,[38] enol ethers,[39] and 2-methoxy-5,6-dihydro-2*H*-pyrans[40] initiated by electrophilic bromine from *N*-bromo-amides have been reported. Bromination and chlorination of 2-pyrone involves addition–elimination.[41]

Because of the electronegativity trend $I < N_3 \sim Br < Cl$, the addition of bromine azide to olefins can be radical or ionic depending on the reaction conditions which can be easily controlled to give either mode exclusively. The additions of chlorine azide and iodine azide can likewise be either radical or ionic but, as expected, forcing conditions are required to make the former ionic (via Cl^+) and the latter radical (via N_3·).[42]

A 3-membered iodonium ion (8) has been suggested to explain the orientation of ionic addition of IN_3 to 1-phenylpropyne, which occurs in the opposite sense to hydration (equation 1).[43] However, since both *cis*- and *trans*-adducts of this orientation are observed, such an intermediate seems to us unnecessary.

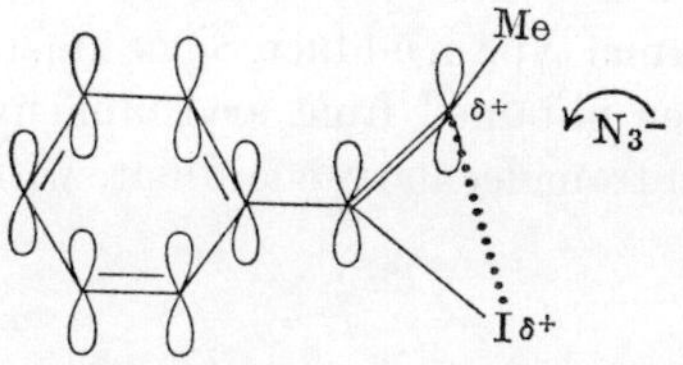

(8)

Greater inductive stabilization
by Me than Ph

$$PhC\equiv CMe \xrightarrow{\text{H}_2\text{O}} Ph\overset{\overset{\displaystyle O}{\|}}{C}CH_2Me \qquad \qquad \cdots (1)$$

The synthetic potential of iodine azide addition to olefins as a route to aziridines[44] has been discussed, and the interesting addition of IN_3 to 1-bromo-2-phenylacetylene[45] has received attention. Anomalies in the order of reactivity of a series of olefins towards electrophilic addition of INCO, depending on whether the reagent is pre-formed or generated *in situ*, are explained by I_2–olefin complexing which occurs during the slow formation of INCO from AgOCN and I_2.[46]

[38] H. E. Smith and R. T. Gray, *Chem. Comm.*, **1968**, 1695.
[39] K. Schrank and W. Pack, *Chem. Ber.*, **102**, 1892 (1969).
[40] M. J. Baldwin and R. K. Brown, *Can. J. Chem.*, **47**, 3098 (1969).
[41] W. H. Pirkle and M. Dines, *J. Org. Chem.*, **34**, 2239 (1969).
[42] A. Hassner and F. Boerwinkle, *Tetrahedron Letters*, **1969**, 3309.
[43] A. Hassner, R. J. Isbister, and A. Friederang, *Tetrahedron Letters*, **1969**, 2939.
[44] A. Hassner, G. J. Matthews, and F. W. Fowler, *J. Am. Chem. Soc.*, **91**, 5046 (1969).
[45] A. Hassner and R. J. Isbister, *J. Am. Chem. Soc.*, **91**, 6126 (1969); J. H. Boyer and R. Selvarajan, *ibid.*, p. 6122.
[46] C. G. Gebelein, S. Rosen, and D. Swern, *J. Org. Chem.*, **34**, 1677 (1969).

The stereochemistry of addition of ICl to trifluorocyclobutenes,[47] the kinetics of I_2 addition to allyl alcohol,[48] and neighbouring group participation in the addition of I_2 to a variety of unsaturated acids,[49] and their derivatives and unsaturated amines[50] have been reported.

Addition of Sulphenyl Halides

The mechanism of sulphenyl halide addition to olefins has been reviewed.[51] The electronic and steric effects of both C- and S-substituents on the reactions of episulphonium ions, which are widely assumed to be intermediates in sulphenyl halide additions, have been reviewed.[52] A cautionary note concerning the normally accepted mechanism for these additions has been raised following the observation that the stable episulphonium salt (9) undergoes attack by nucleophiles at S rather than C,[53] and the detection by NMR at $-5°$ of the intermediate (10) in its reaction with chloride ion.[54] On warming or treatment with excess of chloride ion this gives *trans*-1-chloro-2-methylthio-cyclooctane, the normal type of adduct. Since kinetic evidence for the accepted mechanism has been obtained from arylsulphenyl halides with an *o*-nitro group[55] which can intramolecularly coordinate with S (11), it is suggested that

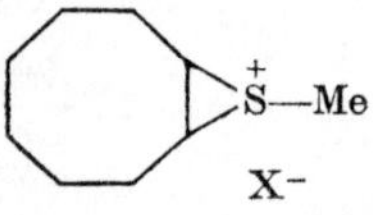

(9) 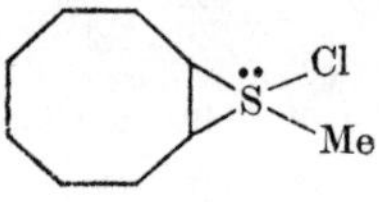   **(10)** 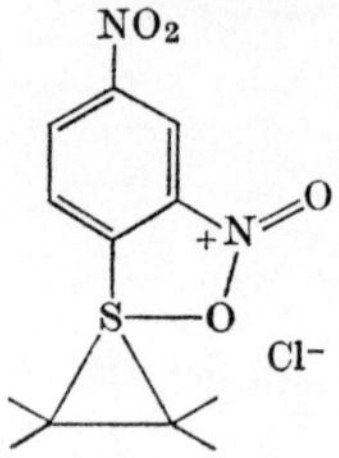   **(11)**

(X = 2,4,6-Trinitrobenzenesulphonate)

the mechanism be reconsidered. In particular, a study of the effect of solvent dielectric constant on the rate and rate laws for the addition to olefins of sulphenyl halides which do not have such *o*-nitro groups may differentiate between free and nucleophile coordinated episulphonium ions.

[47] J. D. Park, R. O. Michael, and R. A. Newmark, *J. Am. Chem. Soc.*, **91**, 5933 (1969).

[48] A. Y. Serguchev and E. A. Shilov, *Ukr. Khim. Zh.*, **34**, 1025 (1968); *Chem. Abs.*, **70**, 86762 (1969).

[49] V. I. Staninets, E. A. Shilov, and E. B. Koryak, *Dokl. Akad. Nauk SSSR*, **183**, 132 (1968); V. I. Staninets, E. A. Shilov, and I. V. Melika, *ibid.*, **187**, 109 (1969).

[50] V. I. Staninets and E. A. Shilov, *Ukr. Khim. Zh.*, **34**, 1132 (1968); *Chem. Abs.*, **70**, 67318 (1969).

[51] D. R. Hogg, *Quart. Rep. Sulfur Chem.*, **2**, 339 (1967).

[52] W. H. Mueller, *Angew. Chem. Intern. Ed. Engl.*, **8**, 482 (1969).

[53] D. C. Owsley, G. K. Helmkamp, and S. N. Spurlock, *J. Am. Chem. Soc.*, **91**, 3606 (1969).

[54] D. C. Owsley, G. K. Helmkamp, and M. F. Rettig, *J. Am. Chem. Soc.*, **91**, 5239 (1969).

[55] D. R. Hogg and N. Kharasch, *J. Am. Chem. Soc.*, **78**, 2728 (1956), and earlier papers.

Reaction of excess of *cis,cis*-cycloocta-1,5-diene with methanesulphenyl chloride surprisingly leads mainly to diadducts (**14**)—(**16**). Transannular overlapping of sulphur orbitals in the monoadduct (**12**) is proposed to explain the enhanced nucleophilicity of the second double bond. The episulphonium ion intermediate (**13**) in the second addition undergoes *trans*-opening by chloride ion to give (**15**) and (**16**) or intramolecular ring opening by the methyl-thio group to give the sulphonium chloride (**14**). In support of this mechanism 5-methylthiocyclooct-1-ene (**17**) is more reactive than cyclooctene and gives closely related bicyclic sulphonium chlorides. Addition of HCl and HI to (**17**) is also unusually rapid and involves transannular neighbouring group participation by the sulphur.[56a] (Recently transannular participation by

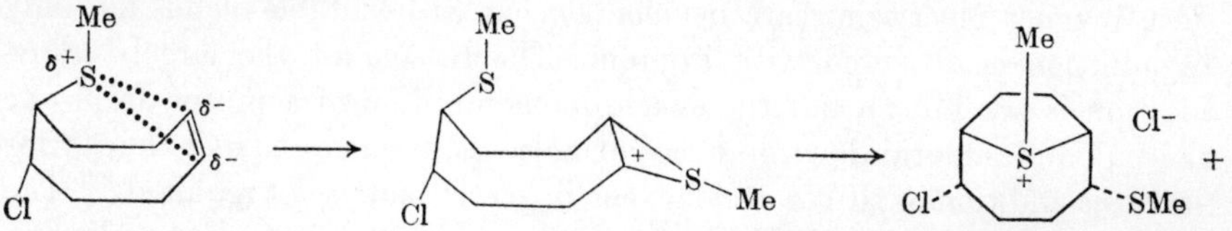

(**12**)   (**13**)   (**14**)   +

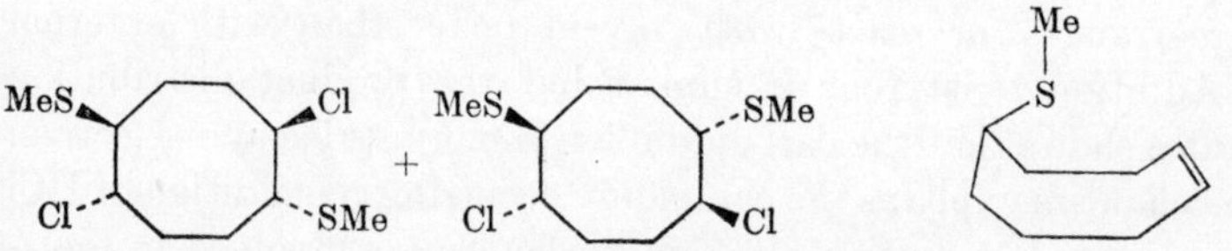

(**15**)   (**16**)   (**17**)

carbon was reported in the addition of methanesulphenyl chloride to *cis,trans*-cyclodeca-1,5-diene.[56b]) In contrast, 2,4-dinitrobenzenesulphenyl chloride, in which such neighbouring group participation is less likely, forms a normal 1,2-addition product with *cis,cis*-cycloocta-1,5-diene. The absence of bicyclic products indicates that an episulphonium ion, rather than an open carbonium ion, is involved.[57] The relative rates of episulphonium ion formation from methanesulphenyl chloride and a series of olefins are quite sensitive to steric factors but less sensitive to the nucleophilicity of the olefin than are other electrophilic additions, e.g. Cl_2 addition, as expected if the transition state for episulphonium ion formation involves positive charge mainly on sulphur.[58] Kinetic studies on the addition of 4-substituted 2-nitrobenzenesulphenyl bromides to cyclohexene in AcOH indicate that they, like the sulphenyl

[56a] W. H. Mueller, *J. Am. Chem. Soc.*, **91**, 1223 (1969).
[56b] J. G. Traynham, G. R. Franzen, G. A. Knesel, and D. J. Northington, *J. Org. Chem.*, **32**, 3285 (1967); see *Org. Reaction Mech.*, **1967**, 136.
[57] G. H. Schmid, *Can. J. Chem.*, **46**, 3757 (1968).
[58] W. A. Thaler, *J. Org. Chem.*, **34**, 871 (1969).

chlorides, add by rate-determining formation of an episulphonium ion.[59] The problems of mechanism and regio-specificity of sulphenyl chloride addition to substituted acetylenes have been further considered[60] and the stereospecifically *trans*-addition of diethoxyphosphorylsulphenyl chloride to cyclohexene[61] has been reported.

Addition of Hydrogen Halides

Addition of HCl to olefins in nitromethane involves irreversible rate-determining proton transfer to the olefin from an undissociated HCl molecule assisted by a second molecule of HCl which hydrogen-bonds with the developing chloride ion.[62] The addition involves tight carbonium ion–HCl_2^- ion-pairs since Wagner–Meerwein rearrangements occur with suitable olefins but only 1,2-addition is observed with isoprene. The reason for the largely *trans*-addition is not known but the synchronous addition of a proton from one molecule and a chloride ion from a second molecule, as suggested by Fahey[63a] for the *trans*-addition of HCl to cyclohexene in acetic acid, is not required.[63b] The addition of HCl to t-butylethylene in acetic acid involves attack of undissociated HCl to give a carbonium ion–chloride intimate ion-pair which collapses to products, rearranges, or reacts with solvent faster than with external nucleophiles. Addition to styrene is similar but the product distribution (chloride:acetate) shows that the carbonium ion is more stable and selective. This type of mechanism explains the commonly occurring *cis*-addition of HCl in ion-pairing solvents but quite a different mechanism is involved in *trans*-addition, e.g. to cyclohexene.[64]

Addition of HCl to 1-benzoylcyclohexene to give *cis*-1-benzoyl-2-chloro-cyclohexane is explained by 1,4-addition and re-ketonization of the most stable enol conformer by protonation from the sterically most favourable direction.[65] Association of alkyl halides with HCl at $-78°$ has been observed; this may complicate studies of HCl addition to alkenes at low temperature.[66] The products from ionic addition of DBr to benzonorbornadiene clearly show that a rearranged classical ion is not involved.[67] Calculations using Hückel

[59] C. Brown and D. R. Hogg, *J. Chem. Soc.*(B) **1969**, 1054.
[60] G. H. Schmid and M. Heinola, *Quart. Rep. Sulfur Chem.*, **2**, 311 (1967); *Chem. Abs.*, **69**, 76163 (1968); V. Caló, G. Scorrano, and G. Modena, *J. Org. Chem.*, **34**, 2020 (1969); L. Di Nunno and G. Scorrano, *Ricerca Sci.*, **38**, 343 (1968); *Chem. Abs.*, **69**, 95605 (1968).
[61] B. Bochivic and A. Kus, *Bull. Acad. Pol. Sci., Ser. Sci. Chim.*, **16**, 463 (1968).
[62] Y. Pocker, K. D. Stevens, and J. J. Champaux, *J. Am. Chem. Soc.*, **91**, 4199 (1969).
[63a] R. C. Fahey and M. R. Monahan, *Chem. Comm.*, **1967**, 936; R. C. Fahey and D. J. Lee, *J. Am. Chem. Soc.*, **89**, 2780 (1967).
[63b] Y. Pocker and K. D. Stevens, *J. Am. Chem. Soc.*, **91**, 4205 (1969).
[64] R. C. Fahey and C. A. McPherson, *J. Am. Chem. Soc.*, **91**, 3865 (1969).
[65] C. Armstrong, J. A. Blair, and J. Homer, *Chem. Comm.*, **1969**, 103.
[66] A. J. Fry, *Tetrahedron Letters*, **1968**, 5853.
[67] L. E. Barstow and G. A. Wiley, *Tetrahedron Letters*, **1968**, 6309.

theory charge densities and localization energies predict the observed direction of ionic addition of hydrogen halides to several dienes.[68]

Studies of the stereochemistry of ionic addition of HBr to *cis*- and *trans*-butene in acetic acid are complicated by rapid isomerization of the olefins by a radical mechanism. Where this is minimized the addition is largely *trans*-stereoselective.[69]

The addition of DCl to norbornene and nortricyclene has been studied in connection with the problem of the norbornyl cation structure in solution.[70] The additions of HCl to 1- and 2-chloronorbornenes[71] and ethynyl ketones,[72] of hydrogen halides to diacetylene and its homologues,[73] of HI to 1,1-difluoroethylene (gas phase),[74] of HBr to alkylidene cyclobutanes,[75] and the kinetics of the copper-catalysed addition of HCl to acetylene have also been studied.[76]

Hydration and Related Reactions

The kinetics of the hydration of fumaric acid to malic acid at pH 0—6 at 118° and 135° have been studied.[77] Although the results do not exclude the intermediate formation of β-malolactonic acid, as proposed by Bender and Connors,[78] this intermediate is considered unlikely in view of the non-stereospecific addition of ammonia to fumaric acid.[79] Fluorofumaric and difluorofumaric acids are hydrated by fumarate hydratase (F is very similar in size to H, therefore steric requirements are similar). The addition is Markovnikov, suggesting that considerable carbonium ion character is developed in this enzyme-catalysed reaction.[80] Kinetic isotope effects in the related fumarate hydratase dehydration of labelled malic acid also support a carbonium ion mechanism.[81]

The hydration of 1-phenylbuta-1,3-diene involves rate-determining protonation to give a carbonium ion.[82] The general-acid catalysis observed in the dehydration of 1-(2,4,6-trimethoxyphenyl)-2-phenylethanol in buffered

[68] M. D. Jordan and F. L. Pilar, *Theor. Chim. Acta*, **10**, 325 (1968); *Chem. Abs.*, **69**, 66792 (1968).
[69] D. J. Pasto, G. R. Meyer, and S. Z. Kang, *J. Am. Chem. Soc.*, **91**, 2163 (1969).
[70] J. M. Brown and M. C. McIvor, *Chem. Comm.*, **1969**, 238.
[71] A. J. Fry and W. B. Farnham, *J. Org. Chem.*, **34**, 2314 (1969).
[72] B. Cavalchi, D. Landini, and F. Montanari, *J. Chem. Soc.*(C), **1969**, 1204.
[73] Y. I. Porfir'eva, L. A. Vasil'eva, E. S. Turbanova, and A. A. Petrov, *Zh. Org. Khim.*, **5**, 591 (1969); *Chem. Abs.*, **71**, 21616 (1969).
[74] T. S. Carlton, A. B. Harker, W. K. Natale, R. E. Needham, R. L. Christensen, and J. L. Ellenson, *J. Am. Chem. Soc.*, **91**, 555 (1969).
[75] S. H. Graham and A. J. S. Williams, *J. Chem. Soc.*(C), **1969**, 391.
[76] K. A. Kurginyan, *Armyan. Khim. Zh.*, **21**, 128 (1968); *Chem. Abs.*, **69**, 66631 (1968).
[77] J. L. Bada and S. L. Miller, *J. Am. Chem. Soc.*, **91**, 3949 (1969).
[78] M. L. Bender and K. A. Connors, *J. Am. Chem. Soc.*, **83**, 4099 (1961), *ibid.*, **84**, 1980 (1962).
[79] J. L. Bada and S. L. Miller, *J. Am. Chem. Soc.*, **91**, 3946 (1969).
[80] W. G. Nigh and J. H. Richards, *J. Am. Chem. Soc.*, **91**, 5847 (1969).
[81] D. E. Schmidt, W. G. Nigh, C. Tanzer, and J. H. Richards, *J. Am. Chem. Soc.*, **91**, 5849 (1969).
[82] Y. Pocker and M. J. Hill, *J. Am. Chem. Soc.*, **91**, 3243 (1969).

60% dioxan indicates that hydration of *trans*-stilbenes proceeds by rate-determining proton transfer. It is considered that intermediate π-complex formation is most unlikely in the protonation of olefinic double bonds.[83] The effect of pressure on the acid-catalysed reversible hydration of acrylic acid has been studied at different concentrations of mineral acid. The data indicate that at least one molecule of water is present in the activated complex and rule out mechanisms involving rate-determining nucleophilic attack of H_2O on protonated acrylic acid. Subtleties of the mechanism are discussed in some detail.[84]

Propene and *p*-bromobenzenesulphonic acid in CF_3CO_2H give isopropyl *p*-bromobenzenesulphonate much faster than the latter solvolyses in CF_3CO_2H. This suggests that in the solvolysis tight ion-pairs (as are formed directly in the addition) return faster than they dissociate.[85] Catalysis of the addition of acetic acid to cyclohexene (in acetic acid) by added salts is paralleled by an increase in the acidity of the acetic acid. This is attributed to specific cationic solvation of acetate.[86]

The Rh(III)-catalysed hydration of acetylenes has also been studied.[87]

The stereoselectivity of the Prins reaction of *cis*- and *trans*-β-methylstyrenes (two-phase reaction) depends both on the acidity and on the concentration of formaldehyde. The maximum stereoselectivity (*trans*-addition) for *trans*-β-methylstyrenes occurs for *meta*-electron-withdrawing substituents. The results fit in with an intermediate which resembles a cyclic oxonium ion which leads to stereoselective *trans*-addition for electron-withdrawing substituents and an open carbonium ion which leads to non-stereospecific addition for electron-releasing substituents.[88] Terminal olefins with paraformaldehyde and HCl in substantially anhydrous media at low temperature give 3-alkyl-4-halogenotetrahydropyrans. The initial step is assumed to be the same as in the conventional Prins reaction to give the carbonium ion (**18**) which can undergo the usual Prins reaction or lose a proton, most likely by a cyclic process, to give the homoallylic alcohol (**19**) which reacts further.[89] In contrast, *cis*- and *trans*-butene give mainly acyclic chloro-alcohols which result from stereoselective *trans*-addition. The high stereoselectivity implies that these are not formed by HCl addition to the homoallylic alcohols since this should lead to loss of stereochemical integrity. The reason for the difference between terminal and internal olefins is not yet apparent.[90] The acid-catalysed addition of

[83] G. M. Loudon and D. S. Noyce, *J. Am. Chem. Soc.*, **91**, 1433 (1969).
[84] S. K. Bhattacharyya and C. K. Das, *J. Am. Chem. Soc.*, **91**, 6715 (1969).
[85] V. J. Shiner and W. Dowd, *J. Am. Chem. Soc.*, **91**, 6528 (1969).
[86] R. Corriu and J. Guenzet, *Tetrahedron Letters*, **1968**, 6083.
[87] B. R. James and G. L. Rempel, *J. Am. Chem. Soc.*, **91**, 863 (1969).
[88] C. Bocard, M. Davidson, M. Hellin, and F. Coussemant, *Tetrahedron Letters*, **1969**, 491.
[89] P. R. Stapp, *J. Org. Chem.*, **34**, 479 (1969).
[90] P. R. Stapp and D. S. Weinberg, *J. Org. Chem.*, **34**, 3592 (1969).

formaldehyde to isobutene[91] and the stereochemistry and mechanism of the Prins reaction of cyclohexene[92] have been studied.

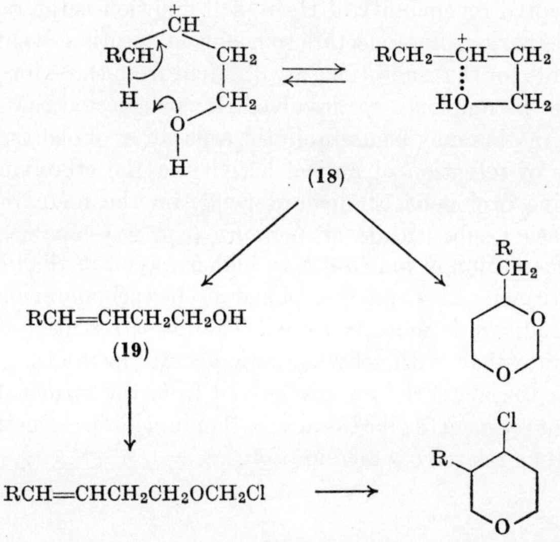

$$RCH \overset{\overset{+}{CH}}{\underset{\underset{\underset{H}{O}}{H}}{\cdots}} CH_2 \quad \rightleftarrows \quad RCH_2 - \overset{+}{\underset{\vdots}{CH}} - CH_2 \\ HO - CH_2$$

(18)

RCH=CHCH₂CH₂OH

$$RCH{=}CHCH_2CH_2OH$$

(19)

$$RCH{=}CHCH_2CH_2OCH_2Cl \longrightarrow$$

The complex problem of the effect of α-acetoxy substituents on hydroboration of olefins has been further studied by two groups. Both mono- and di-ols are formed and the mechanism of mono-ol formation has received particular attention.[93, 94] Hassner and coworkers, from the hydroboration of cholest-2-en-3-ol acetate, favour a mechanism not involving elimination–rehydroboration, at least for this olefin.[93] However, Japanese workers accept elimination–rehydroboration as a major, but not exclusive, route to mono-ols. As regards the directive effect in enol acetates, these workers conclude that addition of boron is mainly at the β-position and that the effect of steric hindrance is comparable to that found in normal olefins, the proportion of α-addition is increased over that in alkyl olefins, and β-addition decreases when two *cis*-alkyl groups are present.[94] Hydroboration of 1-chloronorbornene indicates that the directive effect of chlorine in hydroboration is not merely inductive but more complex.[71] Hydroboration of allylic chlorides has also been studied.[95]

[91] K. L. Friedlin, V. Z. Sharf, V. I. Kheifets, G. K. Oparina, and V. I. Borisova, *Izv. Akad. Nauk SSSR, Ser. Khim.*, **1968**, 514; *Chem. Abs.*, **69**, 66624 (1968).

[92] N. P. Volynskii and A. B. Urin, *Zh. Org. Khim.*, **5**, 344 (1969); *Chem. Abs.*, **70**, 115082 (1969).

[93] A. Hassner, R. E. Barnett, P. Catsoulacos, and S. H. Willen, *J. Am. Chem. Soc.*, **91**, 2632 (1969).

[94] A. Suzuki, K. Ohmori, and M. Itoh, *Tetrahedron*, **25**, 3707 (1969).

[95] H. C. Brown and E. F. Knights, *Israel J. Chem.*, **6**, 691 (1969).

The downfield chemical shift for methoxy group protons when near to a mercury function has been utilized for easy configurational assignment in olefin methoxymercuration products.[96] The first direct demonstration of the generally accepted reversibility of Hg(II) salt addition is the rapid reversible addition of mercuric trifluoroacetate to olefins in aprotic solvents. The equilibrium constants for these additions are different from those for AgNO$_3$–olefin systems where π-complexes are involved. It is suggested that this addition compound formation may be useful in the separation of olefins.[97]

The degree of retention of optical activity in the ethoxymercuration of optically active cyclonona-1,2-diene depends on the mercuric salt used.[98] Electron-releasing substituents on mercury (e.g. Et) capable of stabilizing the bridged mercurinium ion (20) give higher retention than electron-withdrawing substituents where an open planar allylic carbonium ion (21), leading to racemic products, is more favoured. The lower retention observed with mercuric acetate than with ethylmercuric acetate in the case of this allene indicates that the mercurinium ion derived from the former does not completely prevent rotation to give the allyl carbonium ion as suggested by Waters and Kiefer[99] in a related oxymercuration.

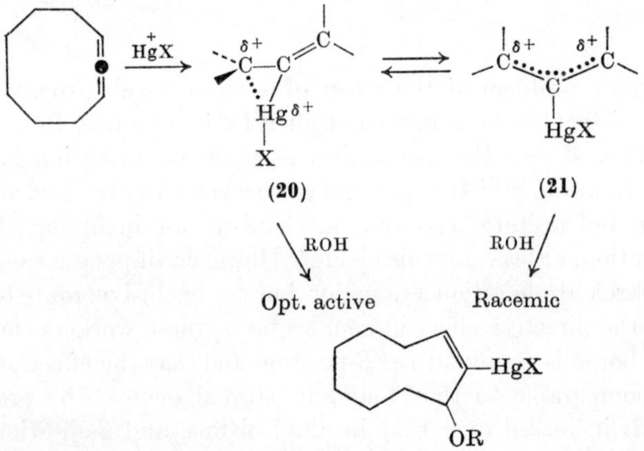

Oxymercuration of methylenecyclopropanes occurs with ring-opening.[100] An n–π complexed mercurinium ion (23) rather than a σ-bridged complex is proposed as an intermediate in the reactions of the bromoallene alcohol (22) with mercuric acetate (e.g. Scheme 1). In the example shown, the epoxide (24)

[96] W. L. Waters, *Tetrahedron Letters*, **1969**, 3769.
[97] H. C. Brown and M. H. Rei, *Chem. Comm.*, **1969**, 1296.
[98] R. D. Bach, *J. Am. Chem. Soc.*, **91**, 1771 (1969).
[99] W. L. Waters and E. F. Kiefer, *J. Am. Chem. Soc.*, **89**, 6261 (1967); see *Org. Reaction Mech.*, **1967**, 141.
[100] R. M. Babb and P. D. Gardner, *Chem. Comm.*, **1968**, 1678.

can be isolated when the reaction is done in acetic acid at room temperature.[101] The stereochemistries of methoxymercuration of acenaphthylene[102] and optically active *trans*-cyclooctene,[103] the solvent-dependent stereoselectivity of the oxymercuration of cyclohex-2-enol,[104] and the influence of silver acetate on the oxymercuration of olefins[105] have received attention. Differences previously reported in the oxymercuration of cyclohex-2-enol have now been explained as due to different work-up procedures.[106]

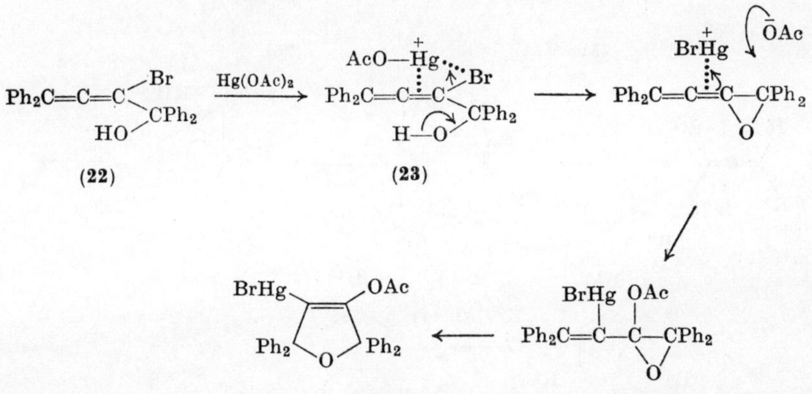

Scheme 1

Aminomercuration in aqueous solution is accompanied by oxymercuration; the aminomercuration products do not arise directly but via oxymercuration.[107] Hydrogen peroxide has been used as a nucleophile in mercuration.[108] Conjugate mercuration of molecules containing cyclopropane rings in place of double bonds has been studied.[109]

Allene oxide (**25**) and dioxaspiropentane (**26**) have been postulated as intermediates in the peracetic acid oxidation of tetramethylallene.[110] Both of these types of intermediate have now been isolated. A dioxaspiropentane derivative was obtained in the buffered peracetic acid oxidation of 2,5,5-trimethylhexa-2,3-diene,[111] and after some difficulty[112] an allene oxide was

[101] F. Toda and K. Akagi, *Tetrahedron*, **25**, 3795 (1969).

[102] V. I. Sokolov, L. L. Troitskaya, and O. A. Reutov, *Zh. Org. Khim.*, **5**, 174 (1969); *Chem. Abs.*, **70**, 86856 (1969).

[103] V. I. Sokolov, L. L. Troitskaya, and O. A. Reutov, *J. Organometal. Chem.*, **17**, 323 (1969).

[104] M. R. Johnson and B. Rickborn, *J. Org. Chem.*, **34**, 2781 (1969).

[105] S. Moon and B. H. Waxman, *J. Org. Chem.*, **34**, 1157 (1969).

[106] S. Moon, C. Ganz, and B. H. Waxman, *Chem. Comm.*, **1969**, 866.

[107] J. J. Perie and A. Lattes, *Tetrahedron Letters*, **1969**, 2289.

[108] V. I. Sokolov and O. A. Reutov, *Zh. Org. Khim.*, **5**, 174 (1969).

[109] V. I. Sokolov, N. B. Rodina, and O. A. Reutov, *J. Organometal. Chem.*, **17**, 477 (1969).

[110] J. K. Crandall and W. H. Machleder, *J. Am. Chem. Soc.*, **90**, 7292 (1968).

[111] J. K. Crandall, W. H. Machleder, and M. J. Thomas, *J. Am. Chem. Soc.*, **90**, 7346 (1968).

[112] J. K. Crandall and W. H. Machleder, *J. Am. Chem. Soc.*, **90**, 7347 (1968).

obtained from 1,3-di-t-butylallene with *m*-chloroperbenzoic acid in hexane.[113] The formation of cyclopentenones by peroxidation of vinylallenes is assumed to involve allene oxides which tautomerize to vinylcyclopropanones which then undergo intramolecular cycloaddition[114] (equation 2) (see also p. 208).

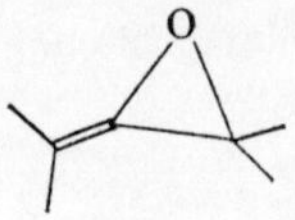

(25) 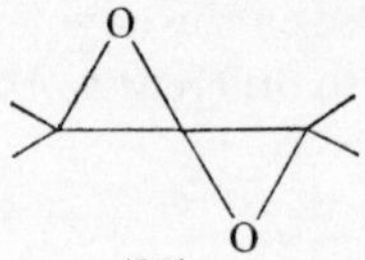   (26)

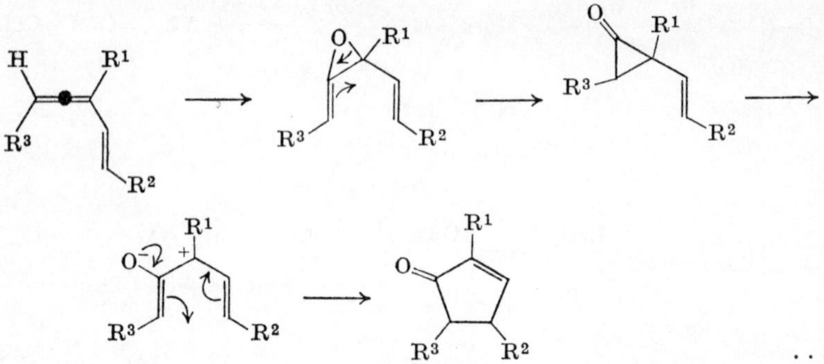

... (2)

MO calculations indicate that the intramolecularly H-bonded and the dipolar forms of peroxy-acids have similar energies. This, together with the π-electron distribution, suggests that both 1,1- and 1,3-addition mechanisms for epoxide formation are theoretically possible, the first being more probable in non-polar and the second in polar media.[115]

A 2-oxabicyclo[1.1.0]butane has been suggested as the most likely intermediate in the peroxidation of cyclopropenes.[116] Asymmetric induction has been observed in the epoxidation of alkenes with optically active peracids.[117] The faster epoxidation of lumisterol than its acetate and 3,5-dinitrobenzoate is thought to be due to 3β-OH–peracid bonding in the transition state.[118] Peroxidations of vinyltrialkylsilanes,[119] unsaturated bicyclic anhydrides,[120] 3,3,6,6-tetramethylcyclohexa-1,4-diene,[121] 3-t-butylcyclohexene,[122] 1,4-di-

[113] R. L. Camp and F. D. Greene, *J. Am. Chem. Soc.*, **90**, 7349 (1968).
[114] J. Grimaldi and M. Bertrand, *Tetrahedron Letters*, **1969**, 3269.
[115] A. Ažman, B. Borštnik, and B. Plesničar, *J. Org. Chem.*, **34**, 971 (1969); see *Org. Reaction Mech.*, **1966**, 136 (equations 5 and 6); **1967**, 142.
[116] J. Ciabattoni and P. J. Kocienski, *J. Am. Chem. Soc.*, **91**, 6534 (1969).
[117] F. Montanari, I. Moretti, and G. Torre, *Chem. Comm.*, **1969**, 135.
[118] K. D. Bingham, G. D. Meakins, and J. Wicha, *J. Chem. Soc.*(C), **1969**, 510.
[119] H. Sakurai, N. Hayashi, and M. Kumadu, *J. Organometal. Chem.*, **18**, 351 (1969).
[120] L. H. Zalkow and S. K. Gabriel, *J. Org. Chem.*, **34**, 218 (1969).
[121] R. W. Gleason and J. T. Snow, *J. Org. Chem.*, **34**, 1963 (1969).
[122] J.-C. Richer and C. Freppel, *Can. J. Chem.*, **46**, 3709 (1969).

bromo-2-methylbut-2-ene,[123] 9-benzylidenefluorenes,[124a] and methyl *exo*-2-methyl-*endo*-bicyclo[2.2.1]hept-5-ene-2-carboxylate[124b] have also been studied.

Various alkyl peroxides in the presence of cyclohexyl metaborate are effective electrophilic epoxidizing agents. A possible structure for the intermediate responsible for peroxidation is (27).[125]

Of wide significance is a caution[126a] concerning the use of relative reactivities of cyclic olefins to indicate whether reactions involve three-membered or larger cyclic transition states.[126b] This follows observations that ozone and osmium tetroxide, which are considered to react to give initially five-membered ring products with olefins, show very much less selectivity than usual. It is pointed out that theoretical calculations correlating these relative rates with strain effects in cyclic olefins depend not only on the ring size of the transition state but also on the position of the transition state on the reaction coordinate (i.e. resemblance to products or reactants.)

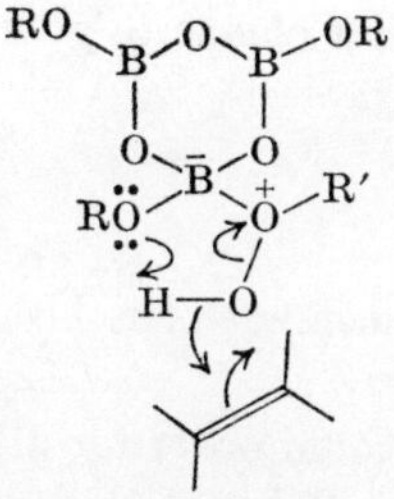

(27) 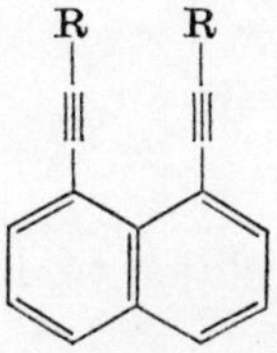 (28) 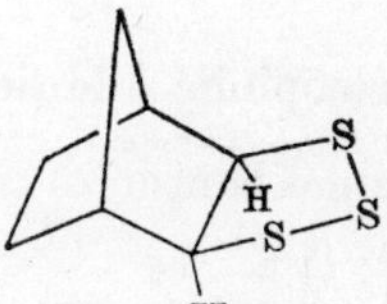 (29)

Miscellaneous Electrophilic and Other Additions

Heterolytic addition of thiocyanogen to alkylolefins proceeds through a cyclic cyanoepisulphonium ion to give *trans*-dithiocyanates and thiocyanato-isothiocyanates. With arylolefins non-stereospecific Markovnikov formation of thiocyanato-isothiocyanates indicates than an open carbonium ion is involved.[127]

Additions of ISCN and 2,4-dinitrobenzenesulphenyl chloride to olefins are

[123] M. S. Malinovskii, V. G. Dryuk, and S. P. Shamrovskaya, *Zh. Org. Khim.*, **4**, 1887 (1968); *Chem. Abs.*, **70**, 28163 (1969).

[124a] B. Muckensturm, *Tetrahedron Letters*, **1968**, 6139.

[124b] P. I. Valov, E. A. Blyumberg, S. L. Knyazhanskii, and L. I. Kas'yan, *Dokl. Akad. Nauk SSSR*, **186**, 120 (1969); *Chem. Abs.*, **71**, 48913 (1969).

[125] P. F. Wolf and R. K. Barnes, *J. Org. Chem.*, **34**, 3441 (1969).

[126a] R. E. Erickson and R. L. Clark, *Tetrahedron Letters*, **1969**, 3997.

[126b] See *Org. Reaction Mech.*, **1966**, 136.

[127] R. G. Guy, R. Bonnett, and D. Lanigan, *Chem. Ind.* (London), **1969**, 1702.

electrophilic but addition of NOCl involves a cyclic four-centre mechanism.[128] Various electrophilic and radical additions to fluorinated enamines and ynamines,[129] and the addition of dinitrogen trioxide to olefins to give nitroso-nitrites,[130] have been studied. Some ionic addition accompanies the radical addition of *N,N*-dichloroethanesulphonamide to terminal olefins in the absence of strong light.[131] Addition of iodobenzene dichloride to olefins is normally radical but an ionic mechanism can intervene in the presence of strong acids.[132]

The following topics have also received attention; electrophilic addition (Br_2, HBr, and HI) to the bisacetylenes (**28**) (R = Me, Ph);[133] electrophilic and nucleophilic additions to *o*-di(phenylethynyl)benzene;[134] addition of isothiocyanic acid to norbornenes;[135] rhodium carbonyl catalysed hydro-formylation of olefins;[136] the kinetics of the reaction of triethylaluminium dimer with olefins;[137] and the Pd(0)-catalysed homogeneous addition of HCN to non-activated olefins.[138] Two reaction schemes have been proposed for $HCo(CO)_4$-catalysed hydroformylation of olefins.[139]

Elemental sulphur, activated by ammonia in the presence of suitable amides, adds stereospecifically to norbornene and similar olefins to give the *cis-exo*-adducts, e.g. (**29**).[140]

Nucleophilic Additions

The mechanism of nucleophilic addition to diynes and enynes has been reviewed.[141]

Aggregation and ion-pair formation leads to a fractional order for alkyl-, vinyl-, and phenyl-lithiums in their additions to 1,1-diphenylethylene in THF, but not for benzyl- and allyl-lithiums which form completely dissociated

128 G. Collin, U. Jahnke, G. Just, G. Lorenz, W. Pritzkow, M. Röllig, L. Winguth, P. Dietrich, C.-E. Döring, H. G. Hauthal, and A. Wiedenhöft, *J. Prakt. Chem.* **311**, 238 (1969); *Chem. Abs.*, **71**, 2717 (1969).
129 J. Freear and A. E. Tipping, *J. Chem. Soc.*(C), **1969**, 1955, 1848, 1963.
130 J. R. Park and D. L. H. Williams, *Chem. Comm.*, **1969**, 332.
131 T. Ohashi, M. Sugie, M. Okahara, and S. Komori, *Tetrahedron*, **25**, 5349 (1969).
132 S. Masson and A. Thuilliere, *Compt. Rend.*, *C*, **266**, 987 (1968).
133 B. Bossenbroek, D. C. Sanders, H. M. Curry, and H. Shechter, *J. Am. Chem. Soc.*, **91**, 371 (1969).
134 H. W. Whitlock, P. E. Sandvick, L. E. Overman, and P. B. Reichardt, *J. Org. Chem.*, **34**, 879 (1969).
135 W. R. Diveley, G. A. Buntin, and A. D. Lohr, *J. Org. Chem.*, **34**, 616 (1969).
136 B. Heil and L. Markó, *Chem. Ber.*, **102**, 2238 (1969).
137 J. N. Hay, P. G. Hooper, and J. C. Robb, *Trans. Faraday Soc.*, **65**, 1365 (1969).
138 E. S. Brown and E. A. Rick, *Chem. Comm.*, **1969**, 112.
139 G. P. Vysokinskii, V. Y. Gankin, and D. M. Rudkovskii, *Karbonilirovanie Nenasyshchennykh Vglvodorodov*, **1968**, 17; *Chem. Abs.*, **71**, 21418 (1969).
140 T. C. Shields and A. N. Kurtz, *J. Am. Chem. Soc.*, **91**, 5415 (1969).
141 E. N. Prilezhaeva, V. N. Petrov, G. S. Vasil'ev, N. S. Nikol'skii, and M. F. Shostakovskii, *Khim. Atsetilena*, **1968**, 113; *Chem. Abs.*, **71**, 2668 (1969).

resonance-stabilized carbanions. This behaviour leads to a concentration-dependent reactivity order, but extrapolation to infinite dilution gives alkyl $> sp^2$-hybridized $>$ resonance-stabilized.[142] Previous work[143] which gave Ph $<$ Bu^n $<$ CH$_2$Ph was carried out at such concentrations that the relative reactivities of n-butyl- and benzyl-lithiums would be inverted. Similar fractional orders for organolithium additions have been observed elsewhere.[144]

Two groups[145, 146] have studied the ready addition of alkyl-lithium reagents to allylic alcohols which proceeds as shown in equation (3). The additions to methallyl alcohol in the presence of N,N,N,N-tetramethylethylenediamine are stereoselective but in the opposite sense to the additions of reactive Grignard reagents. This can be rationalized[145] by a consideration of the different types of coordination which occur in the transition states. 3-Substituted allylic alcohols which cannot form primary organolithium reagents do not react according to equation (3) but react with double-bond migration (Scheme 2 for cyclopent-2-enol) either via a cyclic process or an addition–elimination involving the opposite orientation of addition.[146] The addition of alkyl-lithiums to ethylene requires a catalyst which can solvate Li$^+$, and

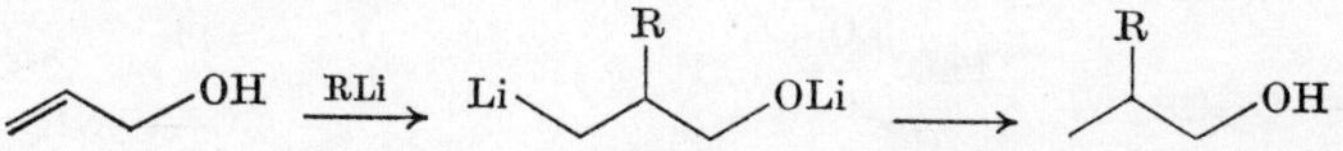

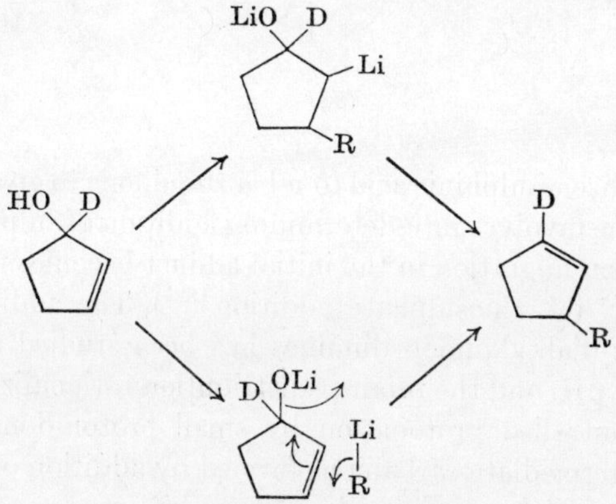

Scheme 2

[142] R. Waack and M. A. Doran, *J. Am. Chem. Soc.*, **91**, 2456 (1969).

[143] E. Grovenstein and G. Wentworth, *J. Am. Chem. Soc.*, **89**, 1852 (1967); *Org. Reaction Mech.*, **1967**, 238.

[144] J. G. Carpenter, A. G. Evans, C. R. Gore, and N. H. Rees, *J. Chem. Soc.*(B), **1969**, 908.

[145] H. Felkin, G. Swierczewski, and A. Tambuté, *Tetrahedron Letters*, **1969**, 707.

[146] J. K. Crandall and A. C. Clark, *Tetrahedron Letters*, **1969**, 325.

only those reagents which produce the "least stable carbanions" will react.[147] Preliminary investigations of isopropyl-lithium addition to unconjugated alkenyl ethers have been reported.[148]

Sodium methoxide adds to allyl p-tolyl sulphoxide via its conjugated propenyl isomer. However, piperidine reacts with the sulphoxide mainly by nucleophilic attack at sulphur on the allyl toluene-p-sulphenamide isomer which must also be in equilibrium with the sulphoxide. Such an equilibrium is in line with the ease of racemization of allyl sulphoxides.[149] Carboxylate anions add to dimethylprop-2-ynylsulphonium bromide through the isomeric allenic salt to give the kinetically controlled non-conjugated adduct (**30**), but more basic nucleophiles give the thermodynamically controlled conjugated products.[150]

Addition of pyrrolidine to *trans*-chalcone in aprotic solvents is second-order in amine and is considered to be a concerted (or nearly so) reaction through a cyclic transition state (**31**). Addition to *cis*-chalcone is first-order in pyrrolidine and involves initial isomerization to *trans*-chalcone through an ionic intermediate (**32**) presumably since a concerted termolecular reaction would not as readily relieve the strain due to the two bulky *cis*-groups.[151]

(**30**)   (**31**)   (**32**)

Addition of benzenesulphinic acid to p-benzoquinone to give 2,5-dihydroxy-diphenyl sulphone involves rate-determining addition of sulphinate ion below pH 3.1, but proton migration in the initial adduct becomes rate-determining at higher pH[152a] (cf. thiosulphate addition[152b]). The addition of aryl sulphinates to N,N-dialkylquinonediimines has been studied as a function of temperature and pH, and the product distribution rationalized.[153]

Kinetically controlled protonation by small proton-donors (e.g. EtOH, AcOH) of the intermediate carbanions formed by addition of nucleophiles to 4-t-butyl-1-cyanocyclohexene is from the least hindered equatorial side when

147 P. D. Bartlett, S. J. Tauber, and W. P. Weber, *J. Am. Chem. Soc.*, **91**, 6362 (1969).

148 A. H. Veefkind, F. Bickelhaupt, and G. W. Klump, *Rec. Trav. Chim.*, **88**, 1058 (1969).

149 D. J. Abbott and C. J. M. Stirling, *J. Chem. Soc.*(C), **1969**, 818.

150 G. D. Appleyard and C. J. M. Stirling, *J. Chem. Soc.*(C), **1969**, 1904.

151 F. M. Menger and J. H. Smith, *J. Am. Chem. Soc.*, **91**, 4211 (1969).

152a Y. Ogata, Y. Sawaki, and M. Isono, *Tetrahedron*, **25**, 2715 (1969).

152b See *Org. Reaction Mech.*, **1968**, 173.

153 K. T. Finley, R. S. Kaiser, R. L. Reeves, and G. Werimont, *J. Org. Chem.*, **34**, 2083 (1969).

the attached nucleophile is equatorial but is from the least hindered axial side when the attached nucleophile is axial.[154]

Introduction of a C-12 carbonyl group, which would be expected to slow down nucleophilic addition to Δ^{16}-20-keto-steroids by sterically disrupting the planarity of the enone system, in fact accelerates such additions. The surprisingly fast addition of methanol for example is explained by hemiketal formation at C-12 followed by intramolecular catalysis as shown (**33**).[155] In the reversible addition of methoxide ion to cycloalk-2-enones the variations of equilibrium constant and rate constant with ring size indicate that conjugation requirements are greater in the transition state, which possesses a high degree of reactant character as regards overlap requirements, than in reactants.[156]

1,3-Oxathioles (**34**) are formed as isolable intermediates in the addition of methanol to 1-acetylthioalk-1-ynes. A similar intermediate is thought to be involved in hydration which occurs in the opposite direction to the hydration of 1-alkylthioalk-1-ynes[157] (equations 4 and 5).

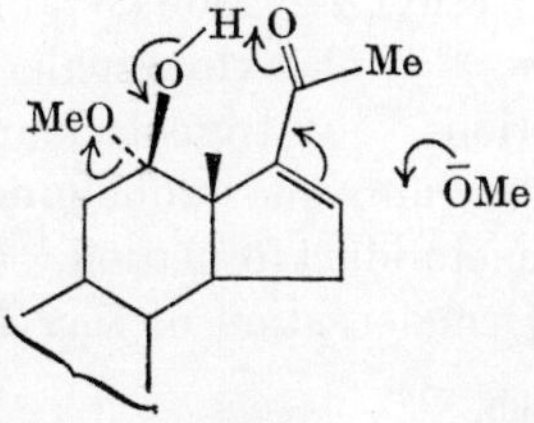

(**33**) 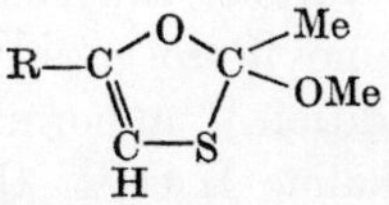   (**34**) 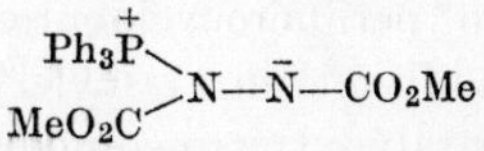   (**35**)

$$RC{\equiv}C-S-\overset{\overset{\displaystyle O}{\|}}{C}-R' \xrightarrow{H_2O} R\overset{\overset{\displaystyle O}{\|}}{C}-CH_2-S-\overset{\overset{\displaystyle O}{\|}}{C}-R' \quad \dots (4)$$

$$RC{\equiv}C-S-R' \xrightarrow{H_2O} RCH_2-\overset{\overset{\displaystyle O}{\|}}{C}-S-R' \quad \dots (5)$$

Exclusive *cis*-addition of phenylacylamines to dimethyl acetylenedicarboxylate even in proton-donating solvents indicates that the zwitterionic intermediates collapse by intramolecular proton transfer (or cyclization to pyrrole derivatives also occurs).[158] *trans*-Addition of dimethylarsine to hexafluorobut-2-yne involves rate-determining nucleophilic attack with

[154] R. A. Abramovitch, M. M. Rogié, S. S. Singer, and N. Venkateswaran, *J. Am. Chem. Soc.*, **91**, 1572 (1969); see also *Org. Reaction Mech.*, **1968**, 170.
[155] G. S. Abernethy and M. E. Wall, *J. Org. Chem.*, **34**, 1606 (1969).
[156] P. Chamberlain and G. H. Whitham, *J. Chem. Soc.*(B), **1969**, 1131.
[157] W. Drenth and G. H. E. Nieuwdorp, *Rec. Trav. Chim.*, **88**, 307 (1969).
[158] S. K. Khetan and M. V. George, *Tetrahedron*, **25**, 527 (1969).

7

subsequent intermolecular proton transfer.[159] Acetone dimethylhydrazone reacts with dimethyl acetylenedicarboxylate to give adducts resulting from initial nucleophilic attack by imino and amino N but not by carbon, the third potentially nucleophilic centre.[160] The kinetics of the base-catalysed addition of ethanol to *p*-substituted acrylanilides indicate that the amide link is an efficient transmitter of electronic activation effects.[161] The hitherto rarely observed addition of carbanions to azobenzene is general in liquid NH_3.[162] Complexing of the *trans*-cinnamylsalicylic acid anion completely prevents addition of sulphite ion presumably because the ligand is close to the reactive region. The effect of complex formation on other reactions was also studied.[163a] The labile adduct from triphenylphosphine addition to dimethyl azodicarboxylate is now believed to have structure (35).[163b]

Other nucleophilic additions which have been studied are those of sulphur nucleophiles to *N*-(β-indol-3-ylacryloyl)imidazole,[163c] HCl, HI, and HCNS to activated acetylenes,[164] alkanethiols and pyrrole to 3,4-methylenedioxy-β-nitrostyrene,[165] piperidine to vinylsilanes,[166] hydrazines to dimethyl acetylenedicarboxylate,[167] alkoxide ion to diacetylenes,[168] ethyl α-cyanobutyrate to the benzene and ferrocenyl analogues of chalcones,[169] triphenylphosphine to perfluorocyclobutene,[170] various Michael additions,[171] intramolecular addition in 1-acyl-1′-cinnamoylferrocenes,[172] potassium and rubidium catalysed nitro-compound addition,[173] abnormal Michael adduct formation,[174] steroid epoxidation with alkaline H_2O_2[175] the self condensation of styryl

[159] W. R. Cullen and W. R. Leeder, *Can. J. Chem.*, **47**, 2137 (1969).
[160] S. F. Nelson, *J. Org. Chem.*, **34**, 2248 (1969).
[161] H. W. Johnson, E. Ngo, and V. A. Pena, *J. Org. Chem.*, **34**, 3271 (1969).
[162] E. M. Kaiser and G. J. Bartling, *Tetrahedron Letters*, **1969**, 4357.
[163a] P. A. Kramer and K. A. Connors, *J. Am. Chem. Soc.*, **91**, 2600 (1969).
[163b] E. Brunn and R. Huisgen, *Angew. Chem. Intern. Ed. Engl.*, **8**, 513 (1969).
[163c] M. F. Dunn and S. A. Bernhard, *J. Am. Chem. Soc.*, **91**, 3274 (1969).
[164] G. F. Dvorko and T. P. Travchuk, *Dopov. Akad. Nauk Ukr. RSR, Ser. B.*, **30**, 1105 (1968); *Chem. Abs.*, **70**, 56903 (1969); G. F. Dvorko and T. P. Travchuk, *Reakts. Sposobnost. Org. Soed.*, **5**, 995 (1968); *Chem. Abs.*, **71**, 21409 (1969).
[165] O. H. Park and T. R. Kim, *Daehan Hwahak Hwoejee*, **12**, 177 (1968); *Chem. Abs.*, **71**, 38053 (1969); T. R. Kim and Y. S. Sang, *ibid.*, p. 170; *Chem. Abs.*, **71**, 38052 (1969); O. H. Park, *ibid.*, p. 106; *Chem. Abs.*, **70**, 36791 (1969).
[166] N. S. Nametkin, V. N. Perchenko, I. A. Grushevenko, and G. L. Kamneva, *Izv. Akad. Nauk SSSR, Ser. Khim.*, **1968**, 2074; *Chem. Abs.*, **70**, 37875 (1969).
[167] N. D. Heindel, P. D. Kennewell, and M. Pfau, *Chem. Comm.*, **1969**, 757.
[168] B. P. Gusev, E. A. El'perina, and V. F. Kucherov, *Khim. Atsetilena*, **1968**, 105; *Chem. Abs.*, **71**, 2892 (1969).
[169] Š. Toma, *Coll. Czech. Chem. Comm.*, **34**, 2771 (1969).
[170] R. F. Stockel, F. Megson, and M. T. Beachem. *J. Org. Chem.*, **33**, 4395 (1968).
[171] E. Winterfeldt, I. Hinz, and P. Strehlke, *Chem. Ber.*, **102**, 310 (1969).
[172] P. Elecko, *Chem. Zvesti*, **23**, 212 (1969); *Chem. Abs.*, **71**, 21594 (1969).
[173] A. Ostaszyński, J. Wielgat, and T. Urbanski, *Tetrahedron*, **25**, 1929 (1969).
[174] A. T. Nielson and T. G. Archibald, *Tetrahedron*, **25**, 2393 (1969); K. Buggle, G. P. Hughes, and E. M. Philbin, *Chem. Ind.* (London), **1969**, 77.
[175] N. Bodor and O. Mantsch, *Rev. Roum. Chim.*, **13**, 1153 (1968); *Chem. Abs.*, **70**, 88055 (1969).

isobutyl ketone,[176] additions to Pt(II) and Pd(II) coordinated olefinic ligands,[177] and hydroplatination of cyclic diolefins.[178] The reaction of nucleophiles with various halogen substituted unsaturated acid derivatives,[179] the mechanism of germanium alkyl and halohydride addition to propargyl chloride,[180] and various other reactions[181] of relevance to nucleophilic addition have been reported. Alkali-metal catalysed reactions of alkylaryls and alkylpyridines with olefins have been extensively studied,[182] and several papers concerning the role of cyclopropane rings in conjugated additions[183] have appeared.

Cycloadditions

The well recognized significance of orbital symmetry in organic chemistry is further underlined in the long awaited review by Woodward and Hoffmann.[184] This includes some significant new advances. In particular, the selection rules for allowed concerted reactions are now reduced to a simple geometric statement: a ground-state pericyclic change is symmetry-allowed when the total number of $(4q + 1)$ suprafacial and $(4r)$ antarafacial components is odd. Cycloadditions receive much attention, and of particular interest is a topical discussion of $2 + 2$-cycloadditions. Concerted thermal $\pi^2 s + \pi^2 s$ cycloaddition (36) is orbital symmetry disallowed; $\pi^2 s + \pi^2 a$ addition (37) is allowed but normally geometrically unfavourable. It has now been pointed out that in certain cases such $\pi^2 s + \pi^2 a$ additions may be possible, e.g. for strained olefins which are initially twisted in such a way that they must twist in the transition state (37) for $\pi^2 s + \pi^2 a$ cycloaddition. A vinyl cation should also be an ideal component for a $\pi^2 s + \pi^2 a$ cycloaddition because of favourable interaction involving the vacant p-orbital in the transition state [(38) and (39) show the two new bonding orbitals]. Striking examples from the literature of both of these types of $2 + 2$-cycloaddition are given. The surprisingly ready formation

[176] A. T. Nielson and D. W. Moore, *J. Org. Chem.*, **34**, 444 (1969).
[177] B. F. G. Johnson, T. Keating, J. Lewis, M. S. Subramanian, and D. A. White, *J. Chem. Soc.* (A), **1969**, 1793; R. Palumbo, A. De Renzi, A. Panunzi, and G. Paiaro, *J. Am. Chem. Soc.*, **91**, 3874, 3879 (1969).
[178] M. A. Schwartz and T. J. Dunn, *J. Am. Chem. Soc.*, **91**, 4007 (1969).
[179] M. Verny and R. Vessière, *Tetrahedron*, **25**, 263 (1969); F. Théron, *Bull. Soc. Chim. France*, **1969**, 278; *Chem. Abs.*, **70**, 105625 (1969); J. C. Chalchat, F. Théron, and R. Vessière, *Compt. Rend.*, **267**, 1864 (1968); *Chem. Abs.*, **70**, 57002 (1969).
[180] M. Sassol, J. Satge, and M. Lesbre, *J. Organometal. Chem.*, **17**, 25 (1969).
[181] S. Danishefsky and B. H. Migdalof, *Tetrahedron Letters*, **1969**, 4331; S. Danishefsky, J. Eggler, and G. Koppel, *Tetrahedron Letters*, **1969**, 4333; W. Schroth, J. Peschel, and A. Zschunke, *Z. Chem.*, **9**, 185 (1969); *Chem. Abs.*, **71**, 30421 (1969); H. O. House and W. F. Fischer, *J. Org. Chem.*, **34**, 3615 (1969).
[182] B. Stipanovic and H. Pines, *J. Org. Chem.*, **34**, 2106 (1969); H. Pines and N. E. Sartoris, *ibid.*, pp. 2113, 2119.
[183] J. M. Stewart and D. R. Olsen, *J. Org. Chem.*, **33**, 4534 (1968); J. M. Stewart and G. N. Pagenkopf, *ibid.*, **34**, 7 (1969); S. Danishefsky, G. Rovnyak, and R. Cavanaugh, *Chem. Comm.*, **1969**, 636.
[184] R. B. Woodward and R. Hoffmann, *Angew. Chem. Intern. Ed. Engl.*, **8**, 781 (1969).

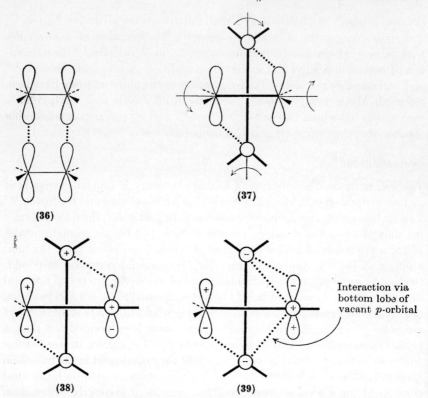

(36)

(37)

(38)

(39)

Interaction via
bottom lobe of
vacant *p*-orbital

of the cyclobutene (**39b**) from the acetylene (**39a**) and HCl, for which a
stepwise mechanism was reasonably proposed,[185] could conceivably be a
further example of a 2 + 2-addition involving a vinyl cation (Scheme 3). The
apparently concerted 2 + 2-additions of ketens are also discussed in the review
(see below).

$$Et_2N-C{\equiv}C-CN \longrightarrow$$

(39a)

Scheme 3

(39b)

[185] T. Sasaki and A. Kojima, *Tetrahedron Letters,* **1969,** 3639.

Salem[186] has improved his PMO approach[187] to interaction between conjugated molecules with overlapping p-orbitals to include the effect of electron repulsion. Also using more sophisticated starting orbitals it has been reapplied to the thermal and photochemical dimerization of unsaturated ketones. The theory now predicts a concerted but asymmetric lower-energy pathway leading to the observed acraldehyde dimer whereas previous calculations[187] indicated that this pathway was of higher energy than that leading to the product of opposite orientation.

Orbital symmetry considerations indicate that the Woodward–Hoffmann rules are generally applicable for thermal but not photochemical cycloadditions to benzene. For the latter, the rules depend on the nature of the excited species and the feasibility of charge-transfer excitation[188] (see also p. 199).

A review of cycloadditions with polar intermediates which includes a valuable critical discussion of the experimental criteria used to differentiate between one- and two-step reactions has been presented.[189] Reviews of the polar cycloaddition of isocyanates to carbon–nitrogen bonds,[190] of reactions of enol ethers including cycloadditions,[191] and of the Woodward–Hoffmann rules[192] have also appeared.

$2 + 4$-*Cycloadditions*

A valuable review of the '-ene' reaction, which is closely related to, and often competes with, the better known Diels–Alder reaction, has appeared.[193]

The extremely rapid dimerization of cyclopentadienones compared with cyclopentadienes is attributed to the low energy difference between the lowest unoccupied and highest occupied diene orbitals.[194] The former symmetric orbital is stabilized by interaction with carbonyl orbitals of the same symmetry, but the latter, being antisymmetric, is unaffected. The rapid dimerization of the ethylene ketal of cyclopentadienone is explained by interactions with oxygen lone-pair orbitals which destabilize the highest occupied diene orbital but do not affect the lowest unoccupied.

In the Diels–Alder reaction of 1,2,3,4,5-pentachlorocyclopentadiene with a variety of dienophiles *endo*-addition predominates as expected. This is attributed to secondary orbital interactions since the *endo*:*exo* ratios do not parallel dipole moments or steric requirements of the dienophiles and are

[186] A. Devaquet and L. Salem, *J. Am. Chem. Soc.*, **91**, 3793 (1969).
[187] L. Salem, *J. Am. Chem. Soc.*, **90**, 543, 553 (1968); see also L. Salem, *Chem. in Britain*, **5**, 449 (1969) for a valuable elementary survey of this approach.
[188] D. Bryce-Smith, *Chem. Comm.*, **1969**, 806.
[189] R. Gompper, *Angew. Chem. Intern. Ed. Engl.*, **8**, 312 (1969).
[190] H. Ulrich, *Accounts Chem. Res.*, **2**, 186 (1969).
[191] F. Effenberger, *Angew. Chem. Intern. Ed. Engl.*, **8**, 295 (1969).
[192] T. L. Gilchrist, *Leicester Chem. Rev.*, **1968**, 7.
[193] H. M. R. Hoffmann, *Angew. Chem. Intern. Ed. Engl.*, **8**, 556 (1969).
[194] E. W. Garbisch and R. F. Sprecher, *J. Am. Chem. Soc.*, **91**, 6785 (1969).

greatest when dienophilic substituents have π- or d-electrons capable of such secondary interactions. The tendency for the bridgehead chlorine in the adducts to be preferentially *anti* to the double bond is opposite to expectations on steric grounds and is assumed to result from dipole–dipole, dipole–induced dipole, and London dispersion forces since the *anti/syn* ratio parallels and increases with the dipole moment of the dienophile.[195] Other workers[196] have shown that the retro-Diels–Alder reaction of norbornylene is faster than that of bornylene. This suggests a concerted reaction since, by assuming a transition state which resembles adduct in its geometry but separated diene and dienophile in its electronic character, simultaneous breaking of the two bonds would increase steric interaction between methyl and potential olefin component and so destabilize the transition state for both forward and reverse reactions, whereas relief of steric strain in a stepwise retro-reaction would lead to bornylene reacting faster. The opposite behaviour of an *anti*-7-chlorine found above[195] is explained by the smaller chlorine lying in the attractive region of van der Waals interaction in the transition state whereas the larger methyl lies in the repulsive zone.

The surprising preponderance of *exo*-addition of 1,2-bistrifluoromethyl-3,3-difluorocyclopropene to cyclopentadiene is attributed to a favourable orientation of dipoles in the transition state for *exo*-addition (**40**) enhanced by attractive van der Waals interactions. Similar considerations explain its exclusive *endo*-addition to furan since in this case the *exo*-transition state (**41**) is destabilized.[197]

It is suggested[198a] that the *endo*-preference (3:1) observed in Diels–Alder additions of pyracycloquinone is much less than that expected if secondary

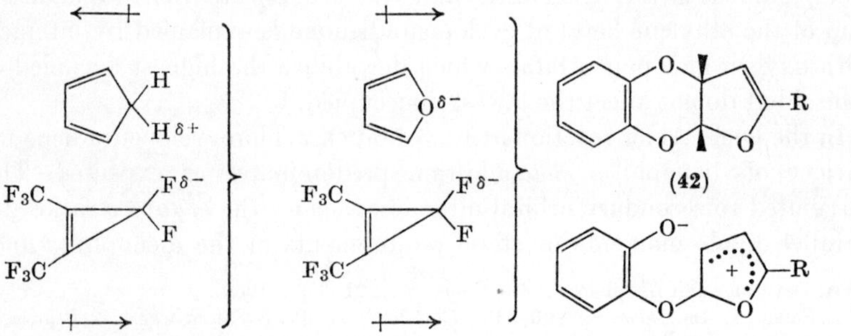

(**40**)   (**41**)   (**43**)

[195] K. L. Williamson, Y. F. L. Hsu, R. Lacko, and C. H. Youn, *J. Am. Chem. Soc.*, **91**, 6129 (1969).
[196] W. C. Herndon and J. M. Manion, *J. Org. Chem.*, **33**, 4505 (1968).
[197] P. B. Sargeant, *J. Am. Chem. Soc.*, **91**, 3061 (1969).
[198a] B. M. Trost, *J. Am. Chem. Soc.*, **91**, 918 (1969).

orbital interactions control *endo*-preference but is not in disagreement with Herndon and Hall's explanation.[198b]

PMO calculations for 2 + 4-cycloadditions of olefins and *o*-quinones suggest that concerted photochemical addition can occur [199a] and that such mechanisms should not be lightly dismissed for reaction of stilbene with tetrachloro-*o*-quinone.[199b] Photochemical addition of but-2-enes to *o*-quinones is predicted to be highly stereoselective. Perfluorocyclopentadiene undergoes Diels–Alder addition (inverse electron demand) to electron-rich dienophiles in high yield, but with normal dienophiles dimerization takes over.[200]

A stepwise mechanism is involved in the reaction between 1,2-benzoquinones and furans to give adducts of type (**42**) since products derived from an intermediate (**43**) formed by electrophilic attack of the quinone on the furan have been isolated.[201] The 'inverse electron demand' Diels–Alder reactions of the acridizinium ion are also stepwise as evidenced [202] by the opposing effects on the reaction rate of 6- and 11-methyl substituents and the previously observed non-stereospecific additions to maleate esters.[203]

The regiospecific and stereospecific polar 1,4-cycloaddition of the benzamidomethylium ion (**44**) to olefins suggests that the addition may be concerted (although not necessarily synchronous) [204a] rather than stepwise with an intermediate as previously suggested.[204b]

(**44**)   (**45**)

The long suspected stereospecific 2 + 4-addition of benzyne to dienes has been proved.[205, 206] This taken together with non-stereospecific 2 + 2-addition is in agreement with a singlet ground-state benzyne molecule with a highest occupied symmetric orbital [206] (see also ref. 267).

[198b] W. C. Herndon and L. H. Hall, *Tetrahedron Letters*, **1967**, 3095; *Org. Reaction Mech.*, **1967**, 155.

[199a] W. C. Herndon and W. G. Giles, *Chem. Comm.*, **1969**, 497.

[199b] See *Org. Reaction Mech.*, **1968**, 176.

[200] R. E. Banks, *J. Chem. Soc.*(C), **1969**, 1372; see also L. P. Anderson, W. J. Feast, and W. K. R. Musgrave, *ibid.*, p. 211.

[201] W. M. Horspool, J. M. Tedder, and Z. U. Din, *J. Chem. Soc.*(C), **1969**, 1694.

[202] C. K. Bradsher and J. A. Stone, *J. Org. Chem.*, **34**, 1700 (1969).

[203] C. K. Bradsher and J. A. Stone, *J. Org. Chem.*, **33**, 519 (1968).

[204a] R. R. Schmidt, *Angew. Chem. Intern. Ed. Engl.*, **8**, 602 (1969).

[204b] W. Seeliger and W. Diepers, *Ann. Chem.*, **697**, 171 (1966).

[205] R. W. Atkin and C. W. Rees, *Chem. Comm.*, **1969**, 152.

[206] M. Jones and R. H. Levin, *J. Am. Chem. Soc.*, **91**, 6411 (1969); see also R. Hoffmann, A. Imamura, and W. J. Hehre, *J. Am. Chem. Soc.*, **90**, 1499 (1968).

The Diels–Alder reactions of cyanoethylenes with cyclopentadiene are believed to involve a transition state which more closely resembles reactants than products.[207] The relative rates of maleic anhydride addition to perylene and its benzo-derivatives correlate with the order of diene character obtained from modified Hückel MO calculations.[208] Diels–Alder addition of 2-substituted tropenoids to acrylonitrile is stereospecific, supporting a one-step reaction.[209] Diels–Alder reactions of 1,2-diphenylcyclobutadiene indicate a rectangular singlet ground-state for this molecule.[210] Some success has been achieved in correlating the observed and calculated orientation of addition of unsymmetrical acetylenes to pyrones.[211] Attempts to obtain Hammett correlations for the Diels–Alder reaction of 2-substituted butadienes have been reported.[212] The high dienophilic activity of perfluoroalkylacetylenes has received further comment.[213] Because of its fused-ring structure, hexafluoro-Dewar benzene unlike other fluoro-olefins is a reactive dienophile.[214] A diradical mechanism has been suggested for the dimerization of 2,4-di-t-butyl-6-ethenylquinone methide.[215] Dienophilic activity of the 4,5-bond of azepines[216] and a tendency for trichloroacetyl isocyanate to act as a diene[217] have been reported. Intramolecular H-transfers occur in the dimerization of phenylpropiolic acid which can be considered as an acid-catalysed dehydro-Diels–Alder reaction.[218]

Other Diels–Alder reactions which have received attention are those of dibenzobicyclo[2.2.0]octatrienes with 2-substituted anthracenes,[219] cyclopentadiene and acraldehyde,[220] α-pyrone with electron-rich olefins,[221] α-bromocrotononitriles with cyclopentadiene,[222] and those of 1,3-dimethyl-cyclohexa-1,3-diene.[223] The reactions of styrenes with activated acetylenes,[224a] benzynes with steroidal dienes and trienes,[224b] the thermal and photo-

207 V. Z. Gabdrakipov and N. K. Altaev, *Izv. Akad. Nauk Kaz. SSR, Ser. Khim.*, **19**, 30 (1969); *Chem. Abs.*, **71**, 38194 (1969).
208 M. Zander, *Ann. Chem.*, **723**, 27 (1969); *Chem. Abs.*, **71**, 38051 (1969).
209 S. Itô, H. Takeshita, and Y. Skoji, *Tetrahedron Letters*, **1969**, 1815.
210 P. Reeves, T. Devon, and R. Pettit, *J. Am. Chem. Soc.*, **91**, 5890 (1969).
211 J. A. Reed, C. L. Schilling, R. F. Tarvin, T. A. Rettig, and J. K. Stille, *J. Org. Chem.*, **34**, 2188 (1969).
212 T. Inukai and T. Kojima, *Chem. Comm.*, **1969**, 1334.
213 R. S. H. Liu and C. G. Krespan, *J. Org. Chem.*, **34**, 1271 (1969).
214 M. G. Barlow, R. N. Haszeldine, and R. Hubbard, *Chem. Comm.*, **1969**, 301.
215 C. D. Cook and L. C. Butler, *J. Org. Chem.*, **34**, 227 (1969).
216 J. R. Wiseman and B. P. Chong, *Tetrahedron Letters*, **1969**, 1619.
217 L. R. Smith, A. J. Speziale, and J. E. Fedder, *J. Org. Chem.*, **34**, 633 (1969).
218 H. W. Whitlock, E. M. Wu, and B. J. Whitlock, *J. Org. Chem.*, **34**, 1857 (1969).
219 S. J. Cristol and W. Y. Lim, *J. Org. Chem.*, **34**, 1 (1969).
220 F. Kasper, *J. Prakt. Chem.*, **311**, 201 (1969).
221 H. Behringer and P. Heckmaier, *Chem. Ber.*, **102**, 2835 (1969).
222 J. Paasivirta and A. Tenhosaari, *Suomen Kemistilehti*, B **42**, 220 (1969).
223 R. N. Mirrington and K. J. Schmalze, *J. Org. Chem.*, **34**, 2358 (1969).
224a E. Ciganek, *J. Org. Chem.*, **34**, 1923 (1969).
224b T. F. Eckhard, H. Heaney, and B. A. Marples, *J. Chem Soc.*(C), **1969**, 2098.

chemical cycloadditions of maleic anhydride,[225] and several retro-Diels–Alder reactions[226] have been reported. The azo-compound (45) undergoes a reverse Diels–Alder reaction to give pyrazole in addition to the expected stepwise loss of nitrogen.[227]

2 + 3-*Cycloadditions*

There have been several interesting but no fundamentally new contributions in the field of 2 + 3-cycloadditions this year. There have been surprisingly few reactions to Firestone's new mechanism for 1,3-dipolar cycloaddition.[228a]

A review on the decomposition and addition reactions of organic azides[228b] has appeared.

Differing reactivity of the stereoisomeric azomethine dicarboxylates arising from conrotatory ring-opening of the aziridines (46) and (47) has been observed

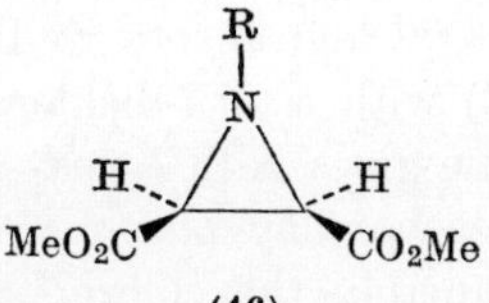

(46) 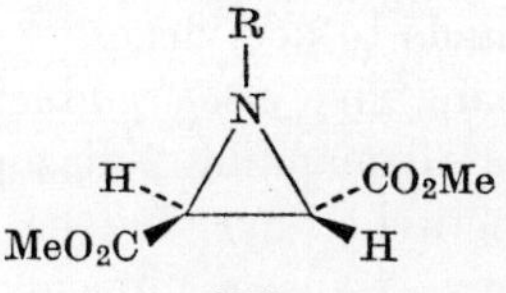   (47)

in their addition to relatively unreactive dipolarophiles. The *trans*-dipole (48) is more reactive and adds stereospecifically, but additions of the less reactive *cis*-dipole (49) are slow enough for rotation to compete, so giving less stereoselective addition. The lower reactivity is possibly due to steric hindrance

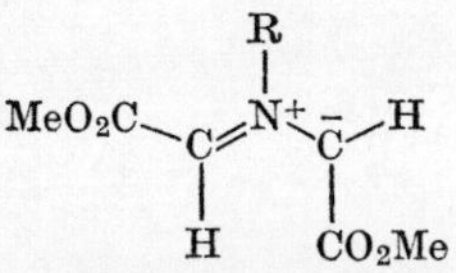

(48) 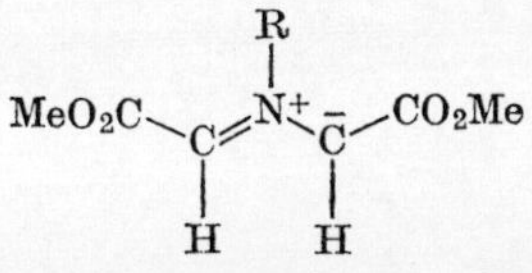   (49) 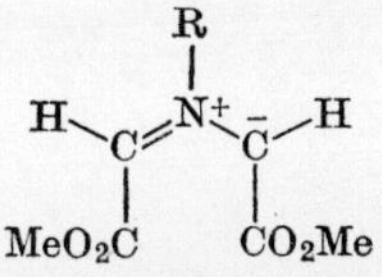   (50)

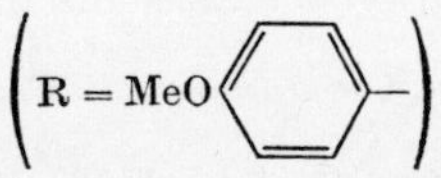

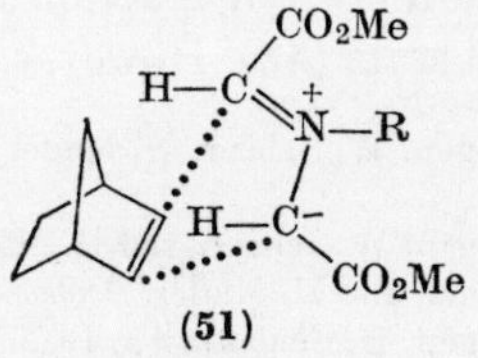

(51)

[225] M. Verbeek, H. D. Scharf, and F. Korte, *Chem. Ber.*, **102**, 2471 (1969).
[226] K. Kirchhoff, F. Boberg, and G. R. Schultze, *Z. Naturforsch.*, **23b**, 1548 (1968); W. C. Herndon and J. M. Manion, *Tetrahedron Letters*, **1968**, 6327.
[227] J. C. Hinshaw and E. L. Allred, *Chem. Comm.*, **1969**, 72.
[228a] See *Org. Reaction Mech.*, **1968**, 180.
[228b] G. L'abbé, *Chem. Rev.*, **69**, 345 (1969).

involving the two ester groups in the transition state. With reactive dipolarophiles addition of both dipoles is stereospecific,[229] and this has been utilized to study the kinetics of isomerization and ring-opening of aziridines (46) and (47).[230] Conrotatory opening of the *trans*-aziridine (47) can give rise to U- (50) or W-shaped (49) *cis*-azomethine ylids. In the addition to norbornene only the W-shaped *cis*-1,3-dipole (49) was able to react via the *exo*-transition state (51). No addition to the U-form (50) was observed, only addition to the 'rotated dipole' (48). The two possible orientations for addition to the U-shaped dipoles are approximately equally unfavourable as shown using a cyclic U-fused 1,3-dipole.[231]

Huisgen and coworkers have reported systematic studies of 1,3-dipolar additions of nitrones. These show the characteristics of the now familiar multicentre mechanism.[232] Day and coworkers have shown that steric effects can outweigh electronic effects in determining the orientation of addition of diazo-compounds to acetylenes,[233] olefins,[234] and allenes.[235] Diazomethane and diazoethane give *cisoid*-adducts (52) with *cis*-3,4-dichlorobutane. The sterically more demanding 2-diazopropane gives both *cisoid*- and *transoid*-adducts. Dimethyl *cis*-cyclobutene-1,2-dicarboxylate only reacts with diazopropane to give *transoid*-adducts. Presumably the chlorine atoms in dichlorocyclobutene interact with the nitrogen atoms in the transition state for the addition, so controlling the orientation and accelerating the addition.[236]

(52)   (53)

The addition of diazomethane to aromatic anils shows the typical characteristics of a concerted reaction in which slight charge build-up occurs in the transition state (53). However, although the addition is largely insensitive to

[229] R. Huisgen, W. Scheer, H. Mäder, and E. Brunn, *Angew. Chem. Intern. Ed. Engl.*, **8**, 604 (1969).

[230] R. Huisgen, W. Scheer, and H. Mäder, *Angew. Chem. Intern. Ed. Engl.*, **8**, 602 (1969).

[231] R. Huisgen and H. Mäder, *Angew. Chem. Intern. Ed. Engl.*, **8**, 604 (1969).

[232] R. Huisgen, H. Seidl, and I. Brüning, *Chem. Ber.*, **102**, 1102 (1969); H. Seidl, R. Huisgen, and R. Knorr, *ibid.*, p. 904; R. Huisgen, H. Hauck, R. Grashey, and H. Seidl, *ibid.*, p. 736; H. Seidl, R. Huisgen, and R. Grashey, *ibid.*, p. 926; R. Huisgen and J. Wulff, *ibid.*, p. 746; R. Huisgen, H. Hauck, H. Seidl, and M. Burger, *ibid.*, p. 1117.

[233] A. C. Day and R. N. Inwood, *J. Chem. Soc.*(C), **1969**, 1065.

[234] S. D. Andrews, A. C. Day, and A. N. McDonald, *J. Chem. Soc.*(C), **1969**, 787.

[235] S. D. Andrews, A. C. Day, and R. N. Inwood, *J. Chem. Soc.*(C), **1969**, 2443.

[236] M. Franck-Neumann, *Angew. Chem. Intern. Ed. Engl.*, **8**, 210 (1969).

solvent dielectric constant, reaction rates are increased in the presence of protic solvents, e.g. H_2O, and often considerably increased in dipolar aprotic solvents, e.g. DMF. These catalytic effects are attributed to the special ability of such solvents to solvate the transition state (particularly the negative site) more efficiently than the reactants. The practical significance of such effects was stressed.[237]

All characteristics of the cyclizations of 1-aryl-3-diazopropenes are typical of intramolecular 1,3-dipolar cycloadditions.[238] The addition of benzonitrile oxide to norbornadiene proceeds without any skeletal rearrangement and is therefore apparently concerted.[239] The formation of open-chain oximes as well as the usual isoxazoles has been observed in the addition of nitrile oxides to arylacetylenes. However, the oximes are formed by a simultaneous independent route.[240] The high orientational selectivity for the intramolecular cycloaddition (equation 6) has been taken as being suggestive of a concerted reaction subject to strict steric requirements rather than a stepwise radical process (Firestone) where only one of the four possible and similar diradicals could have been involved.[241]

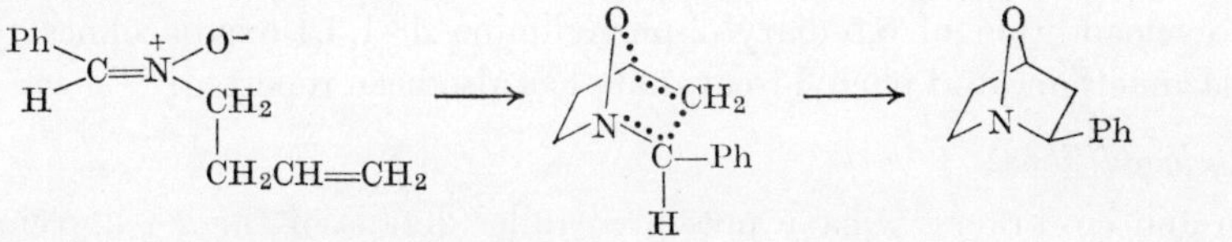

$$\ldots (6)$$

Azides undergo apparently normal concerted dipolar cycloaddition to α-ester phosphorus ylids (**54**) but the ylid acts as dipolarophile in two different ways (Scheme 4) depending on whether R = H or Me.[242] A stepwise mechanism was previously proposed.[243]

Hexafluoro-Dewar benzene, unlike other fluoro-olefins, is a reactive dipolarophile.[244] The products of 1,3-dipolar addition to mesoionic compounds have been observed using the precursors to these systems.[245] Additions of azidobenzenes to activated olefins at 90° lead to pyrrolidines since the first-formed triazolines lose nitrogen to give aziridines which ring-open by C–C

[237] P. K. Kadaba, *Tetrahedron*, **25**, 3053 (1969).

[238] J. L. Brewbaker and H. Hart, *J. Am. Chem. Soc.*, **91**, 711 (1969).

[239] R. Lazar, F. G. Cocu, and N. Barbulescu, *Rev. Chim.* (Roumania), **20**, 3 (1969); *Chem. Abs.*, **70**, 114341 (1969).

[240] S. Morrocchi, A. Ricca, A. Zanorotti, G. Bianchi, R. Gandolfi, and P. Grunanger, *Tetrahedron Letters*, **1969**, 3329.

[241] W. C. Lumma, *J. Am. Chem. Soc.*, **91**, 2820 (1969).

[242] G. L'abbé, P. Ykman, and G. Smets, *Tetrahedron*, **25**, 5421 (1969).

[243] G. R. Harvey, *J. Org. Chem.*, **31**, 1587 (1966).

[244] M. G. Barlow, R. N. Haszeldine, and W. D. Morton, *Chem. Comm.*, **1969**, 931.

[245] K. T. Potts and U. P. Singh, *Chem. Comm.*, **1969**, 66.

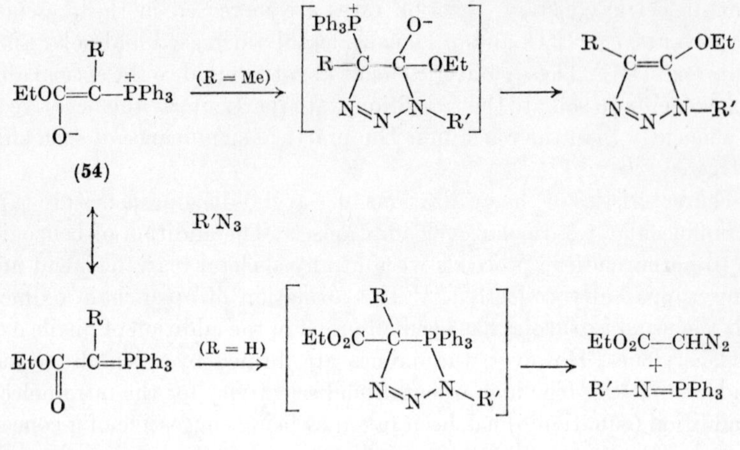

Scheme 4

bond cleavage to give a new 1,3-dipole which adds to a second molecule of olefin.[246]

Other 1,3-dipolar cycloadditions of interest are listed.[247] The retro-1,3-dipolar cycloaddition of 5,5-diaryl-2-phenylimino-Δ^3-1,3,4-oxadiazolines to diaryldiazomethane and phenyl isocyanate has also been reported.[248]

$2 + 2$-*Cycloadditions*

Huisgen and coworkers [249] have presented fuller details of the $2 + 2$-cyclo-additions of ketens. All evidence points to a one-step multicentre mechanism involving partial charge build-up in the transition state for the addition to alkenes and vinyl ethers. This concerted reaction can now be considered as an allowed $\pi^2 s + \pi^2 s + \pi^2 s$ (55) or $\pi^2 s + \pi^2 a + \pi^2 a$ (56) process in which the olefin forms a suprafacial ($\pi^2 s$) component. Alternatively Woodward and Hoffmann[184] have pointed out the similarity of the keten molecule to a vinyl cation (see p. 195). Additional interaction involving the exceptionally low-energy vacant $C{=}O$ π^*-orbital favours the transition state (57) for allowed $\pi^2 s + \pi^2 a$ addition.

246 F. Texier and R. Carrie, *Tetrahedron Letters*, **1969**, 823.
247 S. McLean and D. M. Findlay, *Tetrahedron Letters*, **1969**, 2219; P. Rajagopalan, *ibid.*, p. 311; R. Huisgen, K. V. Fraunberg, and H. J. Sturm, *ibid.*, p. 2589; M. Franck-Neumann and C. Buchecker, *ibid.*, p. 2659; R. Huisgen and J. Wulff, *Chem. Ber.*, **102**, 1833, 1841, 1848 (1969); T. Sasaki and T. Yoshioka, *Bull. Chem. Soc. Japan*, **42**, 258, 826 (1969); T. Sasaki and M. Ando, *ibid.*, **41**, 2960 (1968); T. Sasaki, T. Yoshioka, and I. Izure, *ibid.*, p. 2964; Y. Iwakura, K. Uno, S. Shiraishi, and T. Hongu, *ibid.*, p. 2954; F. Texier and R. Carrie, *Compt. Rend.*, *C*, **268**, 1396 (1969).
248 P. R. West and J. Warkentin, *J. Org. Chem.*, **34**, 3233 (1969).
249 R. Huisgen, L. A. Feiler, and P. Otto, *Chem. Ber.*, **102**, 3444 (1969); R. Huisgen, L. A. Feiler, and G. Binsch, *ibid.*, p. 3460; R. Huisgen and L. A. Feiler, *ibid.*, p. 3391; R. Huisgen, L. A. Feiler, and P. Otto, *ibid.*, p. 3405; L. A. Feiler and R. Huisgen, *ibid.*, p. 3428; R. Huisgen and P. Otto, *ibid.*, p. 3475.

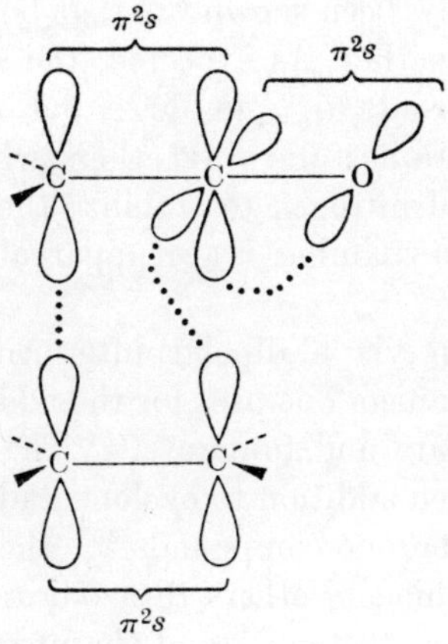

(55) 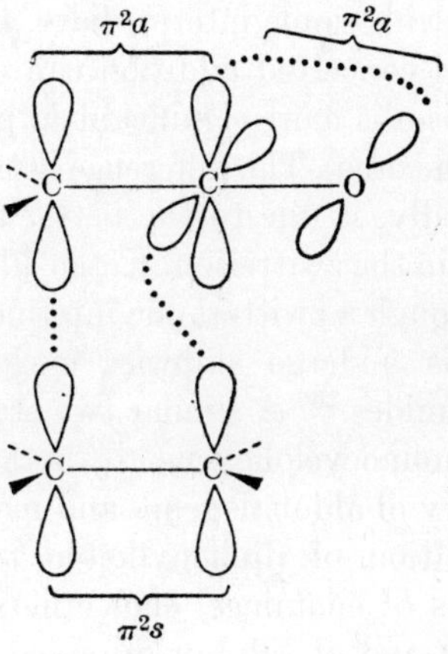　　　　　　　　　　　　(56)

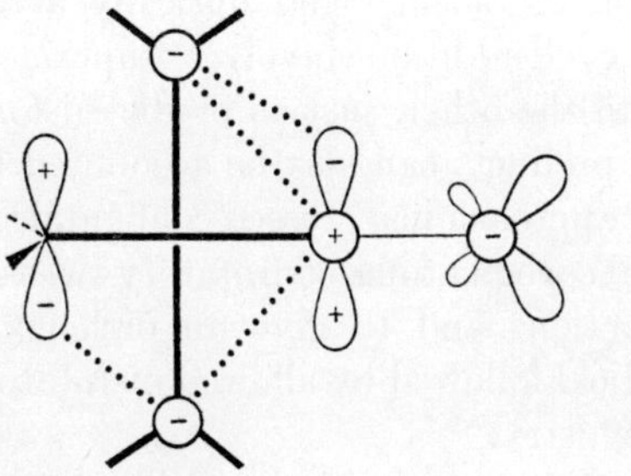

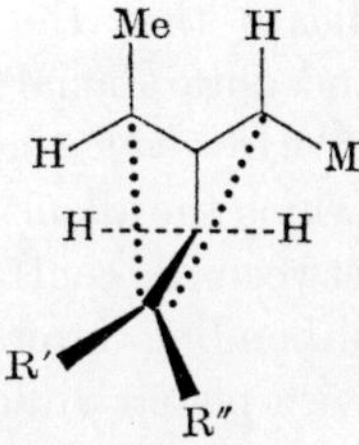

(a, R′ = CN, R″ = H;
b, R′ = H, R″ = CN)

(57)　　　　　　　　　　　　(58)

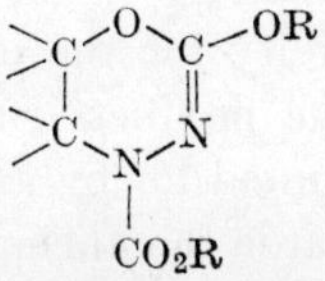

(59)

Also of considerable interest in keten cycloaddition is the demonstration[250] of kinetic secondary deuterium isotope effects at each terminus for styrene addition to diphenylketen. k_H/k_D for the β-position is 0.91, as expected (for $sp^2 \rightarrow sp^3$), but 1.23 for the α-position. Thus significant bond-formation occurs at both α- and β-carbons in the transition state but the nature of the bond-forming processes at each carbon must be different since such isotope effects do not reflect the same process occurring to different extents.

Evidence has been presented[251] that keten addition to enamines is stepwise

[250] J. E. Baldwin and J. A. Kapecki, *J. Am. Chem. Soc.*, **91**, 3106 (1969).
[251] P. Otto, L. A. Feiler, and R. Huisgen, *Angew. Chem. Intern. Ed. Engl.*, **7**, 737 (1968).

with a zwitterionic intermediate. It has now been shown[252a] that both step-wise and concerted addition can occur together. As expected, the stepwise mechanism is more significant in polar solvents (57% in MeCN but only 8% in cyclohexane). The difference between enamines and vinyl ethers, which add concertedly, is due to the better ability of nitrogen to stabilize the formal charges in the zwitterion. Keten addition to ynamines is also apparently step-wise through a zwitterionic intermediate.[252b]

Ketens undergo stepwise cycloaddition via a dipolar intermediate to carbodiimides.[253] A similar two-step mechanism operates for the addition of 1-morpholinocyclohexene to diethyl maleate and fumarate.[254] The stereo-chemistry of aldohalogeno- and methylketen addition to cyclopentadiene,[255] the addition of diphenylketen to vinylborono-compounds,[256] the 2 + 2-additions of enamines, enol ethers, and thioenol ethers to azodicarboxylic esters,[257] and of sulphur diimides to diphenylketen,[258] and the presumably stepwise cycloaddition of sulphenes to ynamines[259] have been reported.

Elegant studies[260] of the dimerization of racemic and optically active cyclonona-1,2-diene indicate that the cycloaddition involves suprafacial addition to one allene and antarafacial to the other, just as predicted for a one-step cycloaddition.[184] However, the products can also be accounted for by a stepwise process in which the allenes approach in a crossed configuration, and as bonding begins between the central carbon atoms conrotatory twisting occurs to minimize non-bonding interactions and to give an orthogonal diradical. Rotation to give a planar diradical followed by allowed disrotatory closure then gives the observed stereochemistry.

Addition of $R(-)$-1,3-dimethylallene to acrylonitrile is highly stereo-selective with respect to the allene component[261a] (see ref. 261b for a similar addition in which a high stereoselectivity for the olefinic component was observed). The stereochemistries of the products rule out any symmetric reactive intermediate but can be accounted for by assuming that the olefin takes the least hindered approach towards the allene and orbital symmetry controlled rotation of a methylene group gives the intermediates (**58a**) and (**58b**) which then collapse to products (dotted lines.)

[252a] R. Huisgen and P. Otto, *J. Am. Chem. Soc.*, **91**, 5922 (1969).
[252b] M. Delaunois and L. Ghosez, *Angew. Chem. Intern. Ed. Engl.*, 8, 72 (1969).
[253] W. T. Brady, E. D. Dorsey, and F. H. Parry, *J. Org. Chem.*, **34**, 2846 (1969); W. T. Brady and E. D. Dorsey, *Chem. Comm.*, **1968**, 1638.
[254] A. Risaliti, E. Valentin, and M. Forchiassin, *Chem. Comm.*, **1969**, 233.
[255] W. T. Brady, E. F. Hoff, R. Roe, and F. H. Parry, *J. Am. Chem. Soc.*, **91**, 5679 (1969).
[256] R. H. Fish, *J. Org. Chem.*, **34**, 1127 (1969).
[257] J. Firl and S. Sommer, *Tetrahedron Letters*, **1969**, 1133, 1137.
[258] T. Minami, O. Aoki, H. Miki, Y. Ohshiro, and T. Agawa, *Tetrahedron Letters*, **1969**, 447.
[259] M. H. Rosen, *Tetrahedron Letters*, **1969**, 647.
[260] W. R. Moore, R. D. Bach, and T. M. Ozretich, *J. Am. Chem. Soc.*, **91**, 5918 (1969).
[261a] J. E. Baldwin and U. V. Roy, *Chem. Comm.*, **1969**, 1225.
[261b] *Org. Reaction Mech.*, **1968**, 179, ref. 186.

Substituted allenes dimerize so that one or both of the larger vinyl sub-stituents in the adduct are inwardly orientated (61). This is explained by a diradical mechanism in which least hindered approach of two allene molecules first leads to the inwardly orientated orthogonal diallyl radical (60). Rotation to a planar structure followed by cyclization gives the required stereochemistry. However, whether this cyclization is non-stereospecific, disrotatory to give *cis*-cyclobutane substituents, or conrotatory to give *trans* is not made clear. Disrotatory closure is symmetry-allowed for the planar singlet diradical but conrotatory for the orthogonal diradical.[262]

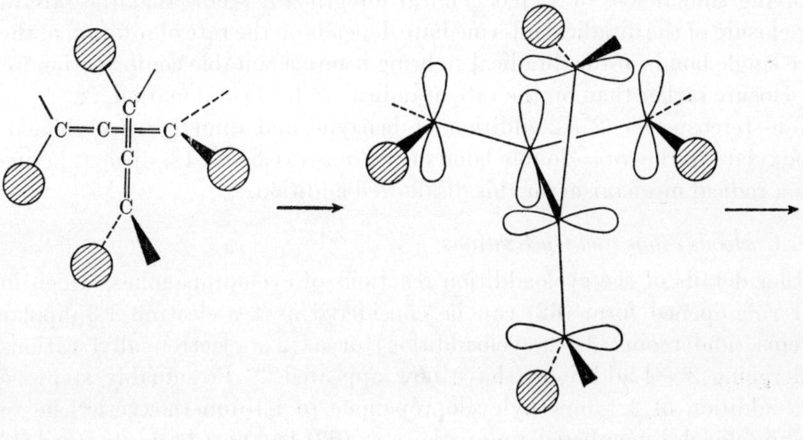

(60)

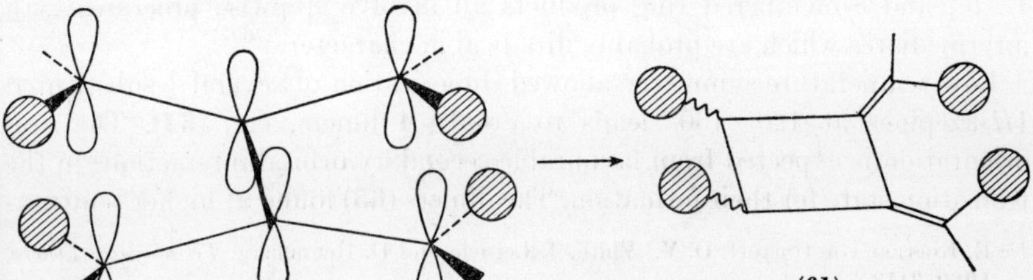

(61)

Vinyl ethers form diazetidines stereospecifically with azodicarboxylates. The rate of reaction is little affected by solvent polarity, and ΔH and ΔS also suggest a concerted addition. However, D-substitution at CH_2 increases the rate ($k_D/k_H = 1.21$) but at CHOR decreases the rate ($k_D/k_H = 0.89$). This,

262 O. J. Muscio and T. L. Jacobs, *Tetrahedron Letters*, **1969**, 2867; T. L. Jacobs, J. R. McClenon, and O. J. Muscio, *J. Am. Chem. Soc.*, **91**, 6038 (1969).

together with the trapping of a dipolar intermediate in diazetidine formation from indene and 4-phenyl-1,2,4-triazoline-3,5-dione, is taken to indicate a stepwise dipolar addition with initial bond formation to CH_2.[263] However, unless the effect of D at CHOR is a β-secondary isotope effect, it seems to us that some form of bonding to CHOR occurs in the transition state. With symmetrical olefins dimethyl azodicarboxylate forms adducts of type (59).[264]

Bartlett has continued his studies of the diradical 2 + 2-addition of fluorinated olefins to dienes.[265, 266] Reversibility of the formation of the intermediate diradical in these additions has been demonstrated. Since tetrafluoroethylene and 1,1-dichloro-2,2-difluoroethylene form diene adducts involving similar loss of stereochemical integrity, it seems that the rate of ring-closure of the diradical intermediate depends on the rate of rotation of the three single bonds of the diradical to bring it into a suitable conformation for ring-closure rather than on the rate of radical–radical combination.[265]

Non-stereospecific 2 + 2-addition of benzyne and dimethyl acetylenedicarboxylate to the *trans*-double bond of *cis,trans*-cycloocta-1,3-diene is in line with a radical mechanism for this disallowed addition.[267]

Other Cycloadditions and Cyclizations

Further details of the cycloaddition reactions of cyclopropanones, which in their ring-opened form (62) can be considered as 4 π-electron 1,3-dipolar systems (undergoing 3 + 2-cycloaddition) or as 2 π-electron allyl cations (undergoing 3 + 4-additions), have now appeared.[268] Presumably stepwise 2 + 2-addition of 2,2-dimethylcyclopropanone to 1,1-dimethoxyethylene to give the novel 4-membered ring orthoester (63) has now been observed.[269] Detailed investigation of chloroprene dimerization suggests that formation of 4-, 6-, and 8-membered ring products all involve stepwise processes with intermediates which are probably diradical in character.[270]

Low-temperature symmetry-allowed dimerization of several 1-substituted 1*H*-azepines at 120—130° leads to *exo*-6 + 4-dimers, e.g. (64). The *exo*-orientation is expected from favourable secondary orbital interactions in the transition state for their formation. The dimers (65) found at higher tempera-

263 E. Koerner von Gustorf, D. V. White, J. Leitich, and D. Henneberg, *Tetrahedron Letters*, **1969**, 3113.

264 E. Koerner von Gustorf, D. V. White, and J. Leitich, *Tetrahedron Letters*, **1969**, 3109.

265 P. D. Bartlett, C. J. Dempster, L. K. Montgomery, K. E. Schueller, and G. E. H. Wallbillich, *J. Am. Chem. Soc.*, **91**, 405 (1969).

266 P. D. Bartlett and G. E. H. Wallbillich, *J. Am. Chem. Soc.*, **91**, 409 (1969).

267 P. G. Gassman, H. P. Benecke, and T. J. Murphy, *Tetrahedron Letters*, **1969**, 1649; see also ref. 206.

268 N. J. Turro, S. S. Edelson, J. R. Williams, T. R. Darling, and W. B. Hammond, *J. Am. Chem. Soc.*, **91**, 2283 (1969).

269 N. J. Turro and J. R. Williams, *Tetrahedron Letters*, **1969**, 321.

270 N. C. Billingham, J. R. Ebdon, R. S. Lehrle, J. L. Markham, and J. C. Robb, *Trans. Faraday Soc.*, **65**, 470 (1969).

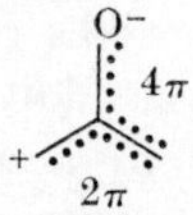

(62)

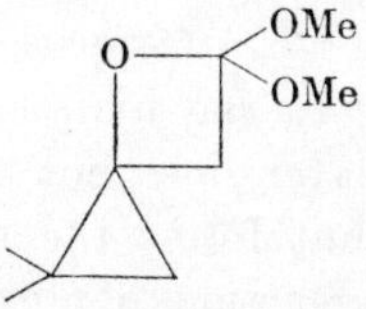

(63)

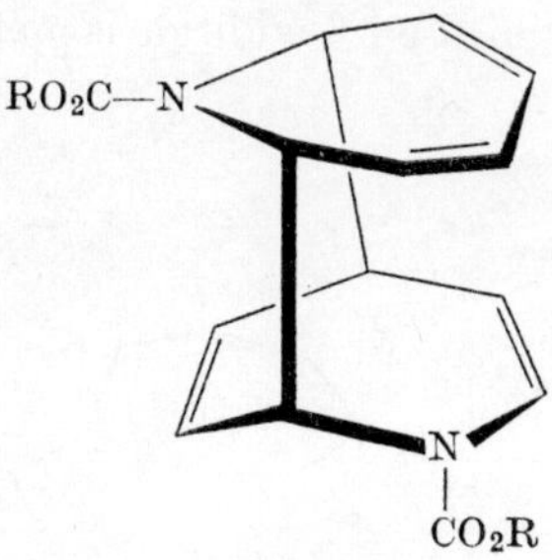

(64)

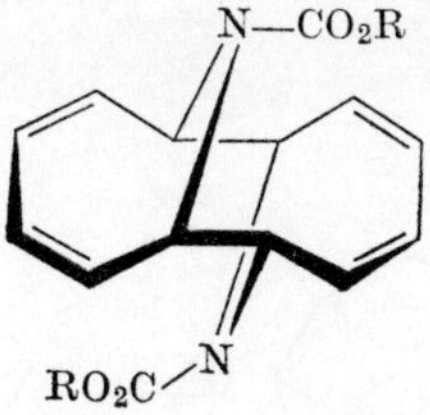

(65)

tures are formed from the low-temperature dimers by a presumably stepwise diradical suprafacial 1,3-bond migration.[271] *N*-Substituted azepines act as dienes in $2 + 4$-addition to dienophiles; *endo*-stereospecificity is observed, in line with orbital symmetry predictions. With dienes, *endo*-$4 + 2$-addition again occurs in which the azepine 4,5-double bond is the dienophile.[272] Addition of chlorosulphonyl isocyanate to bullvalene (Scheme 5) gives 1,2- and 1,6-cyclo-adducts; the 1,2-adducts (67) and (68) isomerize rapidly to (69) and (70). The addition probably involves collapse of an intermediate zwitterion (66) through a 4- or 5-membered transition state, although concerted (or nearly so) addition followed by irreversible conversion of (67) into (66) cannot yet be ruled out.[273] 1,4-Cycloaddition of chlorosulphonyl isocyanate to monosubstituted cyclooctatetraene involves electrophilic attack to give the most stable classical carbonium ion which then collapses via a 1-substituted homotropylium cation to product.[274] Further details of chlorosulphonyl isocyanate addition to hexamethyl-Dewar benzene and its use in the synthesis of the azasemibullvalene system have now appeared.[275]

The formal 1,6-addition of dienophiles to benzocyclooctatetraene involves Diels–Alder reaction of its tricyclic valence tautomer followed by skeletal

[271] I. C. Paul, S. M. Johnson, J. H. Barrett, and L. A. Paquette, *Chem. Comm.*, **1969**, 6; L. A. Paquette, J. H. Barrett, and D. E. Kuhla, *J. Am. Chem. Soc.*, **91**, 3616 (1969).
[272] L. A. Paquette, D. E. Kuhla, J. H. Barrett, and L. M. Leichter, *J. Org. Chem.*, **34**, 2888 (1969); see also ref. 216.
[273] L. A. Paquette, S. Kirschner, and J. R. Malpass, *J. Am. Chem. Soc.*, **91**, 3970 (1969).
[274] L. A. Paquette, J. R. Malpass, and T. J. Barton, *J. Am. Chem. Soc.*, **91**, 4714 (1969).
[275] L. A. Paquette and G. R. Krow, *J. Am. Chem. Soc.*, **91**, 6107 (1969); see *Org. Reaction Mech.*, **1968**, 162.

rearrangement.[276] An apparently symmetry allowed $2\pi + 2\sigma + 2\sigma$-addition of dienophiles to the hydrocarbon (**71**) has been observed.[277] Acetylene-dicarboxylic ester does not undergo $16 + 2\pi$-cycloaddition to the hydro-carbon (**72**) but forms the spiro-products (**73**) by a stepwise process.[278] Cycloaddition reactions of unstable isoindenone derivatives indicate that they function as dienes;[279, 280] the diphenyl derivative readily undergoes $4 + 4$-dimerization.[280] Further proof for the structure of the 1,6-adduct of nitro-sobenzene to cycloheptatriene has appeared. Stepwise 1,6-addition is preferred

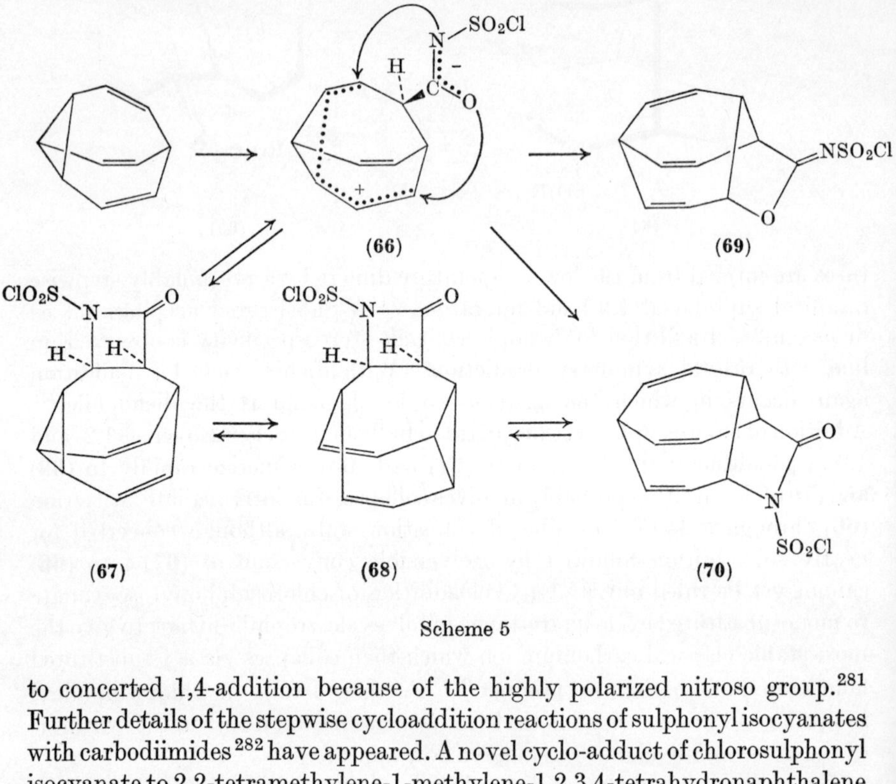

Scheme 5

to concerted 1,4-addition because of the highly polarized nitroso group.[281] Further details of the stepwise cycloaddition reactions of sulphonyl isocyanates with carbodiimides [282] have appeared. A novel cyclo-adduct of chlorosulphonyl isocyanate to 2,2-tetramethylene-1-methylene-1,2,3,4-tetrahydronaphthalene

[276] J. A. Elix and M. V. Sargent, *J. Am. Chem. Soc.*, **91**, 4734 (1969).
[277] E. LeGoff and S. Oka, *J. Am. Chem. Soc.*, **91**, 5665 (1969); see however H. H. Westberg, E. N. Cain, and S. Masamune, *J. Am. Chem. Soc.*, **91**, 7512 (1969).
[278] von H. Prinzbach, L. Knothe, and A. Dieffenbacher, *Tetrahedron Letters*, **1969**, 2093.
[279] K. Blatt and R. W. Hoffmann, *Angew. Chem. Intern. Ed. Engl.*, **8**, 606 (1969).
[280] J. M. Holland and D. W. Jones, *Chem. Comm.*, **1969**, 587.
[281] P. Burns and W. A. Waters, *J. Chem. Soc.*(C), **1969**, 27.
[282] H. Ulrich, B. Tucker, F. A. Stuber, and A. A. R. Sayigh, *J. Org. Chem.*, **34**, 2250 (1969); see also *Org. Reaction Mech.*, **1968**, 185.

has been rationalized.[283a] Cycloaddition reactions of sulphenes[283b] and attempts to isolate 'electronically' stabilized sulphenes[283c] have been reported. Additions of 2-chlorotropone to cyclopentadiene,[284] 2-methoxytropone to cycloheptatriene,[285] benzo[d]tropone to 1,2,3,4 tetrachlorocyclopentadiene,[286] tropone to enamines[287] and benzyne,[288] isonitriles to acetylenedicarboxylic esters,[289] and arylaroyldiazomethane to carbon disulphide[290] have also received attention.

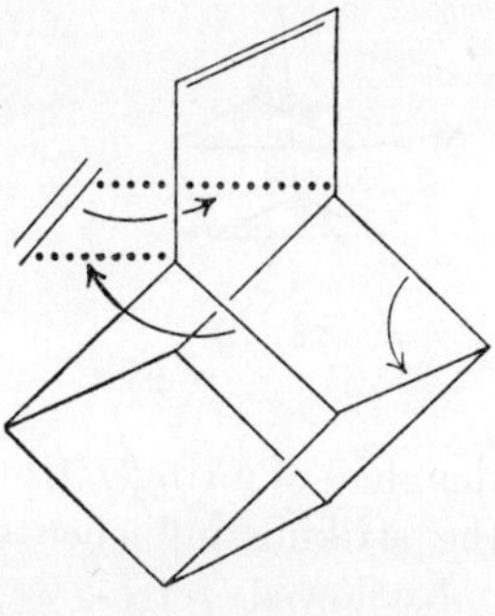

(71) 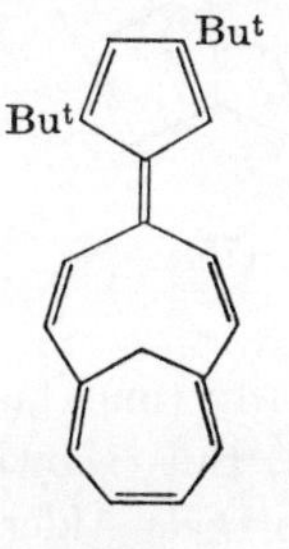   (72) 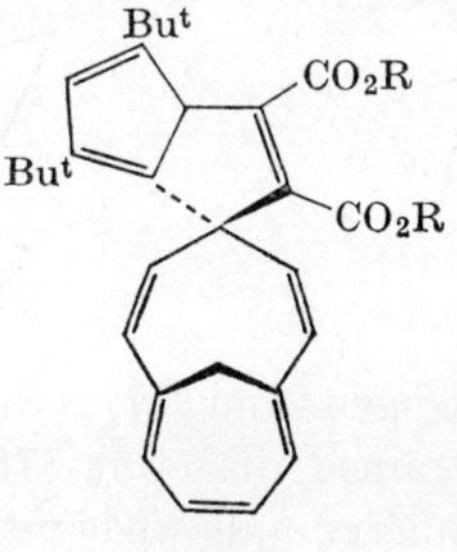   (73)

Full details of the additions of electron-deficient olefins to the strained σ-bond in bicyclo[2.1.0]pentane,[291] and further demonstration[292] of the generality of rear attack on such strained σ-bonds, has appeared. The dimerization of cyclopropene probably involves an '-ene' mechanism[293] (74). Orbital symmetry considerations suggest that conversion of two acetylene ligands on a metal into a cyclobutadiene ligand is not a favourable process.[294] The Rh(I)-catalysed intramolecular cycloaddition (75) to (76) is not concerted.[295] Transition-metal catalysed cyclotrimerization of D-labelled but-2-ynes indicates that no free or complexed tetramethylcyclobutadienes

[283a] T. W. Doyle and T. T. Conway, *Tetrahedron Letters*, **1969**, 1889.
[283b] L. A. Paquette and R. W. Begland, *J. Org. Chem.*, **34**, 2896 (1969).
[283c] L. A. Paquette, J. P. Freeman, and R. W. Houser, *J. Org. Chem.*, **34**, 2901 (1969).
[284] S. Itô, K. Sakan, and Y. Fujise, *Tetrahedron Letters*, **1969**, 775.
[285] S. Itô, Y. Fujise, and M. Sato, *Tetrahedron Letters*, **1969**, 691.
[286] Y. Sasada, H. Shimanouchi, I. Murata, A. Tajiri, and Y. Kitahara, *Tetrahedron Letters*, **1969**, 1185.
[287] M. Oda, M. Funamizu, and Y. Kitahara, *Chem. Comm.*, **1969**, 737.
[288] T. Miwa, M. Kato, and T. Tamano, *Tetrahedron Letters*, **1969**, 1761.
[289] E. Winterfeldt, D. Schumann, and H. J. Dillinger, *Chem. Ber.*, **102**, 1656 (1969).
[290] J. E. Baldwin and J. A. Kapecki, *J. Org. Chem.*, **34**, 724 (1969).
[291] P. G. Gassman, K. T. Mansfield, and T. J. Murphy, *J. Am. Chem. Soc.*, **91**, 1684 (1969).
[292] P. G. Gassman and G. D. Richmond, *Chem. Comm.*, **1968**, 1630.
[293] P. Dowd and A. Gold, *Tetrahedron Letters*, **1969**, 85.
[294] F. D. Mango and J. H. Schachtschneider, *J. Am. Chem. Soc.*, **91**, 1030 (1969).
[295] T. J. Katz and S. A. Cerefice, *J. Am. Chem. Soc.*, **91**, 6519 (1969).

are involved. The product distribution with aluminium chloride as catalyst is, however, consistent with an intermediate with cyclobutadiene-like symmetry. Dichlorobisbenzonitrilepalladium(II) gives an intermediate product distribution.[296] The Ni-catalysed cyclotrimerization of butadiene has also been studied.[297] Four-centre metal-catalysed olefin reactions have been discussed.[298]

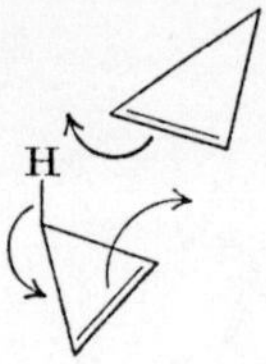

(74) 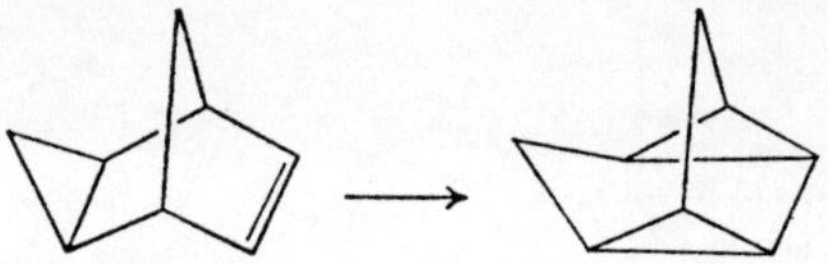   (75)   (76)

Loss of nitrogen from (77) is ca. 10^{22} times faster than loss of nitrogen from the more strained diazetine (78), thus showing the striking influence of orbital symmetry (allowed retro-Diels–Alder *vs.* disallowed retro-2 + 2-addition).[299] The retrohomo-Diels–Alder reaction of (79) and its two isomers proceeds stereospecifically, showing that the cyclopropane ring exerts strict control over the electrocyclic process, surmounting with (79) severe Me–Me repulsion.[300] A concerted retrohomo-Diels–Alder mechanism explains the 10^{11}-fold faster decomposition of (80) than (81) which decomposes by a radical

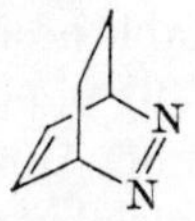

(77) 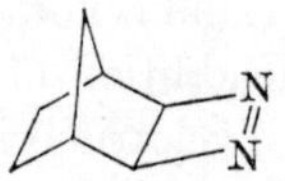   (78) 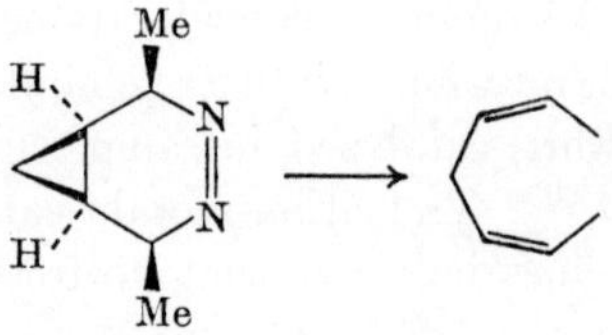   (79)

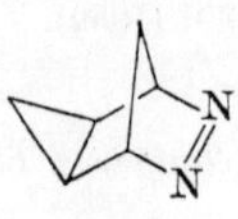

(80) 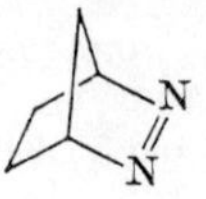   (81) 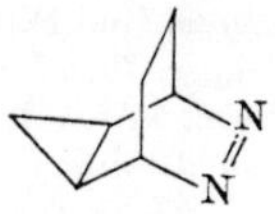   (82) 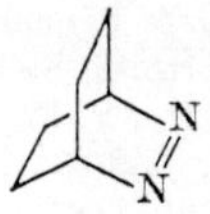   (83)

[296] G. M. Whitesides and W. J. Ehmann, *J. Am. Chem. Soc.*, **91**, 3800 (1969).
[297] B. Bogdanović, P. Heimbach, M. Kröner, G. Wilke, E. G. Hoffmann, and J. Brandt, *Ann. Chem.*, **727**, 143 (1969).
[298] R. L. Banks and G. C. Bailey, *J. Catal.*, **14**, 276 (1969).
[299] N. Rieber, J. Alberts, J. A. Lipsky, and D. M. Lemal, *J. Am. Chem. Soc.*, **91**, 5668 (1969).
[300] J. A. Berson and S. S. Olin, *J. Am. Chem. Soc.*, **91**, 777 (1969).

mechanism.[301] An even more striking rate difference of 10^{17} is observed for the corresponding diazabicyclo[2.2.2]oct-2-enes **(82)** and **(83)**.[302]

The formic acid catalysed cyclization of the allylic isomer of manool involves a cyclooctenyl route.[303] The kinetics and stereochemistry of acid-catalysed pyrazoline ring formation from α,β-unsaturated phenylhydrazones has been studied,[304] and carbonyl participation has been observed in the acid-catalysed cyclization of *endo*-2-acetamido-*exo*-3-methylbicyclo[2.2.1]hept-5-ene.[305] The generation and cyclization of heptatrienyl anions has been reported.[306]

[301] E. L. Allred, J. C. Hinshaw, and A. L. Johnson, *J. Am. Chem. Soc.*, **91**, 3382 (1969).
[302] E. L. Allred and J. C. Hinshaw, *Chem. Comm.*, **1969**, 1021.
[303] S. F. Hall and A. C. Oehlschlager, *Chem. Comm.*, **1969**, 1157.
[304] H. Ferres and W. R. Jackson, *Chem. Comm.*, **1969**, 261.
[305] S. P. McManus, *Chem. Comm.*, **1969**, 235.
[306] R. B. Bates, W. H. Deines, D. A. McCombs, and D. E. Potter, *J. Am. Chem. Soc.*, **91**, 4608 (1969).

CHAPTER 6

Nucleophilic Aromatic Substitution

R. W. ALDER

School of Chemistry, University of Bristol

The S_NAr Mechanism

A book by Miller[1a] is especially concerned with substituent effects, the reactivity of nucleophiles, and mobility of leaving groups. Miller has also extended his semi-empirical reactivity theory to cover the reactions of amines with neutral aromatic substrates.[1b] An interesting short review by Pietra surveys most aspects of nucleophilic aromatic substitution.[2a] Further reviews on the reactions of aromatic nitro-compounds with bases[2b] and the activating effects of the nitro group in nucleophilic aromatic substitution [2c] have appeared.

The rates of reaction of a series of *meta*- and *para*-substituted fluorobenzenes with piperidine in triethylene glycol at 194.5° correlate well[3] with σ^- ($\rho = +4.41$, $r = 0.996$). There is no evidence for benzyne formation. This strongly suggests that the same mechanism (S_NAr) operates for, say, *p*-fluorotoluene, as for *p*-fluoronitrobenzene.

A full paper has appeared [4] on the stopped-flow investigation of the hydrolysis of picryl chloride, reported last year.[5] The suggestion [5] that the intermediate observed was a 3-hydroxide adduct is considered, but rejected. In the normal 2-step mechanism for substitution, either step may be rate-determining, largely depending on the efficacy of the leaving group. Rate-determining formation of the intermediate is, however, normal with better leaving groups than F⁻ (e.g. Cl⁻), and in these cases base catalysis is either absent or quite

[1a] J. Miller, *Aromatic Nucleophilic Substitution*, Elsevier, Amsterdam and New York, 1968.

[1b] J. Miller, *Austral. J. Chem.*, **22**, 921 (1969).

[2a] F. Pietra, *Quart. Rev.*, **23**, 504 (1969).

[2b] P. Buck, *Angew. Chem. Intern. Ed. Engl.*, **8**, 120 (1969).

[2c] T. J. de Boer and I. P. Dirkx, in *The Chemistry of the Nitro and Nitroso Groups* (Ed. H. Feuer), Part I, Wiley, London, 1969, p. 487.

[3] C. L. Liotta and D. F. Pinholster, *Chem. Comm.*, **1969**, 1245.

[4] R. Gaboriaud and R. Schaal, *Bull. Soc. Chim. France*, **1969**, 2683.

[5] See *Org. Reaction Mech.*, **1968**, 187.

modest. An example of this behaviour is the reaction of 2,4-dinitrochlorobenzene with n-butylamine in chloroform.[6] Modest base catalysis by the amine or added nitrate ion is probably associated with the increased nucleophilicity of hydrogen-bonded species such as $H_2RN \cdots H–NHR$. This type of catalysis shows no saturation effect like that observed when catalysis of the second step is involved and when high catalyst concentrations change the rate-determining step back to the first addition step. Modest catalysis of the first step is also observed in the reactions of 1-chloro-4,7-dinitronaphthalene and 4-chloro-3-nitrobenzotrifluoride with piperidine, and there is no isotope effect with [1-²H]piperidine.[7] The result (at 25°) with 4-chloro-3-nitrobenzotrifluoride is in apparent conflict with earlier work of Brieux (done at 35°)[8a] who has however restated his position and provided an example (2-chloronitrobenzene) where the total rate is a linear function of [piperidine] at 75° but a non-linear function at 100°.[8b] Catalysis at both temperatures is small (2-fold increase over a 30-fold change of [piperidine]). With 1-fluoro-4,7-dinitronaphthalene, 1-fluoro-4-nitrobenzene, and 2,4-dinitrophenyl cyclohexyl ether an isotope effect is observed and decomposition of the intermediate is probably rate-determining.[7] It is reported[9] that a similar isotope effect with [1-²H]piperidine in reaction with *o-* and *p*-chloronitrobenzene in benzene is confined to the pyridine-catalysed reaction. The reaction of a series of 4-substituted 2-nitrobenzenes with piperidine in methanol[10a] is not base-catalysed; $\rho = +3.5$. There is also no base catalysis in the reactions of 2,4-dinitrochlorobenzene with anilines in protic–dipolar aprotic solvent mixtures.[10b] Acid catalysis in the reaction of thiophenoxide ion with 2,4-dinitrofluorobenzene could not be detected[11] and it was concluded that the first addition step is rate-determining, since rate-determining loss of fluoride is normally acid-catalysed. An interesting acid-catalysed reaction is the conversion[12] of *p*-nitrosophenol into *p*-nitrosophenetole in EtOH (equation 1).

[6] S. D. Ross, *Tetrahedron*, **25**, 4427 (1969).
[7] F. Pietra, D. Vitali, and S. Frediani, *J. Chem. Soc.*(B), **1968**, 1595.
[8a] See *Org. Reaction Mech.*, **1966**, 162.
[8b] E. Sanhueza, R. L. Toranzo, R. V. Caneda, and J. A. Brieux, *Tetrahedron Letters*, **1969**, 4917.
[9] E. S. Casado, *An. Fac. Quim. Farm. Univ. Chile*, **19**, 105 (1968); *Chem. Abs.*, **70**, 95928 (1969).
[10a] J. F. Bunnett, T. Kato, and N. S. Nudelman, *J. Org. Chem.*, **34**, 785 (1969).
[10b] P. S. Radhakrishnamurti and J. Sahu, *Can. J. Chem.*, **47**, 4499 (1969).
[11] J. F. Bunnett and N. S. Nudelman, *J. Org. Chem.*, **34**, 2038 (1969).
[12] Y. Furuya, I. Urasaki, K. Itoho, and A. Takashima, *Bull. Chem. Soc. Japan*, **42**, 1922 (1969).

There has been a further investigation[13] of electrolyte and micellar effects on the rate of hydrolysis of 2,4-dinitrofluorobenzene, and of salt effects[14] on the reactions of amines with 2,4-dinitrofluorobenzene.

There is an extensive summary of data on S_NAr reactions in Parker's review[15] on protic–dipolar aprotic solvent effects. It appears that the free energies of small anionic nucleophiles are raised in typical dipolar aprotic solvents. However, Haberfield has shown[16] that enthalpies of transfer for these anions from MeOH to DMF are often small, whereas the enthalpies of transfer for the transition states are all highly exothermic. These results are not of course in conflict, but the present writer suggests that the *entropies* of transfer may prove highly significant. Shein[17] has described a number of examples of reactions of *o*- and *p*-substituted chlorobenzenes where, while the reactivity order for the *p*-substituted examples is normal and unaffected by solvent, the reactivity order with *o*-substituents is solvent-dependent. The Russian group has continued[18] its extensive kinetic investigations of activated aromatic substitution in which $-CF_3$, $-SO_2CH_3$, and $-SO_2CF_3$ are used as activating groups. The reactions of aniline with aryl benzenesulphonates containing similar activating groups proceed with C–O bond fission.[19] Other reactions which have been investigated include: a series of cyclic imines, $(CH_2)_nNH$, with *p*-nitrofluorobenzene in DMSO, in which the highest rate was with $n = 3$;[20] *trans*-2-(*p*-X-phenyl)cyclopropylamines with 2,4-dinitrochlorobenzene;[21] 1-halogeno-2,4-dinitrobenzenes and 2-halogeno-5-nitropyridines with amines in acetone;[22] 2-chloronitrobenzene with ethanolamine;[23] 1-[^{14}C]methoxy-2-

[13] C. A. Bunton and L. Robinson, *J. Org. Chem.*, **34**, 780 (1969); see *Org. Reaction Mech.*, **1968**, 189.

[14] F. Naso and A. M. Piepoli, *Ricerca Sci.*, **38**, 427 (1968); *Chem. Abs.*, **70**, 36808 (1969).

[15] A. J. Parker, *Chem. Rev.*, **69**, 1 (1969).

[16] P. Haberfield, L. Clayman, and J. S. Cooper, *J. Am. Chem. Soc.*, **91**, 787 (1969); see also R. Fuchs, J. L. Bear, and R. F. Rodewald, *ibid.*, p. 5797.

[17] S. M. Shein and N. K. Danilova, *Reakts. Spos. Org. Soed.*, *Tartu Gos. Univ.*, **4**, 649 (1967); *Chem. Abs.*, **69**, 85884 (1968); *Zh. Org. Khim.*, **4**, 1940, 1947 (1968); *Chem. Abs.*, **70**, 28118, 28119 (1969); S. M. Shein and A. V. Evstifeev, *ibid.*, **5**, 919 (1969); *Chem. Abs.*, **71**, 38047 (1969); *Reakts. Spos. Org. Soed.*, *Tartu Gos. Univ.*, **5**, 817 (1969); *Chem. Abs.*, **70**, 76945 (1969).

[18] S. M. Shein, K. V. Solodova, and V. I. Govorchenko, *Zh. Obshch. Khim.*, **39**, 1634 (1969); S. M. Shein and A. D. Khmelinskaya, *Zh. Org. Khim.*, **4**, 2160 (1968); *Chem. Abs.*, **70**, 67799 (1969); S. M. Shein, F. V. Pishchugin, V. V. Brovko, and L. F. Chervatyuk, *Izv. Sib. Otd. Akad. Nauk SSSR*, *Ser. Khim.*, **4**, 95 (1968); *Chem. Abs.*, **70**, 67801 (1969); S. M. Shein, A. V. Evstifeev, and V. F. Sturichenko, *Reakts. Spos. Org. Soed.*, *Tartu, Gos. Univ.*, **5**, 384 (1968); *Chem. Abs.*, **70**, 46554 (1969); S. M. Shein and L. A. Suchkova, *ibid.*, **5**, 310 (1968).

[19] R. V. Vizgert, I. M. Ozdrovskaya, V. P. Nazaretyan, and L. M. Yagupolskii, *Zh. Org. Khim.*, **4**, 1511 (1968); R. V. Vizgert, I. M. Ozdrovskaya, and E. N. Ozdrovskii, *Zh. Org. Khim.*, **4**, 1812 (1968); *Chem. Abs.*, **70**, 19403 (1969).

[20] A. Fischer, R. E. J. Hutchinson, R. D. Topsom, and G. J. Wright, *J. Chem. Soc.*(B), **1969**, 544.

[21] A. Fischer and I. J. Miller, *J. Chem. Soc*(B), **1969**, 1135.

[22] T. O. Bamkole, C. W. L. Bevan, and J. Hirst, *Nigerian J. Sci.*, **2**, 11 (1968); *Chem. Abs.*, **70**, 36775 (1969).

[23] F. Hussain and I. U. Khand, *J. Sci. Res.* (Lahore), **1**, 39 (1966); *Chem. Abs.*, **69**, 105564 (1968).

nitronaphthalene with MeO^- where the kinetic order in MeO^- was $3/2$;[24] bromide exchange in 1-bromo-2,4(and 2,6)-dinitronaphthalene;[25] reactions of halogenonaphthoic acids with water and amines;[26] nucleophilic reactivity in a series of 3'- and 4'-substituted 4-chloro-3-nitroazobenzenes;[27] fluoride-ion catalysed methanolysis of picryl chloride;[28] hydrolysis of 1,3-dichloro-4,6-dinitrobenzene;[29] and the displacement of activated NO_2 groups by tervalent phosphorus reagents.[30]

Silver-ion catalysed solvolysis of *N*-t-butyl-*N*-chloroaniline in methanol proceeds[31a] as shown in equation (2). The chloroaniline products possibly

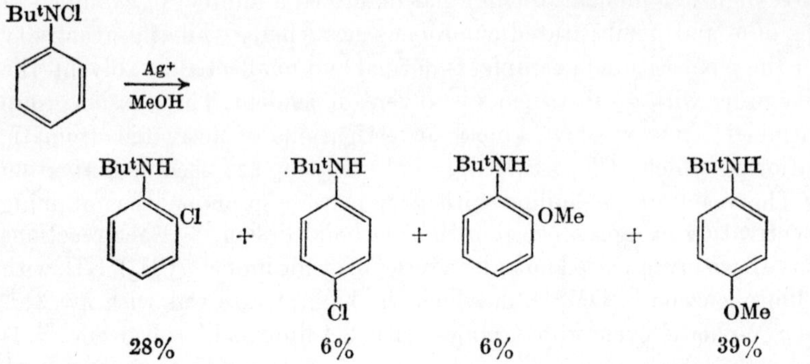

$$\dots (2)$$

result from internal return, but the methoxyanilines do not require a discrete cationic intermediate.[31b] The whole reaction is reminiscent of the acid-catalysed rearrangement of phenylhydroxylamine. Another interesting nucleophilic substitution occurs when complexes of polycyclic hydrocarbons with tetrachloro-*o*-benzoquinone are acidified with HCl and HBr. Aryl halides are formed and halide ion attack on hydrocarbon radical cations is suggested.[32]

In a study of reactions of a series of 3- and 4-substituted 1-chloroanthra-quinones with piperidine in DMF, a fair Hammett plot was obtained but σ^-

[24] N. A. Katsanos, *Z. Phys. Chem.* (Frankfurt), **63**, 168 (1969); *Chem. Abs.*, **71**, 29779 (1969).

[25] M. S. Sharan, *Indian J. Chem.*, **7**, 465 (1969); *Chem. Abs.*, **71**, 29786 (1969).

[26] V. N. Lisitsyn and L. A. Didenko, *Zh. Org. Khim.*, **4**, 1225 (1968); *Chem. Abs.*, **70**, 11392 (1969); **5**, 478 (1969).

[27] J. Kaválek, J. Beneš, J. Socha, F. Brtník, and M. Večeřa, *Coll. Czech. Chem. Comm.*, **34**, 2092 (1969).

[28] V. A. Sokolenko, *Reakts. Spos. Org. Soed.*, *Tartu Gos. Univ.*, **5**, 429 (1968), *Chem. Abs.*, **70**, 19310 (1969).

[29] W. Rauner and F. Wolf, *Z. Chem.*, **8**, 338 (1968).

[30] J. I. G. Cadogan, D. J. Sears, and D. M. Smith, *J. Chem. Soc.*(C), **1969**, 1314.

[31a] P. G. Gassman, G. Campbell, and R. Frederick, *J. Am. Chem. Soc.*, **90**, 7377 (1968).

[31b] See *Org. Reaction Mech.*, **1968**, 37.

[32] M. Wilk and U. Hoppe, *Ann. Chem.*, **727**, 81 (1969).

was not required for the appropriate 4-substituents. Steric inhibition of resonance by the neighbouring carbonyl was suggested to account for this.[33]

Copper-promoted reactions of aryl halides receive increasing attention, although no clear mechanistic picture has emerged even for those reactions occurring in homogeneous solution. Reduction often competes with substitution as in the reaction of various methoxy-substituted bromo- and iodo-benzenes with NaOMe–CuI in 2,4,6-collidine.[34] An *o*-OMe group promotes reduction. Reaction between potassium phthalimide and aryl bromides and iodides occurs smoothly in dimethylacetamide solution with cuprous salt catalysis; in this reaction, ring substituents exert a much attenuated effect in the normal direction for an S_NAr reaction.[35] Bacon has studied several other copper-promoted reactions.[36] A 4-centre transition state is written[37] for the reaction of aryl iodides with cuprous acetylides. Metal–halogen exchange competes with coupling in the reactions of lithium dialkyl- and diaryl-cuprates with aryl iodides.[38] Some other studies of copper-promoted systems are in ref. 39. The reaction of arylthallium bis(trifluoroacetates) with iodide probably proceeds by collapse of an arylthallium diiodide to the aryl iodide.[40] Nucleophilic attack on arene-cyclopentadienyl-iron cations normally occurs on the arene ring and from an *exo*-direction.[41]

The first step in the Smiles rearrangement is internal nucleophilic attack of a benzylic carbanion on a phenylsulphonyl ring. Drozd and his coworkers[42] who are investigating this reaction in detail have shown that in some cases an alternative addition step (1) is reversible. *ortho*-Substituted azobenzenes with aryl Grignard reagents give *o*-arylated products.[43]

[33] J. Šlosar, V. Štěřba, and M. Večeřa, *Coll. Czech. Chem. Comm.*, **34**, 2763 (1969).

[34] R. G. R. Bacon and J. R. Wright, *J. Chem. Soc.*(C), **1969**, 1978.

[35] R. G. R. Bacon and A. Karim, *Chem. Comm.*, **1969**, 578.

[36] R. G. R. Bacon and O. J. Stewart, *J. Chem. Soc.*(C), **1969**, 301; R. G. R. Bacon and S. C. Rennison, *ibid.*, pp. 308, 312.

[37] C. E. Castro, R. Havlin, V. K. Honwad, A. Malte, and S. Mojé, *J. Am. Chem. Soc.*, **91**, 6464 (1969).

[38] G. M. Whitesides, W. F. Fischer, J. San Filippo, R. W. Bashe, and H. O. House, *J. Am. Chem. Soc.*, **91**, 4871 (1969).

[39] F. M. Vainshtein and E. I. Tomilenko, *Zhur. Vses. Khim. Obshch.*, **13**, 709 (1968); *Chem. Abs.*, **70**, 76972 (1969); V. N. Lisitsyn, V. A. Shestakov, and E. D. Sycheva, *ibid.*, **14**, 117 (1969); *Chem. Abs.*, **70**, 114326 (1969); V. N. Lisitsyn and A. D. Tishchenko, *Izv. Vyssh. Ucheb. Zaved., Khim. Tekhnol.*, **10**, 1229 (1967); *Chem. Abs.*, **68**, 113730 (1968); C. Björklund and M. Nilsson, *Acta Chem. Scand.*, **22**, 2581 (1968); I. P. Gragerov and L. F. Kasukhin, *Zh. Org. Khim.*, **5**, 3 (1969); *Chem. Abs.*, **70**, 86736 (1969).

[40] A. McKillop, J. S. Fowler, M. J. Zelesko, J. D. Hunt, E. C. Taylor, and G. McGillivray, *Tetrahedron Letters*, **1969**, 2427.

[41] I. U. Khand, P. L. Pauson, and W. E. Watts, *J. Chem. Soc.*(C), **1969**, 116, 2024.

[42] V. N. Drozd with L. A. Nikonova, K. A. Pak, Y. A. Ustynyuk, and L. I. Zefirova, *Zh. Org. Khim.*, **4**, 1682, 1794, 1961 (1968); *ibid.*, **5**, 320, 325, 933, 1446 (1969); *Izv. Timiryazev Sel'sk. Akad.*, **1969**, 179, 208; *Chem. Abs.*, **70**, 3766, 19723, 28549, 105679, 105680 (1969).

[43] A. Risaliti, S. Bozzini, and A. Stener, *Tetrahedron*, **25**, 143 (1969).

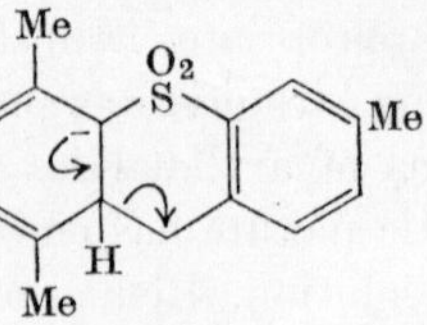

(1)

There have been further reviews of the reactions of hydrated electrons.[44] The reactions of benzene and toluene with lithium and sodium in hexamethyl-phosphoric triamide are strongly dependent on the concentration.[45] At low metal concentrations the major species is the solvated electron, and Birch reduction results. At high concentrations, the presence of metal monomer leads to ring metallation and probably benzyne formation, since with benzene the ESR spectra of naphthalene, biphenyl, and biphenylene radical anions can be detected.

Heterocyclic Systems

3-Bromo-4-nitrothiophen is more than 10^5 times slower in reaction with thiophenoxide than the 2-bromo-3-nitro-, 2-bromo-5-nitro-, and 3-bromo-2-nitro-isomers whose rates vary by only a factor of 2.[46] 3,4-Dinitrothiophen reacts with sodium arylthiolates to give an aryl 2-(4-nitrothienyl) sulphide through a cine-substitution.[47] It is not known if elimination–addition or addition–elimination is involved. However, the same compound with piperidine undergoes ring-scission to yield, 1,4-dipiperidino-2,3-dinitrobutadiene; presumably initial addition is involved.[48]

Nucleophilic attack on 4- and 5-nitroimidazoles results in displacement of the 5- or 4-hydrogen. Meisenheimer complexes were observed in this series in suitable cases.[49] A study of the nucleophilic reactivity of some 4- and 7-substituted 2-chloro-N-methylimidazoles shows, rather surprisingly, that steric as well as electronic effects are operating.[50a] With 2-bromo-5-nitro-1,3,4-thiadiazole, NaSPh replaces $-NO_2$, but AgSPh replaces bromine.[50b] Other studies in 5-ring systems include: reactions of 4-fluoronitro- and 4-fluorodinitro-benzimidazoles with peptide nucleophiles (2-fluoro-3,5-di-nitroaniline was also studied);[51] the hydrolysis of 2,3-dimethylbenzoxazolium

[44] M. Anbar, *Adv. Phys. Org. Chem.*, 7, 115 (1969); E. J. Hart, *Accounts Chem. Res.*, 2, 161 (1969).
[45] J.-E. Dubois and G. Dodin, *Tetrahedron Letters*, 1969, 2325.
[46] D. Spinelli, G. Guanti, and C. Dell'Erba, *Ricerca Sci.*, 38, 1051 (1968); *Chem. Abs.*, 71, 29788 (1969).
[47] C. Dell'Erba, D. Spinelli, and G. Leandri, *Gazzetta*, 99, 535 (1969).
[48] C. Dell'Erba, D. Spinelli, and G. Leandri, *Chem. Comm.*, 1969, 549.
[49] V. Sunjić, T. Fajdiga, M. Japelj, and P. Rems, *J. Heterocyclic Chem.*, 6, 53 (1969).
[50a] A. Ricci, G. Seconi, and P. Vivarelli, *Gazzetta*, 99, 542 (1969).
[50b] H. Newman, E. L. Evans, and R. A. Angier, *Tetrahedron Letters*, 1968, 5829.
[51] K. L. Kirk and L. A. Cohen, *J. Org. Chem.*, 34, 384, 395 (1969).

salts;[52] further studies with benzothiazoles and related systems;[53] methoxy-substitution in halogenonitro-2,1,3-benzoxadiazoles and thiadiazoles;[54] and nucleophilic displacements on some 2-substituted 5-aryl(and arylalkyl)-1,3,4-oxadiazoles.[55] The rate constants for reaction of nine primary and secondary amines with a range of halogenoheterocyclics have been reported.[56] Hydrolytic deamination of some N-heterocyclic amines in aqueous alkali has been studied kinetically.[57]

Methoxide ion displaces methylsulphinyl and methylthio groups from pyridines and related systems. While methylsulphinyl is almost as good a leaving group as methylsulphonyl, methylthio-derivatives are about 10^4 times less reactive. The kinetics are complicated by regeneration of MeO^- during subsequent oxidative reactions of MeS^- and $MeSO^-$.[58] Steric effects in the reactions of 2-RS-, 2-RSO-, and 2-RSO_2-quinoxalines with MeO^- are quite variable. Changing R from Me to Bu^t causes 1.8-, 140-, and 24-fold rate decreases respectively.[59] The reaction of 2-methylsulphonylquinoline with nucleophiles has been studied,[60a] and a $\sigma\rho$ study of reactions of 2-(p-X-phenyl-sulphinyl and -sulphonyl)pyrimidines is reported.[60b] Reaction of 4-nitropyridine with MeO^- is 38 times faster at 50° than reaction of the 2-nitro-isomer,[61] and the kinetics have been reported for reaction of 22 substituted derivatives of 4-nitroquinoline N-oxide with sodium ethoxide.[62] Reactions between 2-halogeno-5-nitropyridines and aniline are base-catalysed in acetone, but not in methanol.[63] The hydrolyses of 2-halogeno-3(and 5)-nitropyridines are acid-catalysed in sulphuric acid. ϕ-Values from a Bunnett–Olsen plot suggested that a slow proton-transfer to water was occurring.[64] Two of the four pathways open to reaction of N-alkoxy-pyridinium and -quinolinium salts with nucleophiles involve attack at the 2-position.[65] Loss of ROH yields a 2-substituted pyridine, but an alternative pathway

[52] L. Oliveros and H. Wahl, *Bull. Soc. Chim. France*, **1969**, 2815.

[53] P. E. Todesco with G. Bartoli, L. DiNunno, S. Florio, and A. Latrofa, *Boll. Sci. Fac. Chim. Ind. Bologna*, **27**, 63, 69, 75, 79 (1969); *Chem. Abs.*, **71**, 49017 (1969).

[54] D. Dal Monte, E. Sandri, L. DiNunno, S. Florio, and P. E. Todesco, *Chim. Ind.* (Italy), **51**, 987 (1969).

[55] R. Madhavan and V. R. Srinivasan, *Indian J. Chem.*, **7**, 760 (1969).

[56] H. Grube and H. Suhr, *Chem. Ber.*, **102**, 1570 (1969).

[57] E. Kalatzis, *J. Chem. Soc.*(B), **1969**, 96.

[58] G. B. Barlin and W. V. Brown, *J. Chem. Soc.*(B), **1968**, 1435; see *Org. Reaction Mech.*, **1967**, 174.

[59] G. B. Barlin and W. V. Brown, *J. Chem. Soc.*(B), **1969**, 333; see also *ibid.*(C), p. 921.

[60a] E. Hayashi and T. Saito, *J. Pharm. Soc. Japan*, **89**, 74 (1969).

[60b] D. J. Brown and P. W. Ford, *J. Chem. Soc.*(C), **1969**, 2720.

[61] A. Dondoni, A. Mangini, and G. Mossa, *J. Heterocyclic Chem.*, **6**, 143 (1969).

[62] T. Okamoto and M. Mochizuki, *Chem. Pharm. Bull.* (Japan), **17**, 987 (1969).

[63] T. Bamkole and J. Hirst, *J. Chem. Soc.*(B), **1969**, 848.

[64] J. D. Reinheimer, J. T. McFarland, R. A. Amos, J. M. Wood, M. Zahniser, and W. Bowman, *J. Org. Chem.*, **34**, 2068 (1969).

[65] A. R. Katritzky and E. Lunt, *Tetrahedron*, **25**, 4291 (1969).

is shown in **(2)**. The factors which influence the choice of pathway were discussed.

(2)

The nature of the small but surprising activation by *o*-methyl groups in nucleophilic substitutions of bromo-pyridines and -picolines by thiophenoxide ion has again been discussed in terms of London forces and ion–dipole effects.[66] However, a simple MO treatment is said to predict higher reactivity at the 2- rather than the 4- or 6-position in 3-substituted pyridines.[67] The reaction of pyridines with ketones and magnesium amalgam leads to 2-substitution,[68] as does the reaction of pyridine *N*-oxides with iminochlorides or nitrilium salts.[69] Competition between attack at the 4-position and at the cyano groups occurs in reactions of 3,5-dicyanopyridines and Grignard reagents.[70] The kinetics of the reactions of piperidine and ammonia with 8-chloro-5-nitroquinoline have been reported.[71] Hydrolyses of 9-aminoacridines have been studied.[72]

van der Plas and his coworkers have continued their studies of ring transformations in reactions of heterocyclic compounds with nucleophiles, especially potassium amide in liquid ammonia.[73] Other studies involving ring-opening are with pyrylium ions;[74] 3-aza-pyrylium[75] and -thiopyrylium ions;[76] and pyridazine *N*-oxides.[77] Methoxide ion attack on 2-trichloromethylpyrazine occurs as shown in equation (3).[78] Other reactions studied include: 1-cyano-2-

[66] R. A. Abramovitch, F. Helmer, and M. Liveris, *J. Org. Chem.*, **34**, 1730 (1969); see *Org. Reaction Mech.*, **1965**, 141; **1967**, 173.
[67] V. N. Drozd, V. I. Minkin, and Y. A. Ostroumov, *Zh. Org. Khim.*, **4**, 1501 (1968).
[68] R. A. Abramovitch and A. R. Vinutha, *J. Chem. Soc.*(C), **1969**, 2104.
[69] R. A. Abramovitch and G. M. Singer, *J. Am. Chem. Soc.*, **91**, 5672 (1969).
[70] J. Kuthan and L. Musil, *Coll. Czech. Chem. Comm.*, **34**, 3173 (1969); J. Paleček, K. Vondra, and J. Kuthan, *ibid.*, p. 2991.
[71] S. M. Shein, N. I. Skrebkova, and N. N. Vorozhtsov, *Khim. Geterotsikl. Soed., Sbl. Azotsoderzhashch. Geterotsikl.*, **1967**, 265; *Chem. Abs.*, **70**, 76977 (1969).
[72] A. Ledochowski and E. Zylkiewicz, *Roczn. Chem.*, **43**, 291 (1969); *Chem. Abs.*, **71**, 12218 (1969).
[73] H. C. van der Plas and H. Jongejan, *Rec. Trav. Chim.*, **87**, 1065 (1968); H. C. van der Plas and B. Zuurdeeg, *ibid.*, **88**, 426 (1969); H. W. van Meeteren, H. C. van der Plas, and D. A. de Bie, *ibid.*, **88**, 728 (1969); H. C. van der Plas, B. Zuurdeeg, and H. W. van Meeteren, *ibid.*, **88**, 1156 (1969).
[74] K. Dimroth and G. Laubert, *Angew. Chem. Intern. Ed. Engl.*, **8**, 370 (1969).
[75] R. R. Schmidt, D. Schwille, and U. Sommer, *Ann. Chem.*, **723**, 111 (1969).
[76] R. R. Schmidt and D. Schwille, *Chem. Ber.*, **102**, 269 (1969).
[77] H. Igeta, T. Tsuchiya, and T. Nakai, *Tetrahedron Letters*, **1969**, 2667.
[78] E. J. J. Grabowski, E. W. Tristram, R. Tull, and P. I. Pollak, *Tetrahedron Letters*, **1968**, 5931.

$$\cdots (3)$$

75%   15%   10%

nitrophenazine with amines;[79] 2-aminopyrimidines with hydroxide;[80] substituted 2-chloropyrimidines with piperidine;[81] thiohydrolysis of 6-methyl-thio-8-(3- or 4-pyridyl)purines;[82] hydration of 4-trifluoromethylpteridines;[83] H–D exchange in uracil derivatives, involving hydration;[84] 2-amination of pyrido[2,3-d]pyridazine and reactions of 5- and 8-chloropyrido[2,3-d]pyridazine with nucleophiles;[85a] substitution in chlorofuro[2,3-d]pyridazines;[85b] hydrolysis of adenosine to inosine by takadiastase adenosine deaminase in which rate-limiting formation of a tetrahedral intermediate is favoured;[86] alkaline hydrolysis of monochloro-s-triazines (a $\sigma\rho$ study of 3- and 5-substituents);[87] reactions of trichloro-s-triazine with nucleophiles in the presence of tertiary amines;[88] monochloro-s-triazines with soluble proteins;[89] C- and N-attack by N,N-dialkylarylamines and C-attack by O-alkyl naphthyl ethers on cyanuric chloride;[90] and reaction of o-aminophenol with hexachlorocyclotriphosphaza-triene in which complete destruction of the ring system occurs.[91]

Nucleophilic substitution occurs in reaction of tris-(3-bromopentane-2,4-dionato)M(III) with thiophenols when M(III) is aluminium, cobalt, and chromium.[92]

[79] S. Pietra, G. Casiraghi, and F. Rolla, *Gazzetta*, **99**, 665 (1969).
[80] S. Munoz, *An. Fac. Quim. Farm.*, *Univ. Chile*, **19**, 76 (1967); *Chem. Abs.*, **70**, 76974 (1969).
[81] V. P. Mamaev, O. A. Zagulyaeva, S. M. Shein, A. I. Shvets, and V. P. Krivopalov, *Reakts. Spos. Org. Soed.*, *Tartu Gos. Univ.*, **5**, 824 (1968); *Chem. Abs.*, **70**, 76976 (1969).
[82] F. Bergmann and M. Rashi, *J. Chem. Soc.*(C), **1969**, 1831.
[83] J. Clark and W. Pendergast, *J. Chem. Soc.*(C), **1969**, 1751.
[84] R. J. Cushley, S. R. Lipsky, and J. J. Fox, *Tetrahedron Letters*, **1968**, 5393.
[85a] D. B. Paul and H. J. Rodda, *Austral. J. Chem.*, **22**, 1745, 1759 (1969).
[85b] M. Robba, M.-C. Zaluski, B. Roques, and M. Bonhomme, *Bull. Soc. Chim. France*, **1969**, 4004.
[86] R. Wolfenden, *Biochemistry*, **8**, 2409 (1969); R. Wolfenden, J. Kaufman, and J. B. Macon, *ibid.*, p. 2412.
[87] T. N. Bykhovskaya and O. N. Vlasov, *Reakts. Spos. Org. Soed.*, *Tartu Gos. Univ.*, **4**, 510 (1967); *Chem. Abs.*, **69**, 43160 (1968).
[88] B. I. Stepanov and G. I. Migachev, *Khim. Geterotsikl. Soed.*, **1969**, 354.
[89] J. Shore, *J. Soc. Dyers Colourists*, **84**, 545 (1968).
[90] R. A. Shaw and P. Ward, *J. Chem. Soc.*(B), **1968**, 1431; **1969**, 596.
[91] H. R. Allcock and R. L. Kugel, *Chem. Comm.*, **1968**, 1606; *J. Am. Chem. Soc.*, **91**, 5452 (1969).
[92] Z. Yoshida, H. Ogoshi, and Y. Shimidzu, *Tetrahedron Letters*, **1969**, 2255.

Meisenheimer and Related Complexes

Interest in this area continues at a high level and a good review has appeared.[93] SCF–CI calculations have been carried out on the 1,3,5-trinitropentadienyl anion, and related cyclohexadienyl anions have been studied by a "composite molecule" method.[94] Full details of two crystal structures containing 6,6-dialkoxy-1,3,5-trinitrocyclohexadienyl anions have appeared.[95] A stopped-flow method has been used to detect unstable Meisenheimer complexes and study their conversion into stable isomers.[96] In the reaction of methoxide with 2-cyano-4,6-dinitroanisole, two transients were observed; the suggested sequence is in equation (4). The kinetics of decomposition of 6-alkoxy-6-methoxy-1,3,5-trinitrocyclohexadienyl anions in water have been studied.[97]

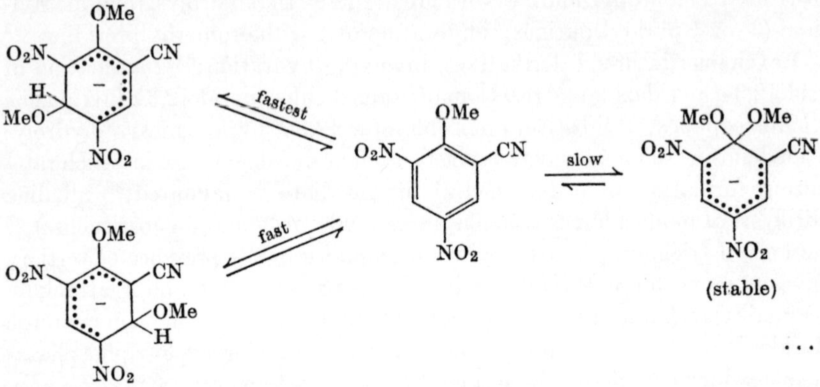

$$\ldots (4)$$

Calorimetry has been used to detect transient Meisenheimer complexes and to find ΔH for several complexes; in the linear relationships of ΔH, ΔG, and $T\Delta S$ with log K, substituents had opposed effects on ΔH and $T\Delta S$.[98] Full kinetic and equilibrium data have been obtained for (3) and (4),[99] and for complexes from methoxide addition to the isomeric 2,4,6-cyanodinitroanisoles.[100] Equilibrium constants for complex-formation between CN^- and 1,3,5-trinitrobenzene vary from 39 to 5×10^5 l mole^{-1} on changing from a methanol to t-butanol solvent.[101] The NMR spectrum at $-30°$ shows cyanide

93 M. R. Crampton, *Adv. Phys. Org. Chem.*, **7**, 211 (1969).

94 H. Hosoya, S. Hosoya, and S. Nagakura, *Theor. Chim. Acta*, **12**, 117 (1968).

95 H. Ueda, N. Sakabe, J. Tanaka, and A. Furusaki, *Bull. Chem. Soc. Japan*, **41**, 2866 (1968); R. Destro, C. M. Gramaccioli, and M. Simonetta, *Acta Cryst.*, B**24**, 1369 (1968).

96 F. Terrier and F. Millot, *Compt. Rend.*, *C*, **268**, 808 (1969); *Bull. Soc. Chim. France*, **1969**, 2692.

97 S. S. Gitis, A. Y. Kaminskii, A. I. Glaz, and Z. N. Kozina, *Reakts. Spos. Org. Soed.*, *Tartu Gos. Univ.*, **4**, 625 (1967); *Chem. Abs.*, **69**, 43164 (1968).

98 J. W. Larsen, J. H. Fendler, and E. J. Fendler, *J. Am. Chem. Soc.*, **91**, 5903 (1969).

99 E. J. Fendler, J. H. Fendler, W. E. Byrne, and C. E. Griffin *J. Org. Chem.*, **33**, 4141 (1968).

100 J. H. Fendler, E. J. Fendler, and C. E. Griffin, *J. Org. Chem.*, **34**, 689 (1969).

101 E. Buncel, A. R. Norris, W. Proudlock, and K. E. Russell, *Can. J. Chem.*, **47**, 4129 (1969).

attachment to either 1- or 3-positions of 2,4,6-trinitroanisole.[102] NMR, IR, and UV data have been obtained at comparable concentrations at −30° for the 6-cyano-1,3,5-trinitrocyclohexadienyl anion,[103] and an attempt made to analyse the IR spectra.[104] Reaction of picrate with aqueous sulphite and hydroxide results in double addition to give multiply charged anions.[105] 2:1-Complexes of methoxide with 1,3-dinitro- and 1,3,5-trinitro-benzene

(3)   (4)   (5)

have been discussed.[106] Addition of chloroform to solutions of 1,3,5-trinitro-benzene and sodium methoxide in MeOH–DMSO leads to an equilibrium between the 6-methoxy- and the 6-trichloromethyl-1,3,5-trinitrocyclo-hexadienyl anions. Trinitroanisole yields the 6-trichloromethyl-2-methoxy anion.[107] The dimsyl anion adduct of 1,3,5-trinitrobenzene has been identi-fied,[108] as have MeO$^-$ and MeCOCH$_2$$^-$ adducts of 3,5-dinitrobenzo-nitrile and -trifluoride,[109a] trinitroanisole,[109b] and the MeCOCH$_2$$^-$ complex of 1,3-dinitrobenzene.[109c] 1,3,5-Trinitrobenzene, triethylamine, and propionaldehyde form the expected Janovsky complex (the first from an aldehyde).[110] The complexes formed from acetone and 1,3,5-trinitrobenzene in the presence of a large excess of alkali[111a] and Janovsky complexes of benzenes containing trifluoromethyl and trifluoromethylsulphonyl groups have been discussed on the basis of UV evidence.[111b] There has been further discussion of the com-plexes with a bicyclo[3.3.1]nonane skeleton formed from (RCH$_2$)$_2$CO and

[102] A. R. Norris, *Can. J. Chem.*, **47**, 2895 (1969).

[103] A. R. Norris, *J. Org. Chem.*, **34**, 1486 (1969).

[104] A. R. Norris and H. F. Shurvell, *Can. J. Chem.*, **47**, 4267 (1969).

[105] M. R. Crampton and M. El-Ghariani, *J. Chem. Soc.*(B), **1969**, 330.

[106] S. S. Gitis, A. Y. Kaminskii, N. A. Pankova, E. G. Kaminskaya, and I. G. L'vovich, *Zh. Org. Khim.*, **4**, 1979 (1968).

[107] S. M. Shein, A. D. Khmelinskaya, and V. V. Brovko, *Chem. Comm.*, **1969**, 1043.

[108] C. A. Fyfe, M. I. Foreman, and R. Foster, *Tetrahedron Letters*, **1969**, 1521.

[109a] M. I. Foreman and R. Foster, *Can. J. Chem.*, **47**, 729 (1969).

[109b] M. Kimura, N. Obi, and M. Kawazoi, *Chem. Pharm. Bull.* (Japan), **17**, 531 (1969).

[109c] M. Kimura, M. Kawata, M. Nakadate, N. Obi, and M. Kawazoe, *Chem. Pharm. Bull.* (Japan), **16**, 634 (1968).

[110] M. J. Strauss, *Tetrahedron Letters*, **1969**, 2021.

[111a] A. Y. Kaminskii and S. S. Gitis, *Zh. Org. Khim.*, **4**, 504 (1968); *Chem. Abs.*, **68**, 113843 (1968).

[111b] A. Y. Kaminskii and S. S. Gitis, *Zh. Org. Khim.*, **4**, 1826; *Chem. Abs.*, **70**, 19697 (1969).

trinitrobenzene, especially of their stereochemistry.[112a] Since the formation of the bicyclic complex with acetone requires diethylamine, and triethylamine is ineffective, an enamine intermediate is suggested.[112b] Picryl phenyl ether and t-butylamine form a Meisenheimer complex which subsequently goes on to the picramide.[113] Reaction of 1,3-dinitrobenzene with alkaline hydroxylamine leads to the 2:1 adduct (5), which subsequently decays to starting 1,3-dinitrobenzene, 2,4-dinitroaniline, and 1,3-diamino-2,4-dinitrobenzene.[114] Acrylonitrile, triethylamine, and trinitrobenzene form a 1:1:1 adduct, for which a zwitterionic structure is suggested,[115] and a Meisenheimer complex may be formed from cyanide and cis-$Pt(Cl)(Picryl)(Ph_3P)_2$.[116]

The Meisenheimer complex from methoxide ion and 2-methoxy-3,5-dinitrothiophen has been isolated and is somewhat more stable than that from trinitroanisole[117]. The methoxide adducts of 2- and 4-methoxy-3,5-dinitropyridine have been discussed further.[118a] With the former,[118b] methoxide ion addition is at the 6-position and, in methanol, there is competition from dealkylation. The crystal structure is reported of the methoxide adduct of 6-methoxy-3,5-dinitrobenzofurazan.[119] Nucleophiles add to the 4-position of 3-nitro-N-methylpyridinium iodide,[120] and NMR evidence is presented for addition of CN^- to the 4-position of several N-methylpyridinium ions.[121] Several 1-lithio-2-phenyl-1,2-dihydropyridines have been isolated and characterized by NMR.[122] Benzyl-lithium, unlike alkyl- and phenyl-lithiums, attacks the 4-position of 3-substituted pyridines.[123]

Substitution in Polyhalogenoaromatic Compounds

Little previous work has been reported on polybromo-compounds, but nucleophilic attack on hexabromobenzene has now been studied.[124] Methoxide ion

[112a] T. Momose, Y. Ohkura, and K. Kohashi, *Chem. Pharm. Bull.* (Japan), **17**, 858 (1969); M. I. Foreman, R. Foster, and M. J. Strauss, *J. Chem. Soc.*(C), **1969**, 2112.

[112b] M. J. Strauss and H. Schran, *J. Am. Chem. Soc.*, **91**, 3974 (1969).

[113] L. B. Clapp, H. Lacey, G. G. Beckwith, R. M. Srivastava, and N. Muhammad, *J. Org. Chem.*, **33**, 4262 (1968).

[114] S. S. Gitis, N. A. Pankova, A. Y. Kaminskii, and E. G. Kaminskaya, *Zh. Org. Khim.*, **5**, 65 (1969); *Chem. Abs.*, **70**, 87178 (1969).

[115] M. J. Strauss and R. G. Johanson, *Chem. Ind.* (London), **1969**, 242.

[116] A. R. Norris and M. C. Baird, *Can. J. Chem.*, **47**, 3003 (1969).

[117] G. Doddi, G. Illuminati, and F. Stegel, *Chem. Comm.*, **1969**, 953.

[118a] P. Bemporad, G. Illuminati, and F. Stegel, *J. Am. Chem. Soc.*, **91**, 6742 (1969).

[118b] C. Abbolito, C. Iavarone, G. Illuminati, F. Stegel, and A. Vazzoler, *J. Am. Chem. Soc.*, **91**, 6746 (1969).

[119] G. G. Messmer and G. J. Palenik, *Chem. Comm.*, **1969**, 470.

[120] T. Severin, H. Lerche, and D. Bätz, *Chem. Ber.*, **102**, 2163 (1969).

[121] R. Foster and C. A. Fyfe, *Tetrahedron*, **25**, 1489 (1969).

[122] C. S. Giam and J. L. Stout, *Chem. Comm.*, **1969**, 142.

[123] R. A. Abramovitch and G. A. Poulton, *J. Chem. Soc.*(B), **1969**, 901.

[124] I. Collins and H. Suschitzky, *J. Chem. Soc.*(C), **1969**, 2337.

causes protodebromination, but with amines nucleophilic substitution is competitive with attack at bromine. The reaction between pentachlorophenyl-lithium and aryl nitriles leads to 2,4-diaryl-5,6,7,8-tetrachloroquinazolines.[125] The unusual attack at the 3-position in octafluorofluorenone, discussed last year, is ascribed to unusual conjugative interaction between this position and the carbonyl, corroborated by ^{19}F shieldings.[126] Methoxide ion attack on octafluorobiphenylene under mild conditions gives a preponderance of 2-substitution.[127] There is competitive *para* and side-chain substitution with perfluorostyrene and methoxide.[128] Other reactions studied include: nucleophilic attack on 3,3′,5,5′-tetrachlorohexafluorobiphenyl;[129] substitution in polychloro-mono- and -di-cyanobenzenes;[130] reactions of pentafluorophenyl-diphenylphosphine, its oxide, and sulphide;[131] attack of amines on polyfluoro-arylamine oxides;[132a] and substitution in pentachloronitrobenzene.[132b] Hexafluorobenzene with potassium fluoride at 550° yields perfluorotoluene and perfluoroxylenes.[133] It is suggested that CF_2 is formed from the heptafluoro-cyclohexadienyl anion. Possibly the heptafluorobicyclo[3.1.0]hexenyl anion is an intermediate.

Musgrave has discussed some interesting examples of acid-catalysed nucleophilic substitutions in perfluoroheterocyclic compounds.[134] Nucleophilic attack on the perfluoropyridazinium cation occurs at C-3 and C-6 in contrast to the C-4 and C-5 attack with the neutral substrate.[135] Kinetic and thermodynamic control of the perfluoroalkylation of perfluoropyridine by perfluoropropene and KF yields the 2,4,5- and 2,4,6-tri(perfluoroisopropyl) isomers respectively.[136] Tetrafluoropyrimidine with iodide gives the 6-iodo-compound; further substitution is in the 4-position.[137] The formation of 2,3,5-trichloro-pyridine and other products from lithium aluminium hydride reduction of pentachloropyridine is rationalized on the basis of 3,4-addition of AlH_3–H^-.[138]

[125] D. J. Berry, J. D. Cook, and B. J. Wakefield, *Chem. Comm.*, **1969**, 1273.
[126] R. D. Chambers and D. J. Spring, *Tetrahedron*, **25**, 565 (1969).
[127] D. V. Gardner, J. F. W. McOmie, P. Albriktsen, and R. K. Harris, *J. Chem. Soc.*(C), **1969**, 1994.
[128] A. E. Pedler, J. C. Tatlow, and A. J. Uff, *Tetrahedron*, **25**, 1597 (1969).
[129] N. Ishikawa and S. Hayashi, *J. Chem. Soc. Japan*, **90**, 913 (1969).
[130] G. Beck and E. Degener, *Ann. Chem.*, **716**, 47 (1968).
[131] J. Burdon, I. N. Rozhkov, and G. M. Perry, *J. Chem. Soc.*(C), **1969**, 2615.
[132a] D. Price, H. Suschitzky, and J. I. Hollies, *J. Chem. Soc.*(C), **1969**, 1967.
[132b] See ref. 196b.
[133] V. E. Platonov, N. V. Ermolenko, G. G. Yacobson, and N. N. Vorozhtsov, *Izv. Akad. Nauk SSSR, Ser. Khim.*, **1968**, 2752; *Chem. Abs.*, **70**, 87166, 114726 (1969).
[134] W. K. R. Musgrave, *Chem. Ind.* (London), **1969**, 943.
[135] R. D. Chambers, J. A. H. MacBride, and W. K. R. Musgrave, *J. Chem. Soc.* (C), **1968**, 2989.
[136] R. D. Chambers, R. P. Corbally, J. A. Jackson, and W. K. R. Musgrave, *Chem. Comm.*, **1969**, 127.
[137] R. E. Banks, D. S. Field, and R. N. Haszeldine, *J. Chem. Soc.*(C) **1969**, 1866.
[138] F. Binns, S. M. Roberts, and H. Suschitzky, *Chem. Comm.*, **1969**, 1211.

Tetrafluorothiophen with sodium methoxide in methanol give 5-methoxy-trifluorothiophen.[139]

Other Reactions

Photoinduced aromatic substitution continues to provide some surprising results. Photolysis of 1-nitronaphthalene in the presence of alkyl chlorides or chloroform produces 1-chloronaphthalene in high yield; the triplet is implicated.[140] Havinga's group have also investigated the photoamination of nitrobenzene and derivatives (apparently p-nitroaniline arises from the nitrobenzene radical anion),[141] and the photosolvolysis of 5-X-3-nitrophenyl phosphates (both O–C and O–P cleavage are observed) and 3-nitrophenyl sulphate.[142] Attack of CN^- on nitroanisoles occurs *meta* to NO_2, and the efficiency of CN^- interception of photoexcited 4-nitroanisole was assessed in competition experiments.[143] Rates and products in these reactions are subject to complex solvent effects,[144] possibly related to solvent effects on $n \rightarrow \pi^*$ and $\pi \rightarrow \pi^*$ bands. Photohydrolysis of 4-nitroanisole is sensitized by benzophenone.[145] m-Halogenophenols and m-halogenoanisoles give both substitution and reduction products on photolysis in alcohols.[146]

A study of the small effects of salts on the rate of disappearance of PhN_2^+ in aqueous solution led to the conclusion that the rate- and product-determining steps are the same.[147] If there is an intermediate, its selectivity is very low. A definite diazirine intermediate is now abandoned for the nitrogen scrambling reaction.[148] It is concluded that there is no common intermediate in the thermal and photochemical hydrolyses of diazonium salts.[149] The following reactions have received attention: decomposition of benzenediazonium chloride in various solvents;[150a] hydride abstraction by diazonium salts from dialkylanilines;[150b] triarylsulphonium salts with phenyl-lithium;[151] and with alkoxides and thioalkoxides, where S_NAr reaction may compete with the

[139] J. Burdon, J. G. Campbell, I. W. Parsons, and J. C. Tatlow, *Chem. Comm.*, **1969**, 27.
[140] G. Fráter and E. Havinga, *Tetrahedron Letters*, **1969**, 4603.
[141] A. van Vliet, J. Cornelisse, and E. Havinga, *Rec. Trav. Chim.*, **88**, 1339 (1969).
[142] R. O. de Jongh and E. Havinga, *Rec. Trav. Chim.*, **87**, 1318, 1327 (1968).
[143] R. L. Letsinger and J. H. McCain, *J. Am. Chem. Soc.*, **91**, 6425 (1969).
[144] R. L. Letsinger and R. R. Hautala, *Tetrahedron Letters*, **1969**, 4205.
[145] R. L. Letsinger and K. E. Steller, *Tetrahedron Letters*, **1969**, 1401.
[146] J. T. Pinhey and R. D. G. Rigby, *Tetrahedron Letters*, **1969**, 1271.
[147] E. S. Lewis, L. D. Hartung, and B. M. McKay, *J. Am. Chem. Soc.*, **91**, 419 (1969).
[148] E. S. Lewis and R. E. Holliday, *J. Am. Chem. Soc.*, **91**, 426 (1969); see also E. S. Lewis and P. G. Kotcher, *Tetrahedron*, **25**, 4873 (1969).
[149] E. S. Lewis, R. E. Holliday, and L. D. Hartung, *J. Am. Chem. Soc.*, **91**, 430 (1969).
[150a] M. Y. Turkina, A. F. Levit, B. V. Chiznov, and I. P. Gragerov, *Zh. Org. Khim.*, **5**, 1238 (1969).
[150b] H. Suschitzky and C. F. Sellars, *Tetrahedron Letters*, **1969**, 1105.
[151] Y. H. Khim and S. Oae, *Bull. Chem. Soc. Japan*, **42**, 1968 (1969); see also *ibid.*, p. 1622.

radical cleavage of tetravalent intermediates.[152] Nesmeyanov's contributions to 'onium chemistry have been reviewed.[153]

Benzyne and Related Intermediates

Aryne chemistry is discussed in a new book.[154] CNDO/2 calculations show *o*- and *m*-benzynes as singlets in their ground state, but *p*-benzyne as a triplet.[155] A study by extended Hückel theory of all possible dehydro-pyridines and -diazines shows 3,4-dehydropyridine as considerably more stable than 2,3-dehydropyridine.[156] The latter, and the 2-pyridyl anion, are destabilized by the neighbouring nitrogen lone-pair. The position of anion capture in unsymmetrical arynes was also considered; for example, a 2,3-dehydropyridine may yield 2-substituted products almost exclusively. SCF calculations on some of these species show triplet states to be only a little higher in energy than the singlets.[157] Another review on reactions of aromatic compounds at high temperatures has appeared,[158] and reactions of benzyne complexes have been discussed.[159]

The reactions of phenyl-lithium with the halogenotoluenes have been carefully studied, including the effects of added piperidine and *N,N,N',N'*-tetramethylethylenediamine.[160] Biaryl formation from fluoro-, chloro-, and bromo-toluenes was exclusively by aryne formation, but displacement apparently competes in the reactions of iodotoluenes. Complications due to metal–halogen exchange occur with the bromo- and iodo-compounds. The reactions of halogenotoluenes with dimsyl anion have also been studied. This time $S_N Ar$ reaction competed with aryne formation with the fluoro-compounds.[161] Studies have been made of the reactions of aryl halides with soda-mide in hexamethylphosphoric triamide (HMPA) and with a combination of sodamide and a sodium t-alkoxide (which appears to be an unusually powerful base) in THF–HMPA in the presence of amines and other nucleophiles.[162] Mechanisms have been suggested for alkaline fusion and copper-catalysed alkaline hydrolysis of halogenophenols.[163]

[152] J. W. Knapczyk and W. E. McEwen, *J. Am. Chem. Soc.*, **91**, 145 (1969).
[153] O. A. Reutov, L. G. Makarova, and T. P. Tolstaya, *Zh. Org. Khim.*, **5**, 1521 (1969).
[154] T. L. Gilchrist and C. W. Rees, *Carbenes, Nitrenes, and Arynes*, Nelson, London, 1969.
[155] M. D. Gheorghiu and R. Hoffmann, *Rev. Roum. Chim.*, **14**, 947 (1969).
[156] W. Adam, A. Grimison, and R. Hoffmann, *J. Am. Chem. Soc.*, **91**, 2590 (1969).
[157] T. Yonezawa, H. Konishi, and H. Kato, *Bull. Chem. Soc. Japan*, **42**, 933 (1969).
[158] E. K. Fields and S. Meyerson, *Accounts Chem. Res.*, **2**, 273 (1969).
[159] I. Tabushi, K. Fujita, K. Okazaki, H. Yamada, and R. Oda, *J. Chem. Soc. Japan, Ind. Chem. Sect.*, **72**, 1677 (1969).
[160] L. Friedman and J. F. Chlebowski, *J. Am. Chem. Soc.*, **91**, 4864 (1969).
[161] D. Lorenz, *Z. Chem.*, **9**, 141 (1969).
[162] P. Caubère and N. Derozier, *Bull. Soc. Chim. France*, **1969**, 1737; P. Caubère and M. F. Hochu, *ibid.*, p. 2854.
[163] S. Oae, N. Furukawa, and T. Asari, *Bull. Chem. Soc. Japan*, **42**, 177 (1969).

Full details of the generation of arynes by aprotic diazotization of anthranilic acids have appeared.[164] Non-benzyne reactions of diazotized anthranilic acid are in ref. 165. Cyclohex-2-en-5-yne-1,4-dione (benzyne quinone) has been generated by lead tetra-acetate oxidation of 1-aminobenzotriazole-4,7-dione and trapped in 40% yield by tetracyclone.[166] Full details of the generation of benzyne by the 1-aminobenzotriazole route have appeared.[167] One possible reason for the high yields of biphenylene obtained by this route is the formation of a lead–benzyne complex. Other oxidizing agents were investigated. Full details have also appeared of the similar generation of 1,8-naphthalyne, and its stereospecific addition to olefins.[168] 1-Nitrosobenzotriazole is deoxygenated by ethyl diphenylphosphinite to yield benzyne, trapped by tetracyclone in 25% yield.[169] o-Phenylene carbonate plus trivalent phosphorus nucleophiles is not a useful source of benzyne.[170]

Benzyne adds to both the A- and B-rings of 1,4-dimethoxyanthracene, and this competition has been utilized to check the identity of benzyne from various precursors. Diphenyliodonium 2-carboxylate gave a different ratio from those of other precursors, but this was shown to be due to the formation of a less reactive iodobenzene complex of the anthracene.[171a] Study of substituent effects in reactions of this type confirms the electrophilic character of benzyne.[171b] The selectivity of benzyne towards aniline and potassium anilide has been examined. The nucleophilicities of these two differed most for attack of benzyne at their 4-positions, and least at their 2-positions.[172] Methoxide ion is 70- and 157-times more reactive than methanol towards the m- and p-positions respectively of 3,4-dehydrochlorobenzene.[173] Orientation and reactivity in nucleophilic additions to benzyne-3- and -4-carboxylates have been studied.[174a] Competition studies of CH_2CN^- and $MeNH_2$ or Me_2NH with various substituted benzynes suggest that both $+I$ and $-I$ substituents decrease the selectivity of benzyne by polarizing the triple bond.[174b]

Diels–Alder adducts of benzyne with 1-methyl-2-pyridone[175] and with

[164] L. Friedman and F. M. Logullo, *J. Org. Chem.*, **34**, 3089 (1969).
[165] G. P. Chiusoli and C. Venturello, *Chem. Comm.*, **1969**, 771; D. C. Dittmer and E. S. Whitman, *J. Org. Chem.*, **34**, 2004 (1969).
[166] C. W. Rees and D. E. West, *Chem. Comm.*, **1969**, 647.
[167] C. D. Campbell and C. W. Rees, *J. Chem. Soc.*(C), **1969**, 742, 748, 752.
[168] C W. Rees and R. C. Storr, *J. Chem. Soc.*(C), **1969**, 756, 760, 765; R. W. Hoffmann, G. Guhn, M. Preiss, and B. Dittrich, *ibid.*, p. 769.
[169] J. I. G. Cadogan and J. B. Thomson, *Chem. Comm.*, **1969**, 770.
[170] C. E. Griffin and D. C. Wysocki, *J. Org. Chem.*, **34**, 751 (1969).
[171a] B. H. Klanderman and T. R. Criswell, *J. Am. Chem. Soc.*, **91**, 510 (1969).
[171b] B. H. Klanderman and T. R. Criswell, *J. Org. Chem.*, **34**, 3426 (1969).
[172] P. Haberfield and L. Seif, *J. Org. Chem.*, **34**, 1508 (1969).
[173] J. F. Bunnett and C. Pyun, *J. Org. Chem.*, **34**, 2035 (1969).
[174a] E. R. Biehl, E. Nieh, H.-M. Li, and C.-I. Hong, *J. Org. Chem.*, **34**, 500 (1969).
[174b] E. R. Biehl, E. Nieh, and K. C. Hsu, *J. Org. Chem.*, **34**, 3595 (1969).
[175] E. B. Sheinin, G. E. Wright, C. L. Bell, and L. Bauer, *J. Heterocyclic Chem.*, **5**, 859 (1968).

o-quinones[176] have been obtained; *o*-quinone diazides yield benzofurans.[176] Nitrile oxides give 1,3-dipolar cycloaddition.[177] A small amount of 1,6-addition occurs with tropone.[178] The anil of *o*-chlorobenzaldehyde cyclized to phenanthroline with potassium amide in ammonia, despite its initially *trans*-geometry.[179]

Benzyne, generated by the aminotriazole route, gives a 60% yield of purely *cis*-1,4-dihydro-1,4-dimethylnaphthalene with *trans,trans*-hexa-2,4-diene.[180] Another report also describes stereospecific addition to dienes but a moderate loss of stereochemistry in 2 + 2-addition to the 1,2-dichloroethylenes.[181] Benzyne addition to *cis*- and *trans*-cyclooctene is also non-stereospecific and, since there is no solvent effect on the product ratio, a diradical is implicated. The *trans*-olefin gives no competing ene reaction.[182] In another study of the reaction of benzyne with olefins,[183] the diradical is suggested as the intermediate yielding both 2 + 2 and ene reaction products. 2 + 2 and ene reaction products are also formed from enamines; here zwitterionic intermediates are more likely.[184]

A careful study of concentration dependence, salt effects, and reversibility, together with trapping experiments, shows that 1-lithio-2-bromocyclopentene yields monomeric cyclopentyne in ether at 20°.[185] Cyclohexyne is implicated in reaction of 1-chlorocyclohexene with $NaNH_2$–Bu^tONa–piperidine in THF.[186]

Tetrafluorobenzyne (trapped by benzene) is formed much more efficiently from pentafluorophenyl-lithium in hexane–benzene than in ether–benzene.[187] Tetrafluorobenzyne undergoes normal 1,4-addition to [2.2]paracyclophane.[188] Reaction with styrene gives mainly 1,2,3,4-tetrafluoro-9,10-dihydrophenanthrene; the H-shift necessary for this is intramolecular, possibly by a conducted-tour mechanism.[189] 1,2,3,4-Tetrafluorophenanthrene is also formed; hydride abstraction by the benzyne is suggested (6). Benzyne itself gives 9-phenyl-9,10-dihydrophenanthrene with styrene. Deuterium labelling shows

[176] W. Reid and J. Tan Sioe Eng, *Ann. Chem.*, **727**, 219 (1969).
[177] T. Sasaki and T. Yoshioka, *Bull. Chem. Soc. Japan*, **42**, 826 (1969).
[178] T. Miwa, M. Kato, and T. Tamano, *Tetrahedron Letters*, **1969**, 1761.
[179] S. V. Kessar and M. Singh, *Tetrahedron Letters*, **1969**, 1155.
[180] R. W. Atkin and C. W. Rees, *Chem. Comm.*, **1969**, 152.
[181] M. Jones and R. H. Levin, *J. Am. Chem. Soc.*, **91**, 6411 (1969).
[182] P. G. Gassman and H. P. Benecke, *Tetrahedron Letters*, **1969**, 1089.
[183] I. Tabushi, K. Okazaki, and R. Oda, *Tetrahedron*, **25**, 4401 (1969).
[184] D. J. Keyton, G. W. Griffin, M. E. Kuehne, and C. E. Bayha, *Tetrahedron Letters*, **1969**, 4163.
[185] G. Wittig and J. Heyn, *Ann. Chem.*, **726**, 57 (1969).
[186] P. Caubère and J. J. Brunet, *Tetrahedron Letters*, **1969**, 3323.
[187] D. D. Callander, P. L. Coe, J. C. Tatlow, and A. J. Uff, *Tetrahedron*, **25**, 25 (1969).
[188] J. P. N. Brewer, H. Heaney, and B. A. Marples, *Tetrahedron*, **25**, 243 (1969).
[189] R. Harrison, H. Heaney, J. M. Jablonski, K. G. Mason, and J. M. Sketchley, *J. Chem. Soc.* (C), **1969**, 1684; see *Org. Reaction Mech.*, **1968**, 123—125.

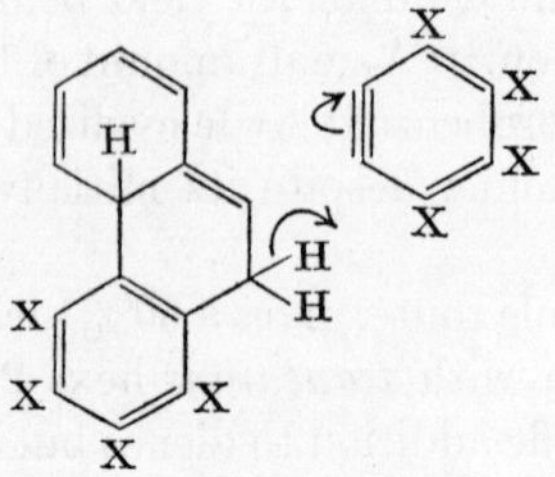

(6)   X = F, Cl 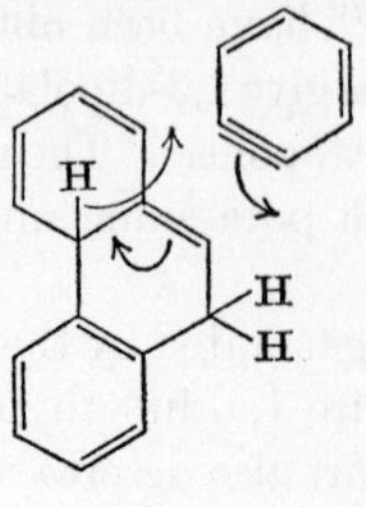   (7)

this to arise by an ene reaction (7). Reactions of tetrahalogenobenzynes with thioanisole give 1,2,3,4-tetrahalogenodiphenyl sulphides. The necessary proton transfer in the intermediate betaine is intramolecular (equation 5).[190]

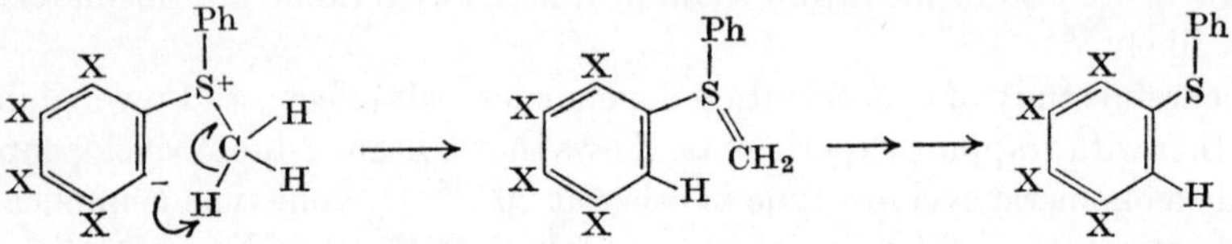

(X = F, Cl)

. . . (5)

Tetrafluorobenzyne reacts with N,N-dimethylaniline to give two 1,4-adducts, a 1,2-adduct, 4a-dimethylamino-5,6,7,8-tetrafluoro-4a,8b-dihydrobiphenyl-ene, and other products from the initial betaine.[191] Tetrafluorobenzyne and pentafluorothiophenolate yield octafluorodibenzothiophen.[192] Benzynes undergo ene reactions with (hindered) steroidal 5,7-dienes,[193] as does tetra-fluorobenzyne with cyclohexene.[194] The generation and reactions of 3-penta-fluorophenyl-4,5,6-trifluorobenzyne,[195] 3,5-dichlorodifluorobenzyne,[196a] and the isomeric alkoxy- and dialkylamino-trichlorobenzynes[196b] are reported. Tetrabromobenzyne and 2,5,6-tribromo-3-pyridyne have been generated and trapped with arenes.[197]

190 J. P. N. Brewer, H. Heaney, and T. J. Ward, *J. Chem. Soc.*(C), **1969**, 355.
191 H. Heaney and T. J. Ward, *Chem. Comm.*, **1969**, 810.
192 R. D. Chambers and D. J. Spring, *Tetrahedron Letters*, **1969**, 2481.
193 I. F. Eckhard, H. Heaney, and B. A. Marples, *J. Chem. Soc.*(C), **1969**, 2098.
194 I. F. Mikhailova and V. A. Barkhash, *Zh. Obshch. Khim.*, **38**, 2117 (1968); *Chem. Abs.*, **70**, 19680 (1969).
195 S. C. Cohen, A. J. Tomlinson, M. R. Wiles, and A. G. Massey, *J. Organometal. Chem.*, **11**, 385 (1968).
196a N. Ishikawa and S. Hayashi, *J. Chem. Soc. Japan*, **89**, 1131 (1968); **90**, 300; *Chem. Abs.*, **70**, 96481, 114727 (1969).
196b D. J. Berry, I. Collins, S. M. Roberts, H. Suschitzky, and B. J. Wakefield, *J. Chem. Soc.* (C), **1969**, 1285.
197 D. J. Berry and B. J. Wakefield, *J. Chem. Soc.* (C), **1969**, 2342.

A variety of routes to dehydrothiophens have been explored.[198] Pyrolysis of either di-2-(3-iodothienyl)mercury or di-3-(4-iodothienyl)mercury in the presence of tetracyclone gave the adduct of 2,3-dehydrothiophen, although the latter should generate 3,4-dehydrothiophen.

Full details have appeared of the double competition method (discussed last year) used to detect concurrent elimination–addition and addition–elimination routes to cine-substitution in pyrimidines.[199] Several 2,3-dehydropyridines have been generated from 3-lithio-4-X-2,5,6-trichloropyridines,[200] but in general the evidence is in favour of the theoretical prediction[156] that 2,3-dehydropyridine is considerably less stable than 3,4-dehydropyridine. The related prediction[156] that the 2-pyridyl anion is destabilized relative to the 4-pyridyl anion is supported by H–D exchange results.[201] Interrupted aryne formation from 3-chloro[2-^{2}H]pyridine in NH_3–$NaNH_2$ shows no loss of deuterium in the starting material.[201a] Similar effects on anion stability are found with the diazines.[201b] MeS^- is competitive with NH_2^- in trapping 3,4-dehydropyridine; a 1:1 ratio of 3- and 4-methylthio-isomers is formed.[202] The *N*-aminotriazole route has been used to generate 2,3- and 3,4-dehydropyridines and -quinolines, which were trapped with tetracyclone; yields were low with the 2,3-isomers.[203] Five reaction pathways have been identified in reactions of 2-X-6-bromopyridines with excess of potassamide in liquid ammonia.[204] These are replacement of bromine, replacement of X, ring-cleavage and formation of 1,3-dicyanopropene, ring-cleavage and recyclization to a 2-methylpyrimidine, and formation of a 2-X-4-aminopyridine. This last pathway, which occurs with 2-alkoxy-6-bromopyridines, may involve a (*meta*) 2,4-dehydropyridine.[205]

[198] G. Wittig and M. Rings, *Ann. Chem.*, **719**, 127 (1968).
[199] T. Kaufmann, R. Nürnberg, and R. Wirthwein, *Chem. Ber.*, **102**, 1161 (1969); T. Kaufmann, R. Nürnberg, and K. Udluft, *ibid.*, p. 1177; see *Org. Reaction Mech.*, **1968**, 200.
[200] J. D. Cook and B. J. Wakefield, *J. Chem. Soc.*(C), **1969**, 1973, 2376; R. A. Fernandez, H. Heaney, J. M. Jablonski, K. G. Mason, and T. J. Ward, *ibid.*, p. 1908.
[201a] J. A. Zoltewicz and C. L. Smith, *Tetrahedron*, **25**, 4331 (1969).
[201b] J. A. Zoltewicz, G. Grahe, and C. L. Smith, *J. Am. Chem. Soc.*, **91**, 5501 (1969).
[202] J. A. Zoltewicz and C. Nisi, *J. Org. Chem.*, **34**, 765 (1969).
[203] G. W. J. Fleet and I. Fleming, *J. Chem. Soc.*(C), **1969**, 1758.
[204] J. W. Streef and H. J. den Hertog, *Tetrahedron Letters*, **1968**, 5945.
[205] H. Boer and H. J. den Hertog, *Tetrahedron Letters*, **1969**, 1943.

CHAPTER 7

Electrophilic Aromatic Substitution

A. R. BUTLER

Department of Chemistry, St. Salvator's College, University of St. Andrews

Full details of a study of electrophilic substitution in substituted [2.2]paracyclophanes have been published by Cram and his coworkers. As reported last year,[1] attack by electrophiles occurs in the unsubstituted ring opposite the most basic site in the substituted ring, and the slow step is a transannular proton shift (Scheme 1).[2] Unfortunately the reaction conditions cause scrambling of the protons and detection of the shift by isotopic labelling proved difficult. However, it was demonstrated in the bromination of deuterated 4-methyl[2.2]paracyclophane. Multiple halogenation and nitration results

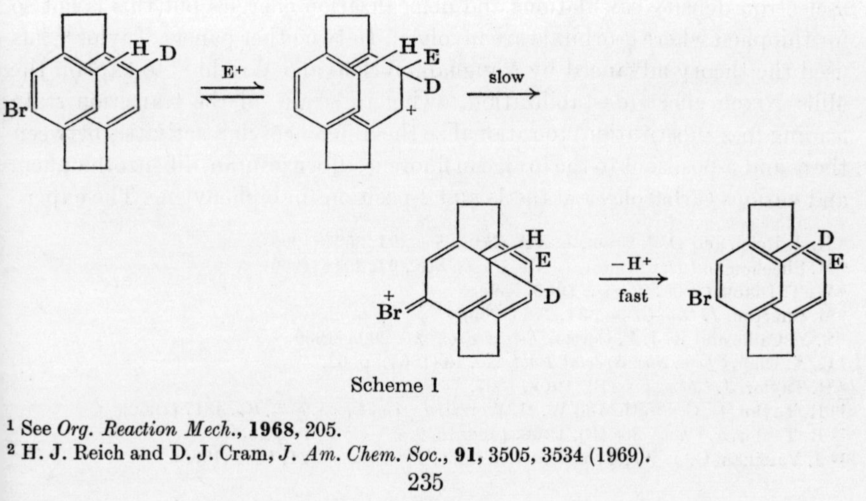

Scheme 1

[1] See *Org. Reaction Mech.*, **1968**, 205.
[2] H. J. Reich and D. J. Cram, *J. Am. Chem. Soc.*, **91**, 3505, 3534 (1969).

in heteroannular substitution.[3] It is only in [2.2]paracyclophane that trans-annular effects are pronounced: in [4.4]paracyclophane there is no evidence for these effects. [3.3]Paracyclophane is intermediate in its properties and the second ring is close enough to enhance the rate of nitration, acetylation, and bromination. Once monosubstitution has occurred both rings are deactivated towards further attack.[4]

The relative reactivities of benzene and cyclobutadiene towards substitution and addition reactions have been compared. Ease of substitution depends upon the acidity of the protons in the intermediate, and molecular orbital calculations show that this is higher in benzene than in cyclobutadiene.[5] By correlation with an extended Hammett equation, Charton[6] has shown that there is no significant steric effect for *ortho*-substituents in a number of cases of electrophilic aromatic substitution: the results may be accounted for by purely electronic effects. Caille and Corriu[7] have discussed Olah's[8] views on the formation of π-complexes during the course of electrophilic substitution and, from a careful study of the relative reactivities of toluene and benzene in bromination and chlorination catalysed by $AlCl_3$, conclude that the low selectivities observed by Olah are not general, thus refuting the evidence for π-complex formation.

Taylor[9] has published details of an interesting study of the gas-phase pyrolysis of 1-arylethyl acetates. This proceeds by partial formation of a carbonium ion on the carbon atom α to the ring and is a model for electrophilic aromatic substitution without the complication of solvation effects. Substituent effects for this reaction correlate well with σ^+ in a Hammett plot, and ρ is -0.66 at $600°K$.[10] From the present data on the rate of reactions of 2- and 3-substituted thiophen and furan compounds, σ^+ constants for thienyl and furyl groups have been determined. The results for furan agree well with π-electron density calculations and delocalization energies but this is not so for thiophen where d-orbitals are involved. In two other papers, Taylor[11] has used the theory advanced by Vaughan, Welch, and Wright[12] to explain the Mills–Nixon effect (destabilization, owing to strain, of the transition state leading to α-substitution) to rationalize the differences in reactivities between the *o*- and *p*-positions to the bridge in fluorene, dibenzofuran, dibenzothiophen, and various carbazoles and the 1- and 2-positions in biphenylene. The experi-

[3] H. J. Reich and D. J. Cram, *J. Am. Chem. Soc.*, **91**, 3527 (1969).

[4] M. Sheehan and D. J. Cram, *J. Am. Chem. Soc.*, **91**, 3544 (1969).

[5] W. T. Dixon, *Chem. Comm.*, **1969**, 559.

[6] M. Charton, *J. Org. Chem.*, **34**, 278 (1969).

[7] S. Y. Caille and R. J. P. Corriu, *Tetrahedron*, **25**, 2005 (1969).

[8] G. A. Olah, *Chem. Soc. Special Publ.* No. 19 (1964), p. 62.

[9] R. Taylor, *J. Chem. Soc.*(B), **1968**, 1397.

[10] R. Taylor, G. G. Smith, and W. H. Wetzel, *J. Am. Chem. Soc.*, **84**, 4817 (1962).

[11] R. Taylor, *J. Chem. Soc.*(B), **1968**, 1402, 1559.

[12] J. Vaughan, G. J. Welch, and G. J. Wright, *Tetrahedron*, **21**, 1665 (1965).

mental results discussed tend to disprove Streitwieser's[13] alternative explanation of the Mills–Nixon effect.

Isomer ratios for electrophilic substitution in a number of compounds have been determined. Nitration of 3-nitro- and 2-acetamido-fluoranthene [see (1)] occurs at the 9- and 3-positions respectively.[14] Halogenation of 3- and 5-halogenoacenaphthene [see (2)] occurs at the 6-position.[15] Östman[16] has

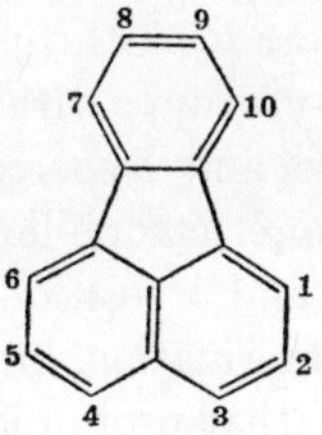

(1) 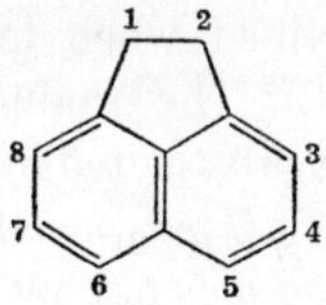   (2) 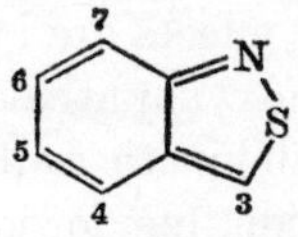   (3)

reviewed electrophilic substitution at the 2- and 3-positions in thiophen and has also reported[17] a study of the relative reactivities of various positions in thiophen substituted by deactivating groups at the 2- ($5 \geqslant 4 \geqslant 3$) and 3-positions ($5 > 2 > 4$). Electrophilic attack on 5-hydroxyindole occurs at the 3- or 4-positions.[18] Attack on 2,1-benzisothiazole (3) by electrophiles occurs mainly at position 5, while nitration of monosubstituted 2,1-benzisothiazole gives products which indicate that the group already present in the benzene ring is the decisive directing influence.[19] Davoll[20] predicted that electrophilic substitutions of pyrrolo[2,3-*d*]pyrimidine (4) should occur at the 5- or 6-positions. Halogenation, nitration, and sulphonation have all been found to occur at the 5-position.[21] Electrophilic substitution in naphtho[2,1-*b*]thiophen

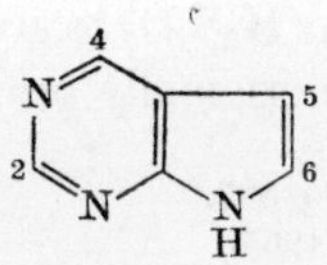

(4) 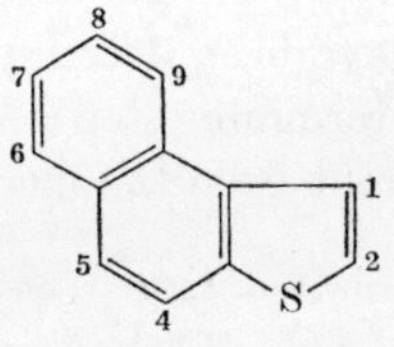   (5) 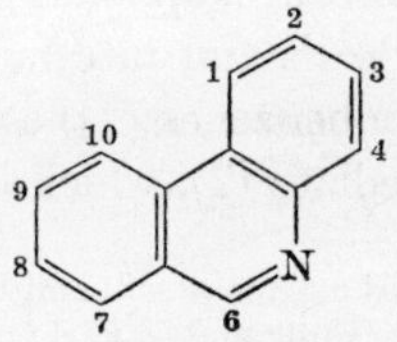   (6)

[13] See *Org. Reaction Mech.*, **1968**, 216.
[14] E. H. Charlesworth and C. U. Lithown, *Can. J. Chem.*, **47**, 1595 (1969).
[15] P. R. Constantine, L. W. Deady, and R. D. Topsom, *J. Org. Chem.*, **34**, 1113 (1969).
[16] B. Östman, *FOA Report*, **1969**, 3; *Chem. Abs.*, **71**, 48927 (1969).
[17] B. Östman, *Acta Chem. Scand.*, **22**, 2754, 2765 (1968).
[18] M. Julia and J.-Y. Lallemand, *Compt. Rend.*, *C*, **267**, 1506 (1968).
[19] M. Davis and A. W. White, *J. Chem. Soc.*(C), **1969**, 2189.
[20] J. Davoll, *J. Chem. Soc.*, **1960**, 131.
[21] J. F. Gerster, B. C. Hinshaw, R. K. Robins, and L. B. Townsend, *J. Heterocyclic Chem.*, **6**, 207 (1969).

(5) and its 1-methyl derivative occurs at the 2-position,[22] a result which does not agree with the molecular orbital calculations of Zahradník and Párkányi.[23] If the 2-position is substituted then attack is at the 5-position.[24] Chandler[25] has determined isomer ratios in the bromination and nitration of phenanthridine (6): the order of reactivities for bromination $(10 > 4 \gg 2)$ is in fairly good agreement with the electron-density calculations of Longuet-Higgins and Coulson.[26]

The best agreement between the partial rate factors for nitration (HNO_3-Ac_2O) of xylenes and the electronic effect of the methyl groups within Pariser–Parr–Pople formalism is obtained when inductive and hyperconjugative mesomeric effects are included.[27] The products from electrophilic attack on 2,6-dimethoxynaphthalene are substituted in the 1- and 5-positions. There is only a small kinetic isotope effect $(k_H/k_D = 1.1)$ for chlorination, bromination, and nitration, but for iodination it is 6.2, indicating that proton loss is rate-determining.[28] Equilibration of isomers produced by electrophilic attack has been studied.[29]

Bonding of a substrate to a metal atom produces changes in its reactivity towards electrophiles. Thus, the phenyl groups in $PhFe(CO)_2Ph$ are more reactive than benzene,[30] and cyclooctatetraene readily undergoes electrophilic substitution if coordinated to iron tricarbonyl.[31]

Sulphonation

Major advances in an understanding of aromatic sulphonation come from Cerfontain and his coworkers: the mechanism depends upon the substrate as well as the concentration of H_2SO_4. For *o*-, *m*-, and *p*-xylene in 72—89% acid the sulphonating species is $H_3SO_4^+$ but at higher acid concentrations *o*- and *m*-xylene are sulphonated by $H_2S_2O_7$.[32] The effect of two methyl groups is additive. Partial rate factors have been determined for the sulphonation of halogenobenzenes:[33] the rate-determining step is attack by $H_2S_2O_7$ to give the intermediate (7), which is different from the mechanism originally proposed

[22] K. Clarke, G. Rawson, and R. M. Scrowston, *J. Chem. Soc.*(C), **1969**, 537.
[23] R. Zahradník and C. Párkányi, *Coll. Czech. Chem. Comm.*, **30**, 195 (1965).
[24] K. Clarke, G. Rawson, and R. M. Scrowston, *J. Chem. Soc.*(C), **1969**, 1274.
[25] G. S. Chandler, *Austral. J. Chem.*, **22**, 1105 (1969).
[26] H. C. Longuet-Higgins and C. A. Coulson, *Trans. Faraday Soc.*, **43**, 87 (1947).
[27] D. T. Clark and D. J. Fairweather, *Tetrahedron*, **25**, 4083 (1969).
[28] J.-C. Richer and A. Rossi, *Can. J. Chem.*, **47**, 3935 (1969).
[29] A. A. Spryskov, *Izv. Vyssh. Ucheb. Zaved., Khim. Khim. Tekhnol.*, **11**, 1349 (1968); *Chem. Abs.*, **71**, 29781 (1969).
[30] E. S. Bolton, G. R. Knox, and C. G. Robertson, *Chem. Comm.*, **1969**, 664.
[31] B. F. G. Johnson, J. Lewis, A. W. Parkins, and G. L. P. Randall, *Chem. Comm.*, **1969**, 595.
[32] A. W. Kaandorp and H. Cerfontain, *Rec. Trav. Chim.*, **88**, 725 (1969); A. J. Prinsen and H. Cerfontain, *ibid.*, p. 833.
[33] C. W. F. Kort and H. Cerfontain, *Rec. Trav. Chim.*, **88**, 860 (1969).

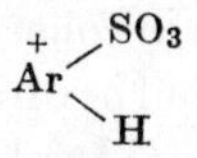

$$\underset{(7)}{Ar\overset{+}{\underset{H}{\diagup}}SO_3} \qquad \underset{(8)}{Ar\overset{+}{\underset{H}{\diagup}}S_2O_6}$$

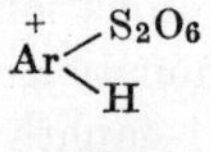

by these authors.[34] Isomerization of *o*- and *m*-xylene sulphonic acids proceeds by slow protiodesulphonation and subsequent sulphonation. The rate of the slow step depends linearly upon the acidity but the slope of a plot of log k against $-H_0$ is less than unity.[35] This is probably due to the occurrence of a side-reaction:

$$ArSO_3^- + H^+ \rightleftharpoons ArSO_3H$$

The mechanism of sulphonation of polyfluorobenzenes is complex. Sulphonation of *p*-difluorobenzene in 100% H_2SO_4 proceeds by attack of $H_2S_2O_7$ but for 1,2,3,4-tetrafluorobenzene in 100—103.5% acid the rate is proportional to $h_0 a_{H_2S_2O_7}$, indicating that the reactive species is $H_3S_2O_7^+$. This applies also to the sulphonation of pentafluorobenzene up to 104% acid but above this attack is by $H_2S_4O_{13}$ to give the intermediate (8).[36] Attack by $H_2S_2O_7$ has been proposed by other workers[37] as the main route in the sulphonation of salicylic acid. Another route is sulphonation of the hydroxy group to give the intermediate (9) which then rearranges. In general, deuterated and tritiated compounds are sulphonated more slowly than the corresponding protiated compounds but Bosscher and Cerfontain[38] report that there is no significant primary isotope effect in the sulphonation of benzene by SO_3 in trichlorofluoromethane or nitromethane.

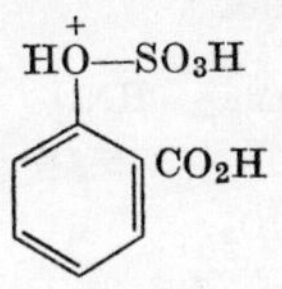

(9) 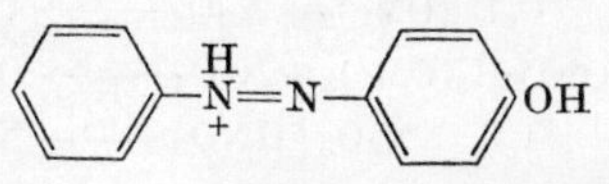   (10)

Sulphonation is known to occur as a subsequent reaction in the Wallach rearrangement of azoxybenzene to 4-hydroxyazobenzene in H_2SO_4 but this reaction has not been studied before. It proceeds by attack of SO_3 on the monoprotonated substrate (10) to give 4-hydroxyazo-4'-sulphonic acid. Above 98% acid there is a second protonation of the phenolic oxygen, and in 100% acid a second sulphonation occurs[39] (see also Chapter 8). Sulphonation of

[34] C. W. F. Kort and H. Cerfontain, *Rec. Trav. Chim.*, **86**, 865 (1967).
[35] A. Koeberg-Telder, A. J. Prinsen, and H. Cerfontain, *J. Chem. Soc.*(B), **1969**, 1004.
[36] C. W. F. Kort and H. Cerfontain, *Rec. Trav. Chim.*, **88**, 1298 (1969).
[37] O. I. Kachurin and L. P. Mel'nikova, *Izv. Vyssh. Ucheb. Zaved., Khim. Khim. Tekhnol.* **11**, 1021 (1968); *Chem. Abs.*, **70**, 36797 (1969).
[38] J. K. Bosscher and H. Cerfontain, *J. Chem. Soc.*(B), **1968**, 1524.
[39] W. M. J. Strachan and E. Buncel, *Can. J. Chem.*, **47**, 4011 (1969).

phenol is retarded by protonation of oxygen in concentrated acid.[40] There has been a study of the sulphonation of toluene and hydrolysis of toluene sulphonic acids.[41] Di-(2-hydroxy-1-naphthyl)methane (11a) has been suggested as an intermediate in the sulphomethylation of 2-naphthol to give 2-hydroxy-1-naphthylmethanesulphonate. An investigation by Ogata, Kawasaki, and Goto[42] has shown that it is really a by-product and the true intermediate is quinomethide (11b).

(11a)   (11b)

Nitration and Nitrosation

The mechanism of aromatic nitration is still under investigation and several workers have produced further evidence for the importance of nitration via nitrosation. This is the mechanism for the nitration of phenacetin (*p*-ethoxy-acetanilide) by dilute nitric acid but the nitroso-compound cannot be oxidized to the corresponding nitro-compound by nitric acid. The authors suggest that the nitro-compound is formed from the transition state resulting from attack of NO^+ without appearance of an isolable nitroso-intermediate.[43] However, it may be that the oxidizing agent is N_2O_4, as has been suggested by Bonner and Hancock.[44] This is thought to be the case for the nitration of dimethoxybenzene by N_2O_4 in carbon tetrachloride:

$$C_6H_4(OMe)_2 + N_2O_4 \xrightarrow{slow} ONC_6H_3(OMe)_2 + HNO_3$$
$$ONC_6H_3(OMe)_2 + N_2O_4 \longrightarrow O_2NC_6H_3(OMe)_2 + N_2O_3$$
$$N_2O_3 + HNO_3 \longrightarrow N_2O_4 + HNO_2$$

Addition of Lewis acids increases the rate of nitrosation, suggesting that the nitrosating species is polarized $NO^+NO_2^-$.[45] Hoggett, Moodie, and Schofield[46] report a dual mechanism for nitration by HNO_3 in acetic anhydride, one being via nitrosation, and the isomer ratios from the two mechanisms are very different. The same workers[47] have concluded that the rate of nitration of activated aromatics in aqueous sulpholan and nitromethane is the encounter

[40] A. A. Spryskov and Z. A. Yakovleva, *Izv. Vyssh. Ucheb. Zaved., Khim. Khim. Tekhnol.*, 12, 439 (1969); *Chem. Abs.*, 71, 48933 (1969).

[41] A. A. Spryskov and V. A. Kozlov, *Izv. Vyssh. Ucheb. Zaved., Khim. Khim. Tekhnol.*, 11, 426 (1966); *Chem. Abs.*, 69, 95576 (1968).

[42] Y. Ogata, A. Kawasaki, and T. Goto, *Tetrahedron*, 25, 2589 (1969).

[43] H. Oelschlaeger, H. Hoffmann, and U. Matthiesen, *Arch. Pharm.*, 302, 45 (1969).

[44] T. G. Bonner and R. A. Hancock, *Chem. Comm.*, 1967, 780.

[45] T. G. Bonner, R. A. Hancock, G. Yousif, and F. R. Rolle, *J. Chem. Soc.(B)*, 1969, 1237.

[46] J. G. Hoggett, R. B. Moodie, and K. Schofield, *Chem. Comm.*, 1969, 605.

[47] J. G. Hoggett, R. B. Moodie, and K. Schofield, *J. Chem. Soc.(B)*, 1969, 1.

rate between the aromatic and NO_2^+. The high selectivities reported by some authors may refer to nitrosation rather than direct nitration. Coombes[48] has investigated the nitration of toluene by nitric acid in carbon tetrachloride and shown that some of the results of Bonner, Hancock, and Rolle[49] for this reaction are due to the reaction mixture becoming heterogeneous towards the end. However, the negative temperature coefficient and the high-order dependence on the nitric acid concentration were confirmed. To accommodate these results the rate-determining step must be breakdown of aggregates of nitric acid molecules, the concentration of which decreases with increasing temperature.

Isomer ratios in the nitration of dimethylnaphthalenes by nitric acid in acetic anhydride are reported,[50] and the reactivities of thiazoles relative to benzene and thiophen have been determined.[51] Nitration of biphenyls and diphenylamines yields products nitrated *ortho* or *para* to the bond between the rings,[52] but with *p*-quaterphenyl only 4-nitro- and 4,4′-dinitro-quaterphenyl are obtained and there are no *o*-substituted products.[53] The main product from the nitration of 2,4,5-triisopropylacetanilide is the 6-nitro-derivative and (12) is thought to be an intermediate. In sulphuric acid the conjugate acid is formed and attack is at the position *meta* to the $-NH_2Ac^+$ group.[54] Nitro-de-t-butylation occurs in the nitration of 2,5-di-t-butyl-benzene.[55]

The reactive species in the nitration of 4-pyridones, 3- and 5-methyl-2-pyridones, and their *O*- and *N*-methyl derivatives is the free base.[56] This is also the case for benzoic acid and acetophenone but it is the conjugate acid for *N,N*-dimethylaniline *N*-oxide and benzamide.[57] Nitration of 6-methyl- and 6-methoxy-quinoline occurs at the 5-position, and that of 7-chloroquinoline at the 8-position.[58] Partial rate factors have been determined for the nitration $(HNO_3-H_2SO_4)$ of benzylidene trifluoride at the *m*-position (6.7×10^{-5}),[59] and of cyclopentyl- and cyclohexyl-benzene.[60] Most positive poles are deactivating

[48] R. G. Coombes, *J. Chem. Soc.*(B), **1969**, 1256.

[49] T. G. Bonner, R. A. Hancock, and F. R. Rolle, *Tetrahedron Letters*, **1968**, 1665.

[50] A. Davies and K. D. Warren, *J. Chem. Soc.*(B), **1969**, 873.

[51] H. J. M. Dou, A. Friedmann, G. Varnin, and J. Metzger, *Compt. Rend.*, *C*, **266**, 714 (1968).

[52] I. F. Falyakhov, G. P. Sharnin, I. E. Moisak, and E. E. Gryazin, *Tr. Kazan. Khim.-Tekhnol. Inst.*, **1967**, No. 36, 564; *Chem. Abs.*, **71**, 21397 (1969).

[53] M. L. Scheinbaum, *Chem. Comm.*, **1969**, 1235.

[54] A. J. Neale, K. M. Davies, and J. Ellis, *Tetrahedron*, **25**, 1423 (1969).

[55] K. H. Bantel and H. Musso, *Chem. Ber.*, **102**, 696 (1969).

[56] P. J. Brignell, A. R. Katritzky, and H. O. Tarhan, *J. Chem. Soc.*(B), **1968**, 1477.

[57] R. B. Moodie, J. R. Penton, and K. Schofield, *J. Chem. Soc.*(B), **1969**, 578.

[58] N. D. Heindel, C. J. Ohnmacht, J. Molnar, and P. D. Kennewell, *J. Chem. Soc.*(C), **1969**, 1369.

[59] R. G. Coombes, R. B. Moodie, and K. Schofield, *J. Chem. Soc.*(B), **1969**, 52.

[60] H. J. M. Dou, G. Vernin, and J. Metzger, *Compt. Rend.*, *C*, **267**, 1515 (1968).

and *m*-directing in aromatic nitration[61] but the $-PPh_3^+$ group deactivates less than $-PMe_3^+$ owing to greater polarization of the former group.[62]

(12)   (13)   (14)

An unexpected product (13) is obtained from the reaction between skatole and tetranitromethane. The mechanism is thought to be nitration at the 3-position followed by attack of $(NO_2)_3C^-$ at the 2-position and elimination of nitrous acid,[63] although previous work has suggested that nitration by tetranitromethane is a free-radical reaction.[64] Nitronium phosphorohexa-fluoride in nitromethane is a useful nitrating agent in that it will mononitrate highly activated aromatics like durene.[65] Tisue and Down[66] report that benzene is nitrated by sodium nitrite or N_2O_4 in the presence of colloidal palladium, while BF_3 is a catalyst in the nitration of *o*-nitrotoluene.[67]

By comparing the rates of the first and second nitrations, and by comparing with 9,10-dihydro-9,10-ethanoanthracene (14), Klanderman and Perkins[68] have shown that there are no transannular effects in the nitration of triptycene. Asano[69] has reported the effect of pressure on the value of the Hammett ρ-factor for nitration (H_2SO_4–HNO_3–AcOH) of six monosubstituted benzenes. The results indicate that solvation of the transition state depends upon the substituent.

Azo Coupling

Kaul and Zollinger[70] report a change of mechanism from heterolytic fission for benzenediazonium tetrafluoroborate to homolytic fission for *p*-nitrobenzene-diazonium tetrafluoroborate in azo coupling with benzene or nitrobenzene. There is good correlation between Hammett σ-constants and the effect of substituents on the rate of coupling of benzenediazonium salts with *N,N*-dimethylaniline.[71] For derivatives of naphthalenesulphonic acids containing

[61] See *Org. Reaction Mech.*, **1968**, 209.
[62] T. A. Modro, *Bull. Acad. Pol. Sci., Ser. Sci. Chim.*, **16**, 585 (1968).
[63] T. F. Spande, A. Fontana, and B. Witkop, *J. Am. Chem. Soc.*, **91**, 6199 (1969).
[64] T. C. Bruice, M. J. Gregory, and S. L. Walters, *J. Am. Chem. Soc.*, **90**, 1612 (1968).
[65] S. B. Hanna, E. Hunziker, T. Saito, and H. Zollinger, *Helv. Chim. Acta*, **52**, 1537 (1969).
[66] T. Tisue and W. J. Down, *Chem. Comm.*, **1969**, 410.
[67] Z. Csuros, G. Deak, L. Fenichel, and A. Torok-Kalmar, *Acta Chim. Acad. Sci. Hung.*, **59**, 381 (1969); *Chem. Abs.*, **71**, 2701 (1969).
[68] B. H. Klanderman and W. C. Perkins, *J. Org. Chem.*, **34**, 630 (1969).
[69] T. Asano, *Bull. Chem. Soc. Japan*, **42**, 2005 (1969).
[70] B. L. Kaul and H. Zollinger, *Helv. Chim. Acta*, **51**, 2132 (1968).
[71] V. Beránek and M. Večeřa, *Coll. Czech. Chem. Comm.*, **34**, 2753 (1969).

both an amino and a hydroxy group the position of coupling depends upon the pH of the reaction medium. A study of the effect of substituents has shown that coupling to the hydroxy group is more selective.[72] The pK_a of the hydroxy group in a number of hydroxynaphthalenesulphonic acids has been determined as part of this study.[73] For the coupling of 6-hydroxynaphthalene-2-sulphonic acid with substituted benzenediazonium chlorides ρ changes from 3.4 in water to 2.9 in 50% aqueous ethanol.[74] The coupling of various *o*- and *p*-quinone diazides with 2-naphthol in buffer solution has been studied by Kazitsyna, Klyueva, and Romanova.[75] The *para*-compound reacts much more rapidly than the *ortho*-compound and electron-attracting substituents increase the rate of reaction.

Friedel–Crafts and Related Reactions

Gallium chloride is a catalyst for the methylation of benzene by methyl chloride. In an excess of methyl chloride the rate equation is

$$\text{Rate} = k[\text{PhH}][\text{GaCl}_3]^2$$

and hexadeuterated benzene is methylated at the same rate as isotopically normal benzene. DeHaan, Brown, and Hill[76] discuss the probable mechanism in detail and conclude that the slow step is attack of the complex $(\text{MeCl})_2\text{GaCl}_2{}^+\text{GaCl}_4{}^-$ on benzene to form a σ-complex. Similar results were obtained for the methylation of toluene and dimethylbenzene.[77] Isotopic exchange occurs between methyl chloride and gallium chloride but the two reactions involve different intermediates. In the latter case a dimethyl-chloronium ion (15) is formed.[78] However, with ethyl chloride exchange involves basically the same mechanism as methylation.[79]

Berg, Jakobsen, and Johansen[80] were unable to confirm the findings of Buu-Hoï and Cagniant[81] that t-butylation of phenanthrene (16) gave the 3,9-disubstituted product. They found, in both phenanthrene and pyrene (17), that attack was at the 2- and 7-positions, and they rationalize this result in terms of the steric effect of the *peri*-hydrogens at the 4-, 5-, 9-, and 10-positions.

[72] J. Panchartek and V. Štěrba, *Coll. Czech. Chem. Comm.*, **34**, 2971 (1969).

[73] K. Kalfus and V. Štěrba, *Coll. Czech. Chem. Comm.*, **34**, 3183 (1969).

[74] E. Kučerová, J. Panchartek, and V. Štěrba, *Coll. Czech. Chem. Comm.*, **33**, 4290 (1968).

[75] L. A. Kazitsyna, N. D. Klyueva, and K. V. Romanova, *Dokl. Akad. Nauk SSSR*, **183**, 105 (1968); *Chem. Abs.*, **70**, 36803 (1969).

[76] F. P. DeHaan, H. C. Brown, and J. C. Hill, *J. Am. Chem. Soc.*, **91**, 4845 (1969).

[77] F. P. DeHaan and H. C. Brown, *J. Am. Chem. Soc.*, **91**, 4850 (1969).

[78] F. P. DeHaan, H. C. Brown, D. C. Conway, and M. G. Gibby, *J. Am. Chem. Soc.*, **91**, 4854 (1969).

[79] F. P. DeHaan, M. G. Gibby, and D. R. Aebersold, *J. Am. Chem. Soc.*, **91**, 4860 (1969).

[80] A. Berg, H. J. Jakobsen, and S. R. Johansen, *Acta Chem. Scand.*, **23**, 567 (1969).

[81] N. P. Buu-Hoï and P. Cagniant, *Ber.*, **77**, 121 (1944).

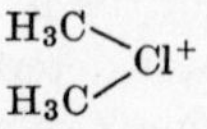

(15) 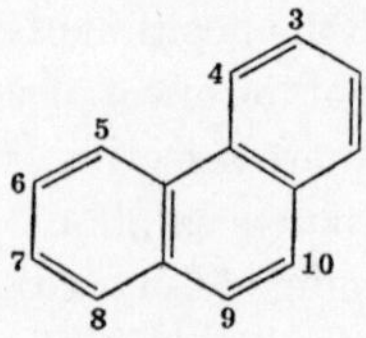   (16) 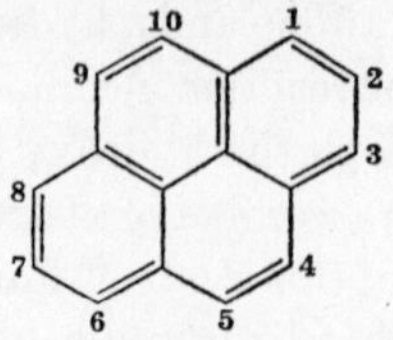   (17)

The alkylation of benzene by 3-chlorobutanoic acid is stereospecific (43% inversion), and it appears that benzene attacks the cyclic intermediate as a nucleophile (18).[82] Nakane, Kurihara, and Natsubori,[83] in an important study, report that in the ethylation of benzene by [2-^{14}C]ethyl iodide catalysed by $AlBr_3$ there is migration of radioactivity from the β- to the α-atom. The authors suggest that the reagent is either a carbonium ion or a non-ionized complex between ethyl halide and metallic halide, the former occurring in non-polar and the latter in basic solvents. Alkylbenzenes, alkyl fluorides, and BF_3 form a complex which was thought to be the σ-complex occurring as an intermediate in Friedel–Crafts alkylations. However, Nakane[84] suggests, from a study of visible, IR, and NMR spectra and the isotope effect, that it is in fact a π-complex.

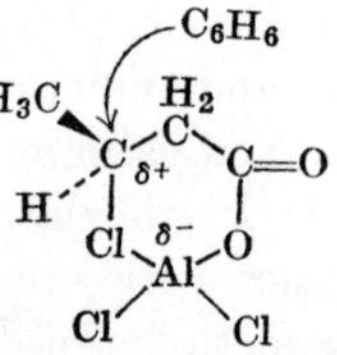

(18) 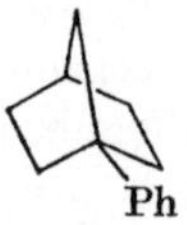   (19)

The relative reactivities of substituted benzenes in Friedel–Crafts alkylations do not always follow an obvious order. For ethylation by ethyl fluoride and BF_3 toluene is *less* reactive than benzene,[83] and for benzhydrylation by benzhydryl acetate the order is Et ∼ Bu^t < Me < Pr^i.[85] The normal order in electrophilic substitution is Me > Et > Pr^i > Bu^t, and steric hindrance to solvation, rather than hyperconjugation, has been suggested as responsible for this. However, Jensen and Smart[86] report that the $AlCl_3$-catalysed benzoylation of phenylnorbornanes is faster than that of isopropylbenzene or even

[82] S. Suga, T. Nakajima, Y. Nakamoto, and K. Matsumoto, *Tetrahedron Letters*, **1969**, 3283.
[83] R. Nakane, O. Kurihara, and A. Natsubori, *J. Am. Chem. Soc.*, **91**, 4528 (1969).
[84] R. Nakane, *J. Chem. Soc. Japan*, **90**, 17 (1969); *Chem. Abs.*, **70**, 86873 (1969).
[85] J. Zabicky, G. Chuchani, and L. Revetti, *Israel J. Chem.*, **7**, 491 (1969).
[86] F. R. Jensen and B. E. Smart, *J. Am. Chem. Soc.*, **91**, 5686 (1969).

toluene, although the norbornyl group is far bulkier than any simple alkyl group. Hyperconjugation must therefore be important, but as the fastest is 1-phenylnorbornane (19), where there are no α-hydrogens, C—C hyperconjugation must be as important as C—H hyperconjugation in toluene.

A second substituent affects the rate of reaction between halogenooctanes and benzene catalysed by HF–BF$_3$ even if separated by seven methylene groups.[87] Anisole is alkylated by propene catalysed by AlCl$_3$ or H$_2$SO$_4$ and eventually gives the triisopropyl derivative.[88]

In very strong acids, acid chlorides are protonated to give chloroacyloxonium ions,[89] and these ions may be involved in Friedel–Crafts acylations. Concurrent second- and third-order kinetics have been reported for two acylation reactions,[90] and the relative rates of the AlCl$_3$-catalysed reaction of methylbenzenes with β-chloropropionyl chloride have been determined.[91] There is a primary isotope effect in the propionylation and benzoylation of benzene and toluene by propionylium and benzoylium hexafluoroantimonate, and deuteration of the methyl group in toluene produces a small secondary effect ($k_H/k_D = 1.04$). This is expected if a σ-complex is formed, as the conjugative effect of the –CD$_3$ group is greater than that of –CH$_3$.[92] Finocchiaro[93] has reported the rates of acetylation of mesitylene, diphenyl ether, dibenzofuran, and dibenzothiophen relative to toluene. A high selectivity was noted.

Arnett and Larsen[94] found that the heats of formation of substituted benzenonium ions in SbF$_5$–HSO$_3$F follow the Baker–Nathan order. A stable carbonium ion results from elimination of fluoride ion from a perhalogenated cycloalka-1,4-diene (Scheme 2).[95] Carbonium ions have been suggested as intermediates in the 1,2-shifts of cyclohexa-2,5-dienones but their existence

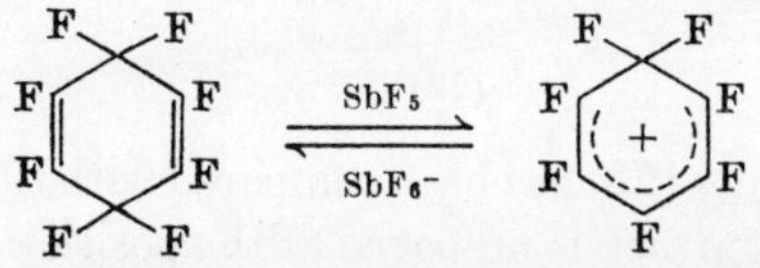

Scheme 2

[87] D. L. Ransley, *J. Org. Chem.*, **34**, 2618 (1969).

[88] E. P. Babin, I. G. Gakh, M. S. Berginina, and L. G. Gakh, *Zh. Prikl. Khim.*, **41**, 342 (1968); *Chem. Abs.*, **69**, 66606 (1968).

[89] F. Carré and R. Corriu, *Bull. Soc. Chim. France*, **1967**, 2898; F. Carré, R. Corriu, and G. Dabosi, *ibid.*, p. 2905.

[90] L. Friedman and R. J. Honour, *J. Am. Chem. Soc.*, **91**, 6344 (1969); G. Hoornaert and P. J. Slootmaekers, *Bull. Soc. Chim. Belges*, **78**, 245, 257 (1969).

[91] P. O. I. Virtanen and H. Ruotsalainen, *Ann. Acad. Sci. Fennicae*, A.II. Chem., No. 143 (1968).

[92] G. A. Olah, J. Lukas, and E. Lukas, *J. Am. Chem. Soc.*, **91**, 5319 (1969).

[93] P. Finocchiaro, *Ann. Chim.* (Italy), **59**, 787 (1969).

[94] E. M. Arnett and J. W. Larsen, *J. Am. Chem. Soc.*, **91**, 1438 (1969).

[95] V. D. Shteingarts, Y. V. Pozdnyakovich, and G. G. Yakobson, *Chem. Comm.*, **1969**, 1264.

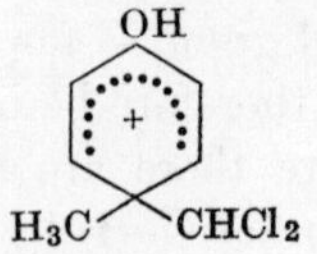

(20)

is transient. However, certain cyclohexadienones such as the 4-dichloromethyl-4-methyl-2,5-dienone form stable cations in H_2SO_4 (20). Protonation was found to follow H_0.[96]

Activation parameters obtained[97] for the cyclodehydration of aromatic ketones to give polynuclear aromatics are in agreement with the mechanism proposed earlier[98] (protonation to give a carbonium ion, cyclization, and elimination of water). The Friedel–Crafts reactions of chlorobenzene with cyanuric acid and 2,4-dichloro-6-methoxy-s-triazine have been examined.[99]

Substituents in tricarbonyl(alkylbenzene)chromium compounds have a very different effect upon the isomer ratios resulting from electrophilic attack than in the free aromatic; e.g. the methoxy group does not activate the *p*-position. In the t-butyl compound the preferred conformation of the CO groups effectively prevents σ-complex formation at the *para*-position (21) but permits it at the *meta*-position (22).[100] The Friedel–Crafts acetylation of a

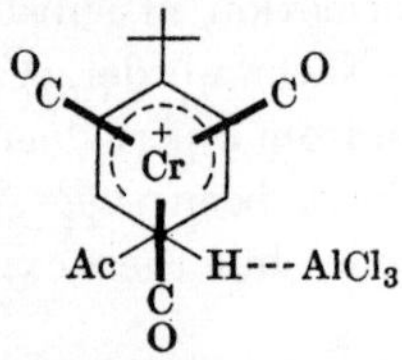

(21) 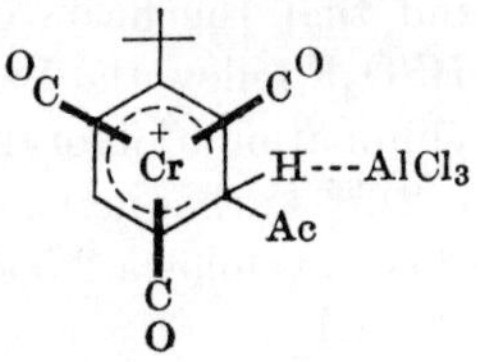   (22) 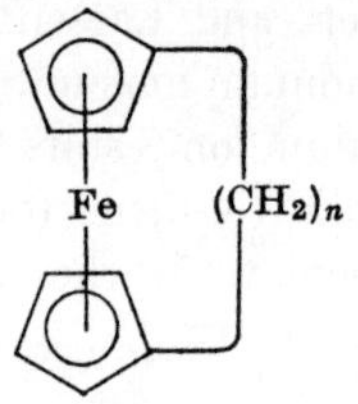   (23)

series of ferrocenophanes (23) has been compared with that of the non-bridged analogues. In general, attack is preferred at the position β to the interannular bridge but the extent of this depends upon the size of n.[101]

Halogenation

The normal course of aromatic bromination is attack by bromine to form a σ-complex in a rate-controlling step and rapid loss of a proton. However,

[96] V. P. Vitullo, *J. Org. Chem.*, **34**, 224 (1969).
[97] M. O. L. Spangler, J. C. Wolford, G. E. Treadwell, and C. Crusenberry, *J. Org. Chem.*, **34**, 892 (1969).
[98] C. K. Bradsher and F. A. Vingiello, *J. Am. Chem. Soc.*, **71**, 1434 (1949).
[99] T. Ishikawa and T. Ishii, *Yuki Gosei Kagaku Kyokai Shi*, **25**, 1042 (1967); *Chem. Abs.*, **68**, 113732 (1968).
[100] W. R. Jackson and W. B. Jennings, *J. Chem. Soc.*(B), **1969**, 1221.
[101] T. H. Barr, E. S. Bolton, H. L. Lentzner, and W. E. Watts, *Tetrahedron*, **21**, 5245 (1969).

Schubert and Gurka[102] report that in the bromination of benzene in aqueous trifluoroacetic acid the first step is reversible. The relative rates of bromination of neopentylbenzene, t-butylbenzene, toluene, and benzene depend upon the water content of the medium. These results are consistent with the views of Schubert and Sweeney[103] on the importance of solvation effects in governing the relative reactivities of alkylbenzenes. However, in a related study by Himoe and Stock[104] on the chlorination of alkylbenzenes it is concluded that a variety of factors (steric inhibition of solvation in the transition state, solvent–substrate interactions, and hyperconjugation) are all important in governing relative reactivities.[105]

The kinetics of bromination are often complicated by high-order terms occurring in the rate equation (generally second-order in bromine and first-order in aromatic). Such terms have been detected in the bromination of quinoline,[106] p-dimethoxybenzene,[107] and p-bromophenol.[108] In the last case there is evidence that acetic acid is involved in the rate-determining process other than as a solvating species. In aqueous solution the bromination of sodium phenolsulphonates results in bromodesulphonation as well as formation of p-benzoquinones. In methanol as solvent the latter reaction does not occur.[109]

In the presence of a base (Na_2CO_3) bromination of 3-phenanthrol (**24**) yields 4-bromo-3-phenanthrol, while without the base present the product is the 9-bromo-compound. The former may be converted into the latter by the

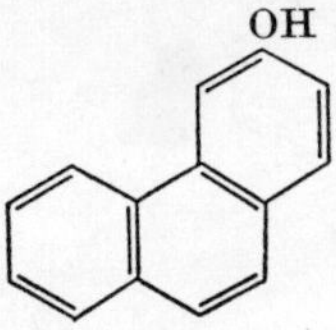

(**24**) 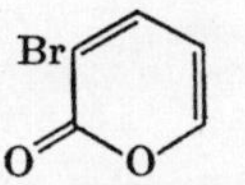   (**25**)

action of HCl in carbon tetrachloride: the mechanism involves protodebromination at the 4-position and subsequent attack at the 9-position. Bromination of 2-phenanthrol occurs at the 1-position.[110] Burgess and

[102] W. M. Schubert and D. F. Gurka, *J. Am. Chem. Soc.*, **91**, 1443 (1969).
[103] W. M. Schubert and W. A. Sweeney, *J. Org. Chem.*, **21**, 119 (1956).
[104] A. Himoe and L. M. Stock, *J. Am. Chem. Soc.*, **91**, 1452 (1969).
[105] See also ref. 86.
[106] V. N. P. Srivastava, B. B. L. Saxena, and B. Krishna, *Indian J. Chem.*, **6**, 306 (1968).
[107] K. V. Seshadri and R. Ganesan, *Z. Phys. Chem.* (Frankfurt), **62**, 29 (1968).
[108] J. Rajaram and J. C. Kuriacose, *Austral. J. Chem.*, **21**, 3069 (1968).
[109] J.-D. Aubort, *Helv. Chim. Acta*, **51**, 2098 (1968).
[110] E. Ota, Y. Iijima, and K. Iwamoto, *Yuki Gosei Kagaku Kyokai Shi*, **26**, 1112 (1968); *Chem. Abs.*, **70**, 76965 (1969).

Latham[111] have described a class experiment in the kinetics of bromination. Pirkle and Dines[112] report that the bromination and chlorination of 2-pyrone go via addition of halogen and elimination of hydrogen halide. The main product of bromination is 3-bromo-2-pyrone (25). The concluding paper[113] in Ridd's study of the effects of positive poles in electrophilic substitution has appeared and is concerned with the orientation of N and As poles in bromination by hypobromous acid. It is concluded that the p/m orientation of N poles is due to electrostatic interaction between the pole and the electrophile in the transition state. For other poles there are π-electrons which overlap with the aromatic ring and orientation is the result of conjugative as well as electrostatic effects.

The use of N-bromosuccinimide as a brominating agent has been reviewed.[114] Its use with various bithienyls results in predominant attack at the α-position, and the results correlate well with extended Pariser–Parr–Pople molecular orbital calculations.[115] Thallium may be used as a catalyst in aromatic bromination.[116] Millington[117] has described the anodic bromination of anthracene: the products depend on the applied potential and must arise from an electrode process involving anthracene and not merely by reaction between anthracene and bromine produced by discharge at the anode. The first step in the bromination of phenothiazine (26) is formation of a cationic free-radical which was detected by ESR.[118]

(26)

The rate of chlorination of p-chloroanisole has been determined,[119] and el Dusouqui, Hassan, and Ibrahim[120] have measured the partial rate factors for the m-nitro group in chlorination by molecular chlorine in dry acetic acid, using acetanilide as the substrate. The result does not agree with previously determined values. Results for the chlorination of t-butylbenzene by chlorine acetate in acetic acid indicate a low selectivity and little steric hindrance for

111 A. E. Burgess and J. L. Latham, *J. Chem. Educ.*, **46**, 370 (1969).
112 W. H. Pirkle and M. Dines, *J. Org. Chem.*, **34**, 2239 (1969).
113 A. Gastaminza, J. H. Ridd, and F. Roy, *J. Chem. Soc.*(B), **1969**, 684.
114 E.-G. Kleinschmidt and H. Bräuniger, *Pharmazie*, **24**, 87, 94 (1969).
115 R. M. Kellog, A. P. Schaap, and H. Wynberg, *J. Org. Chem.*, **34**, 343 (1969).
116 A. McKillop, D. Bromley, and E. C. Taylor, *Tetrahedron Letters*, **1969**, 1623.
117 J. P. Millington, *J. Chem. Soc.*(B), **1969**, 982.
118 P. Alcais and M.-C. Rau, *Bull. Soc. Chim. France*, **1969**, 3390.
119 S. Kalachandra and R. Ganesan, *Current Sci.*, **38**, 64 (1969).
120 O. M. H. el Dusouqui, M. Hassan, and B. Ibrahim, *J. Chem. Soc.*(B), **1969**, 589.

this reagent.[121] Side-chain chlorination of hexamethylbenzene is thought to occur by formation of a σ-complex (27) and then shift of Cl⁺ to a neighbouring methyl group and loss of a proton. This has been elegantly confirmed by the attempted chlorination of the compounds (28) and (29). The former is not

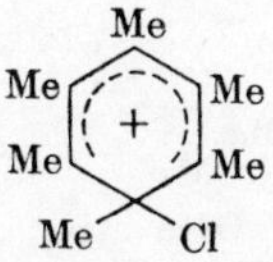

(27) 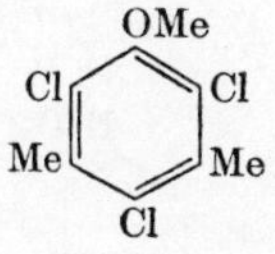   (28) 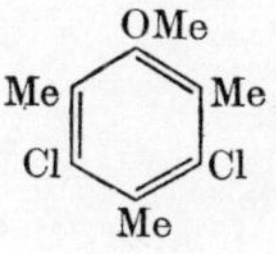   (29) 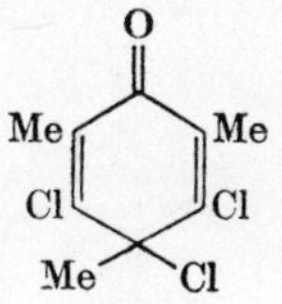   (30)

reactive, indicating that the mechanism is not direct attack on the side-chain, while demethylation to form (30) occurs with (29), indicating nuclear attack by the chlorine.[122] Bromination of hexamethylbenzene has also been studied.[123] In the presence of $AlCl_3$, α,α,α-tribromoacetophenone acts as an electrophilic brominating agent,[124] and Corriu and Coste[125] have shown that chlorination with $SbCl_5$ occurs via a π-complex.

de la Mare[126] has reported further studies of the chlorination of substituted naphthalenes. The product from the molecular chlorination of 1,5-diacetoxy-naphthalene in acetic acid or carbon disulphide is 5-acetoxy-2,4,4-trichloro-4*H*-naphthalen-1-one (31) but use of chloroform yields 5-acetoxy-2,3,4-trichloro-2*H*-naphthalen-1-one (32). The 1-position in triphenylene is the most reactive towards chlorination and does not appear to be sterically hindered.[127] The same group[128] have produced a comprehensive study of the

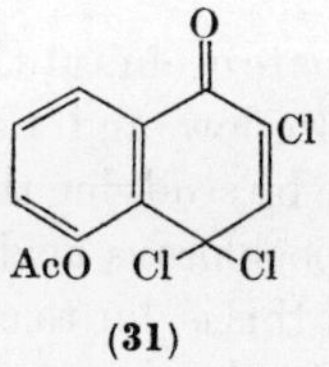

(31) 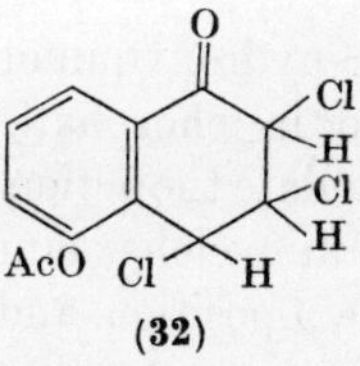   (32)

[121] M. Hassan and G. Yousif, *J. Chem. Soc.*(B), **1969**, 591.

[122] G. Antinori, E. Baciocchi, and G. Illuminati, *J. Chem. Soc.*(B), **1969**, 373.

[123] E. Baciocchi, M. Casula, G. Illuminati, and L. Mandolini, *Tetrahedron Letters*, **1969**, 1275.

[124] B. I. Stepanov and V. F. Traven, *Zh. Org. Khim.*, **5**, 387 (1969); *Chem. Abs.*, **70**, 105613 (1969).

[125] R. Corriu and C. Coste, *Tetrahedron*, **25**, 4949 (1969).

[126] P. B. D. de la Mare, S. de la Mare, and H. Suzuki, *J. Chem. Soc.*(B), **1969**, 429.

[127] P. Bolton, P. B. D. de la Mare, and L. Main, *J. Chem. Soc.*(B), **1969**, 170.

[128] P. B. D. de la Mare, A. Singh, E. A. Johnson, R. Koenigsberger, J. S. Lomas, U. Sanchez de Olmo, and A. M. Sexton, *J. Chem. Soc.*(B), **1969**, 717.

effects of added salts and solvent changes on the products and kinetics of the chlorination of phenanthrene and naphthalene in acetic acid. The bromination of acridone derivatives in methanol leads to addition of bromine and methoxy group rather than substitution (see, for example, Scheme 3).[129] The factors influencing addition to highly oxygenated acridones are discussed in detail.[130]

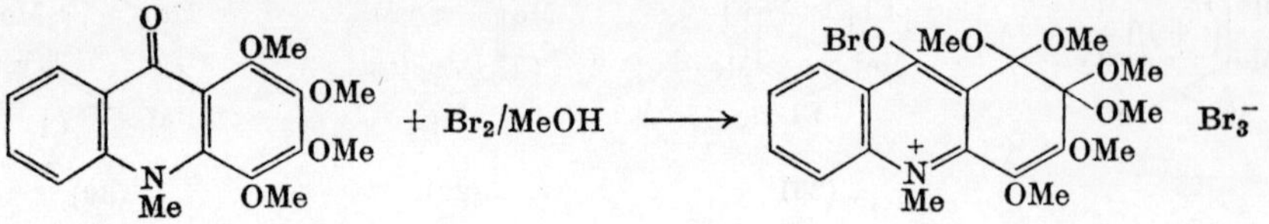

Scheme 3

Bell and De Maria[131] report that, in the presence of perchloric acid, the rate of bromination of various amines varies with the inverse of H_0 or H_0''' depending upon whether the amine is primary or tertiary, showing that the reactive species is the free amine. For 2,6-diethylaniline and 2,4-dimethyl-aniline the rate-determining step is attack of bromine but for 4-bromo-N,N-dimethylaniline the rate varies with the activity of water as well as the acidity, indicating the presence of water in the transition state of the slow step. This must be removal of a proton by the base water from the σ-complex.

Primary isotope effects have been detected in the bromination in dimethyl-formamide of 2-methyl-1,3,5-triethylbenzene and 2,4-dimethyl-1,3,5-triethyl-benzene.[132] They are caused by steric factors. De Fabrizio[133] reports that there is slow proton loss in the iodination of indole and 2-methylindole ($k_H/k_D = 2$—3).

Iodination of 8-hydroxyquinoline-5-sulphonate occurs at the 7-position and is base-catalysed in phosphate buffer.[134] Breslow and Campbell[135] have attempted to simulate the action of an enzyme by studying the chlorination of anisole enclosed in cyclohexaamylose by hypochlorous acid. Substitution is exclusively in the p-position, and it is thought this is due to chlorination of an amylose hydroxy group and subsequent attack of the anisole molecule contained within the cyclohexaamylose.

[129] R. H. Prager and H. M. Thredgold, *Austral. J. Chem.*, **22**, 1477, 1493, 1503 (1969).
[130] R. H. Prager and H. M. Thredgold, *Austral. J. Chem.*, **22**, 1511 (1969).
[131] R. P. Bell and P. De Maria, *J. Chem. Soc.*(B), **1969**, 1057.
[132] A. Nilsson and K. Olsson, *Acta Chem. Scand.*, **23**, 7 (1969).
[133] E. C. R. De Fabrizio, *Ann. Chim.* (Italy), **58**, 1435 (1968).
[134] F. M. Vainshtein, T. A. Degurko, and E. A. Shilov, *Kinet. Katal.*, **9**, 965 (1969); *Chem. Abs.*, **70**, 56888 (1969).
[135] R. Breslow and P. Campbell, *J. Am. Chem. Soc.*, **91**, 3085 (1969).

Hydrogen Exchange

Hydrogen exchange in metallocenes (particularly ferrocenes) has been the subject of a review[136] but there has been no new work in this field apart from a study of the effect of a methoxy group on the rate of hydrogen exchange in anisoletricarbonylchromium in alkali solution. The presence of chromium appears to lessen the effect of the methoxy group.[137]

Hydrogen exchange in hypoxanthines (33) generally occurs at the 8-position but certain substituents direct attack to the 2-position.[138] Acid-catalysed exchange of 2-amino- and 6-amino-2,4-dimethylpyrimidine occurs at the 5-position, and the reactive species is the free base, although the latter can react as the conjugate acid.[139] This is the only reactive species for hydrogen exchange with 2-quinolone, and exchange occurs at the 3-, 8-, 6-, and 5-positions.[140] Hydrogen exchange of pyridine *N*-oxides occurs by formation of the anion, not by addition and deprotonation, and by evaluating the rate relative to benzene it appears that the N^+-O^- group is the most strongly activating group yet reported.[141] From studies of exchange in t-arylphosphines and their oxides, it has been shown that phosphorus is a more powerful electron-attractor than N owing to $p\pi-d\pi$ bonding.[142] The rate is the same at the *m*- and *p*-positions. Barnett and Warkentin[143] report that the species responsible for hydrogen exchange in 2,3-dihydro-5,7-dimethyl-1,4-diazepinium perchlorate (34) in phosphate buffer is $H_2PO_4^-$. The more usual hydronium ion is not effective owing to the charge on the diazepinium ion. Eaborn and Fischer[144] have analysed the effect of substituents on the rate of detritiation of naphthalene, and conclude that the theoretical charge distribution in the Wheland intermediate provides the best simple guide.

(33)   (34)   (35)

[136] V. N. Setkina and D. N. Kursanov, *Uspekhi Khim.*, **37**, 1729 (1968); *Chem. Abs.*, **70**, 20108 (1969).

[137] D. N. Kursanov, V. N. Setkina, N. K. Baranetskaya, E. I. Fedin, K. N. Anisimov, and V. M. Urinyuk, *Dokl. Akad. Nauk SSSR*, **183**, 1340 (1969); *Chem. Abs.*, **70**, 86741 (1969).

[138] F. Bergmann, D. Lichtenberg, and Z. Neiman, *Chem. Comm.*, **1969**, 992.

[139] A. R. Katritzky, M. Kingsland, and O. S. Tee, *J. Chem. Soc.*(B), **1968**, 1484.

[140] G. P. Bean, A. R. Katritzky, and A. Marzec-Pawlowska, *Bull. Acad. Pol. Sci., Ser. Sci. Chim.*, **16**, 453 (1968); *Chem. Abs.*, **70**, 76963 (1969).

[141] J. A. Zoltewicz and G. M. Kauffman, *J. Org. Chem.*, **34**, 1405 (1969).

[142] E. A. Yakovleva, E. N. Tsvetkov, D. I. Lobanov, A. I. Shatenshtein, and M. I. Kabachnik, *Tetrahedron*, **25**, 1165 (1969).

[143] C. Barnett and J. Warkentin, *J. Chem. Soc.*(B), **1968**, 1572.

[144] C. Eaborn and A. Fischer, *J. Chem. Soc.*(B), **1969**, 152.

Base-catalysed and acid-catalysed hydrogen exchange at both the α- and β-positions in thiophen have been compared.[145] The value of $\log k$ for exchange in substituted thiophens at the β-position correlates well with the Hammett σ-constants but for exchange at the α-position there is better correlation with Taft σ^*-constants.[146] It is difficult to draw any inferences from these results and they are in disagreement with the data of other workers.[147]

There has been considerable dispute over the correlation between rate of hydrogen exchange and the acidity function H_0.[148] The slope of a plot of $\log k$ against $-H_0$ is certainly not unity but it may be non-linear. Gold and Adsetts[149] have studied tritium uptake by mesitylene from HCl over a very wide range of acidities (0.1M to 9.16M) with a resultant 10^6 change in rate. A plot of $\log k$ against $-H_0$ is not linear throughout but it is linear at low acidities with a slope of 1.17. They discuss in detail the relevance of H_0 measurements to the transition state for hydrogen exchange and the size of the Brønsted exponent. Failure of the simple linear relationship appears to be due to effects in the activity coefficient of mesitylene and the indicator used to establish the acidity function.

Protonation of $2H$-cyclopenta[d]pyridazine (35) and its 2-methyl derivative occurs at positions 5 and 7 despite the presence of ring nitrogens. With most other 10 π-electron systems (e.g. 5-aza-azulene) protonation does occur on the ring nitrogens.[150] A third dissociation constant for tryptophan ($pK_a = -6.23$ on the H_I scale), corresponding to protonation of the indole ring, has been determined.[151] Modro, Jasinski, and Modro[152] report a study of the protonation of phosphonic and arsonic acids.

Two new acidity scales have been proposed: H_c, based on the protonation of carbon, increases more rapidly than H_0 but slightly less so than $H_R{}'$,[153] and H_m uses the protonation of various substituted azulenes and is identical with $H_0{}'''$ up to 60% H_2SO_4.[154] It is not clear how this scale differs from that proposed by Long and Schulze.[155]

Both aromatic and alkyl hydrogen exchange in alkylbenzenes are catalysed

[145] I. O. Shapiro, L. I. Belen'kii, I. A. Romanskii, F. M. Stoyanovich, Y. L. Gol'dfarb, and A. I Shatenshtein, *Zh. Obshch. Khim.*, **38**, 1998 (1968); *Chem. Abs.*, **70**, 19289 (1969); E. N. Zvyagintseva, L. I. Belen'kii, T. A. Yakushina, Y. L. Gol'dfarb, and A. I. Shatenshtein. *ibid.*, p. 2004; *Chem. Abs.*, **70**, 19290 (1969).

[146] A. Kamrads, *Reakts. Spos. Org. Soed.*, **5**, 701 (1968); *Chem. Abs.*, **70**, 67456 (1969).

[147] A. R. Butler and C. Eaborn, *J. Chem. Soc.*(B), **1968**, 370.

[148] See e.g. C. Eaborn and R. Taylor, *J. Chem. Soc.*, **1960**, 3301.

[149] J. R. Adsetts and V. Gold, *J. Chem. Soc.*(B), **1969**, 950.

[150] A. G. Anderson and D. M. Forkey, *J. Am. Chem. Soc.*, **91**, 924 (1969).

[151] R. C. Armstrong, *Biochem. Biophys. Acta*, **158**, 174 (1968).

[152] A. Modro, T. Jasinski, and T. A. Modro, *Chem. Ind.* (London), **1969**, 381.

[153] M. T. Reagan, *J. Am. Chem. Soc.*, **91**, 5506 (1969).

[154] T. K. Rodima, U. Halda, and E. Varjend, *Reakts. Spos. Org. Soed.*, **5**, 466 (1968); *Chem. Abs.*, **70**, 19468 (1969).

[155] F. A. Long and J. Schulze, *J. Am. Chem. Soc.*, **86**, 327 (1964).

by chloroplatinites in acetic acid.[156] Dimerization is a possible side-reaction during Pt(II)-catalysed exchange under heterogeneous conditions.[157]

Substitution of t-butyl groups in the *peri*-positions of naphthalene should introduce more strain than o-t-butyl groups in benzene. Franck and Leser[158] report that, because of this strain, protode-t-butylation occurs by the action of formic acid on 1,4,6,8-tetra-t-butylnaphthalene. It is the 4-t-butyl group which is lost.

Metalation

There has been little work reported this year on electrophilic metalation. Lithiation of 1,1'-di-(N,N-dimethylaminomethyl)ferrocene occurs at a position *ortho* to the substituent owing to coordination of the lone pair of the $-NMe_2$ group with the electron-deficient lithium (**36**).[159] The mechanism of

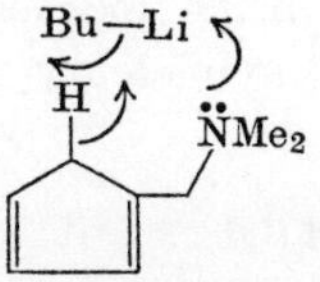

(**36**) 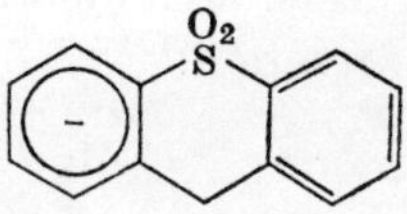   (**37**)

the Smiles rearrangement of aryl sulphones to sulphinic acid by butyl-lithium has been discussed in terms of an anion intermediate (**37**), the presence of which has been demonstrated by NMR.[160]

Metal Cleavage

Eaborn has continued work on a number of metal cleavage reactions. Proto-detrimethylsilylation of *p*-substituted trimethylsilylbenzenes obeys the Yukawa–Tsuno equation with $r = 0.65$; for *o*-substituted compounds $r = 0.3$.[161] This indicates that for *ortho*-substituents the inductive effect is more important than mesomerism compared to *para*-substituents. A study of protodetri-methylsilylation of trimethylsilyl derivatives of benzocyclobutene, indane, and tetralin indicates that both strain and hybridization are operative in the Mills–Nixon effect.[162] Eaborn, Thompson, and Walton[163] have determined

[156] R. J. Hodges and J. L. Garnett, *J. Catal.*, **13**, 83 (1969).
[157] G. E. Calf and J. L. Garnett, *Chem. Comm.*, **1969**, 373.
[158] R. W. Franck and E. G. Leser, *J. Am. Chem. Soc.*, **91**, 1577 (1969).
[159] E. S. Bolton, P. L. Pauson, M. A. Sandhu, and W. E. Watts, *J. Chem. Soc.*(C), **1969**, 2260.
[160] V. N. Drozd, L. A. Nikonova, L. I. Zefirova, and K. A. Pak, *Izv. Timiryazev. Sel'sk. Akad.*, **1969**, 179; *Chem. Abs.*, **71**, 38735 (1969).
[161] C. Eaborn, D. R. M. Walton, and D. J. Young, *J. Chem. Soc.*(B), **1969**, 15; C. Eaborn and P. M. Jackson, *ibid.*, p. 21.
[162] A. R. Bassindale, C. Eaborn, and D. R. M. Walton, *J. Chem. Soc.*(B), **1969**, 12.
[163] C. Eaborn, A. R. Thompson, and D. R. M. Walton, *J. Chem. Soc.*(B), **1969**, 859.

the σ^+-constants for the p- and m-ethynyl groups from a study of protodestan-
nylation of substituted trimethylstannylbenzenes. These groups are electron-withdrawing by induction but capable of electron release. The effect
of positive poles on the rate of protodetrimethylsilylation is similar to that
observed by Ridd[164] in aromatic nitration. The strong electron withdrawal
of the positive charge is enhanced by $p\pi$–$d\pi$ bonding for the –PMe$_3{}^+$ group
and to a lesser extent the AsMe$_3{}^+$ group.[165] Cleavage of the aryl–Si bond in
compounds like Me$_3$SiCH$_2$C$_6$H$_4$SiMe$_3$ indicates hyperconjugative release of
electrons from Me$_3$Si–C.[166]

Rate changes are reflected in the values of $\Delta S^\ddagger$ rather than the size of E_a
in the protodemercuration of diphenylmercury by a number of weak acids.[167]
Substitution in the phenyl ring obeys the Yukawa–Tsuno equation.[168] The
protodemercuration of ferrocenylmercury derivatives has been studied.[169]
Exchange of mercury between phenylmercury bromide and mercury salts is a
bimolecular electrophilic reaction, the rate being increased by electron-
donating substituents.[170] Iodination of diphenylmercury in carbon tetrachloride is an S_Ei reaction:[171]

$$Ph_2Hg + I_2 \rightarrow PhI + PhHgI$$

Nasielski has continued his studies of destannylation reactions. A quantitative treatment of the results for the bromodestannylation of aromatics in
methanol by C/C homogeneous correlations is a direct and satisfactory application of the population-correlation theory. Both the ratio of the rate constants
for breaking of the C(sp^2)–Sn and the C(sp^2)–H bonds and the ρ-values show
differences in the transition states for bromodestannylation and bromo-
deprotonation.[172] Acetolysis of substituted phenyltrimethyltins follows a
Hammett $\rho\sigma^+$ relationship ($\rho = -2.24$). The solvent isotope effect ($k_H/k_D = 10$)
indicates that the transition state must be very different from a σ-complex.[173]

[164] See *Org. Reaction Mech.*, **1968**, 209.
[165] C. Eaborn and J. F. R. Jaggard, *J. Chem. Soc.*(B), **1969**, 892.
[166] A. R. Bassindale, C. Eaborn, D. R. M. Walton, and D. J. Young, *J. Organometal. Chem.*,
     20, 49 (1969).
[167] I. P. Beletskaya, I. L. Zhuravleva, and O. A. Reutov, *Zh. Org. Khim.*, **4**, 729 (1968); *Chem.
     Abs.*, **69**, 18405 (1968).
[168] I. P. Beletskaya, A. L. Kurts, and O. A. Reutov, *Zh. Org. Khim.*, **4**, 542 (1968); *Chem. Abs.*,
     69, 2305 (1968).
[169] A. N. Nesmeyanov, E. G. Perevalova, S. P. Gubin, and A. G. Kozlovskii, *Izv. Akad. Nauk
     SSSR, Ser. Khim.*, **1968**, 654; *Chem. Abs.*, **69**, 66689 (1968).
[170] I. P. Beletskaya, I. I. Zakharycheva, and O. A. Reutov, *Zh. Org. Khim.*, **5**, 793, 798 (1969);
     Chem. Abs., **71**, 38040, 38041 (1969).
[171] I. P. Beletskaya, A. L. Kurts, N. K. Genkima, and O. A. Reutov, *Zh. Org. Khim.*, **4**, 1120
     (1968); *Chem. Abs.*, **69**, 66618 (1968).
[172] P. Alcais and J. Nasielski, *J. Chim. Phys.*, **66**, 95 (1969).
[173] J. Nasielski, O. Buchman, M. Grosjean, and M. Jauquet, *J. Organometal. Chem.*, **19**, 353
     (1969).

Decarboxylation

Decarboxylation in acidic solution has been reviewed by Long.[174] A study of the ^{13}C isotope effect supports the previously proposed mechanism for the decarboxylation of azulene-1-carboxylic acid.[175] Below 0.01M-acid the rate-determining step is slow proton-transfer and, consequently, there is no ^{13}C isotope effect. At 0.3M-acid the slow step has changed to C–C bond fission of the 1-protonated form and the maximum isotope effect (1.035—1.043) is observed.[176] Zieliński[177] has determined the ^{13}C isotope effect for the decarboxylation of a number of pyridinedicarboxylic acids. The effect of the second carboxy group upon the size of the isotope effect depends upon its position. In 1952 Stevens, Pepper, and Lounsberg[178] reported that there is only a very small ^{13}C isotope effect in the decarboxylation of anthranilic acid, making it different from quinaldinic and picolinic acids. This reaction has been re-investigated by Zieliński[179] who found that the size of the isotope effect depends upon the degree of decarboxylation. Extrapolation to zero decarboxylation gives an isotope effect of 1.0105, which is not different from that for other decarboxylation reactions. Thermodynamic data for the thermal decarboxylation of 3-nitro- and 3,5-dinitro-salicylic acid in resorcinol are consistent with an S_E2 mechanism.[180] Decarboxylation of indole-2-carboxylic acid involves protonation at C-3.[181]

Miscellaneous Reactions

Benzyl cations have been shown to be intermediates in the side-chain amination of aryldialkylmethines by trichloroamine, $AlCl_3$, and t-butyl bromide.[182] Adamantane is also aminated at the bridge position by trichloroamine and $AlCl_3$. The first step is attack of Cl^+ and loss of HCl to form the adamantyl carbonium ion (38), and this reacts with NCl_2^-.[183]

Jackson has published a further study of electrophilic attack of 3-substituted indoles: as reported previously[184] initial attack is at the 3-position to form an indolenine intermediate which then rearranges to give a 2,3-disubstituted product. As expected from this mechanism, the same product (2-p-methoxy-benzyl-3-benzylindole) results from attack of p-methoxybenzyl chloride on

[174] F. A. Long, in *Hydrogen-Bonded Solvent Systems* (Ed. A. K. Covington and P. Jones), Taylor and Francis, London, 1968, p. 285.
[175] See *Org. Reaction Mech.*, **1968**, 217.
[176] H. H. Huang and F. A. Long, *J. Am. Chem. Soc.*, **91**, 2872 (1969).
[177] M. Zieliński, *Rocz. Chem.*, **43**, 1547 (1969).
[178] W. H. Stevens, J. M. Pepper, and M. Lounsberg, *Can. J. Chem.*, **30**, 529 (1952).
[179] M. Zieliński, *Rocz. Chem.*, **42**, 1725 (1968).
[180] K. S. Sastry, E. V. Sundaram, and S. S. Muhammod, *Current Sci.* (India), **37**, 643 (1968).
[181] S. Ghosal, *Indian J. Chem.*, **5**, 650 (1967).
[182] P. Kovacic, J. F. Gormish, R. J. Hopper, and J. W. Knapczyk, *J. Org. Chem.*, **33**, 4515 (1968).
[183] P. Kovacic and P. D. Roskos, *J. Am. Chem. Soc.*, **91**, 6344 (1969).
[184] See *Org. Reaction Mech.*, **1967**, 208.

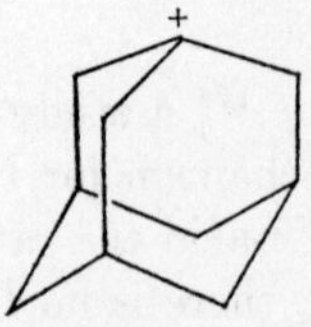

(38) 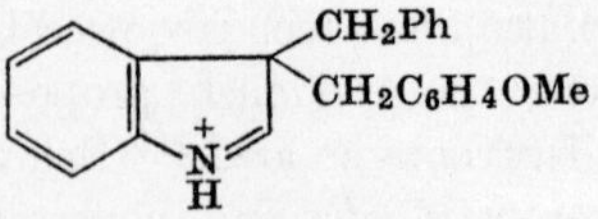   (39) 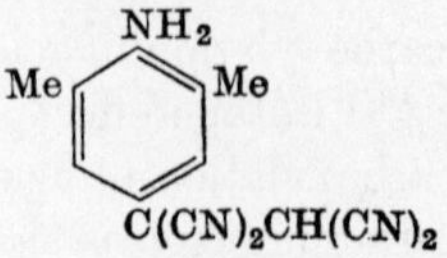   (40)

the Grignard reagent from 3-benzylindole as from the reaction of benzyl
chloride with the Grignard reagent derived from 3-*p*-methoxybenzylindole.
The common intermediate is shown (39): in both cases it is the *p*-methoxy-
benzyl group which migrates.[185] The product of tricyanovinylation by
tetracyanoethylene of 2,6-dimethylaniline is thought to be (40).[186]

[185] K. M. Biswas and A. H. Jackson, *Tetrahedron*, **25**, 227 (1969).
[186] Z. Rappoport and E. Shohamy, *J. Chem. Soc.*(B), **1969**, 77.

CHAPTER 8

Molecular Rearrangements

C. W. Rees

Department of Organic Chemistry, University of Liverpool

In a full and fascinating account of the conservation of orbital symmetry, Woodward and Hoffmann embrace the theory of electrocyclic and sigmatropic reactions and give their generalized selection rules for pericyclic reactions.[1] The first two volumes of a series devoted to molecular rearrangements have appeared[2] and also include an extensive account of orbital symmetry and electrocyclic reactions.[3] Thermal unimolecular reactions of hydrocarbons in the gas phase for which precise quantitative data are available have been reviewed; these include Cope rearrangements, dienyl and homodienyl 1,5-hydrogen shifts, and the rearrangements of cyclopropanes and cyclobutanes.[4] Radical rearrangements are dealt with in Chapter 9.

[1] R. B. Woodward and R. Hoffmann, *Angew. Chem. Internat. Ed. Engl.*, **8**, 781 (1969).
[2] B. S. Thyagarajan (Ed.), *Mechanisms of Molecular Migrations*, Vols. 1 and 2, Interscience, New York, 1968 and 1969.
[3] K. Fukui and H. Fujimoto, in ref. 2, Vol. 2, p. 117.
[4] H. M. Frey and R. Walsh, *Chem. Rev.*, **69**, 103 (1969).

Aromatic Rearrangements

Claisen Rearrangements

Aromatic sigmatropic rearrangements have been reviewed [5] and the stereochemistry of the *ortho*-Claisen rearrangement has been fully analysed. [6]

Makisumi has provided further evidence for the intermediacy of quinoline-thiones, e.g. (1), in the thio-Claisen rearrangement [7] by trapping them in high yield as the *S*-butyryl derivatives in butyric anhydride. [8] The thio-Claisen rearrangement was extended to the propargyl sulphide (2) which on heating at 200° in dimethylaniline gave (4), presumably via the allene (3). [8] The

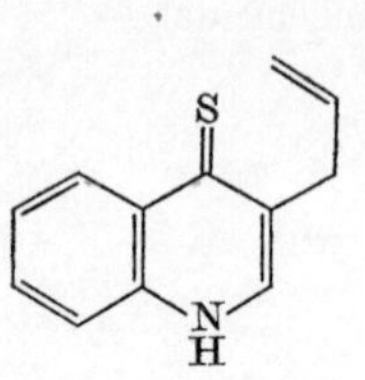

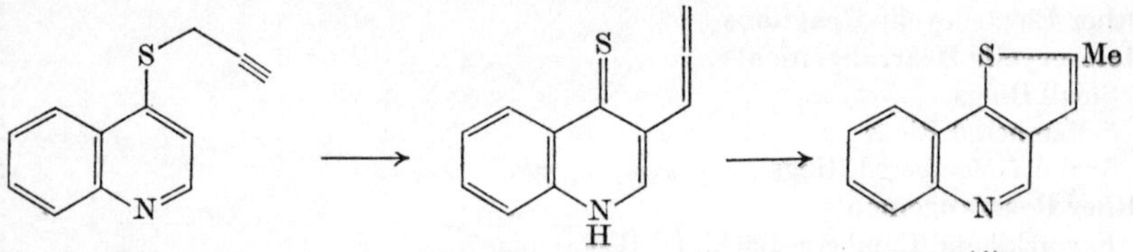

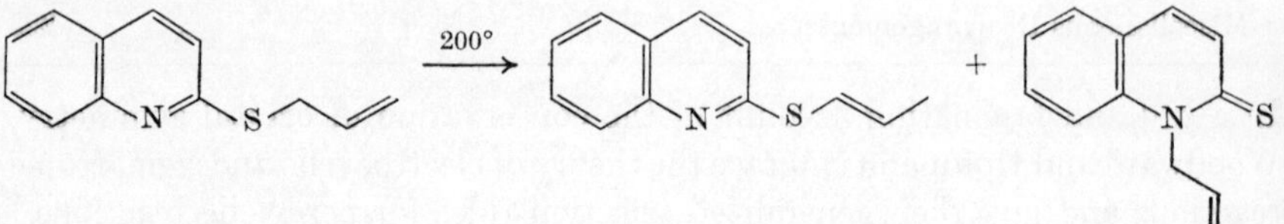

thermal rearrangement of allyl 2-quinolyl sulphides, e.g. (5), was also investigated. The major product (6) has the double bond conjugated with sulphur, but a small amount of the [3,3]-sigmatropic product (7) was detected. The

[5] H. J. Hansen and H. Schmid, *Chem. Brit.*, **5**, 111 (1969).
[6] G. Frater, A. Habich, H. J. Hansen, and H. Schmid, *Helv. Chim. Acta*, **52**, 335 (1969).
[7] *Org. Reaction Mech.*, **1967**, 211.
[8] Y. Makisumi and A. Murabayashi, *Tetrahedron Letters*, **1969**, 1971.

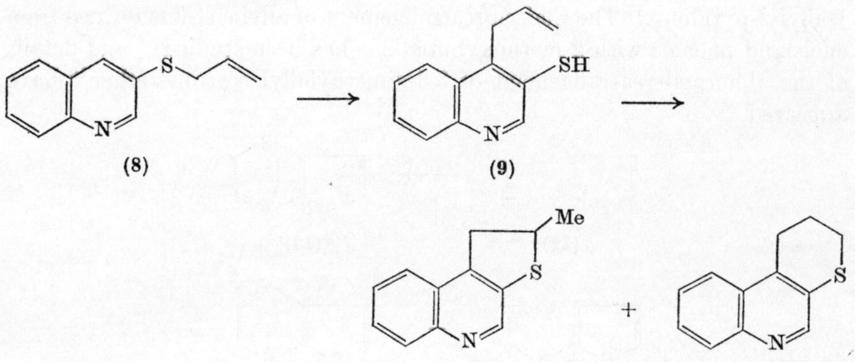

<table>
<tr><td>(8)</td><td>(9)</td></tr>
</table>

(10)   +   (11)

N-allyl compounds (7) also rearranged readily to the more stable S-allyl isomers. The N- and S-benzyl compounds were stable under the same conditions.[9] Allyl 3-quinolyl sulphides, e.g. (8), also rearranged smoothly to the products shown; the same mixture of products in the same yield was obtained from the proposed intermediate (9). Products (10) and (11) were formed in ionic and radical processes, respectively.[10] Evidence for o-allylthiophenol as the intermediate in the rearrangement of allyl phenyl sulphide was provided by trapping of its anion with methyl iodide, and again by formation of the same products in the same ratio from pre-formed o-allylthiophenol.[11]

Thermal rearrangement of (12) to (13) and of (14) to (15) in HMPT at 170—180° also appears to involve acetylenic thio-Claisen migration (see above) on to thiophen.[12] The related acyclic compounds $R^1CH=C(SEt)$-$SCH_2C\equiv CR^2$ rearrange similarly to the allenic dithio-esters $CH_2=C=CR^2$-$CHR^1C(S)SEt$ which in turn cyclize to thiopyrans and thiophens.[13]

The thermal amino-Claisen rearrangement of N-allylpyrazolin-5-ones[14] has been extended to N-allylisoxazolin-5-ones (16); inversion of the allyl group was established by labelling with methyl. With the N-crotyl compound the rearrangement was reversible, giving approximately equal amounts of the isomers in equilibrium at 180°.[15]

Benzyl phenyl ethers do not undergo the Claisen rearrangement; here the bis-allylic double bonds are both part of an aromatic ring.[16] Several Lewis acids catalyse very efficiently the rearrangement of 2-allyloxypyridine to

[9] Y. Makisumi and T. Sasatani, *Tetrahedron Letters*, **1969**, 1975.
[10] Y. Makisumi and A. Murabayashi, *Tetrahedron Letters*, **1969**, 2449, 2453.
[11] H. Kwart and J. L. Schwartz, *Chem. Comm.*, **1969**, 44.
[12] L. Brandsma and H. J. T. Bos, *Rec. Trav. Chim.*, **88**, 732 (1969).
[13] P. J. W. Schuijl, H. J. T. Bos, and L. Brandsma, *Rec. Trav. Chim.*, **88**, 597 (1969).
[14] *Org. Reaction Mech.*, **1967**, 210.
[15] Y. Makisumi and T. Sasatani, *Tetrahedron Letters*, **1969**, 543.
[16] W. J. le Noble and B. Gabrielsen, *Chem. Ind.* (London), **1969**, 378.

1-allyl-2-pyridone.[17] The Claisen rearrangement of allylic ethers formed from enols and phenols with 2-methoxybutadiene has been studied,[18] and details of the abnormal rearrangement of 3,3-dimethylallyl oestrone ether[19] have appeared.[20]

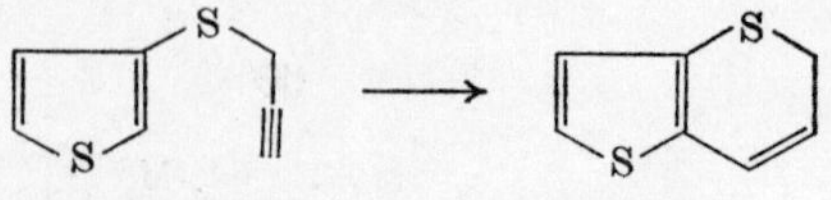

(12)   **(13)**

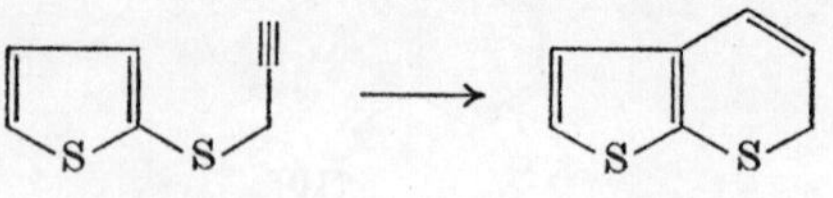

(14)   **(15)**

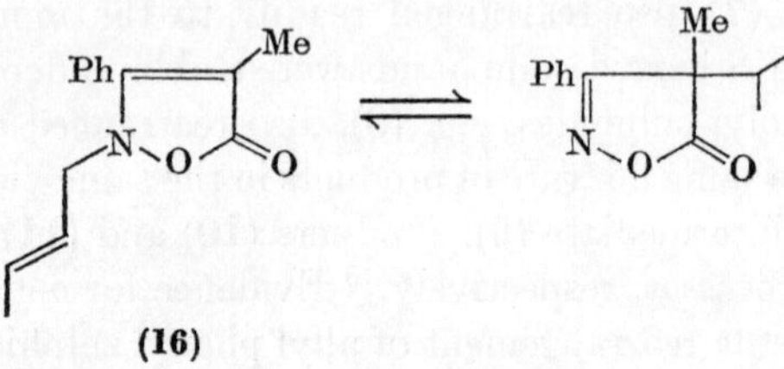

(16)

Dienone–Phenol and Related Rearrangements

The rearrangements of cyclohexadienones have been reviewed.[21] In the phenol to phenol rearrangement of 1-methyl-3-tetralol **(17)** to 3-methyl-1-tetralol **(18)** with 70% perchloric acid at 80°, the ^{14}C label remains largely

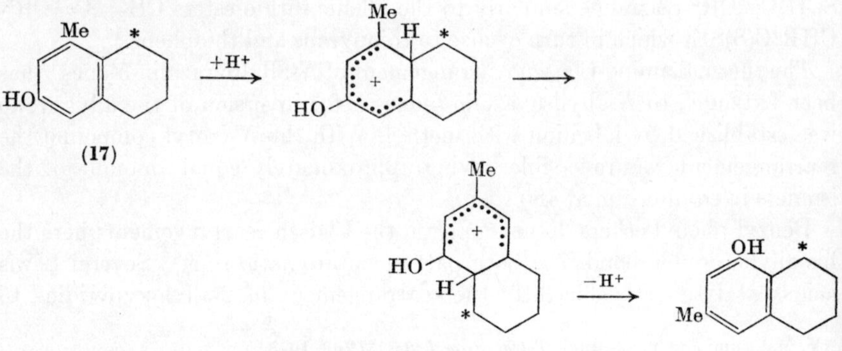

(17)

(18)

[17] H. F. Stewart and R. P. Seibert, *J. Org. Chem.*, **33**, 4560 (1968).

[18] L. J. Dolby, C. A. Elliger, S. Esfandiari, and K. S. Marshall, *J. Org. Chem.*, **33**, 4508 (1968); see also F. Effenberger, *Angew. Chem. Internat. Ed. Engl.*, **8**, 310 (1969).

[19] *Org. Reaction Mech.*, **1966**, 211.

[20] A. Jefferson and F. Scheinmann, *J. Chem. Soc.*(C), **1969**, 243.

[21] B. Miller, in ref. 2, Vol. 1, p. 247.

unscrambled, as shown. Thus the 1,3-alkyl shift predominates over two 1,2-shifts through a spiro-intermediate.[22] The isomerization of hexamethyl-cyclohexadienones in strong acid is greatly accelerated by sulphur trioxide, which also changes the relative stabilities of the isomers.[23]

Work on the mechanism of the Fischer indole synthesis (up to 1967) has been reviewed.[24] The effect of the acid catalyst on the direction of the synthesis with unsymmetrical ketones has been carefully studied and earlier generalizations about such ring-closures have been shown to be incorrect.[25] Further evidence has been provided for dienimine intermediates in the Fischer indole synthesis,[26] and other studies of this reaction have been reported.[27]

Benzidine and Wallach Rearrangements

Benzidine rearrangements have been reviewed.[28] On treatment with acid, 2,2'-hydrazodiphenyl sulphide disproportionates to the corresponding azo-compound and diamine and also undergoes ring-contraction to phenothiazines. The kinetics correspond to the one-proton benzidine rearrangement, and the various products were thought to be formed from a common intermediate (**20**) obtained in the slow step from the conjugate acid (**19**).[29] Attempted rearrange-

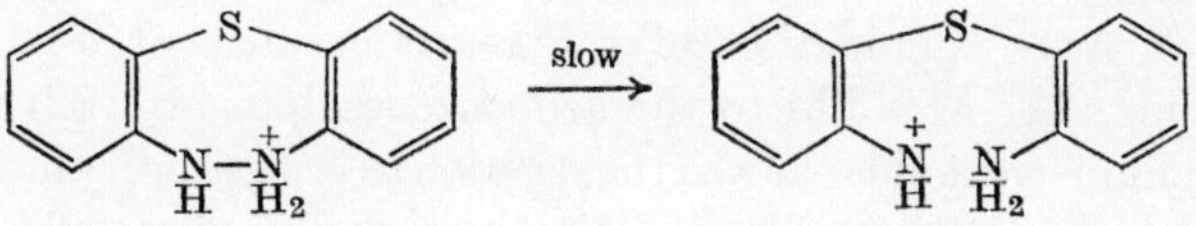

(19)   (20)

ment of 2,2'-hydrazofluorene in alcoholic hydrogen chloride resulted in quantitative disproportionation to the amino- and azo-compounds.[30]

Buncel has reviewed the published work and some of his own unpublished work on the mechanism of the Wallach rearrangement.[31] The kinetics of the rearrangement of azoxybenzene to 4-hydroxyazobenzene have been extended from 96 to 100% sulphuric acid at 25°. The results, coupled with the cryoscopic

[22] R. Futaki, *Tetrahedron Letters*, **1968**, 6245.
[23] R. F. Childs, *Chem. Comm.*, **1969**, 946.
[24] B. Robinson, *Chem. Rev.*, **69**, 227 (1969).
[25] M. H. Palmer and P. S. McIntyre, *J. Chem. Soc.*(B), **1969**, 446.
[26] G. S. Bajwa and R. K. Brown, *Can. J. Chem.*, **47**, 785 (1969).
[27] A. Ebnöther, P. Niklaus, and R. Süess, *Helv. Chim. Acta*, **52**, 629 (1969); R. H. C. Elgersma and E. Havinga, *Tetrahedron Letters*, **1969**, 1735.
[28] H. J. Shine, in ref. 2, Vol. 2, p. 191.
[29] L. D. Hartung and H. J. Shine, *J. Org. Chem.*, **34**, 1013 (1969).
[30] M. J. Namkung and T. L. Fletcher, *Chem. Comm.*, **1969**, 1052; see also S. Hashimoto, J. Sunamoto, I. Shinkai, and S. Nishitani, *Doshisha Daigaku Rikogaku Kenkyu Hokoku*, **8**, 211 (1968); *Chem. Abs.*, **69**, 85968 (1968).
[31] E. Buncel, in ref. 2, Vol. 1, p. 61.

and conductivity behaviour of azoxybenzene and of 4-hydroxyazobenzene, show that rearrangement in 100% H_2SO_4 is very rapid ($t_{\frac{1}{2}} < 0.5$ min) and is followed by two consecutive sulphonation reactions ($t_{\frac{1}{2}} = 4.5$ min. and 28 hr. respectively).[32] No firm evidence was obtained for finite concentrations of intermediates of the type previously postulated.[33] The role of azobenzene-4-hydrogen sulphate, and its hydrolysis, in the Wallach rearrangement has been discussed.[34]

Other Rearrangements

The Fischer–Hepp rearrangement of N-methyl-N-nitrosoaniline in aqueous ethanol containing [^{15}N]nitrous acid gave p-nitroso-N-methylaniline with no ^{15}N enrichment in the presence of hydrochloric acid, and only very little enrichment in the presence of sulphuric acid. Thus the commonly accepted intermolecular mechanism of reversible N-denitrosation followed by C-nitrosation is most unlikely, and this rearrangement may well be intramolecular after all.[35]

Rearrangement of 1-naphthylsulphamic acid to 4-aminonaphthalene-1-sulphonic acid appears from ^{35}S-labelling experiments to be partly intra- as well as inter-molecular, in contrast to the wholly intermolecular nature of the rearrangement of phenylsulphamic to sulphanilic acid under identical conditions.[36] A re-examination of the rearrangement of o-nitrobenzene-sulphenanilide (**21**; Ar = Ph) to the azobenzenesulphinate (**22**) in aqueous alcoholic sodium hydroxide showed the reaction to be first-order in sulphenanilide and in hydroxide ion. The reaction was slightly faster when Ar was p-methoxyphenyl, and was negligibly slow when Ar was p-nitrophenyl. ^{18}O-Labelling showed that both oxygen atoms of the nitro group were transferred to sulphur intramolecularly. The mechanism shown was proposed.[37] Aromatic esters of o-nitrobenzenesulphenic acids rearrange thermally to o- and p-hydroxyphenyl sulphides by heterolysis of the S–O bond to give an ion-pair (**23**).[38]

The reactions of N-aroyldiphenylamines in hot polyphosphoric acid, to give acridines as well as diphenylamines and 4,4-diaroyldiphenylamines, are intermolecular involving complete separation of the aroyl cation.[39] Further proof of the intramolecularity of the thermal isomerization of arylimidates to N,N-diaryl-amides (Chapman rearrangement) is provided by the absence of cross-over products when [^{14}C]-p-bromophenyl N-phenylbenzimidate (**24**;

[32] E. Buncel, W. M. J. Strachan, R. J. Gillespie, and R. Kapoor, *Chem. Comm.*, **1969**, 765.
[33] *Org. Reaction Mech.*, **1968**, 228.
[34] E. Buncel and W. M. J. Strachan, *Can. J. Chem.*, **47**, 911 (1969).
[35] G. Steel and D. L. H. Williams, *Chem. Comm.*, **1969**, 975.
[36] F. L. Scott, C. B. Goggin, and W. J. Spillane, *Tetrahedron Letters*, **1969**, 3675.
[37] C. Brown, *J. Am. Chem. Soc.*, **91**, 5832 (1969).
[38] D. R. Hogg, J. H. Smith, and P. W. Vipond, *J. Chem. Soc.*(C), **1968**, 2713.
[39] J. M. Birchall and D. H. Thorpe, *J. Chem. Soc.*(C), **1968**, 2900.

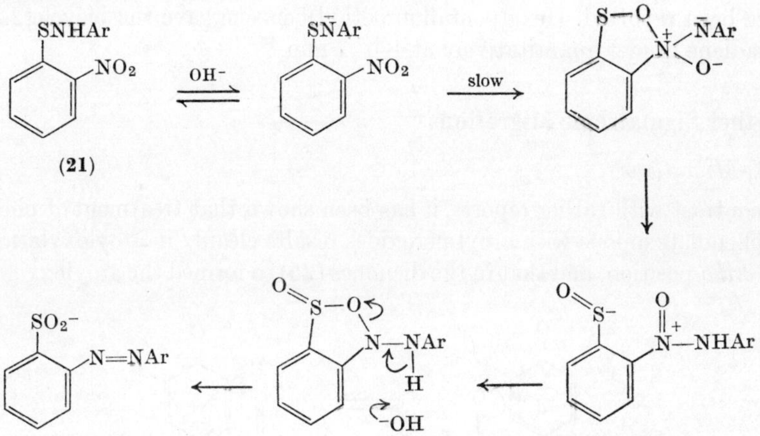

(21)

(22)

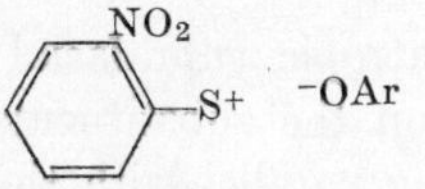

(23)

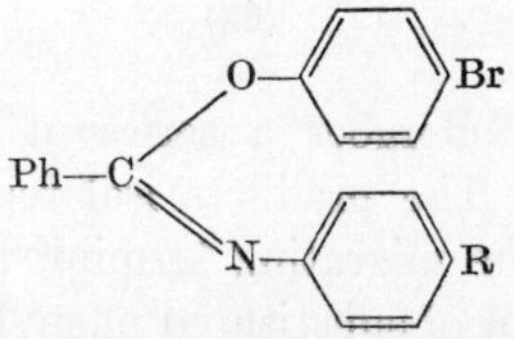

(24)

R = H) and p-bromophenyl N-p-tolylbenzimidate (**24**; R = Me) are rearranged together in boiling tetraglyme. Tritium-labelled allyl N-phenylbenzimidate also rearranged intramolecularly, with inversion of the allyl group.[40] The Fries[41] and Smiles[42] rearrangements have also been investigated. Automerization of the heptamethylbenzenonium ion is considered to proceed by sequential intramolecular 1,2-methyl shifts rather than by random methyl migration in a π-complex, on the basis of nuclear magnetic double and triple resonance experiments.[43] The acid-catalysed isomerization of toluene-o- and -m-sulphonic acids[44] and the isomerization of 1-ethyl-2-methylbenzene over silica–alumina[45]

[40] O. H. Wheeler, F. Roman, and O. Rosado, *J. Org. Chem.*, **34**, 966 (1969).

[41] H. Inone and K. Ishimura, *Seikei Daigaku Kogakubu Kogaku Hokoku*, **7**, 455 (1969); *Chem. Abs.*, **71**, 48989 (1969); R. Martin and J. M. Betoux, *Bull. Soc. Chim. France*, **1969**, 2079.

[42] V. N. Drozd and L. A. Nikonova, *Zh. Org. Khim.*, **4**, 1060 (1968); V. N. Drozd and L. I. Zefirova, *ibid.*, **4** 1961 (1968); V. N. Drozd and L. A. Nikonova, *ibid.*, **5**, 320, 325 (1969); V. N. Drozd, L. I. Zefirova, A. K. Pak, and Y. A. Ustynyuk, *ibid.*, **5** 933 (1969); V. N. Drozd, L. I. Zefirova, and Y. A. Ustynyuk, *ibid.*, p. 1248; V. N. Drozd, K. A. Pak, and Y. A. Ustynyuk, *ibid.*, p. 1267; V. N. Drozd, K. A. Pak, and Y. A. Ustynyuk, *ibid.*, p. 1446; V. N. Drozd and L. A. Nikonova, *ibid.*, p. 1453 .

[43] B. G. Derendyaev, V. I. Mamatyuk, and V. A. Koptyug, *Tetrahedron Letters*, **1969**, 5.

[44] A. Koeberg-Telder, A. J. Prinsen, and H. Cerfontain, *J. Chem. Soc.*(B), **1969**, 1004.

[45] S. M. Csicsery, *J. Org. Chem.*, **34**, 3338 (1969).

have been reported. Hexa(pentafluoroethyl)benzene gave the bicyclo[2.2.0]-hexadiene isomer quantitatively at 400°/1 mm.[46]

Further Sigmatropic Migrations

[3,3]-*Migrations*

In contrast with earlier reports, it has been shown that treatment of phenols or phenolate anions with aroyl peroxides results cleanly in aroyloxylation in the *ortho*-position, and that in the dienones (25) so formed the aroyloxy group

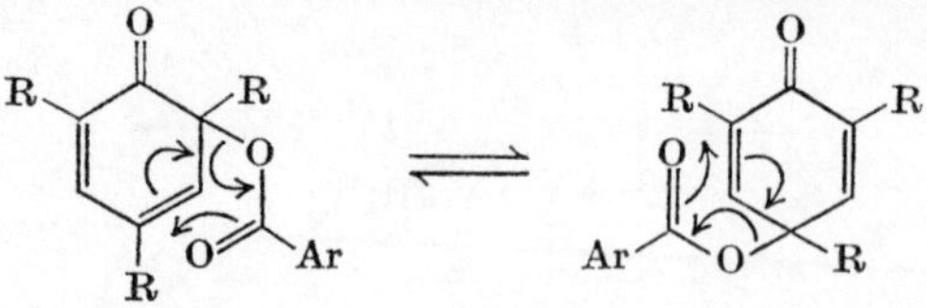

(25)   (26)

smoothly undergoes a series of [3,3]-sigmatropic suprafacial migrations, (25)⇌(26). The products isolated depend on the substituents (R = H or alkyl) at the migrating termini, in satisfying parallel with the Claisen rearrangement of substituted phenyl allyl ethers.[47] The entropies of activation for the allylic migrations would be very interesting (cf. ref. 48).

The transition states for the Cope and Claisen rearrangements resemble the chair conformation of cyclohexane, and the *cis-trans* product ratios in these rearrangements have been explained qualitatively by conformational preferences. This analysis has been strengthened by the demonstration of a quantitative agreement between $RT \ln k_{trans}/k_{cis}$ for the observed product ratios and the free-energy change for conversion of the same substituent from an equatorial to an axial position in cyclohexane. This relationship could possibly be used for estimating *syn*–axial interactions in cyclohexanes which are not available[49] (see also ref. 50). A similar stereoselectivity has been demonstrated[51] in the formation of trisubstituted olefins by the decarboxylative rearrangement of allyl acetoacetates.[52]

Although symmetry-allowed, the degenerate *trans,trans* Cope rearrangement of bicyclo[3.3.0]octa-2,6-diene does not occur, even at 450°, possibly because the rigid molecular geometry is unfavourable for the necessary

[46] E. D. Clifton, W. T. Flowers, and R. N. Haszeldine, *Chem. Comm.*, **1969**, 1216.
[47] D. H. R. Barton, P. D. Magnus, and M. J. Pearson, *Chem. Comm.*, **1969**, 550.
[48] E. S. Lewis, J. T. Hill, and E. R. Newman, *J. Am. Chem. Soc.*, **90**, 662 (1968).
[49] C. L. Perrin and D. J. Faulkner, *Tetrahedron Letters*, **1969**, 2783.
[50] D. J. Faulkner and M. R. Petersen, *Tetrahedron Letters*, **1969**, 3243.
[51] N. Wakabayashi, R. M. Waters, and J. P. Church, *Tetrahedron Letters*, **1969**, 3253.
[52] See *Org. Reaction Mech.*, **1968**, 232.

orbital overlap.[53] The great ease of the Cope rearrangement of cyclopentadiene dimers is ascribed more to ground-state strain than to favourable pre-orientation of the diallylic system.[54] Secondary deuterium isotope effects in the Cope rearrangement are interpreted in terms of relatively little C-1–C-6 bond making and C-3–C-4 bond breaking in the transition state.[55] Gas-phase pyrolysis of (27; $R = CO_2Me$) gave the *cis*-dihydroindenone (29) in high yield; the mechanism shown was proposed and the ketene (28) intercepted with methanol and with cyclohexylamine in xylene at 150°.[56] Gas-phase

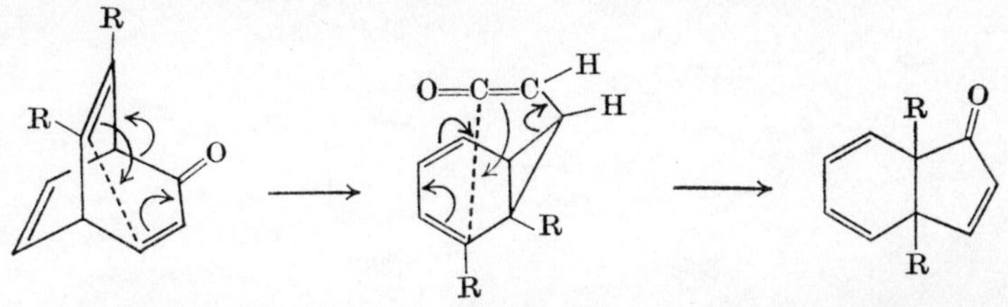

(27)   (28)   (29)

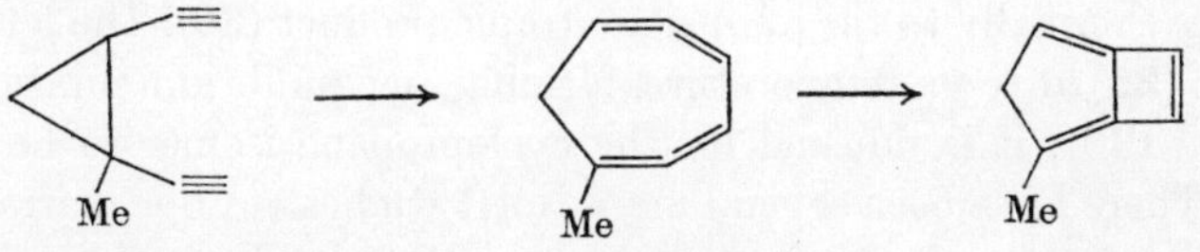

(30)   (31)   (32)

pyrolysis of 1-methyl-1,2-diethynylcyclopropane (30) gave the bicyclo-heptatriene (32) possibly via (31).[57] Other Cope rearrangements include the thermal isomerization of 3,3-dimethylhexa-1,5-diene to 6-methylhepta-1,5-diene,[58] of *cis*-cyclonona-1,2,6-triene to 2,3-divinylcyclopentene,[59] of δ-elemenol and epi-δ-elemenol,[60] and of diethyl allylisopropenylmalonate in different solvents.[61]

There have been further reports on the oxy-Cope rearrangement, particularly of 1-vinylcyclane-1,2-diols,[62] and 1,2-divinylcyclane-1,2-diols,[63] for example

[53] J. E. Baldwin and M. S. Kaplan, *Chem. Comm.*, **1969**, 1354.
[54] I. R. Bellobono, R. Destro, C. M. Gramaccioli, and M. Simonetta, *J. Chem. Soc.*(B), **1969**, 710.
[55] K. Humski, T. Strelkov, S. Borčić, and D. E. Sunko, *Chem. Comm.*, **1969**, 693; see also R. Malojčić, K. Humski, S. Borčić, and D. E. Sunko, *Tetrahedron Letters*, **1969**, 2003.
[56] T. H. Kinstle and P. D. Carpenter, *Tetrahedron Letters*, **1969**, 3943.
[57] M. B. D'Amore and R. G. Bergman, *J. Am. Chem. Soc.*, **91**, 5694 (1969).
[58] H. M. Frey and R. K. Solly, *Trans. Faraday Soc.*, **65**, 1372 (1969).
[59] H. M. Frey and A. M. Lamont, *J. Chem. Soc.*(A), **1969**, 1592.
[60] K. Morikawa and Y. Hirose, *Tetrahedron Letters*, **1969**, 869.
[61] D. C. Berndt, *J. Chem. Eng. Data*, **14**, 112 (1969).
[62] J. M. Conia and J. P. Barnier, *Tetrahedron Letters*, **1969**, 2679.
[63] P. Leriverend and J. M. Conia, *Tetrahedron Letters*, **1969**, 2681.

of *cis-* and *trans-*(33) to *cis-* and *trans-*(34);[64] the cyclooctane analogue rearranges similarly, but the bis-enol does not give the aldol product in this larger ring but ketonizes to cyclododecane-1,6-dione.[65]

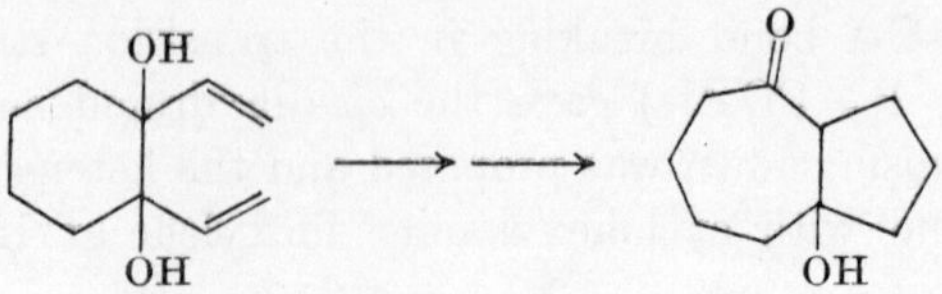

(33)   (34)

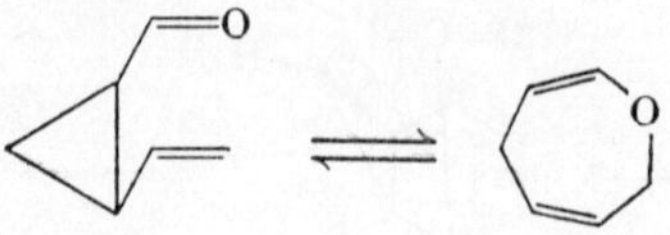

(35)   (36)

Contrary to an earlier report, *cis-*2-vinylcyclopropanecarboxaldehyde (35) rearranges thermally to the [3,3]-sigmatropic product (36). The latter slowly reverts to (35) at room temperature, forming an equilibrium mixture of 95% (35) + 5% (36); it is unusual for the cyclopropane isomer to be the more stable.[66] There have been several more NMR studies of Cope rearrangements in bullvalenes and azabullvalenes.[67] An intriguing synthesis of semibullvalene (39) by rearrangement of the less stable isomer (38), possibly via the doubly allylic diradical shown, has been described. Treatment of dichloro-derivatives

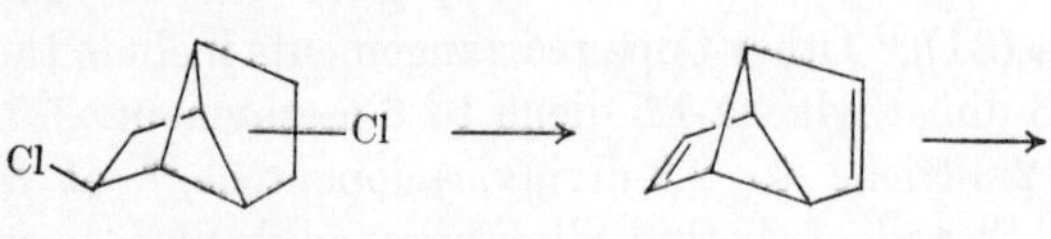

(37)   (38)

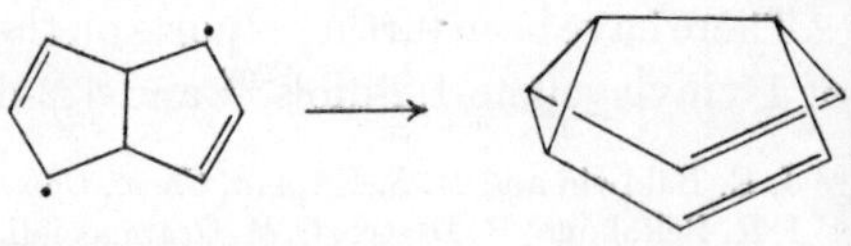

(39)

[64] E. N. Marvell and W. Whalley, *Tetrahedron Letters*, **1969**, 1337.
[65] E. N. Marvell and T. Tao, *Tetrahedron Letters*, **1969**, 1341.
[66] S. J. Rhoads and R. D. Cockroft, *J. Am. Chem. Soc.*, **91**, 2815 (1969).
[67] L. A. Paquette, G. R. Krow, and J. R. Malpass, *J. Am. Chem. Soc.*, **91**, 5522 (1969); L. A. Paquette, J. R. Malpass, G. R. Krow, and T. J. Barton, *ibid.*, p. 5296; L. A. Paquette and G. R. Krow, *ibid.*, p. 6107; H. Klose and H. Günther, *Chem. Ber.*, **102**, 2230 (1969); H. Günther, H. Klose, and D. Wendisch, *Tetrahedron*, **25**, 1531 (1969); H. Röttele, P. Nikoloff, J. F. M. Oth, and G. Schröder, *Chem. Ber.*, **102**, 3367 (1969).

(37) of tricyclo[3.3.0.0^{2,6}]octane with potassium t-butoxide in DMSO at 70° gave **(39)** direct; at lower temperatures **(38)** could be obtained but it rapidly rearranged to **(39)**.[68]

[2,3]-*Migrations*

If 1,2-electrophilic rearrangements, such as the Stevens, are concerted the migrating centre must, from orbital symmetry, undergo inversion of configuration. However, the Stevens rearrangement is intramolecular but proceeds with almost complete retention, so a tight ion-pair mechanism, in which the migrating group retains chirality, has been suggested. The rearrangement of N,N-dimethyl p-nitrobenzylamine acetimide [see **(40)**] has now been shown to involve a radical-pair intermediate. When **(40)** was heated at 180° in [$^2H_5$]nitrobenzene the benzylic proton region in the NMR exhibited a chemically induced dynamic nuclear polarization (CIDNP), showing the presence of a paramagnetic intermediate.[69a] A similar mechanism has been demonstrated, in the same way, for the Stevens rearrangement of benzyldimethylphenacyl ammonium ylid,[69b] and this mechanism therefore appears to be rather

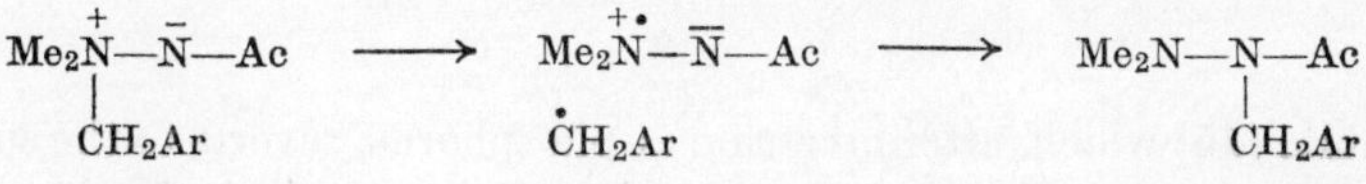

(40)

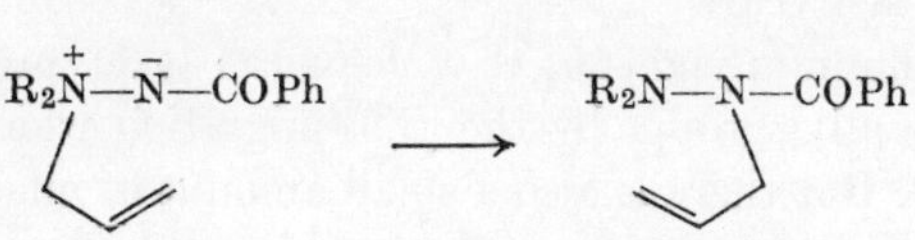

(41)   **(42)** 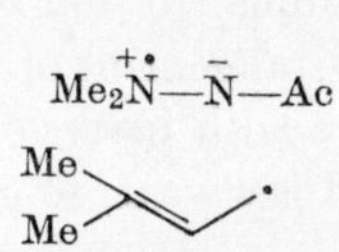   **(43)**

general for this type of 1,2-rearrangement.

Thermal rearrangement of the allylic compounds **(41)** to the hydrazines **(42)** at 120—140° appeared to be a [2,3]-sigmatropic migration.[70] When this reaction was repeated with substituted allyl groups, however, the only products isolated were those in which the allyl group was not inverted. Further, a CIDNP effect was observed in a typical rearrangement and hence the radical-pair intermediate **(43)**, which collapses to the more stable isomer, was proposed. Alternatively, it is possible that these reactions do involve an initial concerted [2,3]-migration, with the product isomerizing via a homolytic cleavage recombination pathway.[71]

[68] J. Meinwald and D. Schmidt, *J. Am. Chem. Soc.*, **91**, 5877 (1969); J. Meinwald and H. Tsuruta, *ibid.*, p. 5877; H. E. Zimmerman, J. D. Robins, and J. Schantl, *ibid.*, p. 5879.

[69a] R. W. Jemison and D. G. Morris, *Chem. Comm.*, **1969**, 1226.

[69b] U. Schöllkopf, U. Ludwig, G. Ostermann, and M. Patsch, *Tetrahedron Letters*, **1969**, 3415.

[70] I. D. Brindle and M. S. Gibson, *Chem. Comm.*, **1969**, 803.

[71] D. G. Morris, *Chem. Comm.*, **1969**, 1345.

The racemization of allyl sulphoxides involves a concerted, reversible, and intramolecular rearrangement.[72] An entirely analogous mechanism has now been demonstrated for racemization of allylmethylphenylphosphine sulphide (**44**) where phosphorus is the chiral centre. A [2,3]-migration gives the thio-

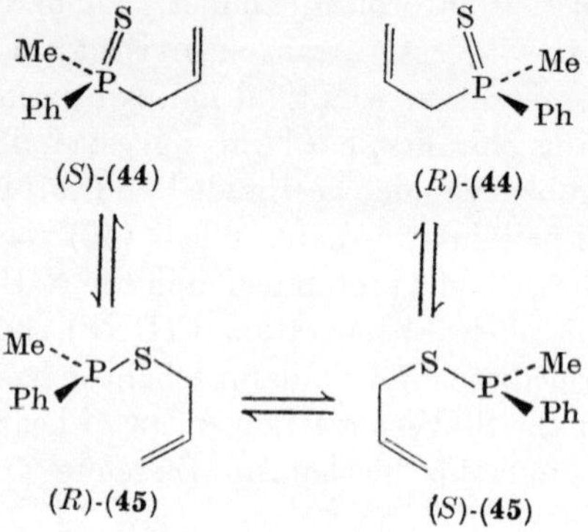

(S)-(**44**)   (R)-(**44**)

(R)-(**45**)   (S)-(**45**)

phosphinite (**45**) which, after inversion at phosphorus, reverts to the sulphide by a second sigmatropic shift. The kinetic and stereochemical evidence was very similar to that reported last year.[73]

Rearrangement of the sulphonium ylid (**46**; R = Me or Et) at 60° proceeds mainly with inversion of the allyl group, by the [2,3]-migration widely reported last year, to give (**47**). But there is also a small amount of migration without inversion to give the Stevens product (**48**), possibly by the homolytic dissociation–recombination mechanism.[74] However, the allylic sulphonium ylids (**49**; R^1 = Ph, p-O_2N-C_6H_4, PhCO; R^2 and R^3 from H, Me, Ph) gave the 5-centred allylic rearrangement products (**50**) to the virtual exclusion of the 3-centred Stevens reaction.[75] Again, only the symmetry-allowed products were formed in the base-catalysed reactions of the allylic ammonium salts (**51**; R^1 and R^2 from H, Me, aryl), via the ylids, to give (**52**) and not (**53**). However, the transformation (**52**)→(**53**) appears to be a general thermal isomerization at 180°, again suggesting the possibility that forbidden Stevens [1,3]-sigmatropic products may be formed in a sequence with this isomerization as the last step.[76]

[72] *Org. Reaction Mech.*, **1968**, 236.
[73] W. B. Farnham, A. W. Herriott, and K. Mislow, *J. Am. Chem. Soc.*, **91**, 6878 (1969).
[74] J. E. Baldwin and R. E. Hackler, *J. Am. Chem. Soc.*, **91**, 3646 (1969).
[75] R. W. C. Cose, A. M. Davies, W. D. Ollis, C. Smith, and I. O. Sutherland, *Chem. Comm.*, **1969**, 293.
[76] R. W. Jemison and W. D. Ollis, *Chem. Comm.*, **1969**, 294; see also G. R. Newkome, *ibid.*, p. 1227.

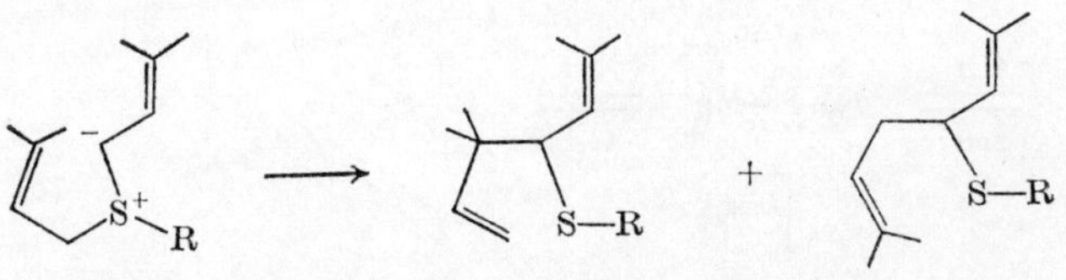

(46)   (47)   +   (48)

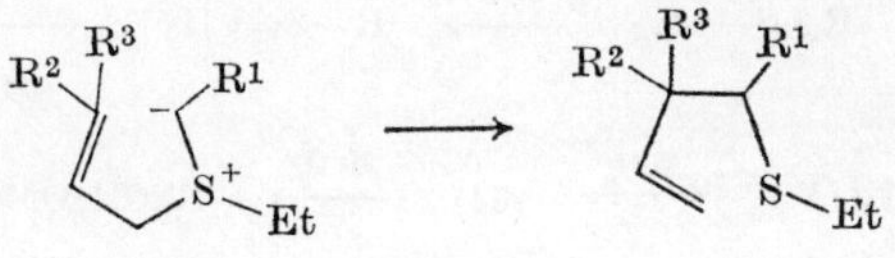

(49)   (50)

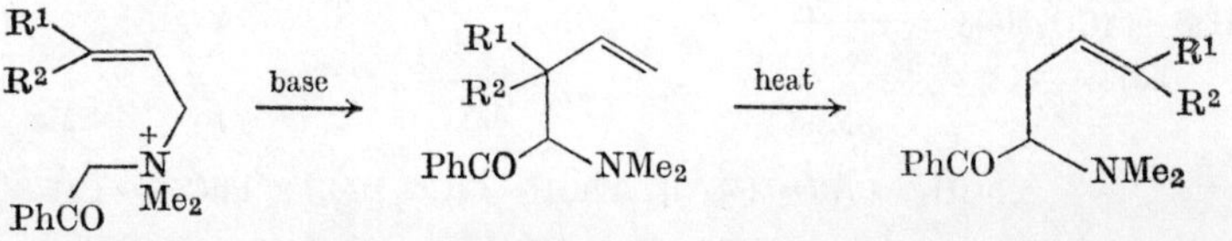

base →   heat →

(51)   (52)   (53)

Details of a study of the base-catalysed rearrangement of alkoxysulphonium salts to α-alkoxy-sulphides and α-acyloxy-sulphides (Pummerer rearrangement) have appeared. The overall mechanism shown was proposed. Evidence for the intermediacy of the sulphur-stabilized carbonium ion (54) was (i) substituents which stabilize a carbonium ion favour the rearrangement, (ii) highly polar solvents facilitate the reaction, and (iii) a mixed rearrangement led to cross-over products indicating that the alkoxy group and sulphur-containing fragment become separated.[77] The acid-catalysed rearrangements of sulphoxides and of amine oxides have been reviewed.[78] Stevens-type rearrangement of the ylid (55) to (56) on heating was accompanied by formation of (57) and (58); the latter products were formed from (56), quantitatively on stronger heating.[79]

An interesting variant on [2,3]-sigmatropic shifts in sulphur ylids is proposed to explain the ready isomerization of allylic di- and poly-sulphides, by equilibration with the isoallylic compounds with branched sulphur chains; e.g. (59)⇌(60).[80] Tertiary allylamine N-oxides rearrange similarly (Meisenheimer rearrangement) but a radical mechanism supervenes.[81] Mechanistic evidence on the Meisenheimer rearrangement has been reviewed and shown to favour a cleavage–recombination radical pathway occurring within a solvent cage,

[77] C. R. Johnson and W. G. Phillips, *J. Am. Chem. Soc.*, **91**, 682 (1969).
[78] G. A. Russell and G. J. Mikol, in ref. 2, Vol. 1, p. 157.
[79] W. Ando, H. Matsuyama, S. Nakaido, and T. Migita, *Tetrahedron Letters*, **1969**, 3825.
[80] D. Barnard, T. H. Houseman, M. Porter, and B. K. Tidd, *Chem. Comm.*, **1969**, 371.
[81] I. Tabushi, J. Hamuro, and R. Oda, *J. Chem. Soc. Japan*, **90**, 197 (1969); *Chem. Abs.*, **70**, 86807 (1969).

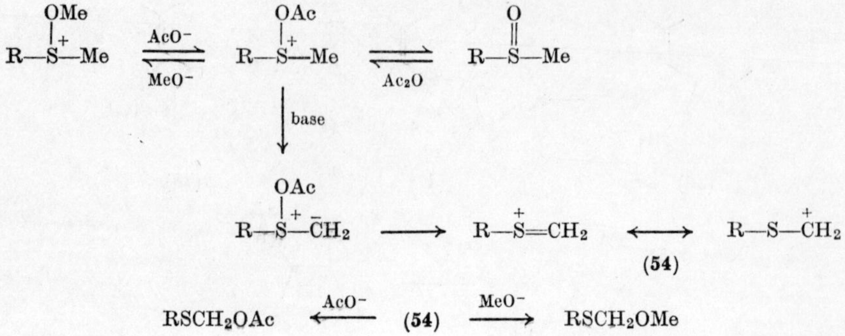

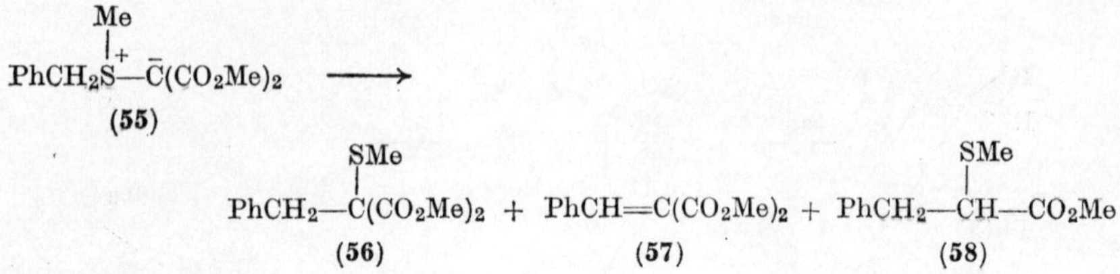

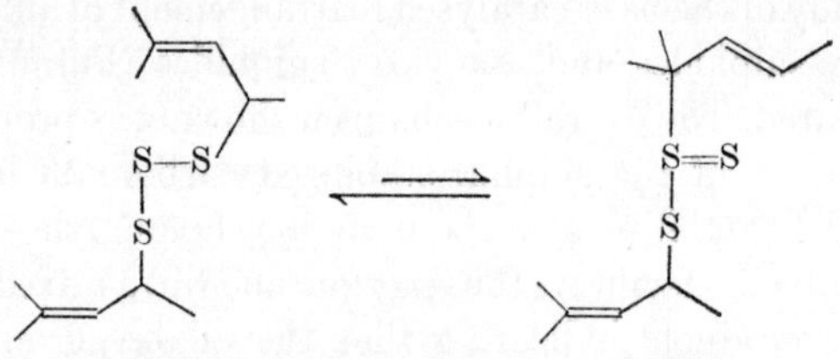

$$R\overset{\overset{\displaystyle OMe}{|}}{\underset{}{S}}\text{—}Me \;\overset{AcO^-}{\underset{MeO^-}{\rightleftharpoons}}\; R\overset{\overset{\displaystyle OAc}{|}}{\underset{}{S}}\text{—}Me \;\underset{Ac_2O}{\rightleftharpoons}\; R\overset{\overset{\displaystyle O}{\|}}{\underset{}{S}}\text{—}Me$$

$$\downarrow \text{base}$$

$$R\overset{\overset{\displaystyle OAc}{|}}{\underset{}{S}}\text{—}\overset{-}{C}H_2 \;\longrightarrow\; R\overset{+}{\text{—}}S{=}CH_2 \;\longleftrightarrow\; R\text{—}\overset{+}{S}\text{—}CH_2$$

$$(54)$$

$$RSCH_2OAc \;\overset{AcO^-}{\longleftarrow}\; (54) \;\overset{MeO^-}{\longrightarrow}\; RSCH_2OMe$$

$$PhCH_2\overset{\overset{\displaystyle Me}{|}}{\underset{}{S}}{}^+\text{—}\overset{-}{C}(CO_2Me)_2 \;\longrightarrow$$

$$(55)$$

$$\underset{(56)}{PhCH_2\overset{\overset{\displaystyle SMe}{|}}{\underset{}{C}}(CO_2Me)_2} \;+\; \underset{(57)}{PhCH{=}C(CO_2Me)_2} \;+\; \underset{(58)}{PhCH_2\overset{\overset{\displaystyle SMe}{|}}{\underset{}{C}}H\text{—}CO_2Me}$$

$$(59) \qquad\qquad\qquad (60)$$

similar to that of the Stevens and Wittig rearrangements.[82] Treatment of a benzyltrimethylammonium salt with n-butyl-lithium in hexane gave (62) together with the Stevens and Sommelet–Hauser products; this was taken as strong support for the *exo*-methylene intermediate (61) in these rearrangements.[83] Stevens rearrangement of (63) was proposed for the formation of (64), one of the products of decomposition of pentafluorophenyl-lithium in the presence of dimethylamine; this unusual migration of an aryl group was ascribed to the strong electron-withdrawing effects of the fluorines.[84] Alkyl allyl ethers have been shown to undergo a 1,4-alkyl shift under Wittig rearrangement conditions, analogous to that known for allylic quaternary ammonium salts in the Stevens rearrangement. Thus (65; R = H, Me) gave (66) with propyl-lithium and tetramethylethylenediamine.[85]

[82] R. A. W. Johnstone, in ref. 2, Vol. 2, p. 249.

[83] S. H. Pine and B. L. Sanchez, *Tetrahedron Letters*, **1969**, 1319.

[84] H. Heaney and T. J. Ward, *Chem. Comm.*, **1969**, 810.

[85] H. Felkin and A. Tambuté, *Tetrahedron Letters*, **1969**, 821.

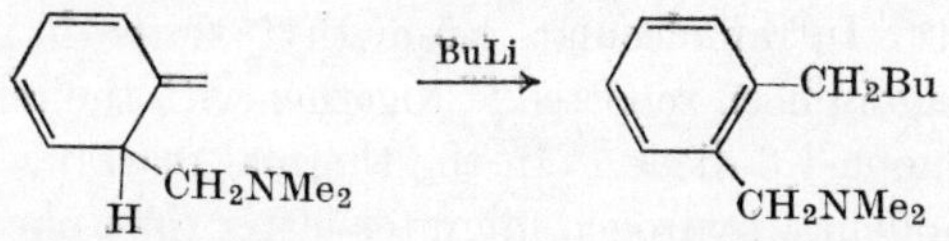

(61) → BuLi → (62)

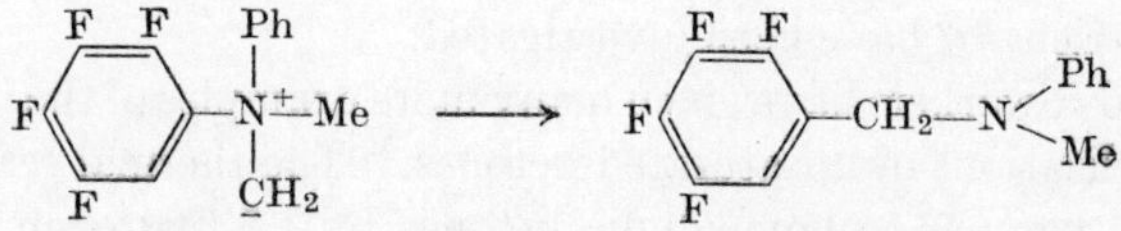

(63) → (64)

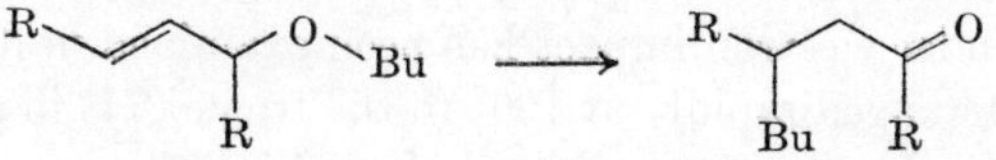

(65) → (66)

[1,5]-*Migrations*

trans-15,16-Dimethyldihydropyrene (67) rearranged at 200—210° to the *trans*-13,15-dimethyl isomer (68) possibly by a concerted [1,5]-suprafacial

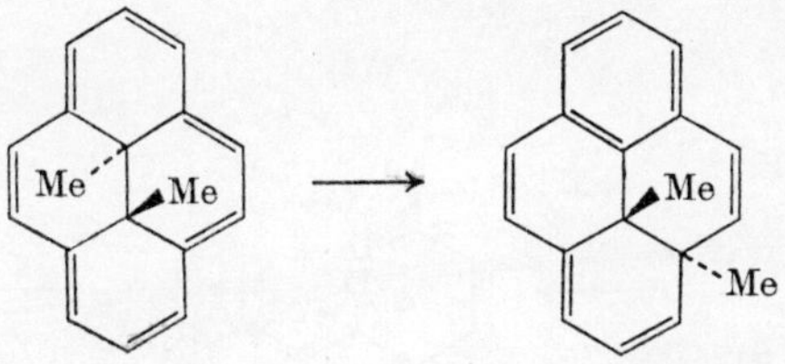

(67) ⟶ (68)

methyl migration. When the optically active 4-carboxy-derivative of (67) was rearranged similarly, recovered starting material had lost none of its optical activity so the methyl groups were not exchanging prior to the rearrangement.[86] The stereochemistry of the thermolysis product of *cis,syn,cis*-tricyclo[8.2.0.0²,⁹]dodeca-3,5,7,11-tetraene, now determined, shows that the formal 1,5-migration of carbon involved is suprafacial.[87] The deactivity effect of the methyl groups in the retro-Diels–Alder reaction of bornylene was explained by the movement of the *anti*-methyl group towards the developing ethylene molecule in the transition state, in a concerted process; the primary product, 1,5,5-trimethylcyclopentadiene rearranged to the 1,2,3-isomers by

[86] V. Boekelheide and E. Sturm, *J. Am. Chem. Soc.*, **91**, 902 (1969).
[87] L. A. Paquette and J. C. Stowell, *Tetrahedron Letters*, **1969**, 4159.

1,5-methyl shifts.[88] Intramolecular 1,5-methyl shifts in trimethylcyclo-
pentadienes have again been reported,[89] together with the related rearrangement of spiro[4.4]nona-1,3-diene.[90] In the thermal rearrangement of 1,1- to
1,2-disubstituted indenes, hydrogen migrates faster than phenyl, and phenyl
faster than methyl.[91] The base-catalysed isomerization of 1-ethyl-3-methyl-
indene [92a] and the thermally induced migration of methoxyl in 7,7-dimethoxy-
cycloheptatriene [92b] have been investigated.

Conia and coworkers have given many more examples of the use in synthesis
of their cyclizations of unsaturated ketones.[93] The thermal rearrangement of
cyclopropyl ketones to homoallylic ketones by 1,5-hydrogen shift has been
extended to cyclopropylmalonic esters (69; R = H or small alkyl).[94] The
structure of the transition state for the γ-hydrogen rearrangement, e.g. in
aliphatic ketones, under electron impact has been examined in detail.[95] The
vinyl allene (70) rearranged rapidly at 100° to the triene (71) in a first-order
process with the rather low activation energy, for a 1,5-hydrogen shift, of 24.6

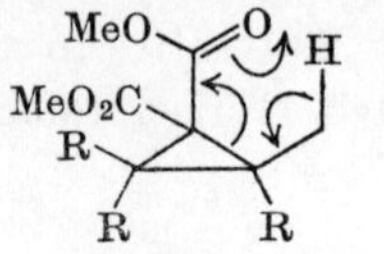

(69) 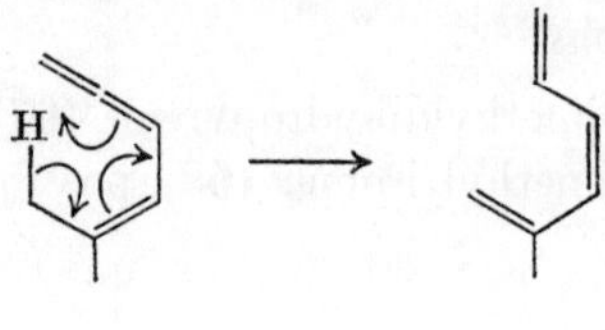   (70)   (71)

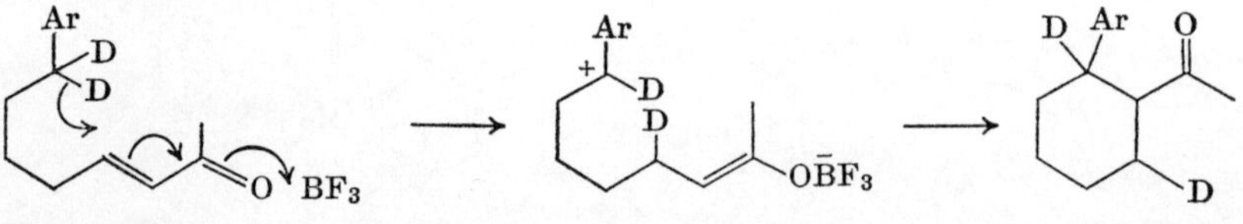

(72)   (73)

[88] W. C. Herndon and J. M. Manion, *J. Org. Chem.*, **33**, 4504 (1968).

[89] V. A. Mironov, A. P. Ivanov, Y. M. Kimelfeld, and A. A. Akhrem, *Tetrahedron Letters*, **1969**, 3985; see *Org. Reaction Mech.*, **1966**, 228; **1968**, 236.

[90] V. A. Mironov, A. P. Ivanov, Y. M. Kimelfeld, L. I. Petrovskaya, and A. A. Akhrem, *Tetrahedron Letters*, **1969**, 3347.

[91] L. L. Miller, R. Greisinger, and R. F. Boyer, *J. Am. Chem. Soc.*, **91**, 1578 (1969).

[92a] G. Bergson and L. Ohlsson, *Acta Chem. Scand.*, **23**, 2175 (1969).

[92b] R. W. Hoffmann, K. R. Eicken, H. J. Luthardt, and B. Dittrich, *Tetrahedron Letters*, **1969**, 3789.

[93] J. M. Conia and P. Beslin, *Bull. Soc. Chim. France*, **1969**, 483; R. Bloch, J. L. Bouket, and J. M. Conia, *ibid.*, p. 489; J. M. Conia and J. L. Bouket, *ibid.*, p. 494; J. M. Conia and G. Moinet, *ibid.*, p. 500; P. Beslin, R. Bloch, G. Moinet, and J. M. Conia, *ibid.*, p. 508; see *Org. Reaction Mech.*, **1965**, 180; **1966**, 218; **1967**, 218; **1968**, 237.

[94] W. Ando, *Tetrahedron Letters*, **1969**, 929.

[95] F. P. Boer, T. W. Shannon, and F. W. McLafferty, *J. Am. Chem. Soc.*, **90**, 7239 (1968).

kcal mole^{-1}.[96] The concerted 1,5-homodienyl hydrogen shift that converts *cis*-1-methyl-2-vinylcyclopropanes into hexa-1,4-dienes at 200° is accompanied by another mechanism, probably homolysis of the cyclopropane, at higher temperatures (300°).[97] Deuterium-labelling shows that 1,5- and 1,7-hydrogen shifts are involved in the cyclization of *cis*- and *trans-o*-(penta-1,3-dienyl)-phenols to 2-ethyl-Δ^3-chromene.[98]

Ketone (**72**) isomerized to (**73**) on heating with boron trifluoride etherate; the intramolecular hydride shift shown was confirmed by rearrangement of the undeuterated and dideuterated compounds together, when only D$_0$- and D$_2$-products were formed.[99]

7-Phenylcycloheptatriene isomerizes at 135° to 3-phenylcycloheptatriene; ^{14}C-labelling at C-7 remains localized, showing that hydrogen and not phenyl migrates. In the photochemical rearrangement at least 18% occurs by phenyl migration, suggesting that there is less discrimination between phenyl and hydrogen in the excited state.[100] The kinetic isotope effect has been determined for the 1,5-hydrogen and 1,5-deuterium shifts in 5-methylcyclopentadiene and 5-methylpentadeuterocyclopentadiene, reactions for which the transition state must be non-linear.[101] Rapid 1,5-hydrogen shifts are involved in the thermal isomerization of benzonorcaradiene, benzonorbornadiene, and 1,2-benzotropylidene.[102] Gas-phase pyrolysis (365°) of allyl formate gave CO$_2$ and propene in a 1,5-transfer of the formyl hydrogen.[103]

o-cis-Butadienylphenol (**74**) was quantitatively isomerized to 2-methyl-2*H*-1-benzopyran (**75**) by heating in DMF; the rearrangement was faster in

(**74**)   (**75**)

less polar solvents, and was unaffected by catalytic amounts of acid or base; deuterium was transferred specifically as shown. A [1,7]-antarafacial sigmatropic shift of hydrogen to the cyclohexadienone, which rapidly cyclized, was proposed.[104]

[96] L. Skattebøl, *Tetrahedron*, **25**, 4933 (1969).
[97] M. J. Jorgenson and A. F. Thacher, *Tetrahedron Letters*, **1969**, 4651.
[98] R. Hug, H. J. Hansen, and H. Schmid, *Chimia*, **23**, 108 (1969).
[99] R. S. Atkinson, *Chem. Comm.*, **1969**, 735.
[100] K. W. Shen, W. E. McEwen, and A. P. Wolf, *Tetrahedron Letters*, **1969**, 827.
[101] S. McLean, C. J. Webster, and R. J. D. Rutherford, *Can. J. Chem.*, **47**, 1555 (1969).
[102] M. Pomerantz and G. W. Gruber, *J. Org. Chem.*, **33**, 4501 (1968).
[103] J. M. Vernon and D. J. Waddington, *Chem. Comm.*, **1969**, 623.
[104] E. E. Schweizer, D. M. Crouse, and D. L. Dalrymple, *Chem. Comm.*, **1969**, 354.

Small-ring Rearrangements

Cyclopropanes

Pyrolysis of cyclopropanes can result in hydrogen shift to give propenes (structural isomerization) and in interconversion of *cis–trans* isomers (geometrical isomerization), and these have been widely studied to clarify the nature of the C–C bond fission. Preliminary reports on "optical" isomerization, which interconverts cyclopropane enantiomers, have appeared this year. There are three limiting sets of molecular motions: simultaneous epimerization at two ring carbons, epimerization exclusively at one, or randomization of stereochemistry at two ring carbons. NMR analysis of product ratios in the degenerate rearrangement of 1-(2-deuterovinyl)-*trans,trans*-2,3-dideuterocyclopropane at 325° in the gas phase accords with the last of these possibilities.[105] The kinetic consequences of the three extreme mechanisms have been elaborated by Berson and Balquist and applied to the racemization and geometrical isomerization of [^{2}H$_6$]tetramethylcyclopropane. The results show that neither the second nor the third mechanism can be the sole process, and that no more than a relatively small fraction of the molecules cleave to produce any planar intermediate or transition state.[106] Bergman and Carter have similarly followed the racemization and geometrical isomerization of optically active *cis*- and *trans*-1-ethyl-2-methylcyclopropanes. The rate of interconversion of enantiomers in both compounds is very close to that for interconversion of geometrical isomers. There is no indication of an electrocyclic process involving "π-cyclopropane" intermediates, and the results are explained by direct cleavage to diradical intermediates in which rotation is faster than cyclization, but not overwhelmingly so.[107]

Thermolysis of the bicyclo[2.1.0]pentane (76) gave the cyclopentene (77) as one of the products,[108] the Ac group undergoing (at a lower temperature) the same 1,2-shift as that reported for the CO$_2$Et group last year.[109]

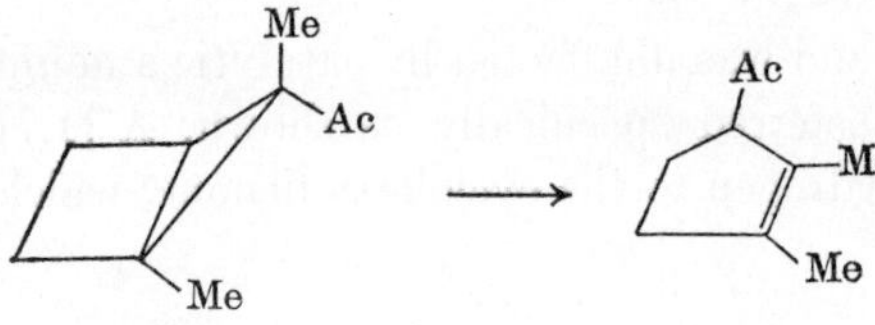

(76)   (77)

105 M. R. Willcott and V. H. Cargle, *J. Am. Chem. Soc.*, **91**, 4310 (1969).
106 J. A. Berson and J. M. Balquist, *J. Am. Chem. Soc.*, **90**, 7343 (1968).
107 W. L. Carter and R. G. Bergman, *J. Am. Chem. Soc.*, **90**, 7345 (1968).
108 M. J. Jorgenson and A. F. Thacher, *Chem. Comm.*, **1969**, 1030.
109 *Org. Reaction Mech.*, **1968**, 238; see also M. J. Jorgenson, T. J. Clark, and J. Corn, *J. Am. Chem. Soc.*, **90**, 7020 (1968).

The reactions of 1,1-dihalogenocyclopropanes have been extensively reviewed.[110] The relative rates of thermal rearrangement of isomeric 1-bromo- and 1,1-dibromo-2,3-dimethylcyclopropanes, in the presence or absence of base, correlate with the stability of the intermediate allylic cations formed by concerted disrotatory opening. In the mass spectra of these bromo-compounds the ratio of peak-heights for $(M - Br)^+$ and M^+ increases in the same sense, suggesting that loss of bromine from the molecular ion may also be a concerted process.[111] Stereochemical and electronic factors influencing the ease of thermal rearrangement of halogenocarbene–cyclic olefin adducts have been investigated further.[112] 9,9-Dibromobicyclo[6.1.0]non-4-ene (78) isomerized under mild conditions to (79) (or its *cis,trans*-isomer) in hot quinoline, i.e. without the loss of HBr; the dichloro-compound behaved similarly. Thus a simple cyclopropyl cation intermediate is unlikely and an ion-pair is proposed for the halogen migration.[113] First-order rate constants for the rearrangement

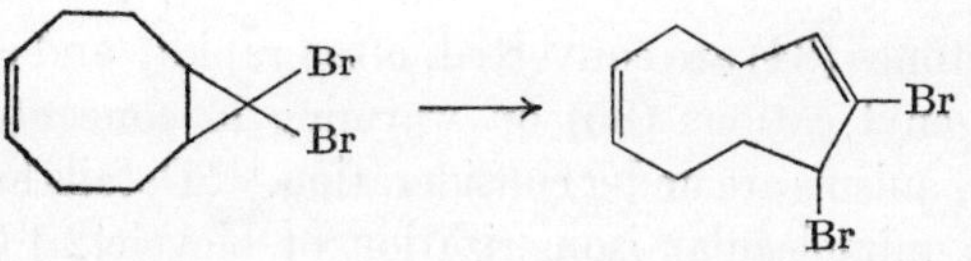

(78)   (79)

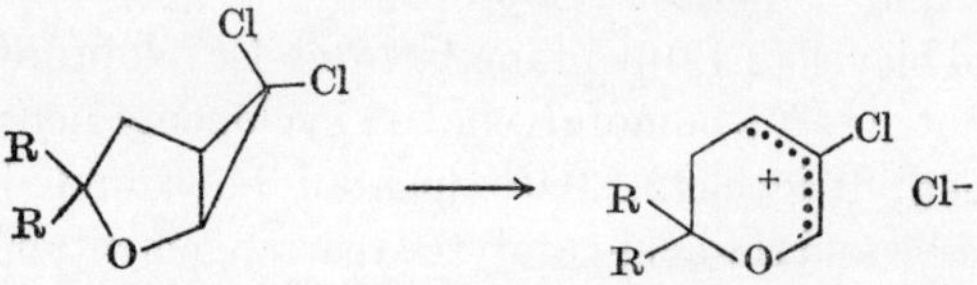

(80)   (81)

in phenyl cyanide at 60° of (80) to (81) are about 1000 times smaller for $R = Me$ than for $R = H$. The most likely explanation is considered to be steric hindrance by methyl to ionization of the *endo*-chlorine in the concerted disrotatory opening to (81).[114] Details of the thermal isomerization of 1,1-dichlorocyclopropanes to alkenes [115] have appeared.[116]

Pyrolysis of the cyclopropylidenecyclopropane (82) at 400° to give mainly the methylenespiropentane (83) was explained by intermediacy of the tri-

[110] R. Barlet and Y. Vo-Quang, *Bull. Soc. Chim. France*, **1969**, 3729.
[111] M. S. Baird and C. B. Reese, *Tetrahedron Letters*, **1969**, 2117.
[112] M. S. Baird, D. G. Lindsay, and C. B. Reese, *J. Chem., Soc.*(C), **1969**, 1173.
[113] M. S. Baird and C. B. Reese, *J. Chem. Soc.*(C), **1969**, 1808.
[114] R. C. De Selms and U. T. Kreibich, *J. Am. Chem. Soc.*, **91**, 3659 (1969).
[115] *Org. Reaction Mech.*, **1967**, 223.
[116] R. Fields, R. N. Haszeldine, and D. Peter, *J. Chem. Soc.*(C), **1969**, 165; K. A. W. Parry and P. J. Robinson, *J. Chem. Soc.*(B), **1969**, 49.

methylenemethane shown.[117] The stereochemistry of the degenerate methylenecyclopropane rearrangement has also been studied.[118]

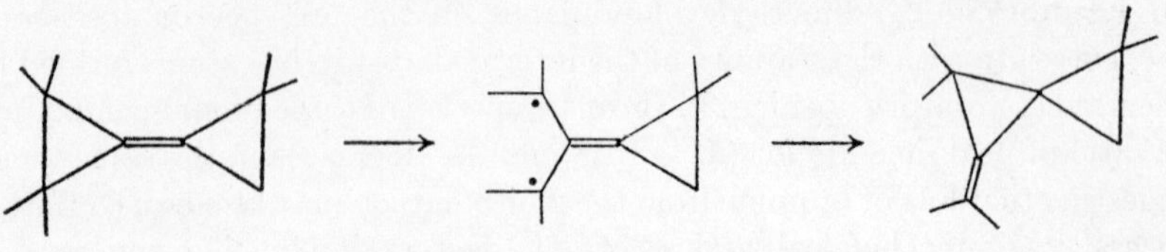

(82)   (83)

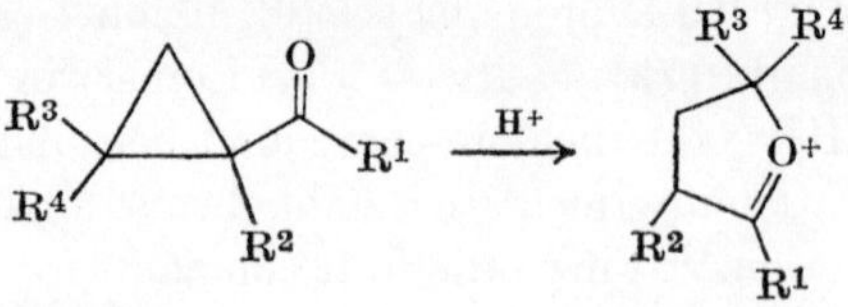

(84)   (85)

Cyclopropyl ketones (84) are converted, often rapidly and completely, into 1-oxacyclopent-1-enyl cations (85) on warming in concentrated sulphuric acid; various mechanisms are under consideration.[119] Details have appeared[120] of the gas-phase unimolecular isomerization of bicyclo[2.1.0]pent-2-ene to cyclopentadiene.[121] Other reactions investigated include the isomerizations of cyclopropene and 1-methylcyclopropene,[122] 3-acetyl-3-methylcyclopropene,[123] a strained bicyclo[2.1.0]pentane,[124] *trans*-1-cyclopropylbuta-1,3-diene to 3-vinylcyclopentene,[125] a homofulvene to a cyclopentadiene,[126] the reaction between 1,5-dimethyltricyclo[2.1.0.0^{2,5}]pentan-3-one and dimethyl acetylenedicarboxylate,[127] and the acid-catalysed rearrangements of 3,3-dimethoxy-1,5-dimethyltetracyclo[3.2.0.0^{2,7}.0^{4,6}]heptane[128] and of quadricyclane and its dimethyl ketal.[129]

Cyclobutanes

Thermolysis of 1-isopropylidene-2-methylene-3,3-dimethylcyclobutane (86) took the course shown, probably by a concerted 1,5-hydrogen shift followed by

117 J. K. Crandall, D. R. Paulson, and C. A. Bunnell, *Tetrahedron Letters*, **1969**, 4217.
118 J. J. Gajewski, *J. Am. Chem. Soc.*, **90**, 7178 (1968).
119 C. U. Pittman and S. P. McManus, *J. Am. Chem. Soc.*, **91**, 5915 (1969).
120 D. M. Golden and J. I. Brauman, *Trans. Faraday Soc.*, **65**, 464 (1969).
121 *Org. Reaction Mech.*, **1968**, 243.
122 R. Srinivasan, *J. Am. Chem. Soc.*, **91**, 6250 (1969).
123 H. Monti and M. Bertrand, *Tetrahedron Letters*, **1969**, 1235.
124 K. Mackenzie, W. P. Lay, J. R. Telford, and D. L. Williams-Smith, *Chem. Comm.*, **1969**, 761.
125 H. M. Frey and A. Krantz, *J. Chem. Soc.*(A), **1969**, 1159.
126 H. Hart and J. D. De Vrieze, *Chem. Comm.*, **1968**, 1651.
127 M. Pomerantz and R. N. Wilke, *Tetrahedron Letters*, **1969**, 463.
128 P. G. Gassman, D. H. Aue, and D. S. Patton, *J. Am. Chem. Soc.*, **90**, 7271 (1968).
129 P. G. Gassman and D. S. Patton, *J. Am. Chem. Soc.*, **90**, 7276 (1968).

electrocyclic opening of the cyclobutene. The isomeric dimethylallene dimer, 1,2-diisopropylidenecyclobutane behaved similarly.[130] A numerical method, based on topological twisting and bending, of by-passing the construction of correlation diagrams is described for the thermal opening of cyclobutenones to ketenes.[131] Details of the rearrangement of *endo*-bicyclo[2.2.0]hex-2-yl

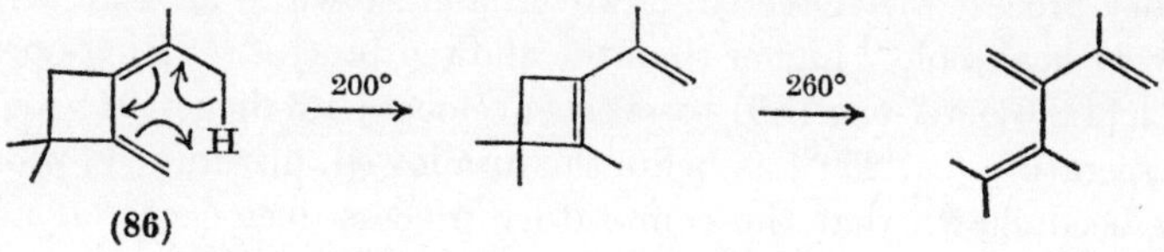

(86)   200°   260°

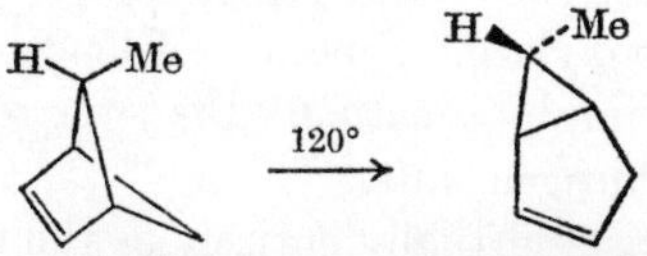

(87)   120°   (88)

acetate[132] have appeared.[133] Thermolysis of (87) gave (88) (98.5%) by a symmetry-controlled 1,3-sigmatropic rearrangement with inversion of the migrating carbon; rearrangement of the *endo*-isomer of (87) was more complex, however.[134] Thermal isomerization of the unmethylated hydrocarbon[135] and its benzo-derivative[136] have also been investigated. Other pyrolyses studied include those of tetramethylcyclobutenes with fused 4- to 8-membered saturated rings,[137] 3,3-diethyl- and 3-ethyl-3-methyl-cyclobutene,[138] 1-n-propyl-, 1-isopropyl-, 1-allyl-, and 1-cyclopropyl-cyclobutene,[139] perdeuterocyclobutene,[140] 1,2-dimethylenecyclobutanes,[141] buta-1,3-dienes to cyclobutenes,[142] and of 1,4-dimethylbicyclo-[2.2.0]hexane, -[2.1.1]hexane, and -[2.1.0]pentane.[143]

[130] E. F. Kiefer and C. H. Tanna, *J. Am. Chem. Soc.*, **91**, 4478 (1969).
[131] C. Trindle, *J. Am. Chem. Soc.*, **91**, 4936 (1969).
[132] *Org. Reaction Mech.*, **1968**, 240.
[133] R. N. McDonald and G. E. Davis, *J. Org. Chem.*, **34**, 1916 (1969).
[134] W. R. Roth and A. Friedrich, *Tetrahedron Letters*, **1969**, 2607.
[135] H. M. Frey, R. G. Hopkins, H. E. O'Neal, and F. T. Bond, *Chem. Comm.*, **1969**, 1069.
[136] H. Tanida and Y. Hata, *J. Am. Chem. Soc.*, **91**, 6775 (1969).
[137] R. Criegee, G. Bolz, and R. Askani, *Chem. Ber.*, **102**, 275 (1969).
[138] H. M. Frey and R. K. Solly, *Trans. Faraday Soc.*, **65**, 448 (1969).
[139] D. Dickens, H. M. Frey, and R. F. Skinner, *Trans. Faraday Soc.*, **65**, 453 (1969).
[140] H. M. Frey and B. H. Pope, *Trans. Faraday Soc.*, **65**, 441 (1969).
[141] J. J. Gajewski and C. N. Shih, *J. Am. Chem., Soc.* **91**, 5900 (1969).
[142] H. A. Brune and W. Schwab, *Tetrahedron*, **25**, 4375 (1969).
[143] R. Srinivasan, *Int. J. Chem. Kinetics*, **1**, 133 (1969).

Other Electrocyclic Reactions

Valence isomerizations have been reviewed.[144] There have been further reports on the thermal isomerization of deca-2,4,6,8-tetraenes and cyclo-octatrienes[145] and related electrocyclic reactions. Activation parameters have been obtained[146] and the very high stereospecificity of this conrotatory cyclization demonstrated.[147] A small amount of the symmetry-forbidden disrotatory process was detected, however, and shown to have an activation enthalpy 11 kcal mole^{-1} higher than the allowed process.[147] Ring-opening of *cis*-bicyclo[4.2.0]oct-7-ene (**89**) to *cis,cis*-cycloocta-1,3-diene (**91**) requires a high temperature (ca. 250°) as befits the disallowed, disrotatory, process. It has now been shown that the conrotatory process does occur, at as low a temperature as 110°, to give the *cis,trans*-diene (**90**) which was trapped with dimethyl acetylenedicarboxylate. Kinetic evidence suggested that the apparent disrotatory opening[148] is actually the two-step process shown, the second step being a 1,5-hydrogen shift.[149]

2-Methoxy-1-azocine exists virtually entirely as the monocyclic tautomer, in striking contrast with the amide, 1,2-dihydroazocin-2-one (**92**), which exists predominantly in the bicyclic form (**93**). This was ascribed to greater

(**89**) (**90**) (**91**)

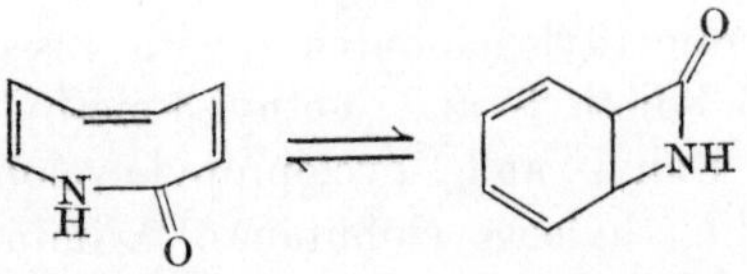

(**92**) (**93**)

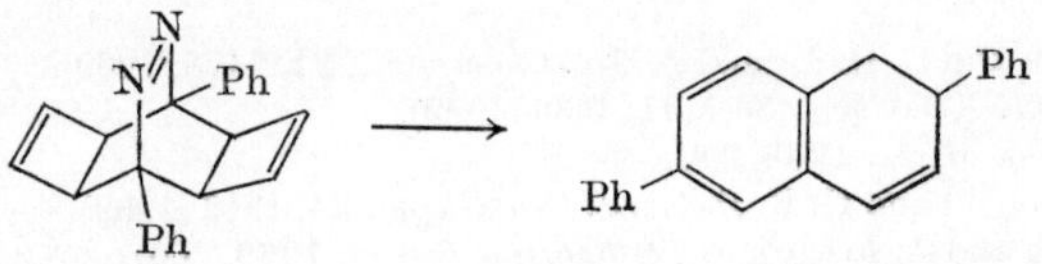

(**94**) (**95**)

144 G. Maier, *Chem. Unserer Zeit.*, **2**, 35 (1968).
145 *Org. Reaction Mech.*, **1968**, 245—246; see also R. Huisgen, *Chim. Ind.* (Milan), **51**, 963 (1969).
146 R. Huisgen, A. Dahmen, and H. Huber, *Tetrahedron Letters*, **1969**, 1461.
147 A. Dahmen and R. Huisgen, *Tetrahedron Letters*, **1969**, 1465.
148 *Org. Reaction Mech.*, **1966**, 223.
149 J. S. McConaghy and J. J. Bloomfield, *Tetrahedron Letters*, **1969**, 3719, 3723.

delocalization in the planar amide function in (**93**) compared with the distorted amide group of (**92**).[150] Thermal decomposition of the cyclic azo-compound (**94**) in *o*-dichlorobenzene gave the 1,2-dihydronaphthalene (**95**) (78.5%) and the 3,4-dihydro-isomer (8%) presumably via 1,6-diphenylcyclodecapentaene and the 9,10-dihydronaphthalene.[151]

Thermal isomerization of the homoazepines (**96** and **98**; $X = NCO_2Et$) is very similar to that of the homotropylidenes ($X = CH_2$). At 220° (**96**) gave (**97**) almost quantitatively in a vinylcyclopropane to cyclopentene rearrangement; (**98**) gave 50% of (**100**; $X = NCO_2Et$) probably by 1,5-hydrogen

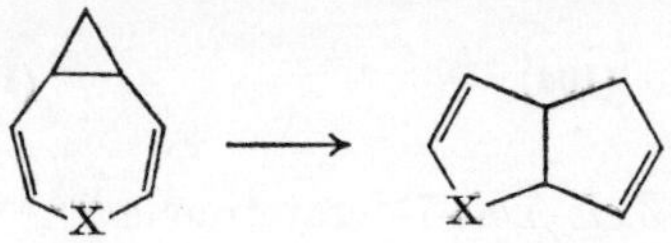

$$(96) \longrightarrow (97)$$

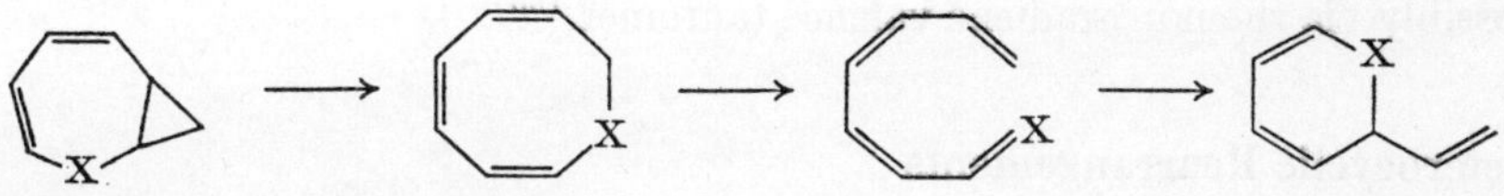

$$(98) \longrightarrow (99) \longrightarrow \longrightarrow (100)$$

shifts to (**99**) followed by the electrocyclic processes shown.[152]

Vapour-phase rearrangement of *anti*- (**101**) and *syn*-9-methyl-*cis*-bicyclo-[6.1.0]nona-2,4,6-triene gave the same five-component mixture, the major products being the epimers (**103**). Compound (**102**; R = Me) was shown to be the common intermediate since methylation of the cyclononatetraenyl anion gave the same five products.[153] Indeed, all-*cis*-cyclonona-1,3,5,7-tetraene (**102**; R = H) rearranges rapidly to *cis*-8,9-dihydroindene at 25°.[154] When the 9,9-dimethyl derivative (**104**) was heated at 151° in nonane the *trans*-dihydro-indene (**105**) was formed stereospecifically; this reversed stereochemistry was attributed to (**104**) being in the *exo*-conformation shown, because of the presence of the methyl groups.[155]

Other examples of valence isomerization include that of bicyclo[4.2.2]deca-2,4,7,9-tetraenes,[156] 3,4-diazabicyclo[4.2.0]octadiene and octatriene,[157]

[150] L. A. Paquette, T. Kakihana, J. F. Kelly, and J. R. Malpass, *Tetrahedron Letters*, **1969**, 1455.
[151] L. A. Paquette and J. F. Kelly, *Tetrahedron Letters*, **1969**, 4509.
[152] W. H. Okamura, *Tetrahedron Letters*, **1969**, 4717.
[153] P. Radlick and W. Fenical, *J. Am. Chem. Soc.*, **91**, 1560 (1969).
[154] A. G. Anastassiou, V. Orfanos, and J. H. Gebrian, *Tetrahedron Letters*, **1969**, 4491.
[155] S. W. Staley and T. J. Henry, *J. Am. Chem. Soc.*, **91**, 1239 (1969).
[156] J. S. McConaghy and J. J. Bloomfield, *Tetrahedron Letters*, **1969**, 1121; J. Altman, E. Babad, M. B. Rubin, and D. Ginsburg, *ibid.*, p. 1125; cf. *Org. Reaction Mech.*, **1968**, 246.
[157] G. Maier, U. Heep, M. Wiessler, and M. Strasser, *Chem. Ber.*, **102**, 1928 (1969).

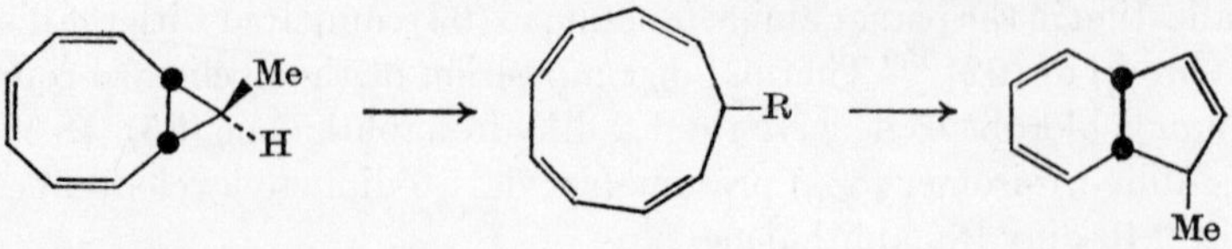

(101)   (102)   (103)

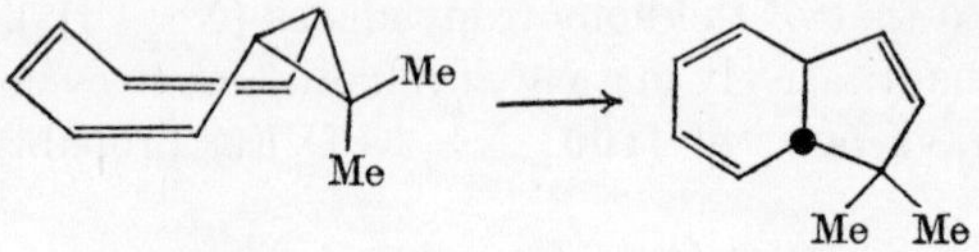

(104)   (105)

1,8-diphenyl-*trans*-1,*cis*-3,*cis*-5,*cis*-7-octatetraene,[158] and spiro[2.6]nona-4,6,8-trienes.[159] When heated at reflux in benzene with ethylene glycol and toluene-*p*-sulphonic acid, 2-bromotropone gave 2-bromoethyl benzoate, possibly via the norcaradiene valence tautomer.[160]

Heterocyclic Rearrangements

Small Rings

Oxidation of the cyclopropenes (**106**; R's = H or Me) with *m*-chloroperbenzoic acid in methylene chloride at 0° gave the mixed α,β-unsaturated aldehydes or ketones shown, probably by rearrangement of the intermediate 2-oxabicyclo-[1.1.0]butane.[161] Thermal rearrangement (80°) of the *exo*-oxide (**107**) of 2-chloronorbornene to *exo*-3-chloronorcamphor gave the enantiomer shown in high optical purity. Thus chloride rather than hydride migrates; an α-ketocarbonium chloride tight ion-pair intermediate[162] was proposed.[163] High selectivity is observed in the conversion of unsymmetrical oxirans into allylic alcohols with lithium diethylamide; for example, pent-2-ene oxide gave exclusively pent-1-en-3-ol. Primary protons are abstracted much more readily than secondary, and secondary more readily than tertiary.[164] Treatment of glycidic esters (**108**) with boron trifluoride in benzene leads to an unusual skeletal rearrangement with migration of ethoxycarbonyl, as shown.[165] Boron trifluoride etherate converted the epoxide (**109**) rapidly and exclusively

158 E. N. Marvell and J. Seubert, *Tetrahedron Letters*, **1969**, 1333.
159 M. Jones and E. W. Petrillo, *Tetrahedron Letters*, **1969**, 3953; C. J. Rostek and W. M. Jones, *ibid.*, p. 3957.
160 J. E. Baldwin and J. E. Gano, *Tetrahedron Letters*, **1969**, 1101.
161 J. Ciabattoni and P. J. Kocienski, *J. Am. Chem. Soc.*, **91**, 6534 (1969).
162 *Org. Reaction Mech.*, **1968**, 247.
163 R. N. McDonald and R. N. Steppel, *J. Am. Chem. Soc.*, **91**, 782 (1969).
164 B. Rickborn and R. P. Thummell, *J. Org. Chem.*, **34**, 3583 (1969).
165 S. P. Singh and J. Kagan, *J. Am. Chem. Soc.*, **91**, 6198 (1969).

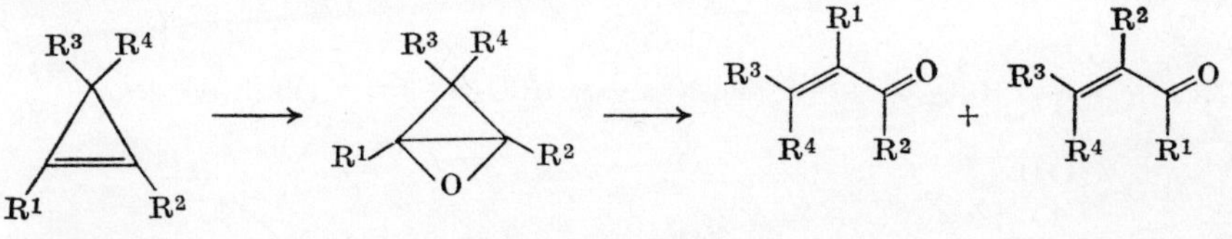

(106)

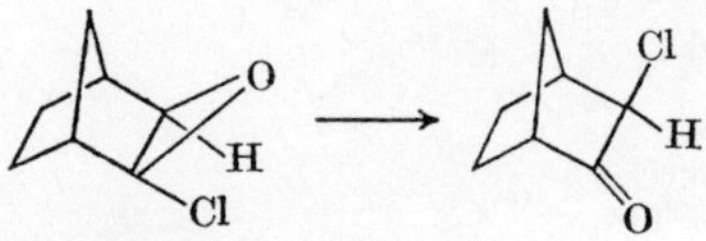

(107) 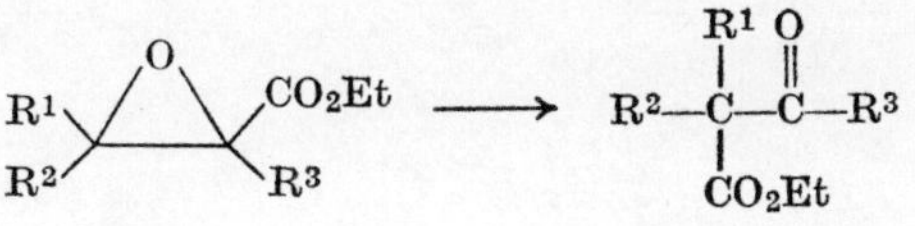   **(108)**

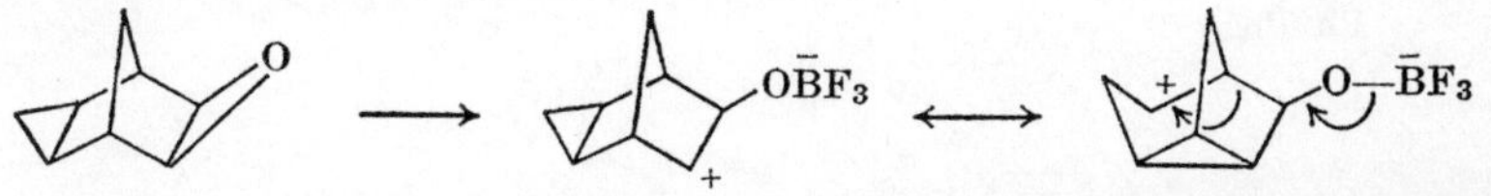

(109)

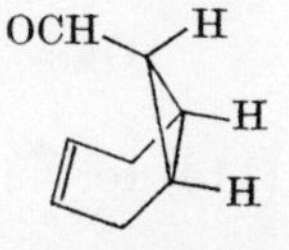

(110)

into (**110**); the mechanism shown was proposed and discussed.[166] Boron trifluoride catalysed rearrangement of exocyclic methylene epoxides to aldehydes appears to involve opening of the complexed oxide ring to a discrete carbonium ion.[167] The rearrangements of 4,5-epoxyeudesmanes[168] and of *cis*- and *trans*-1,2-diphenyl-1,2-ditolyloxiran[169] have also been reported.

Generation of the dipole (**111**) gave products derived from (**112**), presumably formed as shown, rather than dipolar cycloaddition.[170] Pyrolysis of the episulphone (**113**) of *cis*-dibenzoylstilbene is unusual in not giving the stilbene or products derived from it in any quantity; benzil (31%) and diphenylacetylene (31%) are the major products. Their formation is explained by the intriguing (if substantiated) scheme shown.[171] Thiet 1,1-dioxide (**114**) in the vapour phase at 375° or in benzene at 280° gave the sulphinate ester (**115**) (80%),

[166] G. D. Sargent, M. J. Harrison, and G. Khoury, *J. Am. Chem. Soc.*, **91**, 4937 (1969).

[167] B. N. Blackett, J. M. Coxon, M. P. Hartshorn, B. L. J. Jackson, and C. N. Muir, *Tetrahedron*, **25**, 1479 (1969).

[168] H. Hikino, T. Kohama, and T. Takemoto, *Tetrahedron*, **25**, 1037 (1969).

[169] K. Matsumoto, *Tetrahedron*, **24**, 6851 (1968).

[170] M. Takaku, S. Mitamura, and H. Nozaki, *Tetrahedron Letters*, **1969**, 3651.

[171] D. C. Dittmer, G. C. Levy, and G. E. Kuhlmann, *J. Am. Chem. Soc.*, **91**, 2097 (1969).

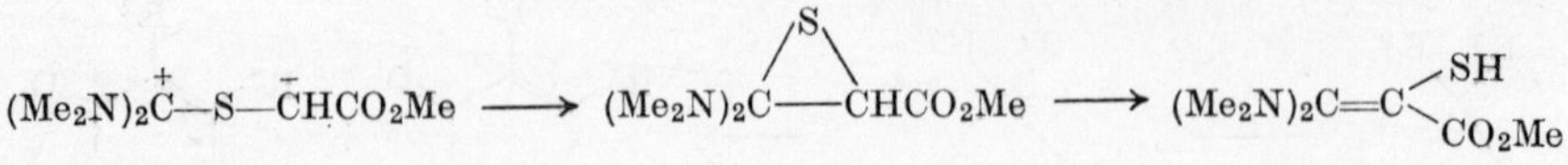

$(Me_2N)_2\overset{+}{C}-S-\overset{-}{C}HCO_2Me \longrightarrow (Me_2N)_2C-\!\!\overset{S}{\triangle}\!\!-CHCO_2Me \longrightarrow (Me_2N)_2C=C\!\!\begin{array}{l}SH\\CO_2Me\end{array}$

(111)   (112)

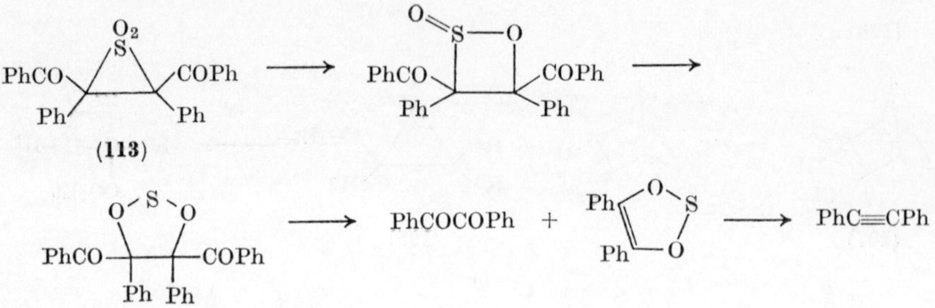

PhCO, Ph $\overset{O_2}{\underset{S}{\triangle}}$ COPh, Ph (113) $\longrightarrow$ PhCO, Ph $\square\!\!\begin{array}{c}O\!=\!S\!-\!O\end{array}$ COPh, Ph $\longrightarrow$

PhCO $\square\!\!\begin{array}{c}O\!-\!S\!-\!O\end{array}$ COPh, Ph Ph $\longrightarrow$ PhCOCOPh $+$ $\overset{Ph}{\underset{Ph}{}}\!\!\begin{array}{c}O\!-\!S\\O\end{array}$ $\longrightarrow$ PhC$\equiv$CPh

$\square\text{—SO}_2$ $\longrightarrow$ $\overset{O}{\underset{O}{S}}$

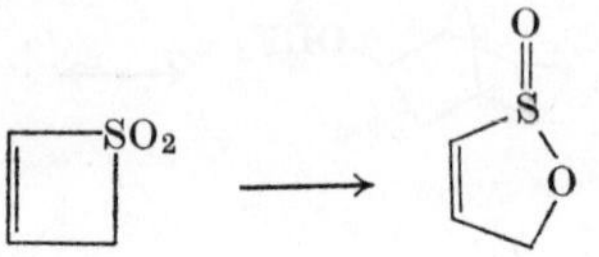

(114)   (115)

possibly by electrocyclic opening of (114) to vinylsulphene and re-closure on to oxygen.[172]

The kinetics of conrotatory ring-opening of dimethyl *cis-* and *trans*-1-(*p*-methoxyphenyl)aziridine-2,3-dicarboxylate to give the *trans-* and *cis*-azomethine ylid, respectively (see ref. 173), are obtained from reaction with tetracyanoethylene; ring-opening is the slow step since the first-order rate constants are independent of the concentration of tetracyanoethylene. This is so effective a dipolarophile that interconversion of the azomethine ylids and their collapse to starting aziridines are suppressed[174] (see also Chapter 5). Details have appeared[175] of the vinylaziridine to 3-pyrroline rearrangement.[173] The contraction of an azetidine ring to an aziridine, and the reverse, have been observed in the reactions of 1-t-butyl-3-chloroazetidine (116) and 1-t-butyl-2-chloromethylaziridine (117); they interconvert on heating, though there was less expansion than contraction. The former was favoured with potassium cyanide in methanol, however; both (116) and (117) gave the cyanoazetidine and no cyanomethylaziridine. Acetolysis rates were greater than for the corresponding carbocyclic systems. These rearrangements and rate enhancements were explained by a relatively slowly equilibrating pair of carbonium

[172] J. F. King, K. Piers, D. J. H. Smith, C. L. McIntosh, and P. de Mayo, *Chem. Comm.*, **1969**, 31, 32.
[173] *Org. Reaction Mech.*, **1967**, 232.
[174] R. Huisgen, W. Scheer, and H. Mäder, *Angew. Chem. Internat. Ed. Engl.*, 8, 602 (1969).
[175] R. S. Atkinson and C. W. Rees, *J. Chem. Soc.*(C), **1969**, 778.

ions stabilized by partial overlap of the empty orbital by the nitrogen lone-pair, as shown.[176] The *N*-chloro-derivative of indano[1,2-*b*]aziridine readily loses HCl to give isoquinoline in an aza-analogue of the cyclopropyl to allyl cation rearrangement.[177]

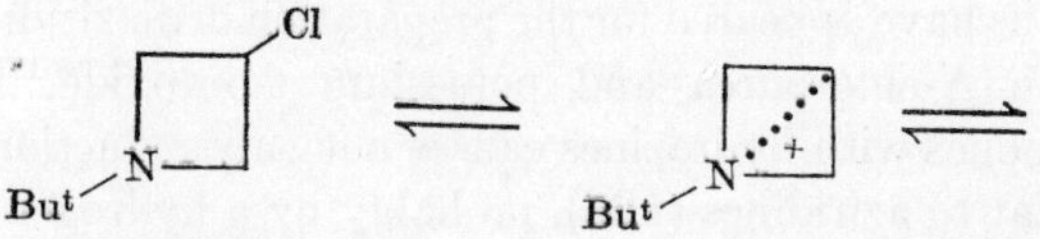

(116)

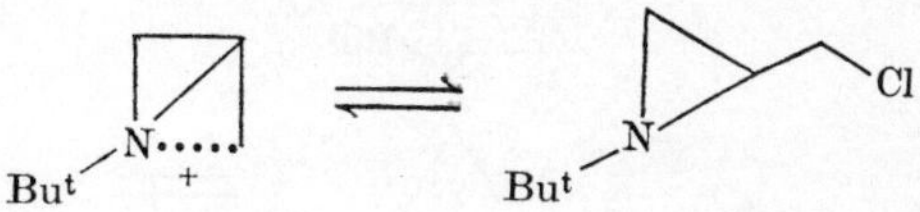

(117)

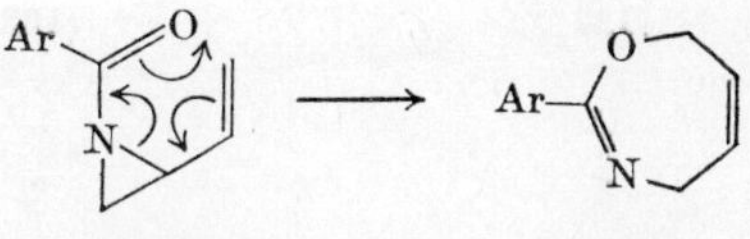

(118)   (119)

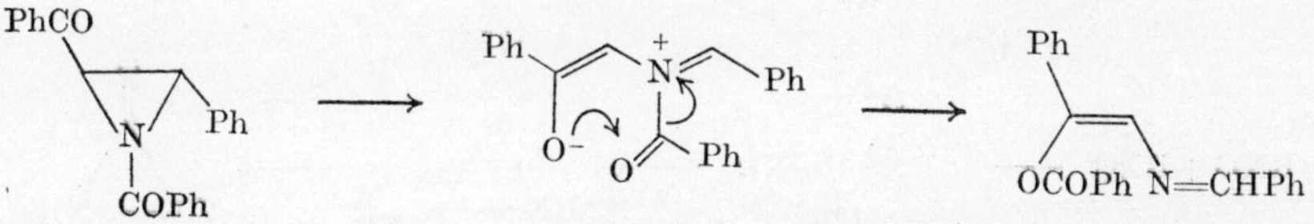

(120)   (121)

Thermal rearrangement in toluene of 1-*p*-nitrobenzoyl-2-vinylaziridine (118) gave a new type of product, 2-*p*-nitrophenyl-4,7-dihydro-1,3-oxazepine (119), probably by a 3,3-sigmatropic shift facilitated by the relief of ring strain.[178] Similar treatment of the *trans*-aziridine (120) gave a high yield of benzoate (121), presumably by initial aziridine C–C bond cleavage.[179] The stereochemistry of the acid-catalysed isomerization of *N*-phenylcarbamoyl and *N*-phenylthiocarbamoyl-aziridines to oxazolines and thiazolines has been investigated with various acids and solvents; S_N1, S_N2, and S_Ni mechanisms were all encountered.[180] Mild heating of 2,2-bistrifluoromethyl-3-methoxy-

[176] V. R. Gaertner, *Tetrahedron Letters*, **1968**, 5919.
[177] D. C. Horwell and C. W. Rees, *Chem. Comm.*, **1969**, 1428.
[178] P. G. Mente, H. W. Heine, and G. R. Scharoubim, *J. Org. Chem.*, **33**, 4547 (1968).
[179] A. Padwa and W. Eisenhardt, *Chem. Comm.*, **1969**, 1215.
[180] T. Nishiguchi, H. Tochio, A. Nabeya, and Y. Iwakura, *J. Am. Chem. Soc.*, **91**, 5835, 5841 (1969).

2*H*-azirine gave α-methylhexafluoroisopropyl isocyanate in a novel rearrangement of unknown mechanism.[181]

Reversible valence isomerization of a heterocyclic analogue of methylenecyclopropane has been demonstrated for the first time, by the interconversion of the diaziridine imines (122) and (123); at 60—90° either gave the equilibrium mixture.[182] Details have appeared for the preparation of diaziridinones, e.g. (124), from the *N*-chlorourea and potassium t-butoxide.[183] Treatment of these diaziridinones with hydrazines causes not only reduction to the urea but rearrangement to aziridines (125), probably by a hydrogen-atom transfer process.[184] Diaziridinones are also formed from isocyanides and

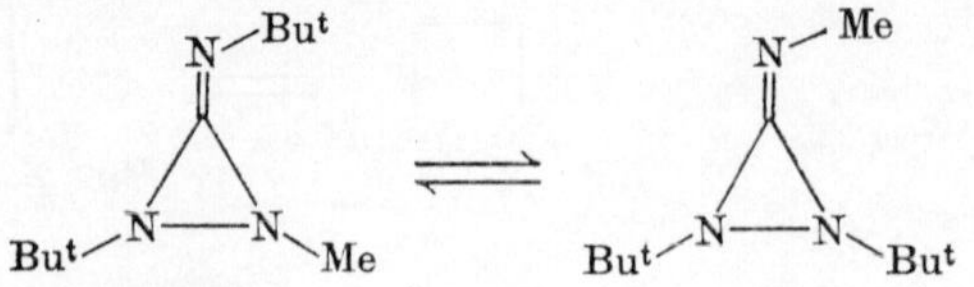

nitrosoalkanes probably via the carbodiimide *N*-oxide (126); (126) was intercepted by the 1,3-dipolarophile phenyl isocyanate. The other products of the reaction, carbodiimide and nitroalkane, could be formed by further reaction of the *N*-oxide and nitrosoalkane.[185] Further evidence for the related conversion of allene oxides into cyclopropanones is provided by the formation of cyclopentenones (127) by the peracid oxidation of vinylallenes.[186]

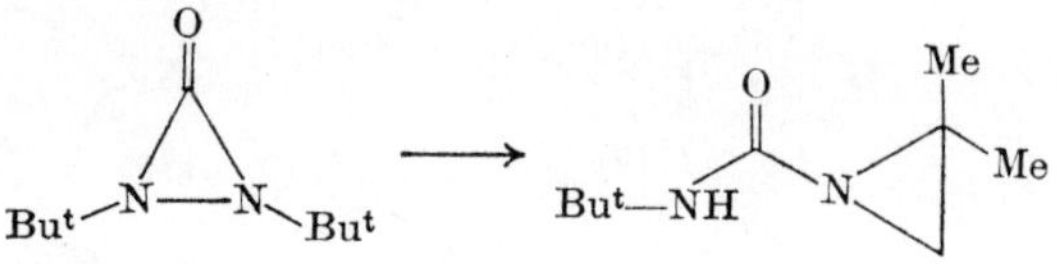

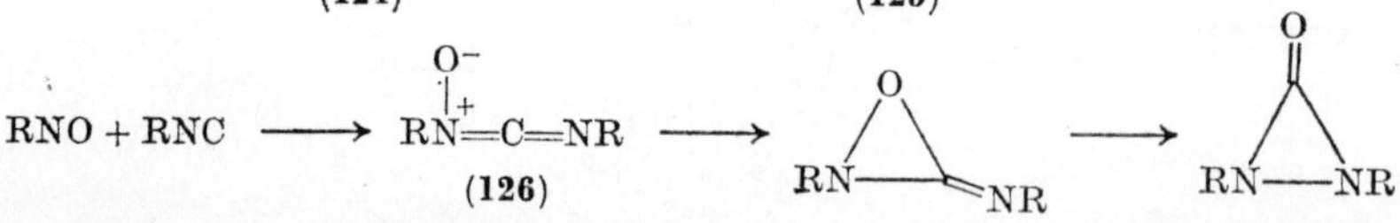

The anion (128) was rapidly formed from the oxaziridine (129) and 1 equivalent of sodium hydride in HMPA; no evidence for the formally aromatic 6 π-electron anion (130) could be obtained.[187]

[181] C. G. Krespan, *J. Org. Chem.*, **34**, 1278 (1969).
[182] H. Quast and E. Schmitt, *Angew. Chem. Internat. Ed. Engl.*, **8**, 449 (1969).
[183] F. D. Greene, J. C. Stowell, and W. R. Bergmark, *J. Org. Chem.*, **34**, 2254 (1969).
[184] F. D. Greene, W. R. Bergmark, and J. G. Pacifici, *J. Org. Chem.*, **34**, 2263 (1969).
[185] F. D. Greene and J. F. Pazos, *J. Org. Chem.*, **34**, 2269 (1969).
[186] J. Grimaldi and M. Bertrand, *Tetrahedron Letters*, **1969**, 3269; see also R. L. Camp and F. D. Greene, *J. Am. Chem. Soc.*, **90**, 7349 (1968).
[187] G. M. Rubottom, *Tetrahedron Letters*, **1969**, 3887.

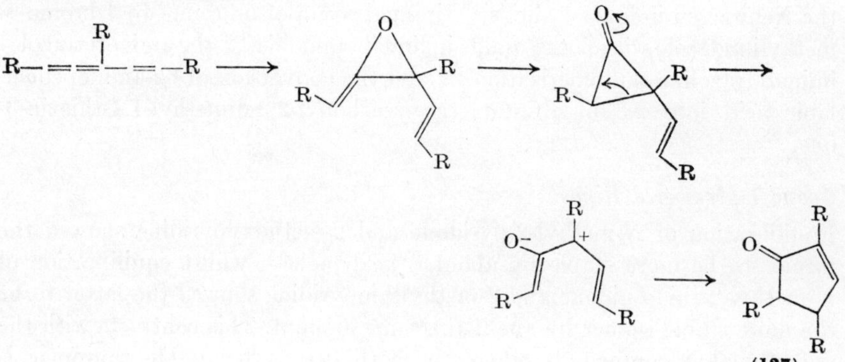

(127)

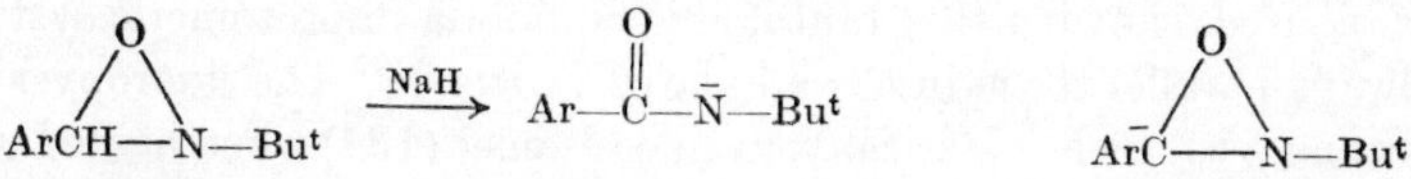

(128) (129) (130)

5-*Membered Rings*

The isopyrazole (131) rearranged smoothly in boiling xylene to the pyrazole (132) by a [3,3]-sigmatropic shift in the enamine; the methyl groups in (131) exchanged deuterium at room temperature in D_2O–NaOD–pyridine. 3-Allyl-2-

(131) (132)

methylindolenines rearrange similarly to indoles.[188] 1-Bromopyrrolo[1,2-*a*]-quinoxaline rearranged to the 3-bromo-isomer on heating with aqueous acid, probably by an intermolecular electrophilic mechanism.[189] The ring-expansions observed when pyrroles, and an imidazole, pyrazole, and 1,2,4-triazole react with dichlorocarbene have been discussed (see Chapter 10). Details have appeared of the base-induced rearrangements of 2-dichloromethyl-2,5-dimethyl-2*H*-pyrrole and related compounds.[190] Other rearrangements of 5-membered ring heterocyclics reported include the Plancher rearrangement,[191]

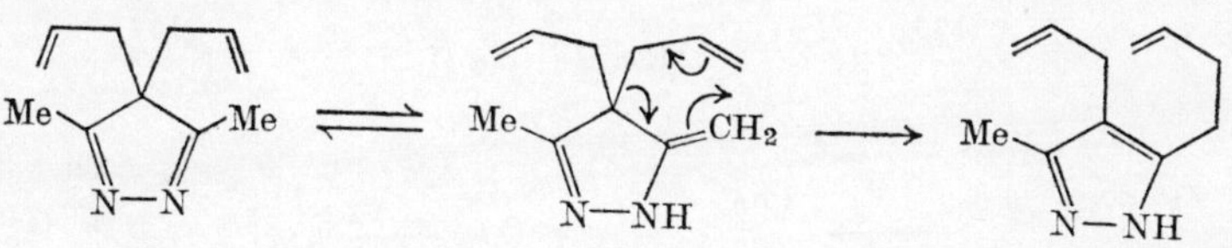

[188] R. K. Bramley and R. Grigg, *Chem. Comm.*, 1969, 99.
[189] G. W. H. Cheeseman and P. D. Roy, *J. Chem. Soc.*(C), 1968, 2848.
[190] R. L. Jones and C. W. Rees, *J. Chem. Soc.*(C), 1969, 2249, 2251, 2255.
[191] Y. Kanaoka, K. Miyashita, and O. Yonemitsu, *Chem. Comm.*, 1969, 1365.

the Nenitzescu indole synthesis,[192] isomerization of 5-bromo- to 4-bromo-1-methylimidazole with potassamide in liquid ammonia,[193] the aminotetrazole–iminotetrazoline tautomerization,[194] and the conversion of 6β-aminopenicillanic acid into 2,3-dihydro-6-methoxycarbonyl-2,2-dimethyl-1,4-thiazin-3-one.[195]

6- and 7-Membered Rings

Equilibration of *N*-methyl-2-pyridone and 2-methoxypyridine showed the former to be more stable by about 8 kcal mole^{-1}, whilst equilibration of *N*-methyl-2-thiopyridone and 2-methylthiopyridine showed the latter to be the more stable isomer by about the same amount. This contrasts with the unmethylated compounds where, in both cases, the amide tautomer is strongly favoured, showing that tautomeric equilibria in protomeric systems do not always parallel those in the alkylated isomers.[196] The fluoropyridine (133) rearranges to the less sterically strained isomer (134) when heated with potassium fluoride. The mechanism proposed is formally a nucleophilic analogue of the acid-catalysed rearrangement of alkylbenzenes.[197] Oxidation of 1-amino-3,4,5,6-tetraphenyl-2-pyridone with lead tetraacetate gave

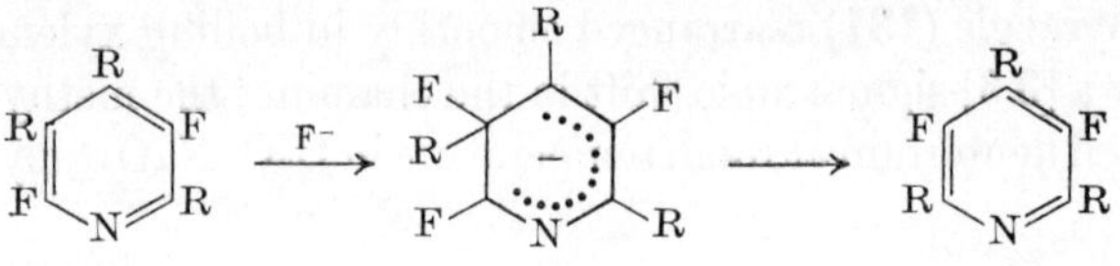

(133) $[R = CF(CF_3)_2]$ (134)

(135)

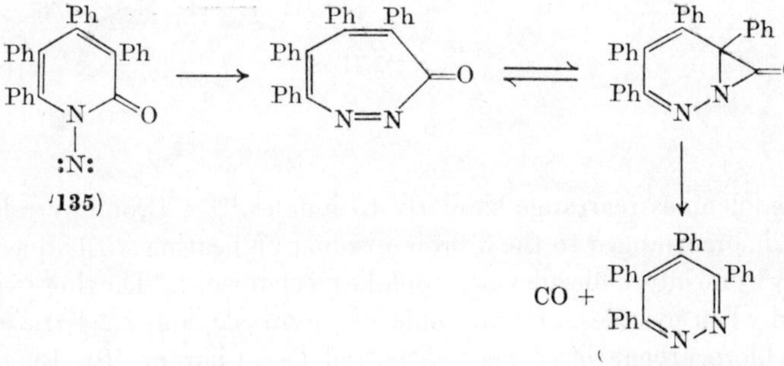

CO +

(136)

[192] R. Littell, G. O. Morton, and G. R. Allen, *Chem. Comm.*, **1969**, 1144.
[193] D. A. de Bie and H. C. van der Plas, *Rec. Trav. Chim.*, **88**, 1246 (1969).
[194] R. N. Butler, *Chem. Comm.*, **1969**, 405.
[195] R. J. Stoodley, *J. Chem. Soc.*(C), **1968**, 2891.
[196] P. Beak and J. T. Lee, *J. Org. Chem.*, **34**, 2125 (1969).
[197] R. D. Chambers, R. P. Corbally, J. A. Jackson, and W. K. R. Musgrave, *Chem. Comm.*, **1969**, 127.

3,4,5,6-tetraphenylpyridazine (**136**) and CO as the major products; formation and rearrangement of the nitrene (**135**), valence isomerization of the α-carbonylazo-intermediate, and concerted elimination of CO was proposed.[198]

The rearrangements of O-acylated N-oxides have been reviewed,[199] and the reactions of pyridine N-oxide with acid anhydrides have been extensively studied.[200] ^{18}O-Labelling experiments suggest that the thermal rearrangement of N- to 4-tosyloxyisocarbostyril and of N- to 8-tosyloxycarbostyril involve a solvent-separated ion-pair formed by N–O bond cleavage.[201]

The Dimroth rearrangement has been reviewed.[202] Treatment of 2- (**137**) and 4 (but not 5)-p-nitrobenzenesulphonamidopyrimidine with strong bases causes elimination of SO_2 and formation of the p-nitrophenylaminopyrimidine.

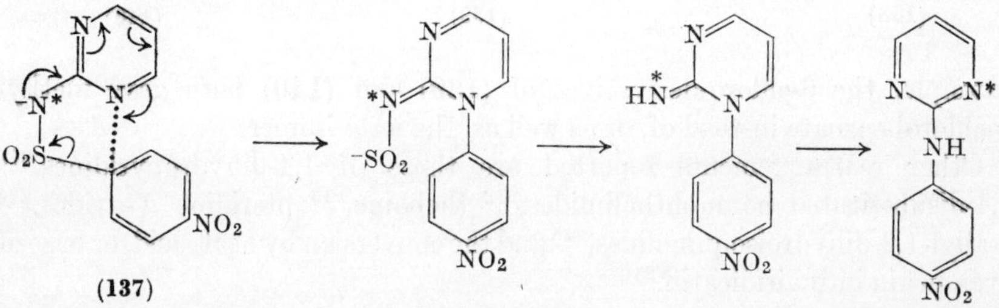

[15]N-Labelling showed that the pyrimidine ring undergoes a Dimroth-type rearrangement and the mechanism shown was proposed.[203] Further intriguing rearrangements and contractions of chloropyrimidines with potassamide in liquid ammonia have been reported.[204] 1-Methyl-2-phenyl-4(1H)-quinazolone was isomerized to the 3-methyl isomer on heating at 260°; the unlikely claim of an *intra*molecular 1,3-methyl shift was made, without experimental evidence.[205]

Pyrolysis (820°) of tetrafluoropyrazine gave tetrafluoropyrimidine probably via the diazabenzvalene isomer, whilst photolysis gave tetrafluoropyrazine probably through the diazaprismane.[206] 2,3-Dihydro-3R-iodomethyl-6-methoxycarbonyl-1,4-thiazine is racemized on heating in various solvents; deuterium-labelling showed that the 2-methylene and exocyclic methylene

[198] C. W. Rees and M. Yelland, *Chem. Comm.*, **1969**, 377.

[199] V. J. Traynelis, in ref. 2, Vol. 2, p. 1.

[200] C. Rüchardt, O. Krätz, and S. Eichler, *Chem. Ber.*, **102**, 3922 (1969).

[201] K. Ogino, S. Kozuka, and S. Oae, *Tetrahedron Letters*, **1969**, 3559.

[202] D. J. Brown, in ref. 2, Vol. 1, p. 209; M. Warren, *Z. Chem.*, **9**, 241 (1969).

[203] G. Malewski, H. Walther, H. Just, and G. Hilgetag, *Tetrahedron Letters*, **1969**, 1057.

[204] H. W. van Meeteren, H. C. van der Plas, and D. A. de Bie, *Rec. Trav. Chim.*, **88**, 728 (1969); H. C. van der Plas, B. Zuurdeeg, and H. W. van Meeteren, *ibid.*, p. 1156.

[205] Y. Hagiwara, M. Kurihara, and N. Yoda, *Tetrahedron*, **25**, 783 (1969).

[206] C. G. Allison, R. D. Chambers, Y. A. Cheburkov, J. A. H. MacBride, and W. K. R. Musgrave, *Chem. Comm.*, **1969**, 1200.

groups become equivalent, possibly in a concerted process (138).[207] Another similar 1,3-migration of sulphur has also been described.[208] Evidence for benzocyclopropenone as a highly reactive intermediate in photolysis of lithium 3-*p*-tolylsulphonylamino-1,2,3-benzotriazin-4(3*H*)-one (139)[209] and in the oxidation of 3-amino-1,2,3-benzotriazin-4(3*H*)-one (140)[210] was provided by the rearrangements observed in the presence of methanol. For

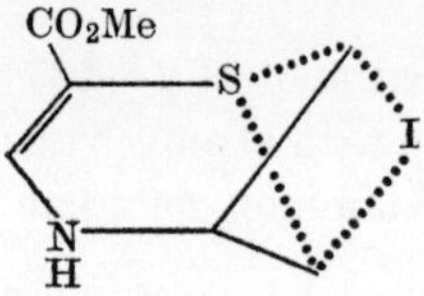

(138) 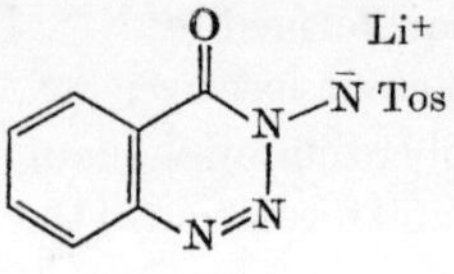   (139) 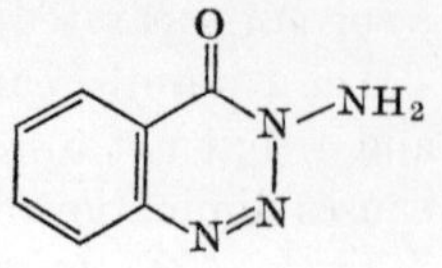   (140)

example, the 6-chloro-derivatives of (139) and (140) both gave methyl *p*-chlorobenzoate instead of, or as well as, the *meta*-isomer.

Other rearrangements reported are those of 1,4-dihydropyridines,[211] 4,4-disubstituted homophthalimides,[212] thebaine,[213] pteridine *N*-oxides,[214] 1-allyl-1,2-dihydroisoquinolines,[215] and the conversion by heat, acid, or base of cyclopenin into viridicatin.[216]

Other Rearrangements

Favorskii and Ramberg–Bäcklund Reactions

Further detailed study of rates and deuterium exchange in the Favorskii rearrangement of ArCH(Cl)COMe[217] and $ArCH_2COCH_2Cl$ in methanolic sodium methoxide has provided additional evidence for a transition state with extensive ionization of the C–Cl bond ($\rho = -5$) aided by π-participation[217] from the neighbouring enolate ion.[218] It follows that change from ArCH= $C(O^-)CH_2X$ to $ArCH=C(O^-)CHXMe$ should have a marked influence on the

[207] A. R. Dunn and R. J. Stoodley, *Chem. Comm.*, **1969**, 1169.

[208] A. R. Dunn and R. J. Stoodley, *Chem. Comm.*, **1969**, 1368.

[209] M. S. Ao, E. M. Burgess, A. Schauer, and E. A. Taylor, *Chem. Comm.*, **1969**, 220.

[210] J. Adamson, D. L. Forster, T. L. Gilchrist, and C. W. Rees, *Chem. Comm.*, **1969**, 221.

[211] J. F. Biellmann and H. J. Callot, *Chem. Comm.*, **1969**, 140; J. F. Biellmann, H. J. Callot, and M. P. Goeldner, *ibid.*, p. 141.

[212] W. J. Gensler, M. Vinovskis, and N. Wang, *J. Org. Chem.*, **34**, 3664 (1969).

[213] R. T. Channon, G. W. Kirby, and S. R. Massey, *J. Chem. Soc.*(C), **1969**, 1215; *Chem. Comm.*, **1969**, 92.

[214] W. Hutzenlaub, G. B. Barlin, and W. Pfleiderer, *Angew. Chem. Internat. Ed. Engl.*, **8**, 608 (1969).

[215] D. W. Brown, S. F. Dyke, R. G. Kinsman, and M. Sainsbury, *Tetrahedron Letters*, **1969**, 1731.

[216] H. W. Smith and H. Rapoport, *J. Am. Chem. Soc.*, **91**, 6083 (1969).

[217] *Org. Reaction Mech.*, **1968**, 254.

[218] F. G. Bordwell, R. G. Scamehorn, and W. R. Springer, *J. Am. Chem. Soc.*, **91**, 2087 (1969).

rearrangement, and this has now been demonstrated. This substitution caused a large (250-fold) increase in rate, a large decrease in the rate of deuterium exchange prior to loss of chloride ion, disappearance of the Br/Cl leaving-group effect (k_{Br}/k_{Cl} from 63 to 0.9), a change in ρ from -5 to $+1$, and a change in product from exclusive Favorskii ester [ArCH$_2$CH(Me)CO$_2$Me] to ArCH$_2$-COCH(Me)OMe in dilute sodium methoxide. These differences are all accom-

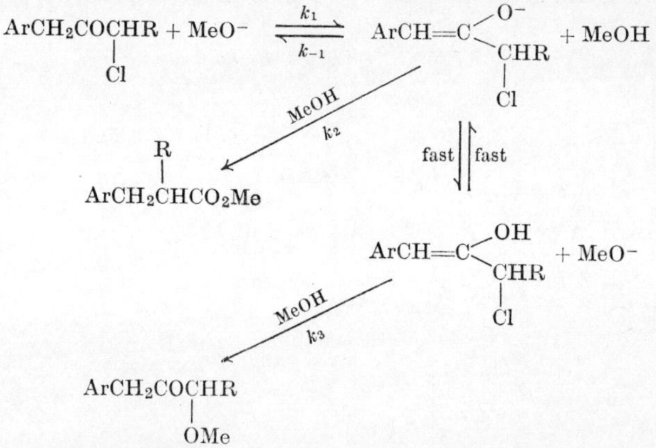

$$\text{ArCH}_2\text{COCHR} + \text{MeO}^- \underset{k_{-1}}{\overset{k_1}{\rightleftharpoons}} \text{ArCH}=\text{C} \overset{\text{O}^-}{\underset{\text{CHR}}{\diagdown}} + \text{MeOH}$$

modated by the mechanism shown. When R = H, formation of the enolate anion is reversible ($k_{-1} > k_2$), making k_2 rate-determining and explaining the extensive deuterium exchange, the large negative ρ, and the larger Br/Cl ratio. When R = Me, the rate of methanolysis is so increased that k_2 and k_3 are greater than k_{-1} and enolate anion formation is effectively irreversible and proton abstraction by methoxide is rate-determining; this explains the absence of deuterium exchange and of a Br/Cl leaving group effect, and the positive ρ. The scheme shown also explains the relative yield of products, the effect of substituents in Ar, and changes in solvent.[219] The distribution of products formed in the reaction of the isomeric pairs of α-bromobutan-2-ones and α-bromo-3-methylbutan-2-ones with sodium methoxide in various solvents was explained by (a) cyclopropanone formation, for formation of the rearranged ester, and (b) attack on the α-halogeno-ketone carbonyl to form epoxy-esters which decomposed to α-methoxy-ketones and/or α-hydroxy-ketones. The results argue against a benzylic-like intermediate, exchange of bromine between the α- and α'-positions, and the occurrence of a dipolar ion prior to rearrangement.[220] In the Favorskii rearrangement of polybromo- and polychloro-ketones in D$_2$O the products were deuterated, indicating a pre-equilibrium step.[221] The chemistry of cyclopropanones, including the Favorskii

[219] F. G. Bordwell, M. W. Carlson, and A. C. Knipe, *J. Am. Chem. Soc.*, **91**, 3949 (1969); see also F. G. Bordwell and M. W. Carlson, *J. Am. Chem. Soc.*, **91**, 3951 (1969).

[220] N. J. Turro, R. B. Gagosian, C. Rappe, and L. Knutsson, *Chem. Comm.*, **1969**, 270.

[221] C. Rappe and L. Knutsson, *Acta Chem. Scand.*, **22**, 2910 (1968).

10

rearrangement, has been reviewed,[222] and an interesting vinylogous re-arrangement of this type described.[223]

The Ramberg–Bäcklund rearrangement (the base-catalysed 1,3-elimination of HX from α-halogeno-sulphones to give alkenes) has been extensively reviewed by Paquette.[224] Further evidence for the W-type transition state in this reaction is provided by the ready conversion of (141) into (142) with aqueous potassium hydroxide at 100°. This must necessarily proceed by a

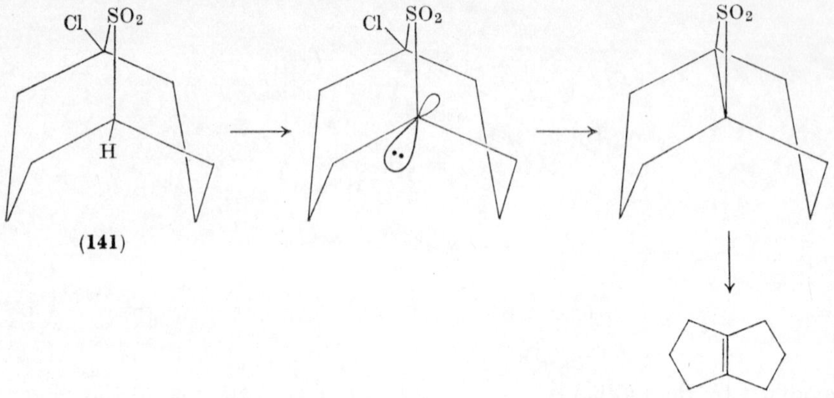

(141)

(142)

W-plan mechanism with inversion of configuration at the two reacting centres. The apparent ease of inversion of the bridgehead carbanion shown is note-worthy.[225] Treatment of the analogous bromo-sulphone with sodium t-pentoxide also gave the bicyclooctene (142) in high yield; this could not be extended to the highly strained bicyclohexene, however.[226]

Beckmann Rearrangements

Treatment of *anti*-2-arylcyclohexanone oximes with toluene-*p*-sulphonyl chloride in pyridine at room temperature gave the pyridinium salts (143) and (144). These salts result from interception of the intermediate carbonium ion before or after its rearrangement, in the Grob[227] mechanism shown.[228] The intermediate observed in the Beckmann rearrangement of 2,4,6-trimethyl-acetophenone oxime in sulphuric acid and in perchloric acid was considered,

[222] N. J. Turro, *Accounts Chem. Res.*, **2**, 25 (1969).
[223] G. M. Iskander and F. Stansfield, *J. Chem. Soc.*(C), **1969**, 669.
[224] L. A. Paquette, in ref. 2, Vol. 1, p. 121.
[225] L. A. Paquette and R. W. Houser, *J. Am. Chem. Soc.*, **91**, 3870 (1969).
[226] E. J. Corey and E. Block, *J. Org. Chem.*, **34**, 1233 (1969).
[227] C. A. Grob, H. P. Fischer, W. Raudenbusch, and J. Zergenyi, *Helv. Chim. Acta*, **47**, 1003 (1964).
[228] A. C. Huitric and S. D. Nelson, *J. Org. Chem.*, **34**, 1230 (1969).

from its NMR spectrum, to be the *N*-arylnitrilium ion (**145**) rather than the bridged phenonium ion.[229] The relative rates of the Beckmann–Chapman rearrangement of the ketoxime picryl ethers (**146**; R and R' = alkyl) are

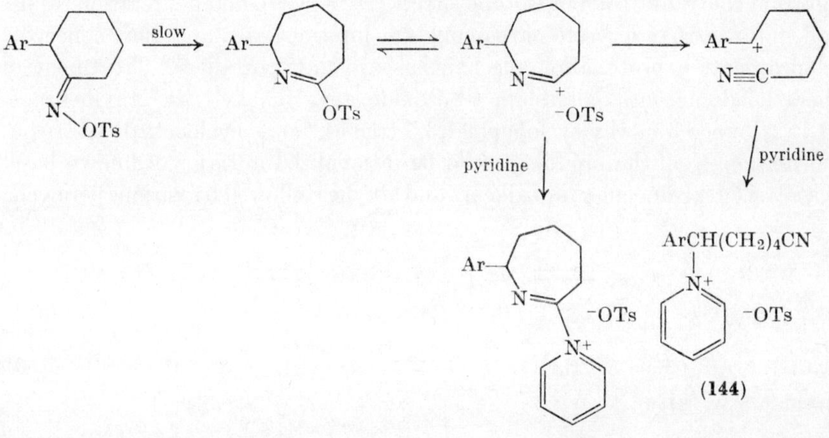

(**144**)

(**143**)

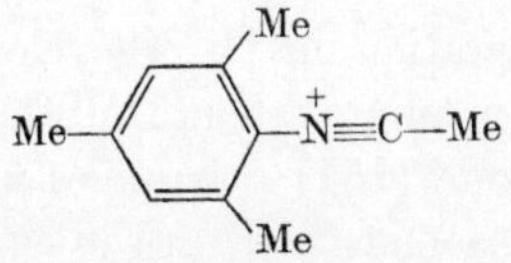

(**145**)

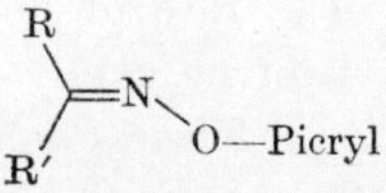

(**146**)

controlled mainly by steric interactions, between R and R' and R' and OPic, in the ground state.[230] Other reactions studied include the abnormal Beckmann rearrangement of $[4\alpha\text{-CD}_3]$-4,4-dimethyl-5α-cholestan-3-one oxime,[231] rearrangement and fission of adamantanone oxime,[232] rearrangement and fragmentation of α-difluoroaminofluorimines,[233] rearrangement of benzyl-ideneacetone oxime,[234] fragmentation of α-keto-oximes to cyanides,[235] rearrangement of chromanone oximes,[236] rearrangement accompanied by cyclization of benzophenone oximes having *ortho*-carboxamide and sulphona-mide groups,[237] and the acid-catalysed Beckmann-type rearrangement of 5-nitronorbornenes.[238]

[229] B. J. Gregory, R. B. Moodie, and K. Schofield, *Chem. Comm.*, **1969**, 645.
[230] H. P. Fischer and F. Funk-Kretschmar, *Helv. Chim. Acta*, **52**, 913 (1969).
[231] G. P. Moss and S. A. Nicolaidis, *Chem. Comm.*, **1969**, 1077.
[232] J. G. Korsloot and V. G. Keizer, *Tetrahedron Letters*, **1969**, 3517.
[233] T. E. Stevens, *J. Org. Chem.*, **34**, 2451 (1969).
[234] W. Zielinski and S. Goszczynski, *Zesz. Nauk. Politech. Slask. Chem.*, 1967, No. 39, 59; *Chem. Abs.*, **69**, 66676 (1968).
[235] M. Green and S. C. Pearson, *J. Chem. Soc.*(B), **1969**, 593.
[236] U. T. Bhalerao and G. Thyagarajan, *Indian J. Chem.*, **7**, 429 (1969).
[237] H. Watanabe, C. L. Mao, and C. R. Hauser, *J. Org. Chem.*, **34**, 1786 (1969).
[238] W. E. Noland, R. B. Hart, W. A. Joern, and R. G. Simon, *J. Org. Chem.*, **34**, 2058 (1969).

Double-bond and Related Rearrangements

Sodium pent-2-ynoate (147), penta-2,3-dienoate (148), and pent-3-ynoate (149) are interconverted in hot aqueous sodium hydroxide. Rates and equilibrium constants, solvent isotope effects (D_2O-H_2O), and deuteration results all indicate that discrete carbanions are intermediates and that concerted deprotonation–protonation mechanisms are unimportant.[239] Treatment of octa-1,7-diyne with potassium t-butoxide gave the xylenes, ethylbenzene, 1-, 2-, and 3-methylcyclohepta-1,3,5-trienes, and cycloocta-1,3,5-triene; formation of all the products could be rationalized in terms of known base-catalysed rearrangements of alkenes and alkynes followed by various pericyclic

$$\text{MeCH}_2\text{C} \equiv \text{C}-\text{CO}_2{}^- \; \rightleftharpoons \; \text{MeCH} = \text{C} = \text{CHCO}_2{}^- \; \rightleftharpoons \; \text{MeC} \equiv \text{CCH}_2\text{CO}_2{}^-$$

$$\quad\quad (147) \quad\quad\quad\quad\quad\quad\quad\quad (148) \quad\quad\quad\quad\quad\quad\quad\quad (149)$$

$$\text{Me(CH}_2)_5\text{COCH}_2\text{CH} = \text{CH(CH}_2)_7\text{CO}_2\text{Me} \; \rightleftharpoons \; \text{Me(CH}_2)_5\text{COCH} = \text{CHCH}_2(\text{CH}_2)_7\text{CO}_2\text{Me}$$

$$\quad\quad\quad\quad (150) \quad\quad\quad\quad\quad\quad\quad\quad\quad\quad\quad\quad\quad (151)$$

processes.[240] The rate of double-bond migration in [1-^{14}C]cyclo-pentene, -hexene, and -heptene, catalysed by potassium t-butoxide in DMSO, increases with ring size.[241] The α,β-unsaturated ketone (151) rearranges at 180° to give nearly 40% of the β,γ-(unconjugated)-isomer (150) in the equilibrium mixture.[242] The disproportionation of the pent-2-enes to but-2-enes and hex-3-enes is stereoselective when a soluble molybdenum catalyst is used.[243]

There are two limiting mechanisms for the thermal geometrical isomerization of compounds about a carbon–nitrogen double bond, involving movement of the group attached to nitrogen sideways in the plane or rotation out of the plane. The latter involves disruption of the C=N. It has now been shown that the rate of isomerization parallels the conjugating ability of the groups (N > S > O > C) attached to the imino carbon rather than their relative electronegatives, suggesting that disruption of C=N is probably involved.[244] Thermal isomerization of an O-alkyl-oxime has been demonstrated for *syn-* and *anti-O*-triphenylmethylbenzaldoxime at 200°; dissociation into triphenyl-methyl and iminoxy radicals, with rapid equilibrium of the latter, was proposed.[245] The azo–hydrazo conversion of 1-tosylazocyclohexene into cyclohex-2-enone tosylhydrazone has been studied under various conditions;

[239] R. J. Bushby and G. H. Whitham, *J. Chem. Soc.*(B), **1969**, 67.
[240] G. Eglinton, R. A. Raphael, and J. A. Zabkiewicz, *J. Chem. Soc.*(C), **1969**, 469.
[241] S. B. Tjan, H. Steinberg, and T. J. de Boer, *Rec. Trav. Chim.*, **88**, 673, 680, 690 (1969).
[242] P. M. Taylor and G. Fuller, *J. Org. Chem.*, **34**, 3627 (1969).
[243] W. B. Hughes, *Chem. Comm.*, **1969**, 431; cf. *Org. Reaction Mech.*, **1968**, 257.
[244] N. P. Marullo and E. H. Wagener, *Tetrahedron Letters*, **1969**, 2555.
[245] E. J. Grubbs and J. A. Villarreal, *Tetrahedron Letters*, **1969**, 1841.

slow abstraction of an allylic proton was suggested in the base-catalysed reaction.[246]

Other investigations include the base-catalysed equilibrium of *trans-Δ^1-* and *-Δ^2-*octalins and of the *cis*-isomers,[247] thermal and nitric oxide catalysed isomerization of *cis-* and *trans-*[1-^{2}H]propene,[248] NMR study of internal rotation in alkenes,[249] gas-phase equilibrium of *cis-* and *trans-*but-2-ene,[250] isomerization of the hexenes,[251] isomerization of *cis*-octadec-9-ene with deuteroperchloric acid,[252] acid- and base-catalysed isomerization of *p*-menthadienes[253] and of alkenesulphonic acids,[254] potassium t-butoxide catalysed rearrangement of methyl linoleate[255] and of hexa-1,5-diyne and hepta-1,6-diyne,[256] the iron pentacarbonyl isomerization of (−)-β-pinene to (−)-α-pinene,[257] isomerization of substituted methylindenes to methyleneindanes,[258] and amine-catalysed geometrical isomerizations.[259]

Base-catalysed rearrangements of acetylenic derivatives have been reviewed.[260]

Miscellaneous Rearrangements

The Wolff rearrangement has been studied further.[261] The kinetics of the thermal Wolff rearrangement of phenylbenzoyldiazomethane and *para*-substituted derivatives in di-n-butyl ether and in aqueous dioxan support slow rate-controlling loss of nitrogen to the α-keto-carbene which then rearranges to the ketene. The rearrangement goes at about the same rate in the presence or absence of nucleophiles.[262] Photolysis of the labelled diazo-ketone (152) gave N_2, CO, and propene (with other minor products); the CO and propene both had nearly 50% each of ^{13}C and ^{12}C isotopes, requiring the intervention of a symmetrical intermediate, the dimethyloxiren. This scrambling, also shown

[246] A. Dondoni, G. Rossini, G. Mossa, and L. Caglioti, *J. Chem. Soc.*(B), **1968**, 1404.

[247] P. Oberhänsli and M. C. Whiting, *J. Chem. Soc.*(B), **1969**, 467.

[248] M. C. Flowers and N. Jonathan, *J. Chem. Phys.*, **50**, 2805 (1969).

[249] Y. Shvo and H. Shanan-Atidi, *J. Am. Chem. Soc.*, **91**, 6683, 6689 (1969).

[250] J. H. Holmes, *J. Chem. Soc.*(A), **1969**, 1924.

[251] R. Maurel, M. Guisnet, and L. Bove, *Bull. Soc. Chim. France*, **1969**, 1975.

[252] J. S. Showell and I. S. Shepherd, *J. Org. Chem.*, **34**, 1097 (1969).

[253] R. B. Bates, E. S. Caldwell, and H. P. Klein, *J. Org. Chem.*, **34**, 2615 (1969).

[254] D. M. Brouwer and J. A. van Doorn, *Rec. Trav. Chim.*, **88**, 1041 (1969).

[255] O. Korver and C. Boelhouwer, *Rec. Trav. Chim.*, **88**, 696 (1969).

[256] D. A. Ben-Efraim and F. Sondheimer, *Tetrahedron*, **25**, 2837 (1969).

[257] P. A. Spanninger and J. L. von Rosenberg, *J. Org. Chem.*, **34**, 3658 (1969).

[258] A. J. Hubert and H. Reimlinger, *J. Chem. Soc.*(C), **1969**, 944.

[259] Y. L. Fan and D. F. Pollart, *J. Org. Chem.*, **33**, 4372 (1968).

[260] I. Iwai, in ref. 2, Vol. 2, p. 73.

[261] R. F. Borch and D. L. Fields, *J. Org. Chem.*, **34**, 1480 (1969); A. L. Wilds, R. L. Von Trebra, and N. F. Woolsey, *ibid.*, p. 2401; Y. Yukawa and T. Ibata, *Bull. Chem. Soc. Japan*, **42**, 802 (1969).

[262] W. Jugelt and D. Schmidt, *Tetrahedron*, **25**, 969 (1969).

for $Me^{13}COCHN_2$, is in contrast with earlier results which had led to the view that oxirens were not involved.[263]

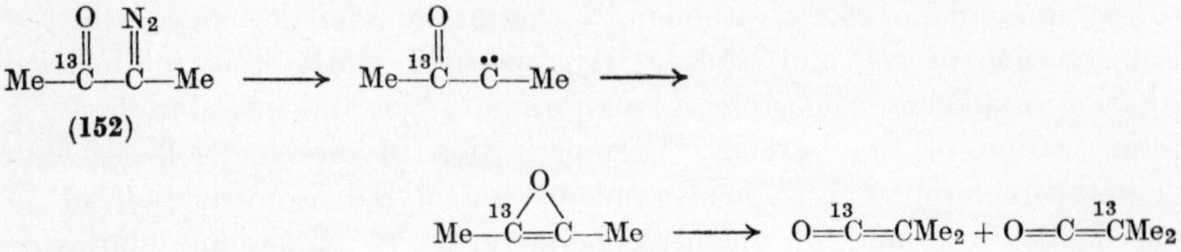

$$Me\overset{O}{\underset{13}{\overset{\|}{-}}}C\overset{N_2}{\underset{}{\overset{\|}{-}}}C\!-\!Me \longrightarrow Me\overset{O}{\underset{13}{\overset{\|}{-}}}C\!-\!\overset{\cdot\cdot}{C}\!-\!Me \longrightarrow$$

(152)

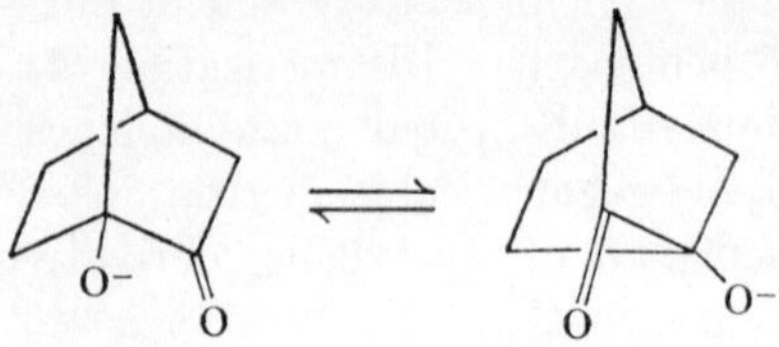

$$Me\overset{O}{\underset{13}{\overset{\diagup}{-}}}C\!=\!C\!-\!Me \longrightarrow O\!=\!\overset{13}{C}\!=\!CMe_2 + O\!=\!C\!=\!\overset{13}{C}Me_2$$

(153)

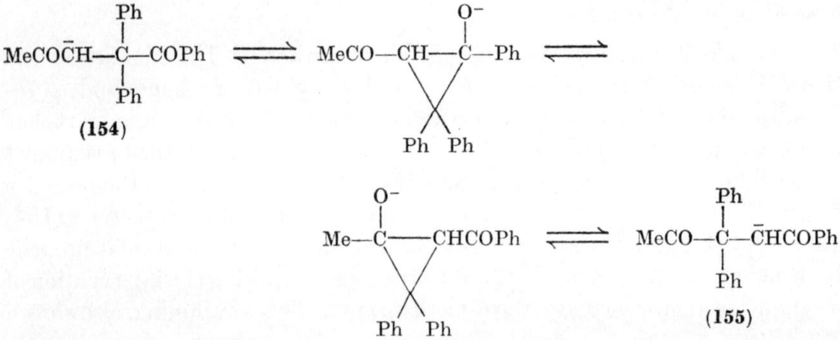

$$MeCO\overset{Ph}{\underset{Ph}{\overset{|}{\underset{|}{C}}}}C\!-\!COPh \rightleftharpoons MeCO\!-\!CH\overset{O^-}{\underset{\diagdown}{\overset{|}{-}}}C\!-\!Ph \rightleftharpoons$$

(154)

$$\underset{Ph\ \ Ph}{\diagup\diagdown}$$

$$Me\overset{O^-}{\underset{\diagdown}{\overset{|}{-}}}C\!-\!\!-\!CHCOPh \rightleftharpoons MeCO\overset{Ph}{\underset{Ph}{\overset{|}{\underset{|}{C}}}}\!-\!\bar{C}HCOPh$$

$$\underset{Ph\ \ Ph}{\diagup\diagdown} \qquad\qquad (155)$$

The bridgehead ketol (153) undergoes an interesting degenerate rearrangement with alkali, the anionic counterpart of the norbornyl cation rearrangement. This was demonstrated by incorporation of *four* atoms of deuterium, when (153) was warmed in D_2O containing potassium carbonate, showing that C-3 and C-7 become equivalent enolizable sites.[264] The base-catalysed rearrangement of diketone (154) to (155) is considered to go through the two homoenolate anions shown.[265] The 3-pyridyl group migrates faster than phenyl or p-methoxyphenyl in the benzilic acid rearrangement.[266]

Pinacol type rearrangements have been studied further.[267]

[263] I. G. Csizmadia, J. Font, and O. P. Strausz, *J. Am. Chem. Soc.*, **90**, 7360 (1968).
[264] A. Nickon, T. Nishida, and Y. Lin, *J. Am. Chem. Soc.*, **91**, 6860 (1969).
[265] P. Yates, G. D. Abrams, and S. Goldstein, *J. Am. Chem. Soc.*, **91**, 6868 (1969).
[266] A. Novelli and J. R. Barrio, *Tetrahedron Letters*, **1969**, 3671.
[267] S. I. Goldberg and W. D. Bailey, *Chem. Comm.*, **1969**, 1059; P. Depovere and R. Devis, *Bull. Soc. Chim. France*, **1969**, 479; K. Bhatia and A. Fry, *J. Org. Chem.*, **34**, 806 (1969).

Heating of N-methylbenzhydroxamic acids (**156**) caused rearrangement to the O-aroyl isomer and formation of some N,O-diaroyl product; on the basis of relative yields of unchanged hydroxamic acid and products, the mechanism shown was proposed.[268] When N,O-bistrimethylsilylhydroxylamine was

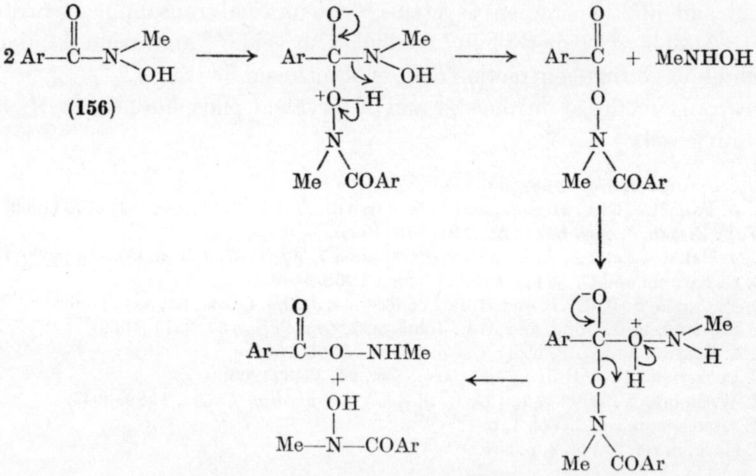

converted into its anion (**157**) with t-butyl-lithium and treated with methyl iodide, the O-methyl derivative (**158**) was formed. The two anions appear to be in equilibrium.[269] New 1,2-migrations of aryl groups from carbon to silicon,[270] from silicon to carbon,[271] and from phosphorus to carbon[272] have been reported. A radical mechanism has been proposed for the thermal rearrangement of isocyanides to cyanides.[273]

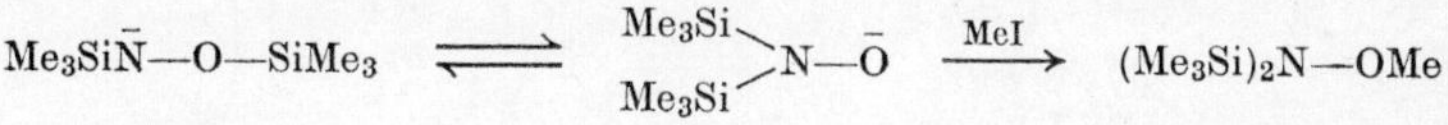

1,3-Alkyl migrations from oxygen to carbon, nitrogen, and oxygen, from sulphur to nitrogen, and from nitrogen to carbon have been reviewed.[274]

[268] W. B. Ankers, D. B. Bigley, R. F. Hudson, and J. C. Thurman, *Tetrahedron Letters*, **1969**, 4539.
[269] R. West, P. Boudjouk, and A. Matuszko, *J. Am. Chem. Soc.*, **91**, 5184 (1969).
[270] H. Sakurai, A. Hosomi, and M. Kumada, *Chem. Comm.*, **1969**, 521.
[271] A. G. Brook and P. F. Jones, *Chem. Comm.*, **1969**, 1324.
[272] J. J. Brophy, K. L. Freeman, and M. J. Gallagher, *J. Chem. Soc.(C)*, **1968**, 2760.
[273] S. Yamada, K. Takashima, T. Sato, and S. Terashima, *Chem. Comm.*, **1969**, 811.
[274] P. S. Landis, in ref. 2, Vol. 2, p. 43.

Tricyclo[5.2.1.0^{4,10}]decane, like its isomer twistane,[275] is smoothly iso-merized to adamantane by aluminium chloride; the shortest pathway involves three 1,2-alkyl shifts.[276]

Other rearrangements investigated include those of chrysanthenone,[277] methylbicyclo[4.3.0]nonanes with aluminium bromide,[278] longifolene with acid,[279] arylazotribenzoylmethanes in the solid state,[280] dibenzobicyclo-[2.2.2]- and -[3.2.1]-octadienyl systems,[281] an unusual transannular rearrange-ment,[282] 3-phenyl-1-methylallyl alcohol with acid,[283] and azido-*p*-quinones which undergo ring-contraction on loss of nitrogen.[284]

Rearrangements of nitrones [285] and of tervalent phosphorus esters [286] have been reviewed.

[275] *Org. Reaction Mech.*, **1968**, 260.
[276] L. A. Paquette, G. V. Meehan, and S. J. Marshall, *J. Am. Chem. Soc.*, **91**, 6779 (1969).
[277] W. F. Erman, *J. Am. Chem. Soc.*, **91**, 779 (1969).
[278] I. M. Makarova and A. A. Petrov, *Neftekhimiya*, **7**, 501 (1967); *Chem. Abs.*, **69**, 18396 (1968).
[279] D. G. Farnum and G. Mehta, *Chem. Comm.*, **1968**, 1643.
[280] D. Y. Curtin, S. R. Byrn, and D. B. Pendergrass, *J. Org. Chem.*, **34**, 3345 (1969).
[281] S. J. Cristol, R. J. Bopp, and A. E. Johnson, *J. Org. Chem.*, **34**, 3574 (1969).
[282] J. E. Baldwin and D. P. Kelly, *Chem. Comm.*, **1968**, 1664.
[283] Y. Pocker and M. J. Hill, *J. Am. Chem. Soc.*, **91**, 3243 (1969).
[284] H. W. Moore, W. Weyler, and H. R. Shelden, *Tetrahedron Letters*, **1969**, 3947.
[285] M. Lamchen, in ref. 2, Vol. 1, p. 1.
[286] V. Mark, in ref. 2, Vol. 2, p. 267.

CHAPTER 9

Radical Reactions

M. J. Perkins

Department of Chemistry, King's College, University of London

Numerous review articles continue to serve the free-radical chemist with digests of various aspects of this enormous field. Reviews published during the past year have dealt, in varying degrees of sophistication, with radical reactivity,[1] electron transfer as an elementary step in organic reactions,[2] radical fragmentation and isomerization,[3] intramolecular hydrogen transfer and related reactions,[4] mechanisms of radical reactions as studied by deuterium labelling,[5] aspects of peroxy radical chemistry,[6] autoxidation[7] and anti-oxidants,[8] the Markovnikov rule and radical addition,[9] kinetics of radical addition and telomerization,[10, 11] cation radicals in polymer synthesis,[12] homoaromaticity in radical ions,[13] chemiluminescence in the reactions of

[1] Y. L. Spirin, *Uspekhi Khim.*, **38**, 1201 (1969).

[2] K. A. Bilevich and O. Y. Okhlobystin, *Uspekhi Khim.*, **37**, 2162 (1968).

[3] J. A. Kew and A. C. Lloyd, *Quart. Rev.*, **22**, 549 (1968).

[4] P. A. Verlrugge, *Chem. Tech.* (Amsterdam), **23**, 286 (1968).

[5] I. P. Gragerov, *Isotopenpraxis*, **3**, 141 (1967); *Chem. Abs.*, **71**, 21489 (1969).

[6] K. U. Ingold, *Accounts Chem. Res.*, **2**, 1 (1969).

[7] I. Seree de Roch, *Ind. Chim. Belge*, **33**, 994 (1968); N. M. Emanuel, Z. K. Maizus, and I. P. Skibida, *Angew. Chem. Internat. Ed. Engl.*, **8**, 97 (1969).

[8] L. R. Mahoney, *Angew. Chem. Internat. Ed. Engl..* **8**, 547 (1969).

[9] N. Isenberg and M. Grdinic, *J. Chem. Educ.*, **46**, 601 (1969).

[10] I. B. Afanas'ev and G. I. Samokhvalov, *Uspekhi Khim.*, **38**, 687 (1969).

[11] W. G. Lloyd, *J. Chem. Educ.*, **46**, 299 (1969).

[12] A. Ledwith, *Ann. N.Y. Acad. Sci.*, **155**, 385 (1969).

[13] S. Winstein, *Quart. Rev.*, **23**, 141 (1969).

radical anions[14] and as a kinetic probe,[15] high-temperature pyrolysis of aromatic nitro-compounds and anhydrides,[16] and evaluation of free-radical thermochemistry by gas-phase reactions with iodine atoms.[17] Also discussed have been methods of generation and study of short-lived free radicals,[18] the use of electron spin resonance (ESR) in conjunction with a fast-flow reactor for precision study of gas kinetics,[19] spin-labelling as a probe for biomolecular structure,[20] substituent effects in ESR spectra coupled with an analysis of radical stability,[21] and applications of NMR to the study of radicals and radical reactions.[22]

The phenomenon of chemically induced dynamic nuclear polarization (CIDNP)—the observation of enhanced absorption, or of emission spectra, when the products of a very fast reaction involving reactive radical intermediates are immediately examined by NMR spectroscopy—was first recorded towards the end of 1967, and the early observations have now been summarized, and a theoretical analysis has been given, by Fischer and Bargon.[23] More recently, radical mechanisms have been inferred for a variety of reactions on the basis of observation of CIDNP phenomena in the products. For example, Baldwin and Brown used this device to demonstrate radical participation in the thermal 1,3-sigmatropic rearrangement of **(1)** into **(2)**, for which orbital symmetry considerations require a non-concerted pathway.[24] Similar observations have been made during the migration of benzyl or allyl groups in the Stevens rearrangement of nitrogen ylids,[25] and in the related rearrangement of the sulphur ylid **(3)** to give **(4)**[26] (see also Chapter 8).

Several new results with lithium alkyls have appeared. Strong nuclear polarization in the exchange reaction:

$$RLi + R'I \rightleftharpoons RI + R'Li$$

has been observed, but the precise mechanistic interpretation seems to be in dispute.[27]

[14] E. A. Chandross, *Trans. N.Y. Acad. Sci.*, **31**, 571 (1969).

[15] L. Matisova, *Chem. Listy*, **62**, 1417 (1968).

[16] E. K. Fields and S. Meyerson, *Accounts Chem. Res.*, **2**, 273 (1969).

[17] D. M. Golden and S. W. Benson, *Chem. Rev.*, **69**, 125 (1969).

[18] S. Kirkiachorian, *Ann. Chim.* (France), **3**, 403 (1968).

[19] A. A. Westenberg, *Science*, **164**, 381 (1969).

[20] O. H. Griffith and A. S. Waggoner, *Accounts Chem. Res.*, **2**, 17 (1969).

[21] E. G. Janzen, *Accounts Chem. Res.*, **2**, 279 (1969).

[22] A. L. Buchachenko and N. A. Sysoeva, *Uspekhi Khim.*, **37**, 1852 (1968).

[23] H. Fischer and J. Bargon, *Accounts Chem. Res.*, **2**, 110 (1969).

[24] J. E. Baldwin and J. E. Brown, *J. Am. Chem. Soc.*, **91**, 3647 (1969).

[25] A. R. Lepley, *J. Am. Chem. Soc.*, **91**, 1237 (1969); U. Schöllkopf, V. Ludwig, G. Ostermann, and M. Patsch, *Tetrahedron Letters*, **1969**, 3415; R. W. Jemison and D. G. Morris, *Chem. Comm.*, **1969**, 1226; D. G. Morris, *ibid.*, p. 1345.

[26] U. Schöllkopf, G. Ostermann, and J. Schossig, *Tetrahedron Letters*, **1969**, 2619.

[27] H. R. Ward, R. G. Lawler, and R. A. Cooper, *J. Am. Chem. Soc.*, **91**, 746 (1969); A. R. Lepley and R. L. Landau, *ibid.*, p. 748; A. R. Lepley, *Chem. Comm.*, **1969**, 64; H. R. Ward, R. G. Lawler, and R. A. Cooper, *Tetrahedron Letters*, **1969**, 527.

The normal method for observing CIDNP phenomena has been to conduct the reaction in the spectrometer probe, where the radical intermediates are in a large magnetic field; electron polarization induced in these intermediates

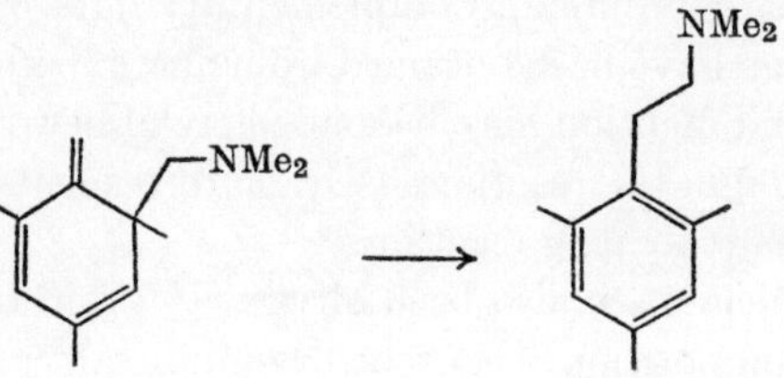

(1)   (2)

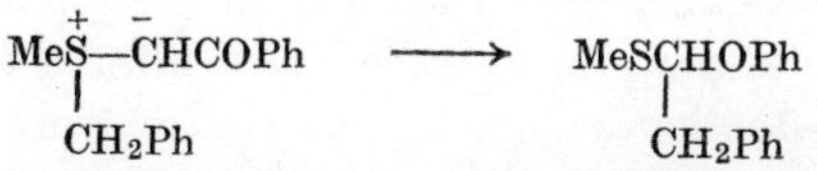

$$\text{Me}\overset{+}{\text{S}}\text{—}\overset{-}{\text{CHCOPh}} \longrightarrow \text{MeSCHOPh}$$

$$\overset{|}{\text{CH}_2\text{Ph}} \qquad\qquad \overset{|}{\text{CH}_2\text{Ph}}$$

(3)   (4)

by the magnetic field appears to be a prerequisite of some interpretations of the spectroscopic data. Closs has devised a system for carrying out photochemical reactions in the spectrometer, and describes examples of nuclear polarization in products of photochemical reactions.[28] Photolysis or thermolysis of diphenyldiazomethane in toluene leads to pairs of benzyl and benzhydryl radicals which, since they are formed from triplet diphenyl-methylene, must initially be "triplets". The spin polarization effect in the 1,1,2-triphenylethane which is produced by radical coupling is different from that observed when the initial radical pairs are obtained by decomposition of the azo-compound $Ph_2CHN=NCH_2Ph$ and are hence "singlet".[29] The results are in accord with a new theoretical analysis of the CIDNP phenomenon[30] and hold promise of providing a valuable technique for probing spin multiplicity in radical pairs. Closs's new theoretical considerations[30] appear to offer an interpretation of the interesting zero-field polarization observed recently, in which several radical reactions were carried to completion in the absence of a magnetic field, and spin polarization was then observed in the spectra of the resulting radical-derived products.[31]

Many of these CIDNP results have been supported by new evidence of other kinds for the existence of radical intermediates in the reaction studied. For

[28] G. L. Closs and L. E. Closs, *J. Am. Chem. Soc.*, **91**, 4549, 4550 (1969).
[29] G. L. Closs and A. D. Trifunac, *J. Am. Chem. Soc.*, **91**, 4554 (1969).
[30] G. L. Closs, *J. Am. Chem. Soc.*, **91**, 4552 (1969).
[31] H. R. Ward, R. G. Lawler, H. Y. Loken, and R. A. Cooper, *J. Am. Chem. Soc.*, **91**, 4928 (1969).

example, benzhydryl radicals have been detected by ESR when diphenyldiazomethane is pyrolysed at 150° in a hydrogen-donor solvent,[32] and alkyl
radicals have been observed after mixing lithium alkyls with alkyl iodides in a
flow system[33] (alkyl radicals were also detected when the alkyl iodide was
replaced by iodine or bromine). Complementary data on the alkyl-lithium–
alkyl halide reaction have been obtained from new experiments with optically
active halides,[34] and mention may be made here of new results which require
radical participation in the reaction of Grignard reagents with trityl esters,[35]
as well as in the Wurtz–Fittig reaction.[36]

CIDNP phenomena have also been observed[37, 38] in the benzene which is
produced from decomposition of benzenediazonium salts under basic conditions
in the presence of hydrogen-donor solvents. Presumably electron transfer
leads to de-diazoniation and the production of phenyl radicals. Again, complementary product studies have been described which support radical participation.[37, 39] These results place a question mark after the interpretation of
another diazonium reduction, reported this year, in terms of aryl cations and
hydride transfer.[40]

Kochi and Krusic last year described the direct observation of the ESR
spectra of several reactive alkyl radicals.[41] Their technique of low-temperature
photolysis of t-butyl peroxide in the cavity of the spectrometer, using a
high-intensity ultraviolet source, has enabled them to generate a wide range
of alkyl radicals in concentrations adequate for direct observation. This
work has been considerably extended and much useful structural information
has been forthcoming. For instance, hydrogen abstraction from methyl-
cyclopropane at −140° gives cyclopropylcarbinyl radicals; the spectra of
these show no evidence of non-classical behaviour, and above −100° isomerization to allylcarbinyl is observed to be very rapid.[42] No estimate of the rate of
this isomerization was made, and it remains to be seen whether the general
technique will be capable of providing useful rate data in addition to structural
and spectroscopic information. One rate constant which did receive comment
was that for geometrical isomerization of methallyl radicals. The radicals
observed were isomerically pure, even at 0°, and it was estimated that the

[32] D. R. Dalton and S. A. Liebman, *J. Am. Chem. Soc.*, **91**, 1194 (1969).
[33] G. A. Russell and D. W. Lamson, *J. Am. Chem. Soc.*, **91**, 3967 (1969).
[34] J. Sauer and W. Braig, *Tetrahedron Letters*, **1969**, 4275.
[35] K. D. Berlin, R. D. Shupe, and R. D. Grigsby, *J. Org. Chem.*, **34**, 2500 (1969).
[36] I. P. Gragerov, *Zh. Obshch. Khim.*, **38**, 2393 (1968); *Chem. Abs.*, **70**, 46575 (1969); I. P.
  Gragerov and L. F. Kasukhin, *Zh. Org. Khim.*, **5**, 9 (1969); *Chem. Abs.*, **70**, 86738 (1969).
[37] A. Rieker, P. Niedener, and D. Leibfritz, *Tetrahedron Letters*, **1969**, 4287.
[38] A. G. Lane, C. Rüchardt, and R. Werner, *Tetrahedron Letters*, **1969**, 3213.
[39] R. Werner and C. Rüchardt, *Tetrahedron Letters*, **1969**, 2407.
[40] H. Suschitzky and C. F. Sellers, *Tetrahedron Letters*, **1969**, 1105.
[41] See *Org. Reaction Mech.*, **1968**, 262, 280.
[42] J. K. Kochi, P. J. Krusic, and D. R. Eaton, *J. Am. Chem. Soc.*, **91**, 1877, 1879 (1969).

rate constant for *cis* to *trans* isomerization must be less than 10^2 sec^{-1}.[43] Hydrogen abstraction from other hydrocarbons, including bicyclobutane,[44] has been studied; no ESR evidence was forthcoming for the formation of

(5)

radical (5) by addition to bicyclobutane. Silyl radicals were observed by abstraction from silanes,[45] and silyl-, germyl-, and stannyl-carbinyl radicals were detected when butoxy radicals reacted with tetraalkyl-silanes, -germanes, or -stannanes[46] (see also p. 326). The possibility of detecting ESR spectra of reactive radicals during di-t-butyl peroxide photolyses was discovered independently by Hudson and his colleagues,[47] and they also observed spectra of silyl, as well as germyl, radicals.[48] Kochi and Krusic have advocated photolysis of aliphatic diacyl peroxides as a means of generating specific alkyl radicals for similar spectroscopic study. In this way they were able to generate the hex-5-enyl radical from hept-6-enoyl peroxide at $-75°$. At $-35°$ the only radical detected was cyclopentylcarbinyl, formed by rapid intramolecular addition.[49] Davies and Roberts showed[50] that specific alkyl radicals could also be generated by photolysis of di-t-butyl peroxide in the presence of a trialkylboron, utilizing a reaction analogous to that which takes place during autoxidation of many organometallic compounds:

$$BuO \cdot + R_3B \rightarrow R_2BOBu + R \cdot$$

Subsequently they demonstrated similar radical displacements from alkyl derivatives of zinc, cadmium, aluminium, antimony, and bismuth.[51] In parallel work, Kochi and Krusic demonstrated displacement of alkyl radicals from boron, aluminium, and gallium,[52] as well as from trivalent phosphorus.[53]

[43] J. K. Kochi and P. J. Krusic, *J. Am. Chem. Soc.*, **90**, 7157 (1968).

[44] P. J. Krusic and J. K. Kochi, *J. Am. Chem. Soc.*, **90**, 7155 (1968); P. J. Krusic, J. P. Jesson, and J. K. Kochi, *ibid.*, p. 4566.

[45] P. J. Krusic and J. K. Kochi, *J. Am. Chem. Soc.*, **91**, 3938 (1969).

[46] P. J. Krusic and J. K. Kochi, *J. Am. Chem. Soc.*, **91**, 6161 (1969).

[47] A. Hudson and H. A. Hussain, *J. Chem. Soc.*(B), **1969**, 793; *Mol. Phys.*, **16**, 199 (1969); A. Hudson and K. D. J. Root, *Tetrahedron*, **25**, 5311 (1969).

[48] S. W. Bennett, C. Eaborn, A. Hudson, A. Hussain, and R. A. Jackson, *J. Organometal. Chem.*, **16**, P36 (1969).

[49] J. K. Kochi and P. J. Krusic, *J. Am. Chem. Soc.*, **91**, 3940 (1969).

[50] A. G. Davies and B. P. Roberts, *Chem. Comm.*, **1969**, 699.

[51] A. G. Davies and B. P. Roberts, *J. Organometal. Chem.*, **19**, P17 (1969).

[52] P. J. Krusic and J. K. Kochi, *J. Am. Chem. Soc.*, **91**, 3942 (1969).

[53] J. K. Kochi and P. J. Krusic, *J. Am. Chem. Soc.*, **91**, 3944 (1969).

In the last of these, direct spectroscopic evidence was obtained in some instances for an intermediate radical adduct, or phosphoranyl radical (e.g.
$Me_3\dot{P}OBu^t$), and phosphoranyl radicals were assumed to participate in all
the displacements from organophosphorus compounds which were examined.
On the other hand, it was argued that displacement of alkyl radicals from
Group III elements probably did not involve a tetracoordinate intermediate.

Perhaps the simplest means of generating specific alkyl radicals in concentrations adequate for ESR study is that reported recently by Hudson and
Jackson.[54] This depends on the high reactivity of silyl radicals in halogen
atom abstraction reactions. Photolysis of di-t-butyl peroxide in the presence
of triethylsilane and a variety of alkyl bromides has yielded good spectra of the
corresponding alkyl radicals produced by the sequence:

$$BuO \cdot + Et_3SiH \rightarrow BuOH + Et_3Si \cdot$$

$$Et_3Si \cdot + RBr \rightarrow Et_3SiBr + R \cdot$$

Finally, in this introductory section, mention might be made of the development of an ingenious wall-less reactor for studying gas-phase reactions. In this
apparatus a laminar flow of inert gas insulates the reactants from contact
with the walls.[55] Use of the apparatus to determine the activation energy for
pyrolysis of neopentane gave a value of 80 kcal mole^{-1}, which is considerably
different from most values cited in the literature.

Structure and Stereochemistry

Last year's report[56] that the 7-norbornenyl radical reacted with trialkyltin
deuteride to give exclusively *anti*-7-deuteronorbornene, from which non-
classical structure (6) was inferred for the radical, has been discredited.[57a,b]
The *anti*-isomer does predominate in the product, possibly because of non-
planar geometry in this strained radical, but it is admixed with *ca.* 20—30%
of the isomeric *syn*-7-deuteronorbornene. There is even less selectivity in the
related deuteration of radical (7).[57b] Nonetheless, the 7-norbornenyl radical

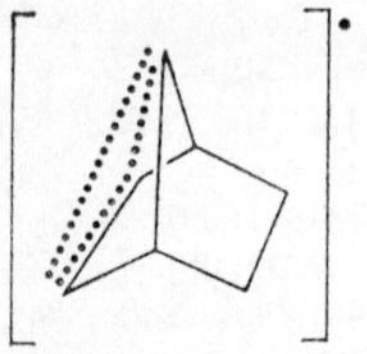

(6) 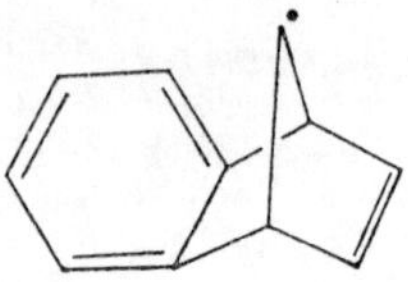   (7)

[54] A. Hudson and R. A. Jackson, *Chem. Comm.*, **1969**, 1323.
[55] J. E. Taylor, D. A. Hutchings, and K. J. Frech, *J. Am. Chem. Soc.*, **91**, 2215 (1969).
[56] *Org. Reaction Mech.*, **1968**, 281.
[57a] G. A. Russell and G. W. Holland, *J. Am. Chem. Soc.*, **91**, 3968 (1969).
[57b] S. J. Cristol and A. L. Noreen, *J. Am. Chem. Soc.*, **91**, 3969 (1969).

must be an attractive candidate for genesis by the Hudson–Jackson procedure mentioned above, since SCF molecular orbital calculations have led to the conclusion that, despite the presence of the unpaired electron in an antibonding orbital, the bridged, non-classical structure should be slightly more stable than the classical structure.[58a,b]

In the general case of an unstrained alkyl radical, Symons has argued the case for planarity at the radical centre which can be made from existing ESR data.[59] Although the evidence is strong, there must remain an element of doubt. Several experimental approaches to this problem depend on the search for strain in bridgehead radicals. Here the evidence again points to a preference for planarity, though this preference may be much less pronounced than is the case with carbonium ions.[60] A new result which provides evidence for strain in the 1-adamantyl radical is the observed reluctance of the radical (**8**; R = 1-adamantyl) to fragment.[61] The results of determination of the product ratio (**9**)/PhCHO from reactions of t-butoxy radicals with a series of alkyl benzyl ethers under standardized conditions are presented in Table 1.[61, 62]

$$\text{BuO} \cdot + \text{ROCH}_2\text{Ph} \rightarrow \text{BuOH} + \text{RO}\dot{\text{C}}\text{HPh}$$
$$(8)$$

$$\text{RO}\dot{\text{C}}\text{HPh} \quad \rightarrow \quad \text{R} \cdot + \text{PhCHO}$$

$$2\text{RO}\dot{\text{C}}\text{HPh} \quad \rightarrow \quad \begin{array}{c} \text{ROCHPh} \\ | \\ \text{ROCHPh} \end{array}$$
$$(9)$$

Table 1. Competition between dimerization and fragmentation for a series of α-alkoxybenzyl radicals, RO$\dot{\text{C}}$HPh

R	Me	Et	Pr^i	Bu^t	PhCH$_2$	1-Adamantyl
Ratio of products, (**9**)/PhCHO	4.4	1.2	0.44	0.33	0	2.4

The ease of fragmentation of acyl radicals to alkyl radicals and carbon monoxide has previously been used to provide information on alkyl radical stability;[63] application of this approach to determining the stability of

[58a] H. O. Ohorodnyk and D. P. Santry, *Chem. Comm.*, **1969**, 510; *J. Am. Chem. Soc.*, **91**, 4711 (1969).

[58b] For direct observation of the radical, see P. Bazukis, J. K. Kochi, and P. J. Krusic, *J. Am. Chem. Soc.*, **92**, 1434 (1970).

[59] M. C. R. Symons, *Nature*, **222**, 1123 (1969).

[60] *Org. Reaction Mech.*, **1968**, 264, 265.

[61] W. H. Chick and S. H. Ong, *Chem. Comm.*, **1969**, 216.

[62] R. L. Huang, Tong Wai-Lee, and S. H. Ong, *J. Chem. Soc.*(C), **1969**, 40.

[63] *Org. Reaction Mech.*, **1965**, 212.

bridgehead triptycenyl radicals leads to the conclusion that these are similar to norbornyl.[64] It was concluded that destabilization was principally by angle strain rather than by inductive electron-withdrawal by the (non-conjugating) benzene rings.[64]

Another fragmentation reaction, that of the radical anions of aliphatic nitro-compounds, does not appear to depend appreciably upon the stability of the incipient alkyl radical, since similar rates were found for the decomposition of radical anions of bridgehead nitroadamantane, nitronorbornane, and nitro-t-butane.[65]

One further observation pertinent to this general discussion (see ref. 60) is that of Rüchardt and coworkers who, having studied the decomposition of a series of peresters of general formula RCO_2OBu^t as sources of alkyl radicals (R·) in mixtures of CCl_4 and $BrCCl_3$, concluded that there was no relationship between the ease of perester decomposition and the selectivity of R· as measured by the relative yields of RCl and RBr.[66a] This probably means that the selectivity in the halogen abstraction reactions is a poor guide to radical stability, because the ease of perester homolysis almost certainly does reflect this stability, albeit to a much smaller degree than the ease of symmetrical azo-compound decomposition.[60] New evidence on this last point has come from an examination of the decomposition rates of a series of peresters (10) as a function of ring-size.[66b] The rate of decomposition responds relatively weakly

(10)   (11)

to changes in ring size when compared with the effects of ring size on the rates of decomposition of the related azo-compounds (11), reported many years ago by Overberger et al. In both series of reactions, however, the pattern of rate dependence shows a close parallel with the ease of carbonium ion formation as a function of ring size in a related series of solvolysis reactions. Once more this points to planar geometry for the alkyl radicals.

Conformational preferences of aryl-substituted methyl radicals have been examined theoretically,[67] and those of cyclopropyl-substituted radicals by experiment.[68] The cyclopropylcarbinyl radicals exhibit a strong preference for

[64] S. F. Nelsen and E. F. Travecedo, *J. Org. Chem.*, **34**, 3651 (1969).
[65] R. H. Gibson and J. C. Crosthwaite, *J. Am. Chem. Soc.*, **90**, 7373 (1968).
[66a] C. Rüchardt, K. Herwig, and S. Eichler, *Tetrahedron Letters*, **1969**, 421.
[66b] P. Lorenz, C. Rüchardt, and E. Schacht, *Tetrahedron Letters*, **1969**, 2787.
[67] R. Hoffman, R. Bissell, and D. G. Farnum, *J. Phys. Chem.*, **73**, 1789 (1969).
[68] N. L. Bauld, J. D. McDermed, C. E. Hudson, Y. S. Rim, J. Zoeller, R. D. Gordon, and J. S. Hyde, *J. Am. Chem. Soc.*, **91**, 6666 (1969); see also *Org. Reaction Mech.*, **1967**, 229, 230.

the bisected conformation (**12**) in cases where substituents effect electron withdrawal from the trigonal carbon, the tendency under these circumstances being towards the situation of a cyclopropylcarbinyl cation.

Secondary cyclopropyl radicals, formed from optically active bromo-

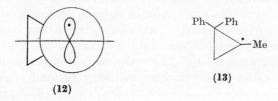

(**12**)

(**13**)

cyclopropanes by reaction with lithium, are further reduced to the configurationally stable anions with partial retention of configuration. It was concluded that this residual asymmetry probably stemmed from the participation of pyramidal (but rapidly inverting) cyclopropyl radicals.[69] Rather better configurational stability was found in some new results with a tertiary cyclopropyl radical (**13**) formed from a bromocyclopropane by dissociative electron capture from naphthalene radical anion (29% net retention). Inversion of the cyclopropyl radical was considered to compete with diffusion-controlled transfer of a second electron to give the cyclopropyl anion.[70] However, scavenging of the same cyclopropyl radical (**13**) with triphenyltin hydride led to hydrocarbon with net inversion of configuration (albeit very small). Possibly in this case there is some interference by the departing triphenyltin bromide molecule, and preferential attack by hydride is from the rear.[71]

From gas-phase studies of decarbonylation of cyclopropanecarboxaldehyde an estimate of 18 kcal mole^{-1} has been made for the activation energy for isomerization of cyclopropyl radicals to allyl radicals,[72] and a new example of the relative difficulty of forming cyclopropyl radicals by fragmentation has been recorded.[73]

In last year's report, conflicting evidence concerning the geometry of 1-phenylvinyl radicals was noted.[74a] Employing relatively low temperatures (60°), a Japanese group reported considerable retention of configuration in reactions of *cis*- or of *trans*-1,2-diphenylvinyl radicals, in contrast to earlier results (at 110°) in which the composition of the stilbene mixture obtained in the reaction appeared not to depend on the geometry of the precursor. (Relevant figures are reproduced in the Errata in this volume. There was an unfortunate

[69] M. J. S. Dewar and J. M. Harris, *J. Am. Chem. Soc.*, **91**, 3652 (1969).
[70] J. Jacobus and D. Pensak, *Chem. Comm.*, **1969**, 400.
[71] L. J. Altman and B. W. Nelson, *J. Am. Chem. Soc.*, **91**, 5163 (1969).
[72] J. A. Kerr, A. Smith, and A. F. Trotman-Dickenson, *J. Chem. Soc.*(A), **1969**, 1400.
[73] M. L. Mihailovic and Z. Cekovic, *Helv. Chim. Acta*, **52**, 1146 (1969).
[74a] See *Org. Reaction Mech.*, **1968**, 265.

misalignment in the original tabulation.) While the Japanese workers considered that they were dealing with isomeric sp^2-hybridized vinyl radicals, the high-temperature results, taken in conjunction with other new information, had led to the suggestion of a linear structure (sp-hybridization) at C-1 of a 1-phenylvinyl radical, and compelling support for this alternative view has now appeared with the publication of data on hydrogen abstraction by 2-methyl-1-phenylvinyl radicals.[74b] The composition of the mixture of isomeric β-methylstyrenes is independent of the geometry of the precursor (t-butyl α-phenylpercrotonate) at 105°, with the *cis*-isomer predominating. More significantly, as the temperature is lowered, the proportion of *cis*-olefin in the product increases when the *cis*-percrotonate (Me and Ph *trans*) is employed as radical precursor. This result can only reasonably be accommodated by assuming an increasing stereoselectivity of hydrogen abstraction by a *linear* radical as the reaction temperature is lowered. The preferred direction of hydrogen abstraction is governed by the bulk of the methyl group on C-2, which results in a preponderance of *cis*-product. These results and the Japanese data appear to be in sharp conflict.

Photolysis of the isomeric 4-iodohept-3-enes in chloroform–pentane gives the isomeric heptenes in proportions which reflect a slight preference for *inversion* of configuration at the vinyl radical centre. The results were discussed in terms of initial production of vibrationally excited radicals.[75] The major reaction product was heptyne, considered to arise by hydrogen transfer from the vinyl radicals to atomic iodine within the solvent cage.

Decomposition of Azo-compounds and Peroxides

In contrast to an earlier result,[76a] it has now been shown by ^{18}O-labelling experiments that cage recombination of benzoyloxy radicals can occur during decomposition of benzoyl peroxide.[76b] A surprising conclusion is that this occurs to a lesser extent than had been found with acetyl peroxide, for which decarboxylation competes with recombination and diffusion from the cage.

Neuman and his colleagues have detailed their results on the effect of pressure on the rates of perester decompositions,[77] and have also attempted to discern the effects of pressure on cage combination and diffusion of butoxy radicals generated from t-butyl hyponitrite.[78]

Koenig's group have extended their work on cage recombination with a study of the rearrangement of *N*-acetyl-*N*-nitroso-*O*-t-butylhydroxylamine

[74b] L. A. Singer and J. Chen, *Tetrahedron Letters*, **1969**, 4849.
[75] R. C. Neuman and G. D. Holmes, *J. Org. Chem.*, **33**, 4317 (1968).
[76a] M. Kobayashi, H. Minato, and Y. Oge, *Bull. Chem. Soc. Japan*, **41**, 2822 (1968).
[76b] J. C. Martin and J. H. Hargis, *J. Am. Chem. Soc.*, **91**, 5399 (1969).
[77] R. C. Neuman and J. V. Behar, *J. Am. Chem. Soc.*, **91**, 6024 (1969).
[78] R. C. Neuman and R. J. Bussey, *Tetrahedron Letters*, **1968**, 5859.

(14) and decomposition of the resulting diazo-ester **(15)**.[79] There was considerably less cage recombination to form t-butyl peracetate than is found in the decomposition of the peracetate itself; presumably this is a consequence

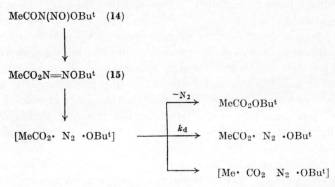

$$MeCON(NO)OBu^t \quad (14)$$

$$MeCO_2N{=\!=}NOBu^t \quad (15)$$

of the initial presence of nitrogen between the caged radical pair. The rate of diffusive separation, k_d, showed inverse dependence on the square-root of the viscosity of the reaction medium, in accord with a new description of the diffusion apart of radical pairs.[80] Experiments with [carbonyl-^{18}O]-labelled **(14)** gave perester with virtually complete scrambling of the label over two oxygens, but the interpretation of this result was not clear. It could have occurred in the caged radical pair, or by an ion-pair mechanism in the diazo-ester **(15)**.

$$MeCO_2N{=\!=}NOBu^t \rightleftarrows MeCO_2^- \overset{+}{N}{\equiv}NOBu^t$$
$$(15)$$

The same group have examined the secondary kinetic isotope effects found during peroxide and perester decompositions. The effect of replacing β-hydrogen atoms by deuterium in peresters which decompose by concerted multiple-bond fission is in good accord with that which would be predicted from a hyperconjugative model;[81] it was also found that earlier estimates of α-deuterium isotope effects were too large. For example, k_H/k_D for decomposition of $PhCD_2CO_2OBu^t$ is ca. 1.06 per deuterium atom. The higher figure quoted previously[82] might have stemmed from an undetected induced decomposition. It was also possible to evaluate an α-deuterium isotope effect on the decarboxylation of acetoxy radicals by means of a detailed examination of the isotopic composition of methyl acetate from the decomposition of acetyl trideutero-acetyl peroxide.[83] The magnitude of this effect is ca. 1.1 for three deuterium

[79] T. Koenig and M. Deinzer, *J. Am. Chem. Soc.*, **90**, 7014 (1968).
[80] T. Koenig, *J. Am. Chem. Soc.*, **91**, 2558 (1969).
[81] T. Koenig and R. Wolf, *J. Am. Chem. Soc.*, **91**, 2568, 2574 (1969).
[82] *Org. Reaction Mech.*, **1965**, 197.
[83] T. Koenig and R. Cruthoff, *J. Am. Chem. Soc.*, **91**, 2562 (1969).

atoms. A consequence of this isotope effect is that cage decarboxylation competes slightly less effectively in the deuterated compound with cage recombination and diffusive separation. Hence there is a tiny, but measurable (by a double-labelling technique), isotope effect on the total rate of peroxide decomposition.

That peroxide decomposition may be induced by one-electron transfer does not appear to be in doubt, but whether this process is a key step in a wide range of peroxide reactions induced by radicals or electron-rich molecules is not clear. It is difficult to differentiate this from the alternative of direct attack by a radical or a molecule on the peroxide either in concert with, or succeeded by, homolysis of the peroxide. For example, decomposition of mixed aliphatic aromatic diacyl peroxides by tributyltin hydride gives both aliphatic and aromatic tributyltin acylates; the presence of electron-withdrawing substituents in the aromatic acyl group leads to an increased proportion of the aromatic stannyl acylate, and has been interpreted in terms of direct attack by nucleophilic stannyl radicals on peroxide oxygen [reactions (1)].[84] It would be a far from simple matter to establish whether or not the electron-transfer sequence (2) intervened.

$$R_3Sn\cdot + R'CO_2OCOAr \rightarrow \begin{array}{c} R_3SnOCOR' + \cdot OCOAr \\ or \\ R_3SnOCOAr + \cdot OCOR' \end{array} \qquad \dots (1)$$

$$R_3Sn\cdot + R'CO_2OCOAr \rightleftharpoons R_3Sn^+ + [R'CO_2OCOAr]^{\overline{\cdot}} \rightarrow$$

$$\begin{array}{ccc} R_3Sn^+ + R'CO_2\cdot + ArCO_2^- & & R_3SnOCOAr + R'CO_2\cdot \\ or & \rightarrow & or & \dots (2) \\ ArCO_2\cdot + R'CO_2^- & & R_3SnOCOR' + ArCO_2\cdot \end{array}$$

Similarly, in the view of the present author, it is unwise to conclude that induced decomposition of benzoyl peroxide by benzophenone ketyl radicals $(Ph_2\overset{\cdot}{C}OH)$ involves a "homolytic hydrogen transfer" because $Ph_2\overset{\cdot}{C}OMe$ radicals failed to react.[85] Formation of products from a reversibly formed ion-pair (16) may be energetically much less favourable than that from (17) in which hydrogen-bonding might facilitate fragmentation of a transient peroxide

$$[Ph_2\overset{+}{C}OMe\ (PhCO_2)_2^{\overline{\cdot}}] \qquad [Ph_2\overset{+}{C}OH\ (PhCO_2)_2^{\overline{\cdot}}]$$
$$\textbf{(16)} \qquad\qquad\qquad \textbf{(17)}$$

radical anion. Again, the isotope effect found in the induced decomposition of di-t-butyl peroxide by 2-deuteroxy-2-propyl radicals[86a] might derive from a proton-transfer step following reversible electron transfer, rather than from rate-determining hydrogen-atom transfer.

---

84 K. Rubsamen, W. P. Neumann, R. Sommer, and U. Frommer, *Chem. Ber.*, **102**, 1290 (1969).
85 W. F. Smith and B. W. Rossiter, *Tetrahedron*, **25**, 2059 (1969); see also W. F. Smith, *ibid.*, p. 2071.
86a K. Ishiza, H. H. Dearman, M. T. Huang, and J. R. White, *Chem. Comm.*, **1969**, 1238.

The oxidation of secondary alcohols by di-t-butyl peroxide is catalysed by the presence of 4-pyridone, probably by the hydrogen transfer mechanism indicated[86b] (distinction from an electron-transfer; proton-transfer mechanism is not implied). Curiously, 2-pyridone retarded the reaction.

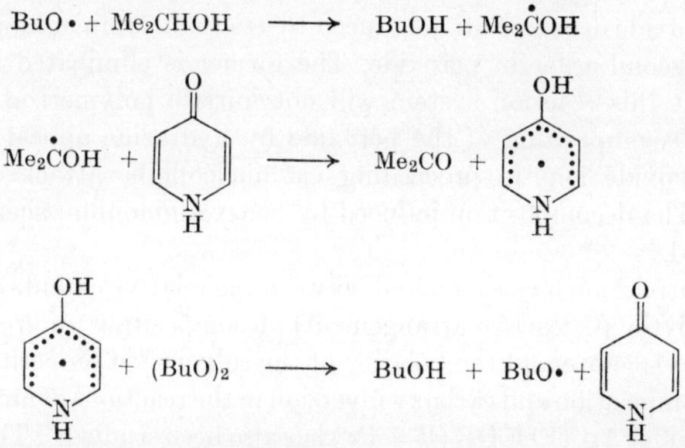

$$BuO\cdot + Me_2CHOH \longrightarrow BuOH + Me_2\overset{\bullet}{C}OH$$

An electron-transfer mechanism has been inferred from an ESR study of reactions of aromatic amines with benzoyl peroxide.[87] When the decomposition of this peroxide is induced by dimethylaniline in the presence of N-phenyl-maleimide, a tetrahydroquinoline derivative (20) is obtained;[88] it was possible to demonstrate that the cation (19) would not add to N-phenyl-maleimide and hence addition of the radical (18) was assumed. Reactions of both the radical (18) and the cation (19) were encountered in γ-radiolysis studies with N,N-dimethylaniline.[89a]

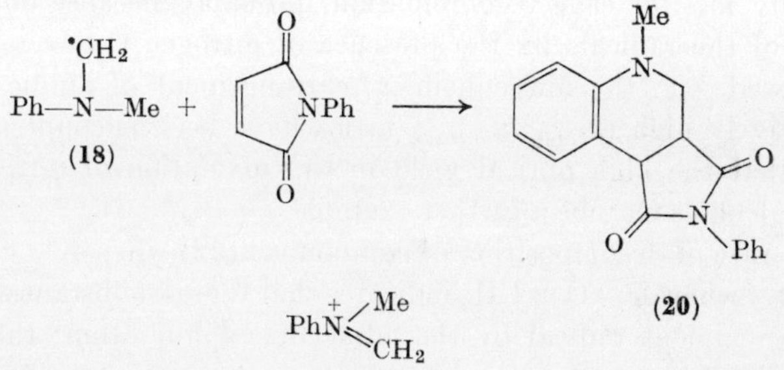

[86b] E. S. Huyser and R. H. C. Feng, *J. Org. Chem.*, **34**, 1727 (1969).

[87] D. G. Pobedimskii, A. L. Buchachenko, and A. L. Neiman, *Zh. Fiz. Khim.*, **42**, 1436 (1968); *Chem. Abs.*, **70**, 67277 (1969); see also B. M. Beileryan and G. K. Dermendzhyan, *Uch. Zap. Erevan. Gos. Univ.*, **1967**, No. 2, 139; *Chem. Abs.*, **70**, 28239 (1969).

[88] R. B. Roy and G. A. Swan, *J. Chem. Soc.*(C), **1969**, 1887; G. A. Swan, *Chem. Comm.*, **1969**, 20; *J. Chem. Soc.*(C), **1969**, 2015.

[89a] J. M. Fayadh and G. A. Swan, *J. Chem. Soc.*(C), **1969**, 1775.

Reactions between benzoyl peroxide and both chloride ions and hydrogen chloride have been studied, in the course of which radical and electrophilic chlorination by benzoyl hypochlorite was encountered.[89b] The decomposition of benzoyl peroxide by acetate ion in aprotic solvents is apparently non-radical, and proceeds via deprotonation of a methyl group.[90] The kinetics of benzoyl peroxide decomposition induced by triethanolamine show terms both first- and second-order in peroxide. The former is eliminated if oxygen is present, but this reaction system will not initiate polymerization of vinyl acetate.[91] Decomposition of the peroxide by hydrazine appears to parallel that by alkoxide ions in proceeding via nucleophilic attack on carbonyl carbons.[75] The decomposition induced by benzylammonium acetate has also been studied.[92]

A good correlation has been found between the relative extents of homolytic and heterolytic (Criegee rearrangement) decomposition of *trans*-9-decalyl peroxyphenylacetate and the polarity of the solvent.[93] Competition between radical decomposition and carboxy inversion in the reactions of unsymmetrical diacyl peroxides $ArCO_2OCOR$ ($R = Pr^i$) has also been studied.[94] The homolytic component of these reactions was a minimum when Ar contained strongly electron-withdrawing substituents. This might be borne in mind in utilization of the carboxy inversion reaction to degrade a carboxylic acid to an alcohol by the sequence:

$$RCO_2H \rightarrow RCO_2OCOAr \rightarrow ROCO_2COAr \rightarrow ROH$$

The cage recombination of radicals formed by thermolysis of optically active (21) is accompanied by 10—15% net retention of configuration, leading to a ratio $(k_{comb}/k_{rac})_{cage}$ of between 0.06 and 0.09. This is considered to be exceptionally low for cage recombination, probably because of the initial separation of the radicals by the presence of nitrogen.[95] Several reactions were discussed, e.g. the Meisenheimer rearrangement of amine oxides, in which relatively high $(k_{comb}/k_{rac})_{cage}$ ratios have been encountered. It was suggested that the high optical yield in Cr^{VI}-oxidation of tertiary hydrocarbons to alcohols may be a further example.

The high rate of decomposition of compounds (22) when $X = S$ compared with the rates when $X = O$ or CH_2 indicates that there is substantial stabilization of the incipient radical by the adjacent sulphur atom; there is little

[89b] N. J. Bunce and D. D. Tanner, *J. Am. Chem. Soc.*, **91**, 6096 (1969).

[90] T. Koenig and R. Wielesek, *J. Am. Chem. Soc.*, **91**, 2551 (1969).

[91] N. M. Beileryan, B. M. Sogomonyan, and O. A. Chaltykyan, *Armyan. Khim. Zh.*, **21**, 551 (1969); *Chem. Abs.*, **70**, 105705 (1969).

[92] N. M. Beileryan, F. D. Karapetyan, and O. A. Chaltykyan, *Uch. Zap. Erevan. Gos. Univ.*, **1968**, No. 2, 44; *Chem. Abs.*, **71**, 38103 (1969).

[93] C. Ruchardt and H. J. Quadbeck-Seeger, *Chem. Ber.*, **102**, 3525 (1969).

[94] R. C. Lamb and J. R. Sanderson, *J. Am. Chem. Soc.*, **91**, 5034 (1969).

[95] K. R. Kopecky and T. Gillan, *Can. J. Chem.*, **47**, 2371 (1969).

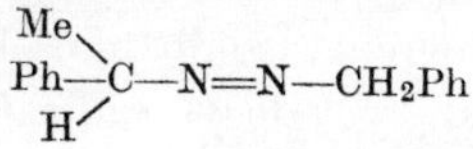

Me
Ph—C—N=N—CH₂Ph
H

$$\text{(21)}$$

Me   Me
PhX—C—N=N—C—XPh
Me   Me

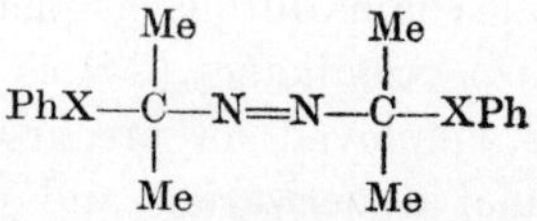

$$\text{(22)}$$

difference between the rates for decomposition of the two compounds with
$X = O$ and $X = CH_2$.[96]

The presence of traces of 2-pyrazoline among the products of photo-decomposition of trifluoromethylazocyclopropane may result from a reaction of the radical $C_3H_5N=N \cdot$;[97] the possible participation of $ArN=N \cdot$ in reactions of aryldiazenes,[98a] and in the polarographic reduction of arenediazonium salts in sulpholan,[98b] has also been discussed.

Hammond and his colleagues have pursued their investigation of radical recombination by photolysing symmetrical azo-compounds in the cavity of an ESR spectrometer, utilizing a rotating-sector technique. Rate constants for bimolecular decay of a variety of tertiary alkyl radicals were estimated to be in the range 0.2—8×10^9 l mole^{-1} sec^{-1} in benzene at room temperature. There was no discernible relationship between rate of decay and radical stability.[99] In view of the formal similarity between di-t-alkyl nitroxides and the hypothetical addition product of a tertiary alkyl radical and an azo-t-alkane, it might be expected that one mode of radical decay in the above reaction system would be addition to undecomposed azo-compounds. Curiously, a stable radical containing two nitrogen atoms *was* observed in the case of di-p-t-butylazocumene but not in any other instance. This type of addition to azo-compound was assumed to be unimportant in a gas-phase study of the photodecomposition of azoisobutane.[100]

The role of secondary reactions during gas-phase photolysis of aliphatic azo-compounds has been discussed,[101] and appreciable stabilization in the radical (23) has been inferred from the observation that it does not propagate

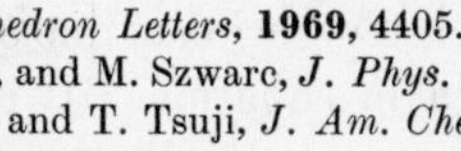

Me   Me
 ⋅C—N=N—C—H
Me   Me

$$\text{(23)}$$

[96] A. Ohno and T. Ohnishi, *Tetrahedron Letters*, **1969**, 4405.
[97] K. Chakravorty, J. M. Pearson, and M. Szwarc, *J. Phys. Chem.*, **73**, 746 (1969).
[98a] E. M. Kosower, P. C. Huang, and T. Tsuji, *J. Am. Chem. Soc.*, **91**, 2324 (1969); cf. *Org. Reaction Mech.*, **1969**, 269, 270.
[98b] R. M. Elafson and F. F. Gadallah, *J. Org. Chem.*, **34**, 855 (1969).
[99] S. A. Weiner and G. S. Hammond, *J. Am. Chem. Soc.*, **91**, 986 (1969).
[100] D. G. L. James and R. D. Suart, *Trans. Faraday Soc.*, **65**, 175 (1969).
[101] O. P. Strausz, R. E. Berkley, and H. E. Gunning, *Can. J. Chem.*, **47**, 3470 (1969).

radical chains during gas-phase decomposition of azoisopropane.[102] Symmetrical azocycloalkanes (24) as well as biscycloalkyl mercury compounds have been employed as precursors for cyclooctyl and cyclodecyl radicals.[103] Normal dimerization and disproportionation were encountered with the

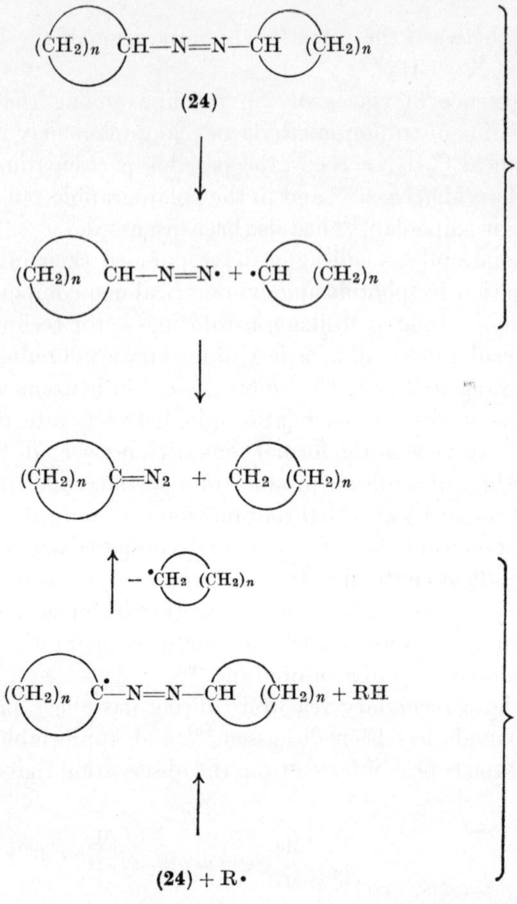

$(CH_2)_n$ CH—N=N—CH $(CH_2)_n$

(24)

$\Biggr\} \cdots (3)$

$(CH_2)_n$ CH—N=N• + •CH $(CH_2)_n$

$(CH_2)_n$ C=N_2 + CH_2 $(CH_2)_n$

$- \text{•}CH_2$ $(CH_2)_n$

$(CH_2)_n$ C•—N=N—CH $(CH_2)_n$ + RH

$\Biggr\} \cdots (4)$

(24) + R•

organometallic precursors, but bicyclic hydrocarbons were by-products of the azo-decompositions, and intervention of diazocycloalkanes was suggested, possibly formed as indicated in equation (3); this could involve geminate disproportionation, following one-bond cleavage of the azo-compound. An attractive alternative might be the route suggested in equation (4), were it

[102] G. Gaseler, *Z. Phys. Chem.* (Frankfurt), **57**, 318 (1968).
[103] A. C. Cope and J. E. Engelhart, *J. Am. Chem. Soc.*, **90**, 7092 (1968).

not for the apparent failure of the radical (**23**) to undergo a similar fragmentation. The radical (**23**) is related to the hydrazonyl radicals which have been generated by oxidation of arylhydrazones.[104]

Radical decompositions of tetrazenes which have received attention[105,106] include reactions in which homolysis to a radical and a radical cation is promoted by prior protonation, alkylation, or acylation.[106]

Diradicals

Some interesting observations have been made on the decomposition of a series of peresters (**25**) and related compounds.[107] The major products are as

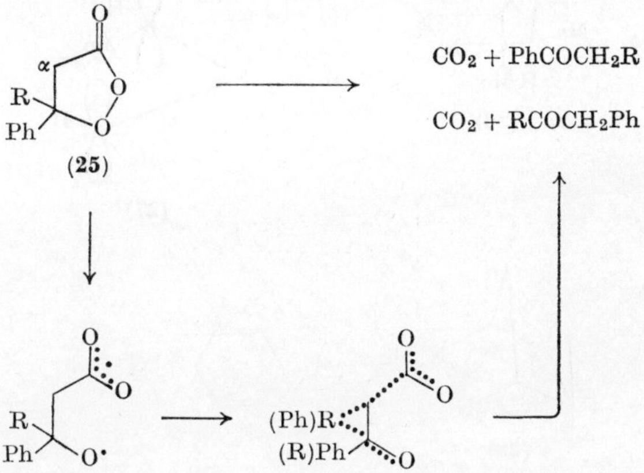

shown, and although the principal reaction path involves alkyl rather than phenyl migration, the rate of decomposition is almost independent of the nature of R ($=$ Me, Pr^i, $PhCH_2$). The proportion of alkyl migration relative to phenyl migration parallels the ease of fragmentation of tertiary alkoxy radicals $RC(Me_2)O\cdot$; if an optically active R-group is attached to the perester by an asymmetric carbon atom there is negligible racemization during the migration step, and if the α-carbon atom is monosubstituted to give an optically active analogue of (**25**), then inversion of configuration at this centre accompanies the rearrangement. This information was interpreted in terms of the mechanistic hypothesis represented as equation (5) in which rearrangement is concerted with decarboxylation in an initially formed diradical.

104 C. Winter and J. Wiecko, *Tetrahedron Letters*, **1969**, 1595; see also *Org. Reaction Mech.*, **1967**, 251.
105 S. F. Nelsen and D. H. Heath, *J. Am. Chem. Soc.*, **91**, 6452 (1969).
106 C. J. Michejda and D. Romans, *Tetrahedron Letters*, **1969**, 4213; S. F. Nelsen, R. B. Metzler, and M. Iwamura, *J. Am. Chem. Soc.*, **91**, 5103 (1969).
107 W. Adam and Y. M. Cheng, *J. Am. Chem. Soc.*, **91**, 2109 (1969); W. Adam, Y. M. Cheng, C. Wilkerson, and W. A. Zaidi, *ibid.*, p. 2111.

Allred and Smith have given[108] a full discussion of their work on the decomposition of 5-methoxy-2,3-diazanorbornene,[109] and Mishra and Crawford have examined the stereochemistry of the products from the acyclic diradicals derived from pyrolysis of the optically active pyrazoline (**26**).[110] The dimethyl cyclopropane produced contained 25% of the *trans*-isomer, and this was 23% optically pure with the geometry expected for the inertial inversion mechanism.[109] The remaining (ca. 94%) of the dimethylcyclopropane was considered to be derived from the planar diradical (**27**). The cyclic azo-compound (**28**)

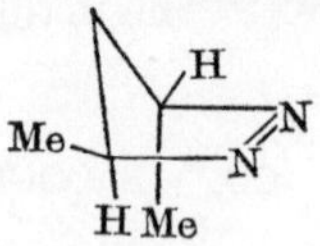

$$
\begin{array}{ccc}
\text{(26)} & & \text{(27)} \\
& & \\
& \text{(28)} & \longrightarrow \quad + \text{N}_2
\end{array}
$$

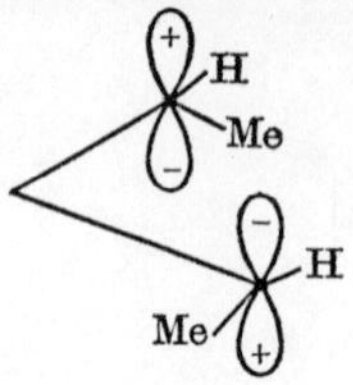

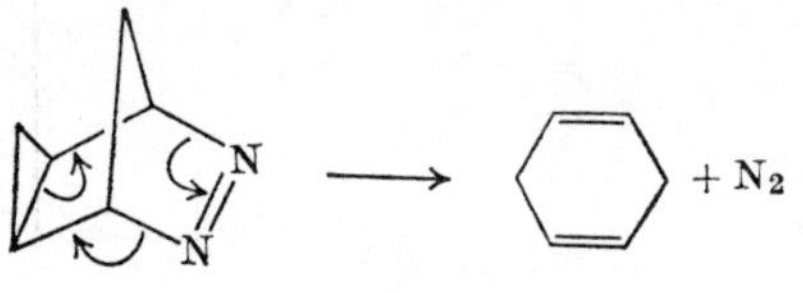

decomposes (to cyclohexa-1,4-diene and nitrogen) some 10^{11} times faster than does 2,3-diazanorbornene, presumably by the pericyclic fragmentation indicated.[111]

Characteristic triplet ESR spectra have been observed for the diradical (**29**),[112] and for the adduct diradicals formed when certain 3,3-disubstituted 3*H*-indazoles, e.g. (**30**), are photolysed in the presence of olefins at liquid-nitrogen temperature.[113] Reactions of diradicals from photolysis of *peri*-naphthotriazines, e.g. (**31**), have been described; addition to vinyl bromide afforded a synthetic route to the new non-benzenoid heteroaromatic compound (**32**).[114]

108 E. L. Allred and R. L. Smith, *J. Am. Chem. Soc.*, **91**, 6766 (1969).
109 See *Org. Reaction Mech.*, **1968**, 267, 268.
110 A. Mishra and R. J. Crawford, *Can. J. Chem.*, **47**, 1515 (1969).
111 E. L. Allred, J. C. Hinshaw, and A. L. Johnson, *J. Am. Chem. Soc.*, **91**, 3382 (1969).
112 D. R. Arnold, A. B. Evnin, and P. H. Kasai, *J. Am. Chem. Soc.*, **91**, 784 (1969).
113 G. L. Closs and L. R. Kaplan, *J. Am. Chem. Soc.*, **91**, 2168 (1969).
114 P. Flowerday and M. J. Perkins, *J. Am. Chem. Soc.*, **91**, 1035 (1969).

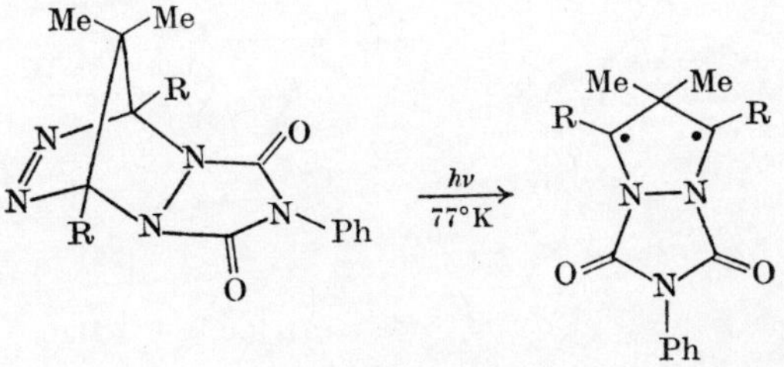

(29)

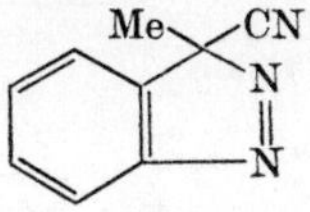

(30)

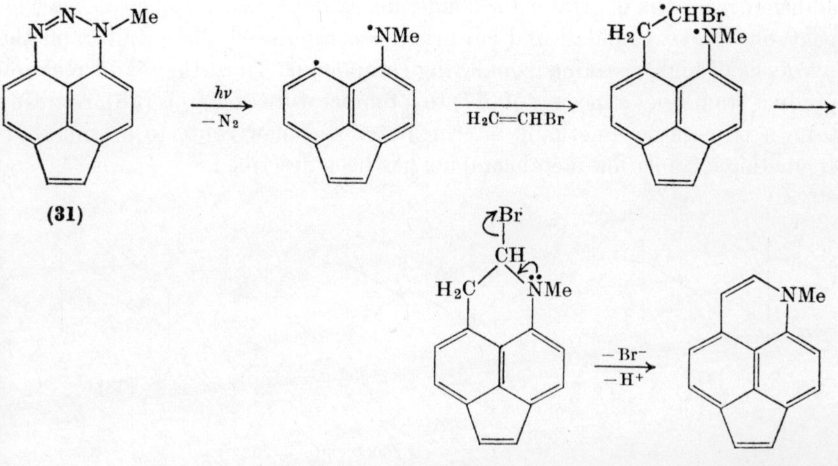

(31)

(32)

Theoretical predictions have been made concerning species which, whilst superficially appearing to be diradicals, should in fact have singlet ground states. Two examples are **(33)** and **(34)**.[115] The facile equilibration of **(35)** and **(36)** presumably involves a diradical mechanism,[116] and details of work on the diradical mechanism for [2.2]paracyclophane isomerization,[117] as well as the description of an elegant stereochemical study of the degenerate vinyl-cyclopropane rearrangement[118] (see p. 274), have appeared.

[115] R. Gleiter and R. Hoffmann, *Angew. Chem. Internat. Ed. Engl.*, **8**, 214 (1969).
[116] R. B. Woodward and D. L. Dalrymple, *J. Am. Chem. Soc.*, **91**, 4612 (1969).
[117] H. J. Reich and D. J. Cram, *J. Am. Chem. Soc.*, **91**, 3517 (1969); see *Org. Reaction Mech.*, **1967**, 256, 257.
[118] M. R. Wilcott and V. H. Cargle, *J. Am. Chem. Soc.*, **91**, 4310 (1969).

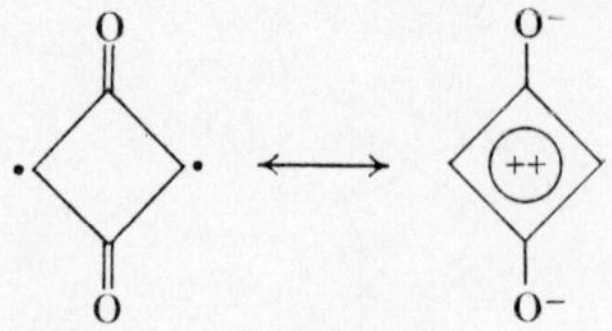

(33) 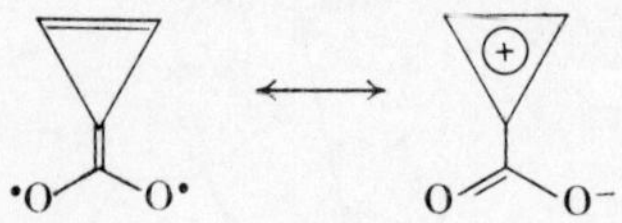   (34)

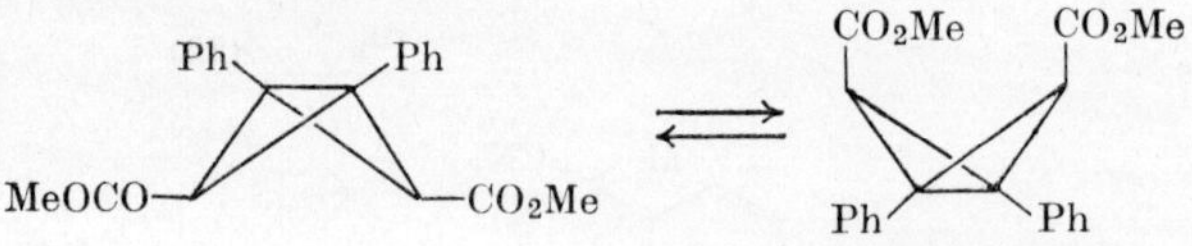

(35)   (36)

New data have been reported on the diradical mechanisms involved in addition reactions of strained σ-bonds, for example the 1,4-bond in bicyclopentane,[119] and Crandall and coworkers have pursued their studies on the pyrolysis of highly strained small-ring compounds. Thus, the major reaction product from brief exposure of (37) to a temperature of 400° is (38), resulting from a trimethylenemethane intermediate.[120] A new route to a trimethylenemethane transition-metal complex has been described.[121]

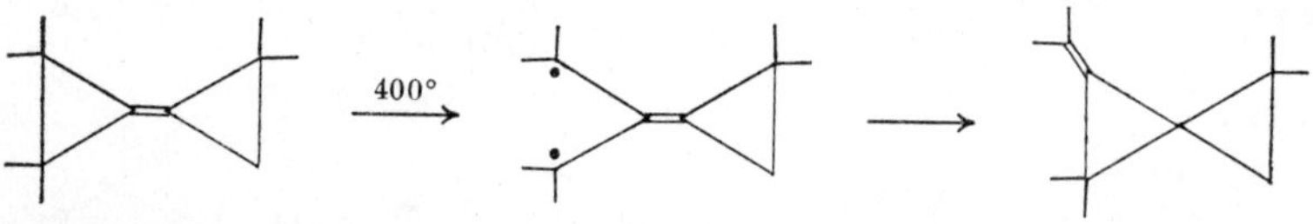

(37)   (38)

(39)

The singlet diradical (39), suggested as an intermediate in certain photo-isomerizations and photoaddition reactions of benzene, appears also to be accessible by reaction of carbon atoms with cyclopentadiene. Benzene and fulvene were encountered among the products of this reaction.[122]

[119] P. G. Gassman and G. B. Richmond, *Chem. Comm.*, **1968**, 1630; P. G. Gassman, K. T. Mansfield, and T. J. Murphy, *J. Am. Chem. Soc.*, **91**, 1684 (1969); see *Org. Reaction Mech.*, **1968**, 275.
[120] J. K. Crandall, D. R. Paulson, and C. A. Bunnell, *Tetrahedron Letters*, **1969**, 4217.
[121] R. Noyori, T. Nishimura, and H. Takaya, *Chem. Comm.*, **1969**, 89.
[122] T. L. Rose, *Chem. Comm.*, **1969**, 95.

Polar effects have been discerned in the behaviour of the 1,4-diradicals formed in Norrish type II photoreactions of ketones (**40**) (see also Chapter 13).[123] The fraction of diradical which reverts to ketone is tabulated, and reveals a marked dependence on the polar nature of substituents in Ar. The results are consistent with a significant dipolar contribution, e.g. (**41**), to the transition state. In view of this it is not surprising that similar hydrogen transfer was not detected in the reactions of 1,4-diradicals not containing a heteroatom.[124]

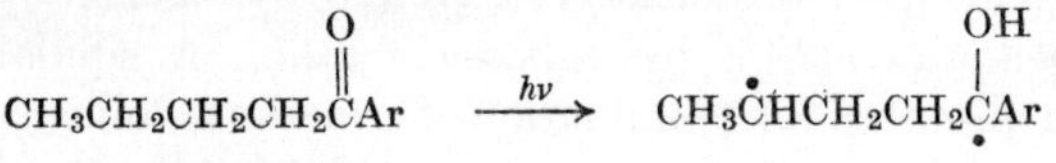

	Ar	Fraction of diradical that reverts to (**40**)
	p-MeOC$_6$H$_4$	31%
	p-MeC$_6$H$_4$	50%
	C$_6$H$_5$	60%
	p-ClC$_6$H$_4$	63%
	3-Pyridyl	71%

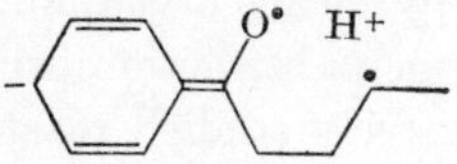

$$CH_3CH_2CH_2CH_2\overset{O}{\overset{\|}{C}}Ar \xrightarrow{h\nu} CH_3\overset{\bullet}{C}HCH_2CH_2\overset{OH}{\overset{|}{C}}Ar$$

$$(40) \qquad\qquad\qquad\qquad\qquad$$

(**41**)

Atom-transfer Processes

Benson and coworkers have discussed the application of kinetic studies to the determination of thermodynamic functions of free radicals,[125] and have described the direct measurement of the rate of reaction of methyl radicals with DBr by a new technique.[126] This allows measurement of rates of radical–molecule reactions without reliance on rate data for termination reactions. Benson's group have also reported thermodynamic data obtained from the gas-phase reaction between iodine and tetrahydrofuran.[127] The α-C–H bond dissociation energy was calculated to be 91.5 kcal mole^{-1}. This is only ca. 4 kcal smaller than the value for C–H in cyclopentane, supporting the earlier conclusion (pp. 310—311) that oxygen atoms do not substantially affect the stability of an adjacent carbon radical. (It should be noted that the ease of hydrogen abstraction α to ether oxygen by electrophilic radicals is a transition-state effect.)

[123] P. J. Wagner and H. N. Schott, *J. Am. Chem. Soc.*, **91**, 5383 (1969).
[124] See *Org. Reaction Mech.*, **1968**, 267.
[125] S. W. Benson and H. E. O'Neal, *Internat. J. Chem. Kinetics*, **1**, 217 (1969).
[126] N. A. Grac, D. M. Golden, and S. W. Benson, *J. Am. Chem. Soc.*, **91**, 3091 (1969).
[127] F. R. Cruickshank and S. W. Benson, *J. Am. Chem. Soc.*, **91**, 1289 (1969).

A formula has been developed for evaluating activation energies of radical displacement reactions in the gas phase, given a knowledge of the dissociation energies of the bond being broken and that being formed.[128]

Quite pronounced solvent effects have been reported on the selectivity of photochlorination reactions of branched aliphatic esters, using either chlorine or trichloromethanesulphenyl chloride as chlorinating agent.[129] The results were obtained by varying the proportion of aromatic component in the solvent mixture, when it was found that the magnitude of the resulting variation in selectivity correlated reasonably well with the $\sigma_p{}^+$ substituent constant of the substituent in the aromatic.

Halogenation by such reagents as N-bromosuccinimide (NBS) is generally believed to involve hydrogen abstraction by a halogen atom. A similar bromine atom chain has been discussed for bromination of α-methoxytoluenes by 4-bromo-2,4,6-tri-t-butylcyclohexa-2,5-dienone, as a result of a measured ρ-value (vs. σ^+) of -0.43, identical with that found in the NBS reaction.[130] (That the formation of benzaldehydes in reactions of methoxytoluenes with NBS does proceed via intermediate bromo-ethers has been clearly demonstrated.[131]) The chlorination of toluenes by means of sulphuryl chloride, on the other hand, appears to involve hydrogen abstraction principally by the $ClSO_2\cdot$ radical.[130] In contrast, the extensively studied reactions of t-butyl hypochlorite may, after all, involve substantial abstraction by chlorine atoms.[132] This is particularly true in the case of benzylic systems, probably because of a relatively slow reaction between the stabilized benzylic radicals and hypochlorite molecules, whereas the reaction, even of benzylic radicals, with molecular chlorine is diffusion-controlled. This result sheds light on some previously discussed anomalies in t-butyl hypochlorite reactions.[133] It seems that the chlorine atom chains in t-butyl hypochlorite reactions may be eliminated by the presence of a trace of olefin. Chlorine atom chains have also been proposed[134] to explain the results of Minisci and Galli and their co-workers, in which unusually high selectivity has been observed in transition-metal salt catalysed chlorination by chloramines in strong acid. The high selectivity claimed for this reaction,[135] and attributed to hydrogen abstraction by $R_2\overset{\cdot+}{N}H$, is at least partly due to the instability of some of the primary chlorination products under the strongly acidic reaction conditions. The products of chlorination of, for example, 1-chlorobutane are stable under these

128 T. Kagiya, Y. Sumida, and T. Inoue, *Bull. Chem. Soc. Japan*, **42**, 2422 (1969).
129 J. P. Soumillion and A. Bruylants, *Bull. Soc. Chim. Belges*, **78**, 425, 435 (1969).
130 K. H. Lee, *Tetrahedron*, **25**, 4357, 4363 (1969); see also *Org. Reaction Mech.*, **1967**, 262.
131 I. Horman, S. S. Friedrich, R. M. Keefer, and L. J. Andrews, *J. Org. Chem.*, **34**, 905 (1969).
132 C. Walling and J. A. McGuinness, *J. Am. Chem. Soc.*, **91**, 2053 (1969).
133 See *Org. Reaction Mech.*, **1965**, 203.
134 D. D. Tanner and M. W. Mosher, *Can. J. Chem.*, **47**, 715 (1969).
135 For example, see *Org. Reaction Mech.*, **1968**, 281.

conditions, for which the isomer distribution of dichlorobutanes has now been found to be independent of chlorinating agent ($h\nu$–Cl_2; Cu^+–Pr^nNCl_2; Cu^+–Et_2NCl; Cu^+–N-chlorosuccinimide).

An example of a radical chain process which almost certainly is carried by succinimidyl radicals is the peroxide-initiated oxidation of 1-phenylethanol to acetophenone by N-iodosuccinimide.[136]

Hydrogen abstraction by Br atoms (using NBS) from a series $PhCH_2X$ shows a rate correlation with σ^+, ρ being -2.46.[137] This is a much greater substituent effect than when X is in the ring, and is consistent with pronounced dipolar character in the transition state, e.g. (42). Reports concerning the

$$PhCHOMe \qquad\qquad PhCH{=}\overset{+}{O}Me$$
$$H\cdot \qquad \longleftrightarrow \qquad H\cdot$$
$$Br \qquad\qquad\qquad Br^-$$

$$(42)$$

reactivities of toluenes[138] and α-substituted toluenes[139] towards trichloromethyl radicals have also appeared, and Pryor and coworkers have found that the reactivity of toluene towards hydrogen abstraction by methyl radicals is almost unaffected by nuclear substitution (ρ ca. -0.1).[140]

The identity reaction of benzyl radicals with toluene has been examined by a deuterium labelling procedure.[141] An isotope effect (k_H/k_D) of about 7 at $150°$ led to the suggestion of tunnelling in the hydrogen transfer step. Tunnelling has also been inferred from the very large isotope effect found in the reactions of methyl radicals, and more especially of trifluoromethyl radicals, with methyl formate and methyl deuteroformate.[142]

Flash photolysis studies with solutions of bromine in toluene revealed charge transfer between atomic bromine and the toluene, and resulted in radical chain bromination of the solvent. Solution spectra of benzyl radicals were not detected.[143]

Ease of hydrogen abstraction from aryl-substituted acetophenones, as measured by their chain transfer constants in the polymerization of methyl methacrylate, does not show any appreciable dependence on the polar

[136] T. R. Beebe and F. M. Howard, *J. Am. Chem. Soc.*, **91**, 3379 (1969).

[137] S. E. Friedrich, E. C. Friedrich, L. J. Andrews, and R. M. Keefer, *J. Org. Chem.*, **34**, 900 (1969).

[138] J. D. Unruh and G. J. Gleicher, *J. Am. Chem. Soc.*, **91**, 6211 (1969).

[139] E. P. Chang, R. L. Huang, and K. H. Lee, *J. Chem. Soc.*(B), **1969**, 878.

[140] W. A. Pryor, V. Tonellato, D. J. Fuller, and S. Jumonville, *J. Org. Chem.*, **34**, 2018 (1969).

[141] R. A. Jackson and D. W. O'Neill, *Chem. Comm.*, **1969**, 1210.

[142] N. L. Arthur and P. Gray, *Trans. Faraday Soc.*, **65**, 424, 434 (1969).

[143] N. Yamamoto, T. Kajikawa, H. Sato, and H. Tsubomura, *J. Am. Chem. Soc.*, **91**, 265 (1969).

character of the substituents, but it does depend on the substituent's resonance parameter E_R.[144]

New attempts to obtain linear free-energy relationships in homolytic reactions of non-conjugated molecules have utilized hydrogen abstraction from substituted adamantanes,[145] from substituted t-butylbenzenes,[146a] and from alkyl acetates.[146b]

It has been demonstrated that intramolecular hydrogen transfer in *N*-halogenoamide photolyses only occurs when there is a hydrogen atom appropriately situated for a six-membered ring cyclic transition state, involving transfer to the amide nitrogen atom. The formal alternative of transfer to

$$\text{CH}_3\text{CMe}_2\text{CH}_2\overset{\overset{\displaystyle Br}{|}}{\underset{\overset{\displaystyle ||}{O}}{C}}{-}\text{N}{-}\text{CMe}_3 \qquad \text{Me}\overset{\overset{\displaystyle Br}{|}}{\underset{\overset{\displaystyle ||}{O}}{C}}{-}\text{N}{-}\text{CMe}_2\text{CH}_2\text{CMe}_2\text{CH}_3$$

(44)   (45)

amide oxygen as, for example, in (43) was not observed;[147] the high yield of t-butylacetamide obtained from this reaction was in sharp contrast to the high yields of intramolecular bromination products obtained on photolysis of (44) and (45). With both of these compounds high quantum yields indicated a chain reaction (involving hydrogen abstraction by amido radicals; cf. NBS). Intramolecular hydrogen abstractions by amido radicals generated by photolysis of nitroso-amides,[148] and by sulphonamido radicals,[149] have also been documented.

Details of a thorough kinetic study of the reactions of halides with trialkyltin hydrides have been presented;[150] this is probably now the best understood of two-step chain reactions. In the course of the work reported, attention was paid to the reactions of radicals known to undergo rapid intramolecular

[144] T. Yamamoto, M. Kasegawa, and T. Otsu, *Bull. Chem. Soc. Japan*, **41**, 2788 (1968); see also *Org. Reaction Mech.*, **1967**, 245.

[145] G. J. Gleicher, J. L. Jackson, P. H. Owens, and J. D. Unruh, *Tetrahedron Letters*, **1969**, 833; see also *Org. Reaction Mech.*, **1968**, 277; I. Tabushi, T. Okada, Y. Aoyama, and R. Oda, *Tetrahedron Letters*, **1969**, 4069.

[146a] L. Harvey, G. J. Gleicher, and W. D. Totherow, *Tetrahedron*, **25**, 5019 (1969).

[146b] J. P. Soumillion and A. Bruylants, *Bull. Soc. Chim. Belges*, **78**, 169 (1969).

[147] Y. L. Chow and T. C. Joseph, *Chem. Comm.*, **1969**, 490.

[148] Y. L. Chow and J. N. S. Tam, *Chem. Comm.*, **1969**, 747.

[149] R. S. Neale and N. L. Marcus, *J. Org. Chem.*, **34**, 1808 (1969).

[150] D. J. Carlsson and K. U. Ingold, *J. Am. Chem. Soc.*, **90**, 7047 (1968).

transformations. Thus, rate constants of ca. 10^5 sec^{-1} at 25° and ca. 5×10^7 sec^{-1} at 100° were determined for the transformation of hex-5-enyl radicals (**46**) into cyclopentylmethyl radicals and for aryl migration in 2,2,2-triphenylethyl radicals respectively. In a similar chain reaction between trisubstituted silanes and alkyl halides, use of an optically active silane gives

$$\xrightarrow[\text{at 25°}]{k \approx 10^5 \text{ sec}^{-1}}$$

(**46**)

optically active silyl halide with ca. 90 % retention of configuration at silicon.[151]

Hydrogen abstraction from PhSiHMe$_2$ by the corresponding silylperoxy radicals is faster than deuterium abstraction from PhSiDMe$_2$ by a factor of only 1.19.[152] This small effect was attributed to a highly unsymmetrical transition state for the hydrogen transfer process. The corresponding value of k_H/k_D for abstraction from silane itself by methyl radicals is ca. 5 at 50°, somewhat *larger* than calculated.[153]

The activation parameters for hydrogen abstraction from neopentane and from tetramethylsilane by methyl radicals are very similar.[154] The same is true for reactions with trifluoromethyl,[154] and thus the silicon atom does not appear to exert any appreciable stabilizing influence on an adjacent carbon radical. Methyl radical reactions with tetramethyl-silane, -germane, -stannane, and -plumbane have also been compared, and whilst no correlation between reactivity and the electronegativity of the central atom in these molecules was evident, there was a correlation between reactivity and the ^{13}C–H coupling constants.[155] It has also been found that the α-position of an alkyltrichlorosilane is activated with respect to hydrogen abstraction by bromine, but deactivated towards attack by chlorine.[156]

Although 1,3-diiodopropane does not give cyclopropane on reaction with triphenyltin hydride, reaction with the corresponding germanium and silicon compounds does. Indeed, with Ph$_3$SiH at 200°, the yield of cyclopropane exceeds that of propane by nearly 2:1.[157] It was believed that the silicon compound was the poorest hydrogen donor of the Group IV hydrides examined, and that the cyclopropane-forming reaction (mechanism unknown) of the intermediate iodopropyl radical could therefore compete most favourably with hydrogen transfer in this case.

[151] H. Sakurai, M. Murakami, and M. Kumada, *J. Am. Chem. Soc.*, **91**, 5191 (1969).
[152] S. Rummel and H. Hübner, *Z. Chem.*, **9**, 150 (1969).
[153] O. P. Strausz, E. Jakubowski, H. S. Sandhu, and H. E. Gunning, *J. Chem. Phys.*, **51**, 552 (1969).
[154] E. R. Morris and J. C. J. Thynne, *J. Organometal. Chem.*, **17**, P3 (1969).
[155] A. U. Chaudhry and B. G. Gowenlock, *J. Organometal. Chem.*, **16**, 221 (1969)
[156] K. W. Michael, H. M. Banks, and J. L. Speir, *J. Org. Chem.*, **34**, 2832 (1969).
[157] L. Kaplan, *Chem. Comm.*, **1969**, 106.

11

Bromination (NBS) of methylenebicyclo[2.2.2]octane **(47)** gives a mixture of **(48)** and **(49)**, in contrast to the analogous reaction with methylenenorbornane which gives almost exclusively methylenebromonorbornane.[158] 2-Norbornyl radicals appear to react in a highly stereoselective fashion in the chain

(47)   (48)   (49)

decarbonylation of *exo*-norbornane-2-carbonyl chloride. At least 95% of the norbornyl chloride produced is the less stable *exo*-isomer.[159]

In contrast to bicyclo[2.1.1]hexane,[160] bicyclo[2.1.0]pentane gives no unrearranged chloro-derivatives on photochlorination.[161] Poutsma has obtained corroborating evidence on the high reactivity of allenic hydrogen atoms towards radical abstraction,[162] and Tedder and coworkers have found that in gas phase chlorination of n-propylbenzene the β-position is deactivated.[163] Formation of β-methylstyrene in this reaction was attributed to elimination of hydrogen chloride from vibrationally excited molecules of 2-chloro-1-phenylpropane; the yield depended on the pressure of added inert gas (SF_6). [In this context, SF_6 has been shown to be inert to attack by methyl or trifluoromethyl radicals up to 365°, despite the calculated exothermicity of the (unobserved) fluorine transfer: $CH_3 \cdot + SF_6 \rightarrow CH_3F + \cdot SF_5$.[164]]

The capture of alkyl radicals by cupric chloride has been employed to study the selectivity in attack on alkanecarboxylic acids by hydroxy and butoxy radicals generated from Fenton's reagent and from t-butyl hydroperoxide/Fe^{2+} respectively.[165] The results indicated that butoxy radicals were slightly more discriminating than were hydroxy. The ease of abstraction of bromine from bromoacetic acid by alkyl radicals gave no evidence for resonance stabilization in the resulting kenoacetic acid.[166, 167]

[158] D. I. Davies and L. T. Parfitt, *Tetrahedron Letters*, **1969**, 293; cf. *Org. Reaction Mech.*, **1968**, 280.
[159] I. Tabushi, T. Okada, and R. Oda, *Tetrahedron Letters*, **1969**, 1605.
[160] R. Srinivasan and F. I. Sonntag, *Can. J. Chem.*, **47**, 1627 (1969).
[161] R. S. Boikess and M. D. Mackay, *Tetrahedron Letters*, **1968**, 5991.
[162] M. L. Poutsma, *Tetrahedron Letters*, **1969**, 2925; see *Org Reaction Mech.*, **1967**, 261.
[163] V. R. Desai, A. Nechvatal, and J. M. Tedder, *J. Chem. Soc.*(B), **1969**, 30.
[164] H. F. Lefevre, J. D. Kale, and R. B. Timmons, *J. Phys. Chem.*, **73**, 1614 (1969).
[165] F. Minisci, R. Galli, and R. Bernardi, *Gazzetta*, **98**, 1113 (1968).
[166] E. D. Safronenko and I. B. Afanas'ev, *Zh. Org. Khim.*, **4**, 2086 (1968); *Chem. Abs.*, **70**, 56958 (1969); see also E. D. Safronenko and I. B. Afanas'ev, *ibid.*, p. 2092; *Chem. Abs.*, **70**, 56959 (1969).
[167] For the use of the prefix "keno" see *Org. Reaction Mech.*, **1968**, 264.

Reaction (6) proceeds by way of a free-radical chain mechanism.[168]

$$PhCH_2C(Ph)\!\!=\!\!NCl \rightarrow PhCN + PhCH_2Cl \qquad \ldots (6)$$

Other results reported during 1969 relate to the kinetics of N-chlorination of amines by t-butyl hypochlorite,[169] the variation of quantum yield in photo-oximation of cyclohexane (with NOCl) as a function of wavelength,[170] photo-oximation of alkanecarboxylic acids,[171] multiple chlorination of 1,2-dichloroethane,[172] abstraction of iodine atoms from aryl iodides by phenyl radicals,[173] hydrogen abstraction from triethylamine by cyanopropyl radicals,[174] and bromination of acenaphthene derivatives.[175] Hydrogen abstraction from aldehydes by $F_2N\cdot$ has been reported,[176] as well as numerous gas phase reactions of methyl,[177, 178] fluoromethyl,[179] trifluoromethyl,[178, 180] and other alkyl radicals.[181]

Additions

The direct observation (ESR) of hex-5-enyl radicals (**46**),[49] and an estimation of their rate of cyclization,[150] have already been mentioned. Separate determination of the rate of cyclization of these radicals utilized dissociative electron transfer of 1-fluorohex-5-ene by sodium naphthalene in dimethoxyethane (the reaction with alkyl fluorides is much slower than that with other alkyl halides); assuming diffusion control for the capture of alkyl radicals by sodium naphthalene, and knowing that hex-5-enyl anions do not cyclize, measurement of the relative yields of hexene and methylcyclopentane permitted estimation of an upper limit (7×10^5 sec^{-1}) for the cyclization rate

[168] M. L. Poutsma and P. A. Ibarbca, *J. Org. Chem.*, **34**, 2848 (1969).
[169] P. Bekiaroglou, *Z. Phys. Chem.* (Frankfurt), **64**, 263 (1969); P. Bekiaroglou and A. Drusas, *ibid.*, p. 288.
[170] H. Miyama, N. Harumiya, Y. Ito, and S. Nakamatsu, *J. Phys. Chem.*, **72**, 4700 (1968).
[171] E. Barale and A. Guillemonat, *Compt. Rend.*, **268**, 1201 (1969).
[172] D. S. Caines, R. B. Paton, D. A. Williams, and P. R. Wilkinson, *Austral. J. Chem.*, **22**, 1177 (1969).
[173] W. C. Danen and D. G. Saunders, *J. Am. Chem. Soc.*, **91**, 5924 (1969).
[174] E. A. Trosman and M. V. Bazilevskii, *Kinet. Katal.*, **9**, 684 (1968); *Chem. Abs.*, **69**, 86013 (1968).
[175] P. R. Constantine, L. W. Deady, and R. D. Topsom, *J. Org. Chem.*, **34**, 1113 (1969).
[176] A. J. White, *Tetracol.*, **7**, 21 (1968).
[177] J. A. Kerr and D. Timku, *J. Chem. Soc.*(A), **1969**, 1241; C. G. Crawforth and D. J. Waddington, *Trans. Faraday Soc.*, **65**, 1334 (1969); P. Gray, A. A. Herod, and L. J. Leyshon, *Can. J. Chem.*, **47**, 689 (1969); P. Gray and L. J. Leyshon, *Trans. Faraday Soc.*, **65**, 780 (1969).
[178] N. L. Arthur, P. Gray, and A. A. Herod, *Can. J. Chem.*, **47**, 1347 (1969).
[179] J. M. Sangster and J. C. J. Thynne, *Can. J. Chem.*, **47**, 2110 (1969).
[180] E. R. Morris and J. C. J. Thynne, *Trans. Faraday Soc.*, **64**, 3021 (1968); J. D. Kale and R. B. Timmons, *J. Phys. Chem.*, **72**, 4239 (1968); P. Gray, N. L. Arthur, and A. C. Lloyd, *Trans. Faraday Soc.*, **65**, 775 (1969).
[181] R. E. Berkley, G. N. C. Woodall, O. P. Strausz, and H. E. Gunning, *Can. J. Chem.*, **47**, 3305 (1969); D. G. L. James and G. E. Troughton, *Trans. Faraday Soc.*, **65**, 763 (1969).

constant.[182] Julia and Maumy have elaborated their ideas on the competition between cyclization to 5- and 6-membered rings in this and related reactions,[183] and other examples of kinetically controlled cyclizations to 5-membered rather than 6-membered rings by Δ^5-free radicals include those observed in radical additions of carbon tetrachloride to substituted hepta-1,6-dienes **(50)**,[184] in the photolysis of pent-4-enyl nitrite **(51)** and related compounds,[185]

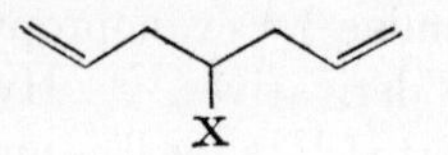

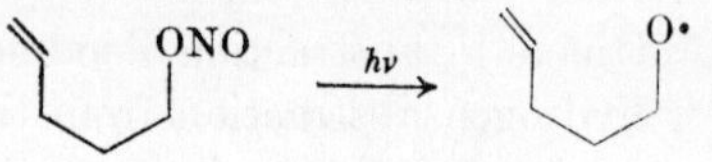

and in the titanous ion catalysed decomposition of diazonium salt **(52;** $n = 1$**)**.[186] In the case of **(52;** $n = 2$**)** a 6-membered ring is obtained. Other sources of the radicals were examined, and a parallel intramolecular hydrogen transfer from the methyl group was also observed in e.g. **(53**; X = O or CH_2**)**,[186] but this did not compete efficiently with hydrogen abstraction from the ethanol solvent. Radical additions of dialkyl phosphonates to the diolefin **(54)**

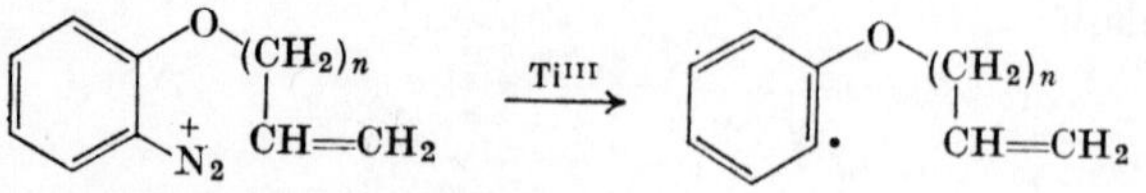

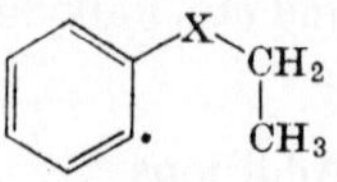

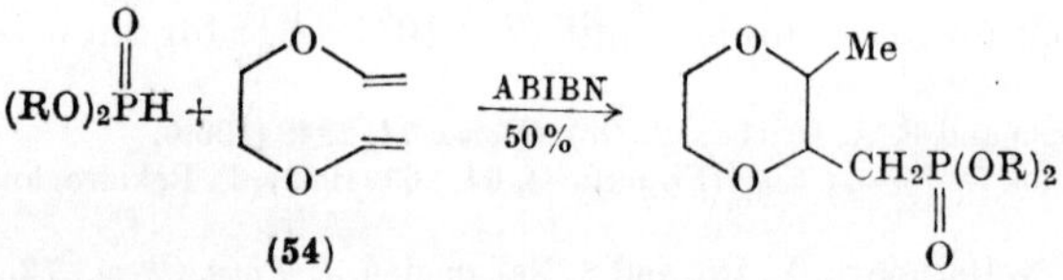

give substituted dioxans in good yield.[187] Examples of intramolecular radical additions to acetylenes have been reported. Thus bromoacetylenes **(55)** react with trialkyltin hydrides to give respectable yields of substituted *exo*-methylene-cycloalkanes **(56)** when $n = 4$ or 5. No internal olefin **(57)** is formed.[188]

[182] J. F. Garst and F. E. Barton, *Tetrahedron Letters*, **1969**, 587.
[183] M. Julia and M. Maumy, *Bull. Soc. Chim. France*, **1969**, 2415, 2427; see *Org. Reaction Mech.*, **1968**, 283.
[184] N. O. Brace, *J. Org. Chem.*, **34**, 2441 (1969); see also *Org. Reaction Mech.*, **1967**, 151.
[185] R. D. Rieke and N. A. Moore, *Tetrahedron Letters*, **1969**, 2035; J. M. Surzur, M. P. Bertrand, and R. Nouguier, *ibid.*, p. 4197.
[186] A. L. J. Beckwith and W. B. Gara, *J. Am. Chem. Soc.*, **91**, 5689, 5691 (1969).
[187] B. A. Trofimov, A. S. Atavin, G. M. Gauvilova, and G. A. Kalabin, *Zh. Obshch. Khim.*, **38**, 2344 (1968).
[188] J. K. Crandall and D. J. Keyton, *Tetrahedron Letters*, **1969**, 1653; see also J. L. Derocque, U. Beisswenger, and M. Hanack, *ibid.*, p. 2149.

More complex reactions were found when the same acetylenes were allowed to react with lithium biphenyl, though the olefins (56) were again the major

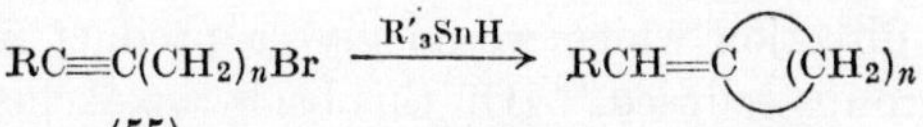

$$RC\equiv C(CH_2)_nBr \xrightarrow{\ R'_3SnH\ } RCH = C\ \ (CH_2)_n$$

(55)

(56)

$$\underset{H}{\overset{R}{>}}\ (CH_2)_n$$

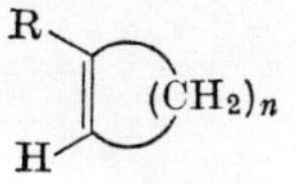

(57)

products. One complication considered important in the latter reaction was electron transfer to the acetylene. For example, with (55; R = Ph and $n = 3$) some cyclization to benzylidenecyclobutane was observed, whereas this was not found among the products of the corresponding stannane reduction. Products of intramolecular thiyl radical addition to acetylenes do include internal cycloalkenes (reaction 7); some representative results are tabulated.[189]

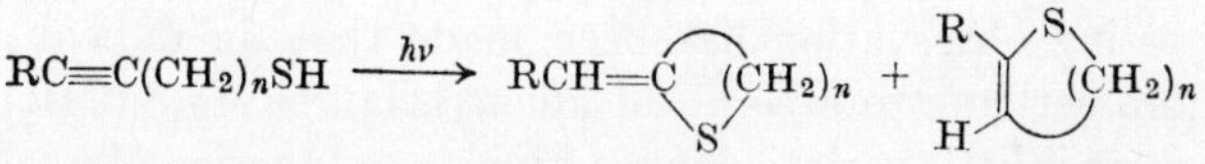

$$RC\equiv C(CH_2)_nSH \xrightarrow{\ h\nu\ } RCH = C\underset{S}{\ }(CH_2)_n\ +\ \underset{H}{\overset{R\ \ S}{>}}(CH_2)_n\ \quad \ldots (7)$$

R	n	%	%
H	3	trace	48
Me	3	17	4
Ph	3	75	—
H	4	trace	38
H	5	—	23

The % yields of cyclized products obtained when initiation is by peroxide are generally lower than those indicated here, which are for photochemical initiation.

Further work on the diradical mechanisms of [2 + 2]- and [2 + 4]-cyclo-addition reactions is discussed in Chapter 5.

The ease of radical additions to cyclic olefins, C_nH_{2n-2}, is markedly dependent on the nature of the initial radical addition. If this is reversible (e.g. addition of RSH or HBr) then the reactivity order $n = 6 > 7 > 5 \gg 8$ is found. If, however, the initial radical addition is not reversible (e.g. addition of BrCCl$_3$, CH$_3$CHO) then the order is $n = 8 > 5 > 7 > 6$. The pronounced change found for cyclooctene was attributed to ring-strain effects in the initial radical adduct.[190] It has been found that trichloromethyl radicals from bromotri-chloromethane appear to add to cycloolefins from an axial-like direction, and

[189] J. M. Surzur, C. Dupuy, M.-P. Crozet, and N. Amor, *Compt. Rend.*, *C*, **269**, 849 (1969).
[190] L. H. Gale, *J. Org. Chem.*, **34**, 81 (1969).

that bromine abstraction which ensues is also preferentially from an axial direction. In flexible systems conformational changes may occur, but the size of the trichloromethyl substituent is such that the bromine abstraction does not occur in the direction which would give a product with bromine and trichloromethyl groups eclipsed.[191] On this basis the exclusive formation of, for example, *trans*-adduct from cyclopentene and (**58**) from *trans*-Δ^2-octalin

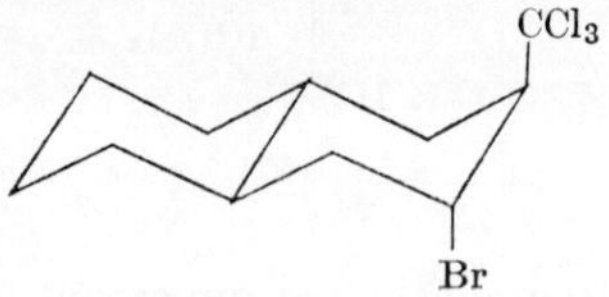

(**58**)

may be rationalized.[192] Le Bel and coworkers have given details[193] of their work[192] on the stereochemistry of the addition of thiols to conformationally rigid cyclohexenes. Again a kinetic preference for axial attack is found. It was considered that sulphur bridging in the initial radical adducts is unimportant.

The interesting observation has been made that, in radical addition of bromotrichloromethane to a series of alkenylsilanes, $Me_3Si(CH_2)_nCH=CH_2$ ($n = 0$—4), there is little variation in reactivity as a function of n, except in the case $n = 1$. This is about eight times more reactive than other members of the series, and it was suggested that there might be homoconjugative stabilization of the intermediate adduct radical by interaction with the d-orbitals on silicon.[194] Unfortunately this result does not appear to be paralleled by especially high reactivity of the β-hydrogens in Kochi's work on abstraction from alkylsilanes, though different transition-state effects would be operating in the two systems.[46]

It has been shown that in many addition reactions of a species HX to an olefin, if HX is not present in very large excess, then the adduct radical abstracts hydrogen not from HX but from an allylic position in a second molecule of olefin.[195]

Non-stereospecific epoxidation of butenes during gas-phase oxidation by oxygen has been construed as evidence for the mechanism shown in equation (8).[196]

[191] J. G. Traynham, A. G. Lane, and N. S. Bhacca, *J. Org. Chem.*, **34**, 1302 (1969).
[192] See *Org. Reaction Mech.*, **1967**, 148, 149.
[193] N. A. LeBel, R. F. Czaja, and A. DeBoer, *J. Org. Chem.*, **34**, 3112 (1969).
[194] H. Sakurai, A. Hosomi, and M. Kumada, *J. Org. Chem.*, **34**, 1764 (1969).
[195] M. Cazaux, G. Bourgeois, and R. Lalande, *Tetrahedron Letters*, **1969**, 3703.
[196] D. J. M. Roy and D. J. Waddington, *J. Am. Chem. Soc.*, **90**, 7176 (1968).

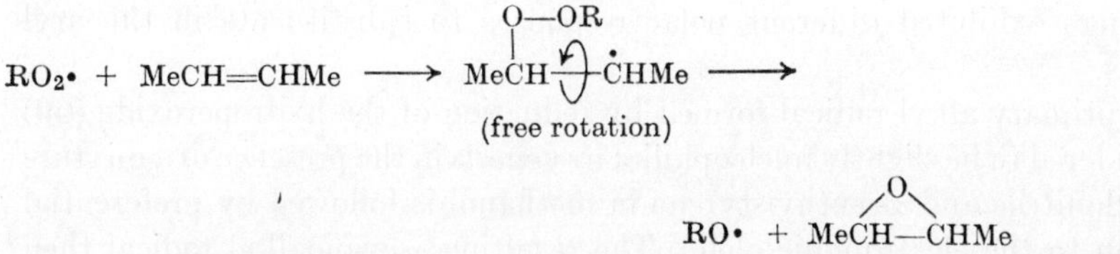

$$RO_2\bullet + MeCH{=}CHMe \longrightarrow \underset{\text{(free rotation)}}{MeCH{-}\overset{O-OR}{\overset{|}{\bigcirc}}{-}\dot{C}HMe} \longrightarrow$$

$$RO\bullet + MeCH{-}\overset{O}{\overset{\triangle}{}}{-}CHMe \qquad \ldots (8)$$

The equilibration of α-methylstyrenes by reversible addition of thiyl radicals gives a mixture containing 97% *trans*-isomer, in contrast to normal thermodynamic equilibrium which, under comparable conditions, gives 78% *trans*-isomer. The difference was attributed to the greater stability of adduct radical than methylstyrene plus thiyl radical, which, it was suggested, could lead to a kinetic control of the composition of the methylstyrene mixture.[197]

New results on the reactivity of vinyl monomers towards phenyl[198, 199] and other aryl radicals have been presented.[199] In the Meerwein arylation reaction of *p*-chlorobenzenediazonium salts with acrylonitrile in acetone, promoted by cuprous chloride, the ratio of chlorobenzene to Meerwein adduct **(59)** is independent of the concentration of added cupric chloride, though the yield of *p*-dichlorobenzene (Sandmeyer product) is increased by added cupric chloride. These results were interpreted in terms of a three-way competition for free aryl radicals.[199] Extension of this work showed that different vinyl

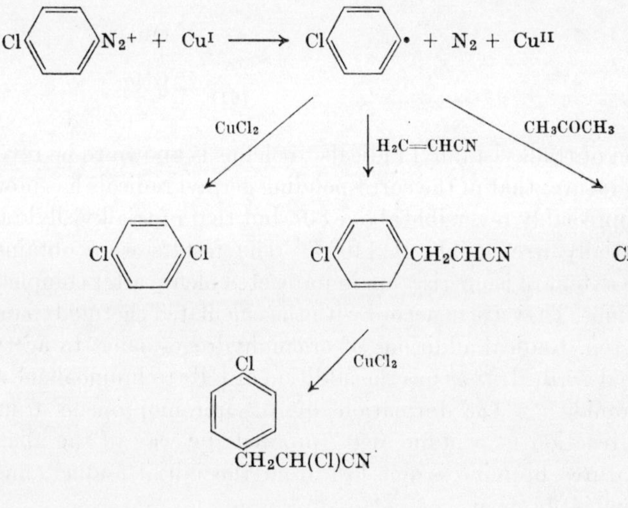

(59)

[197] A. Ohno, Y. Ohnishi, and G. Tsuchihashi, *Tetrahedron Letters*, **1969**, 643.
[198] W. A. Pryor and T. R. Fiske, *Trans. Faraday Soc.*, **65**, 1865 (1969); see *Org. Reaction Mech.*, **1968**, 285.
[199] S. C. Dickerman, D. J. Desouza, and N. Jacobson, *J. Org. Chem.*, **34**, 710 (1969); S. C. Dickerman, D. J. Desouza, M. Fryd, I. S. Megna, and M. M. Skoultchi, *ibid.*, p. 714.

monomers exhibited different polar responses to substituents in the aryl radicals.

The primary alkyl radical formed by reduction of the hydroperoxide (60) is considered to be slightly nucleophilic; its genesis in the presence of a mixture of acrylonitrile and α-methylstyrene in methanol is followed by preferential addition to the electrophilic olefin. The resulting α-cyanoalkyl radical then adds preferentially to α-methylstyrene, giving a new radical with a sufficiently low ionization potential for oxidation by Fe^{III} to be relatively rapid. The result is a good yield of (61).[200]

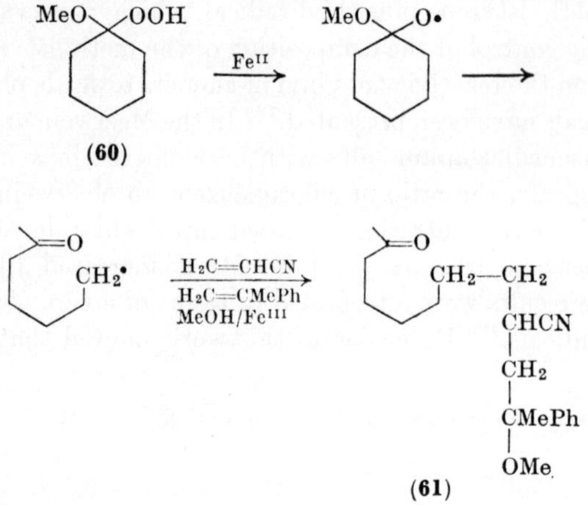

$$MeO\text{—}OOH \xrightarrow{Fe^{II}} MeO\text{—}O\cdot \longrightarrow$$

(60)

$$\begin{array}{c}=O\\ CH_2\cdot\end{array} \xrightarrow[\substack{H_2C=CMePh\\MeOH/Fe^{III}}]{H_2C=CHCN} \begin{array}{c}=O\\CH_2\text{—}CH_2\\|\\CHCN\\|\\CH_2\\|\\CMePh\\|\\OMe\end{array}$$

(61)

The addition of trialkylstannyl radicals to olefins is known to be reversible at room temperature; that of the corresponding germyl radicals has now been found to be appreciably reversible above 80°, but that of trialkylsilyl radicals remains essentially irreversible at 140°.[201] The results were obtained by measuring the extent of isomerization in unreacted olefin, after completion of addition reactions. They are in accord with the calculated thermodynamics of the addition step. Radical additions of organohydrogermanes to acetylenes have also been described,[202] as has the addition of HBr to bromoallene and to propargyl bromide.[203] The formation of 1,2-dibromopropene from the acetylene, by reaction in pentane at 0°, proceeds by way of the allene. At lower temperatures bromine is not lost from the initial adduct, and 1,3-dibromopropene is obtained.

[200] F. Minisci, M. Cecene, R. Galli, and R. Bernardi, *Tetrahedron*, **25**, 2667 (1969).

[201] S. W. Bennett, C. Eaborn, R. A. Jackson, and R. Pearce, *J. Organometal. Chem.*, 15, P17 (1969).

[202] M. Massol, J. Satgé, and Y. Cabadi, *Compt. Rend.*, **268**, 1814 (1969); M. Massol, J. Satgé, and M. Lesbre, *J. Organometal. Chem.*, **17**, 25 (1969).

[203] K. R. Kopecky and S. Grover, *Can. J. Chem.*, **47**, 3153 (1969).

Additions of radicals to unsaturated centres other than carbon–carbon double and triple bonds are represented this year by a new example of phenyl addition to sulphur dioxide (equation 9),[204] additions of alkyl radicals to cyanogen and cyanogen chloride (equations 10 and 11),[205] a new example of addition to carbonyl in biacetyl,[206] and probably also by the efficient γ-induced reduction of esters to ethers by trichlorosilane (equation 12).[207]

$$PhN\!\!=\!\!NCPh_3 \longrightarrow Ph\cdot + N_2 + \cdot CPh_3$$

$$Ph\cdot + SO_2 \longrightarrow PhSO_2\cdot \qquad \cdots (9)$$

$$PhSO_2\cdot + \cdot CPh_3 \longrightarrow PhSO_2CPh_3$$

$$R\cdot + C_2N_2 \longrightarrow RC(CN)\!\!=\!\!N\cdot \xrightarrow{\text{RH}} RC(CN)\!\!=\!\!NH + R\cdot \qquad \cdots (10)$$

$$R\cdot + ClCN \longrightarrow RC(Cl)\!\!=\!\!N\cdot \longrightarrow RCN + Cl\cdot$$
$$\cdots (11)$$

$$Cl\cdot + RH \longrightarrow HCl + R\cdot$$

$$RCOOR' + Cl_3SiH \longrightarrow RCH(OR')OSiCl_3$$
$$\cdots (12)$$
$$RCH(OR')OSiCl_3 \xrightarrow{Cl_3SiH} RCH_2OR'$$

More work on the addition reactions of halogenated norbornenes and related compounds has been discussed,[208] and it has been noted that corresponding olefins in the bicyclo[2.2.2]octane series are appreciably less reactive.[209] The addition of t-butyl and of cyanopropyl radicals to 4-vinylcyclohexene has been compared with the corresponding reaction with butoxy radicals;[210] competition between radical and ionic addition of bromine to butadiene,[211] hydrogen halides to diacetylene,[212] and halogen azides to olefins,[213] has been examined. An indirect method has led to a rate constant of ca. 7×10^7 for the irreversible addition of MeS· to butadiene in the gas phase.[214] Photolytically generated hydroxyl radicals react (at 77° K) with α,β-unsaturated aldehydes, principally by addition to the double bond; abstraction of the aldehyde

[204] H. Takeuchi, T. Nagai, and N. Tokura, *Tetrahedron*, **25**, 2987 (1969).
[205] D. D. Tanner and N. J. Bunce, *J. Am. Chem. Soc.*, **91**, 3028 (1969).
[206] K. J. Hole and M. F. R. Mulcahy, *J. Phys. Chem.*, **73**, 177 (1969).
[207] J. Isurugi, R. Nakao, and T. Fukumoto, *J. Am. Chem. Soc.*, **91**, 4588 (1969).
[208] D. I. Davies and P. J. Rowley, *J. Chem. Soc.*(C), **1969**, 424; D. I. Davies, L. T. Parfitt, C. K. Alden, and J. A. Claisse, *ibid.*, p. 1585.
[209] D. I. Davies and L. T. Parfitt, *J. Chem. Soc.*(C), **1969**, 1401.
[210] J. R. Shelton and E. E. Borchert, *Can. J. Chem.*, **46**, 3833 (1968).
[211] V. L. Heasley and S. K. Taylor, *J. Org. Chem.*, **34**, 2779 (1969).
[212] Y. I. Porfineva, L. A. Vasileva, E. S. Turbanova, and A. A. Petrov, *Zh. Org. Khim.*, **5**, 591 (1969).
[213] A. Hassner and F. Boerwinkle, *Tetrahedron Letters*, **1969**, 3309.
[214] D. H. Graham and J. F. Saltys, *Can. J. Chem.*, **47**, 2529, 2719 (1969).

hydrogen was not detected.[215] Also recorded are additions of SO_3^- to olefins,[216a] of arenesulphonyl iodides to acetylenes,[216b] of trichloromethanesulphenyl chloride to olefins (in which the chain reaction is shown to be carried by $Cl_3CS\cdot$),[217] of N,N-dichloroethanesulphonamide to terminal olefins,[218] of MeN(Cl)CN to tetramethylethylene[149] (the mechanism here is uncertain; the reaction is accelerated by oxygen), $CF_3CF_2CCl_3$ and related compounds to olefins,[219] and of dimethoxymethane to hept-1-ene.[220] The principal products of radical-initiated additions of dioxolan to olefins at 160° are formed by addition of ring-opened radicals (e.g. equations 13).[221] The photochemical

$$R'\overset{\cdot}{C}H(CH_2)_3OCHO \xrightarrow{\text{RH}} R'(CH_2)_4OCHO \quad \dots (13)$$

reaction of chlorine with methylenecyclopropane proceeds mainly by addition to the cyclopropane ring. Addition to the double bond also occurs and gives rise to some cyclopropylcarbinyl→allylcarbinyl rearrangement. In view of this, and the known addition of thiyl radicals to the terminal methylene group of this compound, it was suggested that two modes of chlorine addition to the double bond are possible.[222] These conclusions are summarized in the annexed reaction scheme. Photochemical alkylation of peptides (equation 14) has been noted.[223] Addition reactions of the quinone methide (**62**) have been described,[224] and a radical chain mechanism has been advanced for the addition of aziridine to styrene initiated by sodium.[225] Thiyl radicals add reversibly to acetylene, leading to vinyl mercaptan, which polymerizes.[226] The addition of hydrogen atoms [227] and methyl radicals [228] to butenes has been discussed, as

215 T. Ichikawa and K. Kuwatu, *Bull. Chem. Soc. Japan*, **42**, 2208 (1969).
216a C. J. Norton, N. F. Seppi, and M. J. Reuter, *J. Org. Chem.*, **33**, 4158 (1968).
216b W. E. Truce and G. C. Wolf, *Chem. Comm.*, **1969**, 150.
217 H. Kloosterziel, *Quart. Rep. Sulfur Chem.*, **2**, 353 (1967).
218 T. Ohashi, M. Sugie, M. Okahara, and S. Komari, *Tetrahedron*, **25**, 5349 (1969).
219 P. Tarrant and J. P. Tandon, *J. Org. Chem.*, **34**, 864 (1969).
220 P. Marche and D. Lefort, *Compt. Rend.*, *C*, **269**, 717 (1969).
221 R. Lalande, B. Maillard, and M. Cazaux, *Tetrahedron Letters*, **1969**, 745.
222 A. J. Davidson and A. T. Bottini, *J. Org. Chem.*, **34**, 3642 (1969).
223 J. Sperling, *J. Am. Chem. Soc.*, **91**, 5389 (1969).
224 H. D. Becker, *J. Org. Chem.*, **34**, 2469, 2472 (1969).
225 R. K. Razdan, *Chem. Comm.*, **1969**, 770.
226 J. R. Majer, J. Morton, and J. C. Robb, *J. Chem. Soc.*(B), **1969**, 301.
227 R. D. Kelley, R. Klein, and M. D. Scheer, *J. Phys. Chem.*, **73**, 1160 (1969).
228 N. Yokoyama and R. K. Brinton, *Can. J. Chem.*, **47**, 2987 (1969).

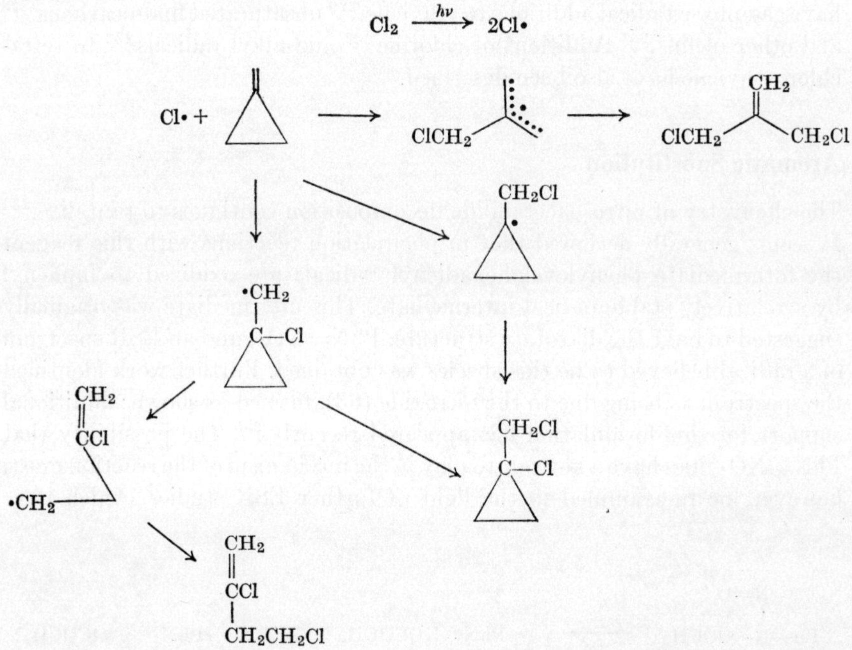

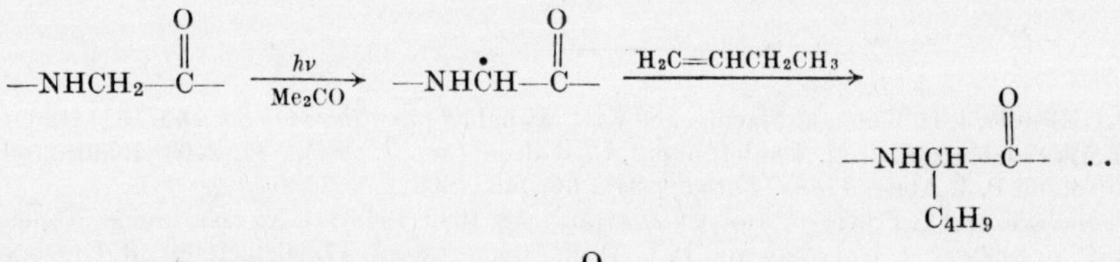

$$Cl_2 \xrightarrow{h\nu} 2Cl\bullet$$

... (14)

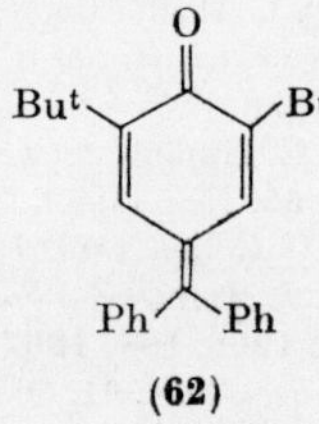

(62)

have gas-phase radical additions to ethylene,[229] unsaturated fluorocarbons,[230] and other olefins.[231] Additions of chlorine [232] and alkyl radicals [233] to tetra-chloroethylene have also been described.

Aromatic Substitution

The chemistry of nitrosoacetanilide decomposition continues to tantalize.[234] It seems generally accepted that in phenylation reactions with this reagent the intermediate phenylcyclohexadienyl radicals are oxidized to biphenyl by a relatively stable radical intermediate. This intermediate was originally suggested to have the diazotate structure, $PhN=NO\cdot$, and an ESR spectrum of a radical believed to be this species was obtained. Further work identified the spectrum as being due to the nitroxide (63), formed as shown; additional support for this formulation has appeared recently.[235] The possibility that $PhN=NO\cdot$ does have a key role to play in the mechanism of the reaction must, however, be re-examined in the light of further ESR studies, which have

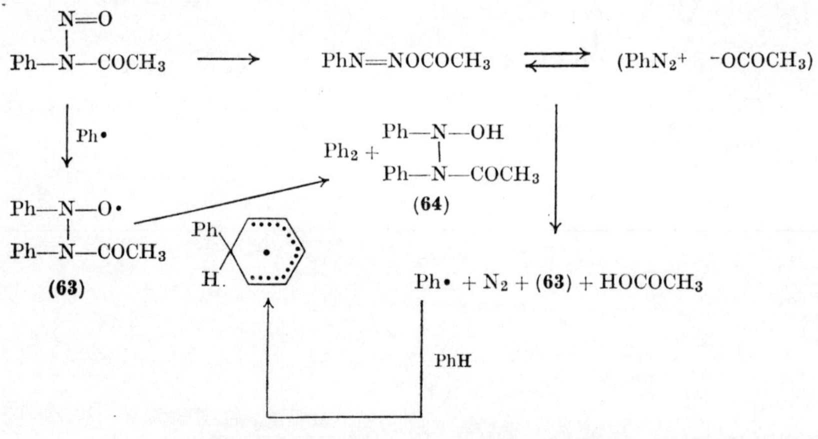

[229] P. C. Beadle, J. H. Knox, F. Placido, and K. C. Waugh, *Trans. Faraday Soc.*, **65**, 1571 (1969); H. W. Sidebottom, J. M. Tedder, and J. C. Walton, *Can. J. Chem.*, **47**, 2103 (1969); R. J. Field and P. I. Abell, *Trans. Faraday Soc.*, **65**, 743 (1969).

[230] G. Navazio and C. Fraccaro, *Ann. Chim.* (Italy), **58**, 1082 (1968); G. Navazio and M. Napoli, *ibid.*, p. 1090; G. J. Karvanos and P. I. Abell, *Can. J. Chem.*, **47**, 2089 (1969); R. Gregory, R. N. Haszeldine, and A. E. Tipping, *J. Chem. Soc.*(C), **1969**, 991; T. S. Carlton and A. B. Harker, *J. Phys. Chem.*, **73**, 3356 (1969).

[231] R. Gregory, R. N. Haszeldine, and A. E. Tipping, *J. Chem. Soc.*(C) **1968**, 3020; J. H. Knox and K. C. Waugh, *Trans. Faraday Soc.*, **65**, 1585 (1969).

[232] J. G. Franklin, G. Huybrechts, and C. Cillien, *Can. J. Chem.*, **47**, 2094 (1969).

[233] A. Horowitz and L. A. Rajbenbach, *J. Am. Chem. Soc.*, **91**, 4626, 4631 (1969).

[234] See *Org. Reaction Mech.*, **1965**, 154; **1966**, 188; **1967**, 184; **1968**, 288.

[235] S. Terabe and R. Konaka, *J. Am. Chem. Soc.*, **91**, 5655 (1969).

revealed the presence of a second relatively stable paramagnetic species in nitrosoacetanilide reactions. This new radical clearly contains two nitrogen atoms, and the larger nitrogen splitting ($a_N = 30.5$ gauss) is consistent with the diazotate formation.[236] The mechanism (outlined) based on structure (63) as the oxidizing radical required that the reduced form (64) be re-oxidized to (63) by a molecule of benzenediazonium acetate (either in its covalent or ion-pair form). The feasibility of such a process seems to have been adequately demonstrated by the model reaction (15) of diphenylhydroxylamine and benzenediazonium fluoroborate.[237] The reaction between these reagents in

$$Ph_2NOH + PhN_2{}^+BF_4{}^- \rightarrow Ph\cdot + N_2 + Ph_2NO\cdot + HBF_4 \qquad \ldots (15)$$

nitrobenzene or acetophenone is very rapid, and fair yields of radical phenylation products (nitrobiphenyls or acetylbiphenyls) are produced. However, this analogy obviously does not prove the mechanism shown for nitrosoacetanilide decomposition, and the competing roles of nitroxide (63) and diazotate radicals remain to be evaluated. The initial step in the model reaction written as (15) was considered to involve electron transfer from the hydroxylamine to the diazonium compound (cf. peroxide reactions p. 308).

An ESR spectrum of a stable radical has also been recorded[238] in the course of the arylation of benzene using homogeneous diazotization of aniline with pentyl nitrite. It was suggested that the radical was $PhN{=}NO\cdot$, but once again the spectrum is that of an aryl nitroxide, with a major nitrogen splitting of ca. 12 gauss.

It has generally been assumed that aryl radical addition to an aromatic ring is essentially irreversible within the normal lifetime of the resulting arylcyclohexadienyl intermediate, and support for this view has been found in the absence of hydrogen isotope effects on the reactivity of aromatic substrates (except in special circumstances in which reversibility of addition is not implicated). New results have discerned an appreciable isotope effect when phenylation of aromatic molecules dissolved in dimethyl sulphoxide is brought about by nitrite-ion induced decomposition of benzenediazonium fluoroborate. The results were not very reproducible, but consistently showed lower reactivity in ArD than in the corresponding ArH. The effect was particularly noticeable at a site adjacent to a nuclear substituent, e.g. *ortho* to a nitro group.[239]

The decomposition of diazonium salts in DMSO containing aromatic substrates has also been noted by Zollinger and Kaul,[240] and Abramovitch and Koleoso have extended their use of sulpholan as a solvent for homolytic

[236] J. I. G. Cadogan, R. M. Paton, and C. Thomson, *Chem. Comm.*, **1969**, 614.
[237] R. M. Cooper and M. J. Perkins, *Tetrahedron Letters*, **1969**, 2477.
[238] A. F. Levit and I. P. Gragerov, *Zh. Org. Khim.*, **5**, 310 (1969); *Chem. Abs.*, **70**, 106094 (1969).
[239] M. Kobayashi, H. Minato, and N. Kobori, *Bull. Chem. Soc. Japan*, **42**, 2738 (1969).
[240] B. L. Kaul and H. Zollinger, *Helv. Chim. Acta*, **51**, 2132 (1968).

arylation to obtain new quantitative results on arylation of anisole.[241] These are in much better accord with the theory of polarized free radicals than existing data for arylation of this substrate.

An intramolecular arylation examined in some detail is the photoinduced cyclization of (65; X = Cl, Br, or I) to fluoroanthene in benzene.[242] Intermolecular arylation of the solvent competes, particularly when X = I, and to

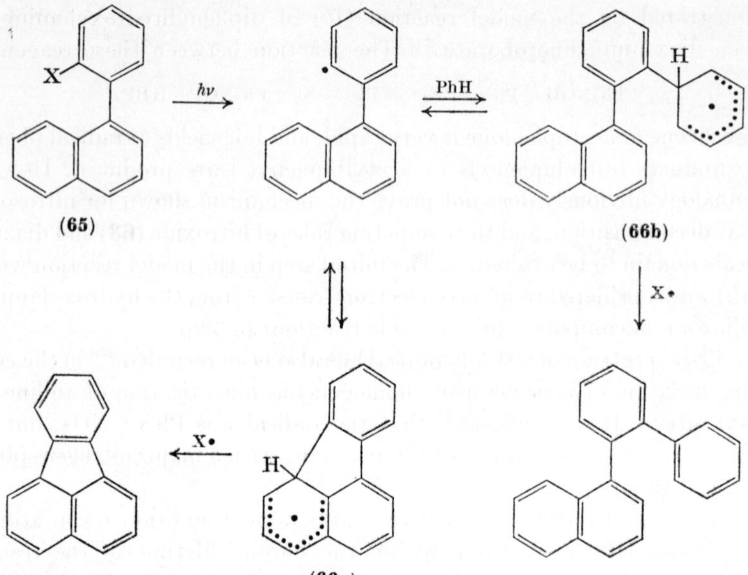

rationalize the different proportions of products obtained with different halogens it was suggested that here, also, cyclohexadienyl radical formation is reversible. This might be a consequence of the strain in (66a) and the steric crowding in (66b). It was argued that the fluoranthene is largely formed by dehydrogenation of (66a) by the halogen atom formed in the photolysis, and that dependence of product ratio on halogen could then be explained by the different reactivities of the different halogens. Other arylation reactions and related reactions of aryl radicals formed by photolysis of aryl halides other than aryl iodides have been discussed.[243] Failure of the reaction was found with certain aryl chlorides whose triplet state was of lower energy than that required to rupture the carbon–chlorine bond; the mechanism of the photo-reaction involves principally a triplet-state process following energy transfer

[241] R. A. Abramovitch and O. A. Koleoso, *J. Chem. Soc.*(B), **1969**, 779.

[242] W. A. Henderson, R. Lopresti, and A. Zweig, *J. Am. Chem. Soc.*, **91**, 6049 (1969).

[243] J. T. Pinhey and R. D. G. Rigby, *Tetrahedron Letters*, **1969**, 1267, 1271; G. E. Robinson and J. M. Vernon, *Chem. Comm.*, **1969**, 977.

from triplet benzene, the benzene solvent absorbing essentially all the radiation.

Other arylations of benzene and its derivatives discussed recently include new examples of transition-metal catalysed reactions of bromobenzene and methylmagnesium iodide with aromatic substrates,[244] γ-radiolysis of benzene and of bromobenzene in benzene,[245] a kinetic study of the thermolysis of benzoyl peroxide in fluorobenzene,[246a] the high-temperature gas-phase pyrolysis of nitrotoluenes in the presence of benzene (the *ortho*-isomer isomerizes to anthranilic acid),[246b] and the photolysis of tetraphenylantimony in benzene.[247] The last reaction gives biphenyl, part of which is derived entirely from the organometallic compound. The extent of this apparent intramolecular reaction is wavelength-dependent; it is also concentration-dependent, and may therefore be the result of a bimolecular rather than an intramolecular process. Also described have been theoretical calculations on the radical reactivities of methylnaphthalenes,[248] of pyridine[249] and the pyrimidines,[250] and experimental studies on the phenylation of isothiazoles,[251] of 2-phenyl-benzothiazole,[252] and of mixtures of biphenyl and the terphenyls,[253] and of phenylation (in which one source of phenyl radicals was photolysis of diphenyl-magnesium) of pyridine[254] and of arylation of methylated pyridines.[255] Thermal decomposition of benzoyl peroxide in furan (under pressure) gives *cis*- and *trans*-2,5-dibenzoyloxy-2,5-dihydrofurans (**68**) in high yield,[256] and adducts (**69**; X = O) were also the major products of decomposition of phenylazotriphenylmethane in furan, though other sources of phenyl radicals give 2-phenylfuran.[257] The adducts (**68**) are presumably the result of scavenging of benzoyloxy radicals by the intermediate (**67**), which is probably very much more stable than is the benzoyloxycyclohexadienyl radical (**70**) (see below). The decomposition of phenylazotriphenylmethane in thiophen gives

[244] D. I. Davies, J. N. Done, and D. H. Hey, *J. Chem. Soc.*(C), **1969**, 1392; see also M. A. Aleksankin, L. I. Fileleeva, and I. P. Gragerov, *Zh. Org. Khim.*, **4**, 1439 (1968).

[245] B. Zimmerli and T. Gäumann, *Helv. Chim. Acta*, **52**, 764 (1969); R. Siegrist and T. Gäumann, *ibid.*, p. 2119; A. Menger and T. Gäumann, *ibid.*, p. 2129.

[246a] P. Lewis and G. H. Williams, *J. Chem. Soc.*(B), **1969**, 120.

[246b] E. K. Fields and S. Meyerson, *J. Org. Chem.*, **33**, 4487 (1968).

[247] K. Shen, W. E. McEwen, and A. P. Wolf, *J. Am. Chem. Soc.*, **91**, 1283 (1969).

[248] J. M. Bonnier, M. Gelus, and J. Renaudo, *J. Chim. Phys.*, **65**, 1326 (1968).

[249] J. Palecek, V. Skala, and J. Kuthan, *Coll. Czech. Chem. Comm.*, **34**, 1110 (1969).

[250] J. N. Herak, *Croat. Chem. Acta*, **40**, 37 (1968).

[251] J. C. Poite, G. Vernin, G. Loridan, H. J. M. Dou, and J. Metzger, *Bull. Soc. Chim. France*, **1969**, 3912; H. J. M. Dou, J. C. Poite, G. Vernin, and J. Metzger, *Tetrahedron Letters*, **1969**, 779.

[252] G. Vernin, H. J. M. Dou, and J. Metzger, *Compt. Rend.*, *C*, **268**, 977 (1969).

[253] H. Stangl and G. Juppe, *Chem. Ber.*, **102**, 2419 (1969).

[254] M. Y. Turkina and I. P. Gragerov, *Zh. Org. Khim.*, **5**, 585 (1969).

[255] G. Vernin, H. J. M. Dou, *Compt. Rend.*, *C*, **266**, 924 (1968).

[256] K. E. Kolb and W. A. Black, *Chem. Comm.*, **1969**, 1119.

[257] L. Berati, N. LaBarba, M. Tiecco, and A. Turdo, *J. Chem. Soc.*(B), **1969**, 1253.

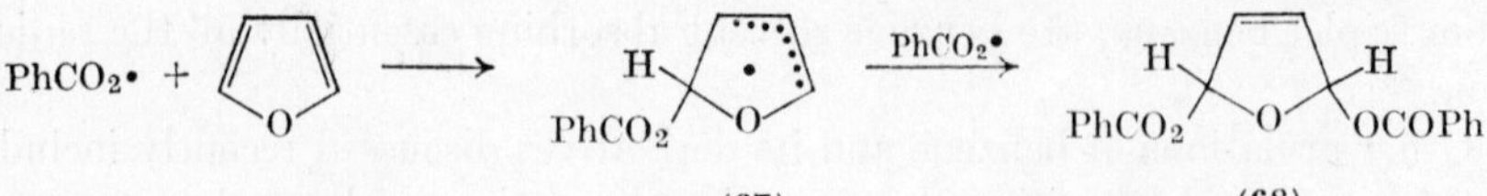

(67)   (68)

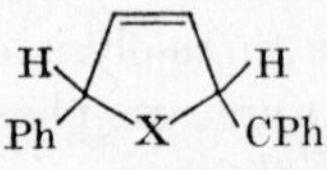

(69) $PhCO_2\cdot + PhH \rightleftharpoons$ (70)

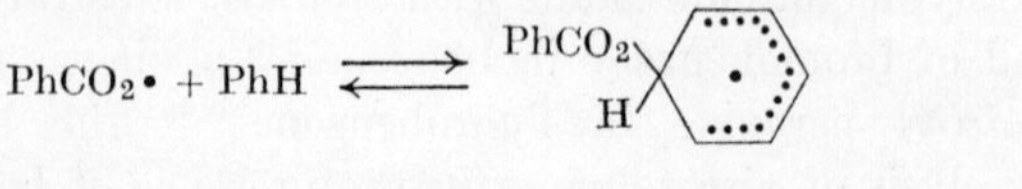

both 2- and 3-phenylthiophen, but the yield of the 2-isomer is unusually low, and this is compensated for by the formation of substantial amounts of the adducts (**69**; X = S).[258]

Arylations of benzene by 2-thiazolyl radicals,[259] by 3-benzothienyl radicals,[260] and by ferrocenyl radicals[261] have been discussed.

Benzoyloxy radicals add reversibly to benzene to give the adduct radical (**70**); in normal circumstances decarboxylation of the benzoyloxy radical is sufficiently fast to preclude significant ester formation by oxidation of (**70**). However, in the presence of excess of a sufficiently powerful oxidizing agent, phenyl benzoate may be formed in high yield.[262] This mechanism has been substantiated for the formation of phenyl benzoate when benzoyl peroxide is photolysed in benzene saturated with oxygen.[263] Kovacic and coworkers have commented further on their work in which cupric chloride is the oxidant,[264] and similar ester formation has now been found for thermal decomposition at 80° in toluene containing iodine.[265] (See also ref. 439 for oxygen-promoted cinnamoylation of benzene.)

Perhaps one of the most interesting recent results in homolytic aromatic substitution is the production of tolan by photolysis of the iodoacetylene (**71**) in benzene.[266] This appears to be the first example of a reaction of an acetylenic free radical in solution. Preliminary results with substituted benzenes suggest that the radical from (**71**) is somewhat more electrophilic in its behaviour than

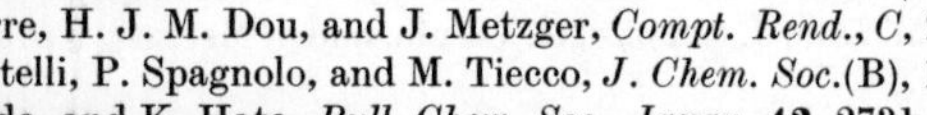

$$PhC\equiv CI \xrightarrow{h\nu} PhC\equiv C\cdot \xrightarrow{PhH} PhC\equiv CPh$$

(**71**)

[258] C. M. Camaggi, R. Leardini, M. Tiecco, and A. Turdo, *J. Chem. Soc.*(B), **1969**, 1251.
[259] G. Vernin, B. Narre, H. J. M. Dou, and J. Metzger, *Compt. Rend., C*, **268**, 2025 (1969).
[260] L. Benati, G. Martelli, P. Spagnolo, and M. Tiecco, *J. Chem. Soc.*(B), **1969**, 472.
[261] T. Sato, S. Shimada, and K. Hata, *Bull. Chem. Soc., Japan*, **42**, 2731 (1969).
[262] See *Org. Reaction Mech.*, **1968**, 291.
[263] M. Kobayashi, H. Minato, and Y. Ogi, *Bull. Chem. Soc. Japan*, **42**, 2737 (1969).
[264] P. Kovacic, C. G. Reid, and M. E. Kurz, *J. Org. Chem.*, **34**, 3302, 3308 (1969).
[265] S. Hashimoto, W. Koike, and M. Yamamoto, *Bull. Chem. Soc. Japan*, **42**, 2733 (1969).
[266] G. Martelli, P. Spagnolo, and M. Tiecco, *Chem. Comm.*, **1969**, 282.

is phenyl, consistent with the *sp*-hybridization of the orbital containing the unpaired electron. In addition, the relatively high yield of *o*-t-butyltolan from the reaction in t-butylbenzene testifies to the small steric requirement of this new radical.

Among alkylation reactions reported during the year has been an attempt to employ the formation of radicals from halides by reaction with methyl-magnesium iodide and a transition-metal salt such as cobaltous chloride.[267] Anomalous behaviour, in the case of 2-phenylethyl radicals, led to the suggestion that these existed in the bridged form (72), but this possibility seems to be discounted by a spectroscopic study of the phenylethyl radical by Kochi

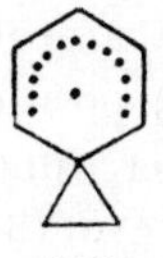

(72)

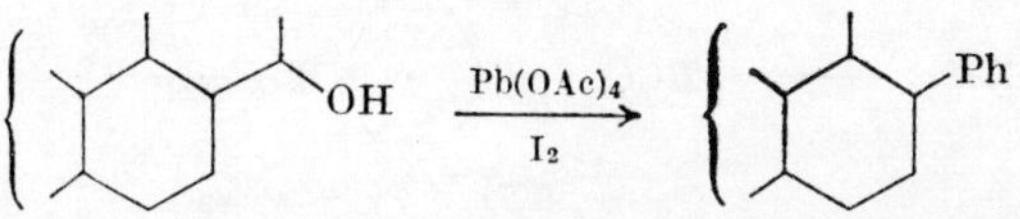

(73) $\xrightarrow[I_2]{Pb(OAc)_4}$ (74)

and Krusic.[49] Also reported have been descriptions of 1-adamantylation of benzene by decomposition of t-butylperoxycarbonyladamantane,[268] n-propylation of benzene by photolysis of tetrapropyl-lead,[269] and benzylation of quinoline and related heterocycles,[270] and of fluorene and pyrene.[271] An unexpected product from the last of these reactions is 1,1'-bipyrenyl. An attempted hypoiodite reaction on a cyclohexane derivative of partial structure (73) gave the product (74), presumably by fragmentation of the alkoxy radical intermediate and alkylation of the benzene solvent by the resulting cyclohexyl radical.[272] Cyclohexyl radicals themselves are known to displace aromatic halogens, and it has now been found that the ease of this reaction is in the order $F > I > Br > Cl$.[273] Clearly, therefore, the mechanism does not involve breaking of the carbon–halogen bond in the rate-determining step.

[267] D. I. Davies, J. N. Done, and D. H. Hey, *J. Chem. Soc.*(C), **1969**, 2021; cf. ref. 244.
[268] G. A. Razuvaev, L. S. Bogulslavskaya, V. S. Etlis, and G. V. Brovkina, *Tetrahedron*, **25**, 4925 (1969).
[269] G. G. Petukhov, Y. A. Kaplin, O. N. Druzhkov, and A. S. Emel'yanova, *Tr. Khim. Khim. Tekhnol.*, **1967**, 152; *Chem. Abs.*, **70**, 77044 (1969).
[270] K. C. Bass and P. Nababsing, *J. Chem. Soc.*(C), **1969**, 388.
[271] K. C. Bass and G. M. Taylor, *J. Chem. Soc.*(C), **1969**, 508.
[272] H. Suginome, H. Ono, and T. Masamune, *Tetrahedron Letters*, **1969**, 2909.
[273] J. R. Shelton and C. W. Uzelmeier, *Rec. Trav. Chim.*, **87**, 1211 (1968).

The ease of displacement of fluorine must reflect the nucleophilic character of the cyclohexyl radical and consequent ease of formation of cyclohexadienyl intermediates such as (75). A reaction which appears to involve homolytic displacement of a nitro group occurs when 1-nitronaphthalene is heated with sulphur to give polysulphides (76).[274]

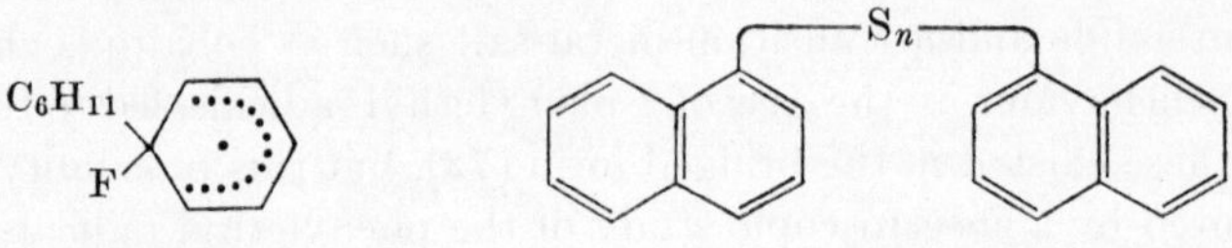

(75)   (76)

The nucleophilic character of acetyl radicals is demonstrated by their selective reactivity at the α- and γ-positions of pyridine heterocycles dissolved in aqueous sulphuric acid.[275] Other aromatic substrates are inert, and a strong contribution to the transition state for the reaction by an acyl cation–pyridinyl radical structure (77) was suggested.

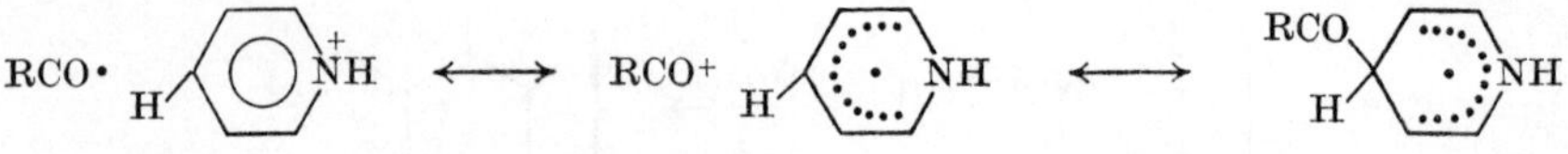

(77)

A homolytic mechanism has been advanced for the fluorination of benzene by xenon difluoride,[276] and aromatic silylation by triethylsilyl radicals has been observed.[277] Norman's group have made some interesting observations on the pH-dependence of aromatic hydroxylation using Fenton's reagent. With toluene, for example, the ratio of cresols to bibenzyl decreases with acidity, and it was suggested that this was a consequence of acid-catalysed degradation of the hydroxycyclohexadienyl intermediates to benzyl radicals as shown.[278]

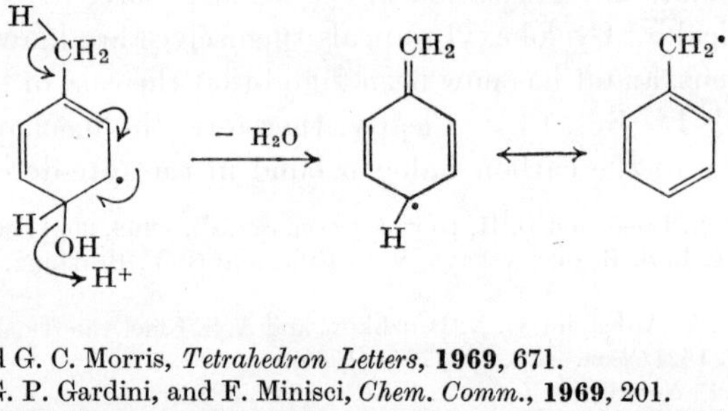

[274] J. A. Elix and G. C. Morris, *Tetrahedron Letters*, **1969**, 671.
[275] T. Caronna, G. P. Gardini, and F. Minisci, *Chem. Comm.*, **1969**, 201.
[276] M. J. Shaw, H. H. Hyman, and R. Filler, *J. Am. Chem. Soc.*, **91**, 1563 (1969).
[277] H. Sakurai, A. Hosomi, and M. Kumada, *Tetrahedron Letters*, **1969**, 1755, 1757.
[278] C. R. E. Jefcoate, J. R. L. Smith, and R. O. C. Norman, *J. Chem. Soc.*(B), **1969**, 1013.

Although new results are in accord with very high rate constants (ca. 10^9 l mole^{-1} sec^{-1}) for the addition of hydrogen atoms to aromatic molecules, mesitylene exchanges nuclear hydrogen for tritium atoms some nine times faster than does benzene.[279]

Rearrangements

Simple 1,2-shifts are relatively rare in free-radical chemistry, with the possible exception of those of aryl groups. Homolytic migration of aryl groups from germanium to oxygen has now been encountered,[280] but no evidence for aryl migration from silicon to carbon could be found, even in the reactions of (78).[281] Several possible reasons for this failure were considered, including the decrease in steric crowding round silicon compared with that round carbon in the 2,2,2-triphenylethyl radical, and the possibility of stabilization of (78) by interaction with silicon *d*-orbitals. Other examples of aryl migrations have

$$Ph_3SiCH_2\bullet \qquad\qquad \overset{\bullet}{C}H_2$$

(78) (79)

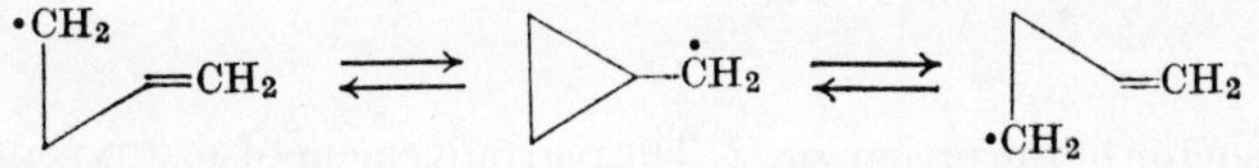

$$\cdot CH_2 \quad \rlap{=\!CH_2}\Big\lfloor\quad \rightleftarrows \quad \triangleright\!\!-\!\overset{\bullet}{C}H_2 \quad \rightleftarrows \quad \underset{\cdot CH_2}{\Big\lceil}\!\!=\!CH_2 \quad \cdots (16)$$

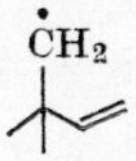

been described, but in related work no vinyl migration was detected in reactions believed to involve (79).[282] Vinyl migrations and the related cyclopropylcarbinyl⇄allylcarbinyl rearrangements (equation 16) have been exemplified by new results involving norbornenyl radicals. Thus biphenyl radical anion reacts at room temperature with either *exo-* or *endo-*5-norbornenyl halides (80) or with nortricyclyl halides (81) to give the same mixture of norbornene and nortricyclene. However, at −58° the radical rearrangement is slowed compared with the electron-transfer reduction of the radicals to carbanions, and the product hydrocarbons show substantial retention of the

[279] J. R. Adsetts and V. Gold, *Chem. Comm.*, **1969**, 353; J. R. Adsetts and V. Gold, *J. Chem. Soc.*(B), **1969**, 1108, 1114; P. Neta and L. M. Dorfman, *J. Phys. Chem.*, **73**, 413 (1969); see also V. D. Shatrov, I. I. Chkheidze, V. N. Shamshev, and N. Y. Buben, *Khim. Vys. Energ.*, **2**, 413 (1968); *Chem. Abs.*, **70**, 28200 (1969); and R. V. Lloyd, F. Magnotta, and D. E. Wood, *J. Am. Chem. Soc.*, **90**, 7142 (1968).
[280] R. L. Dannley and G. C. Farrant, *J. Org. Chem.*, **34**, 2432 (1969).
[281] J. W. Wilt, O. Kolewe, and J. F. Kraemer, *J. Am. Chem. Soc.*, **91**, 2624 (1969).
[282] M. Abramovici and H. Pines, *J. Org. Chem.*, **34**, 266 (1969).

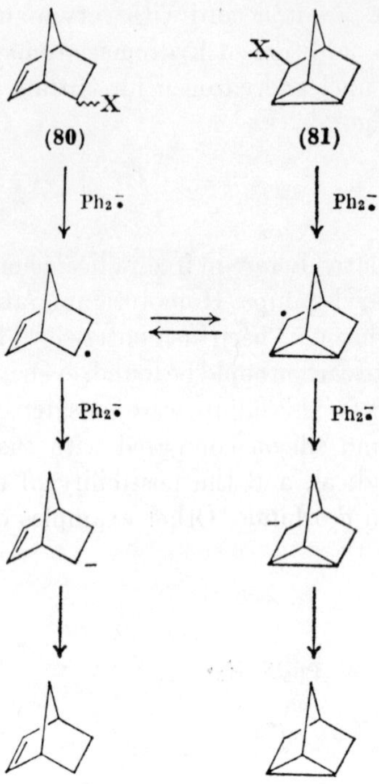

(80)   (81)

$\downarrow$ Ph₂· $\downarrow$ Ph₂·

(structures with Ph₂· arrows, rearrangement equilibrium)

$\downarrow$ Ph₂· $\downarrow$ Ph₂·

carbon skeleton of the precursor.[283] The rearrangement of acetoxynorbornenyl radicals or acetoxynortricyclyl radicals formed by reduction of the organo-mercury compounds (82) and (83) by reaction with sodium borohydride gives products arising from all three isomeric radicals (84)—(86). This is the first instance of identification of products from radical (86), in which the acetoxy substituent is attached to C-7.[284]

Hydrogen abstraction from benzylic or allylic methylene groups which also carry a cyclopropyl substituent constitutes a route to cyclopropylcarbinyl radicals, and several reactions of N-bromosuccinimide with molecules containing this structural component have been shown to furnish products with rearranged (allylcarbinyl) structures.[285] In some of these reactions it is not clear whether carbonium ion intermediates, formed by ionization of allylic bromides, might intervene. However, in the interesting case of bromination of (87), the unrearranged bromides (88) (60% yield) are stable to the reaction

[283] S. J. Cristol and R. W. Gleason, *J. Org. Chem.*, **34**, 1763 (1969).

[284] G. A. Gray and W. R. Jackson, *J. Am. Chem. Soc.*, **91**, 6205 (1969); see also J. J. McCullough and P. W. W. Rasmussen, *Chem. Comm.*, **1969**, 387.

[285] E. C. Friedrich, *J. Org. Chem.*, **34**, 528 (1969); see also D. J. Atkinson, M. J. Perkins, and P. Ward, *Chem. Comm.*, **1969**, 1390.

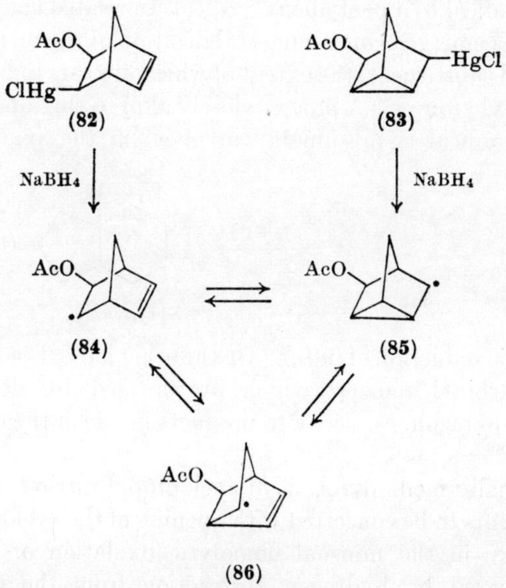

(82)

(83)

NaBH₄

NaBH₄

(84) ⇌ (85)

AcO

(86)

conditions, and hence formation of (89) (27%) and naphthalene (5%) must presumably follow radical rearrangement. The route leading to naphthalene can be considered as involving a symmetry-allowed disrotatory opening of a

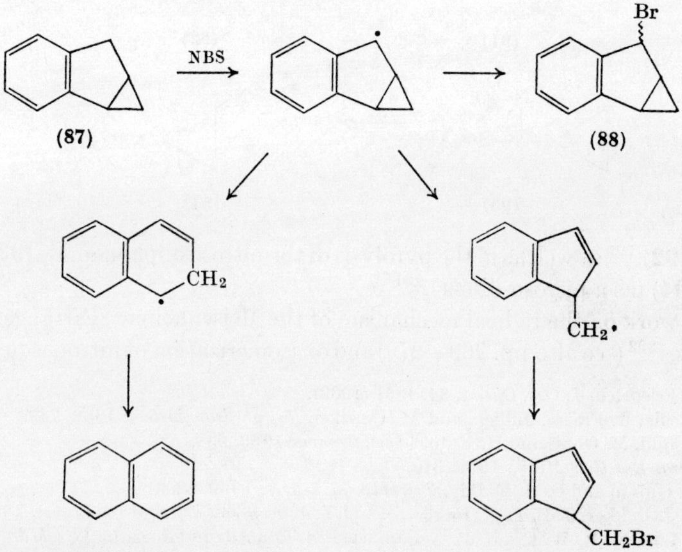

(87) $\xrightarrow{\text{NBS}}$ $\longrightarrow$ (88)

(89)

cyclopentenyl radical to a pentadienyl radical. In related competition studies new evidence has emerged concerning stabilization of the incipient radical by an α-cyclopropyl substituent, the extent of which appears to be comparable to that by an α-vinyl group.[286] A process closely akin to ring-opening of a cyclopropylcarbinyl radical is presumably involved in the synthetically useful

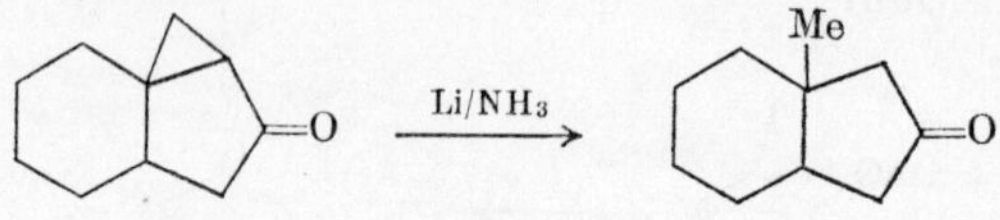

(90)

lithium–ammonia reduction of (**90**).[287] On the other hand diradicals containing a cyclopropylcarbinyl moiety, which are formed by decomposition of 3-cyclopropyl-1-pyrazolines, decay to products in which the cyclopropyl ring remains intact.[288]

The exceptionally facile pyrolysis of cyclopropyl nitrites, in which loss of nitric oxide appears to be concerted with opening of the cyclopropane ring,[289] finds an analogy in the unusual homolytic oxidation of cyclopropanols, apparently initiated by hydrogen abstraction from the OH group, e.g.

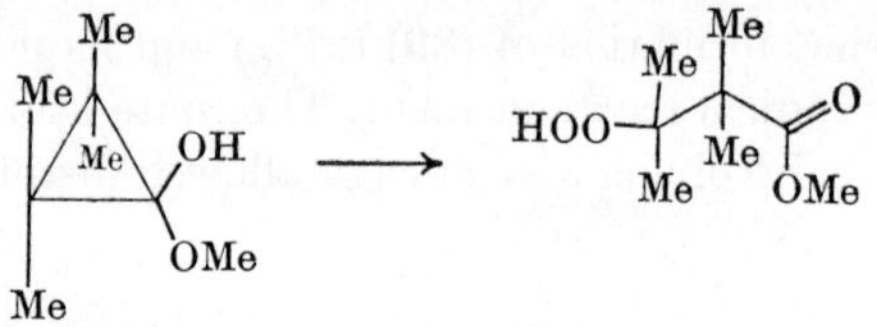

(91)   **(92)**

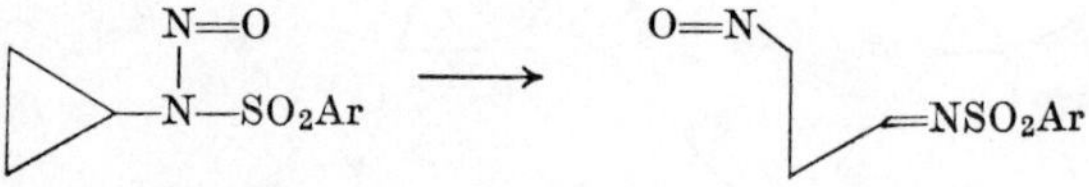

(93)   **(94)**

(**91**)→(**92**),[290] as well as in the pyrolysis of the nitrososulphonamide (**93**) which gives (**94**) (as a nitroso-dimer).[291]

New work on the radical mechanism of the Meisenheimer rearrangement of *N*-oxides[292] (see also pp. 269—270) and on isomerization of nitrones to *O*-alkyl

[286] E. C. Friedrich, *J. Org. Chem.*, **34**, 1851 (1969).
[287] C. Cueille, R. Fraisse-Jullien, and A. Hunziker, *Tetrahedron Letters*, **1969**, 749.
[288] T. Sanjiki, M. Ohta, and H. Kato, *Chem. Comm.*, **1969**, 638.
[289] See *Org. Reaction Mech.*, **1968**, 310.
[290] D. H. Gibson and C. H. DePuy, *Tetrahedron Letters*, **1969**, 2203.
[291] E. E. J. Dekker, J. B. F. N. Engberto, and T. J. de Boer, *Tetrahedron Letters*, **1969**, 2651.
[292] J. P. Lorand, R. W. Grant, P. A. Samuel, E. O'Connell, and J. Zaro, *Tetrahedron Letters*, **1969**, 4087.

oximes has been described;[293] the conversion of optically active alkyl iso-cyanides into nitriles is accompanied by racemization, indicative of a free-radical mechanism,[294] and homolytic mechanisms have been proposed for the pyrolytic rearrangement of (95),[295] and the isomerization of (96).[296]

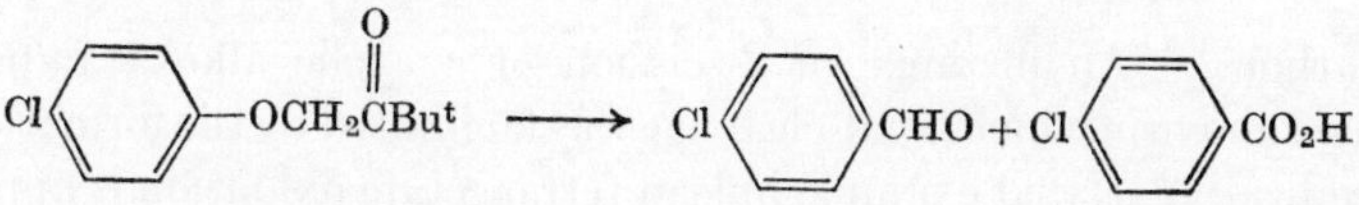

(95)

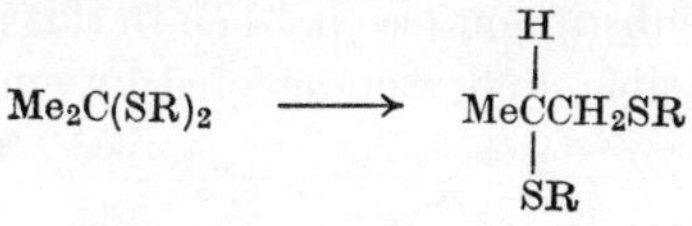

(96)

The complexities of the Ladenburg rearrangement of N-alkylpyridinium salts,[297] and the temperature dependence of 1,3-hydrogen transfer in propyl radicals,[298] have been examined, and an interesting example of the photolysis of an unsaturated nitrite has been encountered in which the intermediate cyclic alkoxy radical undergoes β-scission which results in a stabilized allylic radical; and subsequent processes lead ultimately to a cyclic nitrone.[299]

Reactions Involving Oxidation or Reduction by Metal Salts

(see also Chapter 14)

Several reports have appeared concerning the factors which affect cyclic ether formation during oxidation of aliphatic alcohols by lead tetraacetate.[300] These include a study of the competition between fragmentation of the intermediate alkoxy radical (97) and intramolecular hydrogen transfer. The proportion of fragmentation is increased by alkyl-substitution both on the α-carbon and on

[293] J. S. Vincent and E. J. Grubbs, *J. Am. Chem. Soc.*, **91**, 2022 (1969); see also E. J. Grubbs and J. A. Villarreal, *Tetrahedron Letters*, **1969**, 1841.

[294] S. Yamada, K. Takoshima, T. Sato, and S. Terashima, *Chem. Comm.*, **1969**, 811.

[295] E. Bisugni and C. Rivalle, *Bull. Soc. Chim. France*, **1969**, 3111.

[296] A. B. Terent'ev and G. N. Shvedova, *Izv. Akad. Nauk SSSR, Ser. Khim.*, **1968**, 2231; *Chem. Abs.*, **70**, 28184 (1969).

[297] P. A. Claret and G. H. Williams, *J. Chem. Soc.*(C), **1969**, 146.

[298] O. A. Reutov, G. M. Ostapchuk, K. Uteniyazov, and E. V. Binshtok, *Zh. Org. Khim.*, **4**, 1868 (1968).

[299] H. Suginome, N. Sato, and T. Masamune, *Tetrahedron Letters*, **1969**, 3353.

[300] M. L. Mihailovic et al., *Chem. Comm.*, **1969**, 236; V. M. Micovic, S. Stojcic, M. Bralovic, S. Mladenovic, D. Jeremic, and M. Stefanovic, *Tetrahedron*, **25**, 985 (1969); M. L. Mihailovic, A. Milovanovic, S. Konstantinovic, J. Jankovic, Z. Cekovic, and R. E. Partch, *ibid.*, p. 3205; S. Moon and B. H. Waxman, *J. Org. Chem.*, **34**, 288 (1969).

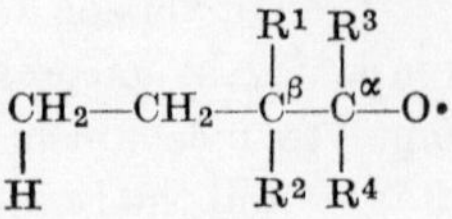

$$CH_2-CH_2-\overset{\overset{R^1}{|}}{\underset{\underset{R^2}{|}}{C}}{}^{\beta}-\overset{\overset{R^3}{|}}{\underset{\underset{R^4}{|}}{C}}{}^{\alpha}-O\cdot$$

$$\underset{H}{|}$$

(97)

the β-carbon.[301] An instance of β-scission of a cyclic alkoxy radical was mentioned above in which bond-cleavage was facilitated by the formation of an allylic radical. A related example in lead tetraacetate oxidation is apparently facilitated by formation of an alkoxyalkyl radical despite the relatively small stabilization by the alkoxy substituent (see p. 317). In this reaction (equation 17) the β-scission was reversible, and was detected by epimerization in the

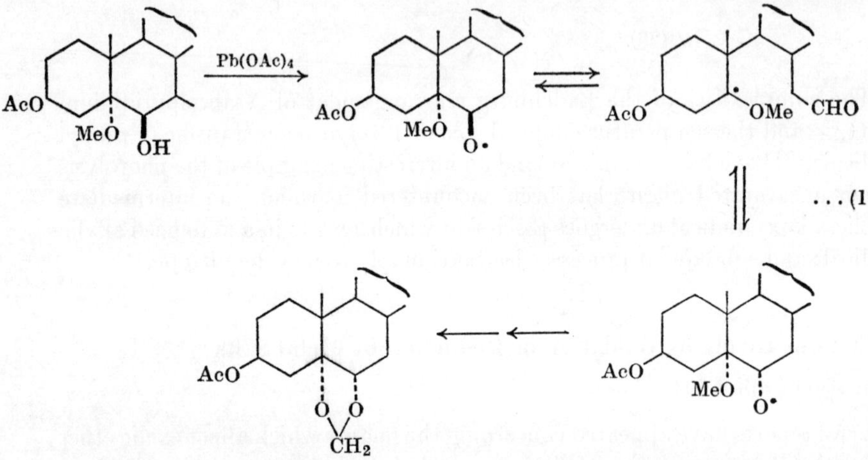

... (17)

product.[302] An attempt to generate butoxy radicals by lead tetraacetate oxidation of t-butanol in the presence of toluene gave some benzyl acetate, possibly formed via benzyl radicals themselves resulting from attack of butoxy radicals on toluene; however, product yields were very low.[303]

Julia's group have used lead tetraacetate to oxidize half-esters of substituted malonic acids, e.g. (98), in a new approach to homolytic ring-closure. Although the intermediate alkoxycarbonyl radicals have a high ionization potential and are consequently difficult to oxidize to carbonium ions, the data pointed to some association of the radical centre with lead (or with added cupric copper).[304]

301 M. L. Mihailovic, M. Jokovljevic, V. Trifunovic, R. Vukov, and Z. Cekovic, *Tetrahedron*, 24, 6959 (1968).
302 P. Morand, *J. Org. Chem.*, 34, 2175 (1969).
303 C. Walling and J. Kjellgren, *J. Org. Chem.*, 34, 1488 (1969).
304 J. C. Chottard, M. Julia, and J. M. Salard, *Tetrahedron*, 25, 4967 (1969).

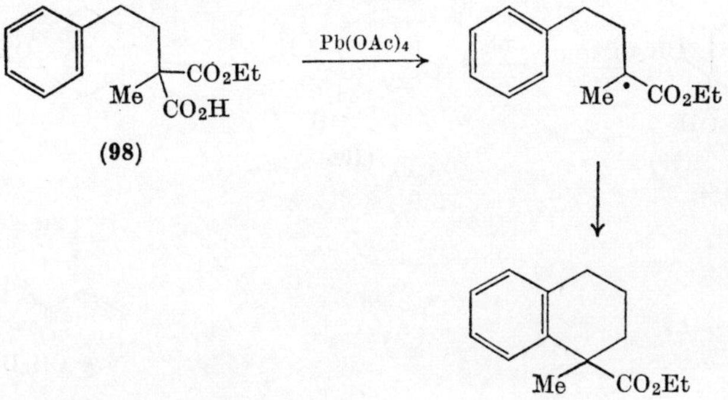

(98)

Oxidation of phenylhydrazine with lead tetraacetate in methylene chloride at low temperature affords benzenediazonium ion, and at higher temperatures in aromatic solvents aromatic phenylation is observed;[305] the possible intermediacy of benzenediazoacetate was considered.

From observations on the relative ease of photoinitiated oxidation of a series of aliphatic carboxylic acids, RCO_2H, Kochi and coworkers concluded that concerted fission of the Pb–O and R–C bonds must occur.[306] ESR studies of photolysis of lead tetraacylates have involved both direct observation of the resulting alkyl radicals (at 77° K)[307] and some interesting applications of the spin trapping techniques described in the following section. Photochemical initiation has also been employed to effect oxidation of alcohols to cyclic ethers, often with greater efficiency than is found in the corresponding thermal reaction.[308]

The formation of alkyl radicals by photolysis of thallium triacylates, unlike the Pb^{IV} reactions, is apparently non-chain and relatively slow.[309]

The Hunsdiecker reaction with (99) gives the anomalous product (101). The formation of a ring-opened product supports a radical mechanism, since the hydroxylated cation (100; · = +) rearranges to give cycloheptanone.[310]

The spectra of acidic oxygen-free solutions of a ceric salt containing a tertiary alcohol show evidence of charge-transfer between ceric ions and the alcohols. If such solutions are frozen to 77° K and irradiated with ultraviolet light, alkyl radicals resulting from fragmentation of the alcohol may be detected by ESR.[311] Similar results are obtained with carboxylic acids.

[305] J. B. Aylward, *J. Chem. Soc.*(C), **1969**, 1663.
[306] J. K. Kochi, R. A. Sheldon, and S. S. Lande, *Tetrahedron*, **25**, 1197 (1969).
[307] K. Heusler and H. Loeliger, *Helv. Chim. Acta*, **52**, 1495 (1969); H. Loeliger, *ibid.*, p. 1516.
[308] M. L. Mihailovic, M. Jakovljevic, and Z. Cekovic, *Tetrahedron*, **25**, 2269 (1969).
[309] N. A. Maier, G. P. Korotyshova, and Y. A. Ol'dekop, *Zh. Obshch. Khim.*, **38**, 2384 (1969); *Chem. Abs.*, **70**, 57071 (1969).
[310] N. G. Kundu and A. J. Sisti, *J. Org. Chem.*, **34**, 229 (1969).
[311] D. Greatorex and T. J. Kemp, *Chem. Comm.*, **1969**, 383.

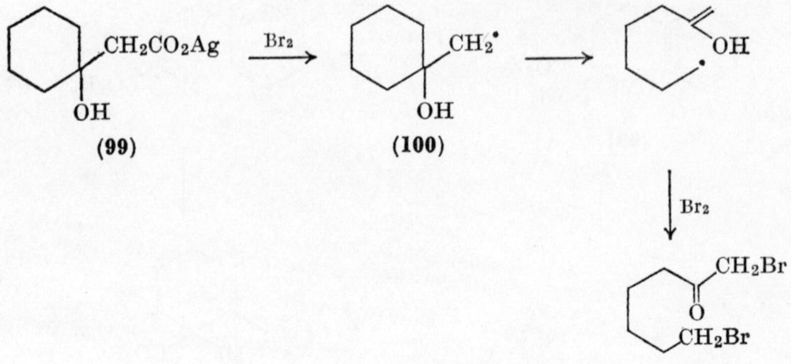

(99)   (100)

(101)

Permanganate oxidation of N-phenyl-2-naphthylamine gives the 1,1'- and 1,N'-dehydro-dimers;[312] no analogue of tetraphenylhydrazine was detected.

In a careful study of the oxidation of toluenes by manganic acetates in acetic acid,[313] it has been found that the principal initial reaction is the formation of carboxymethyl radicals ($\cdot CH_2CO_2H$) which are not readily oxidized to carbonium ions, and which add to the aromatic nucleus or abstract side-chain hydrogen, the former process predominating. When the ionization potential of a substituted toluene is sufficiently low (ca. 8 ev), electron-transfer oxidation may be important as, for example, in the case of p-methoxytoluene. This may be suppressed by addition of potassium acetate, which may lead to a more efficient precursor of carboxymethyl radicals, e.g. $[Mn(OAc)_4]^-$, or by rigorous exclusion of moisture, which again appears to affect the ligands attached to the metal atom. It is possible, by suppressing the electron transfer process in this way, to obtain relative reactivity data for hydrogen abstraction from a series of toluenes by carboxymethyl radicals. These give a good correlation with σ^+, with $\rho = -0.68$. Relative reactivities for the aromatic substitution reaction were also recorded. It is also of interest that the electron transfer process could be suppressed by addition of Mn^{III} ions, which indicates that the electron transfer step is reversible.

A very much stronger response to substituents was found in the side-chain oxidation of toluenes by cobaltic acetate in acetic acid ($\rho = -2.4$).[314] This is consistent with electron transfer followed by proton transfer from toluene radical cation; again the reaction was sensitive to the ligands attached to the metal atom. For example, addition of lithium chloride gave a much more reactive oxidant which was likewise less selective ($\rho = -1.35$).

Some interesting results have also been obtained by Smith and Waters on the

312 R. F. Bridger, D. A. Law, D. F. Bowman, B. S. Middleton, and K. U. Ingold, *J. Org. Chem.*, **33**, 4329 (1968).
313 E. I. Heiba, R. M. Dessau, and W. J. Koehl, *J. Am. Chem. Soc.*, **91**, 138 (1969).
314 E. I. Heiba, R. M. Dessau, and W. J. Koehl, *J. Am. Chem. Soc.*, **91**, 6830 (1969).

Co^{III}-oxidation of cinnamic acid.[315a] Competition between decarboxylation, which in an aqueous medium leads to phenylacetaldehyde, and electron-transfer oxidation is observed. The formation of benzophenone as a product of electron-transfer oxidation of α-phenylcinnamic acid was attributed to aryl migration in the intermediate radical cation.

Mention may also be made here of the publication of the papers presented at a Faraday Society Discussion on homogeneous oxidation.[315b]

The reduction of both isomeric 1-chloro-1-methyl-4-t-butylcyclohexanes by Cr^{II}-ethylenediamine complexes gives the same mixture of 1-methyl-4-t-butylcyclohexanes (60% *cis*, 40% *trans*). The same is true in the presence of a thiol, but now the composition of the product mixture is different (predominantly *trans*). The interpretation is in terms of competition between preferred axial hydrogen abstraction by the radical, and formation of an organochromium intermediate which is more stable with the chromium equatorial, and which is hydrolysed stereospecifically such as to leave the methyl group axial.[316] Bamford's group have demonstrated that electron-transfer reduction of an organic halogen compound may result in a halogen atom (equation 18) instead of the more usual halide anion (equation 19) provided that X is an appropriate grouping.[317]

$$ClX + M^{n+} \rightarrow Cl\cdot + X^- + M^{(n+1)+} \qquad \ldots (18)$$

$$ClX + M^{n+} \rightarrow Cl^- + X\cdot + M^{(n+1)+} \qquad \ldots (19)$$

The formation of traces of ethylene from monoethyl sulphate and ferrous sulphate with t-butyl hydroperoxide may have interesting biochemical implications, the suggested mechanism involving electron transfer across a sigma framework (reaction 20).[318]

$$[R\cdot\text{---}HCH_2CH_2OSO_2O^-\text{---}Fe^{II}] \rightarrow RH + CH_2{=}CH_2 + SO_4^{2-} + Fe^{III} \qquad \ldots (20)$$

The coupling of allyl (or benzyl) groups by pyrolysis of bivalent titanium alkoxides has been shown to involve free radicals by monitoring the

$$Ti(OR)_2 \rightarrow TiO_2 + R_2$$

spectrum of products obtained from an unsymmetrically substituted allyl alcohol. This is typical of coupling of unsymmetrical allyl radicals.[319a] It was also concluded that two allyl groups were probably released simultaneously in view of the failure to obtain radical coupling products (RR) when TiO(OR)Cl was reduced with 1 equivalent of sodium naphthalene, and the product was heated. However, it was apparently not demonstrated that TiO(OR) was

[315a] P. Smith and W. A. Waters, *J. Chem. Soc.*(B), **1969**, 462.

[315b] *Discuss. Faraday Soc.*, **46** (1968).

[316] R. E. Erickson and R. K. Holmquist, *Tetrahedron Letters*, **1969**, 4209.

[317] C. H. Bamford, G. C. Eastmond, and D. Whittle, *J. Organometal. Chem.*, **17**, P33 (1969).

[318] J. Kumamoto, H. H. A. Dollwet, and J. M. Lyons, *J. Am. Chem. Soc.*, **91**, 1207 (1969).

[319a] E. E. van Tamelen, B. Akermark, and K. B. Sharpless, *J. Am. Chem. Soc.*, **91**, 1553 (1969).

present after consumption of the reducing agent was complete. Spontaneous decomposition of TiO(OR) in the presence of naphthalene radical anions would be very unlikely to produce dimer, RR.

Aryl halides are reduced to the corresponding hydrocarbons by copper or cuprous oxide in boiling 2,4,6-collidine containing primary or secondary alkoxide ions,[319b] or suitable phenolic hydrogen-donors.[319c] Aryl radical involvement seems likely with, for example, oxidative coupling of the phenols being observed, though the possibility that this is a reaction of cupric phenoxides is not excluded. Displacement of halogen by hydride was also considered a possibility for the reactions with alkoxide, because other hydride donors such as sodium borohydride were also effective. Cupric phenoxides were invoked to explain oxidative coupling observed (together with dealkylation or deacylation) when 9-alkoxy- or 9-acyloxy-anthracenes reacted with cupric bromide.[319d] Other anthracenes are brominated by this reagent by a ligand-transfer mechanism.

Nitroxides

The utility of *C*-nitroso-compounds, and of nitrones, as diamagnetic radical-scavengers, in conjunction with ESR spectroscopy to identify the nitroxides which result from scavenging experiments with these compounds, was introduced last year as a procedure of some generality for the study of free-radical reactions.[320] New applications of this "spin trapping" technique which have been reported include the scavenging of succinimidyl radicals when solutions of *N*-bromosuccinimide are photolysed in the presence of 2-methyl-2-nitrosopropane (**102**) as well as trapping of solvent-derived radicals when this reaction is conducted in the presence of an olefin.[321] As the authors point out, the trapping of solvent-derived allylic radicals does not permit any definite conclusion to be made concerning the identity of the hydrogen-abstracting species in these reactions. The solvent-derived radicals scavenged by (**102**) were considered to be allylic. However, the possibility that they may have been formed by addition to the olefins, e.g. (**103**), does not appear to be excluded by the spectra of the derived nitroxides.

The more easily accessible nitroso-compound (**104**) has an advantage over (**102**) in that the lines in the spectra of the derived nitroxides are sharper and hence better resolved, and Lagercrantz's group have employed this to study the production of alkoxy and alkyl radicals by lead tetraacetate oxidation of

[319b] R. G. R. Bacon and S. C. Rennison, *J. Chem. Soc.*(C), **1969**, 308, 312.
[319c] R. G. R. Bacon and O. J. Stewart, *J. Chem. Soc.*(C), **1969**, 301.
[319d] D. Mosnaim, D. C. Nonhebel, and J. A. Russell, *Tetrahedron*, **25**, 1591, 3485, 5047 (1969); D. C. Nonhebel and J. A. Russell, *ibid.*, p. 3493.
[320] See *Org. Reaction Mech.*, **1968**, 299.
[321] C. Lagercrantz and S. Forshult, *Acta Chem. Scand.*, **23**, 708 (1969).

Me₃CN=O

(102)

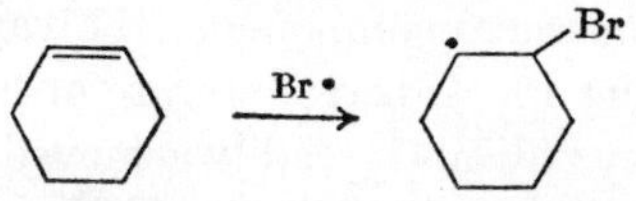

Br •
$\longrightarrow$

Br

(103)

COCH₃
|
Me₂CN=O

(104)

Ph
＼
H ／ =N ↗ O
＼Buᵗ

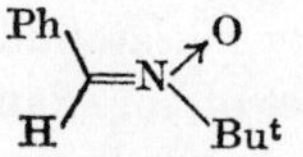

(105)

alcohols and carboxylic acids initiated by ultraviolet light (see p. 343).[322] The same group have demonstrated that photolysis of hydrogen peroxide in the presence of sulphoxides, RSOR, gives alkyl radicals which are also identifiable by scavenging with nitroso-compounds.[323] (In independent work, the direct ESR observation of alkyl radicals formed from sulphoxides and hydroxyl radicals in a flow system has been achieved.[324]) It has been suggested that problems resulting from photolability of aliphatic C-nitroso-compounds, and the consequent formation of symmetrical nitroxides (equations 21 and 22) may

$$\text{Bu}^t\text{N}=\text{O} \rightarrow \text{Bu}^t\cdot + \text{NO} \qquad \ldots (21)$$

$$\text{Bu}^t\cdot + \text{Bu}^t\text{N}=\text{O} \rightarrow \text{Bu}^t_2\text{NO}\cdot \qquad \ldots (22)$$

be avoided in photochemical studies by using light of wavelength ca. 300 nm, to which nitrosobutane is transparent.[325a] In this way it proved possible to detect the first example of a (very short lived) thiyl nitroxide ($a_N = $ ca. 18.5 gauss) by photolysis of a disulphide in the presence of (102). In other experiments, however, it appeared that di-t-butyl nitroxide was formed, and it seems likely that reaction (21) can be photosensitized.

The complex reactions which occur on protracted photolysis of nitrosobenzene have been discussed.[325b]

Radical scavenging by 1-nitrosoadamantane has been recorded,[326] as well as the scavenging by (102) and by nitrosobenzene of radicals formed during nickel peroxide oxidations. These included N-carbazolyl, formed by oxidation of carbazole.[235]

Janzen and Blackburn have given details of their work on spin trapping with the nitrone (105).[327] They show, for example, that in photolyses of

[322] S. Forshult, C. Lagercrantz, and K. Torssell, *Acta Chem. Scand.*, **23**, 522 (1969).

[323] C. Lagercrantz and S. Forshult, *Acta Chem. Scand.*, **23**, 811 (1969).

[324] W. Damerau, G. Lassmann, and K. Lohs, *Z. Chem.*, **9**, 343 (1969).

[325a] I. H. Leaver, G. C. Ramsay, and E. Suzuki, *Austral. J. Chem.*, **22**, 1891 (1969); I. H. Leaver and G. C. Ramsay, *ibid.*, p. 1899.

[325b] R. Tanika, *Bull. Chem. Soc. Japan*, **42**, 210 (1969).

[326] J. W. Hartgerink, J. B. F. N. Engberto, T. A. J. W. Wajer, and T. J. de Boer, *Rec. Trav. Chim.*, **88**, 481 (1969).

[327] E. G. Janzen and B. J. Blackburn, *J. Am. Chem. Soc.*, **91**, 4481 (1969).

organo-lead, -tin, and -mercury compounds, the ease of fragmentation of substituents is in the order Ph > alkyl ≫ acetate or halide. They also report that scavenging of acetoxy radicals is readily achieved, but the observation of nitroxides on mixing certain nitro-amines and (**105**) in the absence of light or radical initiators[328] points to the need for caution in drawing mechanistic conclusions from these scavenging experiments. Other aromatic nitro-compounds do not spontaneously initiate radical reactions in this way, and ultraviolet irradiation of mixtures of, for example, nitrobenzene and (**105**) in hydrogen-donor solvents provides complementary evidence that the radicals once believed to be hydroxy-nitroxides (**106**) are in fact the alkoxy-nitroxides (**107**).[329]

$$ ArNO_2 \xrightarrow{h\nu} ArNO_2{}^* \xrightarrow{RH} ArN(OH)O\cdot + R\cdot $$
$$ (106) $$

$$ R\cdot + ArNO_2 \rightarrow ArN(OR)O\cdot $$
$$ (107) $$

Radicals of the type (**106**) *have* been detected in photochemical reactions of 2,3,5,6-tetrachloronitrobenzene in hydrogen-donor solvents.[330] Photolysis of dinitroprehnitene in diethyl ether gives the corresponding nitro-amine, and

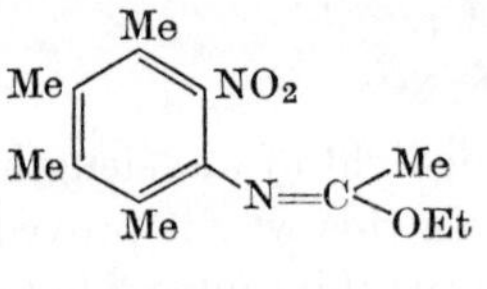

(108) 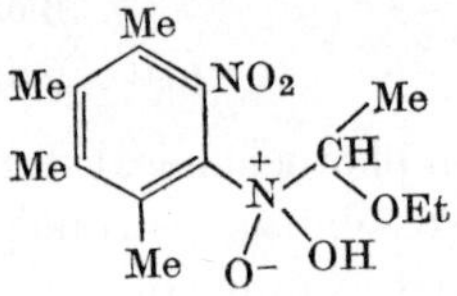　　　　　　　　　　　　　(109)

the imine (**108**).[331] The formation of this product was suggested to involve the intermediate (**109**); this loses water to form a nitrone, which is (subsequently) reduced to (**108**). However, alternative routes can be envisaged involving a nitroso-intermediate.

The literature on nitroxides in which a hydrogen atom is directly bonded to nitrogen (RNHO·) is relatively sparse. New examples have been detected by radical addition to oximes in flow-system experiments (e.g. reaction 23).[332]

$$ R_2C{=}NOH \xrightarrow{\cdot OH} R_2C(OH)\dot{N}OH \rightarrow R_2C(OH)NHO\cdot \qquad \ldots (23) $$

O-Alkyl-oximes give spectra of radicals of general formula (**110**).

$$ R_2C(OH)\dot{N}OR \qquad (110) $$

[328] E. G. Janzen and J. L. Gerlock, *J. Am. Chem. Soc.*, **91**, 3108 (1969).

[329] See *Org. Reaction Mech.*, **1968**, 300; see also J. M. Lynch, P. N. Preston, R. B. Sleight, and L. H. Sutcliffe, *J. Organometal. Chem.*, **20**, 43 (1969).

[330] D. J. Cowley and L. H. Sutcliffe, *Trans. Faraday Soc.*, **65**, 2286 (1969).

[331] H. Hart and J. W. Link, *J. Org. Chem.*, **34**, 758 (1969).

[332] D. J. Edge and R. O. C. Norman, *J. Chem. Soc.*(B), **1969**, 182; P. Smith and W. M. Fox, *Can. J. Chem.*, **47**, 2227 (1969).

Nitroxides with hydrogen attached to nitrogen have also been observed in the reduction of aliphatic nitroso-compounds by hydroxylamines or, photolytically, by thiols.[333]

The relative stabilities of a series of aryl t-butyl nitroxides have been discussed in detail,[334] and the formation of other new nitroxides by oxidation of the corresponding hydroxylamines has been noted.[335] These include a ferrocenyl nitroxide.[336] Other new nitroxides[337] include several designed for use in spin-labelling studies. Products of radical scavenging by tetramethyl-piperidone-N-oxyl (**111**),[338, 339] have been examined, as well as its oxidation of thiols to disulphides[339] and its pyrolytic (80°) breakdown which leads to a polymer of phorone (**112**).[340]

(**111**) $+ R_2NOH$ (**112**)

Other reactions which have been discussed involve the oxidation of secondary amines to nitroxides by ozone,[341] aspects of the chemistry of bistrifluoromethyl nitroxide,[342] and the nitroxide-catalysed isomerization of dimethyl maleate.[343] Stable solutions of protonated di-t-alkyl nitroxides have been obtained by effecting protonation with a strong Brønsted acid, e.g. $AlCl_3$–HCl in methylene chloride;[344a] it seems likely that the radical cations, $R_2\overset{\cdot+}{N}OH$, may be involved as the first stage in oxidations of hydroxylamines to nitroxides. Complexing of nitroxides with Lewis acids has also been described.[344b] Finally,

[333] T. A. J. W. Wajer, A. Mackor, and T. J. de Boer, *Tetrahedron*, **25**, 175 (1969).

[334] A. Calder and A. R. Forrester, *J. Chem. Soc.*(C), **1969**, 1459.

[335] A. Calder, A. R. Forrester, and R. H. Thompson, *J. Chem. Soc.*(C), **1969**, 512; V. S. Griffiths and G. R. Parlett, *J. Chem. Soc.*(B), **1969**, 997.

[336] A. R. Forrester, S. P. Hepburn, R. S. Dunlop, and H. H. Mills, *Chem. Comm.*, **1969**, 698.

[337] D. J. Kosman and L. H. Piette, *Chem. Comm.*, **1969**, 920; see also E. G. Rozantsev and V. I. Suskwa, *Izv. Akad. Nauk SSSR, Ser. Khim.*, **1969**, 1191.

[338] K. Murayama, S. Morimura, and T. Yoshioka, *Bull. Chem. Soc. Japan*, **42**, 1640 (1969).

[339] K. Murayama and T. Yoshioka, *Bull. Chem. Soc. Japan*, **42**, 1942 (1969).

[340] K. Murayama and T. Yoshioka, *Bull. Chem. Soc. Japan*, **42**, 2307 (1969).

[341] S. D. Razumovskii, A. L. Buchachenko, A. B. Shapiro, E. G. Rozantsev, and G. E. Zaikov, *Dokl. Akad. Nauk SSSR*, **183**, 1106 (1969); *Chem. Abs.*, **70**, 95987 (1969).

[342] H. J. Emeléus, *Suomen Kemistilehti*, *B42*, 157 (1969).

[343] A. L. Buchachenko, L. V. Ruban, and E. G. Rozantsev, *Kinet. Katal.*, **9**, 915 (1968); *Chem. Abs.*, **70**, 67353 (1969).

[344a] B. M. Hoffman and T. B. Eames, *J. Am. Chem. Soc.*, **91**, 2169 (1969).

[344b] B. M. Hoffman and T. B. Eames, *J. Am. Chem. Soc.*, **91**, 5168 (1969).

we mention here an interesting discussion of the Linnett double-quartet theory of electronic structure, and its application to the interpretation of phenomena in organic chemistry. It has, for example, been employed to rationalize the ease of the Meisenheimer rearrangement of N-oxides (discussed briefly elsewhere in this chapter), which involves nitroxide formation.[345]

Reactions Involving Radical Ions

The reactions of naphthalene radical anions with alkyl halides have received considerable attention recently. Dissociative electron transfer produces alkyl radicals which may be intercepted by a further equivalent of naphthalene anion, and proton abstraction then gives alkane. The mechanism whereby alkylation of the naphthalene occurs is less clear, but this has now been demonstrated to involve coupling of alkyl radical and naphthalene anion

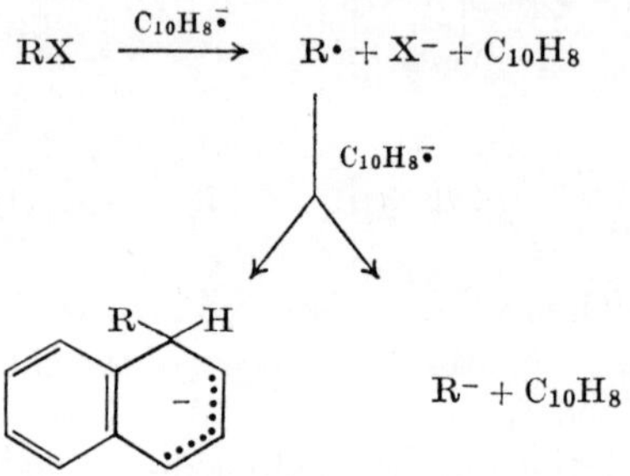

$$RX \xrightarrow{C_{10}H_8\cdot^-} R\cdot + X^- + C_{10}H_8$$

$$R^- + C_{10}H_8 \qquad \qquad \dots (24)$$

(equation 24).[346] Among the factors favouring this mechanism are the increased proportion of alkylation to alkane formation when R is tertiary (reduction $R\cdot \rightarrow R^-$ is most difficult in this case), and that, for a given R, the ratio of alkane to alkylation of naphthalene is independent of the halogen X. When X is I or Br, this reaction also produces some dimer RR; with 1,4-diiodobutane, the only coupling product is cyclobutane. No C_8-products could be detected, and a general cage mechanism for the coupling reaction was advanced which would accommodate this exclusive formation of cyclobutane:[347]

$$RI \xrightarrow{C_{10}H_8\cdot^-} R\cdot \xrightarrow{C_{10}H_8\cdot^- Na^+} R^-Na^+ \xrightarrow{RI} [R\cdot NaI \cdot R] \rightarrow RR$$

In the same study no evidence could be found for the formation of the bridged radical (113);[348] the behaviour of 4-iodobutyl closely parallels that of other

[345] R. A. Firestone, *J. Org. Chem.*, **34**, 2621 (1969).

[346] J. F. Garst, J. T. Barbas, and F. E. Barton, *J. Am. Chem. Soc.*, **90**, 7159 (1968); G. D. Sargent and G. A. Lux, *ibid.*, p. 7160.

[347] J. F. Garst and J. T. Barbas, *Tetrahedron Letters*, **1969**, 3125.

[348] J. F. Garst and J. T. Barbas, *J. Am. Chem. Soc.*, **91**, 3385 (1969); see also P. B. Chock and J. Halpern, *ibid.*, p. 582.

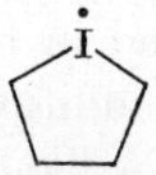

(113)

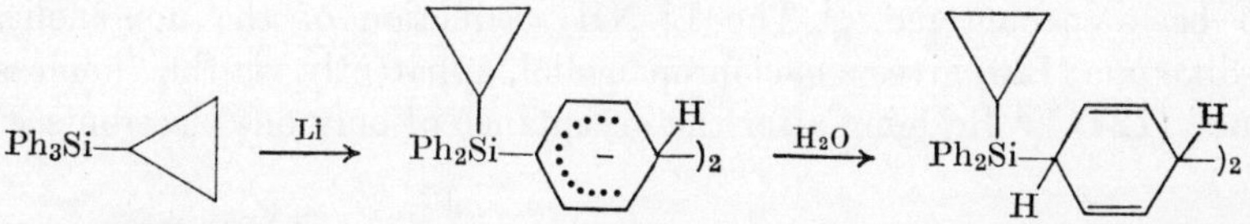

primary radicals in the manner of its distribution between reduction and alkylation of naphthalene anion radicals (see also refs. 27, 182, and 283).

Silicon–carbon bonds are normally cleaved by reaction with lithium, but in THF at −70° a radical anion is formed from the cyclopropyl derivative (114), and this dimerizes.[349]

The interesting electrocyclic transformations of radical anion (115) to (116),[350] and of (117) to (118),[351] have been described; whilst the geometry of the former process is not yet defined, that of the latter is necessarily disrotatory, in violation of symmetry restrictions. The major product from rearrangement of (117) was, however, 9-methylphenanthrene anion radical.

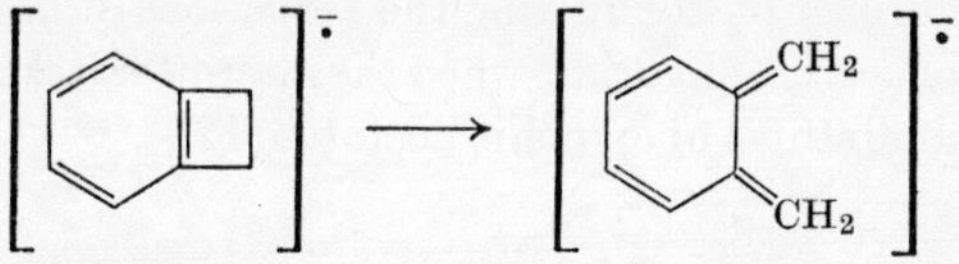

(115)   (116)

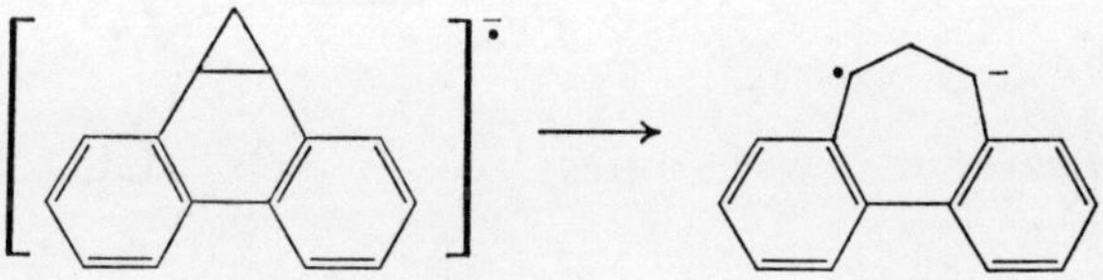

(117)   (118)

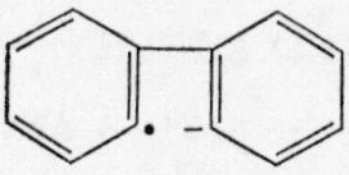

(119)

[349] J. J. Eisch and G. Gupta, *J. Organometal. Chem.*, **20**, P9 (1969).
[350] N. L. Bauld and F. Farr, *J. Am. Chem. Soc.*, **91**, 2788 (1969); a closely related example is given by N. L. Bauld and G. R. Stevenson, *ibid.*, p. 3675.
[351] L. L. Miller and L. J. Jacoby, *J. Am. Chem. Soc.*, **91**, 1130 (1969).

A coupling reaction has been found in the course of experiments with biphenylene radical anion, in which addition to biphenylene leads to dibenzo-[*fg,op*]naphthacene radical anion. A likely first step is bond fission to give **(119)**.[352] A possible example of electrocyclic rearrangement of a semidione has also been encountered.[353] The Li–NH$_3$ reduction of the non-enolizable 1,3-diketone **(120)** gives a cyclopropanediol, apparently via the "homosemidione" **(121)**.[354] Bridging after the acceptance of only one electron seemed

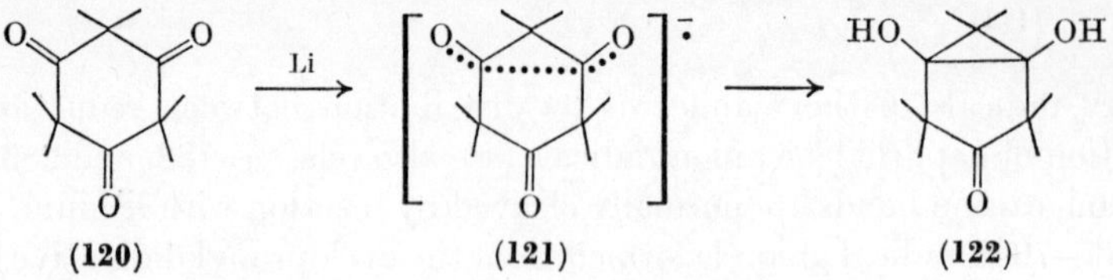

(120)   **(121)**   **(122)**

probable in view of the formation of **(122)** as the only product when a deficiency of lithium was employed. A very similar reaction has been encountered during electrochemical reduction of **(120)**.[355] Cyclopropanol formation during Clemmensen reduction of α,β-unsaturated ketones has been suggested in the past to account for the formation of two isomeric saturated ketones when the supposed cyclopropanol would be unsymmetrically substituted. This mechanism has now been confirmed by conducting the reductions of both **(123)** and **(124)** in the presence of acetic anhydride, when the intermediates were trapped by acetylation, giving mixtures of epimeric acetates **(125)**.[356]

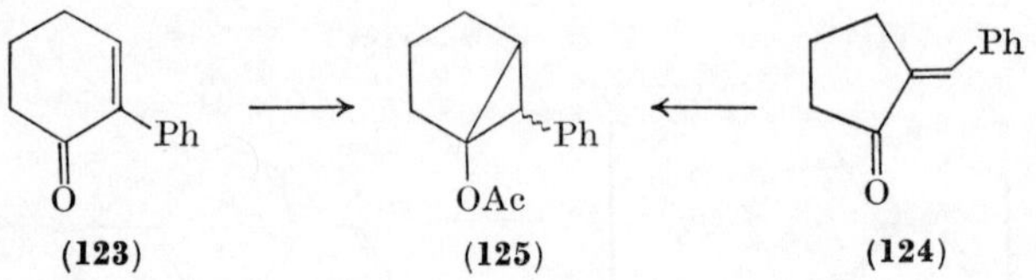

(123)   **(125)**   **(124)**

ESR has been employed to measure the pH-dependence of the decay of nitrobenzene radical anions. The results are in general accord with Kastening's mechanism:

$$ArNO_2^{-} + H^{+} \rightleftarrows Ar\dot{N}O_2H$$

$$Ar\dot{N}O_2H + ArNO_2^{-} \rightarrow ArNO_2 + ArNO + OH^{-}$$

[352] I. B. Goldberg, R. F. Borch, and J. R. Bolton, *Chem. Comm.*, **1969**, 223.
[353] G. A. Russell and P. R. Whittle, *J. Am. Chem. Soc.*, **91**, 2813 (1969).
[354] W. Reusch and D. B. Priddy, *J. Am. Chem. Soc.*, **91**, 3677 (1969).
[355] T. J. Curphey, C. W. Amelotti, T. P. Layloff, R. L. McCartney, and J. H. Williams, *J. Am. Chem. Soc.*, **91**, 2817 (1969).
[356] I. Elphimoff-Felkin and P. Sarda, *Tetrahedron Letters*, **1969**, 3045.

The reaction is sensitive to metal ions, and is catalysed by oxygen, the effects of which were discussed.[357]

Nitrobenzene radical anions in sulpholan initiate polymerization of acrylonitrile.[358] Electron transfer reactions of solutions of nitrobenzene radical anions in acetonitrile have been studied, including transfer to other aromatic nitro-compounds[359] o-Halogenonitrobenzene anions were found to dissociate exceptionally rapidly, both in this work and in a closely related study.[360]

$$[XC_6H_4NO_2]^{\overline{\cdot}} \rightleftarrows X^- + \cdot C_6H_4NO_2$$

When the halide was iodide this dissociation was concluded to be reversible, from the reduction in rate produced by added iodide ions.

The radical anion of 1,1-diphenylethylene adds to diphenylethylene very rapidly in hexamethylphosphoramide, in which it exists as the free ion.[361] The rate of dimerization, on the other hand, is comparable to that in THF where it is almost entirely ion-paired. Disproportionation of acenaphthene-semiquinone has been examined,[362] and the observation that the dimers of radical anions or cations of heptafulvalene are appreciably dissociated points to appreciably greater resonance stabilization in these molecules than in the parent hydrocarbon.[363]

Several pyridines have been reduced to their radical anions in metal–ammonia systems,[364] and the radical anions of arylpyridines in THF behave as very strong bases, readily giving a carbanion from diphenylmethane, for example.[365]

Radical anions of aromatic hydrocarbons donate electrons to sulphur dioxide, leading to the $S_2O_4^{2-}$ ion;[366] it has also been shown, by use of deuterium, that the reduction of molecular hydrogen to hydride by sodium naphthalene is more complex than previously envisaged.[367] The formation of deuterated naphthalene was attributed to interception of deuterium by

[357] D. Kolb, W. Wirths, and H. Gerischer, *Ber. Bunsen. Phys. Chem.*, **73**, 148 (1969); D. Kolb and R. Koopmann, *ibid.*, pp. 284, 624; see also J. Stradins, R. Gavass, V. Gvins, and S. Hillers, *Teor. Eksp. Khim.*, **4**, 774 (1968); *Chem. Abs.*, **70**, 46707 (1969).

[358] J. Martinmoa, *Suomen Kemistilehti*, B**42**, 33 (1969).

[359] A. R. Metcalfe and W. A. Waters, *J. Chem. Soc.*(B), **1969**, 918.

[360] J. G. Lawless and D. M. Hawley, *J. Electroanalyt. Chem. Interfacial Electrochem.*, **21**, 365 (1969).

[361] T. L. Staples, J. Jagur-Grodzinski, and M. Szwarc, *J. Am. Chem. Soc.*, **91**, 3721 (1969); see also K. Höfelman, J. Jagur-Grodzinski, and M. Szwarc, *ibid.*, p. 4645.

[362] A. G. Evans, J. C. Evans, and E. H. Godden, *J. Chem. Soc.*(B), **1969**, 546.

[363] M. D. Sevilla, S. H. Flajser, G. Vincow, and H. J. Dauben, *J. Am. Chem. Soc.*, **91**, 4139 (1969).

[364] A. R. Buick, T. J. Kemp, G. T. Neal, and T. J. Stone, *J. Chem. Soc.*(A), **1969**, 1609.

[365] B. Angelo, *Bull. Soc. Chim. France*, **1969**, 1710; see also M. I. Terekhova, E. S. Petrov, and A. I. Shatenshtein, *Org. Reactivity* (Tartu), **4**, 638 (1967).

[366] S. Bank and D. A. Lloyd, *Tetrahedron Letters*, **1969**, 1413.

[367] S. Bank, T. A. Lois, and M. C. Prislopski, *J. Am. Chem. Soc.*, **91**, 5407 (1969).

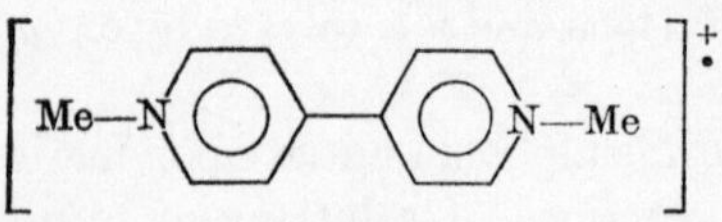

$$\left[\text{Me-N} \bigcirc\bigcirc \text{N-Me} \right]^{+\bullet}$$

(126)

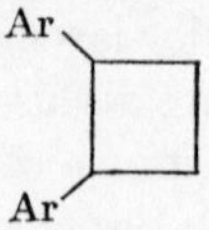

$$\begin{array}{c} \text{Ar} \\ \\ \text{Ar} \end{array}\square$$

(127)

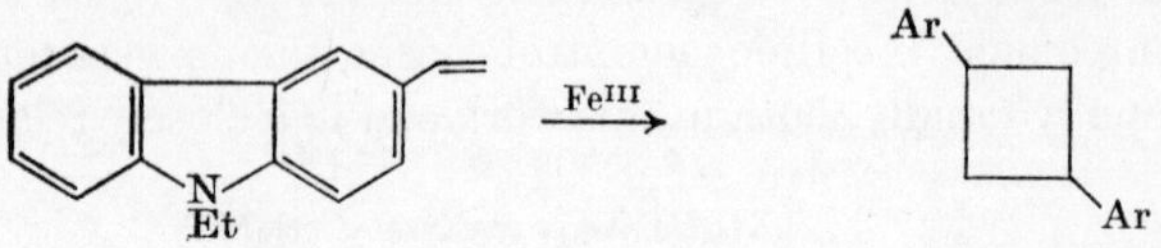

(128) $\xrightarrow{\text{Fe}^{III}}$

$$\text{Ar} \square \text{Ar}$$

(129)

$$\text{ArCH=CH}_2 \quad \underset{\longleftarrow}{\overset{\text{Fe}^{III}}{\rightleftharpoons}} \quad \{[\text{ArCH=CH}_2]^{\ddagger} + \text{Fe}^{II}\}$$

(130)

Path *a* ArCH=CH₂   Path *b*

$$\begin{array}{c} \overset{+}{\text{ArCH--CH}_2} \\ | \\ \text{ArCH--CH}_2 \\ | \\ \text{Fe}^{III} \end{array} \longleftarrow \left\{ \begin{array}{c} \overset{+}{\text{ArCH--CH}_2} \\ | \\ \overset{\bullet}{\text{ArCH--CH}_2} \end{array} + \text{Fe}^{II} \right\} $$

$$\overset{+}{\text{ArCH--CH}_2\text{--Fe}^{III}}$$

$$\downarrow \text{ArCH=CH}_2$$

$$\begin{array}{c} \text{ArCH--CH}_2\text{--Fe}^{III} \\ | \\ \overset{+}{\text{CH}_2\text{--CHAr}} \end{array}$$

$$\begin{array}{c} \text{ArCH--CH}_2 \\ |\qquad| \\ \text{ArCH--CH}_2 \end{array} + \text{Fe}^{II}$$

$$\begin{array}{c} \text{ArCH--CH}_2 \\ |\qquad| \\ \text{CH}_2\text{--CHAr} \end{array} + \text{Fe}^{II}$$

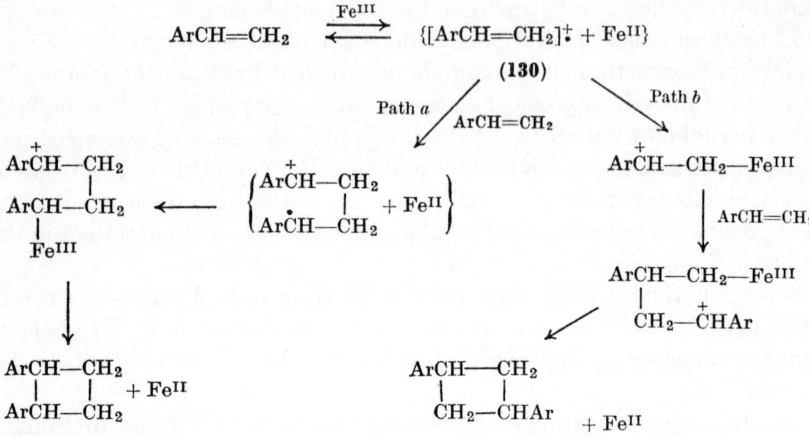

naphthalene radical anions, with the formation of a hydronaphthalene anion:

$$\text{D}_2 \xrightarrow{\text{C}_{10}\text{H}_8^{\bar{\cdot}}} \text{D}\bullet + \text{D}^- + \text{C}_{10}\text{H}_8$$

$$\text{D}\bullet \xrightarrow{\text{C}_{10}\text{H}_8^{\bar{\cdot}}} \text{C}_{10}\text{H}_8\text{D}^-$$

Methoxide anion is oxidized to formaldehyde by two one-electron transfers from paraquat dichloride, giving the radical cation **(126)**.[368]

Photochemical nitration of N-substituted carbazole by tetranitromethane to give 3-nitrocarbazole probably involves the initial photolysis of a charge-transfer complex to give a radical cation, $\text{NO}_2\bullet$, and $^-\text{C}(\text{NO}_2)_3$.[369]

Transition-metal salt oxidations of aromatic enamines such as N-vinyl-carbazole are known to give cyclobutanes **(127)**. C-Vinyl-enamines, e.g. **(128)**, are now reported to give 1,3-diarylcyclobutanes **(129)**.[370] The former system initiates radical polymerization, the latter does not. The difference is ascribed

[368] J. A. Farrington, A. Ledwith, and M. F. Stam, *Chem. Comm.*, **1969**, 259; see also D. J. McClemens, A. K. Garrison, and A. L. Underwood, *J. Org. Chem.*, **34**, 1867 (1969).
[369] D. H. Iles and A. Ledwith, *Chem. Comm.*, **1969**, 364.
[370] F. A. Bell, R. A. Crellin, H. Fujii, and A. Ledwith, *Chem. Comm.*, **1969**, 251.

to radical addition to a second molecule of olefin in the former instance (path *a*) and cationic addition in the latter (path *b*). Similar results were accomplished by photosensitization in the presence of oxygen as the oxidant.[371] The sensitizer acts by accepting an electron from the olefin; subsequent electron-transfer to oxygen regenerates the sensitizer and produces $\{[ArCH{=}CH_2]^{+}$ $O_2^{-}\}$, which then behaves similarly to (**130**). Chloranil sensitizes the reaction in the absence of oxygen, presumably because of its high electron affinity.

One-electron oxidation by the thianthrene cation radical has been discussed.[372] This species has been isolated as its (explosive[373]) perchlorate salt,[374] the hydrolysis of which to a sulphoxide involves an equilibrium concentration of thianthrene di-cation.[375]

Electrochemical Processes

Papers presented at a Faraday Society Discussion on electrode reactions have been published.[376]

Eberson has criticized the use of "relative rates" of anodic oxidation of different carboxylate ions to demonstrate the non-concerted nature of oxidation to acyloxy radical and decarboxylation, though the non-concerted mechanism for this process is not disputed.[377]

Anodic oxidations of carbanions, such as that derived from diethyl malonate, in the presence of olefins give adducts probably formed by addition of malonyl radicals.[378] However, electrolysis of solutions of cyclohexene in acetic acid gives 3-acetoxycyclohexene.[379]

Anodic oxidation of aryl[380] and vinyl[381] ethers, of tertiary amines,[382] and of aromatic hydrocarbons,[383] including benzenes,[384] has received attention, and there have been several studies of aromatic substitution at an anode.[385]

[371] R. A. Carruthers, R. A. Crellin, and A. Ledwith, *Chem. Comm.*, **1969**, 253.
[372] Y. Murata and H. J. Shine, *J. Org. Chem.*, **34**, 3368 (1969).
[373] H. J. Shine and Y. Murata, *Chem. Ind.* (London), **1969**, 782.
[374] See also Y. Murata, L. Hughes, and H. J. Shine, *Inorg. Nucl. Chem. Letters*, **4**, 573 (1968).
[375] H. J. Shine and Y. Murata, *J. Am. Chem. Soc.*, **91**, 1873 (1969).
[376] *Discuss. Faraday Soc.*, **45** (1968).
[377] L. Eberson, *J. Am. Chem. Soc.*, **91**, 2402 (1969); see *Org. Reaction Mech.*, **1968**, 306.
[378] H. Schaefer, *Chem. Ing. Tech.*, **41**, 179 (1969).
[379] T. Shono and T. Kosaka, *Tetrahedron Letters*, **1968**, 6207.
[380] V. D. Parker, *Chem. Comm.*, **1969**, 610.
[381] B. Belleau and Y. K. Au-Young, *Can. J. Chem.*, **47**, 2117 (1969).
[382] N. L. Weinberg, *J. Org. Chem.*, **33**, 4326 (1968); P. J. Smith and C. K. Mann, *ibid.*, p. 1821.
[383] V. D. Parker and L. Eberson, *Chem. Comm.*, **1969**, 340; V. D. Parker, *ibid.*, p. 848; V. D. Parker and R. N. Adams, *Tetrahedron Letters*, **1969**, 1721.
[384] T. Osa, A. Yildiz, and T. Kuwana, *J. Am. Chem. Soc.*, **91**, 3994 (1969).
[385] T. Susuki, K. Koyama, A. Omori, and S. Tsutsumi, *Bull. Chem. Soc. Japan*, **41**, 2663 (1968); K. Koyama, T. Ebara, and S. Tsutsumi, *ibid.*, p. 2668; J. P. Millington, *J. Chem. Soc.*(B), **1969**, 982; V. D. Parker and L. Eberson, *Chem. Comm.*, **1969**, 973; S. Andreades and E. W. Zahnow, *J. Am. Chem. Soc.*, **91**, 4181 (1969); G. Manning, V. D. Parker, and R. N. Adams, *ibid.*, p. 4584; V. D. Parker and L. Eberson, *Tetrahedron Letters*, **1969**, 2839, 2843.

Evidence has been presented that electrolytic reduction of an alkyl iodide involves two one-electron steps,[386] precisely as in the chemical reductions already discussed. Only partial phenyl migration occurred when neophyl halides were reduced electrolytically in dimethylformamide. This is also consistent with participation of a radical intermediate, but not with the alternative possibility that carbonium ions might be involved.[387]

The mechanisms of electrolytic reduction of carbonyl compounds[388] and of reductive dimerization of acrylonitrile have been discussed.[389] Benzene may be reduced electrolytically[390] in hexamethylphosphoramide–ethanol, or by hydrated electrons obtained by γ-radiolysis of aqueous alkaline solutions,[391] but with solvated electrons in pure hexamethylphosphoramide biphenylene is produced possibly by initial formation of benzyne, as occurs under electron-impact in the mass spectrometer.[392]

Autoxidation

(see also Chapter 14)

Evidence was presented last year[393] in support of Russell's rationalization of the relative rapidity of chain termination by pairs of primary or secondary alkylperoxy radicals, when compared with the rate for tertiary alkylperoxy radicals. However, the formation of singlet oxygen in this process has now been reinterpreted as not arising from a simple electrocyclic reorganization of a tetroxide (**131**), but rather from a two-step process in which both oxygen and the carbonyl compound are initially formed in the triplet state, and the two triplets are then mutually quenched with high efficiency within the solvent cage.[394] An unexpected result is Zaikov's observation that the termination reaction between pairs of secondary alkylperoxy radicals from pentan-2-one, and that between pairs of tertiary alkylperoxy radicals from 3-methylbutan-2-one, occur at comparable rates.[395]

The oxygenation of olefins to give, for example, α,β-unsaturated carbonyl compounds, by molecular oxygen catalysed by chlorotris(triphenylphosphine)rhodium(I), apparently involves a radical mechanism. It was shown that (+)-carvomenthene (**132**) gives racemic ketone (**133**), which is consistent with

386 A. J. Fry and M. A. Mitnick, *J. Am. Chem. Soc.*, **91**, 6207 (1969).
387 L. Eberson, *Acta Chem. Scand.*, **22**, 3045 (1968).
388 J. Wiemann, S. L. T. Thuan, D. Lelandais, and M. Dedieu, *Compt. Rend.*, **269**, 30 (1969).
389 T. Asahara, M. Seno, and T. Arai, *Bull. Chem. Soc. Japan*, **42**, 1316 (1969).
390 H. W. Sternberg, R. E. Morkby, I. Wender, and D. M. Mohilner, *J. Am. Chem. Soc.*, **91**, 4191 (1969).
391 M. H. Studier and E. J. Hart, *J. Am. Chem., Soc.* **91**, 4068 (1969).
392 J. E. Dubois and G. Dodin, *Tetrahedron Letters*, **1969**, 2325.
393 See *Org. Reaction Mech.*, **1968**, 316.
394 R. E. Kellogg, *J. Am. Chem. Soc.*, **91**, 5433 (1969).
395 G. E. Zaikov, *Kinet. Katal.*, **9**, 1166 (1968); *Chem. Abs.*, **70**, 46603 (1969).

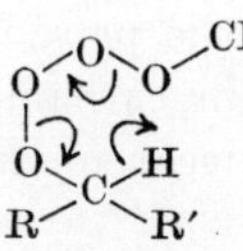

(131)

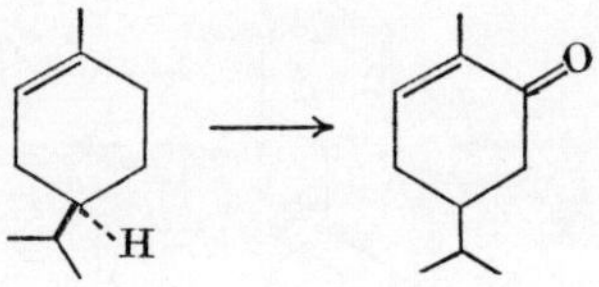

(132)   (133)

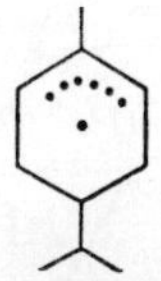

(134) 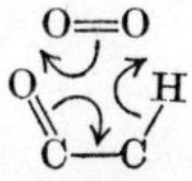   (135)

the participation of a symmetrical intermediate such as (134) rather than with a concerted singlet-oxygen mechanism of which the first step is indicated by (135).[396]

The charge-transfer spectrum observed in solutions of oxygen in tetralin has been employed as a means of monitoring oxygen concentration in this hydrocarbon, and hence of monitoring the rate of autoxidation.[397] Other reports include new data on autoxidation of cumene,[398] octene,[399] cyclohexane,[400] 1,1-diphenylethane,[401a] sorbic acid,[401b] N-butylcaproamide,[402] pyrogallol,[403] 2-arylindanones,[404] and the interesting hydroaromatic compound, 9,10-cyclopenteno-4a,4b-dihydrophenanthrene.[405] Reaction of the

[396] J. E. Baldwin and J. C. Swallow, *Angew. Chem. Internat. Ed. Engl.*, **8**, 601 (1969).

[397] J. Betts and J. C. Robb, *Trans. Faraday Soc.*, **47**, 2144 (1969).

[398] S. Z. Roginskii and T. U. Andrianova, *Dokl. Akad. Nauk SSSR*, **178**, 645 (1968); *Chem. Abs.*, **69**, 58699 (1968); R. V. Kucher and I. P. Shevchuk, *Neftekhimiya*, **8**, 398 (1968); *Chem. Abs.*, **70**, 10726 (1969).

[399] N. Indictor, T. Jochsberger, and D. Kurnit, *J. Org. Chem.*, **34**, 2855, 2861 (1969).

[400] G. E. Zaikov and Z. K. Maizus, *Izv. Akad. Nauk SSSR, Ser. Khim.*, **1969**, 311; *Chem. Abs.*, **70**, 114352 (1969).

[401a] F. A. Guk, V. F. Tsepalov, V. F. Shuvalov, and V. Y. Shlyapintokh, *Izv. Akad. Nauk SSSR, Ser. Khim.*, **1968**, 2250; *Chem. Abs.*, **70**, 28162 (1969).

[401b] L. Pekkarinen, *Suomen Kemistilehti, B42*, 147 (1969).

[402] B. Lanska and J. Sependa, *Coll. Czech. Chem. Comm.*, **34**, 1911 (1969).

[403] N. F. Usacheva, Y. G. Oranskii, G. A. Sedova, E. F. Rul, and M. S. Khaikin, *Zh. Nauch. Prikl. Fotogr. Kinematogr.*, **14**, 201 (1969).

[404] L. P. Zalukaev and G. I. Sorokina, *Dokl. Akad. Nauk SSSR*, **186**, 336 (1969); *Chem. Abs.*, **71**, 48955 (1969).

[405] A. Bromberg and K. A. Muskat, *J. Am. Chem. Soc.*, **91**, 2860 (1969).

aldehyde **(136)** with oxygen in the presence of strong base gives a hydroperoxide **(138)**, apparently via the carbanion **(137)**, salts of which could be isolated.[406] It was proposed that decarbonylation was an ionic process. Studies

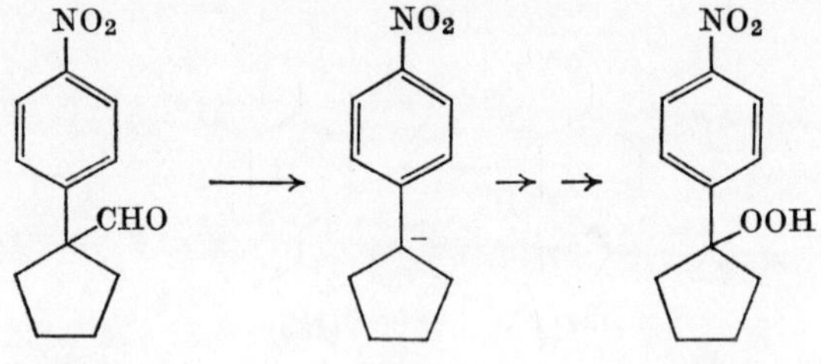

(136) → **(137)** → → **(138)**

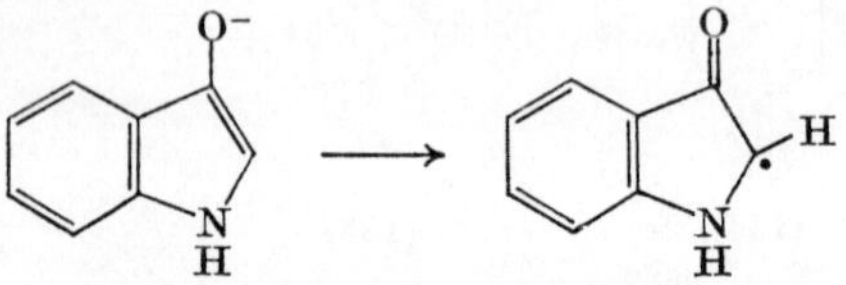

(139) → **(140)**

of autoxidation of other aldehydes[407, 408] include a report that acylperoxy radicals are much more reactive than are alkylperoxy radicals in hydrogen-abstraction, and it is largely for this reason that autoxidation of aldehydes is a relatively facile process.[408]

The oxidation of indoxyl to indigo in basic media is believed to involve oxidation of the indoxyl anion **(139)** to the radical **(140)** which dimerizes [or couples with an anion **(139)**], and the product is further oxidized.[409]

New results on inhibition by amines[410] including gas-phase studies,[411] by phosphites,[412] and by benzoquinone,[413] and synergistic effects with highly hindered phenols[414] have been discussed, and it has been suggested that the

[406] G. Bellucci, B. Macchia, and F. Machia, *Tetrahedron Letters*, **1969**, 3239.
[407] J. P. Franck, I. S. de Roch, and L. Sajus, *Bull. Soc. Chim. France*, **1969**, 1947, 1957; H. Brederech, K. Bühler, K. Posselt, T. Haug, and H. Sonnerborn, *Chem. Ber.*, **102**, 2190 (1969); R. R. Baldwin, R. W. Walker, and D. H. Langford, *Trans. Faraday Soc.*, **65**, 792, 806 (1969).
[408] G. E. Zaikov, J. A. Howard, and K. U. Ingold, *Can. J. Chem.*, **47**, 3017 (1969).
[409] G. A. Russell and G. Kaupp, *J. Am. Chem. Soc.*, **91**, 3851 (1969).
[410] K. Adamic, M. Dunn, and K. U. Ingold, *Can. J. Chem.*, **47**, 287 (1969); K. Adamic and K. U. Ingold, *ibid.*, p. 295; D. F. Bowman, B. S. Middleton, and K. U. Ingold, *J. Org. Chem.*, **34**, 3456 (1969).
[411] P. W. Jones and D. J. Waddington, *Chem. Ind.* (London), **1969**, 492.
[412] D. G. Pobedimskii and A. L. Buchachenko, *Izv. Akad. Nauk SSSR, Ser. Khim.*, **1968**, 1181; *Chem. Abs.*, **69**, 86014 (1968).
[413] E. T. Denisov, *Izv. Akad. Nauk SSSR, Ser. Khim.*, **1969**, 328; *Chem. Abs.*, **70**, 114358 (1969).
[414] L. R. Mahoney, F. C. Ferris, and M. A. DaRooge, *J. Am. Chem. Soc.*, **91**, 3883 (1969).

inhibitor action of many sulphides is due to the formation of sulphur dioxide as the active inhibitor.[415]

Phenylhydroxylamine has been shown to be an active catalyst for autoxidation of systems inhibited by phenols. A key step was suggested to be electron transfer from the hydroxylamine to molecular oxygen.[416] It seems that this observation must be important in the context of amine inhibition, since many amine inhibitors are known to produce nitroxides (e.g. ref. 410) and these might in turn lead to hydroxylamines. Hydroxylamines also induce rapid decomposition of hydroperoxides.

Ingold's group have reported further experiments leading to absolute rate constants in autoxidation systems,[417] and they and others[418] have demonstrated that at temperatures below $-115°$ tertiary alkylperoxy radicals are stable in solution, and are in equilibrium with their dimers (tetroxides). At higher temperatures, irreversible decomposition involves unsymmetrical decomposition of the tetroxide: $(ROOOOR \rightarrow RO\cdot + \cdot OOOR)$. It was also demonstrated that variation in termination rates as a function of R is due to differences in the rate of this decomposition rather than differences in the position of the equilibrium:

$$2RO_2\cdot \rightleftharpoons ROOOOR$$

An attempt to prepare solutions of t-butyl tetroxide by photolysis of t-butyl iodide in the presence of oxygen at $-100°$ was frustrated by competing side-reactions.[419]

A revised value of 4.7 has been given for the isotope effect (k_H/k_D) for abstraction of hydrogen from $PhCD_3$ by hydroperoxy radicals.[420] Chain termination by interaction of two t-butylperoxy radicals occurs more rapidly if the radicals are hydrogen-bonded to solvent than if they are free.[421] Several new papers discuss aspects of hydroperoxide decomposition.[422] The reaction of methyl radicals with oxygen has been studied.[423]

[415] G. Scott, *Chem. Comm.*, **1968**, 1572.

[416] G. T. Knight and B. Saville, *Chem. Comm.*, **1969**, 1262.

[417] J. A. Howard, K. Adamic, and K. U. Ingold, *Can. J. Chem.*, **47**, 3793, 3809; J. A. Howard and K. U. Ingold, *ibid.*, pp. 3797, 3809.

[418] J. E. Bennett, D. M. Brown, and B. Mile, *Chem. Comm.*, **1969**, 504; K. Adamic, J. A. Howard, and K. U. Ingold, *ibid.*, p. 504.

[419] T. Mill and R. Stringham, *J. Phys. Chem.*, **73**, 282 (1969).

[420] H. Hotta, N. Suzuki, and T. Komori, *Bull. Chem. Soc. Japan*, **42**, 2041 (1969).

[421] G. E. Zaikov, Z. K. Maizus, and N. M. Emanuel, *Izv. Akad. Nauk SSSR, Ser. Khim.*, **1968**, 2265; *Chem. Abs.*, **70**, 19386 (1969).

[422] G. E. Zaikov, Z. K. Maizus, and N. M. Emanuel, *Izv. Akad. Nauk SSSR, Ser. Khim.*, **1968**; 53; *Chem. Abs.*, **69**, 35187 (1968); A. I. Prokof'ev, S. P. Solodovnikov, N. N. Bubnov, and N. G. Radzhabov, *ibid.*, p. 1664; A. Y. Valendo and Y. D. Norikov, *ibid.*, p. 275; *Chem. Abs.*, **70**, 114392 (1969); S. Ghosal and S. K. Dutta, *Indian J. Chem.*, **7**, 135 (1969).

[423] N. A. Sokolova, L. V. Nikisha, S. S. Polyak, and A. B. Nalbandyan, *Dokl. Akad. Nauk SSSR*, **185**, 850 (1969); *Chem. Abs.*, **71**, 21490 (1969).

New results are in general accord with a radical chain mechanism for autoxidation of alkylboron compounds,[424-426] and of Grignard reagents.[426] The rate constant k_p for the propagation step shown was estimated to be 1.6×10^4 l mole^{-1} sec^{-1} in isooctane at $30°$.[425]

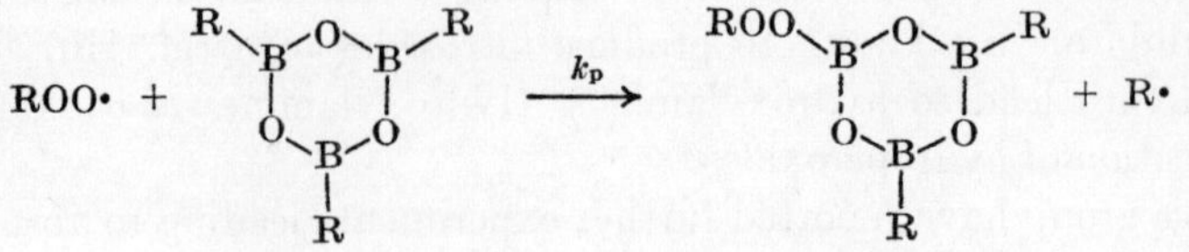

(R = s-butyl)

Miscellaneous

Among the best documented radical oxidations are those of phenols. Efficient oxidative phenol coupling has been reported using manganic tris(acetylacetonate),[427] and using silver carbonate suspended on Celite,[428] as oxidants. A particularly valuable technique for intramolecular oxidative coupling involves conversion of phenoxy groups into vanadium(v) derivatives at $-78°$ using VOCl$_3$ in ether, and subsequently heating the ether solutions to reflux.[429] In this way, diol (**141**) was transformed into the spirodienone (**142**) in 76% yield.

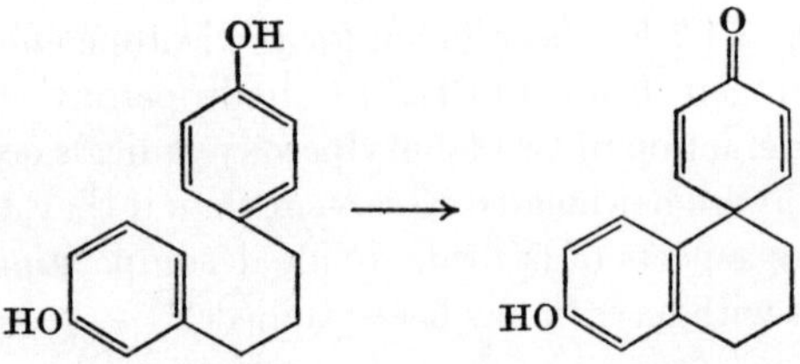

(141)   (142)

The optimum yield with other reagents was 10%. Evidence has been presented suggesting that oxidation of phenols by dichlorodicyanoquinone may in certain circumstances involve phenoxonium ions.[430] Related papers deal with

424 P. G. Allies and P. B. Brindley, *J. Chem. Soc.*(B), **1969**, 1126; J. Grotewold, E. A. Lissi, and J. C. Scaiano, *J. Chem. Soc.*(B), **1969**, 475.
425 K. U. Ingold, *Chem. Comm.*, **1969**, 911.
426 A. G. Davies and B. P. Roberts, *J. Chem. Soc.*(B), **1969**, 311, 317.
427 M. J. S. Dewar and T. Nakaya, *J. Am. Chem. Soc.*, **90**, 7134 (1968).
428 V. Balogh, M. Fetizon, and M. Golfier, *Angew. Chem. Internat. Ed. Engl.*, **8**, 444 (1969).
429 M. A. Schwartz, R. A. Holton, and S. W. Scott, *J. Am. Chem. Soc.*, **91**, 2800 (1969).
430 J. W. A. Findlay, P. Gupta, and J. R. Lewis, *Chem. Comm.*, **1969**, 206.

oxidation of catechin by phenoloxidases,[431] oxidation of hindered phenols,[432] reaction of t-butyl peroxide with phenols,[433] and oxidation of catechol.[434] Data have appeared on the heats of formation of phenoxy radicals, as well as on heats of formation of thiyl and thiophenoxy radicals.[435]

An ingenious procedure for effecting replacement of bridgehead hydroxy groups involves reaction with 1 equivalent of oxalyl chloride, followed by hydrolysis to an oxalic acid half-ester. Oxidation of this with mercuric oxide and iodine gives the bridgehead iodide.[436] The same oxidizing agent cleaves vicinal diols under conditions in which hypoiodites are stable. This suggests that loss of iodine from the intermediate β-hydroxy-hypoiodite is assisted by the adjacent hydroxy group.[437] The silver carbonate reagent [428] converts vicinal diols into α-hydroxy-ketones; other diols may also give hydroxy-ketones but in many cases lactone formation predominates.[438]

2-Phenylvinyl radicals (from cinnamoyl peroxide) react with oxygen to form benzaldehyde, benzoic acid, and some phenylacetaldehyde. These observations were explained in terms of reactions of the peroxy radical (**143**).[439]

The formation of methylphosphorothionates from the reaction between sulphur and alkyl methylphosphinates in the presence of an amine is inhibited by hydroquinone. This and other evidence points to the initial formation of phosphino radicals (**144**) in these reactions[440] (see also ref. 187). Several groups have described reactions of radicals with trivalent organophosphorus compounds, in which phosphoranyl radicals are formed.[53, 441] Powell and Hall have summarized evidence for the formation of triphenylphosphinium cation by one-electron oxidation of triphenylphosphine. An analogous intermediate was considered to be involved in the racemization of an optically active

[431] K. Weinges, W. Ebert, D. Huthwelker, H. Mattauch, and J. Perner, *Ann. Chem.*, **726**, 114 (1969).

[432] H. D. Becker, *J. Org. Chem.*, **34**, 1198, 1203, 1211 (1969); L. M. Strigun, A. I. Prokof'ev, F. N. Pirnazarova, and N. M. Emanuel, *Izv. Akad. Nauk SSSR, Ser. Khim.*, **1968**, 59; L. M. Strigun, L. S. Vartunyan, A. A. Volod'kin, A. I. Prokof'ev, and N. M. Emanuel, *ibid.*, p. 2242; J. Petranek and J. Pilar, *Coll. Czech. Chem. Comm.*, **34**, 79 (1969).

[433] K. M. Johnston, R. E. Jacobson, and G. H. Williams, *J. Chem. Soc.*(C), **1969**, 1424.

[434] N. F. Usacheva, Y. G. Oranskii, R. S. Safiullin, E. F. Rul, and M. S. Khaikin, *Zh. Nauch. Prikl. Fotogr. Kinematogr.*, **13**, 459 (1968); *Chem. Abs.*, **70**, 56990 (1969).

[435] D. H. Fine and J. B. Westmore, *Chem. Comm.*, **1969**, 273.

[436] A. Goosen, *Chem. Comm.*, **1969**, 145.

[437] A. Goosen and H. A. H. Laue, *J. Chem. Soc.*(C), **1969**, 383.

[438] M. Fetizon, M. Golfier, and J. M. Lollis, *Chem. Comm.*, **1969**, 1102, 1119.

[439] K. Tokumaru, *Chem. Ind.* (London), **1969**, 297.

[440] W. A. Mosher and R. R. Irino, *J. Am. Chem. Soc.*, **91**, 756 (1969).

[441] W. G. Bentrude, J. H. Hargis, and P. E. Rusek, *Chem. Comm.*, **1969**, 296; W. G. Bentrude and R. A. Wielesek, *J. Am. Chem. Soc.*, **91**, 2406 (1969); R. S. Davidson, *Tetrahedron*, **25**, 3383 (1969); R. E. Atkinson, J. I. G. Cadogan, and J. T. Sharp, *J. Chem. Soc.*(B), **1969**, 138; W. G. Bentrude and J. J. L. Fu, *Tetrahedron Letters*, **1968**, 6033; K. Terauchi and H. Sakurai, *J. Chem. Soc. Japan, Ind. Chem. Sect.*, **72**, 215 (1969); *Chem. Abs.*, **70**, 86820 (1969); V. F. Drozdovskii, S. M. Kavum, and D. R. Razgon, *Khim. Org. Soed. Fosfora, Akad. Nauk SSSR, Otd. Obshch., Tekh. Khim.*, **1967**, 57; *Chem. Abs.*, **69**, 2303 (1968).

triarylphosphine by tetracyanoquinodimethane.[442] This contrasts with the conformational stability found for the isoelectronic trisubstituted silyl radicals,[151] though the lifetime of the latter intermediates must be much less than that of the phosphinium cation radical. Rate constants for some gas-phase reactions of trimethylsilyl radicals have been reported.[443]

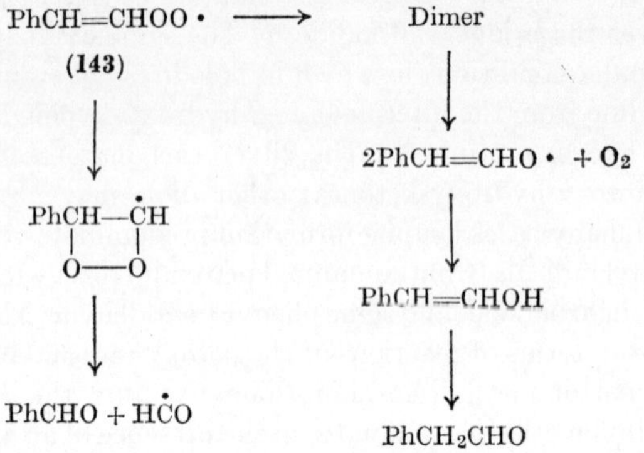

$$PhCH{=}CHOO \cdot \longrightarrow Dimer$$

$$(143)$$

$$\downarrow$$

$$2PhCH{=}CHO \cdot + O_2$$

$$PhCH{-}\overset{\bullet}{CH}$$
$$| \quad |$$
$$O{-}O$$

$$\downarrow$$

$$PhCH{=}CHOH$$

$$\downarrow$$

$$PhCHO + H\overset{\bullet}{C}O$$

$$PhCH_2CHO$$

$$\overset{O}{\underset{|}{\overset{\uparrow}{Me{-}P{-}H}}}\quad\overset{R\cdot}{\longrightarrow}\quad\overset{O}{\underset{|}{\overset{\uparrow}{MeP\cdot}}}$$
$$OR \qquad\qquad\qquad OR$$

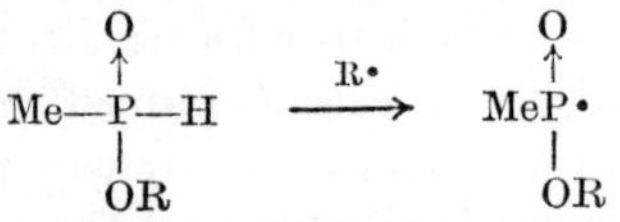

$$(144)$$

Triarylsulphonium salts react with alkoxide ions in alcohols to give aryl alkyl ethers, diaryl sulphides, and products of reactions of free aryl radicals. The reaction is believed to involve initial formation of a tetravalent sulphur derivative, Ar_3SOR, which gives the caged radical pair $[Ar_3S \cdot \; \cdot OR]$. Escape from the cage $(\rightarrow Ar_3S \cdot \rightarrow Ar \cdot + Ar_2S)$ competes with cage formation of diaryl sulphide and the aromatic ether.[444]

Weiner and Hammond have attempted to determine the rate of recombination of t-butoxy radicals by a direct ESR method.[445] However, the radical whose decay they observed was probably not t-butoxy,[446] the spectrum of which should be difficult to detect as a result of exceptionally efficient spin–lattice relaxation which would result in broad-line spectra. A similar technique has, however, been used successfully to monitor the decay of protonated

[442] R. L. Powell and C. D. Hall, *J. Am. Chem. Soc.*, **91**, 5403 (1969).
[443] J. C. J. Thynne, *J. Organometal. Chem.*, **17**, 155 (1969).
[444] J. W. Knapczyk and W. E. McEwen, *J. Am. Chem. Soc.*, **91**, 145 (1969).
[445] S. Weiner and G. S. Hammond, *J. Am. Chem. Soc.*, **91**, 2182 (1969).
[446] M. C. R. Symons, *J. Am. Chem. Soc.*, **91**, 5924 (1969).

semidiones,[447] and in related work rate constants for dimerization of cyclo-heptatrienyl radicals and tri-t-butylcycloheptatrienyl radicals have been measured [448] (see also ref. 99).

Direct measurements on the equilibration of allyl radicals with hexa-1,5-diene have confirmed that the resonance energy of the allyl radical is ca. 10 kcal mole^{-1}.[449] The reactions of allyl radicals have also been studied in the gas-phase pyrolysis of diallyl oxalate [450] and allyl cyclohexa-2,5-dienecarboxylate (which simultaneously produces cyclohexadienyl radicals).[451]

When allyl radicals are generated radiolytically in a methylpentane glass at 77° K, and then irradiated with ultraviolet light, the photoexcited allyl radicals abstract hydrogen atoms from the saturated hydrocarbon.[452]

Norman and West have presented a detailed discussion of the fundamental reactions involved in radical production when titanous ion promoted decomposition of hydrogen peroxide is utilized to produce radicals for ESR study using a flow system.[453] It was not possible to observe phenyl radicals in solution in related experiments involving electron transfer promoted decomposition of benzenediazonium salts. However, these radicals *could* be scavenged by nitrite ions to give nitrobenzene radical anion or by nitric oxide to give diphenyl nitroxide.[454] Reaction of hydroxyl radicals with formate ion in the flow-system experiments gives CO_2^-; this with alkyl halides gives alkyl radicals by dissociative electron-capture, and provides a convenient means of generating specific alkyl radicals.[455] However, the greater convenience of the static photochemical procedures for obtaining solution ESR spectra of alkyl radicals discussed at the beginning of this chapter would, it seems, normally make these the methods of choice for purely spectroscopic studies.

The pyrolysis of 3,3,6,6-tetramethylcyclohexadiene gives *p*-xylene in a non-chain radical process,[456] and not, as previously reported,[457] in a concerted reaction. *trans*-3,6-Dimethylcyclohexadiene (145) gives methane and toluene in a radical chain process on heating to 380°, but the *cis*-isomer (146) gives *p*-xylene and hydrogen in a unimolecular reaction which occurs smoothly at 300°.[456]

[447] S. A. Weiner, E. J. Hamilton, and B. M. Monroe, *J. Am. Chem. Soc.*, **91**, 6350 (1969).
[448] M. L. Morrell and G. Vincow, *J. Am. Chem. Soc.*, **91**, 6389 (1969); see also G. Vincow, H. J. Dauben, F. R. Hunter, and W. V. Volland, *ibid.*, p. 2823.
[449] D. M. Golden, N. A. Gac, and S. W. Benson, *J. Am. Chem. Soc.*, **91**, 2136 (1969).
[450] D. G. L. James, *Trans. Faraday Soc.*, **65**, 1350, 1357 (1969).
[451] D. G. L. James and S. M. Kambamis, *Can. J. Chem.*, **47**, 2081 (1969).
[452] V. A. Roginskii and S. Y. Pshezhetskii, *Khim. Vys. Energ.*, **3**, 140 (1969); *Chem. Abs.*, **70**, 114381 (1969).
[453] R. O. C. Norman and P. R. West, *J. Chem. Soc.*(B), **1969**, 389.
[454] A. L. J. Beckwith and R. O. C. Norman *J. Chem. Soc.*(B), **1969**, 403.
[455] A. L. J. Beckwith and R. O. C. Norman, *J. Chem. Soc.*(B), **1969**, 400.
[456] H. M. Frey, A. Krantz, and I. D. R. Stevens, *J. Chem. Soc.*(A), **1969**, 1734.
[457] W. Reusch, M. Russell, and D. Dzurella, *J. Org. Chem.*, **29**, 2446 (1964).

(145)   (146)

Calculations on the fragmentation reactions of alkanes have been published,[458] as has a report on the effect of inert gas on fragmentation of alkyl radicals.[459]

The formation of benzophenone ketyl radicals ($Ph_2\overset{.}{C}OH$) by thermolysis of benzpinacol was described last year.[460] Base catalysis of this reaction (which gives benzophenone anion radicals) has now been studied,[461] as has its photochemical counterpart.[462] It has been found that pyrolysis of 1,1,2,2-tetraphenylcyclopentane provides less efficient initiation of styrene polymerization than does pyrolysis of 3,3,4,4-tetraphenylhexane.[463]

A correlation between the logarithms of the rates of triethylamine oxidation with the oxidation potentials of a wide variety of one-electron oxidants is consistent with a slow electron-transfer as the rate-determining step in these reactions, which may be summarized by the equation:

$$ Ox + R_3N \underset{fast}{\overset{slow}{\rightleftharpoons}} Ox^{\overline{.}} + R_3N^{\dotplus} \xrightarrow{fast} Products $$

The reversibility of the first step was demonstrated for ferricyanide oxidation by the observation of a reduced rate of oxidation in the presence of added ferrocyanide. The ease of oxidation of other amines proved to be predictable in terms of their ionization potentials.[464]

The role of electron transfer from amines has been discussed for the photoreduction of carbonyl compounds,[465] for oxidation of amines to enamines,[466] for reaction between amines and diaryliodonium salts,[467] and for the oxidation of pyridines by pyrylium salts.[468] The oxidation of tetramethylhydrazine by iodine has been examined in detail.[469]

[458] A. D. Stepukhovich, V. A. Ulitskii, and A. P. Sharaevskii, *Zh. Fiz. Khim.*, **42**, 1276 (1968); *Chem. Abs.*, **69**, 86007 (1968).
[459] D. Done and G. Guiochon, *Can. J. Chem.*, **47**, 3477 (1969).
[460] See *Org. Reaction Mech.*, **1968**, 310.
[461] G. O. Schenck, G. Matthias, M. Pape, M. Cziesla, G. von Bürau, E. Roselino, and G. Koltzenburg, *Ann. Chem.*, **719**, 80 (1968).
[462] R. S. Davidson, F. A. Younis, and R. Wilson, *Chem. Comm.*, **1969**, 826.
[463] E. Borsig, M. Lazav, and M. Capla, *Coll. Czech. Chem. Comm.*, **33**, 4264 (1968).
[464] L. A. Hall, G. T. Davis, and D. H. Rosenblatt, *J. Am. Chem. Soc.*, **91**, 6247 (1969).
[465] E.g. R. S. Davidson, P. E. Lambeth, F. A. Youms, and R. Wilson, *J. Chem. Soc.*(C), **1969**, 2203; see also Chapter 13.
[466] M. Colonna and L. Marchett, *Gazzetta*, **99**, 14 (1969).
[467] O. A. Ptitsyna, O. A. Reutov, and G. C. Lyatiev, *Zh. Org. Khim.*, **5**, 401, 411, 416 (1969).
[468] M. Farcasiu and D. Farcasiu, *Chem. Ber.*, **102**, 2294 (1969).
[469] D. Romans, W. H. Brunning, and C. J. Michejda, *J. Am. Chem. Soc.*, **91**, 3859 (1969).

The dinitrile (**147**) is an electron acceptor comparable with chloranil.[470]
Solvated electrons in liquid ammonia are captured by alkyl cyanides, the anions of which dissociate to give alkyl radicals.[471]

γ-Radiolysis of pyrrolidine at 77° K gives a species which ESR shows to

(**147**)   (**148**)

contain an unpaired electron interacting with three or four nitrogen nuclei—regarded as a solvated electron.[472] Radiolysis of pyridine at the same temperature gives a radical, C_5H_4N (**148**), related to benzyne,[473] and radiolysis of branched-chain alkanes at this temperature gives predominantly *secondary* alkyl radicals.[474]

The participation of radicals has also been discussed for reaction in which pentylamine and iodine are pyrolysed at 500° (giving pyridine),[475] stilbene, cinnamic acid, and related compounds are pyrolysed at 500°,[476] dialkyl-mercury compounds are decomposed thermally or photochemically,[477] and in which *gem*-dinitroalkanes[478] and β-keto-esters are pyrolysed.[479]

The photolysis of mercaptans gives hydrogen and a dialkyl disulphide,[480] and the gas-phase photolysis of dimethyl disulphide gives methane as a major product.[481] Free-radical and caged-radical components of the photo-decarboxylation of N,O-diacylhydroxylamines have been identified.[482]

Thermal hydrogen atoms appear to abstract a hydroxy group from the *gem*-dihydroxy function of solid alloxan;[483] addition at oxygen followed by loss of

[470] S. Chatterjee, *J. Chem. Soc.*(B), **1969**, 725.
[471] P. G. Arapakos, M. K. Scott, and F. E. Huber, *J. Am. Chem. Soc.*, **91**, 2059 (1969).
[472] W. Cronenwett and M. C. R. Symons, *J. Chem. Soc.*(A), **1968**, 2991.
[473] H. J. Bower, J. A. McRae, and M. C. R. Symons, *J. Chem. Soc.*(A), **1968**, 2696.
[474] D. J. Henderson and J. E. Willard, *J. Am. Chem. Soc.*, **91**, 3014 (1969).
[475] W. H. Bell, G. B. Carter, and J. Dewing, *J. Chem. Soc.*(C), **1969**, 352.
[476] T. C. Jones and I. Schmeltz, *J. Org. Chem.*, **34**, 645 (1969).
[477] O. N. Druzhkov, S. F. Zhil'tsov, G. G. Petukhov, *Zh. Obshch. Khim.*, **38**, 2706 (1968); *Chem. Abs.*, **70**, 95958 (1969).
[478] G. M. Nazin, G. B. Manelis, and F. I. Dubovitskii, *Izv. Akad. Nauk SSSR, Ser. Khim.*, **1968**, 2629; *Chem. Abs.*, **70**, 67385 (1969).
[479] B. S. Kirkiacharian, *Compt. Rend.*, *C*, **269**, 721 (1969); B. S. Kirkiacharian, R. Santus, M. Ptak, and M. Boran-Marszak, *ibid.*, p. 842.
[480] B. G. Dzantiev and A. V. Shishkov, *Khim. Vys. Energ.*, **2**, 119 (1968); *Chem. Abs.*, **69**, 66679 (1968).
[481] T. Inaba and H. Ogoro, *J. Chem. Soc. Japan, Ind. Chem. Sect.*, **72**, 114 (1969); *Chem. Abs.*, **70**, 95981 (1969).
[482] B. Danielle, P. Manitto, and G. Russo, *Chem. Ind.* (London), **1969**, 329.
[483] J. N. Herak, *J. Am. Chem. Soc.*, **91**, 5171 (1969).

water may be involved. Evidence for the formation of an oxidizing species, thought to be MeO·, has been found in γ-radiolyses of oxygen-free methanol.[484] Previous work had only identified reducing species. Radiolysis of methanolic sulphuric acid at liquid-nitrogen temperature gives a radical, the ESR spectrum of which is consistent with the structure $\overset{\cdot}{C}H_2\overset{+}{O}H_2$; substantial splitting by the hydrogens on oxygen was attributed to a hyperconjugative mechanism.[485]

Scavenging of carbanions by nitrobenzene has been discussed. This reaction is so efficient that in a triphenylmethane–Bu^{t}OK–Bu^{t}OD system, some of the carbanions (Ph$_3$C$^-$) which would normally revert by internal return to Ph$_3$CH, and not exchange with the solvent, were nonetheless intercepted by electron transfer to added nitrobenzene.[486]

Support for the revised structure (149) for the dimer of the trityl radical comes from the observation that it is aromatized by strong base, giving (150).[487]

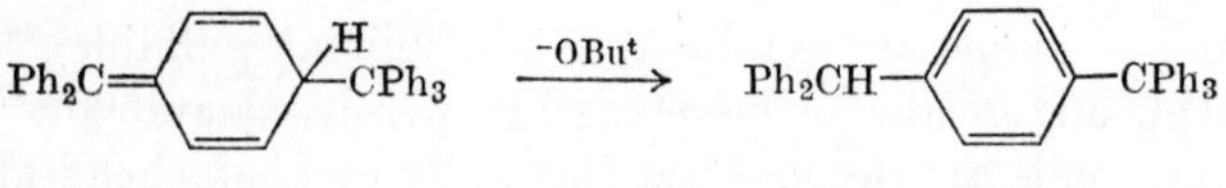

(149)   (150)

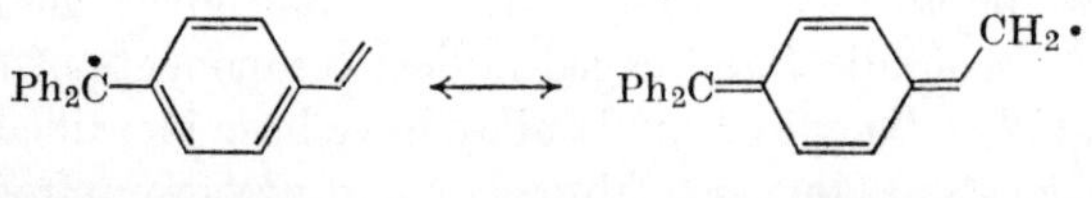

(A)   (151)   (B)

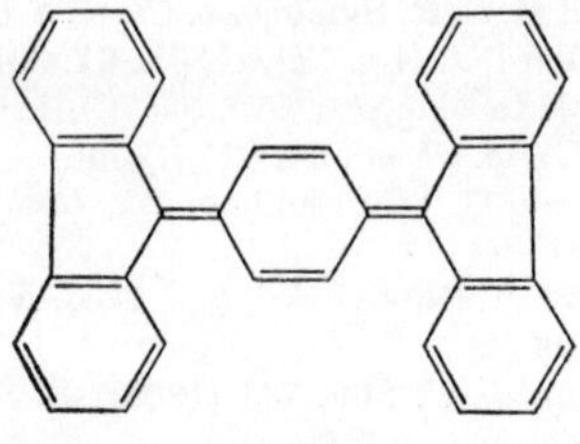

(152)

[484] F. S. Dainton, I. V. Janovsky, and G. A. Salmon, *Chem. Comm.*, **1969**, 335.
[485] D. R. G. Brimage, J. D. P. Cassell, J. H. Sharp, and M. C. R. Symons, *J. Chem. Soc.*(A), **1969**, 2619.
[486] R. D. Guthrie, *J. Am. Chem. Soc.*, **91**, 6201 (1969).
[487] R. D. Guthrie and G. R. Weisman, *Chem. Comm.*, **1969**, 1316.

Approaches to the 4-vinyltriphenylmethyl radical (**151A**) give products arising from reactions of the quinodimethane structure (**151B**).[488] The quinodimethane (**152**) could not be prepared, presumably because of steric crowding; a tetramer was isolated from the attempted synthesis.[489] New work on Tschitschibabin's hydrocarbon has been reported.[490]

The first example of a stable tetrazolinyl radical has been isolated,[491] and a related structure has been demonstrated for the Kuhn–Jerchel radicals.[492]

[488] D. Braun and R. J. Faust, *Monatsh. Chem.*, **100**, 968 (1969).

[489] G. Wittig, E. Dreher, and W. Reuther, *Ann. Chem.*, **726**, 188 (1969).

[490] H. D. Brauer, H. Sheger, and H. Hartmann, *Z. Phys. Chem.* (Frankfurt), **63**, 50 (1969).

[491] F. A. Neugebauer, *Angew. Chem. Internat. Ed. Engl.*, **8**, 520 (1969); see *Org. Reaction Mech.*, **1968**, 313.

[492] F. A. Neugebauer, *Chem. Ber.*, **102**, 1339 (1969).

CHAPTER 10

Carbenes and Nitrenes

T. L. GILCHRIST

Department of Chemistry, The University, Leicester

The emphasis of the year's work has been on the use of carbenes and nitrenes in synthesis, especially in cycloaddition reactions. There is increasing understanding of the structures of the intermediates and on the relationship between their spin state and their reactions. An important review of nitrene chemistry[1] has appeared, as well as two introductory texts on carbenes and nitrenes,[2] a useful introductory review of carbenes,[3] and one on silylenes[4] (bivalent silicon intermediates) which are structurally related to carbenes.

Structure

The structures of carbenes, and the stereochemistry of their additions to olefins, have been discussed in detail by Closs.[5] The most favourable geometries of several states of methylene have been predicted from *ab initio* valence-bond and molecular-orbital calculations, in a paper which also reviews the previous theoretical and spectroscopic evidence on the structure of methylene.[6] The results obtained agree very well with the experimental work, so that the authors are able to predict with confidence the values of various physical

[1] W. Lwowski, *Nitrenes*, Interscience, New York, 1970.
[2] T. L. Gilchrist and C. W. Rees, *Carbenes, Nitrenes, and Arynes*, Nelson, London, 1969; W. Kirmse, *Carbene, Carbenoide, und Carben-Analoge*, Verlag Chemie, Weinheim, 1969.
[3] R. A. Moss, *Chem. Eng. News*, **1969**, June 16th, 60; June 30th, 50.
[4] W. H. Atwell and D. R. Weyenberg, *Angew. Chem. Internat. Ed. Engl.*, **8**, 469 (1969).
[5] G. L. Closs, *Topics Stereochem.*, **3**, 193 (1968).
[6] J. F. Harrison and L. C. Allen, *J. Am. Chem. Soc.*, **91**, 807 (1969).

constants which have so far not been measured, such as the dipole moments. Dicyanocarbene and carbonylcarbene, $:C{=}C{=}O$, have also been the subjects of molecular-orbital calculations.[7] Both are predicted to have a linear triplet ground state, dicyanocarbene having little resistance to bending because of the low π-bond order in the C–C bonds. The Linnett electronic theory has been shown to be capable of giving a rough guide to the structures of singlet and triplet carbenes.[8]

The problems of determining the spin state and the excess energy of methylene generated photochemically have been tackled. Methylene, generated photochemically from diazomethane, reacted with *cis*-butene to give the cyclopropane;[9] the observed and calculated rates of isomerization of the activated product revealed that the methylene carries into the addition reaction only about 30% of the total available excess of energy. A method of determining the proportion of triplet methylene formed in the vapour-phase photolysis of ketene has been described;[10] photolyses were carried out in pairs, one with about 10% added oxygen to remove the triplet, and the other without, in the presence of a substrate and an internal standard which was not a reaction product. By comparing the product distributions the proportion of triplet methylene was determined. Frey and Walsh[11] have reinforced last year's observation[12] that even mercury-sensitized photolysis of ketene may give appreciable amounts of singlet methylene.

An estimate has been made of the lifetimes of aromatic nitrenes, generated by flash photolysis of the azides.[13] Although dimerization and reaction with azido groups are diffusion-controlled processes, hydrogen abstraction is comparatively slow. In hard polymeric matrices the half-life can be as much as 2 sec.

Two groups of workers have pointed out that nitrenium ions, like nitrenes, should be capable of existing as singlets (1) or triplets (2). It is suggested that the *N*-chloro-compound (3) may solvolyse to give a singlet nitrenium ion,[14] which is converted by collisional deactivation into its (ground state) triplet. This then abstracts hydrogen from the solvent, the proportion of abstraction product being increased by a "heavy atom" solvent. On the other hand, the conversion of the oxime (4) into the insertion product (5) is thought to involve a singlet nitrenium ion,[15] despite the heavy atom in the molecule, because the

[7] J. F. Olsen and L. Burnelle, *Tetrahedron*, **25**, 5451 (1969).

[8] R. A. Firestone, *J. Org. Chem.*, **34**, 2621 (1969).

[9] J. W. Simons and G. W. Taylor, *J. Phys. Chem.*, **73**, 1274 (1969).

[10] T. W. Eder and R. W. Carr, *J. Phys. Chem.*, **73**, 2074 (1969).

[11] H. M. Frey and R. Walsh, *Chem. Comm.*, **1969**, 158.

[12] D. C. Montague and F. S. Rowland, *J. Phys. Chem.*, **72**, 3705 (1968).

[13] A. Reiser, F. W. Willets, G. C. Terry, V. Williams, and R. Marley, *Trans. Faraday Soc.*, **64**, 3265 (1968).

[14] P. G. Gassman and R. L. Cryberg, *J. Am. Chem. Soc.*, **91**, 5176 (1969).

[15] P. T. Lansbury and P. C. Briggs, *Chem. Comm.*, **1969**, 1152.

(1) 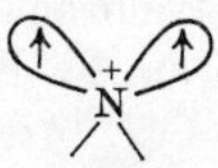　　　　　(2) 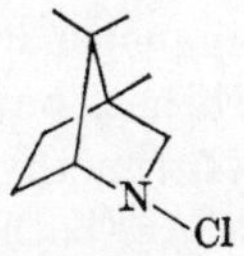　　　　　(3)

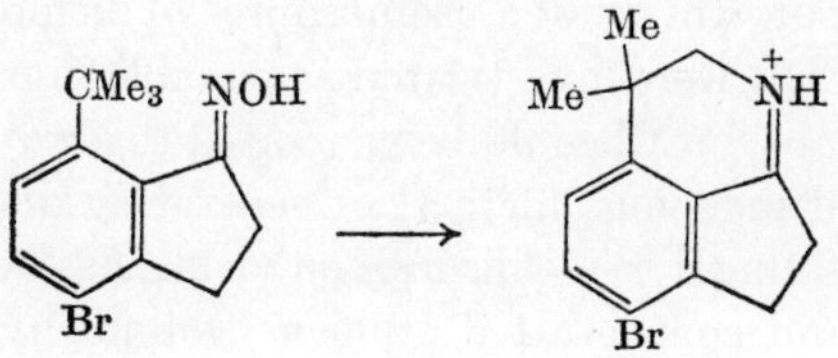

(4)　　　　　　　(5)

isotope effect for the insertion reaction ($k_H/k_D = 1.4$—1.6) is much closer to that expected for a singlet nitrene insertion than for a triplet nitrene abstraction.

Methods of Generation

Carbenes

Seyferth and his group have reported several attempts to extend the use of organomercury and organotin compounds in the generation of carbenes. They have reported an improved synthesis of phenyltrihalogenomethylmercury compounds, $PhHgCX_3$.[16] A curious halogen exchange reaction takes place between these compounds and phenylmercuric fluoride; for example, $PhHgCBr_3$ gives $PhHgCF_3$.[17] The trifluoromethyl compound is thermally stable, but will transfer difluorocarbene to olefins in good yield if sodium iodide is added. Mercury derivatives of the general type (6) have also been synthesized.[18] On thermolysis they appear to give the carbenes (7) which then rearrange, the nature of the product depending on the substituent R.

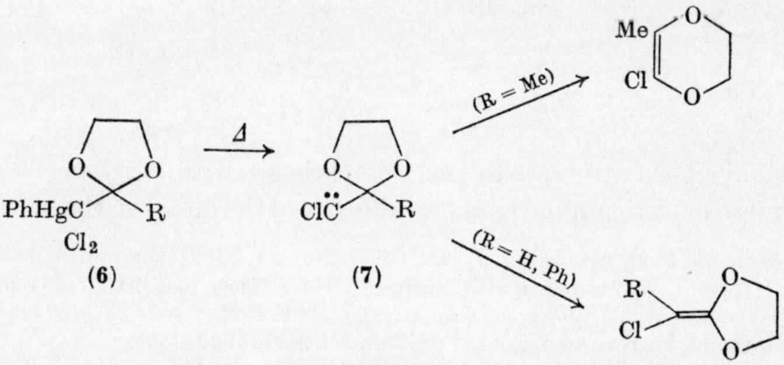

[16] D. Seyferth and R. L. Lambert, *J. Organometal. Chem.*, **16**, 21 (1969).
[17] D. Seyferth, S. P. Hopper, and K. V. Darragh, *J. Am. Chem. Soc.*, **91**, 6536 (1969).
[18] D. C. Mueller and D. Seyferth, *J. Am. Chem. Soc.*, **91**, 1754 (1969).

Various organotin derivatives have been investigated as possible precursors of chlorocarbenes, but none appears to be particularly useful. Thus (trichloromethyl)trimethyltin, Me_3SnCCl_3,[19] and trimethyltin trichloroacetate, $Me_3SnO_2CCCl_3$,[20] both gave some dichlorocarbene on heating, but the precursors offer no advantages over other known sources of the carbene. Similarly (trichloromethyl)tributyltin is not a useful source of dichlorocarbene.[21] Other derivatives Me_3SnCCl_2R were investigated as possible sources of the carbenes $R\ddot{C}Cl$.[19] $Ph\ddot{C}Cl$ and $Me_3Sn\ddot{C}Cl$ could be generated this way, but not $:CHCl$.

A very simple and ingenious method of generating dichlorocarbene from chloroform, and of adding it to olefins in good yields, has been reported.[22] The olefin, chloroform, and concentrated aqueous sodium hydroxide are mixed with a catalytic amount of triethylbenzylammonium hydroxide. The usual hydrolysis of the dichlorocarbene is suppressed because the carbene is generated within the organic layer, probably from the salt $Et_3N^+Bz~^-CCl_3$. Dichlorocarbene has also been generated from chloral and alkoxides in the absence of alcohols, and from trichloroacetophenone.[23] The slow pyrolysis of (dimethoxymethyl)trimethoxysilane is a mild way of generating methoxycarbene:[24]

$$(MeO)_3SiCH(OMe)_2 \xrightarrow[\text{16 hr}]{125°} (MeO)_4Si + :CHOMe$$

Two new precursors of arylcarbenes have been described; both probably give diazoalkanes as intermediates in their decomposition, and the diazoalkanes then give the carbenes. Thermal and photochemical decomposition of phosphazines[25] such as (8), and photolysis of the 5-phenyltetrazole anion (9),[26] are the reported routes; the latter requires a protonation at some stage, probably after the loss of the first molecule of nitrogen.

$$Ph_2C{=}NN{=}PPh_3 \xrightarrow{\Delta,\ h\nu} PPh_3 + Ph_2CN_2 \longrightarrow Ph_2C\textbf{:}$$
(8)

(9)
$$\xrightarrow{h\nu} Ph\bar{C}{=}\overset{+}{N}{=}\bar{N} \xrightarrow{H^+} PhCHN_2 \longrightarrow Ph\ddot{C}H$$

There have been attempts to generate carbenes with ferrocene as a substituent, and to investigate the possible influence of the metal on the nature of the

[19] D. Seyferth and F. M. Armbrecht, *J. Am. Chem. Soc.*, **91**, 2616 (1969).

[20] F. M. Armbrecht, W. Tronich, and D. Seyferth, *J. Am. Chem. Soc.*, **91**, 3218 (1969).

[21] A. G. Davies and W. R. Symes, *J. Chem. Soc.*(C), **1969**, 1892.

[22] M. Makosza and M. Wawrzyniewicz, *Tetrahedron Letters*, **1969**, 4659.

[23] F. Nerdel, H. Dahl, and P. Weyerstahl, *Tetrahedron Letters*, **1969**, 809.

[24] W. H. Atwell, D. R. Weyenberg, and J. G. Uhlmann, *J. Am. Chem. Soc.*, **91**, 2025 (1969).

[25] D. R. Dalton and S. A. Liebman, *Tetrahedron*, **25**, 3321 (1969).

[26] P. Scheiner, *J. Org. Chem.*, **34**, 199 (1969).

ground state and on the reactivity of the species. The action of bases on fluoroborates (10)[27] and the thermal decomposition of tosylhydrazone salts (11)[28] have been studied; of these, the tosylhydrazone route has so far given

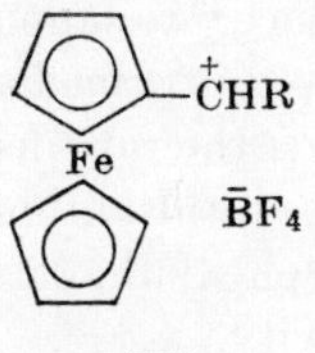

(10) 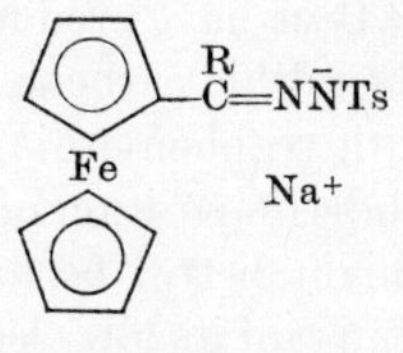   (11)

the more promising results. In this study the products were mainly those of abstraction from the solvent, though some azine was formed and an adduct was obtained with 1,1-diphenylethylene. The authors conclude that the carbenes probably react as triplets, like other arylcarbenes.

A reaction which may involve phenylcarbene is that involving benzaldehyde, zinc, and boron trifluoride in the presence of olefins, which leads to the formation of cyclopropanes. Cyclohexene, for example, gives phenylnorcarane in fair yield. A possible mechanism for the formation of phenylcarbene is shown;[29] it is similar to a mechanism suggested for the Clemmensen reduction several years ago.[30]

$$PhCHO \longrightarrow PhCH{=}\overset{+}{O}\overset{-}{B}F_3 \overset{e^-}{\longrightarrow} Ph\overset{.}{C}HO\overset{-}{B}F_3 \overset{e^-}{\longrightarrow} Ph\overset{..}{C}H + OBF_2{}^- + F^-$$

The carbenes $Br\overset{..}{C}CO_2Et$ and $I\overset{..}{C}CO_2Et$ have been generated by photolysis of the appropriate diazoalkanes;[31] they are more selective than ethoxycarbonylcarbene and may have singlet ground states. The base-induced elimination of HCl from the dichlorocyclopropane (12) gives the allenic carbene (13), but not via a symmetrical "cyclopropyne" intermediate.[32]

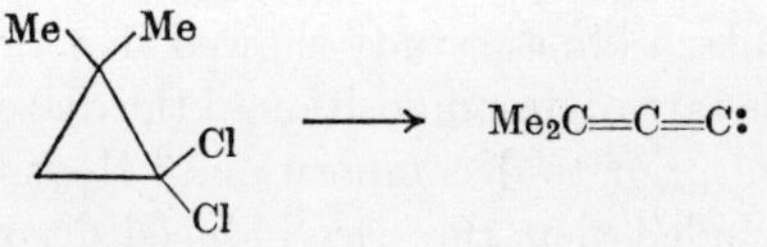

(12)   (13)

Other reactions in which carbenes are probably generated include the pyrolyses of chloroacetonitrile (which may give cyanocarbene) and of trichloroacetonitrile (which may give dichlorocarbene and chlorocyanocarbene),[33]

[27] P. Ashkenazi, S. Lupan, A. Schwarz, and M. Cais, *Tetrahedron Letters*, **1969**, 817.
[28] A. Sonoda, I. Moritani, T. Saraie, and T. Wada, *Tetrahedron Letters*, **1969**, 2943.
[29] I. Elphimoff-Felkin and P. Sarda, *Chem. Comm.*, **1969**, 1065.
[30] D. Staschewski, *Angew. Chem.*, **71**, 726 (1959).
[31] U. Schöllkopf and M. Reetz, *Tetrahedron Letters*, **1969**, 1541.
[32] L. Crombie, P. J. Griffiths, and B. J. Walker, *Chem. Comm.*, **1969**, 1206.
[33] N. Hashimoto, K. Matsumura, and K. Morita, *J. Org. Chem.*, **34**, 3410 (1969).

and the decomposition of diazopropane in the presence of t-butyl hypobromite, in which EtČBr is suggested as an intermediate.[34] 4-Nitrobenzyl bromide and iodide have been shown not to give carbenes with bases, however.[35]

Seebach and Beck have produced kinetic evidence[36] to support their view[37] that free di(phenylthio)carbene (14) is a long-lived intermediate in solution; its reaction with tri(phenylthio)methyl-lithium is the rate-determining step in the thermal decomposition of the latter. Indeed, the authors suggest that the carbene might better be named as the di(phenylthio)acetal of carbon monoxide, because of its low electrophilic reactivity.

$$(PhS)_3CLi \rightleftharpoons (PhS)_2C: + PhSLi$$
$$(14)$$

$$(PhS)_3CLi + (PhS)_2C: \rightarrow (PhS)_2C{=}C(SPh)_2 + PhSLi$$

The mechanisms of the decomposition of diazoalkanes and related compounds in protic solvents have been investigated by several groups of workers. In principle such reactions could involve prior protonation, to give carbonium ions after loss of nitrogen, or they could give carbenes, as in aprotic solvents. The possibilities can be distinguished in the decomposition of diazosuccinic esters by running the reaction in a deuterated solvent: maleic and fumaric esters are formed, and these should incorporate deuterium if the carbonium ion mechanism operates, but not if carbenes are intermediates.[38]

$$EtO_2CCH(N_2^+)CH_2CO_2Et \rightarrow EtO_2C\overset{+}{C}HCH_2CO_2Et \rightarrow EtO_2CCH{=}CHCO_2Et$$

or

$$EtO_2CC(N_2)CH_2CO_2Et \rightarrow EtO_2C\overset{\cdot\cdot}{C}CH_2CO_2Et \rightarrow EtO_2CCH{=}CHCO_2Et$$

The results of such experiments showed that, even in acetic acid, the decomposition is mainly carbenic. In ethanol, about 80% of the reaction went via the carbene intermediate. Similar results were obtained for diaryldiazomethanes with alcohols: a kinetic study showed that the addition of alcohols had little effect on the rate of decomposition of the diazoalkanes, so they were probably not reacting prior to decomposition.[39] However, Newman and his coworkers have concluded that the base-induced decomposition of nitrosooxazolidones (15) goes via a vinyl carbonium ion in protic solvents and through a carbene in aprotic solvents.[40] In the presence of halide ions, the reaction is a useful synthesis of vinyl halides, and in alcohols, of vinyl ethers.

[34] R. J. Bussey and R. C. Neuman, *J. Org. Chem.*, **34**, 1323 (1969).

[35] A. A. Abdallah, Y. Iskander, and Y. Riad, *J. Chem. Soc.*(B), **1969**, 1178.

[36] D. Seebach and A. K. Beck, *J. Am. Chem. Soc.*, **91**, 1540 (1969).

[37] D. Seebach, *Angew. Chem. Internat. Ed. Engl.*, **6**, 443 (1967).

[38] Y. Yamamoto and I. Moritani, *Tetrahedron Letters*, **1969**, 3087.

[39] D. Bethell and R. D. Howard, *J. Chem. Soc.*(B), **1969**, 745.

[40] M. S. Newman and A. O. M. Okorodudu, *J. Org. Chem.*, **34**, 1220 (1969); M. S. Newman and C. D. Beard, *J. Am. Chem. Soc.*, **91**, 5677 (1969).

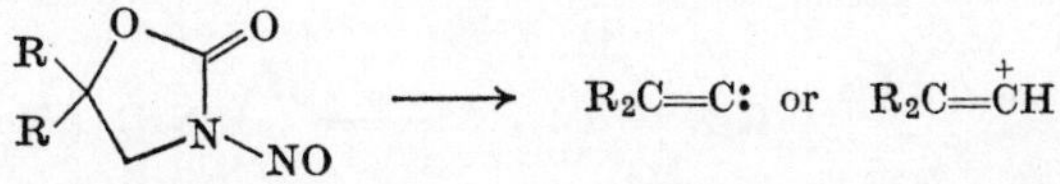

(15)

Nitrenes

A review of the chemistry of organic azides, including their decomposition to give nitrenes, has appeared.[41] Nitrenes have been generated by direct photolysis of alkyl azides,[42] by thermolysis of toluene-p-sulphonyl azide and aliphatic sulphonyl azides,[43] and by the decomposition of benzenesulphinyl azide, $PhSON_3$, at $0°$.[44] The alkyl azides give alkylnitrenes, probably as the singlets, which rearrange faster than they can be trapped. Aliphatic sulphonyl azides decompose both by loss of nitrogen, to give nitrenes, and by loss of sulphur dioxide and nitrogen, to give radicals. The nitrenes insert or add rather than abstract. Benzenesulphinyl azide, the first of its class to be prepared, gives some of the formal nitrene trimer (16). There is the possibility of stabilization of the singlet state of benzenesulphinylnitrene (17) by delocalization of the lone pair of sulphur into the vacant nitrogen orbital, as with the amino-nitrenes.

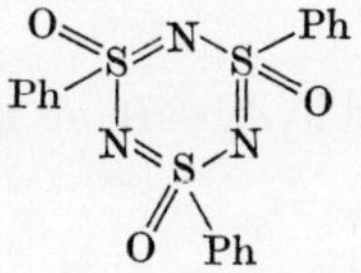

(16) 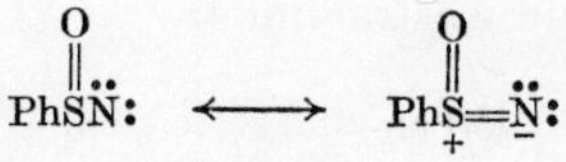   (17)

The thermolysis of 4,6-dimethylpyrimidinyl azide (18) may also give the nitrene.[45] Hydrogen abstraction and CH insertion products were obtained from thermolyses in aliphatic hydrocarbons, and there was no increase in the rate of decomposition in aromatic solvents. With copper acetylacetonate as a catalyst, the nitrene added stereospecifically to *trans*-stilbene[46]—the first time that an arylnitrene has been added to an olefin, though probably a copper complex is involved. 4,6-Dimethylpyrimidinyl azide seems to behave differently from 2-pyridyl azide, which gave no adduct with *trans*-stilbene in comparable conditions.

[41] G. L'Abbé, *Chem. Rev.*, **69**, 345 (1969).
[42] F. D. Lewis and W. H. Saunders, *J. Am. Chem. Soc.*, **90**, 7031 (1968).
[43] D. S. Breslow, M. F. Sloan, N. R. Newburg, and W. B. Renfrow, *J. Am. Chem. Soc.*, **91**, 2273 (1969).
[44] T. J. Maricich, *J. Am. Chem. Soc.*, **90**, 7179 (1968).
[45] R. Huisgen and K. v. Fraunberg, *Tetrahedron Letters*, **1969**, 2595.
[46] K. v. Fraunberg and R. Huisgen, *Tetrahedron Letters*, **1969**, 2599.

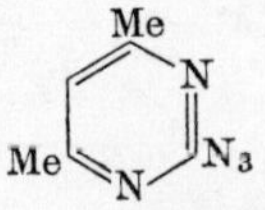

$$R_2NN{=}\overset{\overset{\displaystyle O}{\|}}{S}Me_2 \quad \xrightarrow{h\nu} \quad R_2N\ddot{\ddot{N}}\text{:} + Me_2SO$$

(19)

$$ArSO_2N{=}\overset{\overset{\displaystyle O}{\|}}{S}Me_2 \quad \xrightarrow{h\nu} \quad Ar\bullet$$

(20)

(18)

Very few new methods of generating nitrenes have been reported. Photolysis of sulphoximines (19) appears to be a method of generating aminonitrenes,[47] but the reaction is not a general one: sulphonylsulphoximines (20) give aryl radicals instead.[48]

Sulphonylnitrenes are, however, likely intermediates in the copper-catalysed thermolysis of chloramine-T[49] and in the reaction of sulphonyloxysulphon-amides, $ArSO_2NHOSO_2Ar'$, with triethylamine.[50a] Arylnitrenes have been generated by the solid-state decomposition of acyl carbamoyl peroxides, which are prepared from the aryl isocyanate and a peroxy-acid:[50b]

$$ArNCO + PhCOOOH \rightarrow ArNHCOOOCOPh \rightarrow Ar\ddot{\ddot{N}}\text{:} + PhCO_2H + CO_2$$

Insertions and Abstractions

The reaction between energetic carbon atoms and hydrocarbons is thought to give a carbene as an intermediate:

$$\text{:C: } + R_3CH \rightarrow R_3C{-}\ddot{\ddot{C}}{-}H$$

A study of the pressure dependence of the products of such reactions has shown that it is not the carbene reaction which determines the observed trends, but the stabilization or dissociation of some prior collision complex.[51] The nature of this collision complex is not known, however.

Methylene, generated in the gas phase and in solution by photolysis of diazomethane, inserts exclusively in the CH bonds of acetals and orthoesters. There is a slight preference for insertion in primary CH bonds. With paralde-hyde, the insertion into the ring CH bonds was shown to proceed with retention of configuration.[52a] A study of the gas-phase reaction of methylene with 1-chloropropane has also been reported.[52b]

[47] D. J. Anderson, T. L. Gilchrist, D. C. Horwell, and C. W. Rees, *Chem. Comm.*, **1969**, 146.
[48] R. A. Abramovitch and T. Takaya, *Chem. Comm.*, **1969**, 1369.
[49] D. Carr, T. P. Seden, and R. W. Turner, *Tetrahedron Letters*, **1969**, 477.
[50a] M. Okahara and D. Swern, *Tetrahedron Letters*, **1969**, 3301.
[50b] R. Okazaki and O. Simamura, *Chem. Comm.*, **1969**, 1308.
[51] M. J. Welch and A. P. Wolf, *J. Am. Chem. Soc.*, **91**, 6584 (1969).
[52a] W. Kirmse and M. Buschhoff, *Chem. Ber.*, **102**, 1087, 1098 (1969).
[52b] C. H. Bamford and J. E. Casson, *Proc. Roy. Soc.*, *A*, **312**, 141 (1969).

Dichlorocarbene can insert both into mercury–carbon bonds and into CH bonds. Since $:CCl_2$ does not normally insert into CH bonds, some prior complexing to the mercury seems to be indicated, and an intermediate mercurium ion pair is suggested by Landgrebe and Thurman.[53] Another halogenocarbene insertion which has been observed is that of $CHF_2\ddot{C}F$ into the SiH bonds of trialkylsilanes.[54] The carbene will also insert into SiCl bonds, but not so readily.

Transannular insertion is a major reaction of the cyclooctenylidenes **(21)**—**(23)**. Bicyclo[3.3.0]oct-2-ene **(24)** is the major product from the carbenes **(21)** and **(22)**, and the third isomer **(23)** gives the β-CH insertion product,

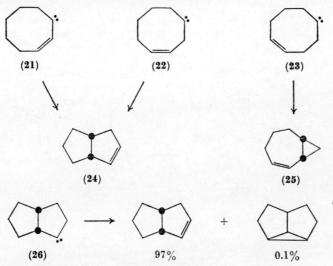

bicyclo[5.1.0]oct-2-ene **(25)**.[55] However, intramolecular insertion is only a minor reaction of the bicyclic carbene **(26)**.[56a] Other transannular insertions have provided useful syntheses of tricyclic ring systems.[56b]

Frey and Walsh have reported an intriguing reaction of triplet methylene with neopentane, in which t-butyl radicals are generated:[57]

$$:CH_2 + CMe_4 \rightarrow Et\cdot + \cdot CMe_3$$

The radicals could be formed either by an abstraction of a methyl group by methylene, or by a displacement of t-butyl by methylene at one of the methyl carbons; both types of reaction are virtually unknown.

[53] J. A. Landgrebe and D. E. Thurman, *J. Am. Chem. Soc.*, **91**, 1759 (1969).
[54] R. N. Haszeldine, A. E. Tipping, and R. O'B. Watts, *Chem. Comm.*, **1969**, 1364.
[55] W. Kirmse and G. Münscher, *Ann. Chem.*, **726**, 42 (1969).
[56a] W. Kirmse and L. Ruetz, *Ann. Chem.*, **726**, 36 (1969).
[56b] T. Sasaki, S. Eguchi, and T. Kiriyama, *J. Am. Chem. Soc.*, **91**, 212 (1969); M. R. Vegar and R. J. Wells, *Tetrahedron Letters*, **1969**, 2565.
[57] H. M. Frey and R. Walsh, *Chem. Comm.*, **1969**, 159.

Dynamic nuclear spin polarization in NMR spectra has been applied for the first time to the study of a triplet carbene reaction.[58] Diphenyldiazomethane, when decomposed in toluene, gives radical pairs generated from the triplet carbene by hydrogen abstraction from the solvent. The crossover from triplet to singlet radical pairs is proposed to cause nuclear spin polarization. This technique promises to be useful for studying other reactions of triplet carbenes.

The synthetically useful reaction in which arylnitrenes insert intra-molecularly into aromatic CH bonds has been extended to other nitrenes. Abramovitch and his coworkers have found such reactions with an alkyl azide (27),[59] sulphonyl azides, such as (28),[60] and ferrocenesulphonyl azide (29).[61]

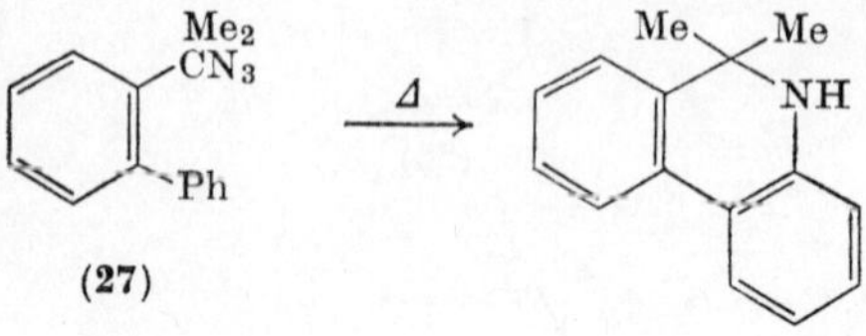

(27)

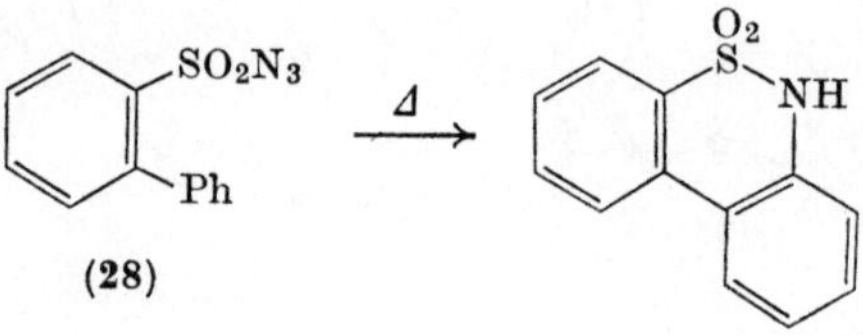

(28)

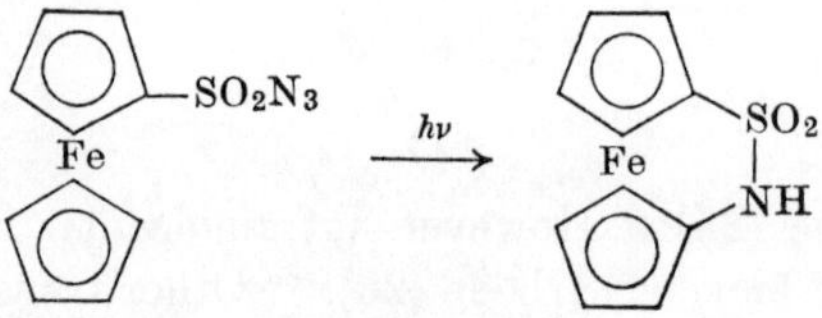

(29)

A re-investigation of the photolysis of 2,2'-diazidobiphenyl[62] has shown that benzo[c]cinnoline is formed, contrary to a previous report,[63] but only in trace amounts. The major product is 4-azidocarbazole, showing that the intermediate nitrene prefers to attack the CH bond rather than the azido group. A kinetic study of the deoxygenation of 2-nitrosobiphenyl by triethyl phosphite did not establish whether a free nitrene is an intermediate; a complex

[58] G. L. Closs and L. E. Closs, *J. Am. Chem. Soc.*, **91**, 4549 (1969); G. L. Closs, *ibid.*, p. 4552.
[59] R. A. Abramovitch and E. P. Kyba, *Chem. Comm.*, **1969**, 265.
[60] R. A. Abramovitch, C. I. Azogu, and I. T. McMaster, *J. Am. Chem.*, *Soc.* **91**, 1219 (1969).
[61] R. A. Abramovitch, C. I. Azogu, and R. G. Sutherland, *Chem. Comm.*, **1969**, 1439.
[62] J. H. Boyer and G. J. Mikol, *Chem. Comm.*, **1969**, 734.
[63] L. Horner and A. Christmann, *Angew. Chem. Internat. Ed. Engl.*, **2**, 599 (1963).

involving both reagents, probably $(EtO)_3P^+-O-N^-Ar$, is formed in the rate-determining step.[64]

There have been important advances in understanding the mechanism of the CH insertion reaction from work by various groups with oxycarbonyl-nitrenes. Lwowski's group has established that only the singlet nitrene is involved in the insertion reaction; the selectivity of the insertion of ethoxy-carbonylnitrene into the CH bonds of 3-methylhexane was unaffected by changes in the concentration of the substrate or by the addition of α-methyl-styrene or dibromomethane, and the reaction with optically active 3-methyl-hexane went with complete retention of configuration.[65] The measurement of the relative reactivities of the different CH bonds in several other hydrocarbons gave the results shown,[66] relative to cyclohexane = 1.0. An independent study of the relative reactivities of the CH bonds in the hydrocarbon (30)

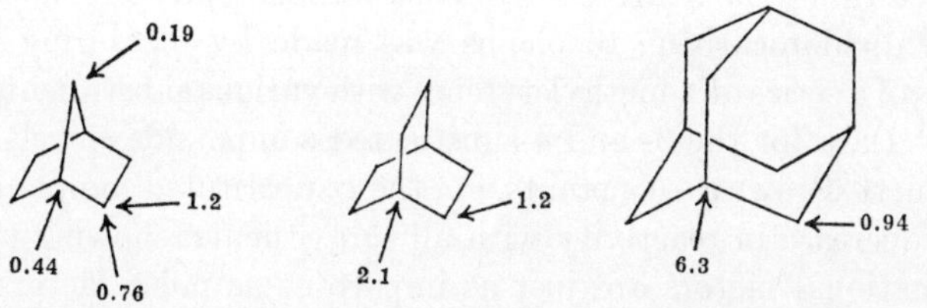

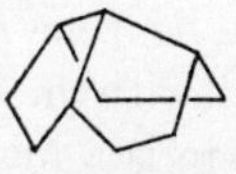

(30)

showed that the tertiary CH bonds (apical and non-apical) were about four times as reactive as the secondary.[67]

The relative reactivities of the hydrocarbons were found to correspond well to the t-butyl perester decomposition rates—a typical free-radical process—and not so well to carbonium ion reactivities. It is probably misleading to think of the transition state as having "free radical character", however; rather, the insertion and radical reactivities are controlled by similar structural features in the substrates.[66] At least, an experimental base for understanding how this structural control operates has now been laid.

Intramolecular equivalents of these oxycarbonylnitrene insertion reactions have also been studied, and show a similar pattern.[68] The reaction has been put to good use in an elegant synthesis of 2-amino-1-adamantanol (31).[69]

[64] J. I. G. Cadogan and A. Cooper, *J. Chem. Soc.*(B), **1969**, 883.
[65] J. M. Simson and W. Lwowski, *J. Am. Chem. Soc.*, **91**, 5107 (1969).
[66] D. S. Breslow, E. I. Edwards, R. Leone, and P. von R. Schleyer, *J. Am. Chem. Soc.*, **90**, 7097 (1968).
[67] L. A. Paquette, G. V. Meehan, and S. J. Marshall, *J. Am. Chem. Soc.*, **91**, 6779 (1969).
[68] S. Yamada and S. Terashima, *Chem. Comm.*, **1969**, 511.
[69] W. V. Curran and R. B. Angier, *J. Org. Chem.*, **34**, 3668 (1969).

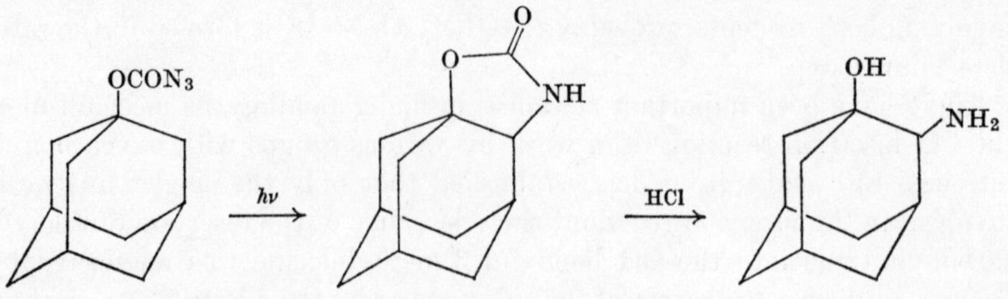

(31)

Cycloadditions

The review by Woodward and Hoffmann on the conservation of orbital symmetry [70] includes a section on cheletropic reactions, of which carbene and nitrene cycloadditions are examples. Closs's review [5] also includes a discussion of the mechanism and stereochemistry of the addition reaction.

An attempt to determine the relative importance of polar and steric factors in the addition of dichlorocarbene to olefins was made by measuring the relative reactivities of a series of α-methylstyrenes with various substituents in the benzene ring.[71] Data for the 3- and 4-substituted compounds correlated well with the Hammett–Brown σ^+-constants. For the 2-substituted compounds there is a marked decrease in reactivity with all substituents, showing that steric and conformational factors are just as important as polar factors in determining relative reactivities. A transition state was suggested for the cycloaddition in which the π-electrons of the olefins are asymmetrically partially delocalized into the vacant p-orbital of the carbene. A similar transition state is proposed for the cycloaddition of dimethylethylidene, $Me_2C=C:$, to olefins:[72] these correspond closely to the transition state suggested last year by Hoffmann for the addition of methylene to ethylene.[73]

Other carbene additions in which steric factors were found to be important were the reaction of triphenylcyclopentadienylidene (**32**) with olefins,[74] where the proportion of addition to insertion decreases with a more hindered olefin; and the reactions of diphenylcarbene, fluorenylidene, and dichlorocarbene with olefins $Me_3Si(CH_2)_nCH=CH_2$.[75] The order of reactivity towards diphenylcarbene ($n = 0 > 1 > 2$) was different from that for dichlorocarbene ($n = 1 > 2 > 0$); the enhanced reactivity of $Me_3SiCH=CH_2$ with diphenylcarbene may arise because the carbene reacts as a triplet.

[70] R. B. Woodward and R. Hoffmann, *Angew. Chem. Internat. Ed. Engl.*, 8, 781 (1969).
[71] I. H. Sadler, *J. Chem. Soc.*(B), **1969**, 1024.
[72] M. S. Newman and T. B. Patrick, *J. Am. Chem. Soc.*, **91**, 6461 (1969).
[73] R. Hoffmann, *J. Am. Chem. Soc.*, **90**, 1475 (1968); see *Org. Reaction Mech.*, **1968**, 321.
[74] H. Dürr and L. Schrader, *Chem. Ber.*, **102**, 2026 (1969).
[75] I. A. D'yakonov, I. B. Repinskaya, and T. D. Marinina, *Zh. Obshch. Khim.*, **39**, 717 (1969); I. A. D'yakonov, V. P. Dushina, and G. V. Golodnikov, *ibid.*, p. 923.

Several synthetically useful intramolecular carbene cycloadditions to double bonds have been reported.[76] Other additions to olefins include the reactions of dihalogenocarbenes with fluorinated olefins[77] and with steroids,[78] and of substituted diazocyclopentadienes (33) with norbornadiene.[79] Unlike the reaction of diazofluorene with norbornadiene, the last reaction is thought

(32)   (33)   (34)

not to involve an intermediate pyrazoline, but to go via the carbene, though the evidence is equivocal.

The chemistry of cycloheptatrienylidene (34) has been reviewed.[80] The carbene adds stereospecifically to fumarates and maleates, and to fumaronitrile and maleonitrile, contrary to a previous report, but does not add to nucleophilic olefins like cyclohexene and *cis*-butene.[81] The carbene probably has a nucleophilic singlet ground state.

Dihalogenocarbenes have been added to large cyclic polyenes capable of accommodating *trans* as well as *cis* double bonds. The *trans* double bonds are preferentially attacked.[82] With 1-chlorobutadiene, dichlorocarbene adds exclusively to the unsubstituted double bond;[83] acetylcarbene gives both possible cyclopropanes with unsymmetrical dienes, however.[84] Competitive alkyne additions by ethoxycarbonylcarbene gave a confused picture. With enynes, the carbene, generated thermally in the presence of copper, added to the double bonds and to triple bonds with alkyl substituents.[85] With olefin–acetylene pairs such as oct-4-yne and *trans*-oct-4-ene, the carbene added

---

[76] H. Tsuruta, K. Kurabayashi, and T. Mukai, *J. Am. Chem. Soc.*, **90**, 7167 (1968); G. Cannic, G. Linstrumelle, and S. Julia, *Bull. Soc. Chim. France*, **1968**, 4913; D. J. Beames and L. N. Mander, *Chem. Comm.*, **1969**, 498; S. K. Dasgupta, R. Dasgupta, S. R. Ghosh, and U. R. Ghatak, *ibid.*, p. 1253.

[77] M. L. Deem, *Chem. Comm.*, **1969**, 993.

[78] P. Crabbé, *Ind. Chim. Belge*, **34**, 15 (1969).

[79] H. Dürr, G. Scheppers, and L. Schrader, *Chem. Comm.*, **1969**, 257.

[80] T. Mukai, H. Tsuruta, T. Nakazawa, K. Isobe, and K. Kurabayashi, *Sci. Rep. Tohoku Univ.*, First Ser., **51**, 113 (1968); *Chem. Abs.*, **70**, 87155 (1969).

[81] W. M. Jones and C. L. Ennis, *J. Am. Chem. Soc.*, **91**, 6391 (1969); W. M. Jones B. N. Hamon, R. C. Joines, and C. L. Ennis, *Tetrahedron Letters*, **1969**, 3909.

[82] J. Graefe and M. Mühlstädt, *Z. Chem.*, **9**, 23 (1969); *Tetrahedron Letters*, **1969**, 3431; M. Mühlstädt, *Z. Chem.*, **9**, 303 (1969).

[83] I. A. D'yakonov and T. A. Kornilova, *Zh. Org. Khim.*, **5**, 178 (1969).

[84] V. A. Kalinina and Y. I. Kheruze, *Zh. Org. Khim.*, **4**, 1347 (1968).

[85] I. A. D'yakonov, L. P. Danilkina, and R. N. Gmyzina, *Zh. Org. Khim.*, **5**, 1026 (1969).

more readily to the olefins when generated photochemically, but surprisingly, added more readily to the acetylenes when generated thermally with a copper catalyst.[86]

The gas-phase addition of dideuteromethylene to allene gave methylenecyclopropane having a random distribution of deuterium.[87] A symmetrical intermediate seems to be involved, possibly singlet trimethylenemethane; or a vibrationally excited methylenecyclopropane may be formed which then undergoes the known degenerate rearrangement. Methylenecyclopropane and dideuteromethylene similarly gave methylenecyclobutane, again with a statistical distribution of deuterium. If an excited spiropentane is an intermediate, the excess of vibrational energy must be evenly distributed within the molecule prior to rearrangement, and not localized at the carbene carbon, otherwise the deuterium distribution would be uneven.

A carbene addition to an acetylene may be involved in the formation of the methylenecyclobutene derivative (35); the mechanism shown for its production from the acetylene, benzylidene chloride, and potassium t-butoxide is one of those suggested by the author.[88]

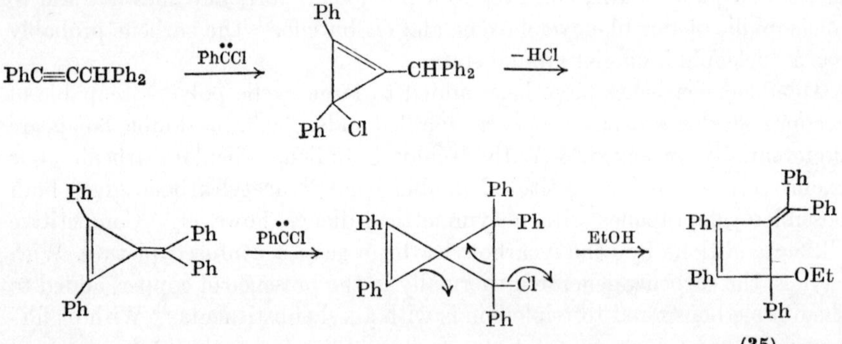

PhC≡CCHPh₂ → PhC̈Cl → [structure] CHPh₂ → − HCl

(35)

The photochemical addition of the diazo-compounds (36)—(38) to dimethyl acetylenedicarboxylate[89] and to benzene[90] may involve carbenes, but 1,3-dipolar addition followed by loss of nitrogen seems equally likely. The main interest of the additions to benzene is that the products have the norcaradiene structure rather than the cycloheptatriene. This is ascribed to the small angle, α, at the spiro carbon, although the corresponding adduct with hexafluorobenzene has the cycloheptatriene structure.[91]

[86] I. A. D'yakonov, M. I. Komendantov, L. P. Danilkina, R. N. Gmyzina, T. S. Smirnova, and A. G. Vitenberg, *Zh. Org. Khim.*, **5**, 383 (1969).
[87] W. von E. Doering, J. C. Gilbert, and P. A. Leermakers, *Tetrahedron*, **24**, 6863 (1968).
[88] B. Föhlisch, *Tetrahedron Letters*, **1969**, 3009.
[89] H. Dürr and L. Schrader, *Angew. Chem. Internat. Ed. Engl.*, **8**, 446 (1969).
[90] D. Schönleber, *Angew. Chem. Internat. Ed. Engl.*, **8**, 76 (1969); M. Jones, *ibid.*, p. 76.
[91] M. Jones, *J. Org. Chem.*, **33**, 2538 (1968); see *Org. Reaction Mech.*, **1968**, 336.

The synthetic value of the reaction between dichlorocarbene and pyrroles or imidazoles has been enhanced by the work of two groups, reported this year. Jones and Rees found that dichlorocarbene reacts with alkyl-substituted pyrroles in basic media in two ways:[92] by addition to the double bonds, to

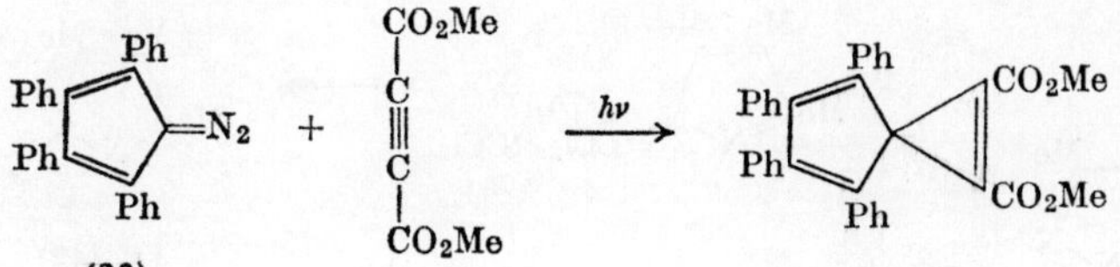

(36)

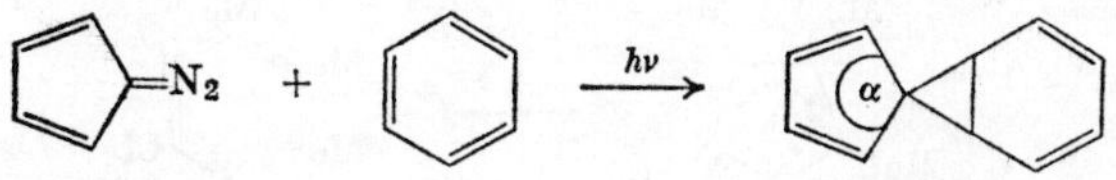

(37)

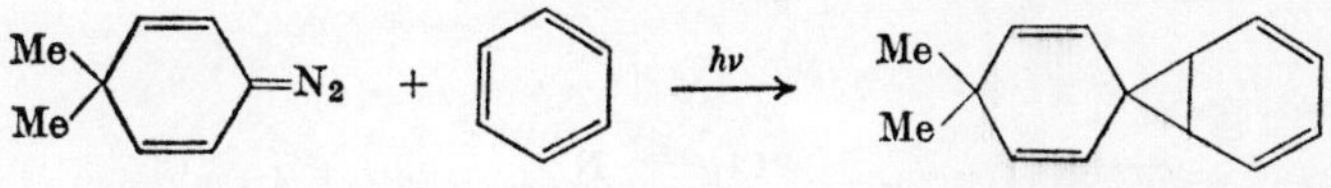

(38)

give 3-chloropyridines, and by electrophilic attack on the pyrrole anions [(39)→(40)]. The latter reaction was suppressed, however, when the carbene was generated in neutral aprotic solutions, and 3-chloropyridines could be obtained in good yields. Similar reactions were also found with methyl-substituted imidazoles and pyrazoles,[93] but the product yields were lower. The investigation revealed some interesting side-reactions; for example 3,4,5-trimethylpyrazole (41) gave the trichloroethylene (42) and the chloropyrimidine (43) as minor products: the mechanisms shown were suggested for their formation.

The other group of workers studied very similar systems, but in the gas phase, the dichlorocarbene being generated by pyrolysis of chloroform. The reactions appear to be cleaner and to go in higher yield than in solution. Pyrrole gave a mixture of 2- and 3-chloropyridines in an overall yield of 86%.[94] The 2-chloropyridine is thought to be formed by an electrophilic attack of the carbene either at nitrogen or at the α-carbon of pyrrole. Imidazole gave mainly 5-chloropyrimidine (44) with some chloropyrazine (45) which is

[92] R. L. Jones and C. W. Rees, *J. Chem. Soc.*(C), **1969**, 2249.

[93] R. L. Jones and C. W. Rees, *J. Chem. Soc.*(C), **1969**, 2251.

[94] F. S. Baker, R. E. Busby, M. Iqbal, J. Parrick, and C. J. G. Shaw, *Chem. Ind.* (London), **1969**, 1344.

13

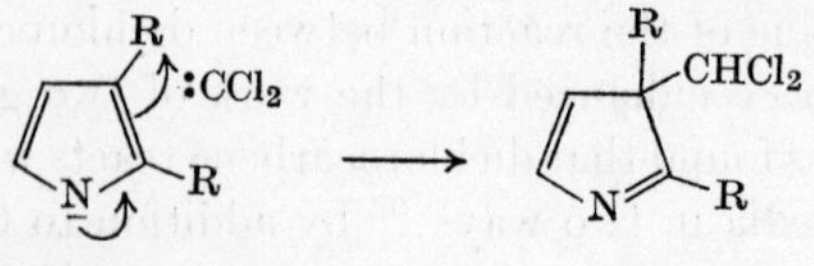

(39)   (40)

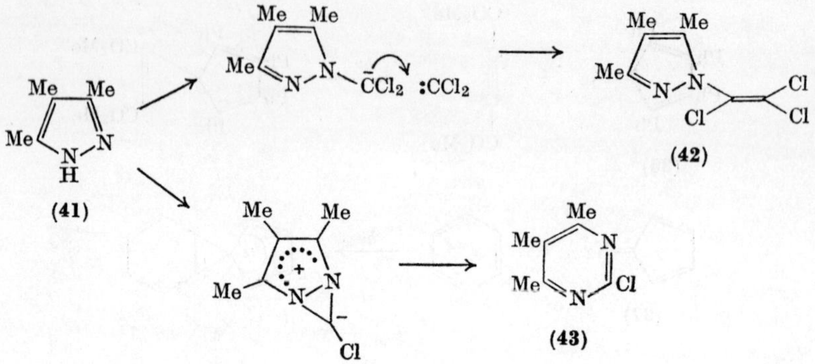

(41)   (42)

(43)

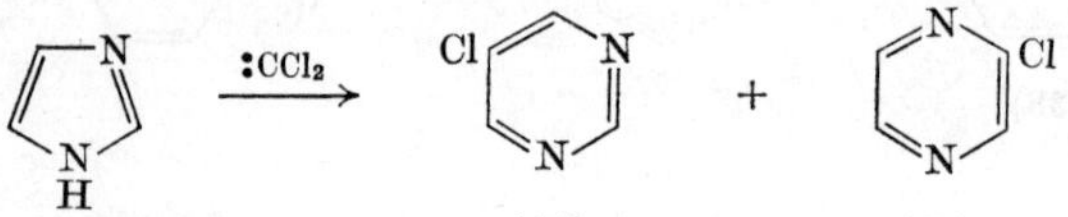

(44)   (45)

probably formed by addition of the carbene to the C=N bond, followed by ring expansion.[95]

The first isolable carbene adducts (46) and (47) of a pyrrole derivative have also been obtained.[96]

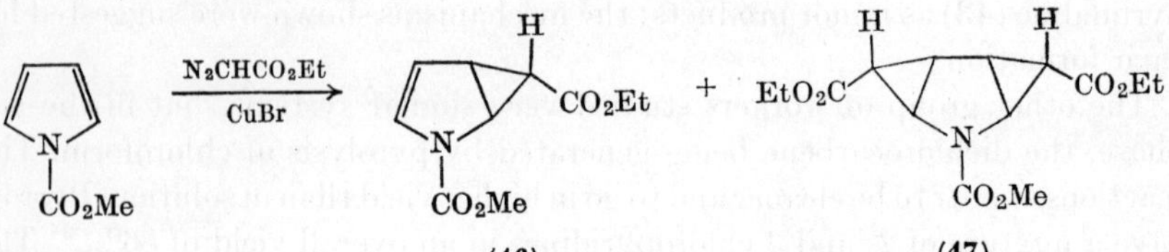

(46)   (47)

Other carbene additions to aromatic systems include the reaction of energetic carbon atoms with benzene at low temperature, which leads to the production of methylene, phenylcarbene, and cycloheptatrienylidene (34) as inter-

[95] R. E. Busby, M. Iqbal, J. Parrick, and C. J. G. Shaw, *Chem. Comm.*, **1969**, 1344.
[96] F. W. Fowler, *Chem. Comm.*, **1969**, 1359.

mediates,[97] the addition of ethoxycarbonylcarbene to a porphyrin,[98] and the reaction of aromatic ethers with dihalogenocarbenes.[99]

The photolysis of vinyl azides to form 1-azirines, which may involve an intramolecular nitrene addition, is establishing itself as the best general synthesis of these compounds; several new examples have been reported,[100, 101] including the first synthesis of a 1-azirine unsubstituted at the 2-position, (48).[101]

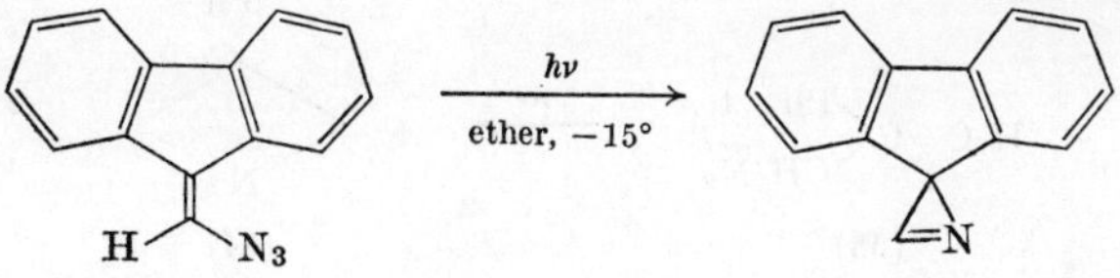

(48)

However, some vinyl azides can apparently react differently. Indirect attempts to form the nitrene (50) by photolysis of the vinyl azide (49) gave no products derived from the nitrene, but instead, dicyanostilbene was formed;[102] this could have come from the isomeric phenylcyanocarbene (51). The mechanism by which this is formed is not established, but a possible sequence is loss of halogen from the vinyl azide (49) to give the ethynyl azide and then the nitrene (50). Thus, it appears that there is an alternative reaction path for halogenated vinyl azides.

$$PhC\!\!=\!\!CBrI \xrightarrow{h\nu} PhC\!\!\equiv\!\!CN_3 \xrightarrow{h\nu} PhC\!\!\equiv\!\!C\ddot{N}: \longrightarrow Ph\ddot{C}C\!\!\equiv\!\!N$$
$$\underset{(49)}{\overset{|}{N_3}} \qquad\qquad\qquad (50) \qquad\qquad (51)$$

An attempt to form azirines by the photolysis of 2-azidovinyl ketones, such as (52), failed;[103] nitriles were formed instead.

$$PhCOCH\!\!=\!\!CHN_3 \xrightarrow{h\nu} PhCOCH_2CN$$
(52)

A similar reaction has previously been observed in the photolysis of β-styryl azides.[104] However, it may be that these reactions do involve the azirines as

[97] H. M. Pohlit, T.-H. Lin, and R. M. Lemmon, *J. Am. Chem. Soc.*, **91**, 5425 (1969).

[98] H. J. Callot and A. W. Johnson, *Chem. Comm.*, **1969**, 749.

[99] S. D. Saraf, *Can. J. Chem.*, **47**, 1169 (1969); M. Rabinovitz and H. Bregman, *Israel J. Chem.*, **6**, 933 (1968).

[100] K. Isomura, M. Okada, and H. Taniguchi, *Tetrahedron Letters*, **1969**, 4073; A. Hassner, R. J. Isbister, R. B. Greenwald, J. T. Klug, and E. C. Taylor, *Tetrahedron*, **25**, 1637 (1969).

[101] W. Bauer and K. Hafner, *Angew. Chem. Internat. Ed. Engl.*, **8**, 772 (1969).

[102] J. H. Boyer and R. Selvarajan, *J. Am. Chem. Soc.*, **91**, 6122 (1969); A. Hassner and R. J. Isbister, *ibid.*, p. 6126.

[103] S. Sato, *Bull. Chem. Soc. Japan*, **41**, 2524 (1968).

[104] J. H. Boyer, W. E. Krueger, and G. J. Mikol, *J. Am. Chem., Soc.* **89**, 5504 (1967).

intermediates, which then rearrange to the nitriles on further irradiation. Significantly, further photolysis of the azirine (48) caused it to rearrange to the nitrile, 9-cyanofluorene.

Allyl azides may also have potential as precursors of small heterocyclic systems; the azabicyclobutane (54) is formed when the allyl azide (53) is irradiated.[105] Intramolecular nitrene cycloaddition may also occur in the

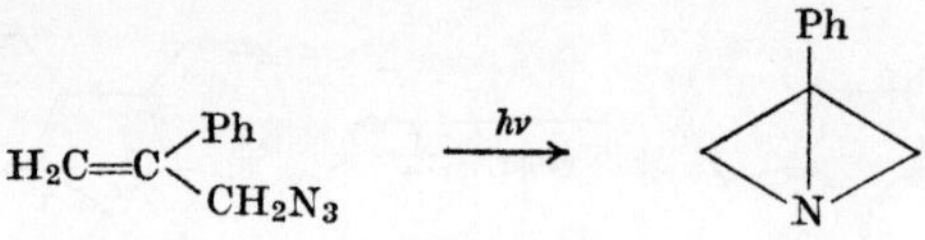

$$H_2C = C \begin{matrix} Ph \\ CH_2N_3 \end{matrix} \xrightarrow{h\nu}$$

(53)   (54)

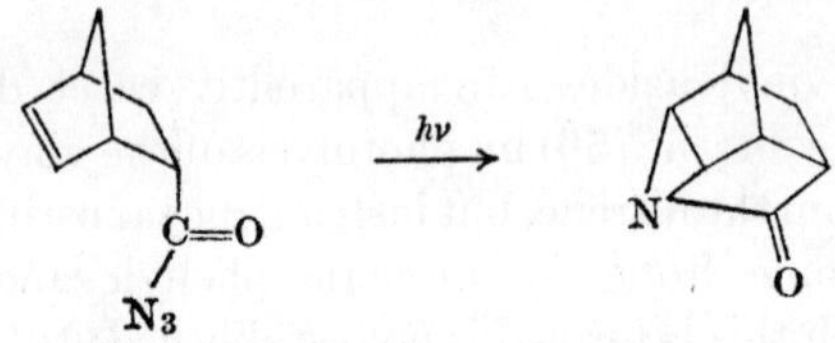

(55)

photolysis of acyl azides, such as (55), with a suitably placed double bond. The aziridines were not isolated, but were detected indirectly as their hydrolysis products, and by solution IR spectroscopy.[106]

Several heterocyclic *N*-amino-compounds, such as *N*-aminophthalimide (56), are oxidized by lead tetra-acetate, probably to give the *N*-nitrenes, which can be trapped in good yield by a wide range of electrophilic and nucleophilic olefins.[47, 107] Attempts were made to trap *N*-phthalimidonitrene with mono- and di-alkylacetylenes, to give the 2-azirines (57). The products isolated were instead the corresponding 1-azirines (58).[108] The mechanism by

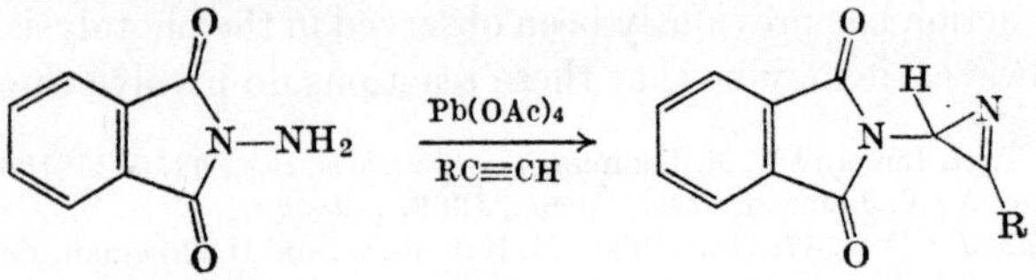

$$\xrightarrow[RC\equiv CH]{Pb(OAc)_4}$$

(56)   (58)

[105] A. G. Hortmann and J. E. Martinelli, *Tetrahedron Letters*, **1968**, 6205.

[106] I. Brown, O. E. Edwards, J. M. McIntosh, and D. Vocelle, *Can. J. Chem.*, **47**, 2751 (1969).

[107] R. S. Atkinson and C. W. Rees, *J. Chem. Soc.*(C), **1969**, 772.

[108] D. J. Anderson, T. L. Gilchrist, and C. W. Rees, *Chem. Comm.*, **1969**, 147.

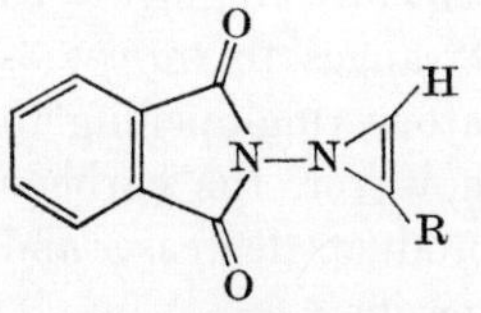

(57) 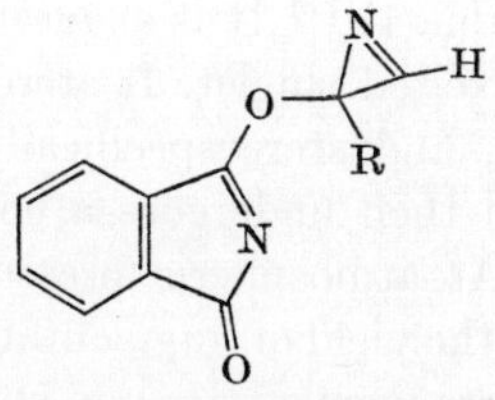 (59) 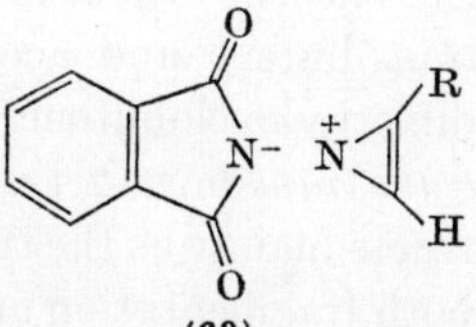 (60)

which they are formed is unknown; assuming that the 2-azirines are indeed the primary products, they could rearrange to 1-azirines via the *O*-azirines (**59**) or via the nitrenium ions (**60**). The driving force for the rearrangement is presumably the antiaromaticity of the 2-azirines.

Rearrangements and Fragmentations

A survey of the rearrangements of various classes of nitrenes has been published.[109]

The reaction in which allenes are formed from diazocyclopropanes and their precursors has been reinvestigated.[110] In the presence of an olefin, the formation of a spiropentane competes with collapse to the allene. As the olefin concentration is increased, the ratio of spiropentane to allene also increases, but not linearly. This originally led to the suggestion that the allene must have two precursors, the diazocyclopropane and the carbene; but the reinvestigation has shown that the non-linearity is due to a change in the polarity of the solvent, so it is not necessary to have two precursors for the allene.

A study of the effect of pressure on the gas-phase photolysis of (*trans*-2,3-dimethylcyclopropyl)diazomethane (**61**) revealed some significant trends in

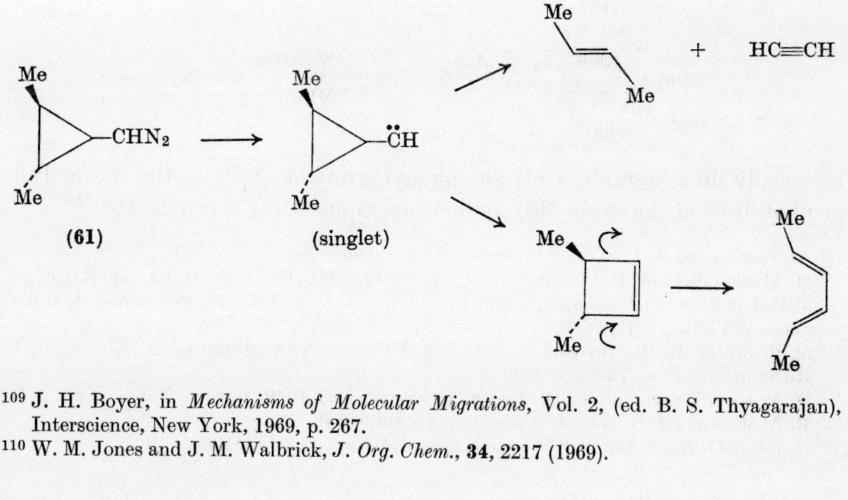

(61) (singlet)

[109] J. H. Boyer, in *Mechanisms of Molecular Migrations*, Vol. 2, (ed. B. S. Thyagarajan), Interscience, New York, 1969, p. 267.
[110] W. M. Jones and J. M. Walbrick, *J. Org. Chem.*, **34**, 2217 (1969).

the product distribution (Table 1).[111] It is suggested that, at low pressures, the carbene reacts as the excited singlet. It stereospecifically fragments to *trans*-butene and acetylene, and stereospecifically rearranges to *trans*-2,3-dimethylcyclobutene, which then undergoes a conrotatory ring-opening to *trans,trans*-hexa-2,4-diene. At atmospheric pressure, however, the carbene reacts mainly as the triplet; the yield of fragmentation products decreases and both fragmentation and rearrangement are non-stereospecific.

Table 1. Products of the gas-phase photolysis of
(*trans*-2,3-dimethylcyclopropyl)diazomethane (**61**)

Pressure (mm)	Acetylene (%)	Butene		Hexa-2,4-diene	
		trans	*cis*	*trans,trans*	*cis,trans*
11	41	43	< 2	20	< 4
23	40	41	< 2	25	6
21 + 739 N_2	21	10	9	8–15	7–13

Other examples of the Wolff rearrangement[112] and of 1,2-hydrogen shifts in carbenes[113] have been noted. There is evidence that sulphonylnitrenes can undergo Curtius-type rearrangements.[114] A 1,2-alkyl shift is the major mode of reaction of norbornan-7-ylidene (**62**);[115] intramolecular insertion is geo-

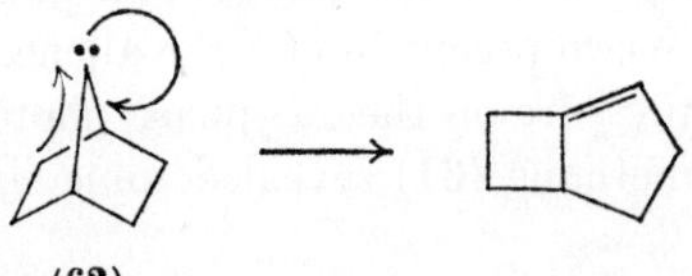

(**62**)

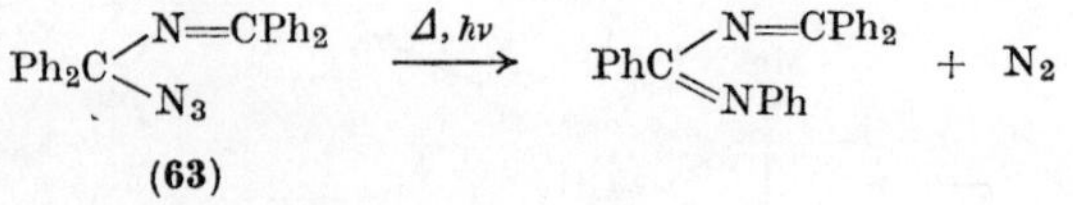

(**63**)

metrically unfavourable. Only the phenyl groups migrate in the thermolysis or photolysis of the azide (**63**); no benzophenone azine was detected.[116]

[111] A. Guarino and A. P. Wolf, *Tetrahedron Letters*, **1969**, 655.
[112] H. Veschambre and D. Vocelle, *Can. J. Chem.*, **47**, 1981 (1969); W. D. Barker, R. Gilbert, J.-P. Lapointe, H. Veschambre, and D. Vocelle, *ibid.*, p. 2853; M. Regitz and J. Rüter, *Chem. Ber.*, **102**, 3877 (1969).
[113] M. R. Bridge, H. M. Frey, and M. T. H. Liu, *J. Chem. Soc.*(A), **1969**, 91; W. Kirmse and L. Ruetz, *Ann. Chem.*, **726**, 30 (1969).
[114] R. A. Abramovitch and W. D. Holcomb, *Chem. Comm.*, **1969**, 1298.
[115] R. A. Moss and J. R. Whittle, *Chem. Comm.*, **1969**, 341.
[116] N. Koga, G. Koga, and J.-P. Anselme, *Can. J. Chem.*, **47**, 1143 (1969).

There have been further studies of the known ring expansions of aryl-nitrenes and their conversion into pyridines,[117, 118] and of the thermal conversion of 2-pyridylnitrene into 2- and 3-cyanopyrroles.[119] Crow and Wentrup have shown that in pyridylnitrene the two nitrogens become equivalent prior to ring contraction, and propose a symmetrical carbene (64) as the intermediate.[119] A similar ring expansion is proposed to explain the conversion of the arylnitrene (65) into 2-methyl-6-vinylpyridine in the gas phase; although

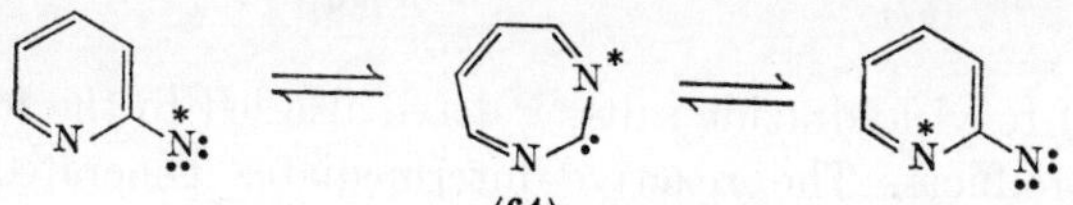

(64)

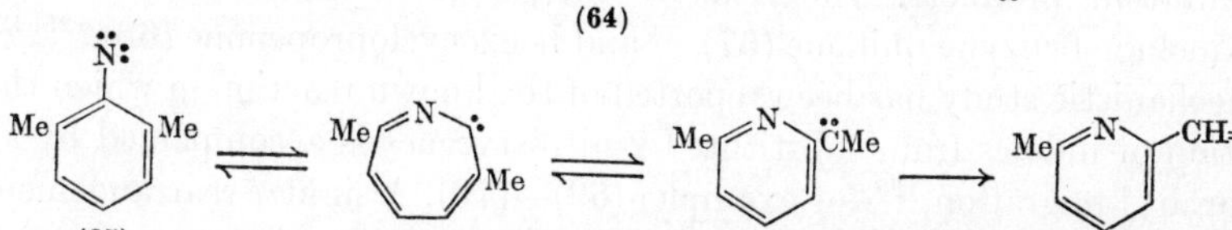

(65)

the nitrene–carbene equilibrium lies to the left, it is pushed to the right by the isomerization of the carbene.[118]

2-Quinolylnitrene (66) undergoes a different type of rearrangement in the gas phase, one product being homophthalonitrile; again, however, a ring-contraction–ring-expansion sequence may be involved in the conversion.[120]

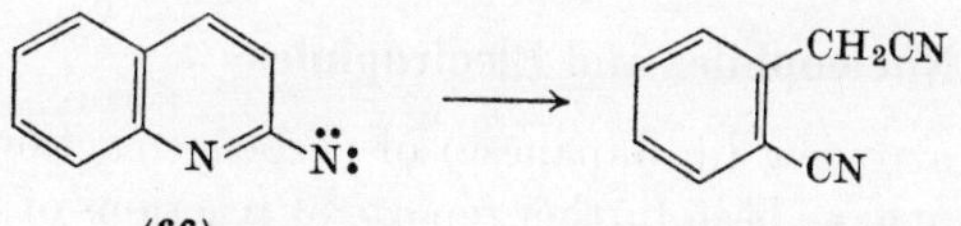

(66)

There have been several reports of fragmentation and rearrangement of 1,1-disubstituted hydrazines on oxidation,[121–125] reactions which may involve aminonitrenes as intermediates, though the intermediacy of free nitrenes in at

[117] R. J. Sundberg, B. P. Das, and R. H. Smith, *J. Am. Chem. Soc.*, **91**, 658 (1969).
[118] C. Wentrup, *Chem. Comm.*, **1969**, 1386.
[119] W. D. Crow and C. Wentrup, *Chem. Comm.*, **1969**, 1387.
[120] R. F. C. Brown and R. J. Smith, *Chem. Comm.*, **1969**, 795.
[121] L. A. Carpino, J. Ferrari, S. Göwecke, and S. Herliczek, *J. Org. Chem.*, **34**, 2009 (1969); G. Koga and J.-P. Anselme, *J. Am. Chem. Soc.*, **91**, 4323 (1969); C. D. Campbell and C. W. Rees, *J. Chem. Soc.*(C), **1969**, 742, 752; C. W. Rees and R. C. Storr, *ibid.*, p. 760; C. W. Rees and M. Yelland, *Chem. Comm.*, **1969**, 377; L. A. Carpino, *J. Org. Chem.*, **34**, 461 (1969).
[122] H. E. Baumgarten, W. F. Wittman, and G. J. Lehmann, *J. Heterocyclic Chem.*, **6**, 333 (1969).
[123] C. G. Overberger, M. Valentine, and J.-P. Anselme, *J. Am. Chem. Soc.*, **91**, 687 (1969).
[124] C. W. Rees and D. E. West, *Chem. Comm.*, **1969**, 647.
[125] J. Adamson, D. L. Forster, T. L. Gilchrist, and C. W. Rees, *Chem. Comm.*, **1969**, 221.

least one such reaction has been questioned.[122] Alternative methods of generating aminonitrenes, such as reduction of *N*-nitroso-compounds[123, 126]

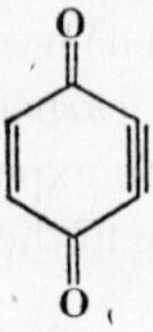

(67) 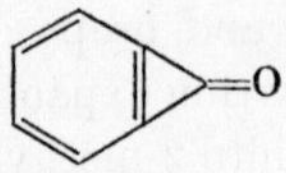   (68)

and photolysis of tosyl hydrazide salts,[127] have also led to the formation of fragmentation products. The reactive intermediates generated by these routes include benzyne quinone **(67)**[124] and benzocyclopropenone **(68)**.[125, 127]

A mechanistic study has been reported of the known reaction in which the formation of indoles from substituted 2-nitrostyrenes is accompanied by an alkyl or aryl migration,[128] for example, **(69)**→**(70)**. A similar rearrangement has been observed in another deoxygenation.[129]

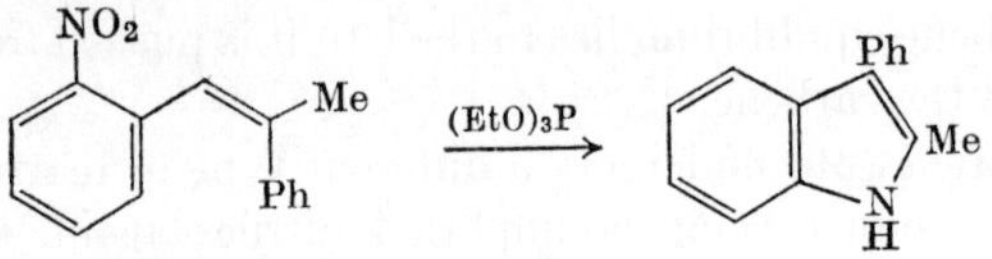

(69)   (70)

Reactions with Nucleophiles and Electrophiles

There has been a review (in Japanese) of carbene reactions which involve ylids,[130] and there have been further reports of reactions of this sort. Careful studies by Ando, Migita, and their coworkers have established that the spin state of the carbene is important in determining the way in which it reacts with a potential nucleophile. For di(methoxycarbonyl)carbene, :$C(CO_2Me)_2$, they showed that the singlet carbene, generated photochemically or with a copper catalyst, adds readily to sulphides,[131, 132] sulphoxides,[131] and allyl halides,[133] the major reaction in each case being the formation of an ylid. With simple sulphides and sulphoxides the ylids are stable and can be isolated, but

126 J. I. G. Cadogan and J. B. Thomson, *Chem. Comm.*, **1969**, 770.
127 M. S. Ao, E. M. Burgess, A. Schauer, and E. A. Taylor, *Chem. Comm.*, **1969**, 220.
128 R. J. Sundberg and G. S. Kotchmar, *J. Org. Chem.*, **34**, 2285 (1969).
129 T. Kametani, T. Yamanaka, and K. Ogasawara, *J. Org. Chem.*, **33**, 4446 (1968).
130 H. Nozaki, *Yuki Gosei Kagaku Kyokai Shi*, **27**, 125 (1969); *Chem. Abs.*, **70**, 105568 (1969).
131 W. Ando, T. Yagihara, S. Tozune, S. Nakaido, and T. Migita, *Tetrahedron Letters*, **1969**, 1979; W. Ando, T. Yagihara, S. Tozune, and T. Migita, *J. Am. Chem. Soc.*, **91**, 2786 (1969).
132 W. Ando, K. Nakayama, K. Ichibori, and T. Migita, *J. Am. Chem. Soc.*, **91**, 5164 (1969).
133 W. Ando, S. Kondo, and T. Migita, *J. Am. Chem. Soc.*, **91**, 6516 (1969).

with the allyl halides and with sulphides $RSCH_2CH=CH_2$, the ylids spontaneously rearrange, for example $(71)\rightarrow(72)$. The formation of the ylids competes favourably with cycloadditions, either to double bonds within the same molecule, or to other added olefins. When the triplet carbene was produced, by sensitized photolysis,[132, 133] the ylids were no longer formed.

$$MeCH=CHCH_2Cl + :C(CO_2Me)_2 \longrightarrow MeCH \overset{CH-CH_2}{\underset{\underset{MeO_2C \quad CO_2Me}{\overset{|}{C}}}{\diagdown Cl^+}} \longrightarrow \underset{ClC(CO_2Me)_2}{\overset{MeCHCH=CH_2}{|}}$$

(71) (72)

It is clear from other work that electrophilic attack by a singlet carbene to form an ylid is always an important primary reaction with substrates containing a heteroatom. Such reactions probably take place between dichlorocarbene and sulphur[134] or N,N-dimethylaniline,[135] between methylene[136] or ethoxycarbonylcarbene[137] (both produced as singlets by copper catalysis) and ethers, between diphenylcarbene and water or alcohols,[138] and between monatomic[139] or triatomic[140] carbon and alcohols. In the case of the diphenylcarbene in alcohols or water, abstraction of a proton to give a carbonium ion and direct insertion into the OH bond can be excluded: a primary formation of an oxygen ylid is indicated. A similar pattern may apply with nitrenes: it was found that direct, unsensitized, photolysis of p-azidobenzonitrile in dimethylamine gave the hydrazine (73) as the major product, with only a little

$$NC\!\!-\!\!\langle\bigcirc\rangle\!\!-\!\!N_3 \quad \xrightarrow[Me_2NH]{h\nu} \quad Me_2NNH\!\!-\!\!\langle\bigcirc\rangle\!\!-\!\!CN \; + \; NC\!\!-\!\!\langle\bigcirc\rangle\!\!-\!\!NH_2$$

(73)

p-aminobenzonitrile; but in the presence of a triplet sensitizer the yields were reversed.[141] The formation of the hydrazine, presumably via an ylid, therefore appears to be a reaction only of the singlet nitrene.

[134] D. Seyferth and W. Tronich, *J. Am. Chem. Soc.*, **91**, 2138 (1969).
[135] S. D. Saraf, *Can. J. Chem.*, **47**, 1173 (1969).
[136] M. Kapps and W. Kirmse, *Angew. Chem. Internat. Ed. Engl.*, **8**, 75 (1969).
[137] S. T. Murayama and T. A. Spencer, *Tetrahedron Letters*, **1969**, 4479.
[138] D. Bethell, A. R. Newall, G. Stevens, and D. Whittaker, *J. Chem. Soc.*(B), **1969**, 749.
[139] P. S. Skell and R. F. Harris, *J. Am. Chem. Soc.*, **91**, 4440 (1969).
[140] P. S. Skell and R. F. Harris, *J. Am. Chem. Soc.*, **91**, 699 (1969).
[141] R. A. Odum and A. M. Aaronson, *J. Am. Chem. Soc.*, **91**, 5680 (1969).

The reaction of dibromocarbene with an isocyanide to form a ketenimine (74) has been reported;[142] a similar reaction of diphenylcarbene has been suggested by the other workers to explain the formation of a ketenimine.[143] In these reactions, the isocyanide presumably plays the role of nucleophile. Boyer and Beverung found that irradiation of a mixture of cyclohexyl isocyanide and diphenyldiazomethane also gave a ketenimine, (75);[144] presumably an electrophilic attack by diphenylcarbene on the isocyanide

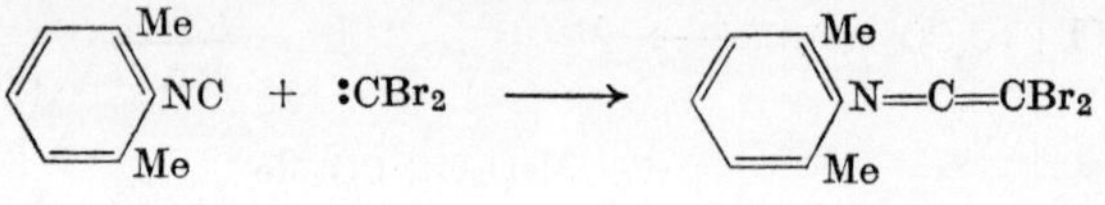

$$\text{(xylyl)}NC \ + \ :CBr_2 \ \longrightarrow \ \text{(xylyl)}N{=}C{=}CBr_2$$

(74)

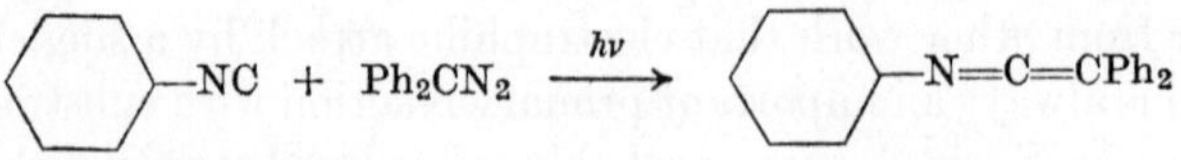

$$\text{(cyclohexyl)}{-}NC \ + \ Ph_2CN_2 \ \xrightarrow{h\nu} \ \text{(cyclohexyl)}{-}N{=}C{=}CPh_2$$

(75)

$$\xrightarrow{h\nu}$$

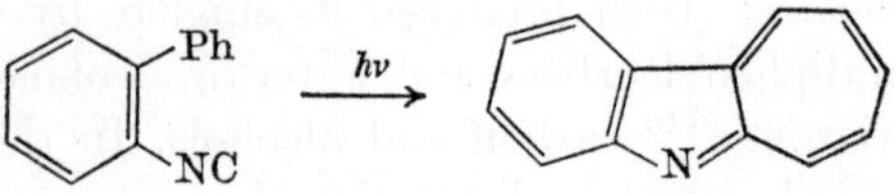

(76)

would account for the formation of the product, but the authors prefer an alternative explanation, that an excited electrophilic isocyanide attacks the diazoalkane. In a previous study they did show that a photochemically excited isocyanide could have electrophilic carbene properties;[145] irradiation of *o*-biphenyl isocyanide gave the insertion product (76).

An intramolecular nucleophilic attack on a carbene may be involved in the conversion of 1,3-bisdiazopropane into pyrazole:[146]

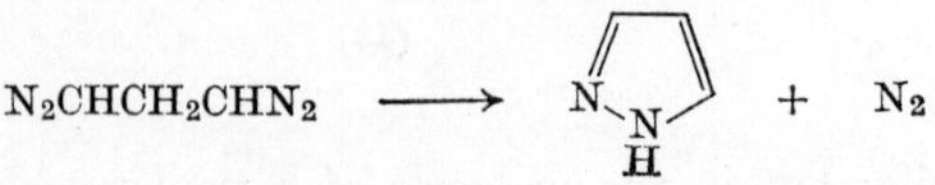

$$N_2CHCH_2CHN_2 \ \longrightarrow \ \text{(pyrazole)} \ + \ N_2$$

There have also been reports of intramolecular deoxygenation of carbonyl groups in reductions of aromatic nitro-compounds[147] and in aromatic azide

[142] T. Takizawz, N. Obata, Y. Suzuki, and T. Yanagida, *Tetrahedron Letters*, **1969**, 3407; see footnote 6.

[143] P. R. West and J. Warkentin, *J. Org. Chem.*, **34**, 3233 (1969).

[144] J. H. Boyer and W. Beverung, *Chem. Comm.*, **1969**, 1377; see also J. A. Green and L. A. Singer, *Tetrahedron Letters*, **1969**, 5093.

[145] J. H. Boyer and J. De Jong, *J. Am. Chem. Soc.*, **91**, 5929 (1969).

[146] H. Hart and J. L. Brewbaker, *J. Am. Chem. Soc.*, **91**, 706 (1969).

[147] D. G. Saunders, *Chem. Comm.*, **1969**, 680; T. Kametani, T. Yamanaka, and K. Ogasawara, *J. Chem. Soc.*(C), **1969**, 385.

decompositions.[148] It is unlikely that free nitrenes are involved in the deoxygenations, however.

The products of the reaction between organoboranes and methoxycarbene can be rationalized by postulating the formation of an ylid (**77**). In this reaction the borane acts as an electrophile.[149] Similarly, the boron trichloride catalysed decomposition of acyl azides may involve zwitterions (**78**) as intermediates.[150]

$$\overset{-}{R_3B}-\overset{+}{C}HOMe \qquad\qquad Cl_3\overset{-}{B}-O-C(R)\!\!=\!\!\overset{+}{N}$$
$$(\textbf{77}) \qquad\qquad\qquad (\textbf{78})$$

It has also been suggested that acetic acid may protonate arylnitrenes to give nitrenium ions; the products of the deoxygenation of aromatic nitro- and nitroso-compounds were different when 5% acetic acid was added.[151] This observation raises the point that nitrenium ions, and not nitrenes, might be involved in other reactions, such as the oxidation of hydrazines by lead tetra-acetate, where acetic acid is formed.

Carbenoids and Metal Complexes

There have been several comparative studies of carbenoids and free carbenes, and a comparative review (in Japanese) has appeared.[152] Skell and Cholod have shown that dichlorocarbene, generated by a variety of routes, including the action of butyl-lithium on chloroform, is always "free", and carbenoids are not involved.[153] The paper also contains a useful summary of previous work on free and complexed carbenes. The picture is not so clear with phenylchlorocarbene, which was generated by the action of potassium t-butoxide on benzylidene chloride and by photolysis of phenylchlorodiazirine.[154] The selectivities of the intermediates from the two sources in additions to olefins were very similar, except for the competitive addition to isobutene and tetramethylethylene, where a considerable difference in selectivity was observed; surprisingly, the photochemically generated intermediate was the more selective. The stereoselectivity of the additions to olefins of phenylfluorocarbene (or -carbenoid),[155] and of phenylcarbenoid,[156] have also been studied. In a study of the effect of catalysts (RO)₃PCuCl on the carbenoid

[148] S. Bradbury, C. W. Rees, and R. C. Storr, *Chem. Comm.*, **1969**, 1428.
[149] A. Suzuki, S. Nozawa, N. Miyaura, and M. Itoh, *Tetrahedron Letters*, **1969**, 2955.
[150] E. Fahr and L. Neumann, *Ann. Chem.*, **721**, 14 (1969).
[151] R. J. Sundberg, R. H. Smith, and J. E. Bloor, *J. Am. Chem. Soc.*, **91**, 3392 (1969).
[152] I. Moritani and S. Murahashi, *Kagaku No Ryoiki*, **22**, 777 (1968); *Chem. Abs.*, **70**, 36723 (1969).
[153] P. S. Skell and M. S. Cholod, *J. Am. Chem. Soc.*, **91**, 6035 (1969).
[154] R. A. Moss, J. R. Whittle, and P. Freidenreich, *J. Org. Chem.*, **34**, 2220 (1969).
[155] R. A. Moss and J. R. Przybyla, *Tetrahedron*, **25**, 647 (1969).
[156] M. Schlosser and G. Heinz, *Angew. Chem. Internat. Ed. Engl.*, **8**, 760 (1969).

addition of ethoxycarbonylcarbene to cyclohexene,[157] it was concluded that the final transition state leading to products must contain the olefin, metal, and carbene, and must be asymmetric. A tetrahedral copper complex (79) was proposed.

Two studies of the Simmons–Smith reaction have clarified details of the mechanism. With cyclic allylic alcohols (80), the assumption that the *syn*-cyclopropane is always formed, because of coordination of the reagent to the oxygen atom, has been shown to be invalid;[158] for $n > 3$, the *anti*-cyclopropane is formed almost exclusively. However, models show that for these rings, *anti* addition of the reagent, coordinated to the oxygen, is more favourable. Thus,

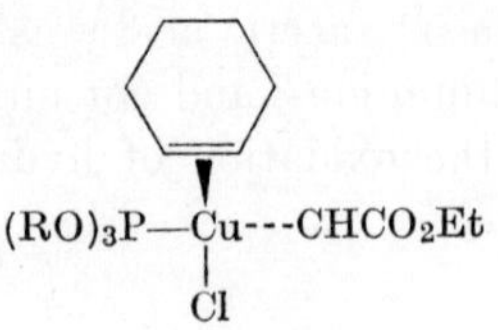

$$(\text{RO})_3\text{P}\!-\!\overset{|}{\underset{|}{\text{Cu}}}\!-\!-\!-\text{CHCO}_2\text{Et}$$

(79)

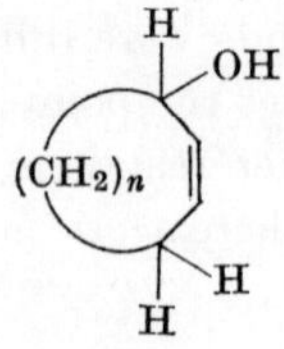

(80)

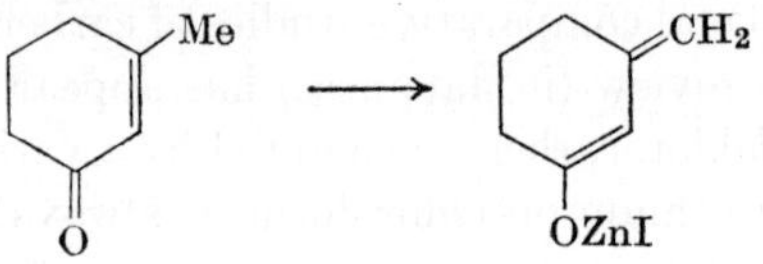

(81)

the reaction seems to involve prior complexing to the oxygen in all cases, but this does not necessarily lead to formation of the *syn*-cyclopropane. A curious anomaly in the ease of addition of the Simmons–Smith reagent to α,β-unsaturated ketones has been convincingly explained;[159] it was found that β,β-dialkyl substituted ketones gave no cyclopropanes. The suggestion is that such ketones can readily enolize and form a complex, for example (81), which effectively "ties up" the reagent and prevents addition. There has also been a report that the addition of the reagent to an olefin was non-stereospecific,[160] but in fact the olefin probably isomerized prior to the addition. Reactions related to the Simmons–Smith reaction have also been reported.[161, 162] Other

[157] W. R. Moser, *J. Am. Chem. Soc.*, **91**, 1135, 1141 (1969).
[158] C. D. Poulter, E. C. Friedrich, and S. Winstein, *J. Am. Chem. Soc.*, **91**, 6892 (1969).
[159] J.-C. Limasset, P. Amice, and J.-M. Conia, *Bull. Soc. Chim. France*, **1969**, 3981.
[160] J. A. Donnelly and P. O'Boyle, *Chem. Comm.*, **1969**, 1060.
[161] J. Nishimura, N. Kawabata, and J. Furakawa, *Tetrahedron*, **25**, 2647 (1969); S. Sawada and Y. Inouye, *Bull. Chem. Soc. Japan*, **42**, 2669 (1969).
[162] S. H. Goh, L. E. Closs, and G. L. Closs, *J. Org. Chem.*, **34**, 25 (1969).

reactions probably involving carbenoids include examples of insertion,[163] rearrangement,[164] and addition.[165]

A series of very stable complexes of nucleophilic carbenes, for example (**82**) and (**83**), have been prepared by Öfele.[166] The crystal structure of the cyclopropenylidene complex (**82**) showed that the C-2–C-3 bond was slightly shorter than the others,[167] so that the π-electrons are not completely delocalized in the ring.

(**82**)   (**83**)

The researches of Fischer's group on the related methoxymethyl- and (methylamino)methyl-carbene complexes have revealed several more cases of *cis–trans* isomerism.[168] Pyrolysis of the phenylmethoxycarbene complex (**84**) gave *cis-* and *trans-α,α*-dimethoxystilbene:[169]

$$\text{PhC(OMe)}:\text{Cr(CO)}_5 \rightarrow \text{PhC(OMe)}{=}\text{C(OMe)Ph}$$
(**84**)

A ruthenium–nitrene complex (**85**) has been suggested as an intermediate in the reaction sequence shown, since the addition of nucleophiles such as iodide prevents the formation of the dimer (**86**).[170]

$$[(\text{NH}_3)_5\text{RuN}_3]^{2+} \xrightarrow{\text{H}^+} [(\text{NH}_3)_5\text{RuNHN}{\equiv}\text{N}]^{3+} \longrightarrow$$

$$\text{N}_2 + [(\text{NH}_3)_5\text{RuNH}]^{3+} \longrightarrow [(\text{NH}_3)_5\text{RuN}{=}\text{NRu(NH}_3)_5]^{4+} + 2\text{H}^+$$
(**85**)   (**86**)

Nitrene–metal complexes are also probably involved in the reaction of hexafluoroazomethane with $(\text{Ph}_2\text{PMe})_2\text{Ir(CO)Cl}$,[171] and in the decomposition of aryl azides catalysed by di-iron nonacarbonyl.[172] In the latter study, the original observation [173] of the striking catalysis of azide decompositions by iron

[163] B. Fraser-Reid, R. L. Sun, and J. T. Brewer, *Tetrahedron Letters*, **1969**, 2775, 2779.

[164] P. G. Gassman, J. P. Andrews, and D. S. Patton, *Chem. Comm.*, **1969**, 437; A. Zurquiyah and C. E. Castro, *J. Org. Chem.*, **34**, 1504 (1969).

[165] R. M. Carlson and P. M. Helquist, *Tetrahedron Letters*, **1969**, 173; G. Köbrich and H. Heinemann, *Chem. Comm.*, **1969**, 493.

[166] K. Öfele, *Angew. Chem. Internat. Ed. Engl.*, **7**, 950 (1968); *ibid.*, **8**, 916 (1969).

[167] G. Huttner, S. Schelle, and O. S. Mills, *Angew. Chem. Internat. Ed. Engl.*, **8**, 515 (1969).

[168] E. Moser and E. O. Fischer, *J. Organometal. Chem.*, **16**, 275 (1969); C. G. Kreiter and E. O. Fischer, *Angew. Chem. Internat. Ed. Engl.*, **8**, 761 (1969); see *Org. Reaction Mech.*, **1968**, 344.

[169] E. O. Fischer, B. Heckl, K. H. Dötz, J. Müller, and H. Werner, *J. Organometal. Chem.*, **16**, P29 (1969).

[170] L. A. P. Kane-Maguire, F. Basolo, and R. G. Pearson, *J. Am. Chem. Soc.*, **91**, 4609 (1969).

[171] J. Ashley-Smith, M. Green, N. Mayne, and F. G. A. Stone, *Chem. Comm.*, **1969**, 409.

[172] C. D. Campbell and C. W. Rees, *Chem. Comm.*, **1969**, 537.

[173] M. Dekker and G. R. Knox, *Chem. Comm.*, **1967**, 1243.

carbonyls was confirmed, but the reactions are complex and the nature of the products is unpredictable. The reaction of aromatic nitro-compounds with ferrous oxalate might involve nitrene–iron complexes,[174] but most of the products can be accounted for without invoking nitrenes.

[174] R. A. Abramovitch, B. A. Davis, and R. A. Brown, *J. Chem. Soc.*(C), **1969**, 1146.

CHAPTER 11

Reactions of Aldehydes and Ketones and their Derivatives

B. CAPON

Chemistry Department, Glasgow University

Formation and Reactions of Acetals and Ketals

Several examples of general-acid catalysis in acetal hydrolysis have been reported.[1-3] The important factor if general-acid catalysis is to occur appears to be that carbon–oxygen bond fission shall occur easily. This may be achieved by having an acetal or ketal which forms a stable carbonium ion and/or has a good leaving group. Benzaldehyde aryl methyl acetals (1) have both of these features and their hydrolyses show marked general-acid catalysis.[1] The ρ-value for the acetic acid-catalysed reaction is $+0.89$. This was interpreted as indicating that proton transfer is concerted with carbon–oxygen bond breaking with the transition state as (2). It is interesting that the ρ-value is very similar to that found in the enzymically catalysed hydrolyses of aryl glycosides which are thought to involve general-acid catalysis within the enzyme–substrate complex.[4] If the intermediate carbonium ion is made sufficiently stable, general-acid catalysis is also found in the hydrolysis of dialkyl acetals. A nice example is the hydrolysis of tropone diethyl ketal (3) which is general-acid catalysed in tris, phosphate, and bicarbonate buffers.[2] It was also claimed that the hydrolysis of benzophenone ketals is general-acid catalysed[3] but other

[1] E. Anderson and B. Capon, *Chem. Comm.*, **1969**, 390; *J. Chem. Soc.*(B), **1969**, 1033.
[2] E. Anderson and T. H. Fife, *J. Am. Chem. Soc.*, **91**, 7163 (1969).
[3] R. H. De Wolfe, K. M. Ivanetich, and N. F. Perry, *J. Org. Chem.*, **91**, 848 (1969).
[4] B. Capon, *Chem. Rev.*, **69**, 434 (1969).

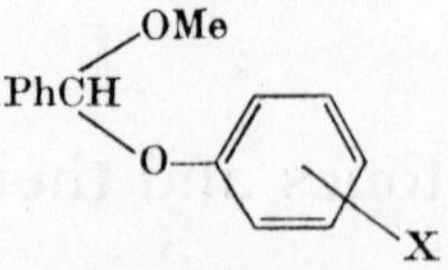

OMe
PhCH
O—⟨benzene ring⟩—X

(1)

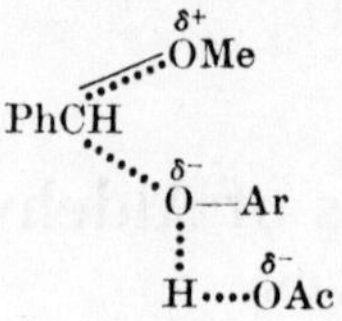

$\delta+$
..........OMe
PhCH
$\delta-$
O—Ar
$\delta-$
H····OAc

(2)

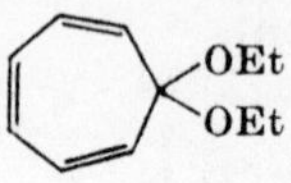

OEt
OEt
(cycloheptatriene ring)

(3)

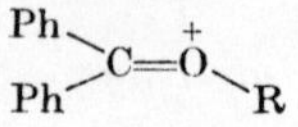

Ph
   C=$\overset{+}{O}$
Ph   R

(4)

Ph
   C=$\overset{+}{O}$
H   R

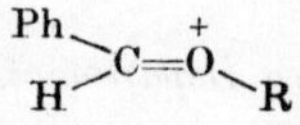

(5)

workers[2, 5] were unable to observe any rate increase on increasing the buffer concentration. It is possible that ion (4), which is an intermediate in the hydrolysis of benzophenone ketals, is less stable than ion (5) which is an intermediate in the hydrolysis of benzaldehyde acetals. Both phenyl groups of (4) cannot lie in the same plane as the carbon–oxygen bond; one of them must be twisted, and if the angle of twist were sufficiently large the mesomeric stabilization would be less than the inductive destabilization. An effect of this type in the transition state could explain the slower hydrolyses of benzophenone ketals compared to benzaldehyde acetals.

The entropies of activation for the hydrolysis of 2-(2,2,2-trichloroethoxy)- and 2-(2,2,2-trifluoroethoxy)-tetrahydropyran in hydrochloric acid are −13.0 and −3.8 e.u., and the solvent isotope effect, $k(H_2O)/k(D_2O)$, is 1.59 and 1.28 respectively. A mechanism involving a rate-determining proton transfer was proposed.[6]

Details of Capon and his coworkers' investigation of intramolecular catalysis in the hydrolysis of 2-(methoxymethoxy)benzoic acid and 2-carboxyphenyl glucosides have been published.[7a]

The rate of the acid-catalysed hydrolysis of benzaldehyde diethyl acetal is enhanced by sodium dodecyl sulphate at concentrations above the critical micelle concentration. The effect is larger with the more reactive acetals, and the acid-catalysed and acid- and micelle-catalysed reactions yield ρ-values of −3.3 and −4.1 respectively. This behaviour is similar to that found with methyl orthobenzoates which yield ρ-values of −2.0 and −2.5 respectively.[7b]

The acid-catalysed hydrolyses of *cis*- and *trans*-2,5-dimethoxytetrahydrofuran are 7—8 times slower than that of 2-methoxytetrahydrofuran. This was interpreted in terms of the deactivating effect of the second methoxy group

[5] B. Capon and M. C. Smith, *J. Chem. Soc.*(B), **1969**, 1031.
[6] A. Kankaanperä, *Suomen Kemistilehti*, *B42*, 460 (1969).
[7a] B. Capon, M. C. Smith, E. Anderson, R. H. Dahm, and G. H. Sankey, *J. Chem. Soc.*(B), **1969**, 1038; see *Org. Reaction Mech.*, **1965**, 238.
[7b] R. B. Dunlap, G. A. Ghanim, and E. H. Cordes, *J. Phys. Chem.*, **73**, 1898 (1969).

on the formation of the cyclic carbonium ion. The hydrolyses of the *cis-* and *trans*-2,6-dimethoxytetrahydropyrans were also studied. The rates are approximately 0.85 and 0.5 times that for the hydrolysis of 2-methoxytetrahydropyran.[8] The equilibration of the dimethoxytetrahydrofurans and the diethoxytetrahydropyrans was also studied.[9]

The solvent isotope effect for the hydrolysis of 2-ethoxy-2,3-dihydropyran, $k(D_3O^+)/k(H_3O^+) = 1.26$, lies between those for the hydrolyses of 2-methoxytetrahydropyran (2.94) and 2,3-dihydropyran (0.453), and the reaction was thought to occur partly with protonation on carbon and partly with protonation on the exocyclic oxygen.[10]

Two investigations of the acid-catalysed hydrolysis of 2-aryl-1,3-oxathiolans (6) have been reported.[11, 12a] Although it is not certain if the rate-limiting step involves carbon–sulphur or carbon–oxygen bond breaking, the former was preferred because the solvent isotope effect for the reaction of the *p*-methoxy-compound, $k(D_2O)/k(H_2O) = 1.93$, is smaller than that normally found in acetal hydrolysis.[12a] If this is correct the 1330-fold smaller rate of reaction of the *p*-methoxy-compound compared to that for 2-*p*-methoxyphenyldioxolan is the result of the lower basicity of the sulphur which leads to a lower concentration of the conjugate acid. Although $\Delta S^{\ddagger}$ for the hydrolysis of 2-phenyl-1,3-oxathiolan is −13.2 e.u., an *A*2 mechanism was considered unlikely since 2-phenyl-2-methyl-1,3-oxathiolan reacts twice as fast as 2-phenyl-1,3-oxathiolan.[12a] The ρ-value for the hydrolysis of the 2-(substituted phenyl)-1,3-oxathiolans in water is −2.81. The mercuric-ion promoted hydrolysis was also studied.[11] Equilibration of the hemithioketals of 3,3,5-trimethylcyclohexanone has been studied.[12b]

The hydrolysis of methyl pseudo-8-benzoyl-1-naphthoate (7) in aqueous sulphuric acid shows all the characteristics of an *A*1 mechanism; viz. a linear plot of log k_{obs} against H_0 with slope of ca. 1, a Bunnett *w*-value of −0.50, a Bunnett ϕ-value of −0.083, an entropy of activation of +0.7 e.u., and a ρ-value for substituents in the phenyl ring of −2.1. The ρ-value indicates that ketone–oxygen fission rather than acyl–oxygen fission is occurring in the rate-limiting step, and two mechanisms are possible (equations 1 and 2). At present it is not possible to distinguish between them. The ρ-value for the hydrolysis of methyl pseudo-2-(*para*-substituted)benzoylbenzoates is −1.0. This suggests that, of the two possible bimolecular mechanisms for the hydrolysis of these

[8] A. Kankaanperä, *Suomen Kemistilehti*, *B*42, 208 (1969); cf. *Org. Reaction Mech.*, **1968**, 350.
[9] A. Kankaanperä and K. Miikki, *Acta Chem. Scand.*, **23**, 1471 (1969).
[10] A. Kankaanperä, *Acta Chem. Scand.*, **23**, 1465 (1969); cf. *Org. Reaction Mech.*, **1968**, 351, and ref. 144 on p. 419 of this volume.
[11] N. C. De and L. R. Fedor, *J. Am. Chem. Soc.*, **90**, 7266 (1968).
[12a] T. H. Fife and L. K. Jao, *J. Am. Chem. Soc.*, **91**, 4217 (1969).
[12b] M. P. Mertes, H.-K. Lee, and R. L. Schowen, *J. Org. Chem.*, **34**, 2080 (1969).

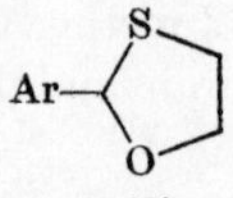

(6) 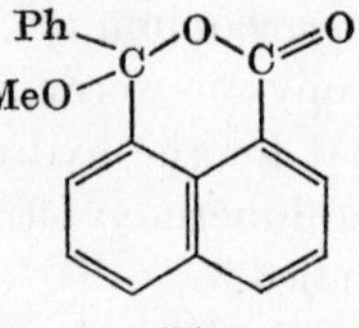   (7) 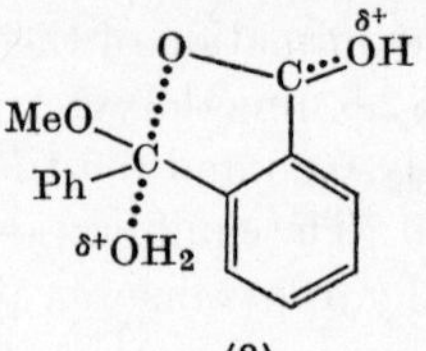   (8)

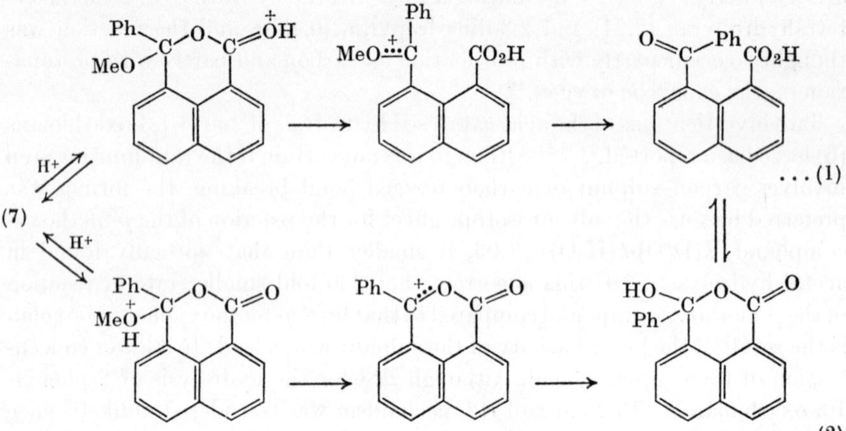

... (1)

... (2)

compounds,[13] the one which involves attack at the ketone-carbon is the more likely, and that the transition state is as (8).[14a]

The formation of acetals from aldehydes and orthoformates has been studied.[14b]

Other topics which have been investigated include the hydrolyses of 2-t-butoxy-tetrahydrofuran and -tetrahydropyran,[15a] of the tetrahydro-pyranyl derivatives of 1,3,3-trimethyl-*endo*-2-norbornanol (α-fenchol),[15b] and of the isopropylidene acetals of some 2-pentuloses,[15c] the formation of 1,3-dioxolans,[15d] and the ethylidenation[16a] and acetonation[16b] of sorbose.

The basicities of the oxygen atoms of symmetrical and unsymmetrical acetals have been determined.[17]

The hydrolysis of 1,3-dioxans has been reviewed.[18]

[13] See *Org. Reaction Mech.*, **1968**, 349—350.
[14a] D. P. Weeks and G. W. Zuorick, *J. Am. Chem. Soc.*, **91**, 477 (1969).
[14b] J. W. Scheeren, J. E. W. van Melick, and R. J. F. Nivard, *Chem. Comm.*, **1969**, 1175.
[15a] A. Kankaanperä and K. Miiki, *Suomen Kemistilehti*, B**42**, 430 (1969).
[15b] J. Korvola and P. J. Mälkönen, *Suomen Kemistilehti*, B**42**, 430 (1969).
[15c] R. S. Tipson, B. F. West, and R. F. Brady, *Carbohydrate Res.*, **10**, 181 (1969).
[15d] S. L. T. Thuan and J. Wiemann, *Bull. Soc. Chim. France*, **1968**, 4550.
[16a] T. Maeda, M. Kiyokawa, and K. Tokuyama, *Bull. Chem. Soc. Japan*, **42**, 492 (1969).
[16b] T. Maeda, Y. Miichi, and K. Tokuyama, *Bull. Chem. Soc. Japan*, **42**, 2648 (1969).
[17] A. Kankaanperä, *Acta Chem. Scand.*, **23**, 1723, 1728, 2211 (1969).
[18] A. V. Bogatskii and N. L. Garkovik, *Uspekhi Khim.*, **37**, 581 (1968) [*Russ. Chem. Rev.*, **37**, 264 (1968)].

Hydrolysis and Formation of Glycosides[19, 20]

Non-enzymic Reactions

When compound (9) is treated with acidic methanol it yields a non-equilibrium mixture of (10) and (11) which is converted into the equilibrium mixture at a slightly lower rate. No products in which there had been anomerization at C-1, or furanosides, or open-chain compounds could be detected. It was concluded that the anomerization at C-5, (10)⇌(11), proceeded via the cyclic ion (12), since it was thought that formation of the open ion (13) would

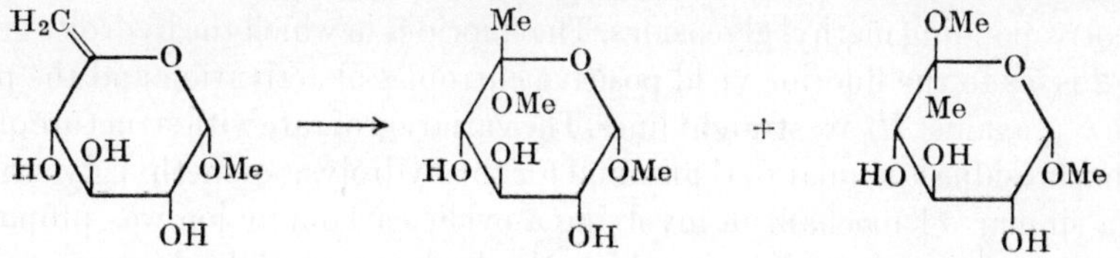

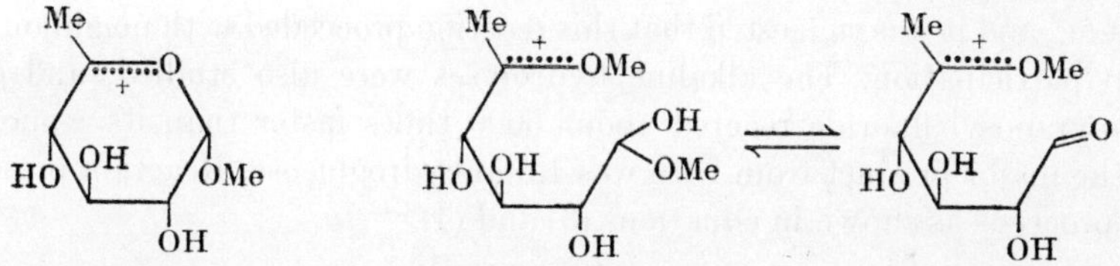

lead to anomerization at C-1.[21] In our view this is by no means certain since anomerization at C-1 would require dissociation to (14) and recombination which could be much slower than recyclization.

The ρ-value for the hydrolysis of substituted phenyl xylopyranosides in 0.1M-hydrochloric acid is −0.146.[22]

The acid-catalysed hydrolyses of α-D-ribofuranose-1-phosphate (15) and α-D-glucopyranose-1-phosphate (16) proceed with carbon–oxygen bond

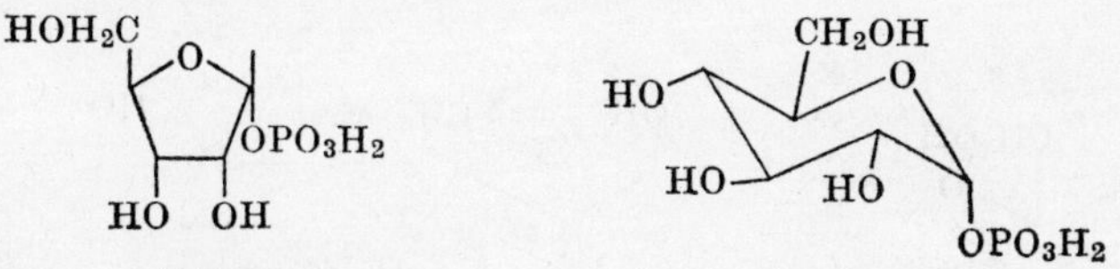

[19] B. Capon, *Chem. Rev.*, **69**, 415 (1969).
[20] J. Szejtli, *Kem. Kozlem.*, **31**, 83 (1969); *Chem. Abs.*, **71**, 13286 (1969); *Ernährungsforsch.*, **13**, 371 (1968).
[21] J. Lehmann, E. Pfeiffer, and H. Reinshagen, *Chem. Ber.*, **102**, 2745 (1969).
[22] F. Van Wijnendaele and C. K. De Bruyne, *Carbohydrate Res.*, **9**, 277 (1969).

fission by an $A1$ mechanism which probably involves a cyclic carbonium ion. The ribofuranose phosphate reacts several hundred times faster than the glucopyranose phosphate and the entropies of activation are $+7.4$ and $+14.9$ e.u. respectively. It was suggested that overlap between the lone-pair orbital of the ring oxygen and the developing p-orbital at C-1 in the transition state occurred more easily in a five- than in a six-membered ring. The less strongly positive entropy of activation for the hydrolysis of the ribofuranose phosphate was attributed to a higher initial-state entropy which arose from the greater flexibility of the five-membered ring.[23]

The acid-catalysed hydrolyses of glycosyl fluorides are faster than those of the corresponding methyl glycosides. The fluorides in which the hydroxy group at C-2 is *cis* to the fluorine yield positive entropies of activation and the plots of $\log k_{obs}$ against H_0 are straight lines. The variation of rate with structure of the glycosyl residue is similar to that found for the hydrolysis of methyl glycosides, and a similar $A1$ mechanism involving a cyclic carbonium ion was proposed. The one compound studied in which the hydroxy at C-2 was *trans* to the fluorine, β-D-glucopyranosyl fluoride, yielded a negative entropy of activation, -5.4 e.u., and it was suggested that this reaction proceeded with neighbouring group participation. The alkaline hydrolyses were also studied, and β-D-glucopyranosyl fluoride reacted about 5000 times faster than its α-anomer and the major product from both was 1,6-anhydroglucose. Reaction presumably proceeds as shown in equations (3) and (4).[24]

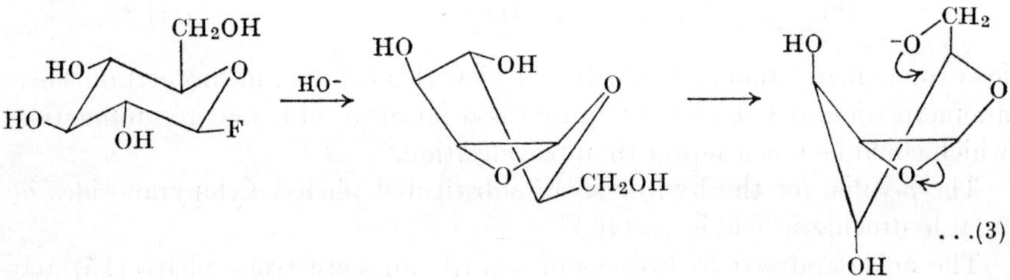

$$\dots (3)$$

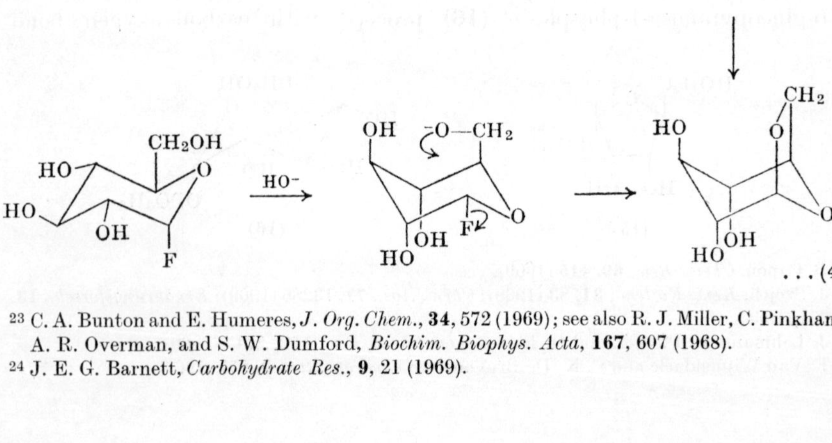

$$\dots (4)$$

[23] C. A. Bunton and E. Humeres, *J. Org. Chem.*, **34**, 572 (1969); see also R. J. Miller, C. Pinkham, A. R. Overman, and S. W. Dumford, *Biochim. Biophys. Acta*, **167**, 607 (1968).
[24] J. E. G. Barnett, *Carbohydrate Res.*, **9**, 21 (1969).

The hydrolyses of sucrose,[25] cellotriose,[26] oligomannuronic and oligogulur-onic acids,[27] and of glycosides with a 2,4-dinitrophenylamino substituent at C-2,[28] and the methanolysis of methyl xylo- and arabino-pyranosides[29] have been studied.

Enzymic Reactions [30-37]

The hydrolysis of compounds (17) and (18) catalysed by lysozyme occurs wholly with loss of the *p*-nitrophenoxy group, but compound (19) is hydrolysed partly by cleavage between the *N*-acetylglucosamine residues. This makes it difficult to compare the kinetics for the hydrolysis of (19) with those for the hydrolysis of (17) and (18). The hydrolyses of (17) and (18) catalysed by lysozyme obey the Michaelis–Menten law, and the values of K_m are similar to the values of K_s for productive binding determined by NMR. The values of V_{max}, 1.7×10^{-9} and 2.2×10^{-8} mole sec^{-1} respectively, are very low, and this suggests that the mechanism of breakdown of the enzyme–substrate complex may be different from that for good substrates.[38a] The lysozyme-catalysed hydrolysis of several other β-aryl di-*N*-acetyl chitobiosides has also been studied.[38b]

The ratio of the rates of hydrolysis of compounds (20a) and (20b) catalysed by lysozyme has been determined by working with (20a) in which the phenyl group is tritium-labelled and (20b) with the phenyl group ^{14}C-labelled. The hydrolysis of a mixture of them was allowed to proceed to 2—5% and the deuterium isotope effect was determined from the ^{3}H to ^{14}C ratio in the released phenol. The value of k_H/k_D was 1.13 and taken to indicate that reaction proceeded via a carbonium ion. The α-deuterium isotope effect for the hydrolysis of phenyl β-D-glucoside catalysed by almond-emulsin β-glucosidase is $k_H/k_D = 1.01$, which suggests that the reaction involves a direct substitution.[39]

Hen-egg-white lysozyme and human lysozyme catalyse the conversion of di-*N*-acetylchitobiose and tri-*N*-acetylchitotriose into higher molecular weight

[25] A. D. Pethybridge, *J. Chem. Soc.*(A), **1969**, 1345.

[26] A. Meller, *Carbohydrate Res.*, **10**, 313 (1969).

[27] O. Smidrød, B. Larsen, T. Painter, and A. Haug, *Acta Chem. Scand.*, **23**, 1573 (1969).

[28] P. F. Lloyd and B. Evans, *J. Chem. Soc.*(C), **1969**, 2753.

[29] M. Szymczyk and A. Temeriusz, *Rocz. Chem.*, **43**, 1227 (1969).

[30] B. Capon, *Chem. Rev.*, **69**, 433 (1969).

[31] W. P. Jencks, *Catalysis in Chemistry*, McGraw-Hill, New York, 1969.

[32] M. R. Hollaway, *Ann. Reports Progr. Chem.*, *B*, **65**, 601 (1968) (published 1969).

[33] S. M. Hopkinson, *Quart. Rev.*, **23**, 98 (1969).

[34] A. Williams, *Introduction to the Chemistry of Enzyme Action*, McGraw-Hill, London, 1969.

[35] S. Doonan, *The Chemistry and Physics of Enzyme Catalysis*, R.I.C. Reviews, **2**, 117 (1969).

[36] P. Jolles, *Angew. Chem. Internat. Ed. Engl.*, **8**, 227 (1969).

[37] D. M. Chipman and N. Sharon, "Mechanism of lysozyme action", *Science*, **165**, 458 (1969).

[38a] T. Rand-Meir, F. W. Dahlquist, and M. A. Raftery, *Biochemistry*, **8**, 4206 (1969); cf. *Org. Reaction Mech.*, **1968**, 354.

[38b] C. S. Tsai, J. Y. Tang, and S. C. Sabbarao, *Biochem. J.*, **114**, 529 (1969).

[39] F. W. Dahlquist, T. Rand-Meir, and M. A. Raftery, *Biochemistry*, **8**, 4214 (1969).

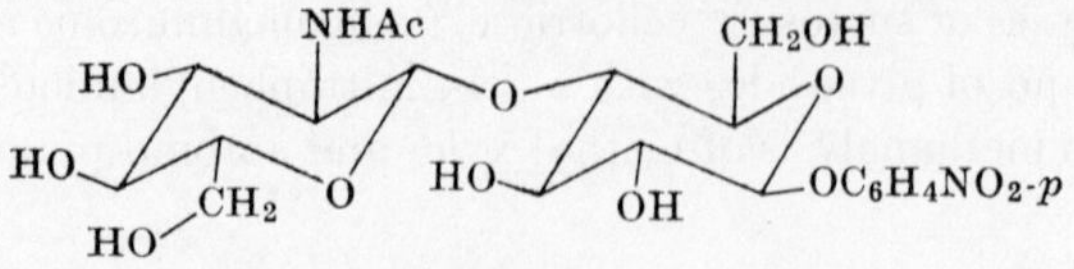

NHAc CH_2OH

HO   O   O

HO CH_2 O   HO   OH $OC_6H_4NO_2$-*p*

HO

(17)

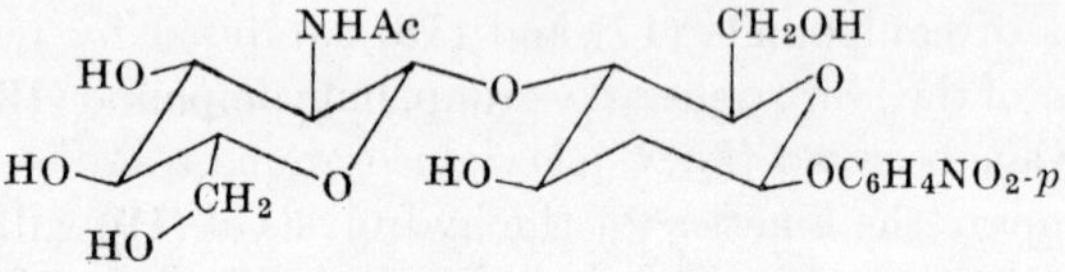

NHAc CH_2OH

HO   O   O

HO CH_2 O   HO $OC_6H_4NO_2$-*p*

HO

(18)

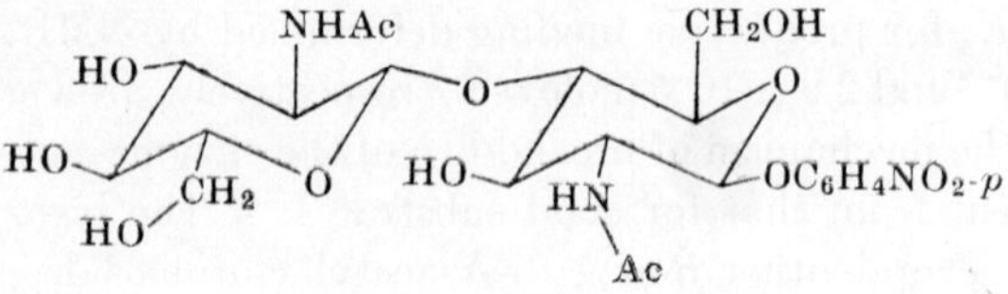

NHAc CH_2OH

HO   O   O

HO CH_2 O   HO   HN $OC_6H_4NO_2$-*p*

HO   Ac

(19)

NHAc CH_2OH

HO   O   O

HO CH_2 O   HO   HO OC_6H_5

HO   R

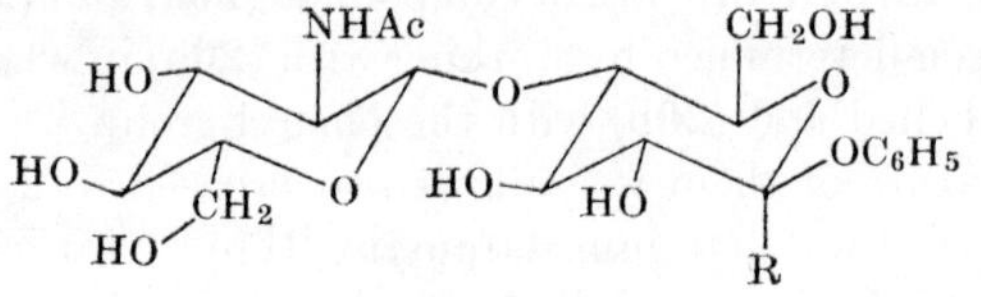

(20a) R = H
(20b) R = D

oligosaccharides but papaya lysozyme does not. In the presence of methanol di-*N*-acetylchitobiose is converted partly by hen-egg-white and human lysozyme into methyl 2-acetamido-2-deoxyglucoside. This is at least 99.7 and 99.9% of the β-configuration, i.e. formed with retention of configuration. Papaya lysozyme does not catalyse this reaction. The hydrolysis of tetra-*N*-acetylchitotetraose catalysed by papaya lysozyme yields a product which mutarotates downward and must therefore be predominantly of the α-configuration.[40]

Lysozyme has one carboxy group which reacts preferentially with triethyloxonium fluoroborate at pH 4. The esterified enzyme binds tri-*N*-acetylchitotriose twenty times less efficiently than the free enzyme but retains 57% of its specific activity. It was suggested that the carboxy group which is esterified is one which is close to the strong (non-productive) binding site for tri-*N*-acetylchitotriose rather than at the catalytic site. The ester is hydrolysed easily at pH 7 presumably with intramolecular catalysis. At pH 4.7 two

[40] F. W. Dahlquist, C. L. Borders, G. Jacobson, and M. A. Raftery, *Biochemistry*, **8**, 694 (1969).

carboxy groups are esterified by triethyloxonium fluoroborate. If the labile ester is then hydrolysed, a second monoester is obtained which binds tri-N-acetylchitotriose 10 times less effectively than the free enzyme and has no catalytic activity.[41] It was shown that the carboxy group which is esterified in this derivative is that of aspartic acid 52.[42]

It has been shown that Biebrich Scarlet binds in the catalytic site of lysozyme. Use has been made of this to determine the equilibrium constants for the productive binding of tri-N-acetylchitotriose and hexa-N-acetylchitohexaose at pH 7.6 which are 2×10^{-2}M and 5×10^{-6}M respectively. The dissociation constant for the non-productive binding of tri-N-acetylchitotriose had previously been determined to be 5×10^{-6}M. Hexa-N-acetylchitohexaose therefore binds much more strongly than tri-N-acetylchitotriose in the productive mode. It binds about as strongly in the productive mode as tri-N-acetylchitotriose in the non-productive mode.[43]

The variation of the relaxation time with concentration of solutions of di-N-acetylchitobiose, lysozyme, and p-nitrophenol at pH 6 and 7 has been interpreted in terms of a binding mechanism involving a diffusion-controlled formation of a first enzyme-inhibitor complex followed by a slower protein isomerization.[44] The values of the rate constants were in good agreement with those obtained by an NMR method.[45, 46] Equilibrium constants for the binding of N-acetylglucosamine oligosaccharides to lysozyme have also been determined.[45-48]

The 2′,3′-epoxypropyl glycoside of N-acetylglucosamine is an irreversible inhibitor for lysozyme.[49]

The NMR spectra of hen-egg-white[50] and human[50, 51] lysozyme have been reported.

The titration curve of hen-egg-white lysozyme is consistent with six of the eight carboxy groups having pK_a ca. 4.3 and the remaining two having pK_a 3.5 and 6.0.[52]

[41] S. M. Parsons, L. Jao, F. W. Dahlquist, C. L. Borders, T. Groff, J. Racs, and M. A. Raftery, *Biochemistry*, **8**, 700 (1969).

[42] S. M. Parsons and M. A. Raftery, *Biochemistry*, **8**, 4199 (1969); see also T. Y. Lin and D. E. Koshland, *J. Biol. Chem.*, **244**, 505 (1969).

[43] G. L. Rossi, E. Holler, S. Kumar, J. A. Rupley, and G. P. Hess, *Biochem. Biophys. Res. Comm.*, **37**, 757 (1969).

[44] E. Holler, J. A. Rupley, and G. P. Hess, *Biochem. Biophys. Res. Comm.*, **37**, 423 (1969).

[45] B. D. Sykes, *Biochemistry*, **8**, 1110 (1969).

[46] B. D. Sykes and C. Parravano, *J. Biol. Chem.*, **244**, 3900 (1969).

[47] M. A. Raftery, F. W. Dahlquist, S. M. Parsons, and R. G. Wolcott, *Proc. Nat. Acad. Sci.*, **62**, 44 (1969).

[48] F. W. Dahlquist and M. A. Raftery, *Biochemistry*, **8**, 713 (1969).

[49] E. W. Thomas, J. F. McKelvy, and N. Sharon, *Nature*, **222**, 485 (1968).

[50] C. C. McDonald, J. D. Glickson, and W. D. Phillips, *Biochem. Biophys. Res. Comm.*, **35**, 43 (1969).

[51] J. S. Cohen, *Nature*, **223**, 43 (1969).

[52] R. Sakakibara and K. Hamaguchi, *J. Biochem.* (Tokyo), **64**, 613 (1968).

Chitin, with about 70% of the 6-hydroxy groups substituted, is still bound by lysozyme. The rate of hydrolysis of chitin and glycol-chitin catalysed by lysozyme is reduced to about 20% of its original value when 50% of the *N*-acetyl groups are removed, but the ability to bind appears to be unchanged.[53]

The catalytic actions of hen- and goose-egg-white lysozyme have been compared.[54] Papaya lysozyme[55] and the lysozyme from Bacteriophage[56] have been studied. The sequence of human lysozyme has been determined.[57] Other investigations on lysozyme are described in ref. 58.

The steric courses of hydrolyses catalysed by several other glycosidases have been determined this year, and the results are given in Table 1.[59-61]

Phenol is formed from a mixture of maltose and phenyl α-D-glucoside in the presence of α-amylase from *B. subtilis*. This was thought to arise from the formation of a small amount of phenyl maltoside produced from maltose and phenyl α-D-glucoside by glucosyl transfer.[62]

The hydrolyses of *p*-nitrophenyl maltoside[63] and of a mixture of *O*-methylated derivatives of phenyl maltoside[64] catalysed by Taka-Amylase A have been studied. There has been a detailed kinetic study of the hydrolysis of phenyl maltoside catalysed by the α-amylase of *B. subtilis*.[65] Other investigations on amylases are given in ref. 66.

Pullullanase from *Acetobacter aerogenes* hydrolyses an α-glucan consisting of maltotriose units linked by α-1,6-glycosidic bonds exclusively to maltotriose. It was shown that hydrolysis occurs at the end of the substrate.[67]

[53] K. Hayashi, N. Fujimoto, M. Kugimiya, and M. Funatsu, *J. Biochem.* (Tokyo), **65**, 401 (1969).

[54] A. C. Dianoux and P. Jolles, *Helv. Chim. Acta*, **52**, 611 (1969).

[55] J. B. Howard and A. N. Glazer, *J. Biol. Chem.*, **244**, 1399 (1969).

[56] L. W. Black and D. S. Hogness, *J. Biol. Chem.*, **244**, 1968, 1976, 1982 (1969).

[57] R. Canfield, *Brookhaven Symposium Biol.*, No. 21, 136 (1969).

[58] J. B. Cole, M. C. Bryan, and W. P. Bryan, *Arch. Biochem. Biophys.*, **130**, 86 (1969); R. C. Davies and A. Neuberger, *Biochim. Biophys. Acta*, **178**, 306 (1969); R. C. Davies, A. Neuberger, and B. M. Wilson, *ibid.*, p. 294; K. C. Aune and C. Tanford, *Biochemistry*, **8**, 4572, 4579 (1969); B. Bonavida, A. Miller, and E. E. Secarz, *ibid.*, p. 968; T. Miyazaki and Y. Matsushima, *Bull. Chem. Soc. Japan*, **41**, 2754 (1968); J. J. Pollock and N. Sharon, *Biochem. Biophys. Res. Comm.*, **34**, 673 (1969); A. Marzotto and G. Kollin, *Z. Physiol. Chem.*, **350**, 427 (1969).

[59] D. E. Eveleigh and A. S. Perlin, *Carbohydrate Res.*, **10**, 87 (1969).

[60] G. Semeza, H. C. Curtius, O. Raunhardt, P. Hore, and M. Müller, *Carbohydrate Res.*, **10**, 417 (1969).

[61] K. Hiromi, T. Shibaoka, H. Fukube, and S. Ono, *J. Biochem.* (Tokyo), **66**, 63 (1969).

[62] H. Yoshida, K. Hiromi, and S. Ono, *J. Biochem.* (Tokyo), **66**, 183 (1969).

[63] N. Suetsugu, K. Hiromi, M. Takagi, and S. Ono, *J. Biochem.* (Tokyo), **64**, 619 (1968).

[64] Y. Isemura, T. Ikenaka, and Y. Matsushima, *J. Biochem.* (Tokyo), **66**, 77 (1969).

[65] H. Yoshida, K. Hiromi, and S. Ono, *J. Biochem.* (Tokyo), **65**, 741 (1969).

[66] H. Yamaguchi, T. Mega, T. Ikenaka, and Y. Matsushima, *J. Biochem.* (Tokyo), **66**, 441 (1969); A. Yutani, K. Yutani, and T. Isemura, *ibid.*, **65**, 201 (1969); K. Kainuma and D. French, *FEBS Letters*, **5**, 257 (1969); J. Wakim, M. Robinson, and J. A. Thoma, *Carbohydrate Res.*, **10**, 487 (1969).

[67] K. Wallenfels, I. R. Rached, and F. Hucho, *European J. Biochem.*, **7**, 231 (1969).

Table 1. The steric course of some reactions catalysed by glycosidases

Enzyme	Source	Substrate	Major product	Predominant steric course	Method	Reference
α-D-Glucosidase	*Saccharomyces cerevesiae*	Phenyl α-D-glucoside	α-D-Glucose	Retention	NMR	59
β-D-Glucosidase	Almond emulsin	Salicin	β-D-Glucose	Retention	NMR	59
Glucamylase	*A. niger*	Starch	β-D-Glucose	Inversion	NMR	59
exo-Laminaranase	Basidiomycete sp. QM 806	Laminarin	α-D-Glucose	Inversion	NMR	59
β-Amylase	Sweet potato	Amylose	β-Maltose	Inversion	NMR	59
β-Amylase	Barley	Amylose	β-Maltose	Inversion	GLC	60
α-Amylase	Hog pancreas	Amylose	α-Maltose	Retention	NMR	59
α-Amylase	Hog pancreas	Amylose	α-Maltose and α-Maltotriose	Retention	GLC	60
α-Amylase	*B. subtilis*	Phenyl α-maltoside	α-Maltose	Retention	Optical rotation	61
endo-Laminaranase	*Rhizopus arrhizus* QM 1032	Laminarin	β-Oligosaccharides	Retention	NMR	59
endo-Laminaranase	*Cytophaga* sp.	Laminarin	β-Oligosaccharides	Retention	NMR	59
Lactase	Rat intestine	Lactose	β-Galactose	Retention	GLC	60
Sucrase	Rat intestine	Sucrose	α-Glucose	?[a]	GLC	60

[a] The position of bond fission appears to be unknown.

Several β-galactofuranosides have been reported to be hydrolysed by almond-emulsin β-galactosidase but not by that from *E. coli* K12 and bovine liver.[68] Galactosidases from a wide variety of sources are inhibited by D-galactal.[69] Other investigations on galactosidases are given in ref. 70.

The following glycosidases have also been studied: β-glucosidase from almond emulsin,[71] *Aspergillus oryzae*,[72] and the liver of *Charonia lampas*,[73] β-xylosidase from the liver of *Charonia lampas*,[73] cellulase,[74] yeast invertase,[75] and mammalian α-acetylgalactosamidase.[76]

Hydration of Aldehydes and Ketones and Related Reactions[77]

The hydroxide-ion-catalysed decompositions of the hemithioacetals of acetaldehyde with benzenethiol, *p*-nitrobenzenethiol, and thioacetic acid have rate constants of ca. 10^{10} M^{-1} sec^{-1} and the rates are largely controlled by the rates of diffusion together of the reactants. The mechanism is therefore as shown in equations (5)—(7) with k_d the slow step. It follows that the rate-

$$\overset{|}{\underset{|}{R\dot{S}COH}} + {}^-OH \underset{k_{-d}}{\overset{k_d}{\rightleftharpoons}} \overset{|}{\underset{|}{R\dot{S}COH}}\cdots{}^-OH \qquad \dots (5)$$

$$\overset{|}{\underset{|}{R\dot{S}COH}}\cdots{}^-OH \rightleftharpoons RS^- + \overset{}{\underset{}{{>}C{=}O}} + H_2O \qquad \dots (6)$$

$$RS^- + H^+ \rightleftharpoons RSH \qquad \dots (7)$$

limiting step for the formation of the hemithioacetal is k_{-d}, the diffusion away of the hydroxide ion. Weaker bases than ^-OH catalyse the breakdown more slowly and yield a β-value of 0.8. This indicates that proton transfer and C–S bond breaking must be "in some sense concerted".[78]

The equilibrium constants for the reaction of propane-1-thiol with carbonyl compounds in methylene chloride solution have been determined.[79]

[68] K. Yoshida, N. Iino, T. Kamota, and K. Kato, *Chem. Pharm. Bull.* (Japan), **17**, 1123 (1969).
[69] Y. C. Lee, *Biochem. Biophys. Res. Comm.*, **35**, 161 (1969).
[70] S. Shifrin and G. Hunn, *Arch. Biochem. Biophys.*, **130**, 530 (1969); P. M. Dey and J. B. Pridham, *Biochem. J.*, **115**, 47 (1969); K. K. Mäkinen, *Acta Chem. Scand.*, **22**, 3339 (1968).
[71] J. P. Horwitz, C. V. Easwaran, and L. S. Kowalczyk, *Carbohydrate Res.*, **9**, 305 (1969).
[72] G. Legler and L. M. O. Osama, *Z. Physiol. Chem.*, **349**, 1488 (1968).
[73] M. Fukuda and F. Egami, *J. Biochem.* (Tokyo), **66**, 157 (1969).
[74] K.-E. Eriksson and B. H. Hollmark, *Arch. Biochem. Biophys.*, **133**, 233 (1969).
[75] M. M. Tong and R. E. Pincock, *Biochemistry*, **8**, 908 (1969).
[76] B. Weissmann and D. F. Hinrichsen, *Biochemistry*, **8**, 2034 (1969).
[77] P. Le Henaff, *Bull. Soc. Chim. France*, **1968**, 4687.
[78] R. E. Barnett and W. P. Jencks, *J. Am. Chem. Soc.*, **91**, 6758 (1969).
[79] L. Field and B. J. Sweetman, *J. Org. Chem.*, **34**, 1799 (1969).

In D_2O solution glutaraldehyde exists mainly as compounds (21)—(23).[80] 1,1'-Oxydiethanol is present in solutions of acetaldehyde in aqueous hydrochloric acid.[81]

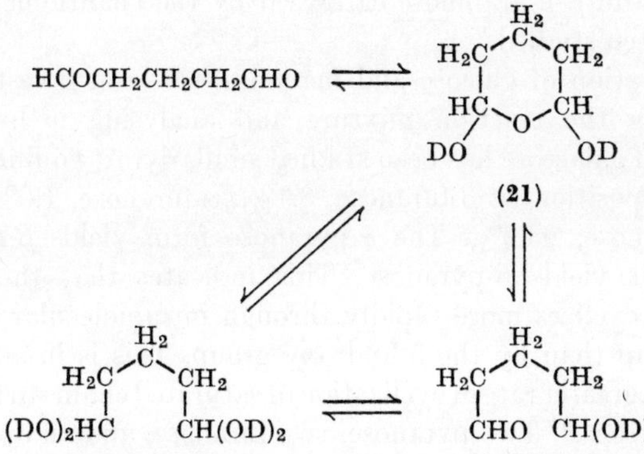

$HCOCH_2CH_2CH_2CHO \rightleftharpoons$

(21)

(22)   (23)

The hydration of the following compounds has also been studied: formaldehyde,[82] pyruvic acid,[83] α-ketoglutaric acid,[84] glyoxalate, mesoxalate, and ethyl pyruvate,[85] and pyridinecarboxaldehydes.[86, 87]

It has been demonstrated that 1,4-diazabicyclo[2.2.2]octane is an effective catalyst for the mutarotation of tetramethylglucose in benzene, and it was suggested that the catalytic effect of a mixture of pyridine and phenol is the result of general-base catalysis by the phenoxide ion rather than concerted acid–base catalysis by the un-ionized pyridine and phenol.[88] It was also suggested that 2-pyridone is an effective catalyst for the mutarotation, not because it is bifunctional but because it is tautomeric. Other tautomeric catalysts are pyrazole, 1,2,4-triazole, and benzoic acid. 2-Aminophenol, which is bifunctional but not tautomeric, is not a catalyst.[89] Catalysis of the mutarotation of tetramethylglucose in benzene is very insensitive to the strength of the acid ($\alpha = 0.3$), which was interpreted as being consistent with bifunctional tautomeric catalysis. Thioacetic acid is a very poor catalyst. It exists mainly in the form CH_3COSH and catalysis would involve conversion into the unstable form CH_3CSOH, and so it is not very effective.[90]

[80] P. M. Hardy, A. C. Nicholls, and H. N. Rydon, *Chem. Comm.*, **1969**, 565.
[81] G. Socrates, *Chem. Comm.*, **1969**, 702; *J. Org. Chem.*, **34**, 2958 (1969).
[82] H.-G. Schecker and G. Schulz, *Z. Phys. Chem.* (Frankfurt), **65**, 221 (1969).
[83] H. Patting and H. Strehlow, *Ber. Bunsenges. Phys. Chem.*, **73**, 534 (1969).
[84] J. Jen and W. Knoche, *Ber. Bunsenges. Phys. Chem.*, **73**, 539 (1969).
[85] M. L. Ahrens, *Ber. Bunsenges. Phys. Chem.*, **72**, 691 (1968).
[86] Y. Pocker and J. E. Meany, *J. Phys. Chem.*, **73**, 1857 (1969).
[87] S. Cabani, G. Conti, and P. Gianni, *J. Chem. Soc.*(A), **1969**, 1363.
[88] P. E. Rony, W. E. McCormack, and S. W. Wunderly, *J. Am. Chem. Soc.*, **91**, 4244 (1969); *Org. Reaction Mech.*, **1968**, 358.
[89] P. E. Rony, *J. Am. Chem. Soc.*, **91**, 6090 (1969).
[90] A. Kergomard and M. Renard, *Tetrahedron*, **24**, 6643 (1968).

(S)-α-(2-Naphthyloxy)propionic acid is a slightly less effective catalyst than its (R)-enantiomer for the mutarotation of tetramethylglucose in benzene.[91] The mutarotation of α-L-rhamnose catalysed by the enantiomers of 1-phenyl-ethylamine and of β-D-arabinose catalysed by the enantiomers of mandelic acid has also been studied.[92]

The mutarotation of glucose and mannose has been investigated by tri-methylsilylating the reaction mixture and analysing it by GLC.[93] The mutarotation of galactose has been studied similarly. At equilibrium in water at 25° the composition is: β-furanose, 3.1%; α-furanose, 1.0%; α-pyranose, 32.0%; β-pyranose, 63.9%. The α-pyranose form yields β-furanose more rapidly than it yields β-pyranose. This indicates that the intermediate aldehydo-form cyclizes more rapidly through intramolecular attack by the 4-hydroxy group than by the 5-hydroxy group. This is in accord with the normally found greater rate of cyclization of saturated chains to form five- than six-membered rings.[94] Two pyranose, two furanose, and an open-chain form have been shown to be present in solutions of fructose.[95]

The mutarotation of tetramethylglucose in water is subject to steric hindrance since 2-picoline and 2,6-lutidine are less effective catalysts than expected from their pK_a values.[96] It was already known that the mutarotation of glucose is subject to steric hindrance.[97] The occurrence of steric hindrance suggests that catalysis does not occur through a chain of water molecules.

Other studies of mutarotation are reported in ref. 98.

The mutarotatase from bovine kidney cortex has been purified and studied.[99]

The composition and conformation of sugars in solution has been reviewed.[100]

Reactions with Nitrogen Bases

The formation of p-chlorobenzaldehyde thiosemicarbazone is similar to the formation of semicarbazones in showing a change in rate-limiting step from attack by thiosemicarbazide at low pH to dehydration of the intermediate

[91] A. Kergomard and M. Renard, *Tetrahedron, Letters* **1969**, 3041.

[92] V. A. Pavlov, E. I. Klabunovskii, and A. A. Balandin, *Zh. Fiz. Khim.*, **42**, 2475 (1968); *Chem. Abs.*, **70**, 58160, 58161 (1969).

[93] C. Y. Lee, T. E. Acree, and R. S. Shallengberger, *Carbohydrate Res.*, **9**, 356 (1969).

[94] T. E. Acree, R. S. Shallenberger, C. Y. Lee, and J. W. Einset, *Carbohydrate Res.*, **10**, 355 (1969).

[95] H. C. Curtius, J. Voellmin, and M. Mueller, *Z. Anal. Chem.*, **243**, 341 (1968).

[96] H. H. Huang, A. N. H. Yeo, and L. H. L. Chia, *J. Chem. Soc.*(B), **1969**, 836.

[97] F. Covitz and F. H. Westheimer, *J. Am. Chem. Soc.*, **85**, 1773 (1963).

[98] H. S. Isbell, H. L. Frush, C. W. R. Wade, and C. E. Hunter, *Carbohydrate Res.*, **9**, 163 (1969); A. de Granchamp-Chaudin, *Ann. Pharm. Fr.*, **26**, 115 (1968); K. D. Dzhundubaev and R. I. Kozhakhmetonva, *Izv. Akad. Nauk Kirg. SSR*, **1968**, 85; *Chem. Abs.*, **69**, 77638 (1968); E. I. Klabunovskii, V. A. Pavlov, and A. A. Balandin, *Zh. Fiz. Khim.*, **42**, 2487 (1968).

[99] J. M. Bailey, P. H. Fishman, and P. G. Pentchev, *J. Biol. Chem.*, **243**, 4827 (1968); **244**, 781 (1969).

[100] S. J. Angyal, *Angew. Chem. Internat. Ed. Engl.*, **8**, 157 (1969).

carbinolamine at high pH. Nucleophilic attack by thiosemicarbazide is general-acid catalysed with $\alpha = 0.15$. The main difference between thiosemicarbazone and semicarbazone formation is that the dehydration step in the former is general-base catalysed. A good Brønsted plot was obtained with bases of several types to yield a β-value of 0.71. Two mechanisms were considered (equations 8 and 9). That of equation (8) was preferred since if that of equation (9) were followed the rate constant for the reaction of the ionized carbinolamine with BH^+ would have to be greater than that for diffusion.[101]

$$\text{B} \cdots \text{H} - \text{N} - \text{C} - \text{OH} \rightleftharpoons \text{B} - \text{H}^+ \quad {}_{\diagdown}\text{N} \doteq \text{C}{\diagup} \quad \bar{\text{O}}\text{H} \overset{\text{fast}}{\rightleftharpoons} \text{B} + {}_{\diagdown}\text{N}{=}\text{C}{\diagup} + \text{H}_2\text{O}$$

$$\cdots (8)$$

$$\text{H} - \text{N} - \text{C} - \text{OH} + \text{B} \overset{\text{fast}}{\rightleftharpoons} {}_{\diagdown}\bar{\text{N}} - \text{C} - \text{O} \quad \text{H} - \overset{+}{\text{B}} \rightleftharpoons {}_{\diagdown}\overset{\diagdown}{\text{N}}{=}\text{C}{\diagup} \quad \text{O} - \text{H} \cdots \text{B}$$
$$\overset{\text{H}}{} \qquad\qquad\qquad \overset{\text{H}}{} \qquad \cdots (9)$$

The reaction of formaldehyde with urea is general-acid and general-base catalysed with α- and β-coefficients 0.31 and 0.37 respectively. The catalytic constants for HCO_3^-, HPO_4^{2-}, and $H_2PO_4^-$ are greater than expected from the Brønsted plot for the other acids. The entropies of activation for catalysis by these ions were 15—20 e.u. lower than those for catalysis by water and ^-OH, and it was suggested that they acted as bifunctional tautomeric catalysts with a transition state (**24**). This was visualized as being formed via hydrogen-bonded complexes between the catalyst and the formaldehyde or urea.[102]

The solvent isotope effect for the hydrolysis of cyclohexanone, cyclopentanone, and acetaldehyde oxime in hydrochloric acid is $k(D_2O)/k(H_2O) = 1.45$.[103]

The *syn-* and *anti-*forms of several semicarbazones have been isolated.[104]

Ring–chain tautomerism of imides and ketones with *o*-carboxyamide and sulphonamido groups has been studied.[105]

There have been several investigations on the formation of oximes,[106]

[101] J. M. Sayer and W. P. Jencks, *J. Am. Chem. Soc.*, **91**, 6353 (1969).

[102] B. R. Glutz and H. Zollinger, *Helv. Chim. Acta*, **52**, 1976 (1969); P. Eugster and H. Zollinger, *ibid.*, p. 1985.

[103] N. W. Ikonomov, *Glas. Hem. Drus. Beograd.*, **32**, 295 (1969); *Chem. Abs.*, **71**, 48917 (1969).

[104] V. I. Stenberg, P. A. Barks, D. Bays, D. D. Hammagren, and D. V. Rao, *J. Org. Chem.*, **33**, 4402 (1968).

[105] H. Watanabe, C. L. Mao, I. T. Barnish, and C. R. Hauser, *J. Org. Chem.*, **34**, 919 (1969); see also H. J. Roth and G. Langer, *Arch. Pharm.*, **301**, 736 (1968).

[106] V. A. Komarov and L. P. Ivanova, *Zh. Org. Khim.*, **5**, 49 (1969); *Chem. Abs.*, **70**, 86838 (1969); L. P. Ivanova, *ibid.*, p. 52; *Chem. Abs.*, **70**, 86835 (1969); A. Heymes and J. Jacques, *Ann. Chim.* (France), **3**, 555 (1968); A. Heymes, M. Dvolaitzky, and J. Jacques, *ibid.*, p. 543.

hydrazones,[107] and osazones.[108] The mutarotation of sugar osazones has been reviewed.[109]

Dehydration of the formaldehyde adduct of the tetrahydroquinoxaline (25) to yield (27) has a pH–rate profile which shows a break indicative of a change in rate-determining step from dehydration of the carbinolamine intermediate

[107] H. Dorn, H. Dilcher, and K. Walter, *Ann. Chem.*, **720**, 111 (1969); S. Tagami, K. Sasayama, and D. Shiho, *Chem. Pharm. Bull.* (Japan), **17**, 5 (1969).

[108] H. Simon, W. Moldenhauer, and A. Kraus, *Chem. Ber.*, **102**, 2777 (1969); H. Simon and W. Moldenhauer, *ibid.*, pp. 1191, 1198; J. Buckingham and R. D. Guthrie, *J. Chem. Soc.*(C), **1968**, 3079.

[109] L. Mester, *Chimia*, **23**, 133 (1969).

(26) at high pH to ring-closure at low pH. The mechanism of equation (10) was proposed. The reaction shows kinetic general-acid catalysis which was interpreted as resulting from the kinetically equivalent general-base catalysed cyclization of the iminium cation (28). This led to a reasonable β-value of 0.65.[110]

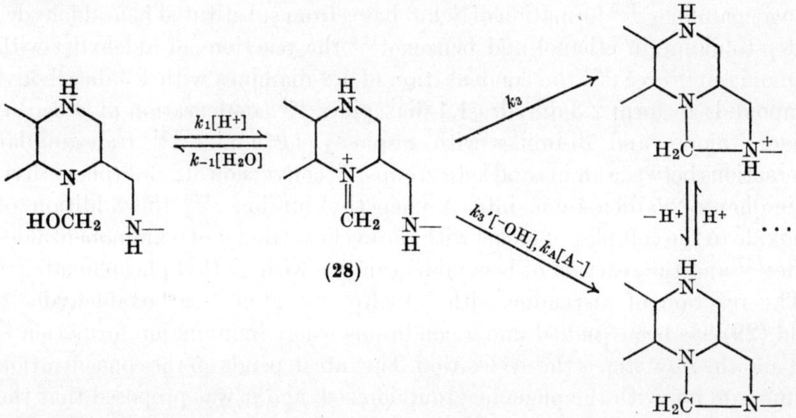

$$
\text{HOCH}_2 \quad \underset{k_{-1}[\text{H}_2\text{O}]}{\overset{k_1[\text{H}^+]}{\rightleftharpoons}} \quad \text{CH}_2 \quad (28) \quad \overset{k_2}{\longrightarrow} \quad \overset{k_3[^-\text{OH}],\, k_A[\text{A}^-]}{\longrightarrow} \quad \cdots (10)
$$

The rate of formation of 2-hydroxybenzylideneanilines in acetone, ether, and benzene is reported to be significantly greater than that of 4-hydroxybenzylideneanilines. This was attributed to intramolecular hydrogen-bonding.[111] The hydrolysis of salicylideneanilines has been studied [112] and the equilibrium constant for the formation of N-(2-hydroxyethyl)salicylideneimine in aqueous solution measured.[113] The metal-ion promoted condensation of salicaldehyde with glycine [114] and the condensation of pyridoxal with glycine and glycine amide in methanol [115] have been studied.

cis–trans-Isomerization of α-methylbenzylideneaniline is inhibited on complexing with triethylaluminium. It was therefore concluded that isomerization involved inversion of the nitrogen lone pair, not rotation about the C=N bond.[116]

An X-ray crystal structure determination has shown that benzylideneaniline has a structure in which the aniline ring is twisted out of the plane of the C–N=C–C bonds by 40—45°.[117]

[110] S. J. Benkovic, P. A. Benkovic, and D. R. Comfort, *J. Am. Chem. Soc.*, **91**, 1860, 5270 (1969).
[111] P. Nagy and Z. Molnar, *Szeged. Pedagog. Foiskola Evk. Masodik Resz.*, **1966**, 145; *Chem, Abs.*, **69**, 43199 (1968); P. Nagy, *ibid.*, **1967**, 61; *Chem. Abs.*, **69**, 43200 (1968).
[112] Y. A. Davydovskaya and T. I. Vainshtein, *Azometing*, **1967**, 234; *Chem. Abs.*, **69**, 76115 (1968).
[113] R. W. Green and R. J. Sleet, *Austral. J. Chem.*, **22**, 917 (1969).
[114] D. Hopgood and D. L. Leussing, *J. Am. Chem. Soc.*, **91**, 3740 (1969).
[115] Y. Matsushima, *Chem. Pharm. Bull.* (Japan), **16**, 2151 (1968).
[116] E. A. Jeffery, A. Meisters, and T. Mole, *Tetrahedron*, **25**, 741 (1969).
[117] H. B. Bürgi and J. D. Dunitz, *Chem. Comm.*, **1969**, 472.

N-Pyruvoylanthranilic acid does not exist in a cyclic form and is not converted into 1-acetyl-3-methylene-4,1-benzoxazepine-2,5-dione via such a structure as suggested previously.[118]

Other reactions involving Schiff bases which have been studied include the following: hydrolysis of substituted benzylideneanilines [119] and benzylidene-halogenoanilines;[120] formation of Schiff bases from substituted benzaldehydes and *p*-toluidine in ethanol and benzene;[121] the reactions of aldehydes with ammonia and urea;[122] the condensation of 1,2-diamines with 1,3-dicarbonyl compounds to form 2,3-dihydro-1,4-diazepines;[123] condensation of *p*-benzo-quinone mono- and di-imines with *m*-phenylenediamine;[124] transannular interactions between amino and keto groups;[125] conversion of 2,6-di(phenethyl-amino)hepta-2,5-dien-4-one into *N*-phenethyl-lutidone;[126] the addition of ethoxide to the complex of nickel with the cyclic tetramer of *o*-aminobenzalde-hyde;[127] and the reaction of benzylideneaniline with diethyl phosphonate.[128]

The reaction of histamine with 3-hydroxypyridine-4-carboxaldehyde to yield (29) has been studied under conditions where iminium ion formation is fast and the slow step is the cyclization. The rate depends on the concentration of iminium ion with the phenolic group ionized, and it was proposed that the reaction involves cyclization of the kinetically equivalent form with the imidazole group ionized and the un-ionized phenolic group providing intra-molecular general-acid catalysis as shown in equation (11).[129]

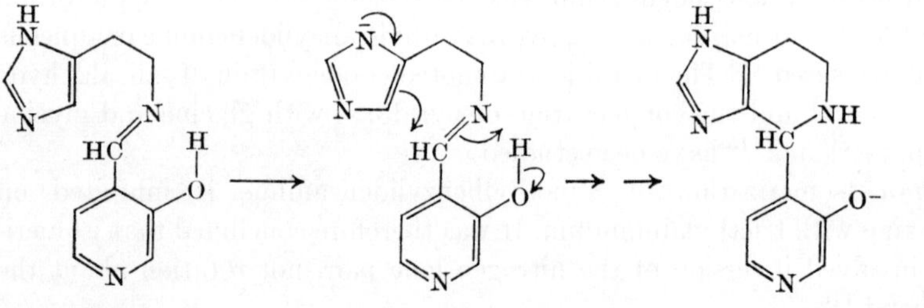

$$\cdots (11)$$

(29)

[118] P. F. Wegfahrt and H. Rapoport, *J. Org. Chem.*, **34**, 3035 (1969).
[119] A. Mesli and J. Tirouflet, *Compt. Rend.*, *C*, **267**, 838 (1968).
[120] I. R. Bellobono and G. Favini, *Tetrahedron*, **25**, 57 (1969).
[121] P. Nagy, *Szeged. Pedagog. Foiskola Evk. Masodik Resz.*, **1966**, 153; *Chem. Abs.*, **69**, 26632 (1968).
[122] A. Kawasaki and Y. Ogata, *Mem. Fac. Eng. Nagoya Univ.*, **19**, 1 (1967); *Chem. Abs.*, **69**, 58669 (1968).
[123] C. Barnett, D. R. Marshall, and D. Lloyd, *J. Chem. Soc.*(B), **1968**, 1536.
[124] J. F. Corbett, *J. Chem. Soc.*(B), **1969**, 818, 823, 827.
[125] N. J. Leonard and T. Sato, *J. Org. Chem.*, **34**, 1066 (1969).
[126] S. Goto, A. Kono, and S. Iguchi, *J. Pharm. Sci.*, **57**, 791 (1968).
[127] L. T. Taylor, F. L. Urbach, and D. H. Busch, *J. Am. Chem. Soc.*, **91**, 1072 (1969).
[128] W. M. Henderson and W. H. Shelver, *J. Pharm. Sci.*, **58**, 106 (1969).
[129] T. C. Bruice and A. Lombardo, *J. Am. Chem. Soc.*, **91**, 3009 (1969).

3,5-Di-t-butyl-1,2-benzoquinone reacts with amines of the type R_2CHNH_2 in a transamination reaction (equation 12).[130]

$$\text{But–benzoquinone} + R_2CHNH_2 \longrightarrow \text{But–}NCHR_2 \text{ quinone}$$

$$\cdots (12)$$

$$\text{But–}\overset{+}{N}H_3\text{–OH} + R_2CO \longleftarrow \text{But–}N{=}CR_2\text{–OH}$$

There have been many investigations of transamination.[131] The racemization of L-glutamic acid catalysed by salicaldehydes in the presence of cupric ions[132] and the vitamin-B_{12}-catalysed exchange of the β-protons of α-amino-acids in the presence of metal ions[133] have been studied.

The kinetics of the reactions of monosaccharides with aniline in 80% aqueous ethanol have been determined. A mechanism involving reaction of the aniline with the carbonyl form of the sugar was proposed.[134] Other investigations on glycosylamines are described in ref. 135.

The hydrolysis of deoxyuridine, thymidine, and 5-bromodeoxyuridine is independent of pH in the range 3—7 and also independent of the nature and the concentration of the buffer. The entropies of activation at 75° are 8.7,

[130] E. J. Corey and K. Achiwa, *J. Am. Chem. Soc.*, **91**, 1429 (1969).
[131] T. L. Fisher and D. E. Metzler, *J. Am. Chem. Soc.*, **91**, 5323 (1969); O. A. Gansow and R. H. Holm, *ibid.*, pp. 573, 5984; D. L. Leussing and L. Anderson, *ibid.*, p. 4698; V. M. Doctor and J. Oró, *Biochem. J.*, **112**, 691 (1969); P. Hermann and I. Willhardt, *Z. Physiol. Chem.*, **349**, 395 (1968); B. von Kerékjártó and U. Gebert, *ibid.*, **350**, 118 (1969); U. Gebert and B. von Kerékjártó, *Ann. Chem.*, **718**, 249 (1968); D. J. Whelan and G. J. Long, *Austral. J. Chem.*, **22**, 1779 (1969); H. C. Dunathan, L. Davis, P. G. Kury, and M. Kaplan, *Biochemistry*, **7**, 4532 (1968); J. E. Ayling, H. C. Dunathan, and E. E. Snell, *ibid.*, p. 4537; H. C. Dunathan, L. Davis, P. G. Kury, and M. Kaplan, *ibid.*, p. 4532; J. E. Churchich and J. G. Famelly, *J. Biol. Chem.*, **244**, 72 (1969); Y. Matsushima, *Chem. Pharm. Bull.* (Japan), **16**, 2046 (1968); see also Y. Matsushima and T. Hino, *ibid.*, p. 2277.
[132] M. Ando and S. Emoto, *Bull. Chem. Soc. Japan*, **42**, 2624, 2628 (1969).
[133] E. H. Abbott and A. E. Martell, *Chem. Comm.*, **1968**, 1501.
[134] V. A. Afanas'ev and V. A. Kharmats, *Zh. Fiz. Khim.*, **43**, 500 (1969); *Chem. Abs.*, **70**, 106790 (1969); see also V. A. Kharmats and V. A. Afanas'ev, *ibid.*, **42**, 2078 (1968); *Chem. Abs.*, **70**, 68665 (1969).
[135] J. Jasinska and J. Sokolowski, *Rocz. Chem.*, **42**, 2121 (1968); Z. Pawlak and E. Góvska, *ibid.*, **43**, 1237 (1969); T. Jasinski, K. Smiataczowa, T. Sokolowski, and J. Sokolowska, *Zesz. Nauk., Mat. Fiz. Chem., Wyzsz. Szk. Pedagog. Gdansk.*, **7**, 137 (1967); *Chem. Abs.*, **70**, 4463 (1969); B. I. Stepanov, B. A. Korolev, and N. A. Rozanelskaya, *Zh. Obshch. Khim.*, **39**, 2105 (1969); T. N. Matsevich, E. P. Trailina, I. A. Savich, and V. I. Spitsyn, *Dokl. Akad. Nauk SSSR*, **188**, 601 (1969); B. N. Stepanenko, E. S. Volkova, and M. G. Chentsova, *ibid.*, **183**, 1353 (1968); B. N. Stepanenko, E. S. Volkova, and M. G. Chentsova, *ibid.*, p. 1356; *Chem. Abs.*, **70**, 88163 (1969).

3.5, and 10.3 e.u. respectively and the relative rates at pH 4 are 2.85:1.05:52.8. An S_N1 mechanism to form a cyclic 2-deoxyribosyl cation was proposed.[136] It was suggested that the acid-catalysed hydrolysis proceeded by an analogous *A*1 mechanism.[136, 137]

The acid-catalysed hydrolysis of a large series of *N*-tetrahydropyranyl and tetrahydrofuryl pyridones and pyridazones has been studied. The rate is increased strongly when the oxygen of the pyridone or pyridazone is replaced by sulphur, but decreased when the tetrahydropyran or furan oxygen is replaced by sulphur. It was concluded that the reaction proceeded via ring-opening. If this is correct it is surprising that the rate of reaction of compounds (30) increases when X is changed along the series H < CN < NO$_2$ < Cl < Br.[138]

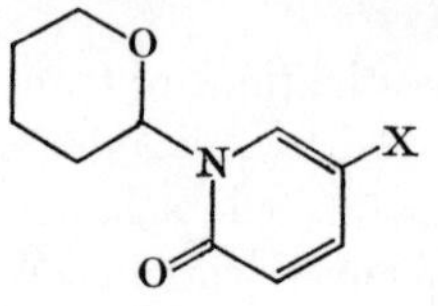

(30)

The Doebner–Miller reaction has been studied.[139]

Hydrolysis of Enol Ethers and Esters[140,141]

The isotope effect $k(\text{HF})/k(\text{DF})$ for the hydrogen fluoride catalysed hydrolysis of ethyl vinyl ether is 3.35. This is less than one-quarter of the theoretical maximum. It was considered that this was probably not the result of a large stretching vibration in the transition state with a frequency dependent on the mass of the hydrogen being transferred, since the α-value is 0.6. This was taken to mean that in the transition state this "hydrogen is held with approximately equal forces between the two bases between which it is moving" and the low isotope effect for the hydrogen fluoride catalysed reaction was ascribed to the absence of a bending vibration in the initial state and the presence of one in the transition state.[142]

Boric acid (pK_a 9.2) is approximately a ten-fold less effective catalyst for the hydrolysis of methyl isopropenyl ether than is bicarbonate (pK_a 10.2)

[136] R. Shapiro and S. Kang, *Biochemistry*, **8**, 1806 (1969).

[137] See also B. Capon, *Chem. Rev.*, **69**, 449 (1969).

[138] H. Kuhmstedt and G. Wagner, *Arch. Pharm.*, **302**, 213 (1969).

[139] A. B. Turner, *Chem. Comm.*, **1968**, 1659; T. P. Forrest, G. A. Dauphinee, and W. F. Miles, *Can. J. Chem.*, **47**, 2121 (1969); Y. Ogata, A. Kawasaki, and S. Suyama, *J. Chem. Soc.*(B), **1969**, 805.

[140] A. F. Rekasheva, *Uspekhi Khim.*, **37**, 2272 (1968) [*Russ. Chem. Rev.*, **37**, 1009 (1968)].

[141] F. Effenberger, *Angew. Chem. Internat. Ed. Engl.*, **8**, 295 (1969).

[142] A. J. Kresge and Y. Chiang, *J. Am. Chem. Soc.*, **91**, 1025 (1969).

despite the fact that it is a stronger acid. It was suggested that this is because the ionization of boric acid occurs as shown in equation (13).[143]

$$B(OH)_3 + 2H_2O \rightleftharpoons {}^-B(OH)_4 + H_3O^+ \qquad (13)$$

The hydrolyses of compounds (31) and (32) proceed with rate-determining

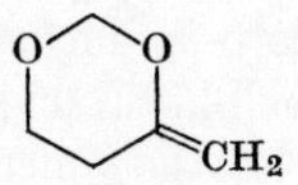

(31) 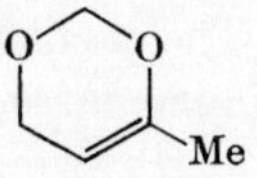   (32)

proton transfer to carbon since they are general-acid catalysed, have $k_D/k_H = 0.39$ and 0.41, and proceed much more rapidly than if the compounds were reacting as acetals; (31) reacts about twenty times faster than (32).[144]

General-acid catalysis could not be detected in the hydrolysis of 4-methoxy-but-3-en-2-one; the isotope effect $k(D_2O)/k(H_2O)$ was 2.08, and the entropy of activation was -26.0 e.u. A mechanism involving reversible protonation of the carbonyl oxygen was proposed (equation 14).[145]

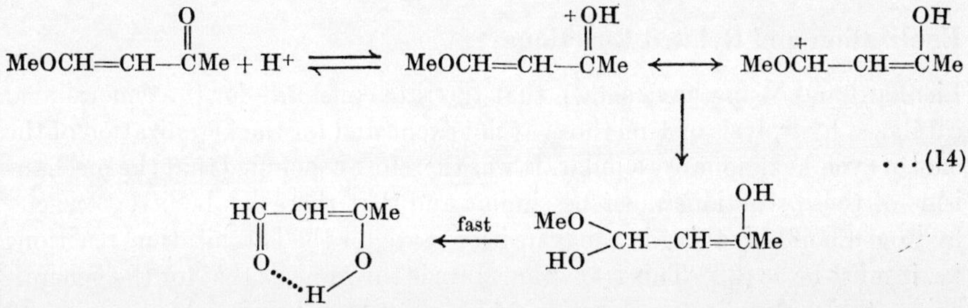

$$\cdots (14)$$

The temperature variation of the solvent isotope effect $k(EtOH)/k(EtOD)$ for the HCl-catalysed addition of ethanol to 2-chloroethyl vinyl ether has been interpreted in terms of concurrent catalysis by un-ionized HCl molecules and ion pairs.[146]

[143] A. Kankaanperä and P. Salomaa, *Acta Chem. Scand.*, **23**, 712 (1969).
[144] P. Salomaa and L. Hautonieme, *Acta Chem. Scand.*, **23**, 709 (1969); cf. *Org. Reaction Mech.*, **1968**, 351, and K. Pihlaja and J. Heikkilä, *Suomen Kemistilehti*, *B42*, 338 (1969).
[145] L. R. Fedor and J. McLaughlin, *J. Am. Chem. Soc.*, **91**, 3594 (1969).
[146] B. A. Trofimov, A. S. Atavin, and O. N. Vylegzhanin, *Izv. Akad. Nauk SSSR, Ser. Khim.*, **1968**, 927; *Chem. Abs.*, **69**, 43191 (1968); *Org. Reactivity* (Tartu), **5**, 450 (1968); *Chem. Abs.*, **70**, 28102 (1969).

The mercuric-ion catalysed hydrolyses of vinyl butyl ether and vinyl acetate have been studied.[147]

The mechanism of hydrolysis of α-acetoxy-4-nitrostyrene changes from A_{Ac}-2 at low acid concentrations to an A-S_E2 mechanism involving protonation of the double bond at high acidities ($> 55\%$ H_2SO_4). In 6% H_2SO_4 the solvent isotope effect, k_H/k_D, is 0.75 and the rate is similar to that for isopropyl acetate, but in 69% H_2SO_4 k_H/k_D is 3.25 and the reaction is faster than the hydrolysis of isopropyl acetate. The *p*-chloro-, *m*-chloro-, *p*-nitro-, and unsubstituted compounds react at similar rates in dilute sulphuric acid, which is consistent with their reacting via an A_{Ac}-2 mechanism, but the *p*-methoxy- and *p*-methyl-compounds react faster and presumably partly by an A-S_E2 mechanism under these conditions.[148] In contrast to these conclusions the acid-catalysed hydrolyses of several vinyl esters are reported to involve vinyl–oxygen fission. The entropies of activation are negative and the solvent isotope effect for the hydrolysis of vinyl formate is $k_H/k_D = 0.84$.[149] The latter result seems to be more in accord with an A_{Ac}-2 mechanism. It has also been reported that in acidic solutions the vinyl group of methyl vinyl esters of dicarboxylic acids is hydrolysed more rapidly than the methyl group.[150]

Other investigations on the hydrolysis of vinyl ethers are described in ref. 151.

Enolization and Related Reactions

Lienhard and Wang have shown that the rate constants for the general-acid catalysed hydrolysis of 1-methoxycyclohexene and for the ketonization of the enol of cyclohexanone are similar. It was therefore concluded that the mechanisms of these reactions must be similar and that there can be little oxygen–hydrogen bond breaking in the transition state for the ketonization reaction; i.e. it must be as (**33**). This transition state is the same as that for the general-acid catalysed enolization reaction which must therefore involve a rapid and reversible proton transfer followed by abstraction of the α-proton by the conjugate base of the catalysing acid (see equations 15 and 16).[152]

[147] I. P. Samchenko and A. F. Rekasheva, *Ukr. Khim. Zh.*, **34**, 450 (1968); *Chem. Abs.*, **69**, 85862 (1969).

[148] D. S. Noyce and R. M. Pollack, *J. Am. Chem. Soc.*, **91**, 119 (1969).

[149] L. F. Kulish and O. I. Korol, *Ukr. Khim. Zh.*, **34**, 495 (1968); *Chem. Abs.*, **69**, 85861 (1969).

[150] P. M. Zaitsev, G. R. Freidlin, and Z. V. Zaitseva, *Ukr. Khim. Zh.*, **35**, 380 (1969); *Chem. Abs.*, **71**, 48923 (1969).

[151] B. A. Trofimov, M. F. Shostakovskii, A. S. Atavin, B. V. Prokop'ev, V. I. Lavrov, and N. M. Denglazov, *Khim. Atsetilena*, **1968**, 259; B. A. Trofimov, A. S. Atavin, T. S. Emel'-yanov, B. V. Prokop'ev, and V. I. Lavrov, *Org. Reactivity* (Tartu), **4**, 778 (1967); *Chem. Abs.*, **69**, 43277 (1969); M. F. Shostakovskii, B. V. Prokop'ev, N. M. Denglazov, A. K. Filippova, and E. I. Dubinskaia, *Dokl. Akad. Nauk SSSR*, **187**, 835 (1969).

[152] G. E. Lienhard and T.-C. Wang, *J. Am. Chem. Soc.*, **91**, 1146 (1969).

$$\underset{\underset{\underset{B}{\overset{\delta-}{\vdots}}}{\overset{H}{\vdots}}}{\overset{\delta+}{>C\cdots\cdots C\cdots}}\overset{\delta+}{\text{OH}}$$

$$(33)$$

$$\underset{\underset{H}{\overset{|}{}}}{>C-C}\overset{O}{\diagdown} + HB \quad \rightleftharpoons \quad \underset{\underset{H}{\overset{|}{}}}{>C-C}\overset{+}{\diagup}\overset{\overset{+}{O}H}{} + B^- \qquad \dots (15)$$

$$\underset{\underset{H}{\overset{|}{}}}{>C-C}\overset{\overset{+}{O}H}{\diagup} + B^- \quad \longrightarrow \quad >C=C\overset{OH}{\diagdown} + BH \qquad \dots (16)$$

The ratio of the rate constants for ketonization by proton transfer from water and reaction with bromine of the enol of acetone has been determined. This ratio is 3.4 times faster in H_2O than in D_2O. On the assumption that the rate of reaction of the enol with bromine is the same in H_2O as in D_2O, the ketonization of acetone enol has an isotope effect $k(H_2O)/k(D_2O) = 3.4$. This is close to the isotope effect for the hydrolysis of ethyl isopropenyl ether (3.5) and is therefore additional evidence that the mechanism of ketonization is similar to the mechanism of hydrolysis of enol ethers.[153]

The relative rates of ketonization and reaction with bromine of enols in aqueous acetic acid have been measured. They vary from 4.30 for the enol of Bu^tCOMe to 0.26 for the enols of Bu^tCOR where R is primary alkyl. This was attributed to a decrease in the ketonization rate constant as a result of hyperconjugation.[154]

The rates of iodination of pyruvic acid, methylpyruvic acid, and dimethylpyruvic acid vary only slightly with acid concentration in the range 0.02—0.1M-HCl.[155] This is not the result of intramolecular catalysis by the carboxy group since pyruvic acid dimethylamide and diethylamide behave similarly.[156]

The DCl-catalysed deuterium exchange of several unsymmetrical ketones in D_2O has been studied (Table 2). It was again[157] found that a substituent in one branch affects the rate of exchange in both branches.[158]

The ρ-value for iodination of substituted acetophenones in 1.388M-HClO$_4$ is -1.785 at $30°$,[159] and for the pyridine-catalysed bromination in 75% acetic acid it is 0.75 at $30°$.[160]

[153] J. E. Dubois and J. Toullec, *Chem. Comm.*, **1969**, 478.
[154] J. E. Dubois and J. Toullec, *Chem. Comm.*, **1969**, 292, 478.
[155] A. Schellenberger, H. Lehmann, and G. Oehme, *Z. Chem.*, 8, 144 (1968).
[156] A. Schellenberger and G. Fisher, *Z. Chem.*, 8, 460 (1968).
[157] *Org. Reaction Mech.*, **1967**, 318; **1968**, 365.
[158] M. Chevallier, J. Jullien, and T. L. Nguyen, *Bull. Soc. Chim. France*, **1969**, 3332; see also J. Jullien and N. Thoi-Lai, *ibid.*, **1968**, 4669.
[159] S. Mishra, P. L. Nayak, and M. K. Rout, *J. Indian Chem. Soc.*, **46**, 645 (1969).
[160] D. N. Nanda, P. L. Nayak, and M. K. Pont, *Indian J. Chem.*, **7**, 469 (1969).

Table 2. The rates of deuterium exchange in
each branch of some unsymmetrical ketones in
DCl–D$_2$O at 41.8° [158]

$10^6 k$	Ketone	$10^6 k$
—	H–CH$_2$COMe	51.2
0.28	F–CH$_2$COMe	2.97
15	Cl–CH$_2$COMe	2.56
136.3	CH$_3$–CH$_2$COMe	42.9
60	CH$_3$–CH$_2$COEt	—

Other investigations on the enolization of ketones under acidic conditions are described in ref. 161.

The racemization of (+)-2-carboxybenzylindan-1-one (**34**) is proportional to the concentration with the carboxy group ionized. Intramolecular catalysis was proposed.[162] The rate of enolization of several amino-ketones, (**35**)—(**37**), depends on the concentration with the amino group unprotonated, which suggests that the mechanism involves intramolecular general-base catalysis. The pH–rate profiles for enolization of ketones (**38**) and (**39**) are not bell-shaped, which suggests that there is no concerted acid–base catalysis.[163]

The iodination of acetone in the presence of amines and diamines has been studied.[164, 165]

The deuterium isotope effect for iodination of CH$_3$CD$_2$COCD$_2$CH$_3$ is 4.0 and 6.8 for catalysis by pyridine and 2,6-lutidine respectively.[166] Proton tunnelling in the base-catalysed detritiation of 2-ethoxycarbonylcyclopentanone has been discussed.[167] There has been a review on proton tunnelling.[168]

More examples of slow α-hydrogen exchange by cyclopropyl ketones have been reported.[169, 170]

4-Homoadamantanone underwent a facile base-catalysed deuterium

[161] U. L. Haldna, H. Kuura, M. Tamme, and V. Palm, *Org. Reactivity* (Tartu), **5**, 1009 (1968); *Chem. Abs.*, **71**, 12244 (1969); A. Talvik and S. Hiidmaa, *ibid.*, pp. 309, 297; *Chem. Abs.*, **70**, 19401 (1969); E. S. Lewis, J. D. Allen, and E. T. Wallick, *J. Org. Chem.*, **34**, 255 (1969); M Gaudry and A. Marquet, *Compt. Rend.*, *C*, **268**, 1174 (1969).
[162] C. Rappe and H. Bergander, *Acta Chem. Scand.*, **23**, 214 (1969).
[163] J. K. Coward and T. C. Bruice, *J. Am. Chem. Soc.*, **91**, 5339 (1969).
[164] A. A. Yasnikov, E. A. Shilov, L. P. Koshechkina, and N. V. Volkova, *Ukr. Khim. Zh.*, **33**, 1315 (1967); *Chem. Abs.*, **69**, 35160 (1968).
[165] L. P. Koshechkina, E. A. Shilov, and A. A. Yasnikov, *Ukr. Khim. Zh.*, **35**, 55 (1969); *Chem. Abs.*, **70**, 76964 (1969).
[166] J. P. Calmon, M. Calmon, and V. Gold, *J. Chem. Soc.*(B), **1969**, 659.
[167] J. R. Jones, *Trans. Faraday Soc.*, **65**, 2430 (1969).
[168] E. F. Caldin, *Chem. Rev.*, **69**, 135 (1969).
[169] H. W. Amburn, K. C. Kauffman, and H. Schechter, *J. Am. Chem. Soc.*, **91**, 530 (1969); *Org. Reaction Mech.*, **1968**, 366.
[170] C. Agami and M. Audouin, *Compt. Rend.*, *C*, **268**, 1267 (1969).

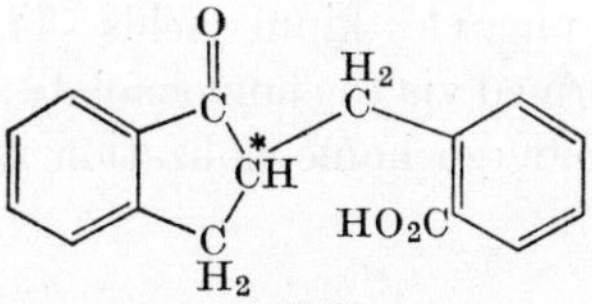

(34)

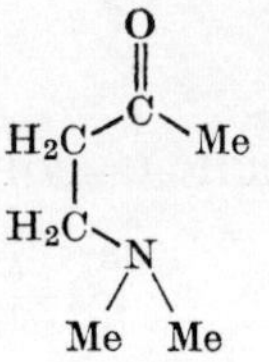

(35) 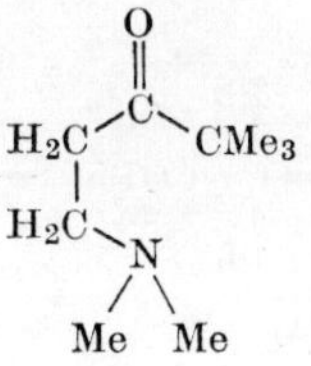 (36) 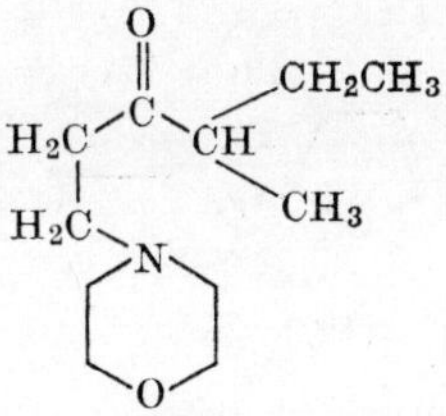 (37)

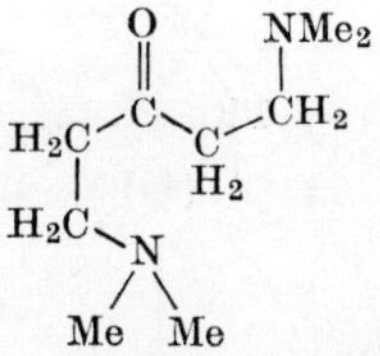

(38) 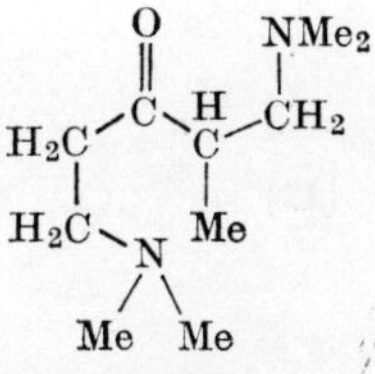 (39)

exchange at the α-methylene group but no exchange could be observed at the bridgehead position.[171]

Other examples of enolization are reported in ref. 172.

There have been several investigations of the alkylation of enolate anions,[173] and many measurements of the equilibrium constants for enolization reactions.[174]

[171] P. von R. Schleyer, E. Funke, S. H. Liggero, *J. Am. Chem. Soc.*, **91**, 3965 (1969).

[172] M. Gaudry and A. Marquet, *Bull. Soc. Chim. France*, **1969**, 4160; W. J. Albery and B. H. Robinson, *Trans. Faraday Soc.*, **65**, 980 (1969); P. Alcais and J. E. Dubois, *J. Chim. Phys.*, **65**, 1800 (1968); *Chem. Abs.*, **70**, 56906 (1969); P. Alcais, *ibid.*, **65**, 1794 (1968); *Chem. Abs.*, **70**, 67305 (1969).

[173] I. T. Harrison, E. Kimura, E. Bohme, and J. H. Fried, *Tetrahedron Letters*, **1969**, 1589; B. J. L. Huff, F. N. Tuller, and D. Caine, *J. Org. Chem.*, **34**, 3070 (1969); P. A. Tardella, *Tetrahedron Letters*, **1969**, 1117.

[174] M. Regnitz and H.-J. Geelhaar, *Ann. Chem.*, **728**, 108 (1969); H. Wamhoff, G. Hoffer, H. Lander, and F. Korte, *ibid.*, **722**, 12 (1968); Z. Bánkowska and I. Zadrozna, *Rocz. Chem.*, **42**, 1591 (1968); D. J. Sardella, D. H. Heinert, and B. L. Shapiro, *J. Org. Chem.*, **34**, 2817 (1969); D. P. Venter and J. Dekker, *ibid.*, p. 2224; P. Courtot, J. Le Saint, and N. Platzer, *Compt. Rend.*, *C*, **267**, 1332 (1968); J. P. Calmon, *ibid.*, **268**, 1435 (1969); W. Hansel and R. Haller, *Arch. Pharm.*, **302**, 147 (1969); K. K. Babievskii, V. M. Belikov, P. V. Petrovskii, and E. I. Fedin, *Dokl. Akad. Nauk SSSR*, **186**, 1079 (1969); O. Bohman and S. Allenmark, *Acta Chem. Scand.*, **22**, 2716 (1968); W. Rubaszenstia and Z. R. Grabowski, *Tetrahedron*, **25**, 2807 (1969); D. D. Mahajan and R. K. Chaturvedi, *Z. Phys. Chem.* (Leipzig), **241**, 33 (1969).

Treatment of (40) with phenyl-lithium yields (44) as well as (41). The former was thought to be formed via the homoenolate anions (42) and (43).[175] Adamantanone does not undergo homoenolization under conditions where camphenilone does.[176]

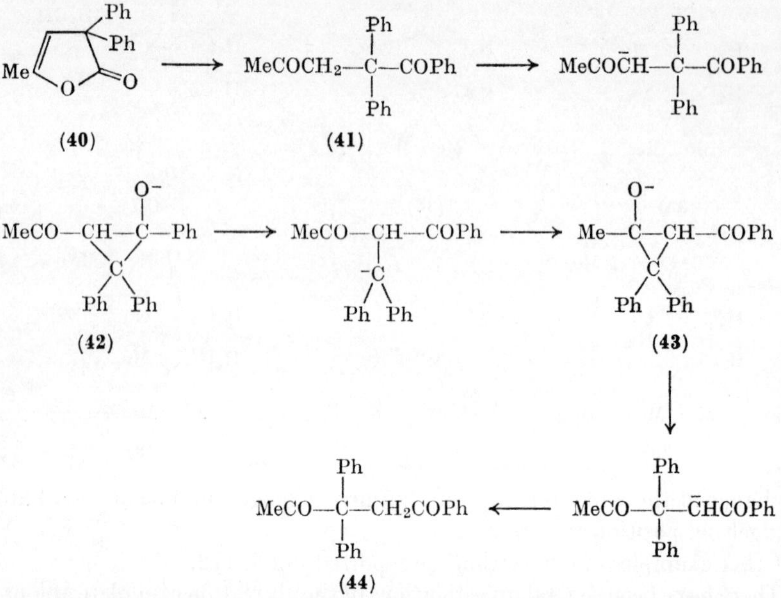

$$\text{(40)} \quad\longrightarrow\quad \text{MeCOCH}_2\text{—C—COPh} \quad\longrightarrow\quad \text{MeCO}\bar{\text{C}}\text{H—C—COPh}$$

(40)   (41)

(42) $\longrightarrow$ (42) $\longrightarrow$ (43)

(44)

Aldol Reaction

The aldol reaction of cyclopentanone and isobutyraldehyde catalysed by lithium or potassium hydroxide in cyclopentanone itself as solvent yields more than 95% of the *threo*-isomer. When tetramethylammonium hydroxide is the catalyst only 30% of the *threo*-isomer is formed, and when methanol is the solvent only 30% of the *threo*-isomer is formed with potassium, lithium, or tetramethylammonium hydroxide as catalyst. It was suggested that the undissociated enolate reacts as shown in (45) and (46), and that the former is strongly preferred.[177]

[175] P. Yates, G. D. Abrams, and S. Goldstein, *J. Am. Chem. Soc.*, **91**, 6868 (1969).
[176] J. E. Nordlander, S. P. Jindal, and D. J. Kitko, *Chem. Comm.*, **1969**, 1136.
[177] J. E. Dubois and M. Dubois, *Chem. Comm.*, **1968**, 1567; *Bull. Soc. Chim. France*, **1969**, 3120, 3126.

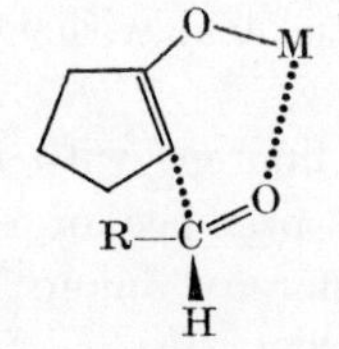

(46) 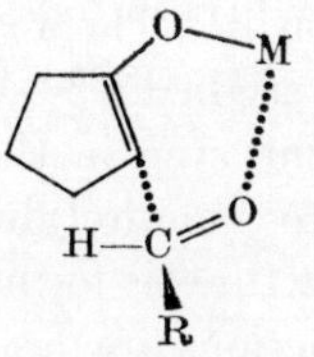   (45)

Equilibration of the ketols (47) and (49) occurs via ion (48), not via a retro-aldol reaction and re-aldolization, since in the presence of cyclohexanone no ketol of cyclohexanone and isobutyraldehyde is formed. It was shown by independent experiment that cyclohexanone reacts more than ten times faster than cyclopentanone with isobutyraldehyde under these conditions.[178]

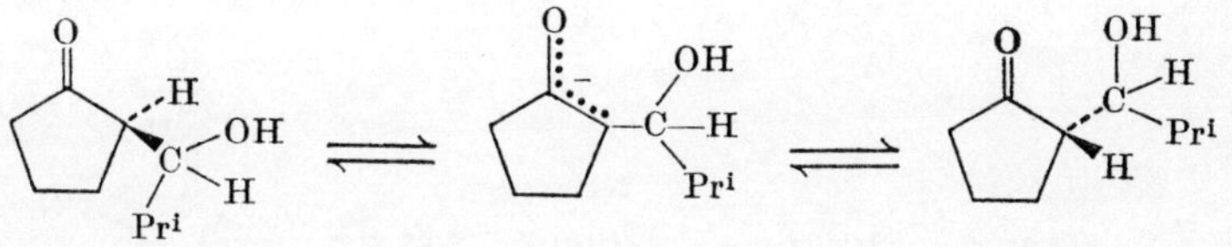

(47)   (48)   (49)

The reaction of butan-2-one and formaldehyde in chloroform and nitro-methane in the presence of boron trifluoride to yield 5-acetyl-5-methyl-1,3-dioxan (51) shows an isotope effect $k_H/k_D = 2$—3 when pentadeuterobutan-2-one is used. It was suggested that the rate-limiting step is enolization, and, consistent with this, it was shown that (50) yields (51) rapidly under the

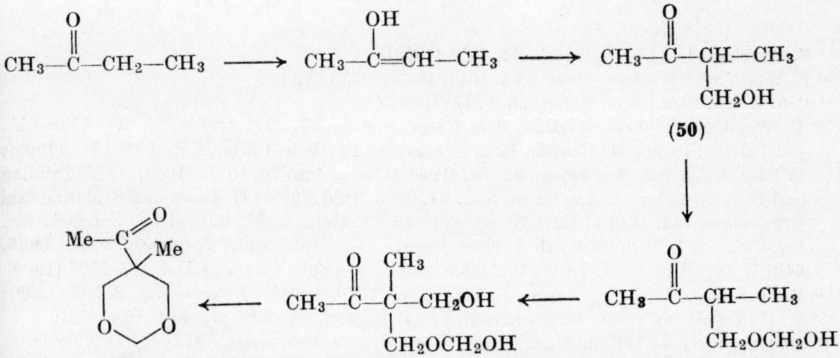

(50)

(51)

[178] J. E. Dubois and M. Dubois, *Bull. Soc. Chim. France*, **1969**, 3553.

reaction conditions.[179] The reaction in acetic acid with sulphuric acid as catalyst was also studied.[180]

The rate-limiting step in the reaction of fluorene with formaldehyde to yield 9,9-di(hydroxymethyl)fluorene in the presence of ethoxide ions in DMSO–EtOH mixtures is formation of the fluorenyl anion.[181]

Other aldol reactions are described in ref. 182*a*.

There have been several investigations of the mechanism of action of aldolase.[182*b*]

Reactions of Enamines[183]

The hydrolysis of enamines (**52**) and (**53**) follows a rate law of the form of equation (17), where K_a is the ionization constant for non-productive *N*-

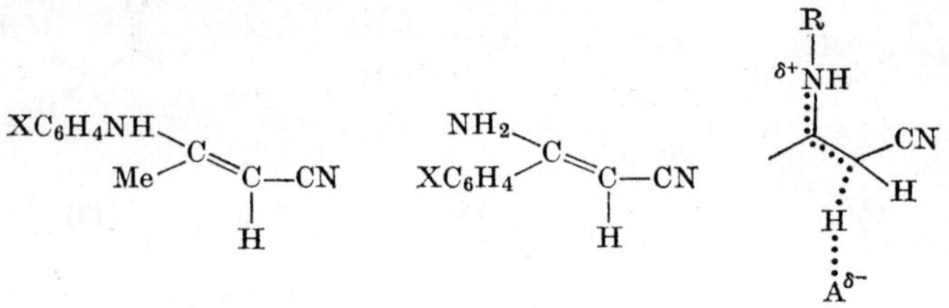

(**52**)   (**53**)   (**54**)

protonation and k_r is the rate constant for the hydrogen-ion catalysed hydrolysis of the unprotonated enamine (see equations 18 and 19). The ρ-values for the variation of k_r and K_a with structure are -0.612 and $+2.80$ for compounds (**52**) and -0.489 and $+0.84$ with compounds (**53**). It also appears that the rate

[179] B. Wesslén, *Acta Chem. Scand.*, **22**, 2085 (1968).

[180] B. Wesslén, *Acta Chem. Scand.*, **23**, 1017, 1023, 1033 (1969).

[181] B. Wesslén, *Acta Chem. Scand*, **23**, 1247 (1968).

[182*a*] B. Wesslén, and L-O. Ryrfors, *Acta Chem. Scand.*, **22**, 2071 (1968); B. Wesslén, *ibid.*, p. 2993; W. Broser, J. Reusch, H. Kurreck, and P. Siegle, *Chem. Ber.*, **102**, 1715 (1969); K. Sasaki, *J. Chem. Soc. Japan*, **89**, 797 (1968); *Chem. Abs.*, **70**, 10778 (1969); D. E. Tallman and D. L. Leussing, *J. Am. Chem. Soc.*, **91**, 6253, 6256 (1969); H. Hoser and S. Malinowski *Rocz. Chem.*, **43**, 1211 (1969); W. Kiewlicz and S. Malinowski, *Bull. Acad. Pol. Sci.*, *Ser. Sci. Chim.*, **17**, 259 (1969); B. J. Kurtev and C. G. Kratchanov, *J. Chem. Soc.*(B), **1969**, 649; E. Lee-Ruff, N. J. Turro, P. Amice, and J. M. Conia, *Can. J. Chem.*, **47**, 2797 (1969).

[182*b*] R. D. Kobes, R. T. Simpson, B. L. Vallee, and W. J. Rutter, *Biochemistry*, **8**, 585 (1969); D. R. Trentham, C. H. McMurray, and C. I. Pogson, *Biochem. J.*, **114**, 19 (1969); J. F. Biellmann, E. L. O'Connell, and I. A. Rose, *J. Am. Chem. Soc.*, **91**, 6484 (1969); D. P. Hanlon and E. W. Westhead, *Biochemistry*, **8**, 4247, 4255 (1969); T. H. Gawronski and E. W. Westhead, *ibid.*, p. 4261; K. Brand, O. Tsolas, and B. L. Horecker, *Arch. Biochem. Biophys.*, **130**, 521 (1969); J. F. Riordan and P. Christen, *Biochemistry*, **8**, 2381 (1969).

[83] See *Org. Reaction Mech.*, **1965**, 254; **1966**, 332; **1967**, 324—326; **1968**, 367.

is very sensitive to the substituents directly attached to the double bond. The reaction is general-acid catalysed and a rate-determining proton transfer

$$k_{\text{obs}} = k_r a_{\text{H}}(K_a/K_a + a_{\text{H}}) \tag{17}$$

$$\text{E} + \text{H}^+ \underset{}{\overset{K_a}{\rightleftharpoons}} \text{EH}^+ \tag{18}$$

$$\text{E} + \text{H}^+ \overset{k_r}{\longrightarrow} \text{Product} \tag{19}$$

to form the rapidly hydrolysed protonated Schiff base was proposed with the transition state as (**54**). At low pH there appears to be a change in rate-determining step to hydrolysis of the protonated Schiff base.[184]

The reactions of enamines with Schiff bases[185] and the alkylation of enamines[186] have also been studied.

Other Reactions

The equilibrium constant for the methoxide-catalysed addition of methanol to cycloalk-2-enones varies with ring size as follows: 5 (0), 6 (1.18), 7 (100), 8 (100), 9 (21.7). The high values for the cycloheptenone and cyclooctenone were attributed to poor conjugation between the olefinic and carbonyl double bonds. The rates of addition did not vary greatly with ring size and were in the reverse order to the equilibrium constants. This suggests that conjugation is more important in the transition state than in the initial state. Two mechanisms were considered. The first, equation (20), involves formation of the enolate anion and is the reverse of an *E1cB* mechanism, and the second, equation (21), is concerted and is the reverse of an *E2* elimination.[187] At present it is not possible to decide which is correct.

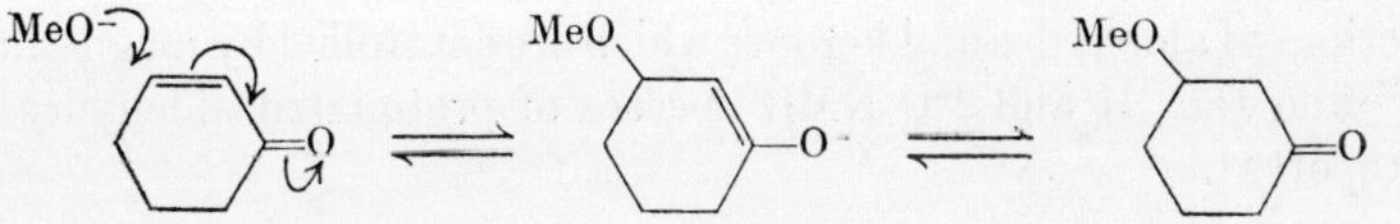

$$\dots (20)$$

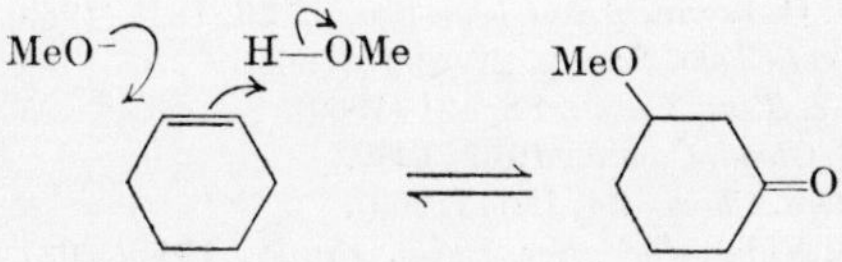

$$\dots (21)$$

[184] J. K. Coward and T. C. Bruice, *J. Am. Chem. Soc.*, **91**, 5329 (1969); see also P. Y. Sollenberger and R. B. Martin, Abstracts of Papers, ORGN 66, 156th Am. Chem. Soc. National Meeting, Atlantic City, September, 1968.

[185] S. Tomoda, Y. Takeuchi, and Y. Nomura, *Tetrahedron Letters*, **1969**, 3549.

[186] N. F. Firrell and P. W. Hickmott, *Chem. Comm.*, **1969**, 544.

[187] P. Chamberlain and G. H. Whitham, *J. Chem. Soc.*(B), **1969**, 1131.

The cleavage of β-keto-sulphides by HBr in F_3CCO_2H takes place with non-enolizable ketones and so the first step cannot be enolization. It was suggested that the reaction involved protonation of the ketone oxygen and nucleophilic attack by Br$^-$ on sulphur.[188] The reaction of ketones with 2,4-dinitrobenzenesulphenyl chloride has been studied.[189, 190]

The rate of hydrolysis of cyclohexanone bisulphite compound is 1.10 times as fast as that of the analogous α-tetradeutero-compound. The corresponding isotope effects for hydrolysis of the bisulphite compounds of 4-t-butylcyclo-hexanone are 1.05 (bisulphite equatorial) and 1.10 (bisulphite axial).[191]

exo-6-Hydroxybicyclo[3.3.1]nonan-2-one reacts with sodium in D_2O–dioxan to incorporate up to six atoms of deuterium. No deuterium was incorporated at position 6, which excludes the occurrence of homoenolization and a redox process. A transannular hydride shift (equation 22) provides the most likely explanation.[192]

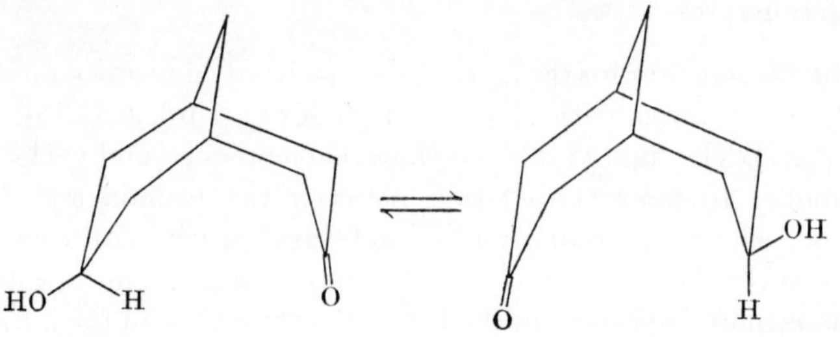

$$\cdots (22)$$

The relative rates of reduction with sodium borohydride in propan-2-ol of acetophenone, cyclopropyl phenyl ketone, cyclobutyl phenyl ketone, cyclo-pentyl phenyl ketone, and cyclohexyl phenyl ketone are respectively 1, 0.1, 0.23, 0.36, and 0.25.[193] Other investigations of the reduction of ketones are reported in ref. 194.

More reactions of aldehydes and ketones which are controlled by magnesium chelation[195] and the ^1H and ^{13}C NMR spectra of protonated aldehydes[196] have been reported.

[188] C. Rappe and R. Gustafsson, *Acta Chem. Scand.*, **22**, 2915 (1968).
[189] R. Gustafsson, C. Rappe, and J. O. Levin, *Acta Chem. Scand.*, **23**, 1843 (1969).
[190] C. Rappe and R. Gustafsson, *Acta Chem. Scand.*, **22**, 2927 (1968).
[191] G. Lamaty and J. P. Roque, *Rec. Trav. Chim.*, **88**, 131 (1969).
[192] W. Parker and J. R. Stevenson, *Chem. Comm.*, **1969**, 1289.
[193] S. F. Sun and P. R. Neidig, *J. Org. Chem.*, **34**, 1854 (1969).
[194] P. Geneste, G. Lamaty, and B. Vidal, *Bull. Soc. Chim. France*, **1969**, 2027; P. Geneste, G. Lamaty and B. Vidal, *Compt. Rend., C*, **266**, 1387 (1968); P. Geneste, G. Lamaty, and L. Moreau, *ibid.*, p. 1012; W. L. Dilling, C. E. Reineke, and R. E. Plepys, *J. Org. Chem.*, **34** 2605 (1969); A. Colombeau and A. Rassat, *Chem. Comm.*, **1968**, 1587; E. J. Denney and B. Mooney, *J. Chem. Soc.*(B), **1969**, 1410.
[195] T. M. Harris, M. P. Wachter, and G. A. Wiseman, *Chem. Comm.*, **1969**, 177; cf. *Org. Reaction Mech.*, **1966**, 332.
[196] A. M. White and G. A. Olah, *J. Am. Chem. Soc.*, **91**, 2943 (1969).

The following reactions have also been studied: addition of Grignard reagents to 2-t-butylcyclohexanone,[197] benzophenone,[198] camphor,[199] and sterically hindered ketones;[200] the Reformatsky reaction;[201] the Wittig reaction;[202] Haller–Bauer cleavage of ketones;[203] acid-catalysed cleavage of tertiary alkyl ketones;[204] H_2SO_4- and BF_3-catalysed reaction of aromatic aldehydes with aryl acetonitriles to yield aryl bis-arylacetamides;[205] decomposition of Mannich bases;[206] Tischenko reaction of benzaldehyde with aluminium t-butoxide to yield benzyl benzoate;[207] thiamine-catalysed condensation of furfural to furoin;[208] and the reactions between formaldehyde and phenols,[209] *cis-* and *trans-*4-methoxy-3-t-butylcyclohexanone and acetylene,[210] 2,4,6-trinitrotoluene and methylene bis-piperidine,[211] picoline and benzaldehyde,[212] 2,3-dimethylquinoxaline and benzaldehyde,[213] and 7-ketonorbornane and dimethyloxosulphonium methylid.[214]

[197] J. C. Richer and D. Eugène, *Can. J. Chem.*, **47**, 2387 (1969).
[198] T. Holm, *Acta Chem. Scand.*, **23**, 579 (1969).
[199] K. Suga, S. Watanabe, and Y. Yamaguchi, *Austral. J. Chem.*, **22**, 669 (1969).
[200] R. A. Benkeser, W. G. Young, W. E. Broxterman, D. A. Jones, and S. J. Piaseczynski, *J. Am. Chem. Soc.*, **91**, 132 (1969).
[201] J.-L. Luche and H. B. Kagan, *Bull. Soc. Chim. France*, **1969**, 1680.
[202] W. P. Schneider, *Chem. Comm.*, **1969**, 785.
[203] C. L. Bumgardner and K. G. McDaniel, *J. Am. Chem. Soc.*, **91**, 6821 (1969).
[204] P. Bauer and J. E. Dubois, *Chem. Comm.*, **1969**, 229.
[205] Z. Csurös, G. Deak, M. Haraszthy-Papp, and I. Hoffmann, *Acta Chim. Acad. Sci. Hung.*, **59**, 119 (1969).
[206] M. M. Dolgaya, Y. N. Belokon, and V. M. Belikov, *Izv. Akad. Nauk SSSR, Ser. Khim.*, **1969**, 74; *Chem. Abs.*, **70**, 114397 (1969).
[207] Y. Ogata and A. Kawasaki, *Tetrahedron*, **25**, 929 (1969).
[208] R. Hüttenrauch and U. Olthoff, *Pharmazie*, **24**, 238 (1969).
[209] M. F. Sorokin and O. I. Kulikovskii, *Tr. Mosk. Khim. Tekhnol. Inst.*, **1968**, No. 57, 88; *Chem. Abs.*, **70**, 95953 (1969).
[210] A. A. Akhren, A. V. Kamernitskii, and A. M. Prokhoda, *Khim. Atsetilena*, **1968**, 76; *Chem. Abs.*, **71**, 2716 (1969).
[211] J. E. Fernandez, J. D. Mones, M. L. Schwartz, and R. E. Wulff, *J. Chem. Soc.*(B), **1969**, 506.
[212] J. Beneš, J. Kaválek, J. Bartošek, and J. Churáček, *Coll. Czech. Chem. Comm.*, **34**, 819 (1969).
[213] J. Kaválek, *Coll. Czech. Chem. Comm.*, **34**, 1819 (1969).
[214] R. K. Bly and R. S. Bly, *J. Org. Chem.*, **34**, 304 (1969); cf. *Org. Reaction Mech.*, **1968**, 369—371.

CHAPTER 12

Reactions of Acids and their Derivatives

B. Capon

Chemistry Department, Glasgow University

Carboxylic Acids [1,2]

Tetrahedral Intermediates

Barnett and Jencks have shown that the rate-limiting step in the reaction of a carboxylic acid derivative can be proton transfer to a tetrahedral intermediate. The reaction studied was the *S*- to *N*-acetyl transfer of *S*-acetyl-mercaptoethylamine which was formulated as shown in Scheme 1. Above pH 2.3 and at low buffer concentrations the predominant rate-determining step is general-base catalysed (rate constant defined in terms of protonated ester SH^+). At lower pH and higher buffer concentrations the reaction is not general-base catalysed and there is a decrease in rate with increasing acidity. The predominant rate-limiting step at high acidities is nucleophilic attack of the unprotonated amino group on the unprotonated thiolester group. The rate of this step should decrease with increasing acidity as the concentration of species S decreases. At lower acidities the predominant rate-limiting step

[1] W. P. Jencks, *Catalysis in Chemistry and Enzymology*, McGraw-Hill, New York, 1969.
[2] S. Patai (Ed.), *The Chemistry of Carboxylic Acids and Esters*, Wiley, London, 1969.

becomes protonation of the zwitterionic tetrahedral intermediate, $I^\pm$, which is general-base catalysed with $\beta = 0.97$ when the rate constant is expressed in terms of the concentration of protonated aminothiolester, SH^+. This is equivalent to general-acid catalysis of the reaction of the free amine with $\alpha = 0.03$. Under these conditions $I^\pm$ is formed rapidly and reverts to starting aminothiolester, S, faster than it is protonated. The latter step should be diffusion-controlled and hence exhibit general acid catalysis with an α-value of zero, as found. Scheme 1 also explains the variation with pH of the products

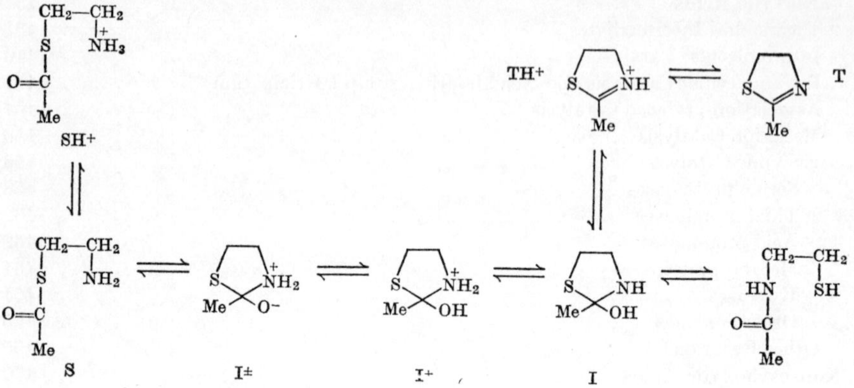

Scheme 1

of hydrolysis of thiazoline T. The rate of hydrolysis of this compound follows the ionization curve of the protonated form down to about pH 2 but at higher acidities there is a change in rate-limiting step with mid-point at about pH 0.85. At low acidities the rate-limiting step is attack of water on protonated thiazoline but this changes to breakdown of the tetrahedral intermediate at high acidities. Under conditions where the rate-limiting step remains attack of water on protonated thiazoline, the proportion of thiolester changes from 0.37 at pH 2 to 0.01 at pH 4. This is explained if there is an additional barrier to formation of thiolester which according to Scheme 1 is deprotonation of the tetrahedral intermediate I^+; i.e. the reverse of the step proposed to be rate-limiting in the S- to N-acetyl migration. As required by this scheme the mid-point in the change in rate-limiting step in the latter occurs at the same pH (2.30) as that (2.37) at which the yield of thiolester in the hydrolysis of thiazoline is reduced to half its maximum value.

   The independence of k_{cat} on K_a for general-acid catalysis of the S- to N-migration was not found with acids of pK_a greater than ca. 7. These acids were less effective catalysts than those with pK_a less than 7. This behaviour would be expected if the pK_a of $I^\pm$ were about 7, which is reasonable since proton transfer from a weak acid to the conjugate base of a stronger acid should no longer be diffusion-controlled. When the pK_a of I$^\pm$ is changed by working with the chloroacetyl compound this break occurred at pK_a about

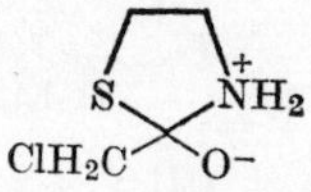

(1)

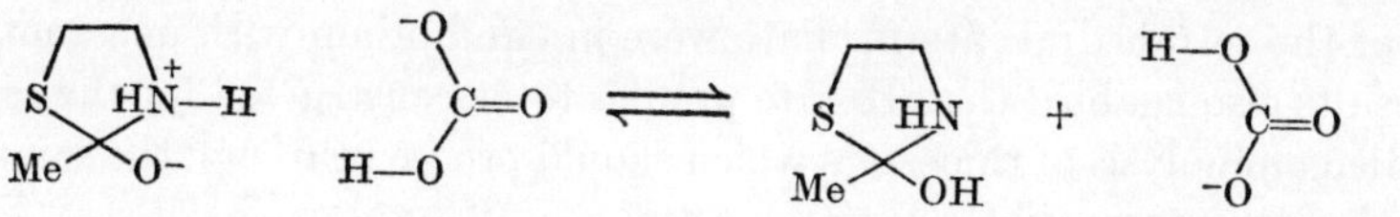

... (1)

5.9, as would be expected since it was estimated that the pK_a of (1) is about 1.5 units less than that of I$^\pm$. Bicarbonate and water are more effective catalysts than expected from their pK_a values and the Brønsted plot for catalysis by substituted ammonium ions, and it was suggested that they act as bifunctional catalysts (e.g. equation 1).[3]

Chaturvedi and Schmir have extended their investigation of tetrahedral intermediates generated in the hydrolysis of imidate esters to compounds (2)

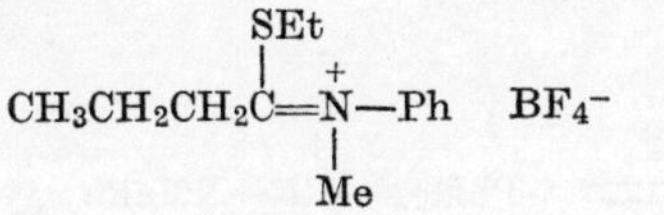

(2) 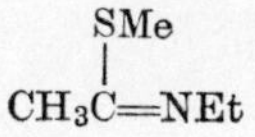   (3)

and (3). The pH–rate profiles for the hydrolysis of (2) and (3) follow equations (2) and (3) respectively. The products also depend on pH, with thiol and amide

$$k_{obs} = k_5 + k_7[^-OH] \qquad \qquad \ldots(2)$$

$$k_{obs} = [H^+](k_5 + k_7[^-OH])/([H^+] + K_a) \qquad \ldots(3)$$

predominating at high pH and amine and thiol ester at low pH. The variation in the percentage of amine with pH for both reactions can be explained in terms of a hypothetical di-acidic tetrahedral intermediate existing in three forms, TH$^+$, TH, and T$^-$, which yield 97, 14.2, and 0.6% respectively with (2) and 84, 6.5, and 0.1% amine with (3). Two reaction schemes were considered in which the forms were either in equilibrium with one another or in which one step in their interconversion was rate-limiting. In the light of the work of Barnett and Jencks [3] this was thought most likely to be step k_2''' of equation (4). The variation of the percentage amine with buffer concentration enabled relative values of k''' for catalysis by a series of bases to be obtained. These

[3] R. E. Barnett and W. P. Jencks, *J. Am. Chem. Soc.*, **91**, 2359 (1969).

yielded an α-value of 0.94 which was interpreted as indicating that this step is diffusion-controlled. The fact that the reaction scheme with step k''' rate-limiting enabled the effect of buffer concentration to be interpreted in this

$$
\begin{array}{ccccc}
& \text{SMe} & & \text{SMe} & & \text{SMe} \\
& | & \text{fast} & | \quad + & k_2''' & | \quad + \\
\text{Me}-\text{C}-\text{NHEt} & \rightleftharpoons & \text{Me}-\text{C}-\text{NH}_2\text{Et} & \rightleftharpoons & \text{Me}-\text{C}-\text{NH}_2\text{Et} & \quad \dots (4) \\
& | & & | & & | \\
& \text{OH} & & \text{OH} & & \text{O}^-
\end{array}
$$

way led to its being favoured over the alternative scheme in which the various forms of the tetrahedral intermediate were in equilibrium with one another. The results also enabled the pH–rate profiles to be constructed for the as yet unstudied aminolysis of thioesters which should proceed through the same set of tetrahedral intermediates.[4]

More work on the hydrolysis of acetimidate ester is reported in ref. 5.

The base-catalysed hydrolysis of a series of N,N'-dimethyl-N,N'-diphenyl-amidinium salts (4) has been studied. The rate increases as the group R is changed along the series Ph < Me < H, and substituted phenyl compounds yield a ρ-value of $+1.57$. The reaction is general-base catalysed, and a mechanism involving reversible formation of a tetrahedral intermediate which undergoes a rate-limiting breakdown to products was proposed (see equations 5—7). The positive ρ-value presumably results from the dominating effect of substituents on K_1.[6]

$$
\text{PhMeN}\!\cdots\!\overset{+}{\text{C}}\text{R}\cdots\text{NMePh} + {}^-\text{OH} \quad \overset{K_1}{\rightleftharpoons} \quad \text{PhMeN}-\overset{\overset{\displaystyle \text{OH}}{|}}{\text{C}}\text{R}-\text{NMePh} \qquad \dots (5)
$$

(4)

$$
\text{PhMeN}-\overset{\overset{\displaystyle \text{OH}}{|}}{\text{C}}\text{R}-\text{NMePh} + \text{H}_2\text{O} \quad \overset{K_2}{\rightleftharpoons} \quad \text{PhMeN}-\overset{\overset{\displaystyle \text{OH}}{|}}{\text{C}}\text{R}-\overset{+}{\underset{\text{H}}{\text{N}}}\text{MePh} + {}^-\text{OH} \qquad \dots (6)
$$

$$
\text{PhMeN}-\overset{\overset{\displaystyle \text{OH}}{|}}{\text{C}}\text{R}-\overset{+}{\underset{\text{H}}{\text{N}}}\text{MeR} \quad \overset{k_2}{\underset{\text{slow}}{\longrightarrow}}
$$

$+ \text{B}$

$$
\left[\begin{array}{c} \text{PhMeN}-\text{CR}\overset{\delta+}{\cdots}\text{NHMePh} \\ \overset{\delta-}{|} \\ \text{O}\cdots \\ \quad\quad \text{H}\cdots\overset{\delta+}{\text{B}} \end{array} \right] \longrightarrow \quad \begin{array}{c} \text{PhNMeCR} \\ \| \\ \text{O} \\ + \text{PhNHMe} + \text{BH}^+ \end{array} \qquad \dots (7)
$$

[4] R. K. Chaturvedi and G. L. Schmir, *J. Am. Chem. Soc.*, **91**, 737 (1969); cf. *Org. Reaction Mech.*, **1968**, 375.

[5] T. C. Pletcher, S. Koehler, and E. H. Cordes, *J. Am. Chem. Soc.*, **90**, 7072 (1968); cf. *Org. Reaction Mech.*, **1967**, 333—334; **1968**, 373—377.

[6] R. H. DeWolfe and M. W.-L. Cheng, *J. Org. Chem.*, **34**, 2595 (1969).

The pH–rate profile for hydrolysis of ethyl S-trifluoroacetylmercaptoacetate (5) is similar to that for hydrolysis of S-ethyl trifluorothioacetate,[7] except that the rate does not start to decrease with increasing acidity until higher acidities, i.e. with $-\log[H^+] < 0.5$. This is the behaviour expected if breakdown of the tetrahedral intermediate (6) to yield reactants is acid-catalysed and to yield

$$
\begin{array}{ccccc}
\overset{\displaystyle O}{\underset{\displaystyle \|}{}}\!\!\text{CF}_3\text{—C—SR} & \underset{k_{-1}(\text{H}^+)}{\overset{k_1(\text{H}_2\text{O})}{\rightleftharpoons}} & \text{CF}_3\!\!\underset{\displaystyle \text{OH}}{\overset{\displaystyle O^-}{\underset{\displaystyle |}{\overset{\displaystyle |}{\text{—C—SR}}}}} & \xrightarrow{k_2} & \text{CF}_3\text{COO}^- + \text{HSR}
\end{array}
$$

(5)   **(6)**

(R = CH$_2$CO$_2$Et)

products is not. The mercaptoacetate ion is a better leaving group than the ethylmercapto ion and hence a higher concentration of acid has to be reached before the acid-catalysed loss of $^-$OH from the tetrahedral intermediate becomes competitive. No decrease in rate was found in the hydrolysis of phenyl thioformate at high acidities. Here the phenylthio group is an even better leaving group, and hence a very high acid concentration would be needed before OH loss became competitive with phenylthio loss. The rate constant, k_2, for loss of mercaptoacetate ion from the anionic form of the tetrahedral intermediate (6) was estimated to be 4×10^9 sec^{-1}.[8]

The pH–rate profile for cyanolysis of S-ethyl thioacetate is sigmoid with pK_a(app) (8.57 at 25°) lower than the pK_a of HCN (9.27). A mechanism involving reversible formation of a tetrahedral intermediate and acid-catalysed expulsion of the ethylthio group was proposed (see equation 8).[9]

$$
\begin{array}{ccc}
\overset{\displaystyle O}{\underset{\displaystyle \|}{}}\!\!\text{MeCSEt} + {}^-\text{CN} & \rightleftharpoons & \text{MeC}\underset{\displaystyle \text{CN}}{\overset{\displaystyle O^-}{\underset{\displaystyle |}{\overset{\displaystyle |}{\text{SEt}}}}} \xrightarrow{\quad} \text{Products}
\end{array}
$$

$$\Big\updownarrow \text{H}^+$$

$$
\left[
\text{MeC}\underset{\displaystyle \overset{+}{\text{CN}}}{\overset{\displaystyle O^-}{\underset{\displaystyle |}{\overset{\displaystyle |}{\text{—SEt}}}}}\!\!\!{}^{\text{H}} \rightleftharpoons \text{MeC}\underset{\displaystyle \text{CN}}{\overset{\displaystyle \text{OH}}{\underset{\displaystyle |}{\overset{\displaystyle |}{\text{SEt}}}}}
\right] \xrightarrow{\quad} \text{Products} \quad \dots (8)
$$

The variation with pH of the rate of the bicarbonate-catalysed hydrolysis of the trimethylammonioacetanilide ion has been interpreted as resulting from the intervention of a complex between the tetrahedral intermediate and

[7] See *Org. Reaction Mech.*, **1965**, 260; **1967**, 330.
[8] R. Barnett and W. P. Jencks, *J. Org. Chem.*, **34**, 2777 (1969).
[9] T. Maugh and T. C. Bruice, *Chem. Comm.*, **1969**, 1056; cf. *Org. Reaction Mech.*, **1968**, 381.

the bicarbonate ion.[10a] An extensive investigation of general acid–base catalysis of the hydrolysis of N-(p-nitrophenyl)dichloroacetamide has been reported.[10b]

The alkaline hydrolysis of 3-methyldihydrouracil (7) follows a rate law of the form:

$$k_{obs} = [^-OH]k_a(k_1 + k_2[^-OH])/(k_a + k_1 + k_2[^-OH])$$

which was interpreted in terms of the reaction scheme of equation (9).[11]

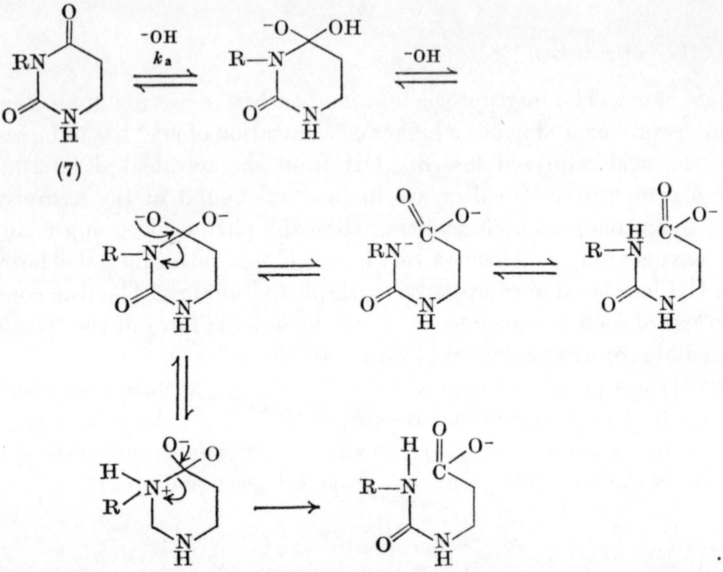

$$\cdots (9)$$

The hydrolysis of substituted barbituric acids has been studied.[12]

The lactonization of acids (8) shows general-acid and general-base catalysis. It was considered that reaction involved reversible formation of a tetrahedral intermediate followed by rate-limiting loss of ^-OH with transition states (9) and (10) for catalysis by HA and A^- respectively. The ρ-values were -0.55 for catalysis by water, -0.80 and -0.75 for catalysis by AcO^- and HCO_2^-, -1.50 for catalysis by AcOH and HCO_2H, and -1.68 for catalysis by H_3O^+.[13]

10a S. O. Eriksson, U. Meresaar, and U. Wahlberg, *Acta Chem. Scand.*, **22**, 2773 (1968); see also S. O. Eriksson and M. Källrot, *Acta Pharm. Suec.*, **6**, 63 (1969); *Chem. Abs.*, **70**, 105596 (1969).

10b R. F. Pratt and J. M. Lawlor, *J. Chem. Soc.*(B), **1969**, 230.

11 E. G. Sander, *J. Am. Chem. Soc.*, **91**, 3629 (1969).

12 S. O. Eriksson and C. G. Regardh, *Acta Pharm. Suec.*, **5**, 457 (1968); S. O. Eriksson, *ibid.*, **6**, 321 (1969).

13 S. Milstien and L. A. Cohen, *J. Am. Chem. Soc.*, **91**, 4585 (1969).

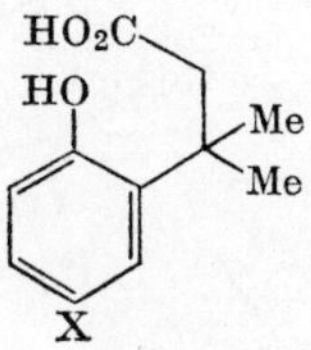

HO2C
HO
Me
Me
X

(8)

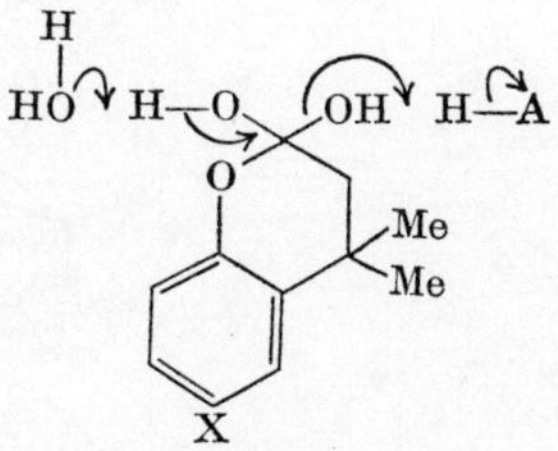

H
HO
H—O
OH
H—A
O
Me
Me
X

(9)

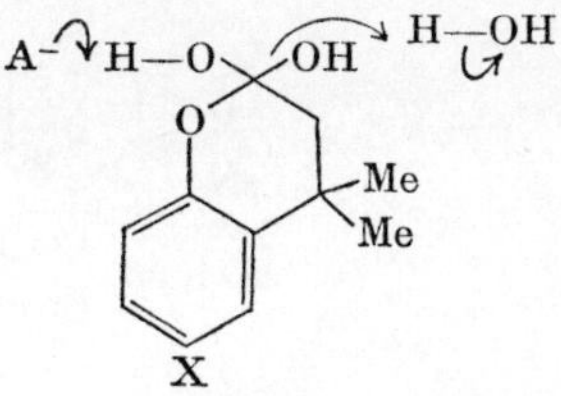

A⁻
H—O
OH
H—OH
O
Me
Me
X

(10)

On treatment with alkali, α-nitroisobutyramide yields the 2-nitropropane anion, cyanate ion, and ammonia. Presumably the tetrahedral intermediate expels the carbanion and ammonia at competitive rates as shown in equation (10).[14]

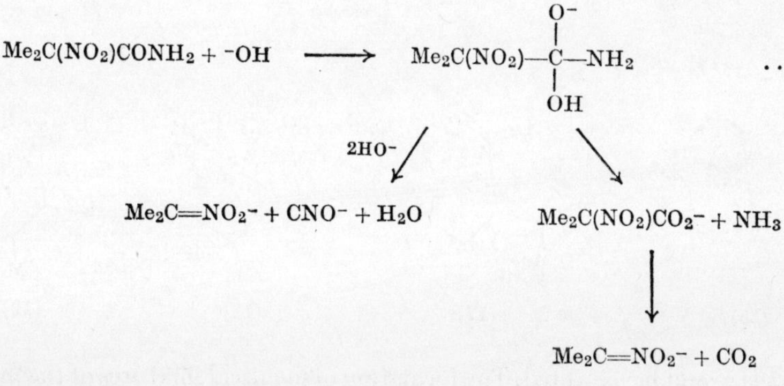

Me2C(NO2)CONH2 + ⁻OH
O⁻
Me2C(NO2)—C—NH2
OH
2HO⁻
Me2C=NO2⁻ + CNO⁻ + H2O
Me2C(NO2)CO2⁻ + NH3
Me2C=NO2⁻ + CO2

. . . (10)

Cleavage of ketones (**11**) and (**15**) with potassium hydroxide yields different mixtures of acids (**12**) and (**14**), which indicates that reaction cannot proceed via the kinetically free anion (**13**). It was suggested that proton transfer was partly intramolecular. If this were so, the two transition states for reaction of (**11**) would be (**16**) and (**17**), leading to (**14**) and (**12**) respectively. It was considered that (**17**) was favoured and that there was therefore a predominance of (**12**) in the product. Similarly (**15**) would lead to (**18**) and (**19**), and it was proposed that (**19**), analogous to (**17**), was disfavoured because of the presence of the *gem*-dimethyl group and hence formation of similar quantities of (**12**)

[14] M. Masui, H. Sayo, H. Ohmori, and T. Minami, *Chem. Comm.*, **1969**, 404.

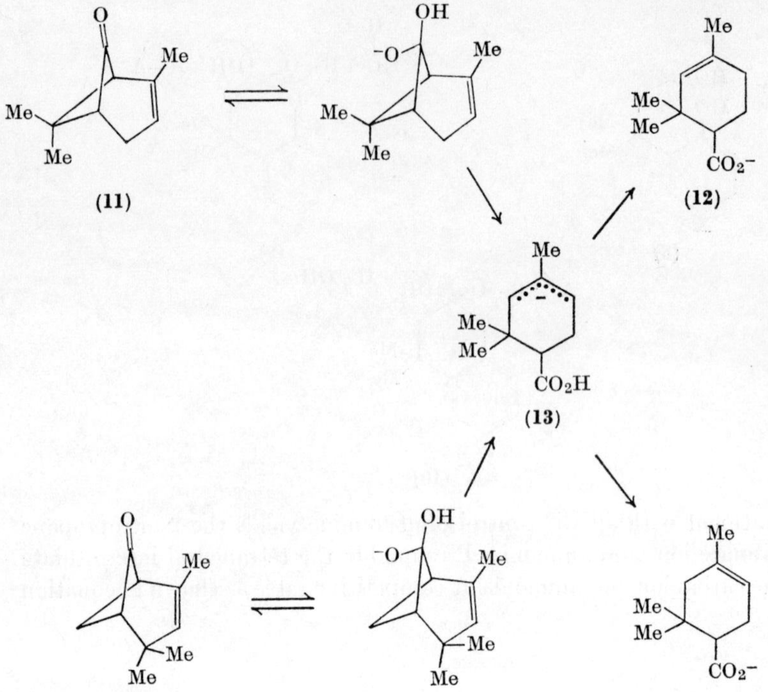

(11)   (12)

(13)

(15)   (14)

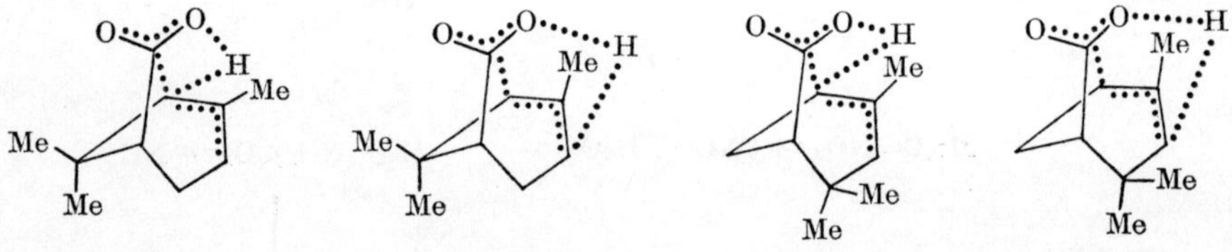

(16)   (17)   (18)   (19)

and (14) would be expected. The formation of identical mixtures of the methyl esters of (12) and (14) from (11) and (15) on treatment with methoxide ions is consistent with these proposals.[15]

The kinetic law for the alkaline cleavage of acetonylacetone has a term which is second-order in hydroxyl ions. The solvent isotope effect on this term is $k(H_2O)/k(D_2O) = 3.82$. This was interpreted in terms of a mechanism which involves reversible formation of a tetrahedral intermediate which undergoes a general-base catalysed breakdown to products as shown in equation (11).[16]

[15] W. F. Erman, E. Wenkert, and P. W. Jeffs, *J. Org. Chem.*, **34**, 2196 (1969).
[16] J. P. Calmon, *Compt. Rend., C*, **268**, 1256 (1969).

$$\underset{\underset{O}{\parallel}\ \underset{O}{\parallel}}{MeCCH_2CMe} \overset{-OH}{\rightleftharpoons} \underset{\underset{O^-}{\mid}\ \underset{O}{\parallel}}{\overset{\overset{OH}{\mid}}{MeCCH_2CMe}} \overset{-OH}{\longrightarrow} \underset{\underset{O^-}{\mid}}{\overset{\overset{\overset{\delta^-}{O}\cdots H\cdots \overset{\delta^-}{OH}}{\parallel}}{MeC\cdots CH_2\cdots C\overset{Me}{\underset{\overset{\delta^-}{O}}{\diagdown}}}}$$

$$\downarrow$$

$$MeCO_2{}^- + CH_2\cdots\underset{=}{C}\overset{Me}{\underset{O}{\diagdown}} \quad \cdots(11)$$

The base-catalysed cleavage of cyclopropenones has been studied.[17, 18]

It has been suggested that the second-order dependence on methanol of the methanolysis of *p*-nitrobenzoyl chloride in acetonitrile at 25° results from a mechanism involving reversible formation of a tetrahedral intermediate whose breakdown with loss of chloride ion is catalysed by proton removal by a second methanol molecule. The reaction is strongly accelerated by tetramethylammonium chloride probably because chloride ion removes a proton from the tetrahedral intermediate.[19]

Schowen and his coworkers have carried out an intensive investigation of the methoxide-promoted methanolysis of aryl acetates and aryl carbonates. Plots of the free energies of activation against the free energies of ionization of the corresponding phenols are curves for both series. This is not the result of a change in the rate-limiting step from breakdown to formation of the tetrahedral intermediate when the substituent becomes more electron-withdrawing, since no methoxide exchange of the labelled carbonate (20) prior to loss of the aryloxy residue could be detected. It probably results either from the contribution of the resonance interaction being greater in the equilibrium than in the rate process or from the length of the transition-state nucleophile bond increasing with electron withdrawal in the leaving group. Both the aryl acetates and carbonates react 1.5—2.1 times faster in CH_3OD than in CH_3OH.[20, 21]

$$\underset{p\text{-}XC_6H_4O}{\overset{{}^{13}CH_3O}{\diagdown}}C{=}O$$

$(X = CH_3O, H, CHO)$

(20)

$$\underset{Me}{\overset{Me}{\diagdown}}N\diagdown\diagup\underset{OH}{\overset{\overset{O}{\parallel}}{N}}{-}CMe$$

(21)

[17] C. W. Bird and A. F. Harmer, *J. Chem. Soc.*(C), **1969**, 959.
[18] F. G. Bordwell and S. C. Crooks, *J. Am. Chem. Soc.*, **91**, 2084 (1969).
[19] D. N. Kevill and F. D. Foss, *J. Am. Chem. Soc.*, **91**, 5054 (1969).
[20] C. G. Mitton, R. L. Schowen, M. Gresser, and J. Shapley, *J. Am. Chem. Soc.*, **91**, 2036 (1969).
[21] C. G. Mitton, M. Gresser, and R. L. Schowen, *J. Am. Chem. Soc.*, **91**, 2045 (1969).

Intermolecular Catalysis

The hydrolysis of 2-naphthyl chloroacetate $(1.94 \times 10^{-3}\text{M})$ at pH 7.35 is increased 4.4-fold in the presence of N-methylacetohydroxamic acid $(6.33 \times 10^{-4}\text{M})$. The reaction is first-order for at least 90%. The solvent isotope effect, $k(\text{H}_2\text{O})/k(\text{D}_2\text{O})$, is 1.05 and nucleophilic catalysis was proposed (equation 12). The effectiveness of this catalysis is controlled by the rate of hydrolysis of the intermediate (**22**) which is about 80 times the rate of the spontaneous hydrolysis of 2-naphthyl chloroacetate. The hydrolysis of p-nitrophenyl acetate in the presence of N-methylacetohydroxamic acid occurs with rapid release of p-nitrophenol equal in concentration to that of the catalyst, followed by a slow turnover controlled by the hydrolysis of the N,O-diacetyl-N-methylhydroxylamine. To obtain more effective catalysis, a catalyst (**21**) containing a dimethyl-

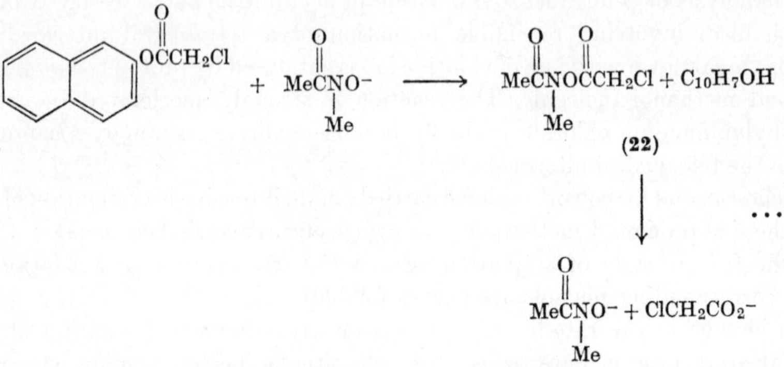

$$\text{(22)}$$

$$\ldots \text{(12)}$$

amino group capable of catalysing the hydrolysis of the intermediate intramolecularly was studied. With the concentration of (**21**) $6.34 \times 10^{-4}\text{M}$ and of p-nitrophenyl acetate $2.01 \times 10^{-3}\text{M}$ there was a 25-fold increase in the rate of hydrolysis and the kinetics were first-order.[22]

Benzamidoxime is a 700—800-fold better nucleophile than benzaldoxime towards ethyl chloroformate and benzoyl fluoride at pH 7 but only about 3-fold better towards p-nitrophenyl acetate.[23] The reactions of p-nitrophenyl benzoate and p-nitrobenzoate with benzohydroxamic acids have been studied.[24]

The acetyl-pyridinium ion has been detected as an intermediate in the pyridine-catalysed hydrolysis of acetic anhydride.[25]

2-Pyridone, benzoic acid, pyrazole, and 1,2,4-triazole are effective catalysts for the reaction of p-nitrophenyl acetate with butylamine and glycine methyl

[22] W. O. Gruhn and M. L. Bender, *J. Am. Chem. Soc.*, **91**, 5883 (1969).
[23] J. D. Aubort and R. F. Hudson, *Chem. Comm.*, **1969**, 1342.
[24] H. Kwart and H. Omura, *J. Org. Chem.*, **34**, 318 (1969).
[25] A. R. Fersht and W. P. Jencks, *J. Am. Chem. Soc.*, **91**, 2124 (1969).

ester in benzene. A mechanism involving concerted tautomeric catalysis (equations 13 and 14) was proposed[26] (see also p. 411). 2-Pyridone has been

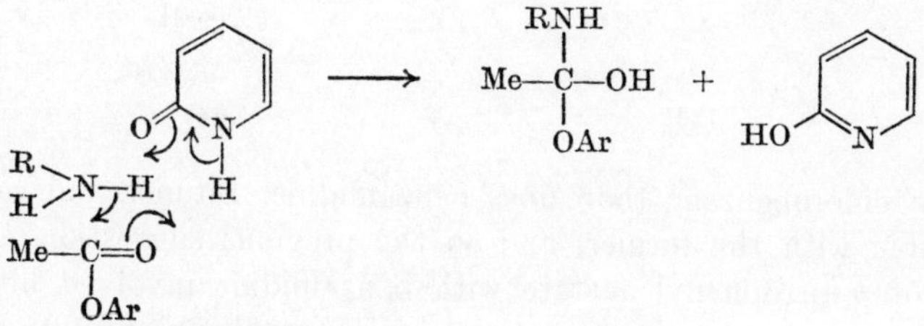

$$
\text{... (13)}
$$

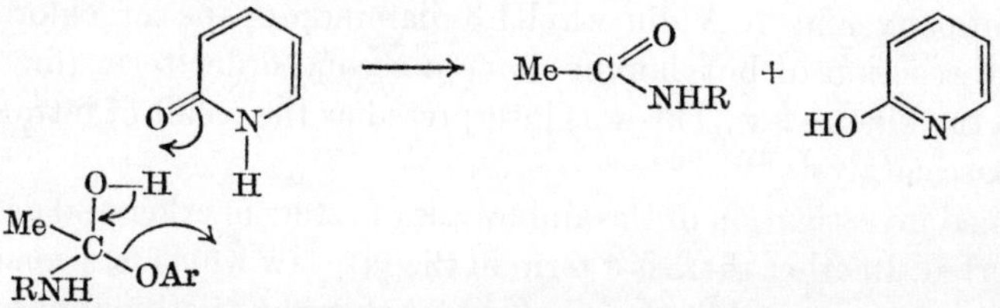

$$
\text{... (14)}
$$

recommended as a catalyst in the synthesis of amides from unreactive esters and amines.[27] The effectiveness of catalysts for the reaction of benzyloxy-carbonyl-L-phenylalanine *p*-nitrophenyl ester with glycine *t*-butyl ester in dioxan decreases in the order trimethylacetic acid > acetic acid > 2-pyridone > monochloroacetic acid.[28]

It has been proposed that hydrolysis of *p*-nitrophenyl acetate catalysed by mercaptomethyl- and mercaptoethyl-imidazole proceeds via concerted catalysis by the imidazole and thiol groups.[29a]

Pyridine-catalysis of the methanolysis of *o*-nitrophenyl acetate has been suggested to be nucleophilic catalysis since 2,6-lutidine is a ten-fold less effective catalyst than expected from its pK_a and the Brønsted plot for non-sterically hindered pyridines.[29b]

1,4-Diazabicyclooctane, although 150 times less basic than n-butylamine *in water*, is a slightly better catalyst than n-butylamine for the n-butyl-aminolysis of *p*-nitrophenyl acetate in chlorobenzene. It was therefore concluded that the catalysis by n-butylamine was general-base catalysis with transition state (23) rather than bifunctional catalysis with transition state (24).[30] 1,4,5,6-Tetrahydropyrimidone reacts more rapidly with *p*-nitrophenyl

[26] P. R. Rony, *J. Am. Chem. Soc.*, **91**, 6090 (1969).
[27] H. T. Openshaw and N. Whittaker, *J. Chem. Soc.*(C), **1969**, 89.
[28] N. Nakamizo, *Bull. Chem. Soc. Japan*, **42**, 1071 (1969).
[29a] F. Schneider, E. Schaich, and H. Wenck, *Z. Physiol. Chem.*, **349**, 1525 (1968).
[29b] A. Kirkien-Konasiewicz, R. J. Simkin, and R. Murphy, *Chem. Ind.* (London), **1968**, 1842.
[30] See *Org. Reaction Mech.*, **1966**, 343—344, and ref. 67 of this chapter.

$$R-NH_2\cdots\overset{\overset{\displaystyle Me}{\overset{\displaystyle R\,\delta+}{|}}}{\underset{\overset{\displaystyle \parallel}{\underset{\displaystyle O\,\delta-}{}}}{\underset{\displaystyle H_2}{N}}}\cdots C-OAr$$

(23)

$$\begin{array}{c} Me \quad OAr \\ R-N\cdots C \\ H \qquad O \\ H \\ \qquad N\cdots H \\ H \quad R \end{array}$$

(24)

acetate in chlorobenzene than does benzamidine. Bifunctional catalysis is not possible with the former, and so the previous suggestion[30] that the reaction of p-nitrophenyl acetate with benzamidine involved bifunctional catalysis was rejected. The reactions of p-nitrophenyl acetate with 1,3-diaminopropane and N,N-dimethyl-1,3-diaminopropane in chlorobenzene, unlike the reaction of butylamine, have a second-order term (first-order in amine) in the kinetic law. This was interpreted as the result of intramolecular general-base catalysis.[31]

A detailed investigation of the aminolysis of esters in ether and acetone has been reported. In ether there is a term in the rate law which is second-order in amine. On the grounds that there was no catalysis by triethylamine this term was thought to arise not from general-base catalysis but from a hydrogen-bonded cyclic transition state.[32]

The reactions of p-nitrophenyl benzoates with ammonia in 33% acetonitrile–water show only a first-order dependence on ammonia but those of p-chlorophenyl benzoates show both a second- and a first-order dependence. The ρ-values are 1.426 for the reactions of p-nitrophenyl benzoates and 1.08 and 1.876 for the reactions of p-chlorophenyl benzoates which are first- and second-order in ammonia respectively.[33]

Transamidation of 1,3-diphenylurea with n-butylamine catalysed by 1,4-diazabicyclooctane in dioxan was thought to involve nucleophilic catalysis by the 1,4-diazabicyclooctane.[34] Nucleophilic catalysis by imidazole and o-mercaptobenzoic acid in the hydrolysis of p-nitrophenyl N-benzyloxy-carbonylglycinate has been studied.[35]

It has been suggested that ethyl acetate reacts with hydroxylamine at pH 9.5 (40°) by N- and O-attack, and that the product of O-attack (O-acetyl-hydroxylamine) is hydrolysed rapidly under the reaction conditions.[36]

The reactions of pyromellitic anhydride with alcohols in DMF catalysed by tertiary amines,[37] p-nitrophenyl benzoate with carboxylic acid hydrazides

[31] H. Anderson, C.-W. Su, and J. W. Watson, *J. Am. Chem. Soc.*, **91**, 483 (1969).
[32] D. P. N. Satchell and I. I. Secemskii, *J. Chem. Soc.*(B), **1969**, 130.
[33] J. F. Kirsch and A. Kline, *J. Am. Chem. Soc.*, **91**, 1841 (1969).
[34] Y. Furuya, K. Itoho, and H. Miyagi, *Bull. Chem. Soc. Japan*, **42**, 2348 (1969).
[35] R. W. Hay and R. J. Trethewey, *Austral. J. Chem.*, **22**, 109 (1969).
[36] R. E. Notari, *J. Pharm. Sci.* **58**, 1069 (1969).
[37] B. H. M. Kingston, J. J. Garey, and W. A. Hellmig, *Analyt. Chem.*, **41**, 86 (1969).

in DMF,[38] succinimide with hydroxylamine to give N-hydroxysuccinimide,[39] and propionic anhydride with 2-dimethylaminoethanethiol to give the S-ester,[40] and the aminolysis of benzoxazolines,[41] have also been investigated.

The ρ-values of the carboxylic-acid catalysed benzoylation of aromatic amines by benzoyl chloride, bromide, and fluoride are 2.72, 2.98, and 2.95 respectively, and the ratios of the rate constants for catalysed and uncatalysed reactions are 31, 141, and 294,000 respectively when p-methoxyaniline is the amine.[42] Catalysis by carboxylic acids and amides of the reaction of benzoyl chloride with m-chloroaniline and with phenols has been compared.[43,44]

2-Methoxy-2-imidazoline (25) reacts with p-nitrophenyl acetate whereas imidazolone (26) is inert. It was therefore suggested that the enzymically catalysed reaction of the imidazolone ring of biotin with carbon dioxide proceeds via the enol form.[45]

(25)   (26)

The α-effect appears to be unimportant in substitution reactions at saturated carbon[46] (see also p. 121).

The hydrolysis of phenyl orthoformate is very slow for an orthoester, and can be studied in 1—6M-HCl in 40% dioxan. In our view the low rate probably results from the weakened ability of a phenoxy group relative to an alkoxy group to stabilize the intermediate ion and transition state for its formation. The isotope effect $k(D_2O)/k(H_2O)$ in 1M-hydrochloric acid is 1.07.[47]

[38] A. P. Grekov and M. I. Shandruk, *Zh. Org. Khim.*, **4**, 1077 (1968); *Chem. Abs.*, **69**, 51295 (1968).

[39] R. E. Notari, *J. Pharm. Sci.*, **58**, 1064 (1969).

[40] A. Hussain and P. Schurman, *J. Pharm. Sci.*, **58**, 684 (1969).

[41] D. Simov, V. Kalcheva, M. Arnaudov, and B. Galabov, *Dokl. Bogl. Akad. Nauk*, **21**, 377 (1968); *Chem. Abs.*, **70**, 36785 (1969).

[42] L. M. Litvinenko, N. M. Oleinik, and G. V. Semenyuk, *Ukr. Khim. Zh.*, **35**, 278 (1969); *Chem. Abs.*, **71**, 2694 (1969).

[43] G. V. Semenyuk, N. M. Oleinik, and L. M. Litvinenko, *Zh., Obshch. Khim.*, **38**, 2009 (1968); *Chem. Abs.*, **70**, 19475 (1969); *Ukr. Khim. Zh.*, **35**, 278 (1969); L. M. Litvinenko, G. D. Titskii, and V. A. Tarasov, *Org. Reactivity* (Tartu), **5**, 325 (1968); *Chem. Abs.*, **90**, 19338 (1969).

[44] L. V. Koshkin, R. M. Basaev, N. T. Fedorina, N. N. Baszeva, *Zh. Org. Khim.*, **4**, 2175 (1968); *Chem. Abs.*, **70**, 67300 (1969).

[45] A. F. Hegarty, T. C. Bruice, and S. J. Benkovic, *Chem. Comm.*, **1969**, 1173; see also M. Caplow, *Biochemistry*, **8**, 2656 (1969); R. B. Huston and P. Cohen, *ibid.*, p. 2658; H. Sigel, D. B. McCormick, R. Griesser, B. Prijs, and L. D. Wright, *ibid.*, p. 2687.

[46] S. Oae, Y. Kadoma, and Y. Yano, *Bull. Chem. Soc. Japan*, **42**, 1110 (1969); cf. *Org. Reaction Mech.*, **1967**, 335, 365; **1968**, 361.

[47] M. Price, J. Adams, C. Lagenaur, and E. H. Cordes, *J. Org. Chem.*, **34**, 22 (1969); see also E. H. Cordes, *Rev. Fac. Quim., Univ. Nac. Mayor San. Marcos*, **18**, 43 (1966); *Chem. Abs.*, **69**, 85824 (1968); p. 623 of ref. 2.

The ρ-value for the hydrolysis of methyl benzoates in 95% sulphuric acid is
−3.6, consistent with an $A_{Ac}1$ mechanism.[48] Other studies of ester hydrolysis
in concentrated sulphuric acid are reported in refs. 49 and 50.

The acid-catalysed hydrolysis of t-butyl formate occurs with 3% alkyl–
oxygen fission at 35° and 40% at 76°.[51] There have been several investigations
of the hydrolysis of t-butyl acetate.[52-56] An investigation of the relative
importance of acyl–oxygen and alkyl–oxygen fission in the hydrolysis of
esters of allyl alcohols has been reported.[57] α,α-Dimethyl-β-propionolactone
undergoes acyl–oxygen fission in HF–BF$_3$ to yield the protonated hydroxy-
methyldimethylacetyl cation.[58] The ring-opening of β-propionolactone by
trimethyltin compounds has been studied.[59]

The hydrolysis of actic anhydride in H_2O–D_2O mixtures has been studied.[60]

Protonation of the carbonyl groups of peptides, amino-acids, and amides
has been studied by NMR spectroscopy.[61]

Intramolecular Catalysis and Neighbouring-group Participation

The rates of hydrolysis of imides (27) and (28) in alkaline buffers are proportional to the concentrations of the species with the hydroxy group ionized,
and reach constant values at pH values ca. 2 units greater than the pK_a
values. This plateau rate is about 6000 times greater for (27) than for (28).
These reactions could proceed by an intramolecularly general-acid catalysed
attack of hydroxyl ion on the species with the hydroxy group un-ionized
(equation 15) or by an intramolecularly general-base catalysed reaction of
water with the species with the hydroxy group ionized (equation 16). The
expressions for the plateau rate constants for these two mechanisms are
respectively:

$$k(\text{plateau}) = kK_w/K_a$$

$$k(\text{plateau}) = k'$$

[48] H. van Bekkum, H. M. A. Buurmans, B. M. Wepster, and A. M. van Wijk, *Rec. Trav. Chim.*,
    88, 301 (1969).
[49] A. C. Hopkinson, *J. Chem. Soc.*(B), **1969**, 861.
[50] A. C. Hopkinson, *J. Chem. Soc.*(B), **1969**, 203.
[51] R. A. Fredlein and I. Lauder, *Austral. J. Chem.*, **22**, 19, 33 (1969).
[52] P. Salomaa, *Suomen Kemistilehti*, *B42*, 134 (1969).
[53] G. Costeanu, O. Landauer, and C. Mateescu, *Rev. Roum. Chim.*, **14**, 845 (1969).
[54] P. Salomaa, *Acta Chem. Scand.*, **23**, 713 (1969).
[55] H. Saimiya and S. Terazawa, *Koatsu Gasu*, **5**, 427 (1968); *Chem. Abs.*, **69**, 95573 (1968).
[56] J. Vuori and J. Koskikallio, *Suomen Kemistilehti*, *B42*, 136 (1969).
[57] G. Meyer, P. Viout, and R. Rumpf, *Bull. Soc. Chim. France*, **1968**, 4436.
[58] H. Hogeveen, *Rec. Trav. Chim.*, **87**, 1303 (1969).
[59] K. Itoh, Y. Kato, and Y. Ishii, *J. Org. Chem.*, **34**, 459 (1969).
[60] B. D. Batts and V. Gold, *J. Chem. Soc.*(A), **1969**, 984; see also V. Gold, *Adv. Phys. Org.
    Chem.*, **7**, 259 (1969); P. Salomaa, *Suomen Kemistilehti*, *A42*, 17 (1969).
[61] J. L. Sudmeier and K. E. Schwartz, *Chem. Comm.*, **1968**, 1646; G. A. Olah and P. J. Szilagyi,
    J. Am. Chem. Soc., **91**, 2949 (1969); M. Liler, *J. Chem. Soc.*(B), **1969**, 385.

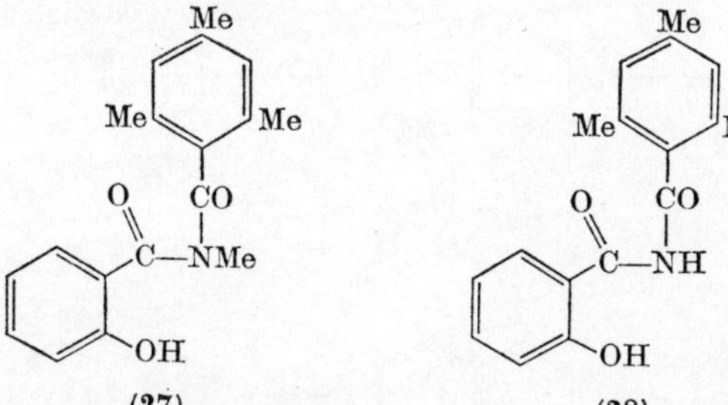

(27)   (28)

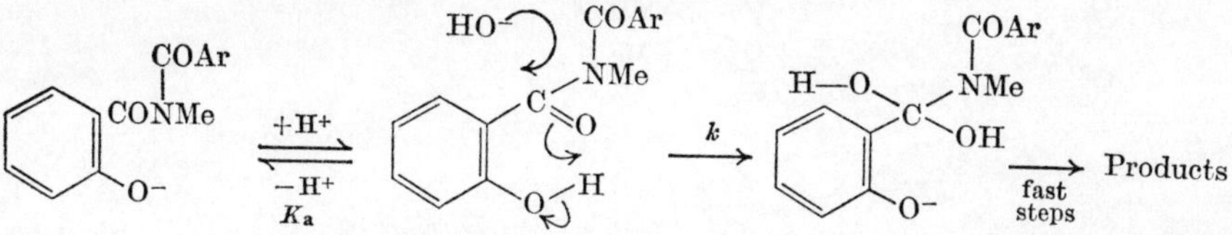

... (15)

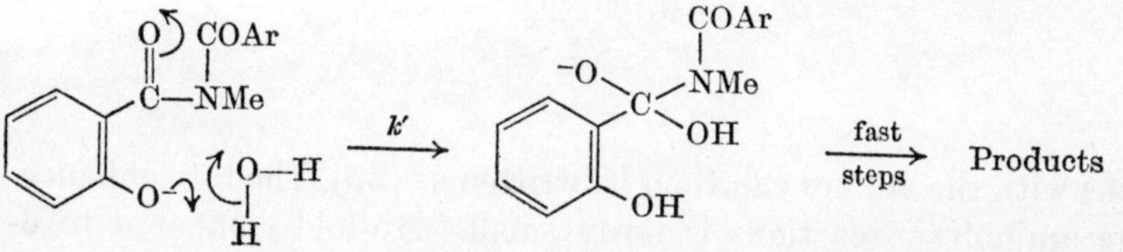

... (16)

The mechanism involving intramolecular acid catalysis was preferred since this enabled the large difference in k(plateau) to be attributed to a difference in K_a. pK_a for (28) is very low (6.38), and on the assumption that pK_a for (27) is normal (8—9), the 6000-fold difference in k(plateau) could be explained.[62]

The rate of hydrolysis of 8-acetoxyquinoline at pH ca. 4—7 is about 500 times greater than that of 6-acetoxyquinoline.[63] The solvent isotope effect $k(H_2O)/k(D_2O)$ is 2.35, which suggests that the reaction involves a rate-limiting proton transfer. Two mechanisms were considered (equations 17 and 18), and that of equation (18) was favoured on the basis of the solvent isotope effect.[63] The plot of log k for the reactions of 8-acetoxyquinoline with water and with primary and secondary amines against log k for the reactions of 6-acetoxyquinoline with the same nucleophiles is a straight line. A different line is obtained when the rate constants for the reaction with tertiary amines are plotted, which suggests that the reactions of 8-acetoxyquinoline with water and with primary and secondary amines follow a similar mechanism.

[62] R. M. Topping and D. E. Tutt, *J. Chem. Soc.*(B), **1969**, 106; cf. *Org. Reaction Mech.*, **1968**, 386.

[63] Cf. *Org. Reaction Mech.*, **1966**, 349.

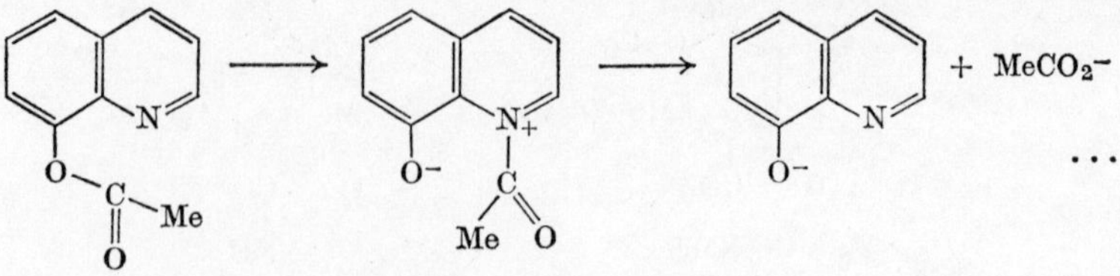

... (17)

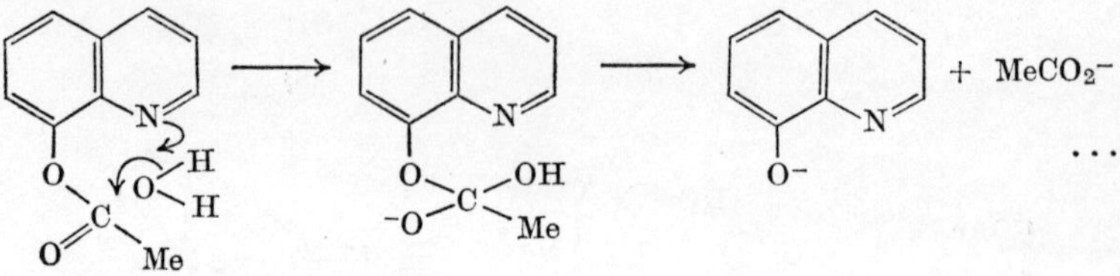

... (18)

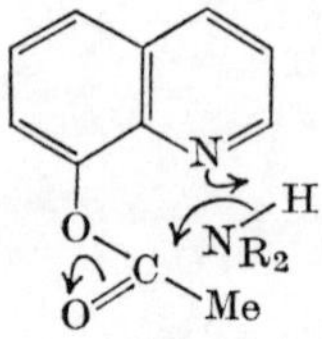

(29)

The reactions with the amines can then be written as (29). The rate enhancement of the aminolysis reactions is fairly small (<15-fold compared to 6-acetoxyquinoline).[64] Details of Felton and Bruice's investigation of the aminolysis of 4-(2-acetoxyphenyl)imidazole were also reported.[65]

The reaction of 2-pyridyl p-nitrobenzoate in chlorobenzene is faster than expected on the basis of its carbonyl stretching frequency. The reaction shows mixed second- and third-order kinetics with $k_{obs} = k_2[\text{Amine}] + k_3[\text{Amine}]^2$, and the rate enhancement appears to result from increased values of both k_2 and k_3.[66]

The reaction of phenyl salicylate with n-butylamine in chlorobenzene is second-order in amine, and the third-order constant is 132 times greater than that for aminolysis of phenyl o-methoxybenzoate. Both reactions are catalysed by 1,4-diazabicyclooctane but only the reaction of phenyl o-methoxybenzoate is catalysed by n-butylamine hydrochloride. A mechanism involving general-base catalysis and intramolecular general-acid catalysis was proposed for the reaction of phenyl salicylate (see equation 19).[67]

The pH–rate profile for hydrolysis of methyl 2,6-dihydroxybenzoate is bell-shaped. The rate constant at the maximum is slightly less than the rate

[64] T. C. Bruice and S. M. Felton, *J. Am. Chem. Soc.*, **91**, 2799, 6721 (1969).
[65] See *Org. Reaction Mech.*, **1968**, 385.
[66] F. R. Smith and J. W. Watson, *Chem. Comm.*, **1969**, 786.
[67] F. M. Menger and J. H. Smith, *J. Am. Chem. Soc.*, **91**, 5346 (1969).

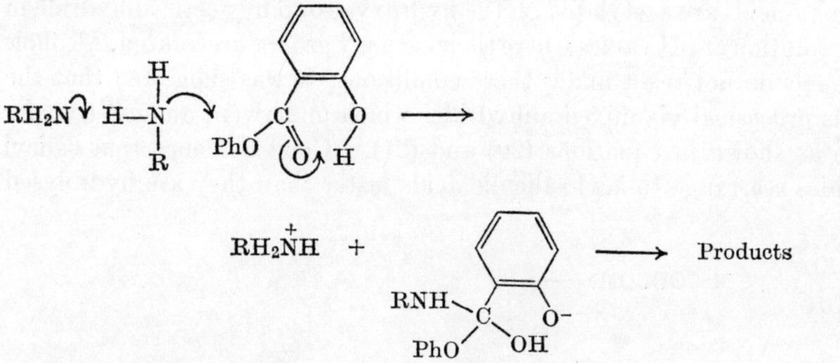

$$RH_2\overset{+}{N}H \quad + \quad \text{[structure]} \quad \longrightarrow \quad \text{Products} \quad \cdots (19)$$

constant for the hydrolysis of methyl salicylate at the plateau of its pH–rate profile. This was interpreted as indicating that the hydrolysis of methyl 2,6-dihydroxybenzoate is strongly accelerated since the hydrolysis of a 2,6-disubstituted benzoate should be very slow. Possible mechanisms are those depicted in (30) and (31).[68]

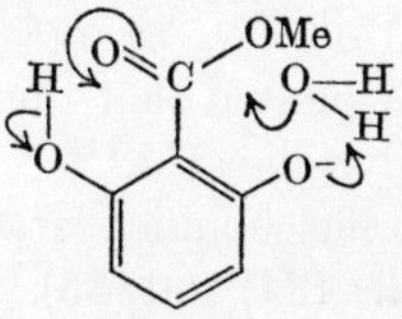

(30) 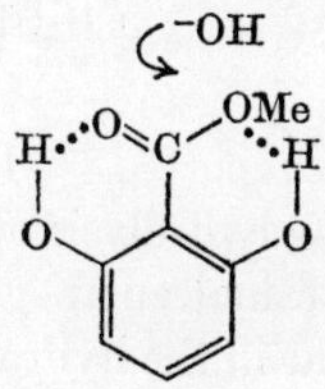   (31)

The labelled salicyl salicylate (32) has been shown to undergo equilibration of the label about 10^4 times faster than it is hydrolysed at pH 8.2. It was shown that hydrazinolysis proceeds via direct attack on the salicyl salicylate rather than by trapping of the salicylic anhydride (33).[69]

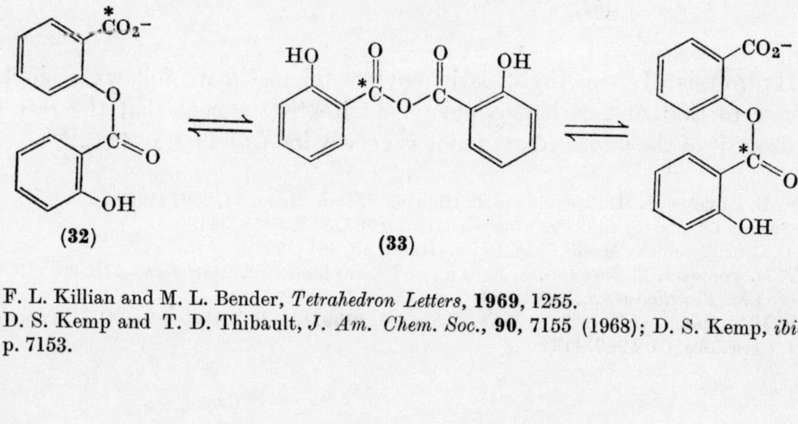

(32)   (33)

[68] F. L. Killian and M. L. Bender, *Tetrahedron Letters*, **1969**, 1255.
[69] D. S. Kemp and T. D. Thibault, *J. Am. Chem. Soc.*, **90**, 7155 (1968); D. S. Kemp, *ibid.*, p. 7153.

Hydroxy-acids are acetylated on the hydroxy group by acetic anhydride in aqueous solution at pH values where their carboxy groups are ionized. Alcohols and phenols do not react under these conditions. It was suggested that the reactions proceeded via mixed anhydrides which underwent an acyl transfer reaction as shown in equations (20) and (21).[70] Good evidence that salicyl anhydrides rearrange to acyl salicylic acids faster than they are hydrolysed

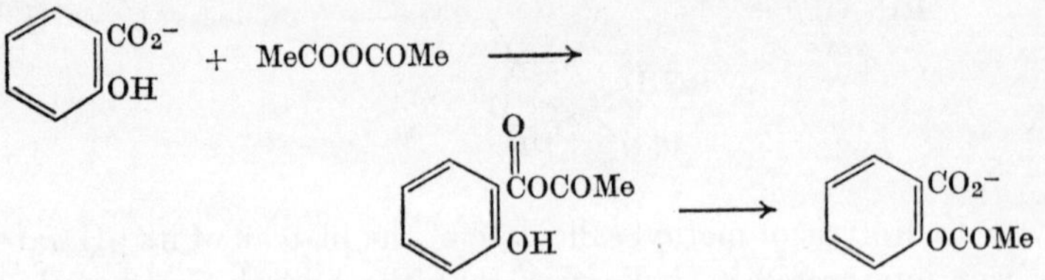

$$\cdots (20)$$

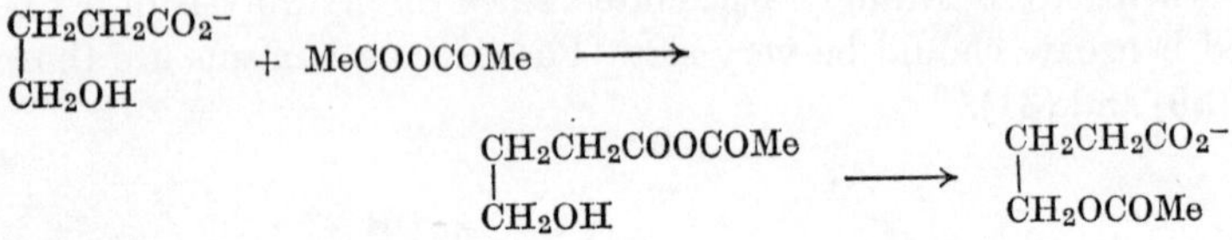

$$\cdots (21)$$

has already been reported.[71a] It has been suggested that the carboxy group participates in the hydrolysis of O-benzoylglycollic acid.[71b]

Concerted intramolecular general-base nucleophilic catalysis has been suggested to occur in the hydrolysis of amides (34) and (35).[72]

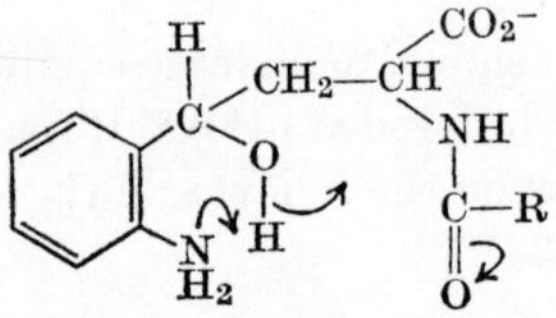

(34) 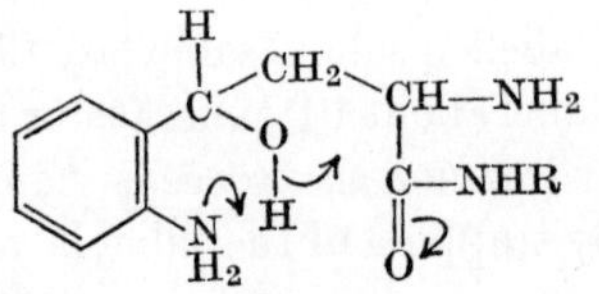   (35)

Hydrolysis of N-methyl-2-carbamoylphenyl mesitoate follows a similar course to that of 2-carbamoylphenyl mesitoate,[73] except that the rate of hydrolysis of the intermediate imide is greatly increased (cf. p. 444).[74]

[70] R. B. Paulssen, I. H. Pitman, and T. Higuchi, *J. Org. Chem.*, **34**, 2097 (1969).
[71a] See ref. 69 (above) and *Org. Reaction Mech.*, **1967**, 337; **1968**, 383.
[71b] C. Concillo and A. Arcelli, *Ann. Chim.* (Italy), **58**, 881 (1968).
[72] F. M. Veronese, E. Boccu, C. A. Benassi, and E. Scoffone, *Z. Naturforsch.*, **24b**, 294 (1969).
[73] See *Org. Reaction Mech.*, **1968**, 386.
[74] R. M. Topping and D. E. Tutt, *J. Chem. Soc.*(B), **1969**, 164; P. L. Russell and R. M. Topping, *J. Chem. Soc.*(C), **1969**, 1134.

The hydrolyses of peptides with structure (**36**) are much faster than those of peptides with structure (**37**). Presumably the benzimidazole group provides

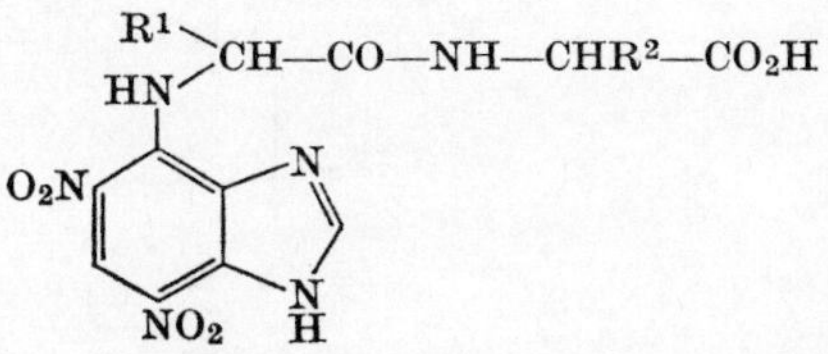

$$
\begin{array}{cc}
\text{R}^1\!\!>\!\!\text{CH—CO—NH—CHR}^2\text{—CO}_2\text{H} & \text{R}^1\!\!>\!\!\text{CH—CO—NH—CHR}^2\text{—CO}_2\text{H} \\
\text{(36)} & \text{(37)}
\end{array}
$$

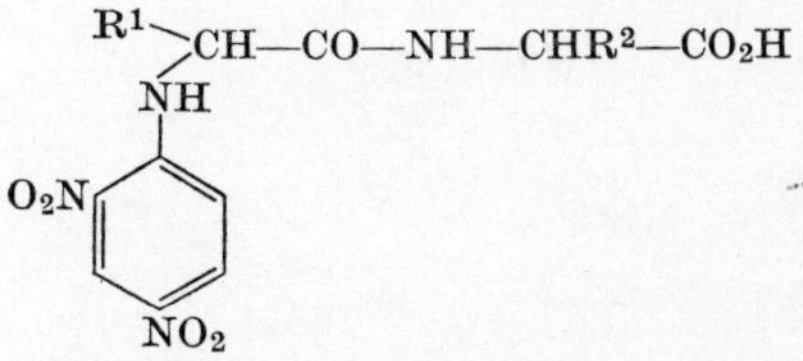

intramolecular catalysis which, in view of the large rate enhancements found, is probably nucleophilic catalysis.[75]

The acetates with structures (**38**) do not appear to undergo hydrolysis with participation by the pyrimidine nitrogen.[76]

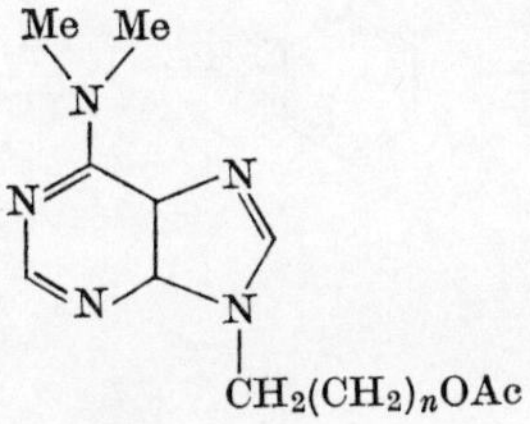

$$\text{(38)} \quad n = 1, 2$$

Intramolecular catalysis in the hydrolysis of 5-nitro-2-pyridyl-β-alanyl-glycine[77] and ring-closure of glycylglycine ethyl ester[78] has been studied.

The rate of conversion of (**39**) into (**41**) depends on the concentration of ionized form and is independent of pH in the range 5—12. The reaction probably involves the form with the amide group ionized and the carboxy group un-ionized as shown. The ester (**40**) cyclizes very rapidly to (**41**).[79]

The alkaline cyclization of N-(o-carboxyphenyl)urea is proportional to the concentration of the species with the carboxy group ionized multiplied by the concentration of hydroxyl ions. Reaction was thought to involve intramolecular nucleophilic attack by the ionized urea group on the ionized carboxy group as shown in equation (22).[80]

[75] K. L. Kirk and L. A. Cohen, *J. Org. Chem.*, **34**, 390, 395 (1969).
[76] H. Determan and D. Merz, *Ann. Chem.*, **728**, 209 (1969).
[77] E. Bordignon, A. Rastrelli, G. Rigatti, and A. Signor, *Ricerca Sci.*, **38**, 715 (1968); cf. *Org. Reaction Mech.*, **1965**, 270; **1966**, 349; **1967**, 344.
[78] U. Meresaar and A. Agren, *Acta Pharm. Suec.*, **5**, 85 (1968).
[79] P. J. Taylor, *J. Chem. Soc.*(B), **1968**, 1554.
[80] A. F. Hegarty and T. C. Bruice, *J. Am. Chem. Soc.*, **91**, 4924 (1969).

15

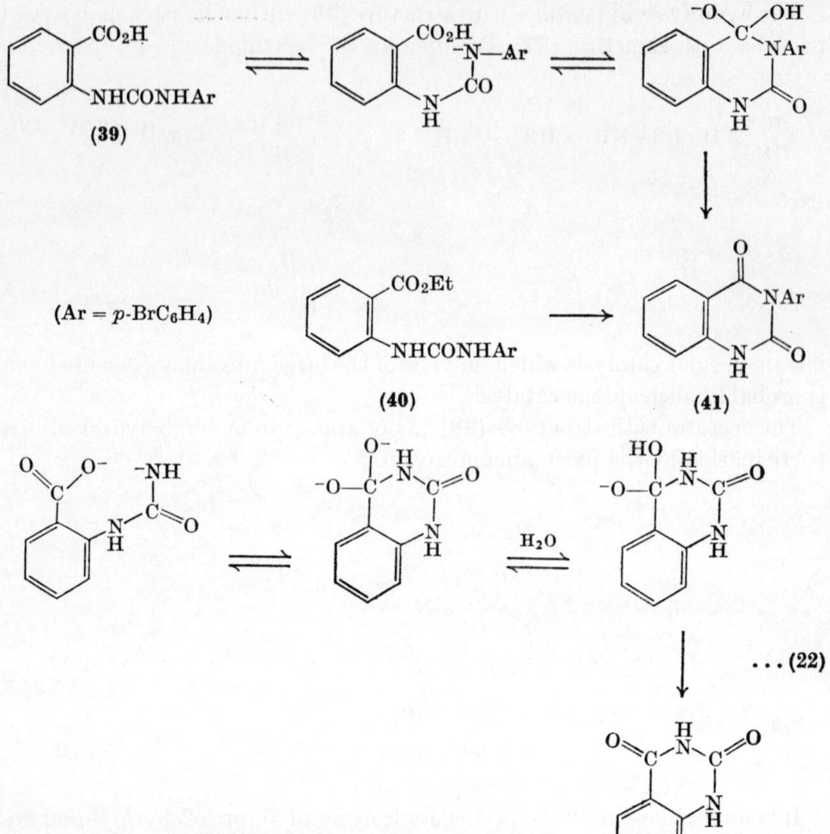

(39) (Ar = p-BrC$_6$H$_4$) (40) (41) . . . (22)

The rates of hydrolysis of amides of structure $H_2N(CH_2)_nCONMe_2$ vary with n in the order $4 > 3{,}5 > 6 > 7{,}1{,}2$. It seems likely that the first three react with neighbouring-group participation.[81]

The rearrangement of (**42**) into (**44**) proceeds via the pseudothiohydantoic acid (**43**). The ρ-value is 0.25.[82]

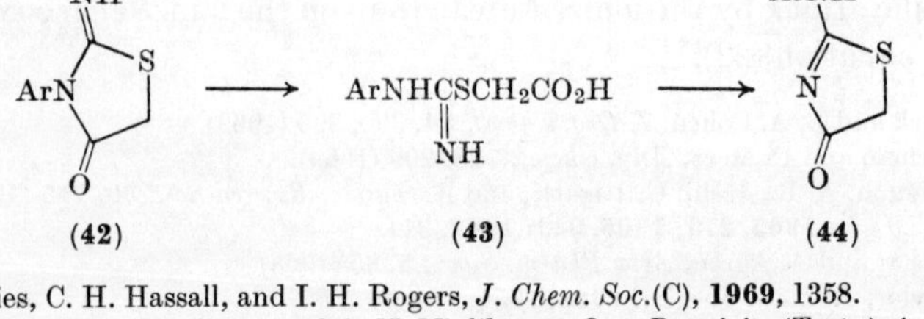

(42) (43) (44)

[81] J. A. Davies, C. H. Hassall, and I. H. Rogers, *J. Chem. Soc.*(C), **1969**, 1358.
[82] Y. V. Svetkin, S. A. Vasil'eva, and A. N. Minlibaeva, *Org. Reactivity* (Tartu), **4**, 705 (1967).

Hydrolysis of (45) proceeds via the anhydride (46) at pH 7 since, in the presence of glycine, both (47) and (48) are formed.[83]

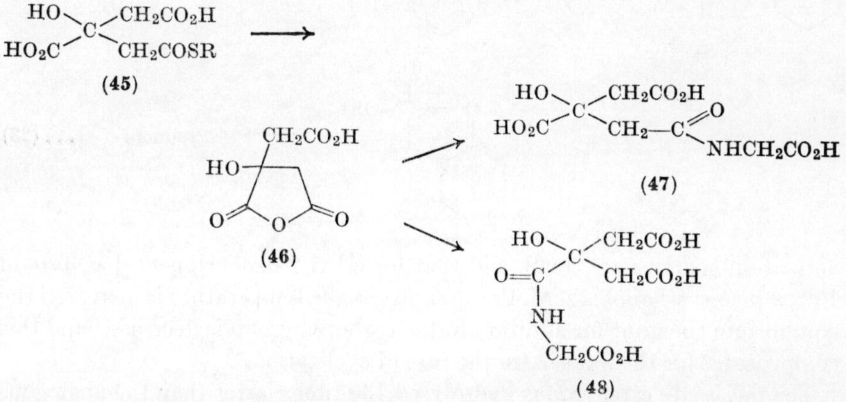

4-Bromo-4′-nitrobenzhydryl hydrogen phthalate does not react with participation by the carboxylate group in *alkaline solution*.[84]

The major initial products of the dehydration of N-arylmaleamic acids (49) in sodium acetate–acetic anhydride are isoimides (50). These rearrange to the

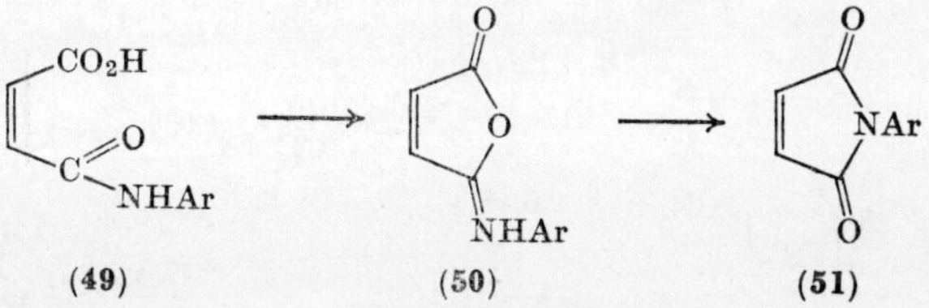

imides (51). This reaction has a ρ-value of 1.7. There is also some acetolysis to yield maleic anhydride and the acetanilide.[85]

Ethyl 2-oxocyclohexylacetate and 2-oxocyclopentylacetate are hydrolysed 66 and 199 times faster than ethyl cyclohexylacetate and ethyl cyclopentylacetate respectively under alkaline conditions.[86] Reaction presumably involves addition to the ketone carbonyl group as shown in equation (23). The apparent

[83] W. Buckel and H. Eggerer, *Z. Physiol. Chem.*, **350**, 1367 (1969).
[84] K. G. Rutherford, O. A. Mamer, and B. K. Tang, *Can. J. Chem.*, **47**, 1487 (1969).
[85] C. K. Sauers, *J. Org. Chem.*, **34**, 2275 (1969).
[86] K. C. Kemp and M. L. Mieth, *Chem. Comm.*, **1969**, 1260.

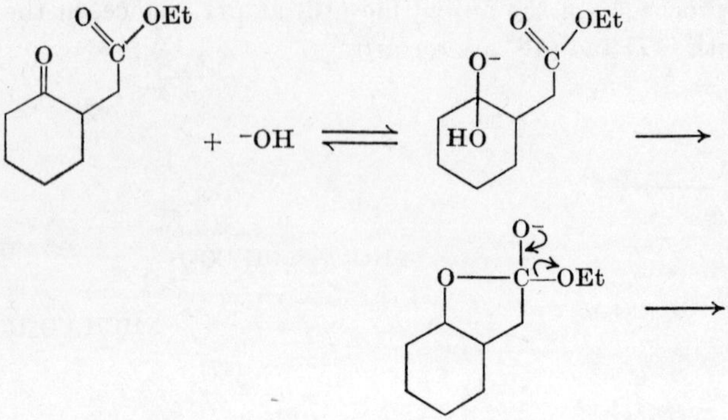

activation energies are small, and that for ethyl 2-oxocyclopentyl acetate in 10% aqueous ethanol is zero. Presumably, as the temperature is increased the equilibrium constant for addition to the carbonyl group is decreased and this compensates for the increase in the rate of cyclization.[86]

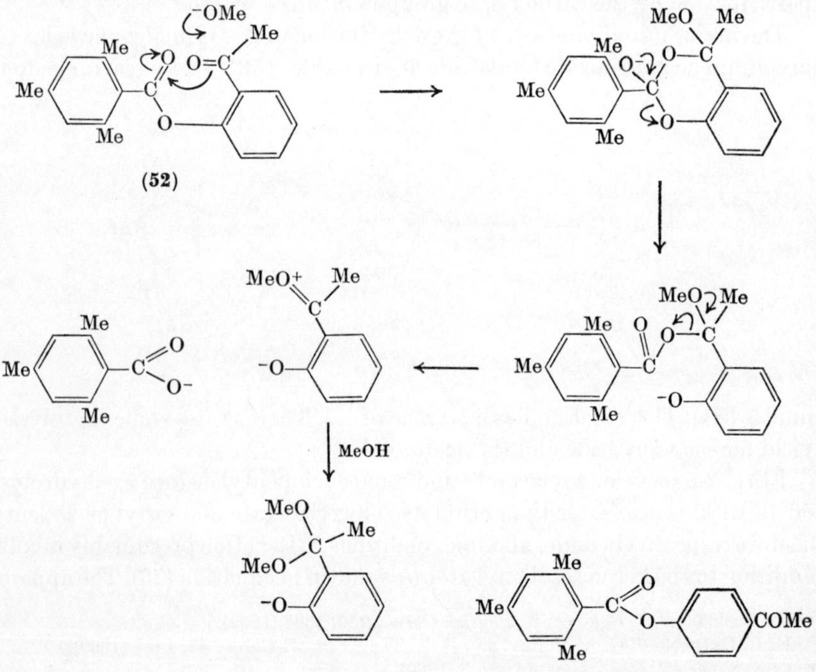

The mesitoate ester (52) is hydrolysed 130 times faster than the analogous ester (54) with the acetyl group in the *para*-position. In methanolic methoxide, compound (52) was shown to react with simultaneous disappearance of the ketone and ester carbonyl groups and formation of ketal (53) as shown. The high rate of hydrolysis presumably results from an analogous mechanism.[87]

In the pH range 4—8, 2-dimethylaminoethyl thiopropionate (55) is hydrolysed about 100 times faster than propionylthiocholine (56). Above pH 11 (56) is hydrolysed 32 times faster than (55). It was suggested that the protonated amino group stabilizes the transition state through hydrogen-

$$\text{Et}\overset{\text{O}}{\overset{\|}{\text{C}}}\text{SCH}_2\text{CH}_2\text{NMe}_2 \qquad\qquad \text{Et}\overset{\text{O}}{\overset{\|}{\text{C}}}\text{SCH}_2\text{CH}_2\overset{+}{\text{N}}\text{Me}_3$$

$$(55) \qquad\qquad\qquad\qquad (56)$$

bonding or electrostatically, and that a quaternary amino group also has some stabilizing effect.[88]

Details of Zerner and his coworkers' investigation of the hydrolysis of aryl esters of *N*-acyl-amino-acids and of the intermediate oxazolinones have been published.[89] The racemization of *N*-benzoyl-L-leucine 4-(methylsulphonyl)-phenyl ester[90] and the cyclization of hydrazides of unsaturated *N*-benzoyl-amino-acids[91] have been studied.

The hydroxy-nitrile (57) is converted rapidly into the lactone (59), probably via the imido-lactone (58).[92]

$$(57) \qquad\qquad\qquad (58) \qquad\qquad\qquad (59)$$

Other intramolecular reactions which have been investigated include: hydrolysis of 1,2,5-thiadiazole-3,4-dicarboxylic acid bishydrazide,[93] *N*-[2-(4-imidazolyl)ethyl]phthalimide,[94] glucuronamide,[95] 5,6-dialkyl-2-pyridyl-1-fumaric esters,[96] and arabinosylcytosine to arabinosyluracil;[97] acetylation of

[87] H. D. Burrows and R. M. Topping, *Chem. Comm.*, **1969**, 904.
[88] A. Hussain and P. Schurman, *J. Pharm. Sci.*, **58**, 687 (1969).
[89] J. de Jersey, P. Willadsen, and B. Zerner, *Biochemistry*, **8**, 1959 (1969); J. de Jersey and B. Zerner, *ibid.*, pp. 1967, 1974; see *Org. Reaction Mech.*, **1966**, 344; J. O. Branstad, *Acta Pharm. Suec.*, **6**, 49 (1969); *Chem. Abs.*, **70**, 114325 (1969).
[90] B. J. Johnson and P. M. Jacobs, *J. Org. Chem.*, **33**, 4524 (1968).
[91] C. Bodea and I. Oprean, *Rev. Roum. Chim.*, **13**, 1647 (1968).
[92] J. F. Biellmann, H. J. Callot, and M. P. Goeldner, *Chem. Comm.*, **1969**, 141; see also M. A. Davis, T. A. Dobson, and J. M. Jordan, *Can. J. Chem.*, **47**, 2827 (1969).
[93] I. Sekikawa, *J. Heterocyclic Chem.*, **6**, 129 (1969).
[94] S. C. K. Su and J. A. Shafer, *J. Org. Chem.*, **34**, 2911 (1969).
[95] T. Yamana, Y. Mizukami, and F. Ichimura, *J. Pharm. Soc. Japan*, **89**, 173 (1969); *Chem. Abs.*, **70**, 10680 (1969).
[96] N. P. Shusherina, O. V. Slavyanova, R. Y. Levina, *Zh. Obshch. Khim.*, **39**, 1182 (1969).
[97] R. E. Notari, M. L. Chin, and A. Cardoni, *Tetrahedron Letters*, **1969**, 3499.

6-epi-mesembranol,[98] and of the 7-hydroxy group of methyl cholate;[99] cyclization of phenylglyoxal aldoxime-(carboxylic acid benzyl ester)-hydrazone,[100] enol esters,[101] and of *o*-aminoacetanilides to benzimidazoles;[102] rearrangement of 2-methyl-1,3($2H,3H$)-dioxoisoquinoline-4-carboxylic acid anilide;[103] reduction of the methyl ester of *o*-nitrophenylpyruvic acid monoxime;[104] and the reaction of *N*-acylanilines with dicarboxylic acids in the molten state to yield cyclic imides.[105]

Several *N→O*-acyl migrations [106, 107] and a *C→O*-acyl migration [108] have been reported.

Association-prefaced Catalysis

Knowles and Parsons have reported another striking example of catalysis involving the pre-association of catalyst with substrate through hydrophobic interactions. The *N*-ethylimidazole-catalysed hydrolysis of *p*-nitrophenyl acetate is about thirty times faster than that of *p*-nitrophenyl decanoate, but the *N*-decylimidazole-catalysed hydrolysis of *p*-nitrophenyl decanoate is about 45 times faster than that of *p*-nitrophenyl acetate. *N*-Decylimidazole is a 740 times more effective catalyst for the hydrolysis of *p*-nitrophenyl decanoate than is *N*-ethylimidazole.[109]

Polyethyleneimines are 2—4 times more effective catalysts for the hydrolysis of *p*-nitrophenyl acetate, caproate, and laurate than is propylamine. Polyethyleneimine of molecular weight 600 with 10% of the imine residue lauroylated is a much more effective catalyst especially towards *p*-nitrophenyl laurate when the catalytic efficiency is 10,000 times that of propylamine. Presumably the laurate ester is bound to the lauroylated polyethyleneimine through hydrophobic interactions.[110]

The hydrolysis of 3-nitro-4-acetoxybenzoic acid is catalysed by copolymers of 1-vinyl-2-methylimidazole and 1-vinylpyrrolidone. When the polymer has

98 P. W. Jeffs, R. L. Hawks, and D. S. Farrier, *J. Am. Chem. Soc.*, **91**, 3831 (1969).
99 R. T. Bickenstaff and B. Orwig, *J. Org. Chem.*, **34**, 1377 (1969).
100 I. Lalezari, A. Shafiee, and M. Yalpani, *Tetrahedron Letters*, **1969**, 3059; see also F. J. Lalor and F. L. Scott, *J. Chem. Soc.*(C), **1969**, 1034.
101 T. Széll, G. Schöbel, and L. Balaspiri, *Tetrahedron*, **25**, 707 (1969); T. Széll, L. Dózsai, M. Zarandy, and K. Menyhárth, *ibid.*, p. 715.
102 K. J. Morgan and A. M. Turner, *Tetrahedron*, **25**, 915 (1969).
103 S. B. Kadin, *J. Org. Chem.*, **34**, 3178 (1969).
104 R. T. Coutts, G. Mukherjee, R. A. Abramovitch, and M. A. Brewster, *J. Chem. Soc.*(C), **1969**, 2207.
105 M. Krasnoselsky and S. Patai, *J. Chem. Soc.*(B), **1969**, 24.
106 S. Suzuki and T. Ishimaru, *J. Chem. Soc. Japan*, **90**, 795, 798, 800 (1969).
107 E. Varga, B. I. Kurtev, and A. Orakhovats, *Izv. Otd. Khim. Nauk.*, *Bulg. Akad. Nauk*, **1**, 79 (1968); *Chem. Abs.*, **71**, 2753 (1969).
108 D. Y. Curtin and J. E. Richman, *Tetrahedron Letters*, **1969**, 3081.
109 J. R. Knowles and C. A. Parsons, *Nature*, **221**, 53 (1969); cf. *Org. Reaction Mech.*, **1967**, 344.
110 G. P. Royer and I. M. Klotz, *J. Am. Chem. Soc.*, **91**, 5885 (1969).

3.5—15.7% imidazole content Michaelis–Menten kinetics were followed but with polymers of higher imidazole content (40 and 60%) more complex behaviour was found. Electrostatic forces do not appear to be important in binding substrate to catalyst since the rate increases when the charge on the catalyst decreases at high pH. Hydrophobic interactions involving the pyrrolidone residue were thought to be more important.[111]

The hydrolysis of ester (60) is catalysed by compound (61). At pH 8 the rate is increased with increasing substrate concentration at low concentrations but reaches a maximum value at high concentrations. The pH dependence of the rate shows that the neutral and monopronated forms of (61) have different catalytic activities, with the former the more active. This suggests that

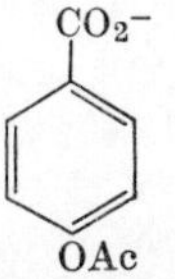

(60) 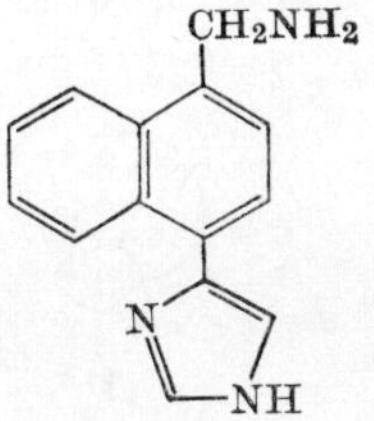   (61)

hydrophobic interactions are more important than electrostatic forces for binding substrate to catalyst.[112]

The hydrolysis of a range of carboxylic acid derivatives in the presence of substituted theophyllines has been studied by Kramer and Connors. It was shown *inter alia* that the reaction of *p*-nitrophenyl benzoate with sulphite is completely inhibited on complex-formation with theophylline but that the reactions of hydroxylamine, hydrazine, and hydrogen peroxide are only partly inhibited. The intermolecular reaction of cinnamoylsalicylic acid with hydroxide ion is strongly inhibited on complex-formation but the intramolecularly catalysed hydrolysis is hardly influenced at all. The hydrolysis of phthalamic acid, which involves intramolecular catalysis, is slightly accelerated in the presence of caffeine, the rate of hydrolysis of methyl hydrogen phthalate is decreased slightly, and that of *p*-nitrophenyl hydrogen glutarate is not affected.[113] The alkaline hydrolysis of aromatic amides in the presence of alkylxanthines has been studied.[114] Imidazole decreases the rate of the

[111] T. Kunitake, F. Shimada, and C. Aso, *J. Am. Chem. Soc.*, **91**, 2716 (1969).
[112] C. Aso, T. Kunitake, and S. Shinkai, *Chem. Comm.*, **1968**, 1483.
[113] P. A. Kramer and K. A. Connors, *J. Am. Chem. Soc.*, **91**, 2600 (1969); see also K. A. Connors, M. H. Infeld, and B. J. Kline, *ibid.*, p. 3597.
[114] K. Kakemi, H. Sezaki, M. Nakano, K. Ohsuga, and T. Mitsunaga, *Chem. Pharm. Bull.* (Japan), **17**, 901 (1969).

hydroxyl-ion catalysed hydrolysis of N-methylphthalimide through complex-formation. It was suggested that the complex might have the structure of a tetrahedral intermediate.[115]

The hydrolyses of methyl orthobenzoate,[116] phenyl orthoformate[117] (see p. 443), and ethyl acetate[118] in the presence of micelles have been investigated.

There have been several other investigations of hydrolyses catalysed by synthetic polymers,[119] and of hydrolyses in the presence of ion-exchange resins.[120]

Metal-ion Catalysis[121]

The hydrolysis of the peptide bond of the glycylglycine complex of β-(triethyl-enetetramine)cobalt(III) **(62)** is about 6.5 times faster than that of glycyl-glycine.[122]

(62)

In contrast is the behaviour found with the analogous nickel complex,[123] and the rate of hydrolysis of L-histidine methyl ester in the presence of Cu(II)-L-histidine is the same as that of D-histidine methyl ester.[124]

Details of Hay and Porter's work on the metal-ion catalysed hydrolysis of cysteine methyl ester have been published.[125]

[115] S. C. K. Su and J. A. Shafer, *J. Org. Chem.*, **34**, 926 (1969).

[116] R. B. Dunlap and E. H. Cordes, *J. Phys. Chem.*, **73**, 361 (1969); cf. *Org. Reaction Mech.*, **1968**, 389, and pp. 217, 400, 473, and 475 of this volume.

[117] M. Price, J. Adams, C. Lagenaur, and E. H. Cordes, *J. Org. Chem.*, **34**, 22 (1969).

[118] Y. Saheki, H. Yoshizane, and K. Negoro, *J. Chem. Soc. Japan*, **89**, 1183 (1968).

[119] M. A. Schwartz, *J. Medicin. Chem.*, **12**, 36 (1969); S. K. Pluzhnov, I. E. Kirsh, V. A. Kabanov, and V. A. Kargen, *Dokl. Akad. Nauk*, **185**, 843 (1969); C. G. Overberger, H. Maki, and J. C. Salamone, *Svensk Kem. Tidskr.*, **80**, 156 (1968); C. G. Overberger, J. C. Salamone, I. Cho, and M. Maki, *Ann. N.Y. Acad. Sci.*, **155**, 431 (1969); C. G. Overberger, *Accounts Chem. Res.*, **2**, 217 (1969); H. Morawetz and B. Vogel, *J. Am. Chem. Soc.*, **91**, 563 (1969).

[120] M. B. Ordyan, T. Y. Eidus, R. K. Bostandzhyan, and A. E. Akopyan, *Armyan. Khim. Zh.*, **20**, 873 (1967); *Chem. Abs.*, **69**, 76123 (1968); *ibid.*, **21**, 728 (1968); *Chem. Abs.*, **70**, 114321 (1969).

[121] A. E. Martell, *Pure Appl. Chem.*, **17**, 129 (1968).

[122] R. W. Hay and P. J. Morris, *Chem. Comm.*, **1969**, 1208.

[123] See *Org. Reaction Mech.*, **1968**, 392.

[124] R. W. Hay and P. J. Morris, *Chem. Comm.*, **1969**, 18; B. E. Leach and R. J. Angelici, *J. Am. Chem. Soc.*, **91**, 6297 (1969).

[125] L. J. Porter, D. D. Perrin, and R. W. Hay, *J. Chem. Soc.*(A), **1969**, 118; R. W. Hay and L. J. Porter, *ibid.*, p. 127; see *Org. Reaction Mech.*, **1967**, 348.

The rate of hydrolysis of esters (**63**) is independent of R but is increased about four-fold when X is changed from Cl to Br. This indicates that loss of halide is rate-limiting. When the reaction was carried out in ^{18}O-enriched water the label was incorporated in approximately equal amounts into each oxygen of

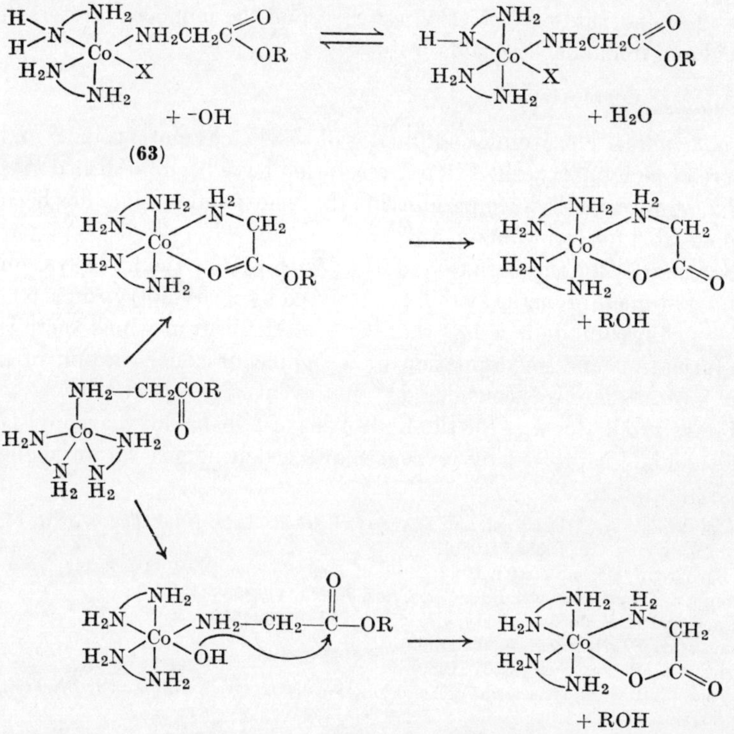

(**63**)

... (24)

... (25)

the product. This was interpreted as indicating that reaction occurred equally by the pathways of equations (24) and (25).[126]

The hydrolyses of thiolbenzoic acid, thiobenzoic anhydride, and *S*-ethyl thiobenzoate are promoted by mercuric ions.[127] The rate of silver-ion catalysed hydrolysis of *S*-methyl thiobenzoates, p-XC_6H_4COSEt, decreases as X is changed in the order $NO_2 > H > MeO$, but the rate of the mercuric-catalysed follows the order $MeO > H > NO_2$. It was suggested that the latter reactions might proceed via a unimolecular decomposition of the complexed ester.[128]

The cobalt-oxalato-complex $[Co(C_2O_4)_3]^{3-}$ exchanges only six of its oxygens with labelled water.[129]

[126] D. A. Buckingham, D. M. Foster, and A. M. Sargeson, *J. Am. Chem. Soc.*, **91**, 4102 (1969).
[127] D. P. N. Satchell and I. I. Secemski, *Chem. Ind.* (London), **1969**, 1632.
[128] D. P. N. Satchell and I. I. Secemski, *Tetrahedron Letters*, **1969**, 1991.
[129] J. A. Broomhead, I. Lauder, and P. Nimmo, *Chem. Comm.*, **1969**, 652.

Other reactions which have been studied include hydrolysis of *p*-nitrophenyl acetate catalysed by copper complexes of dipeptides,[130] Cu(II)-promoted hydrolysis of 2-cyanopyridine to pyridine-2-carboxamide,[131] copper-catalysed formation of peptide bonds,[132] opening of the carbonate ring of $[(en)_2CoCO_3]^+$,[133] and the Cu(II)-promoted hydrolysis of asparagine ethyl ester,[134] *N*-(2-aminoethyl)glycine ethyl ester,[134] and the monoethyl esters of aspartic acid and iminodiacetic acid.[135]

Enzymic Catalysis [1, 136a−142]

(a) *Serine proteinases.* The atomic coordinates of tosyl α-chymotrypsin[143] and the structure of γ-chymotrypsin at 5.5 Å resolution have been published.[144]

1-Oxy-2,2,6,6-tetramethyl-4-piperidinylmethyl phosphofluoridate has been used as a spin label for serine enzymes.[145]

The apparent pK_a for the variation of V_{max} with pH for the hydrolysis of *N*-acetyl-L-tryptophan *p*-methoxyanilide catalysed by α-chymotrypsin is 6.6, and for the *p*-chloroanilide it is less than 6. A mechanism in which there is reversible formation and accumulation of a significant concentration of a tetrahedral intermediate was considered[146] but eventually rejected.[147]

The pH–rate profile for k_{cat} for the hydrolysis of *N*-benzyloxycarbonyl-L-arginine *p*-toluidide catalysed by trypsin shows a dependence on an acidic group of $pK_a(app)$ 6.38.[148]

130 V. I. Salakhutdinov, A. P. Borisova, and I. A. Savich, *Dokl. Akad. Nauk Tadzh. SSR*, **11**, 34 (1968); *Chem. Abs.*, **70**, 106855 (1969).
131 P. F. B. Barnard, *J. Chem. Soc.*(A), **1969**, 2140; cf. *Org. Reaction Mech.*, **1967**, 347.
132 C. Shinomuja and T. Kishikawa, *Bull. Chem. Soc. Japan*, **41**, 3028 (1968).
133 D. J. Francis and R. B. Jordan, *J. Am. Chem. Soc.*, **91**, 6626 (1969).
134 A. Pilbrant, *Acta Pharm. Suec.*, **6**, 469 (1969).
135 A. Pilbrant, *Acta Pharm. Suec.*, **6**, 37 (1969).
136a M. F. Perutz, "X-Ray analysis, structure, and function of enzymes", *European J. Biochem.*, **8**, 455 (1969).
136b A. C. T. North and D. C. Phillips, "X-Ray studies of crystalline proteins", *Progr. Biophys. Mol. Biol.*, **19**, Pt. 1, 5 (1969).
137 M. R. Hollaway, "Enzyme mechanisms", *Ann. Rep. Progr. Chem.*, B, **65**, 601 (1968).
138 A. Williams, "Mechanism of action and specificity of proteolytic enzymes", *Quart. Rev.* (London), **23**, 1 (1969).
139 A. Williams, *Introduction to the Chemistry of Enzyme Action*, McGraw-Hill, London, 1969.
140 S. Doonan, *The Chemistry and Physics of Enzyme Catalysis*, R.I.C. Reviews, **2**, 117 (1969).
141 B. L. Vallee and J. F. Riordan, "Chemical approaches to the properties of active sites of enzymes", *Ann. Rev. Biochem.*, **38**, 733 (1969).
142 S. T. Singer, "Covalent labelling of active sites", *Adv. Protein Chem.*, **22**, 1 (1967).
143 J. J. Birktoft, B. W. Matthews, and D. M. Blow, *Biochem. Biophys. Res. Comm.*, **36**, 131 (1969); see also D. M. Blow, *Biochem. J.*, **112**, 261 (1969).
144 G. M. Cohen, E. W. Silverton, B. W. Matthews, H. Braxton, and D. R. Davies, *J. Mol. Biol.*, **44**, 129 (1969).
145 J. D. Morrisett, C. A. Broomfield, and B. E. Hackley, *J. Biol. Chem.*, **244**, 5758 (1969).
146 M. Caplow, *J. Am. Chem. Soc.*, **91**, 3639 (1969); see also T. Inagami, A. Patchornik, and S. S. York, *J. Biochem.* (Tokyo), **65**, 809 (1969).
147 Vishnu and M. Caplow, *J. Am. Chem. Soc.*, **91**, 6754 (1969).
148 T. Inagami, *J. Biochem.* (Tokyo), **66**, 277 (1969).

Proton uptake on binding of α-chymotrypsin to acetonitrile, benzyl alcohol, and N-acetyl-D- and -L-tryptophan amide has been studied. On binding these compounds the apparent pK of a basic group of the enzyme is changed from 8.8 to a value greater than 10. Dissociation constants for the benzyl alcohol–chymotrypsin complex calculated from the proton uptake agreed quite well with those determined from inhibition studies. These results support the view that the decrease in k_2/K_s at alkaline pH results from an increase in K_s.[149]

K_m for the δ-chymotrypsin-catalysed hydrolysis of N-acetyl-L-tryptophan methyl ester and N-furylacryloyl-L-tryptophan amide increases much less in alkaline solution than does K_m for reactions catalysed by α-chymotrypsin. The apparent pK_a for the relatively small variation in K_m is 9.25. This suggests that the ability of α-chymotrypsin to bind substrates depends only slightly on the state of ionization of isoleucine-16. It was therefore suggested that the reversible inactivation of α-chymotrypsin at high pH depends on the state of ionization of tyrosine-146 or alanine-149 which are end-groups in α- but not in δ-chymotrypsin.[150]

The change in conformation of α-chymotrypsin at alkaline pH has been studied by observing the change in the CD and ORD spectra which takes place.[151] The CD spectrum of chymotrypsinogen has been reported.[152]

It has been suggested from acetylation studies that the N-terminal α-amino group of isoleucine controls the activity of trypsin at high pH.[153] Acylation of the exposed tyrosine residues of trypsin increases its activity towards specific low molecular weight substrates.[154] Trypsin has been allowed to react with ethyl p-nitrophenyl diazomalonate. The resulting acyl enzyme was photolysed so that the carbene formed reacted with an alanine residue which was converted on further reaction into glutamic acid.[155a]

The mechanisms of action of trypsin and chymotrypsin have been compared.[155b]

α-Chymotrypsin reacts with methyl p-nitrobenzenesulphonate to yield a catalytically inactive product in which the imidazole group of histidine-57 is methylated.[156] Phenacyl bromides alkylate methionine-192 of chymotrypsin. The alkylated enzyme has a strong optically active absorption at 290—365 nm which was attributed to ylid formation.[157]

[149] F. C. Wedler and M. L. Bender, *J. Am. Chem. Soc.*, **91**, 3894 (1969); *Org. Reaction Mech.*, **1967**, 349—350.
[150] P. Valenzuela and M. L. Bender, *Proc. Nat. Acad. Sci.*, **63**, 1214 (1969).
[151] J. McConn, G. D. Fasman, and G. P. Hess, *J. Mol. Biol.*, **39**, 551 (1969).
[152] E. H. Strickland, J. Horwitz, and C. Billups, *Biochemistry*, 8, 3205 (1969).
[153] J. Chevallier, J. Yon, and J. Labouesse, *Biochem. Biophys. Acta*, **181**, 73 (1969); see also A. Nureddin and T. Inagami, *Biochim. Biophys. Res. Comm.*, **36**, 999 (1969).
[154] H. L. Trendholm, W. E. Spomer, and J. F. Wootton, *Biochemistry*, 8, 1741 (1969).
[155a] R. J. Vaughan and F. H. Westheimer, *J. Am. Chem. Soc.*, **91**, 217 (1969).
[155b] H. P. Kasserra and K. J. Laidler, *Can. J. Chem.*, **47**, 4021, 4031 (1969).
[156] Y. Nakagawa and M. L. Bender, *J. Am. Chem. Soc.*, **91**, 1566 (1969).
[157] D. S. Sigman, P. A. Torchia, E. R. Blout, *Biochemistry*, 8, 4560 (1969).

A number of active-site directed irreversible inhibitors for trypsin[158] and α-chymotrypsin[159] have been reported.

The lactones (**64**) and (**65**) react with α-chymotrypsin to form acyl enzymes. That derived from (**65**) undergoes deacylation to yield the acid but that derived from (**64**) re-forms the lactone.[160]

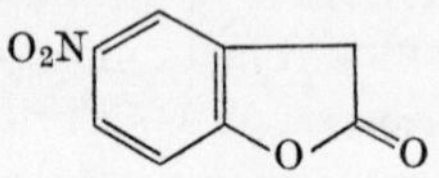

(**64**) 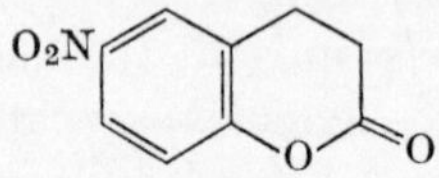   (**65**)

The rate of acylation of chymotrypsin by the *p*-nitrophenyl esters of aliphatic acids follows the order expected from the operation of a steric effect; i.e. acetate > propionate > isobutyrate > trimethylacetate. However, *p*-nitrophenyl hexanoate reacts nearly ten times faster than *p*-nitrophenyl acetate.[161] Deacylation of trifluoromethyl-substituted-acyl chymotrypsins[162] and the trypsin-catalysed hydrolysis of *p*-nitrophenyl esters of *N*-(α)-tosyl derivatives of lysine and arginine[163] have been studied.

k_{cat}/K_m for the chymotrypsin-catalysed hydrolysis of (D-**66**) is about 34,000 times greater than that for (L-**66**) in DMSO–water mixtures. (D-**67**) is less reactive than (D-**66**) and k_{cat}/K_m is only about 4 times greater than for (L-**67**).[164] The active-site specificity of α-chymotrypsin has been discussed.[165]

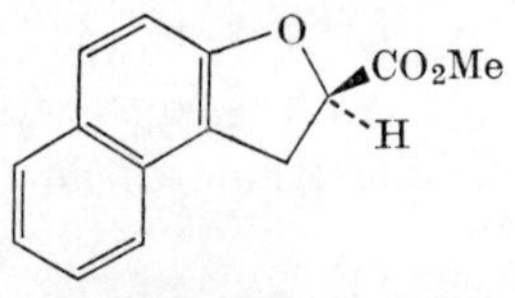

(D-**66**) 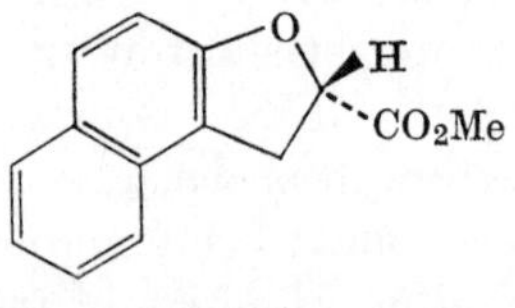   (L-**66**)

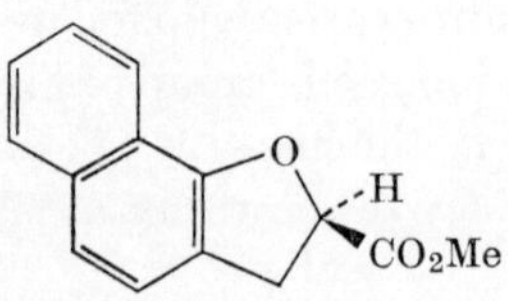

(D-**67**) 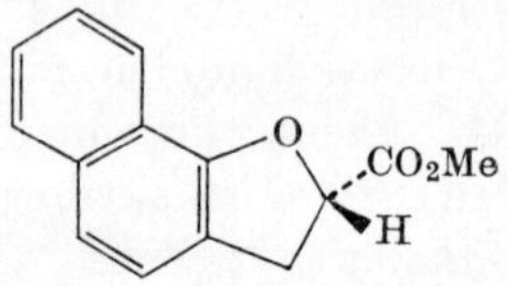   (L-**67**)

[158] B. R. Baker and E. H. Erickson, *J. Medicin. Chem.*, **12**, 112 (1969).
[159] B. R. Baker and J. A. Hurlbut, *J. Medicin. Chem.*, **12**, 118 (1969).
[160] P. Tobias, J. H. Heidema, K. W. Lo, E. T. Kaiser, and F. J. Kezdy, *J. Am. Chem. Soc.*, **91**, 202 (1969).
[161] J. B. Milstien and T. H. Fife, *Biochemistry*, **8**, 623 (1969).
[162] P. M. Enriquez and J. T. Gerig, *Biochemistry*, **8**, 3156 (1969).
[163] L. G. Yurganova, *Biokhimiya*, **34**, 55 (1969); *Chem. Abs.*, **70**, 106830 (1969).
[164] Y. Hayashi and W. B. Lawson, *J. Biol. Chem.*, **244**, 4158 (1969).
[165] S. G. Cohen, *Trans. N.Y. Acad. Sci.*, **31**, 705 (1969); S. G. Cohen, A. Milovanović, R. M. Schultz, and S. Y. Weinstein, *J. Biol. Chem.*, **244**, 2664 (1969).

p-Nitrophenyl *p*-guanidinobenzoate has been used to titrate trypsin, thrombin, plasmin,[166] and cocoonase,[167] and 2-phenyloxazolin-5-one and 4,4-dimethyl-2-phenyloxazolin-5-one have been used to titrate α-chymotrypsin, trypsin, and papain.[168]

The kinetics of binding of *N*-trifluoroacetyl-D-phenylalanine to chymotrypsin and DFP-chymotrypsin have been measured by an NMR method.[169]

Although thiol-subtilisin[170] binds substrates as well as subtilisin, it is without catalytic activity except for a very small effect on the hydrolysis of *p*-nitrophenyl esters. It was suggested that in subtilisin itself there must be a critical alignment of catalytic groups which is necessary for activity.[171]

The specificity of α-chymotrypsin and subtilisin has been compared.[172]

Other investigations on chymotrypsin[173] and trypsin[174] have been reported.

The plots of k_{cat}/K_m for the hydrolysis of *N*-furylacryloyl-L-alanine methyl ester and *N*-benzoylalanine methyl ester catalysed by elastase appear to be bell-shaped with apparent pK_a values 6.85 and 11.4.[175] It seems likely that the fall in rate at high pH is the result of deprotonation of an *N*-terminal-valyl α-amino group, since the activity of the enzyme is lost when this group is acetylated.[176a] The specificity of elastase has been investigated.[176b]

[166] T. Chase and E. Shaw, *Biochemistry*, **8**, 2212 (1969).

[167] J. F. Hruska, J. H. Law, and F. J. Kezdy, *Biochem. Biophys. Res. Comm.*, **36**, 272 (1969).

[168] J. de Jersey and B. Zerner, *Biochemistry*, **8**, 1967, 1975 (1969).

[169] B. D. Sykes, *J. Am. Chem. Soc.*, **91**, 949 (1969); *Biochem. Biophys. Res .Comm.*, **33**, 727 (1968).

[170] See *Org. Reaction Mech.*, **1966**, 352—353.

[171] K. E. Neet, A. Nanci, and D. E. Koshland, *J. Biol. Chem.*, **243**, 6392 (1968); see also L. Polgar and M. L. Bender, *Biochemistry*, **8**, 136 (1969).

[172] K. Morihara, T. Oka, and H. Tsuzuki, *Biochem. Biophys. Res. Comm.*, **35**, 210 (1969).

[173] T. Inagami, A. Patchornik, and S. S. York, *J. Biochem.* (Tokyo), **65**, 809 (1969); W. H. K. Martinek, Z. A. Stvel'tsova, and I. V. Beresin, *Molekulyarnaya Biologiya*, **3**, 554 (1969); K. L. Carraway, P. Spoerl, and D. E. Koshland, *J. Mol. Biol.*, **42**, 133 (1969); J. P. Abita and M. Lazdunski, *Biochem. Biophys. Res. Comm.*, **35**, 707 (1969); A. J. Hymes, C. C. Cuppett, and W. J. Canady, *J. Biol. Chem.*, **244**, 637 (1969); B. H. J. Hofstee, *ibid.*, **243**, 6306 (1968); R. Biltonen and R. Lumry, *J. Am. Chem. Soc.*, **91**, 4251, 4256 (1969); S. N. Timasheff, *Arch. Biochem. Biophys.*, **132**, 165 (1969); F. Friedberg and S. Bose, *Biochemistry*, 8, 2564 (1969); V. K. Antonov and L. D. Rumsh, *Dokl. Akad. Nauk SSSR*, **185**, 821 (1969); A. Matsushima, Y. Inada, and K. Shibata, *J. Biochem.* (Toyko), **66**, 423 (1969); A. A. Aaviksaar, I. G. Nolvak, and V. A. Palm, *Org. Reactivity* (Tartu), **5**, 1053 (1969); A. A. Aaviksaar, E. V. Rozengart, P. F. Sikk, and R. A. Herbst. *ibid.*, p. 1059; M. G. Gonikberg, N. I. Prokhorova, and B. S. El'yanov, *Izv. Akad. Nauk SSSR, Ser. Khim.*, **1968**, 2841; B. Meloun, I. Frič, and F. Šorm, *Coll. Czech. Chem. Comm.*, **34**, 3127 (1969).

[174] T. M. Radhakrishnan, K. A. Walsh, and H. Neurath, *Biochemistry*, **8**, 4020 (1969); V. Holeyšovsky, B. Keil, and F. Sorm, *FEBS Letters*, **3**, 107 (1969); L. G. Yurganova, *Biokhimiya*, **34**, 55 (1969); J. P. Abita, M. Delaage, M. Lazdunski, and J. Savada, *European J. Biochem.*, **8**, 314 (1969); M. A. Coletti-Previero, A. Previero, and E. Zuckerkandl, *J. Mol. Biol.*, **39**, 493 (1969); T. Inagami and H. Hatano, *J. Biol. Chem.*, **244**, 1176 (1969); F. M. Pohl, *European J. Biochem.*, **7**, 146 (1968); D. E. Lenz and W. P. Bryan, *Biochemistry*, 8, 1123 (1969).

[175] P. Geneste and M. L. Bender, *Proc. Nat. Acad. Sci.*, **64**, 683 (1969).

[176a] H. Kaplan and H. Dugas, *Biochem. Biophys. Res. Comm.*, **34**, 681 (1969).

[176b] A. S. Narayanan and R. A. Amwar, *Biochem. J.*, **114**, 11 (1969).

Streptococcal proteinase[177] and kallilrein from porcine pancreas[178] have been investigated.

(b) *Thiol proteinases.* It has been shown that K_I for inhibition by α-N-benzoyl-D-arginine ethyl ester for the hydrolysis of α-N-benzoyl-L-arginine ethyl ester catalysed by ficin is independent of pH over the range 4.0—9.5. This suggests that K_s for binding of the substrate is also independent of pH, and that the bell-shaped curve for the variation of $k_{cat}/K_m(app) = k_2/K_s$ results from variation of k_2 the acylation rate constant. This then appears to depend on the ionization of two groups of pK_a 4.46 and 8.37.[179]

Further work on the sequence of papain has been reported.[180]

Other investigations on ficin,[181] papain,[182] and bromelain[183] have been described.

(c) *Acid proteinases.* When benzyloxycarbonyl-[^{3}H]-L-tyrosyl-[^{3}H]-L-tyrosine is allowed to react with pepsin, and the reaction is terminated by the addition of trichloroacetic acid, the precipitated protein is radioactive. When [^{3}H]-L-tyrosyl-[^{3}H]-L-tyrosine is added immediately before precipitation, the protein is not radioactive. The radioactivity could not be removed from the protein by dialysis in the presence or absence of hydroxylamine. There was no incorporation of radioactivity when benzyloxycarbonyl-[^{3}H]-L-tyrosyl-L-tyrosine was used. The radioactivity that was incorporated must therefore have been that which was in the carboxy terminal residue. This result provides good evidence for the incursion of a covalent amino-protein. The mechanism shown in the scheme was proposed. It was suggested that pepsin has a potential anhydride group which cleaves peptides with formation of an intermediate which is an acyl and an amino enzyme (**69**). This is subsequently hydrolysed via (**70**) and (**71**). The method of isolation used in the above experiments would yield only (**71**).[184] When benzyloxycarbonyl-L-tyrosyl-L-tyrosine is allowed to react with pepsin in the presence of [^{3}H$_3$]methanol, tritium is incorporated into the enzyme but no methyl esters are formed when dipeptides are hydrolysed in the presence of radioactive methanol. This supports the

[177] T. Y. Liu, N. Nomura, E. K. Jonsson, and B. G. Wallace, *J. Biol. Chem.*, **244**, 5745 (1969).

[178] F. Fiedler and E. Werk, *European J. Biochem.*, **7**, 27 (1968).

[179] J. R. Whitaker, *Biochemistry*, **8**, 1896, 4591 (1969).

[180] S. S. Husain and G. Lowe, *Biochem. J.*, **114**, 279 (1969).

[181] M. R. Hollaway, E. Antonini, and M. Brunori, *FEBS Letters*, **4**, 299 (1969).

[182] B. G. Wolthers, *FEBS Letters*, **2**, 143 (1969); L. A. A. Sluyterman and M. J. M. De Graaf, *Biochim. Biophys. Acta*, **171**, 277 (1969); L. A. A. Sluyterman, J. Wijdenes, and B. G. Wolthers, **178**, 392 (1969); E. Shapira and R. Arnon, *J. Biol. Chem.*, **244**, 1004, 1026, 1033 (1969); I. B. Klein and J. F. Kirsch, *ibid.*, p. 5928; I. B. Klein and J. F. Kirsch, *Biochem. Biophys. Res. Comm.*, **34**, 575 (1969); E. L. Smith and I. M. Chaikin, *J. Biol. Chem.*, **244**, 5087, 5095 (1969).

[183] C. W. Wharton, E. M. Crook, and K. Brocklehurst, *European J. Biochem.*, **6**, 572 (1968).

[184] M. Akhtar and J. M. Al-Janabi, *Chem. Comm.*, **1969**, 859.

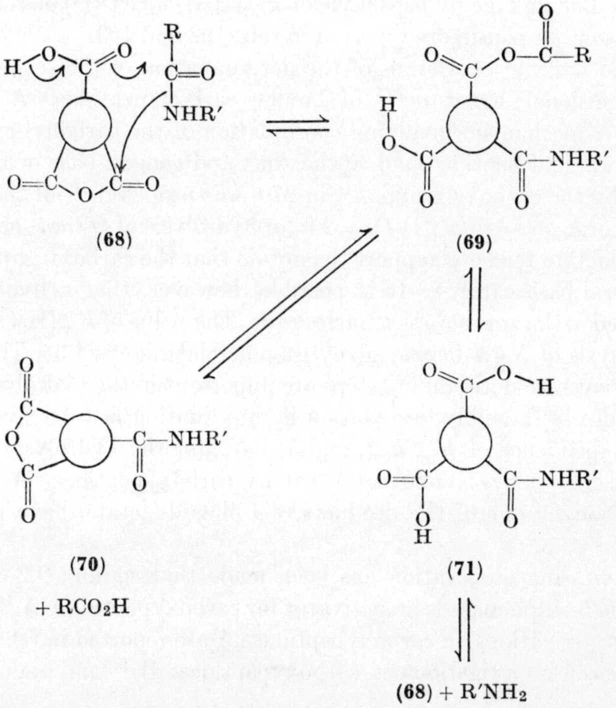

(68)

(69)

(70)

+ RCO$_2$H

(71)

(68) + R'NH$_2$

view that there is an anhydride intermediate, e.g. **(69)**, which is cleaved by attack at the enzyme carboxy group.[185, 186]

The binding of the uncharged inhibitor for pepsin, N-acetyl-D-phenylalanyl-L-phenylalanine amide is independent of pH over the range 0.22—5.84. The value of K_i for binding of N-acetyl-D-phenylalanyl-L-phenylalanine and its L-D-isomer increases at pH about 3.5. This was attributed to the ionization of their carboxy groups. The value of the ratio k_{cat}/K_m for the hydrolysis of N-acetyl-L-phenylalanyl-L-phenylalanine amide depends on the ionization of two groups of apparent pK_a 1.05 and 4.75. In view of the pH independence of the binding constant of the uncharged inhibitor, it was considered that these groups were involved in the catalytic action.[187—189] Pepsin reacts with N-diazoacetyl-L-phenylalanine methyl ester with loss of catalytic activity. The reagent reacts with an aspartic acid carboxy group in the sequence Ile-Val-Asp-Thr-Gly-Thr-Ser.[190] Cleavage of Gly-Gly-Gly-Phe(NO$_2$)-Phe-OMe at the

[185] M. Akhtar and J. M. Al-Janabi, *Chem. Comm.*, **1969**, 1002; see, however, M.Akhtar *ibid.*, **1970**, 361; T. M. Kitson and J. R. Knowles, *ibid.*, p. 361.

[186] A. J. Cornish-Bowden, P. Greenwell, and J. R. Knowles, *Biochem. J.*, **113**, 369 (1969).

[187] J. R. Knowles, H. Sharp, and P. Greenwell, *Biochem. J.*, **113**, 343 (1969).

[188] A. J. Cornish-Bowden and J. R. Knowles, *Biochem. J.*, **113**, 353 (1969).

[189] P. Greenwell, J. R. Knowles, and H. Sharp, *Biochem. J.*, **113**, 363 (1969).

[190] R. S. Bayliss, J. R. Knowles, and G. B. Wybrandt, *Biochem. J.*, **113**, 377 (1969).

Phe(NO$_2$)–Phe linkage by pepsin yields $k_{cat}(H_2O)/k_{cat}(D_2O)$ of 1.97.[191] Other investigations on pepsin are reported in refs. 192 and 193.

(d) *Metallo-proteinases*. Details of the determination of the sequence[194] and three-dimensional structure[195] of bovine carboxypeptidase A have been reported. A mechanism involving coordination of the carbonyl group of the carboxy terminal peptide bond to the zinc, and general-base or nucleophilic catalysis by the carboxy group of Glu-270, was proposed.[195] If this is correct the value of $k_{cat}(H_2O)/k_{cat}(D_2O) = 2.0$ for hydrolysis of *O-trans*-cinnamoyl-L-α-phenyl-lactate tends to support the notion that the carboxy group is acting as a general-base catalyst. It is possible, however, that a hydroxy group coordinated to the zinc acts as a nucleophile. The value of $k_{cat}(H_2O)/k_{cat}(D_2O)$ for hydrolysis of *N*-(*N*-benzoylglycyl)-L-phenylalanine is 1.33. This suggests that different rate-determining steps are important in the hydrolysis of esters and peptides.[196] Possibly deacylation is rate-limiting in ester hydrolysis.[195] The pH dependence of K_m, k_{cat}, and k_{cat}/K_m for the hydrolysis of *O-trans*-cinnamoyl-L-α-phenyl-lactate catalysed by carboxypeptidase A, and of K_i for inhibition by one of the products, L-α-phenyl-lactate, have been determined.[197]

The interesting observation has been made that sodium 2,2-dimethyl-2-silapentane-5-sulphonate is an activator for carboxypeptidase A.[198]

Other investigations on carboxypeptidase A are reported in ref. 199. There have also been investigations on carboxypeptidase B[200] and aminopeptidase B.[201]

(e) *Esterases*. There have been several investigations of liver esterases. The

191 T. R. Hollands and J. S. Fruton, *Proc. Nat. Acad. Sci.*, **62**, 1116 (1969); T. R. Hollands, I. M. Voynick, and J. S. Fruton, *Biochemistry*, **8**, 515 (1969); G. E. Trout and J. S. Fruton, *ibid.*, p. 4183; G. P. Sachdev and J. S. Fruton, *ibid.*, p. 4231.

192 R. L. Lundblad and W. H. Stein, *J. Biol. Chem.*, **244**, 154 (1969); A. M. Tamburro, A. Scattarm, and R. Rocchi, *Gazzetta*, **98**, 1256 (1968).

193 S. W. May and E. T. Kaiser, *J. Am. Chem. Soc.*, **91**, 6491 (1969).

194 R. A. Bradshaw, L. H. Ericsson, K. A. Walsh, and H. Neurath, *Proc. Nat. Acad. Sci.*, **63**, 1389 (1969); P. H. Pétra and H. Neurath, *Biochemistry*, **8**, 2466 (1969); M. Nomoto, N. G. Srinivasan, R. A. Bradshaw, R. D. Wade, and H. Neurath, *ibid.*, p. 2755; P. H. Petra, R. A. Bradshaw, K. A. Walsh, and H. Neurath, *ibid.*, p. 2762; R. A. Bradshaw, D. R. Babin, M. Nomoto, N. G. Srinivasin, L. H. Ericsson, K. A. Walsh, and H. Neurath, *ibid.*, p. 3859; R. A. Bradshaw, *ibid.*, p. 3871.

195 W. N. Lipscomb, J. A. Hartsuck, F. A. Quiocho, and G. N. Reeke, *Proc. Nat. Acad. Sci.*, **64**, 28 (1969).

196 B. L. Kaiser and E. T. Kaiser, *Proc. Nat. Acad. Sci.*, **64**, 36 (1969).

197 P. L. Hall, B. L. Kaiser, and E. T. Kaiser, *J. Am. Chem. Soc.*, **91**, 485 (1969).

198 M. Epstein and G. Navon, *Biochem. Biophys. Res. Comm.*, **36**, 126 (1969).

199 B. L. Vallee, *Ann. N.Y. Acad. Sci.*, **158**, 377 (1969); H. M. Kagan and B. L. Vallee, *Biochemistry*, **8**, 4223 (1969); R. C. Adelman and A. G. Lacko, *Biochem. Biophys. Res. Comm.*, **33**, 596 (1968).

200 J. H. Seeley and N. L. Benoiton, *Biochem. Biophys. Res. Comm.*, **37**, 721 (1969); T. H. Plummer, *J. Biol. Chem.*, **244**, 5246 (1969).

201 K. K. Makinen, *Suomen Kemistilehti*, *B42*, 129, 246 (1969).

sequences around the active-site serines are identical in the pig, sheep, and horse enzymes and differ by one residue in the ox and chicken enzymes.[202] The hydrolysis of p-nitrophenyl dimethylcarbamate catalysed by the pig-liver enzyme and of o-nitrophenyl dimethylcarbamate catalysed by the ox-liver enzyme show an initial rapid burst followed by a slow zero-order turnover of substrate. This indicates the intervention of acyl-enzyme intermediates. The rates of deacylation of acyl-enzymes derived from the liver esterases are 10^5 times greater than those of the corresponding acyl-chymotrypsins.[203, 204] The pig-liver enzyme shows substrate activation and is also activated by benzene and acetone.[203, 205] Catalysis of acyl transfer to amino-acid esters by the beef-liver esterase has been studied.[206]

There have been several investigations of choline-esterases.[207]

(f) *Other enzymes.* Carbonic anhydrase[208] and jack bean urease[209] have been studied.

Other Reactions

Bruice and Holmquist have shown that the esters $NCCH_2CO_2Ar$ and $EtO_2CCH_2CO_2Ar$ with Ar o- or p-nitrophenyl are hydrolysed via an elimination–addition mechanism with a ketene intermediate. The rates of these reactions reach steady values in alkaline solutions where the methylene groups are ionized, and at pH 7 the rates are about 40 and 10^4 times greater than expected from the rates of the water-catalysed reactions on the basis of a linear free-energy relationship for the ^-OH- and H_2O-catalysed reactions of other esters. The hydrolyses of α,α-dimethylcyanoacetic and dimethylmalonic esters are normal.[210] Similar behaviour is found with $Me_2S^+–CH_2CO_2Ar$ when Ar is o-nitrophenyl but the rate enhancement is only about 10, so although

[202] R. C. Augusteyn, J. de Jersey, E. C. Webb, and B. Zerner, *Biochim. Biophys. Acta*, **171**, 128 (1969).

[203] D. J. Morgan, J. K. Stoops, E. C. Webb, and B. Zerner, *Biochemistry*, **8**, 2000 (1969); D. J. Morgan, J. R. Dunstone, J. K. Stoops, E. C. Webb, and B. Zerner, *ibid.*, p. 2006; M. T. C. Runnegar, K. Scott, E. C. Webb, and B. Zerner, *ibid.*, p. 2013; M. T. C. Runnegar, E. C. Webb, and B. Zerner, *ibid.*, p. 2018; J. K. Stoops, D. J. Morgan, M. T. C. Runnegar, J. de Jersey, E. C. Webb, and B. Zerner, *ibid.*, p. 2026.

[204] B. Zerner, *Proc. Austral. Biochem. Soc.*, **1969**, 8.

[205] D. L. Barker and W. P. Jencks, *Biochemistry*, **8**, 3879, 3890 (1969).

[206] M. I. Goldberg and J. S. Fruton, *Biochemistry*, **8**, 86 (1969).

[207] J. Patočka and J. Tulach, *Coll. Czech. Chem. Comm.*, **34**, 3191 (1969); A. H. Beckett, C. L. Vaughan, and M. Mitchard, *Biochem. Pharmacol.*, **17**, 1591 (1968); W. Leuzinger, M. I. Goldberg, and E. Chauvin, *J. Mol. Biol.* **40**, 217 (1969); R. D. O'Brien, *Biochem. J.*, **113**, 713 (1969); F. Iverson and A. R. Main, *Biochemistry*, **8**, 1889 (1969).

[208] R. L. Ward, *Biochemistry*, **8**, 1879 (1969); E. T. Kaiser and K.-W. Lo, *J. Am. Chem. Soc.*, **91**, 4912 (1969); S. L. Bradbury, *J. Biol. Chem.*, **244**, 2002, 2010 (1969); R. W. Henkens, G. D. Watt, and J. M. Sturtevant, *Biochemistry*, **8**, 1874 (1969).

[209] R. L. Blakeley, E. C. Webb, and B. Zerner, *Biochemistry*, **8**, 1984 (1969); R. L. Blakeley, J. A. Hinds, M. E. Kunze, E. C. Webb, and B. Zerner, *ibid.*, p. 1991.

[210] T. C. Bruice and B. Holmquist, *J. Am. Chem. Soc.*, **90**, 7136 (1968); **91**, 2993 (1969).

reaction via a ketene intermediate is possible it is less certainly established.[211] Alkaline hydrolysis of $Me_2S^+-CH_2CO_2Me$ was thought not to proceed via a ketene intermediate.[212] The hydrolysis of 5-nitrocoumarone shows a dependence of rate on pH similar to the cyanoacetic and malonic esters but here, on the basis of isotope effects, formation of the carbanion was thought to be a parasitic equilibrium[213] (see also ref. 340, p. 479).

The methanolysis of acyl halides with MeOD in pentane in the presence of trimethylamine yields an ester which is partly monodeuterated, and it was suggested that reaction proceeds partly via a ketene and partly via an acyl quaternary derivative.[214]

A plot of log $k(H_2O)$ for the water-catalysed reactions of a series of esters of structure $XO_2CC_6H_4NO_2$-o against log $k(OH)$ for the reactions with hydroxide ions is a straight line. Esters in which X is positively or negatively charged show no significant deviations, and it was concluded that electrostatic effects are unimportant in the reactions of these with ^-OH. When log k for reactions of other nucleophiles was plotted against log $k(H_2O)$ there were significant deviations from straight-line plots. Positively charged esters react abnormally rapidly with acetate, phosphate, and carbonate but not with hydroxide and trifluoroethoxide, and abnormally slowly with ethylenediamine, methoxylamine, and glycine ethyl ester. The high rates shown by acetate, phosphate, and carbonate were attributed to electrostatic stabilization of the transition state, and the low rates for the other nucleophiles to electrostatic destabilization.[215]

The mechanism of hydrolysis of 3-t-butylsydnones and 3-furfurylsydnones appears to be $A1$.[216]

The effects of substituents on the rates of the alkaline hydrolysis of ethyl nicotinates[217] and methyl and ethyl acetates[218] have been determined. The alkaline hydrolyses of the esters from polycyclic hydrocarbon carboxylic acids[219] and from benzofuran, benzothiophen, and indole carboxylic acids[220a] have been studied.

A linear free-energy relationship has been established between the rate of solvolysis of substituted alkyl esters and the pK_a of the alcohol.[220b]

[211] B. Holmquist and T. C. Bruice, *J. Am. Chem. Soc.*, **91**, 3003 (1969).
[212] J. Casanova and D. A. Rutolo, *J. Am. Chem. Soc.*, **91**, 2347 (1969).
[213] P. S. Tobias and F. J. Kézdy, *J. Am. Chem. Soc.*, **91**, 5171 (1969).
[214] W. E. Truce and P. S. Bailey, *J. Org. Chem.*, **34**, 1341 (1969).
[215] B. Holmquist and T. C. Bruice, *J. Am. Chem. Soc.*, **91**, 2982, 2985 (1969).
[216] S. Aziz and J. G. Tillett, *Tetrahedron Letters*, **1969**, 2855; cf. *Org. Reaction Mech.*, **1967**, 357; **1968**, 399.
[217] Y. Veno and E. Imoto, *J. Chem. Soc. Japan*, **88**, 1210 (1967); *Chem. Abs.*, **69**, 66782 (1968).
[218] J. Barthel, G. Bäder, and G. Schmeer, *Z. Phys. Chem.* (Frankfurt), **62**, 63 (1968).
[219] J. F. Corbett, A. Feinstein, P. H. Gore, G. L. Reed, and E. C. Vignes, *J. Chem. Soc.*(B), **1969**, 974.
[220a] A. Feinstein, P. H. Gore, and G. L. Reed, *J. Chem. Soc.*(B), **1969**, 205.
[220b] J. R. Robinson and L. E. Matheson, *J. Org. Chem.*, **34**, 3630 (1969).

The hydrolyses of the following esters have also been investigated: *o*- and *p*-nitrophenyl alkanoates;[221] ethyl esters of methylenedioxy-, ethylenedioxy-, and triethylenedioxy-benzoic acid;[222] atropine;[223] aminoacyl derivatives of 2-acetamido-2-deoxy-D-glucose, 2-amino-2-deoxy-D-glucose, and *N*-acetyl-neuramic acid;[224] dimethyl oxalate;[225] acetates, propionates, and butyrates of steroid alcohols;[226] α,β-unsaturated amino-acid esters;[227] acrylic esters of phenols with phosphine oxide substituents;[228] diethyl carbonate;[229] esters of thiamine;[230a] positively charged esters;[230b] alkyl adipates;[231] methyl,[232, 233] ethyl,[233, 234] and pentyl[235] acetate; acetates of 4-hydroxyazobenzenes and 2-hydroxy-5-methylazobenzenes;[236] and Meldrum's acid.[237]

There have been several investigations of the reactions of carboxylic acids with diphenyldiazomethane,[238] and of other esterification reactions.[239]

The rates of the acid-catalysed hydrolyses of aliphatic amides have been correlated by the Taft equation,[240] and those of substituted benzanilides by the Hammett equation.[241]

[221] V. Rangacah and B. S. Rao, *Current Sci.* (India), **37**, 71 (1968).
[222] M. N. Byrne and N. H. P. Smith, *J. Chem. Soc.*(B), **1968**, 809.
[223] W. Lund and T. Waaler, *Acta Chem. Scand.*, **22**, 3085 (1968).
[224] V. A. Derevitskaya and V. M. Kalinevich, *Khim. Prir. Soed.*, **4**, 28 (1968); *Chem. Abs.*, **69**, 36387 (1968).
[225] J. Vuori and J. Koskikallio, *Suomen Kemistilehti*, B**42**, 136 (1969).
[226] Z. Vesetý, J. Pospíšek, and J. Trojánek, *Coll. Czech. Chem. Comm.*, **34**, 1801 (1969).
[227] Y. I. Turyan, F. K. Ignateva, and M. A. Korshuvov, *Zh. Obshch. Khim.*, **38**, 2405 (1968).
[228] V. E. Bel'skii, M. S. Lapin, A. A. Muslinkin, and S. Z. Shengurova, *Izv. Akad. Nauk SSSR, Ser. Khim.*, **1968**, 1732; *Chem. Abs.*, **70**, 10735 (1969).
[229] A. Kivinen and A. Viitala, *Suomen Kemistilehti*, B**41**, 372 (1968).
[230a] K. Inazu, S. Shigeyo, and R. Yamamoto, *J. Pharm. Soc. Japan*, **88**, 1552 (1968); *Chem. Abs.*, **70**, 76947 (1969).
[230b] M. R. Wright, *J. Chem. Soc.*(B), **1969**, 707.
[231] C. Kimura, K. Murai, and H. Yamamoto, *J. Chem. Soc. Japan, Ind. Chem. Sect.*, **71**, 1569 (1968); *Chem. Abs.*, **70**, 36764 (1969).
[232] R. S. Roy, *Analyt. Chem.*, **40**, 1724 (1968).
[233] R. Vilcu and I. Ciocazanu, *Rev. Roum. Chim.*, **14**, 429, 561 (1969); R. S. Roy and H. N. Al-Jallo, *Analyt. Chem.*, **40**, 1725 (1968).
[234] R. N. O'Brien and A. Glasel, *Can. J. Chem.*, **47**, 223 (1969).
[235] E. S. M. De Ponce and C. A. Ponce, *An. Quim.*, **64**, 379 (1968); *Chem. Abs.*, **69**, 76121 (1969).
[236] J. Socha, J. Horska, and M. Večeřa, *Coll. Czech. Chem. Comm.*, **34**, 3178 (1969).
[237] K. Pihlaja and M. Seilo, *Acta Chem. Scand.*, **22**, 3053 (1968).
[238] A. Buckley, N. B. Chapman, and J. Shorter, *J. Chem. Soc.*(B), **1969**, 195; N. B. Chapman, J. R. Lee, and J. Shorter, *ibid.*, p. 769; K. Bowden and D. C. Parkin, *Can. J. Chem.*, **47**, 177, 185, 3909 (1969); K. Bowden and R. C. Young, *ibid.*, p. 2775.
[239] A. L. Markham, M. Mirzabaeva, and A. I. Glushenkova, *Uzb. Khim. Zh.*, **12**, 28 (1968); *Chem. Abs.*, **70**, 36763 (1969); F. Mareš, V. Bazant, and J. Krupička, *Coll. Czech. Chem. Comm.*, **34**, 2208 (1969); M. F. Sorokin, Z. A. Kochnova, and I. S. Krivopatova, *Tr. Mosk. Khim. Tekhnol. Inst.*, **57**, 61 (1968); *Chem. Abs.*, **70**, 105605 (1969).
[240] P. D. Bolton and G. L. Jackson, *Austral. J. Chem.*, **22**, 527 (1969).
[241] V. L. Levashova, N. P. Lushina, and V. F. Mandyuk, *Zh. Org. Khim.*, **4**, 1846 (1968); *Chem. Abs.*, **70**, 19311 (1969).

The hydrolyses of the following amides have also been investigated:
phenylacetamide;[242] *N*-benzylnicotinamide;[243] 1-formyl-1,2,3,4-tetrahydro-
isoquinoline;[244] *o*-nitroformanilide;[245] diketopiperazines;[246] peptides;[247] and
N-acetylsulphonamides.[248]

The acid-catalysed hydrolysis of carbamic esters appears to involve the
O-protonated form and a rate-limiting C–O bond cleavage.[249]

The hydrolysis of ϵ-caprolactam[250] and its reaction with benzoic acid
hydrazide[251] have been investigated.

The hydrolyses of the following compounds have also been studied:
peresters;[252] D-glucuronolactone;[253] nitriles;[254] Antimycin A_1;[255] phthali-
mides;[256] anhydrides;[257] and 3,5-disubstituted tetrahydro-1,3,5-thiadiazine-
2-thiones.[258]

The rate of hydrolysis of benzoyl chloride in mixtures of water and organic
solvents has been correlated with the chemical shift of the water protons.[259]
The rate of hydrolysis of benzoyl chloride in dioxan–water mixtures has been
measured at 1—1000 atmospheres.[260] The reaction of benzoyl chloride in

[242] H. Rokeya and M. M. Huque, *Pakistan J. Sci. Ind. Res.*, **10**, 147 (1967); *Chem. Abs.*, **69**, 66593 (1968).
[243] A. B. Uzienko and A. A. Yasnikov, *Org. Reactivity*, **4**, 616 (1967); *Chem. Abs.*, **69**, 43168 (1968).
[244] N. G. Zarakhani, V. M. Promyslov, M. I. Vinnik, and L. G. Yudin, *Zh. Fiz. Khim.*, **43**, 637 (1969); *Chem. Abs.*, **71**, 12230 (1969).
[245] I. M. Medvetskaya and M. I. Vinnik, *Zh. Fiz. Khim.*, **43**, 630 (1969); *Chem. Abs.*, **71**, 12229 (1969).
[246] O. Grahl-Nielsen, *Tetrahedron Letters*, **1969**, 2827.
[247] R. G. Lee, D. A. Long, and T. G. Truscott, *Trans. Faraday Soc.*, **65**, 503, 820 (1969).
[248] F. Muzalewski, *Rocz. Chem.*, **43**, 1157 (1969).
[249] V. C. Armstrong and R. B. Moodie, *J. Chem. Soc.*(B), **1969**, 934.
[250] M. M. Mhala and M. M. Jagdale, *Indian J. Chem.*, **6**, 711 (1969).
[251] A. P. Grekov and S. A. Sukhorukova, *Zh. Org. Khim.*, **4**, 2155 (1968); *Chem. Abs.*, **70**, 67308 (1969).
[252] V. L. Antonovskij, Z. S. Frolova, and O. N. Romancova, *Zh. Org. Khim.*, **5**, 44 (1969); V. L. Antonovskij, Z. S. Frolova, E. V. Skorobogatova, and O. K. Ljašenko, *ibid.*, p. 46; V. N. Nosan and T. T. Yurzhenko, *Zh. Org. Khim.*, **5**, 458 (1969); *Chem. Abs.*, **71**, 2686 (1969); T. I. Yurshenko and V. N. Nosan, *Ukr. Khim. Zh.*, **33**, 1283 (1967).
[253] T. Yamana, Y. Mizukami, and F. Ichimura, *J. Pharm. Soc. Japan*, **88**, 1504 (1968).
[254] D. Zavoianu and F. Cocou, *Rev. Roum. Chim.*, **12**, 1141 (1967); *Chem. Abs.*, **69**, 26413 (1968); D. Zavoianu, *An. Univ. Bucuresti Ser. Sti. Nat. Chim.*, **16**, 21 (1967); *Chem. Abs.*, **70**, 19300 (1969).
[255] A. Hussain, *J. Pharm. Sci.*, **58**, 316 (1969).
[256] E. Brode, *Arzneim.-Forsch.*, **18**, 1317 (1968); *Chem. Abs.*, **70**, 36743 (1969).
[257] T. F. Fagley and R. L. Oglukian, *J. Phys. Chem.*, **73**, 1438 (1969); I. I. Bardyshev and O. D. Strizhakov, *Dokl. Akad. Nauk Beloruss. SSSR*, **12**, 1100 (1968); *Chem. Abs.*, **70**, 56881 (1969); G. Calvaruso and F. P. Cavasino, *Ann. Chim.* (Italy), **58**, 1039 (1968); T. Oshima and T. Iwai, *J. Chem. Soc. Japan*, **89**, 1036 (1968); *Chem. Abs.*, **70**, 36754 (1969); M. A. Beg and H. N. Singh, *Z. Phys. Chem.* (Leipzig), **237**, 128 (1968); J. Konecny, *Helv. Chim. Acta*, **52**, 2151 (1969).
[258] P. B. Talukdar and S. Banerjee, *J. Indian Chem. Soc.*, **44**, 1060 (1967).
[259] P. Frölich, W. Köhler, and R. Radeglia, *Z. Chem.*, **8**, 467 (1969).
[260] D. Buttner and H. Heydtmann, *Ber. Bunsenges. Phys. Chem.*, **73**, 640 (1969).

benzene–nitrobenzene mixtures[261] and the alcoholysis of acyl chlorides have also been studied.[262]

The exchange reactions of acetyl fluoride with acetyl hexafluoroarsenate [263] and of benzoyl iodides with iodine [264] have also been studied.

Other reactions which have been investigated include: methanolysis of sugar lactones;[265] acetyl exchange of acetylated methyl glycosides;[266] hydrogen exchange of amides;[267] addition of pentan-3-ol to ketones in n-hexane;[268] and the reaction of cyanamides with amines.[269]

The effects of *ortho*-substituents on the rates of the acid-catalysed hydrolysis of benzoic esters and on the esterification of benzoic acids have been discussed.[270]

There have been several mechanistic investigations of the reactions of isocyanates,[271] and of decarboxylation reactions.[272a] Decarboxylation reactions have been reviewed.[272b]

[261] E. Tommila and T. Vihavainen, *Suomen Kemistilehti*, **42**, 89 (1969).

[262] F. Akiyama and N. Tokura, *Bull. Chem. Soc. Japan*, **41**, 2690 (1968); I. G. Murgulescu and I. Demetrescu, *Rev. Roum. Chim.*, **14**, 149 (1969); L. V. Kuritsyn and N. K. Vorob'ev, *Izv. Vyssh. Ucheb. Zaved., Khim. Khim. Tekhnol.*, **12**, 416 (1969).

[263] L. Lunazzi and S. Brownstein, *J. Am. Chem. Soc.*, **91**, 3034 (1969).

[264] D. W. Hamilton and R. M. Noyes, *J. Am. Chem. Soc.*, **91**, 1740 (1969).

[265] I. Matsunaga and Z. Tamura, *Chem. Pharm. Bull.*, **17**, 1383 (1969).

[266] J. Strucínski and J. Swiderski, *Rocz. Chem.*, **43**, 1236 (1969).

[267] C. Y. S. Chen and C. A. Swenson, *J. Am. Chem. Soc.*, **91**, 234 (1969).

[268] W. T. Brady, W. L. Vaughn, and E. F. Hoff, *J. Org. Chem.*, **34**, 843 (1969).

[269] V. G. Golov, M. G. Ivanov, and Y. I. Mushkin, *Zh. Vses. Khim. Obshch.*, **13**, 231 (1968); *Chem. Abs.*, **69**, 66628 (1968); M. G. Ivanov, V. G. Golov, and Y. I. Mushkin, *Kinet. Katal.*, **9**, 681 (1968); *Chem. Abs.*, **69**, 95612 (1969).

[270] M. Charton, *J. Am. Chem. Soc.*, **91**, 619, 624, 6648 (1969).

[271] G. Ostrogovich, F. Kerek, A. Buzás, and N. Doca, *Tetrahedron*, **25**, 1875 (1969); S. G. Entelis, O. V. Nesterov, and R. P. Tiger, *Dokl. Akad. Nauk SSSR*, **178**, 661 (1968); *Chem. Abs.*, **69**, 76176 (1968); F. A. Kryuchkov and M. V. Shoshtaeva, *Khim. Polim.*, **1968**, No. 5, 147; A. K. Zhitinkina and M. V. Shoshtaeva, *ibid.*, pp. 117, 129; *Chem. Abs.*, **70**, 2974, 2975, 2976 (1969); N. N. Zolotarevskaya, E. Z. Zhuravlev, Z. V. Gerega, and I. I. Konstantinov, *Kinet. Katal.*, **10**, 427 (1969); *Chem. Abs.*, **71**, 21411 (1969).

[272a] G. S. Pande, *J. Sci. Ind. Res.* (India), **26**, 393 (1967); D. B. Bigley and R. W. May, *J. Chem. Soc.*(B), **1969**, 994; M. Zielinski, *Rocz. Chem.*, **42**, 1725 (1968); **43**, 1547 (1969); H. Zweifel and T. Völker, *Chimia*, **23**, 389, 391 (1969); G. W. Kosicki, *Biochemistry*, **7**, 4299, 4310 (1968); G. W. Kosicki and F. H. Westheimer, *ibid.*, p. 4303; S. Oae and K. Uneyama, *Kagaku* (Kyoto), **21**, 132 (1966); *Chem. Abs.*, **70**, 19276 (1969); G. Gambaretto and G. Trollo, *Ann. Chim.* (Italy), **59**, 690 (1969); K. G. Klaus and J. V. Rund, *Inorg. Chem.*, **8**, 59 (1969); A. Schellenberger, G. Hübner, and H. Lehmann, *Angew. Chem. Internat. Ed. Engl.*, **7**, 886 (1968); S. Allenmark and O. Bohman, *Arkiv Kemi*, **31**, 305 (1969); V. W. Joshi and G. Bagavant, *Current Sci.* (India), **37**, 165 (1968); A. N. Bourns, J. Buccini, G. E. Dunn, and W. Rodewald, *Can. J. Chem.*, **46**, 3915 (1968); E. Bamann, V. S. Sethi, and G. Laskawy, *Arch. Pharm.*, **301**, 78 (1968); M. H. O'Leary, *J. Am. Chem. Soc.*, **91**, 6886 (1969); H. H. Huang and F. A. Long, *ibid.*, p. 2872; K. S. Sastry, E. V. Sundaram, and S. S. Muhammad, *Current Sci.*, **37**, 643 (1968); S. Ghosal, *Indian J. Chem.*, **5**, 650 (1967).

[272b] L. W. Clark, p. 589 in ref. 2.

Non-carboxylic Acids

Phosphorus-containing Acids

The hydrolysis of methyl ethylene phosphate can occur by two pathways (endo- and exo-cyclic cleavage; equations 26 and 27). The proportion of

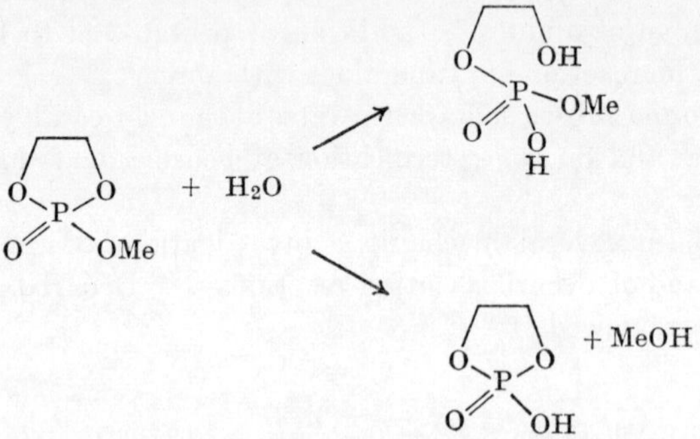

$$\cdots (26)$$

$$+ \text{ H}_2\text{O}$$

$$+ \text{ MeOH} \qquad \cdots (27)$$

reaction occurring by each of these depends in a complicated way on the pH and has been explained on the following assumptions: (*i*) that formation and breakdown of the pentacovalent intermediate occurs only through attack and expulsion from an apical position; (*ii*) that the five-membered ring in the pentacovalent intermediate is constrained to occupy apical and equatorial positions; (*iii*) that sometimes pseudorotation can be a rate-limiting step. At low

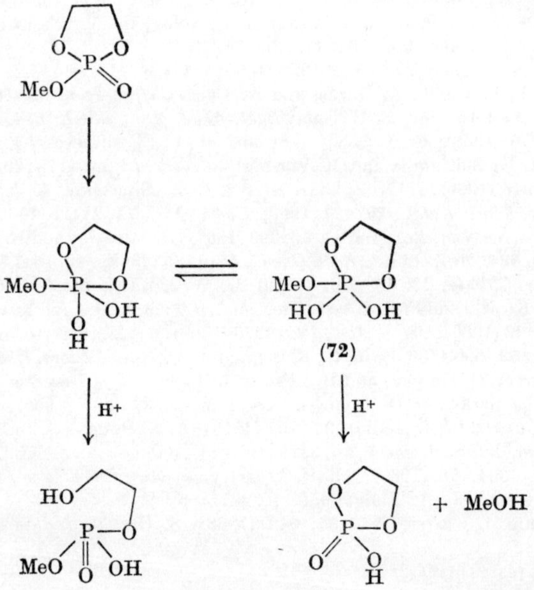

$$\text{H}^+$$

(72)

$$\text{H}^+$$

$$+ \text{ MeOH}$$

pH (<4) the proportion of exocyclic cleavage drops from ca. 50% to zero with increasing acidity. It was assumed that formation and breakdown of the pentacovalent intermediate was acid-catalysed but that pseudorotation was not and occurred in the unprotonated intermediate. Then at high acidities pseudorotation becomes slower than the acid-catalysed breakdown of the pentacovalent intermediate and so the amount of exocyclic cleavage which must proceed through conformation (**72**) would fall with increasing acidity.

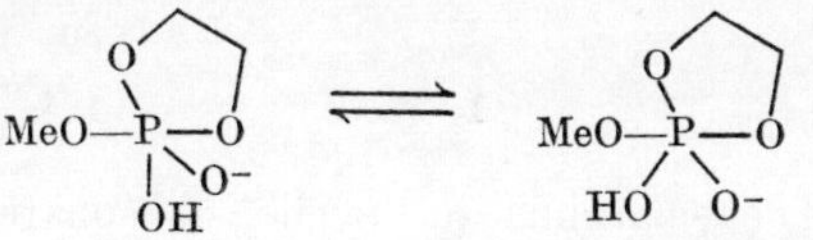

(**73**) (**74**)

In moderately alkaline solution (pH 12) the proportion of exocyclic cleavage again drops to zero, and this was attributed to reaction via the mono-anionic pentacovalent intermediate (**73**) which reacts with ring-opening faster than it pseudorotates to (**74**), the conformation necessary for exocyclic cleavage.

In 10M-alkali the proportion of exocyclic cleavage rises to 10%, which was attributed to rapid pseudorotation of the di-anionic pentacovalent intermediate (**75**) in which the negative charge is in an unfavourable apical position. This leads to conformation (**76**) which is a favourable conformation for expulsion of methoxide from an apical position.[273]

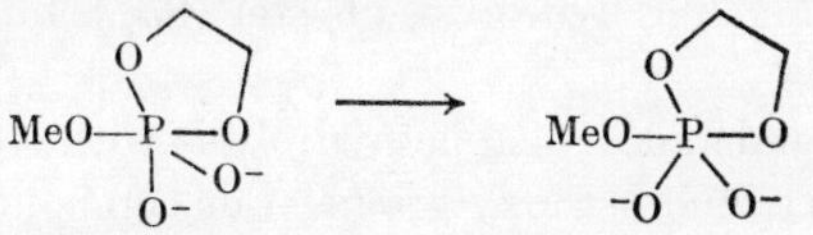

(**75**) (**76**)

Kluger and Westheimer have reported details of their investigation of the hydrolysis of the bicyclic phosphinate esters now known[274] to be (**77**) and (**79**). Both esters are hydrolysed preferentially at the bridgehead phosphorus at rates larger than those for the analogous monocyclic esters, (**78**) and (**80**). The high rates were attributed to release of strain on going to the pentacovalent intermediate. It was proposed that the origin of this strain was the very small C–P–C bond angle of 87° at the bridgehead.[274] At high acidities the pH–rate profiles for (**77**) and (**79**) both show breaks which were attributed to a change in rate-limiting step from attack of water on the protonated ester to pseudorotation.[275]

[273] R. Kluger, F. Covitz, E. Dennis, L. D. Williams, and F. H. Westheimer, *J. Am. Chem. Soc.*, **91**, 6066 (1969).
[274] Y. H. Chiu and W. N. Lipscomb, *J. Am. Chem. Soc.*, **91**, 4150 (1969).
[275] R. Kluger and F. H. Westheimer, *J. Am. Chem. Soc.*, **91**, 4143 (1969); cf. *Org. Reaction Mech.*, **1967**, 363.

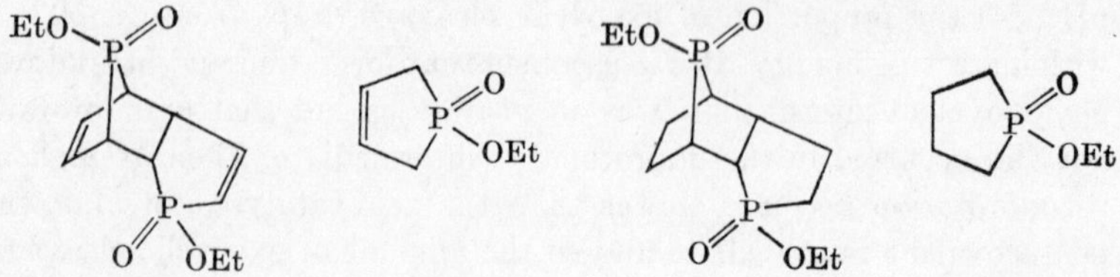

(77)   (78)   (79)   (80)

Rel. ⎰ Acid 10^5 1   50   1
rate ⎱ Alkali 4×10^4 1 10^4 1

Molecular-orbital calculations on cyclic phosphate esters have been reported.[276]

On the basis of the greater rate of reaction of (81) than of (82) with phenyl isocyanate it has been suggested that (81) has greater ring-strain. This was attributed to conjugation between the phenyl group and the phosphorus which leads to a change in hybridization of the latter.[277]

(81)   (82)

The neutral and alkaline hydrolysis of ester (83) is much faster than that of ester (84).[278,279]

Compound (85) undergoes alkaline hydrolysis with stepwise, non-geminal replacement of the trifluoroethoxy groups, but compound (86) reacts much more rapidly and yields catechol phosphate as the first identifiable product. The hexaphenoxy-compound is stable to alkali.[280]

The hydrolyses of compound (87)[281] and of 3-hydroxy-2-oxo-2-phenyl-3,5,5-trimethyl-1,2-oxaphospholan[282] have been studied.

The enthalpy of hydrolysis of the 3′-bond of adenosine-3′,5′-monophosphate is −14,100 cal mole^{-1} and of guanosine-3′,5′-monophosphate is −10,500 cal mole^{-1}.[283]

[276] D. R. Boyd, *J. Am. Chem. Soc.*, **91**, 1200 (1969).
[277] R. Greenhalgh, J. E. Newbery, R. Woodcock, and R. F. Hudson, *Chem. Comm.*, **1969**, 22; R. F. Hudson and A. Mancuso, *ibid.*, p. 522.
[278] V. E. Bel'skii and I. P. Gozman, *Zh. Obshch. Khim.*, **37**, 2730 (1967); *Chem. Abs.*, **69**, 58646 (1968).
[279] V. E. Bel'skii, N. N. Bezzubova, and I. P. Gozman, *Zh. Obshch. Khim.*, **38**, 1330 (1968); *Chem. Abs.*, **69**, 85866 (1968).
[280] H. R. Allcock and E. J. Walsh, *J. Am. Chem. Soc.*, **91**, 3102 (1969).
[281] J.-P. Majoral, J. Devillers, and J. Navech, *Compt. Rend.*, *C*, **268**, 1077 (1969).
[282] K. Bergensen, *Acta. Chem. Scand.*, **23**, 696 (1969).
[283] P. Greengard, S. A. Rudolph, and J. M. Sturtevant, *J. Biol. Chem.*, **244**, 4798 (1969).

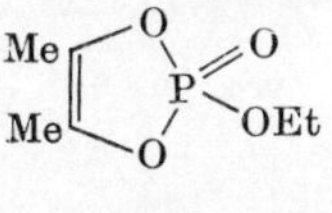

Me—⟨O⟩P(=O)—O—OEt

(83)

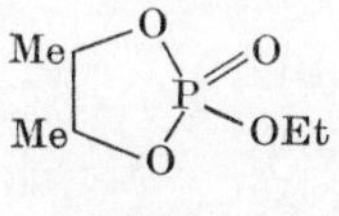

Me—⟨O⟩P(=O)—O—OEt

(84)

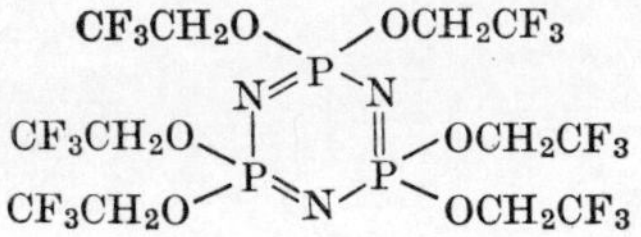

(85)

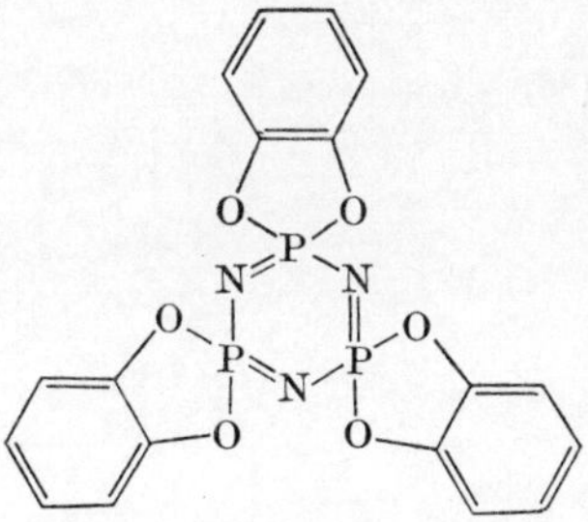

(86)

EtO—P(=O)(—O—CH₂—)₂C(Me)(Me)

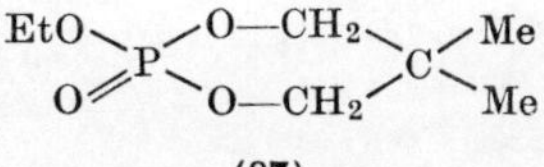

(87)

Plots of k_{obs} against acid concentration for the hydrolysis of *p*-nitrophenyl, 2,4-dinitrophenyl, bis-2,4-dinitrophenyl, and *p*-nitrophenyl diphenyl phosphate generally show maxima which cannot be explained by complete protonation; $k(H_3O^+)/k(D_3O^+)$ falls in the range 0.7—1.1. It was suggested that "proton transfers are partially rate limiting and that initial-state effects are also important".[284] The acid hydrolysis of 3-chloro-4-methylphenyl phosphate[285] and of *p*-chloro- and *p*-bromo-phenyl phosphate,[286] and the hydrolysis of bis-dinitrophenyl phosphate in the pH range 1—14[287] have also been studied. The reactions between *p*-nitrophenyl diphenyl phosphate and HO⁻ and F⁻ ions are catalysed by micelles of cetyltrimethylammonium bromide.[288, 289] It has been confirmed by tracer studies that the hydrolysis of 2,4-dinitrophenyl phosphate at pH 6 and in 1M-potassium hydroxide occurs with phosphorus–oxygen cleavage.[290]

The hydrolysis of (88) is 10^5 times faster than that of (93) and 8×10^7 times faster that that of its *para*-isomer. Compound (90) was shown to be an intermediate in the hydrolysis of (88), and to be itself hydrolysed at an enhanced

[284] C. A. Bunton and S. J. Farber, *J. Org. Chem.*, **34**, 3396 (1969).
[285] M. M. Mhala and M. D. Patwardhan, *Indian J. Chem.*, **6**, 704 (1969).
[286] M. M. Mhala, M. D. Patwardhan, and G. Kasturi, *Indian J. Chem.*, **7**, 145 (1969).
[287] C. A. Bunton and S. J. Farber, *J. Org. Chem.*, **34**, 767 (1969).
[288] C. A. Bunton and L. Robinson, *J. Org. Chem.*, **34**, 773 (1969).
[289] C. A. Bunton, L. Robinson, and L. Sepulveda, *J. Am. Chem. Soc.*, **91**, 4813 (1969).
[290] C. A. Bunton and J. M. Hellyer, *J. Org. Chem.*, **34**, 2798 (1969).

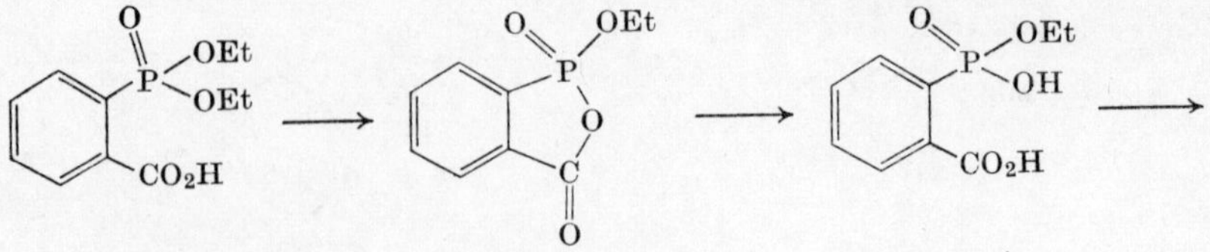

(88)   (89)   (90)

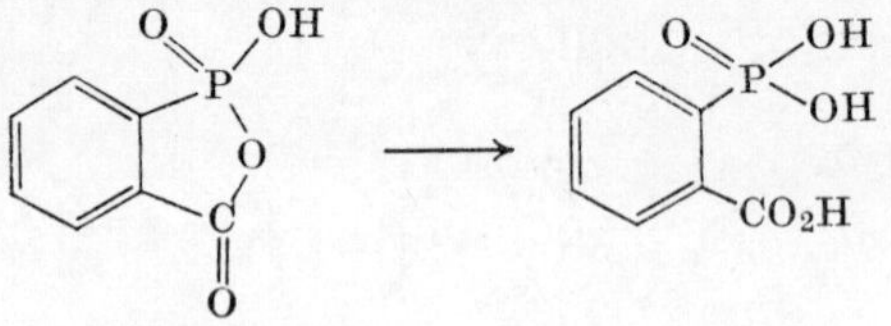

(91)   (92)   ... (28)

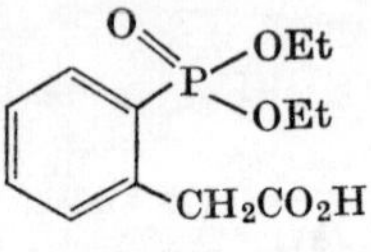

(93)

rate.[291] In the presence of hydroxylamine, (88) yields a hydroxamic acid which is possibly formed by way of an intermediate (89) or (91). Reaction was therefore considered to proceed as shown in equation (28).[291]

The hydrolyses of *syn-* and *anti-p*-nitrophenyl phenacyl methylphosphonate are especially fast but the *syn*-isomer reacts fastest. It was suggested that both react by intramolecular general-base catalysis as shown in (94) and (95).[292]

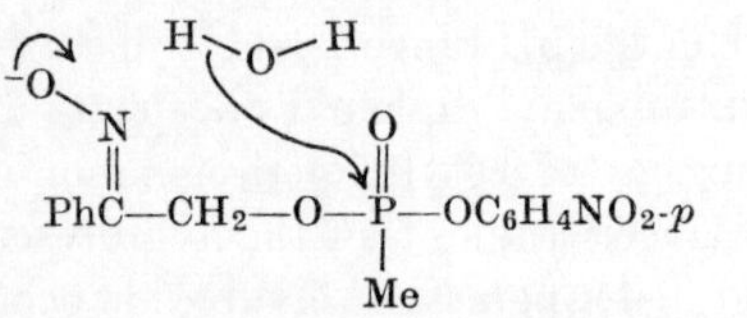

$$PhC—CH_2—O—P—OC_6H_4NO_2\text{-}p$$

(94) 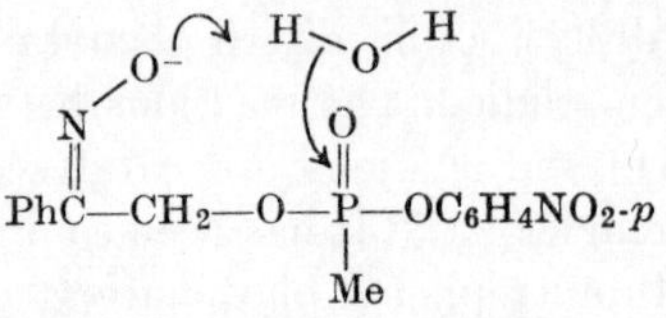   (95)

The hydrolysis of the phosphates of glyceraldehyde and dihydroxyacetone,[293] glycerol,[294] and myoinositol[295] has been studied.

[291] G. M. Blackburn and M. J. Brown, *J. Am. Chem. Soc.*, **91**, 525 (1969); see also p. 9 of ref. 1.
[292] C. N. Lieske, J. W. Hovanec, and P. Plumbergs, *Chem. Comm.*, **1969**, 976.
[293] N. Y. Kozlova, I. V. Mel'nichenko, and A. A. Yasnikov, *Ukr. Khim. Zh.*, **34**, 1041 (1969); *Chem. Abs.*, **70**, 86743 (1969); N. Y. Kozlova, I. V. Mel'nichenko, and A. A. Yasniko, *Ukr. Khim. Zh.*, **34**, 1145 (1968); *Chem. Abs.*, **70**, 68724 (1969).
[294] V. E. Bel'skii, Z. V. Lustina, N. I. Rizpolozhenskii, and L. V. Stepashkina, *Izv. Akad. Nauk SSSR, Ser. Khim.*, **1968**, 2813; *Chem. Abs.*, **70**, 95929 (1969).
[295] Cf. J. K. Raison and W. J. Evans, *Biochim. Biophys. Acta*, **170**, 448 (1968).

P^1,P^1-Diethyl pyrophosphate has been synthesized from P^1,P^1-diethyl P^2-*m*-nitrophenyl pyrophosphate by a photochemical aromatic substitution.[296] It has pK_a values of 3 and 5.2. The di-anion is 50 times more reactive than the mono-anion. This is unlike the behaviour of alkyl phosphates but similar to that of 2,4-dinitrophenyl phosphate.[297] Both the latter and the pyrophosphate have good leaving groups which can depart without undergoing protonation.[298] Other investigations on pyrophosphates are reported in refs. 299 and 300.

The rates of hydrolysis of the mono- and di-anions of acyl phosphates decrease with increasing branching of the acyl-chain. The reactions proceed with phosphorus–oxygen bond cleavage, probably as shown in equations (29) and (30). It seems likely that the rate-decreasing effect of methyl substitution in the acyl group is probably a steric effect.[301]

$$CH_3-\overset{\overset{\displaystyle O}{\|}}{C}-O-\overset{\overset{\displaystyle O}{\|}}{\underset{\displaystyle O^-}{P}}{=}O \longrightarrow CH_3-\overset{\overset{\displaystyle O}{\|}}{C}-OH + [PO_3]^- \qquad \ldots (29)$$

$$CH_3-\overset{\overset{\displaystyle O}{\|}}{C}-O-\overset{\overset{\displaystyle O^-}{\|}}{\underset{\displaystyle O^-}{P}}{=}O \longrightarrow CH_3-\overset{\overset{\displaystyle O}{\|}}{C}-O^- + [PO_3]^- \qquad \ldots (30)$$

$$[PO_3]^- + H_2O \longrightarrow H_2PO_4^-$$

The hydrolyses of 2-, 3-, and 4-pyridylmethyl phosphates have been studied. The rate of hydrolysis of the 2-pyridylmethyl compound is strongly increased by metal ions.[302a] Thorium ions increase the rate of hydrolysis of 3-pyridyl phosphate and 8-quinolyl phosphate.[302b] The metal-ion catalysed hydrolysis of trimetaphosphate has been investigated.[303]

The acid-catalysed hydrolysis of 1-phenylvinyl diethyl phosphate shows solvent isotope effect $k(H_2O)/k(D_2O) = 2.5$ and an entropy of activation of -8.8 e.u. An $A-S_E2$ mechanism was proposed. The rate is increased in the presence of sodium lauryl sulphate at concentrations above the critical micelle concentration.[304]

[296] Cf. *Org. Reaction Mech.*, **1967**, 360, 414.
[297] See *Org. Reaction Mech.*, **1967**, 359.
[298] D. L. Miller and T. Ukena, *J. Am. Chem. Soc.*, **91**, 3050 (1969).
[299] V. E. Bel'skii, M. V. Efremova, and Z. V. Lustina, *Izv. Akad. Nauk SSSR, Ser. Khim.*, **1967**, 1236; E. I. Budovskii and N. N. Shibaev, *Khim. Prir. Soed.*, **4**, 233 (1968); *Chem. Abs.*, **70**, 78303 (1969).
[300] G. M. Kosolapoff and H. G. Kirksey, *Dokl. Akad. Nauk. SSSR*, **176**, 1339 (1967); *Chem. Abs.*, **69**, 18452 (1968).
[301] D. R. Phillips and T. H. Fife, *J. Org. Chem.*, **34**, 2710 (1969).
[302a] T. Murakami and T. Takagi, *J. Am. Chem. Soc.*, **91**, 5130 (1969).
[302b] Y. Murakami, J. Sunamoto, and H. Sadamori, *Chem. Comm.*, **1969**, 983.
[303] D. L. Miller, G. I. Krol, and U. P. Strauss, *J. Am. Chem. Soc.*, **91**, 6882 (1969).
[304] C. A. Bunton and L. Robinson, *J. Am. Chem. Soc.*, **91**, 6072 (1969); cf. ref. 116, p. 456.

The hydrolysis of dibenzyl phosphoenolpyruvic acid occurs with loss of benzyl alcohol, and the rate depends on the concentration of the form with an undissociated carboxy group. In the presence of hydroxylamine the product is dibenzyl phosphate and pyruvic acid hydroxamate. It was suggested that hydroxylamine traps a pentacovalent intermediate or an acyl phosphate as shown in equation (31). An acyl phosphate would have to be formed from the pentacovalent intermediate after pseudorotation.[305]

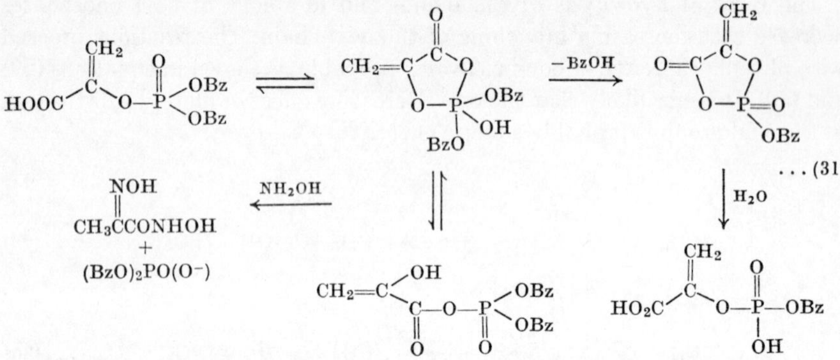

$$\dots (31)$$

The relative rates of reaction of compounds $(Pr^iO)_2POX$ with H_2O, HO^-, and n-butylamine have been determined with $X = CN$, Cl, $-OP(O)(OPr^i)_2$, and F.[306]

Acidic hydrolysis of a series of dialkyl methylphosphonates by alkyl–oxygen fission has been studied and the rates have been compared with the relative rates of ageing of phosphonylate cholinesterases.[307]

The hydrolyses of the following compounds have also been studied: phosphinate esters;[308] tetraethyl methylene phosphonate;[309] dialkyl S-methyl thiolphosphates;[310] and dialkyl phosphites.[311]

Other investigations include those of the solvolyses of phosphoramidic

305 S. J. Benkovic and K. J. Schray, *J. Am. Chem. Soc.*, **91**, 5653 (1969).

306 R. F. Hudson and R. Greenhalgh, *J. Chem. Soc.*(B), **1969**, 325.

307 J. I. G. Cadogan, D. Eastlick, F. Hampson, and R. K. Mackie, *J. Chem. Soc.*(B), **1969**, 144.

308 V. E. Bel'skii and M. V. Efremova, *Zh. Fiz. Khim.*, **42**, 2926 (1968); *Chem. Abs.*, **70**, 86748 (1969); V. E. Bel'skii, M. V. Efremova, and Z. V. Lustina, *Izv. Akad. Nauk SSSR, Ser. Khim.*, **1969**, 1293; V. E. Bel'skii, M. V. Efremova, and A. R. Panteleeva, *ibid.*, **1968**, 2278; *Chem. Abs.*, **70**, 28109 (1969); V. E. Bel'skii, M. V. Efremova, I. M. Shermergorn, and A. N. Pudovick, *ibid.*, **1969**, 307; *Chem. Abs.*, **71**, 2682 (1969); V. E. Bel'skii, A. N. Pudovick, M. V. Efremova, V. N. Eliseenkov, and A. R. Panteleeva, *Dokl. Akad. Nauk SSSR*, **180**, 351 (1968); H. Christol and C. Marty, *J. Organometal. Chem.*, **12**, 471 (1968).

309 V. E. Bel'skii and L. A. Kudryavtseva, *Izv. Akad. Nauk SSSR, Ser. Khim.*, **1968**, 2160; *Chem. Abs.*, **70**, 10734 (1969).

310 V. E. Bel'skii, N. N. Bezzubova, and Z. V. Lustina, *Zh. Obshch. Khim.*, **39**, 181 (1969); *Chem. Abs.*, **70**, 95931 (1969); V. E. Bel'skii, N. N. Bezzubova, V. N. Eliseenkov, and A. N. Pudovick, *ibid.*, p. 1011.

311 V. E. Bel'skii, G. Z. Motygullin, V. N. Eliseenkov, and A. N. Pudovick, *Izv. Akad. Nauk SSSR, Ser. Khim.*, **1969**, 1297.

chlorides,[312] orthophosphoramidates,[313] allylic diphenyl phosphates,[314] and of halides of type R_2POCl and $RPOCl_2$,[315] transesterification of phosphinate esters,[316] and of dialkyl phosphites,[317] the reaction of isopropyl methyl-phosphonofluoridate with 2-hydroxyiminomethyl-1-methylpyridinium iodide,[318] and oxidative phosphorylation.[319]

Hückel and extended Hückel calculations on the bonding in phosphate diesters have been reported.[320]

The mechanism of action of ribonucleases has been reviewed.[321]

Uridine-2′,3′-cyclic phosphonate (**96**) is hydrolysed by ribonuclease more slowly than is uridine-2′,3′-cyclic phosphate. It was suggested that this could

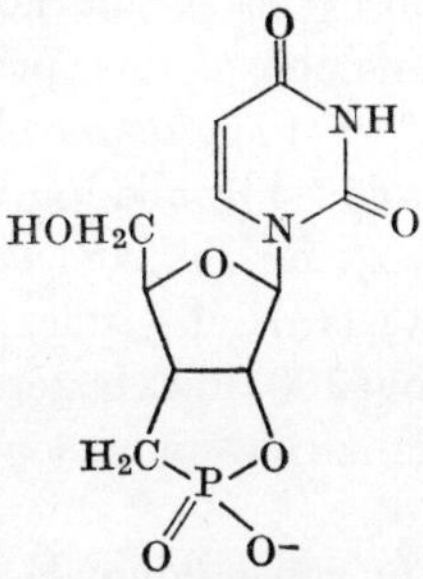

(96) 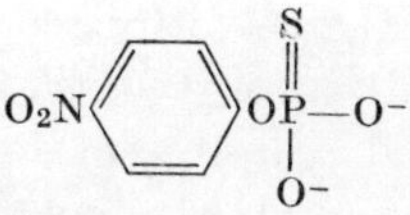   (97)

be explained by a mechanism involving pseudorotation.[322] On the other hand, a linear mechanism has been preferred on the basis of the results of X-ray and NMR studies of the binding of cytidine-3′-monophosphate.[323]

Titration curves for the four histidine residues of ribonuclease have been

[312] A. F. Gerrard and N. K. Hamer, *J. Chem. Soc.*(B), **1969**, 369.

[313] C. J. Peacock, *Z. Naturforsch.*, **24b**, 391 (1969).

[314] J. A. Miller, *J. Chem. Soc.*(B), **1968**, 1427.

[315] N. N. Mel'nikov, A. F. Vasil'ev, B. A. Khaskin, N. N. Tuturina, and T. M. Ivanova, *Zh. Obshch. Khim.*, **38**, 1745 (1968); A. A. Neimysheva and I. L. Knunyants, *ibid.*, p. 595; *Chem. Abs.*, **69**, 66629 (1968).

[316] A. N. Pudovick and G. I. Evstof'ev, *Dokl. Akad. Nauk SSSR*, **183**, 842 (1968); *Chem. Abs.*, **70**, 67280 (1969).

[317] F. K. Samigulin, I. M. Kafengauz, and A. P. Kafengauz, *Kinet. Katal.*, **9**, 898 (1968); *Chem. Abs.*, **70**, 76960 (1969).

[318] J. H. Blanch, *J. Chem. Soc.*(B), **1969**, 1172.

[319] D. O. Lambeth and H. A. Lardy, *Biochemistry*, **8**, 3395 (1969); C. A. Bunton and J. Hellyer, *Tetrahedron Letters*, **1969**, 187.

[320] R. L. Collin, *Ann. N.Y. Acad. Sci.*, **158**, 50 (1969).

[321] E. A. Barnard, *Ann. Rev. Biochem.*, **38**, 677 (1969).

[322] M. R. Harris, D. A. Usher, H. P. Albrecht, G. H. Jones, and J. G. Moffatt, *Proc. Nat. Acad. Sci.*, **63**, 247 (1969); D. A. Usher, *ibid.*, **62**, 661 (1969).

[323] G. C. K. Roberts, E. A. Dennis, D. H. Meadows, J. S. Cohen, and O. Jardetzky, *Proc. Nat. Acad. Sci.*, **62**, 1151 (1969).

obtained by studying the variation of the chemical shifts of the C-2 hydrogens with pH.[324] The NMR spectrum of ribonuclease is also discussed in ref. 325.

When the free radical 2,2,6,6-tetramethyl-4-hydroxypiperidinyl-1-oxy monophosphate is bound to ribonuclease the NMR signals of the protons at C-2 of His-12 and His-119 are broadened, which suggests that binding occurs at the active site.[326]

Other investigations on ribonuclease are described in ref. 327. Ribonuclease-T_1 has also been investigated.[328]

Initial rapid-burst experiments which demonstrate the intervention of a phosphoryl enzyme in the hydrolysis of 4-methylumbelliferyl phosphate catalysed by alkaline phosphatase have been described.[329] The intervention of a phosphoryl-enzyme is also suggested by the observation that the reactions of a wide range of phosphates catalysed by alkaline phosphatase in aqueous tris buffer all yield the same ratio of phosphorylated tris to inorganic phosphate.[330]

p-Nitrophenyl phosphorothioate (**97**) is hydrolysed by alkaline phosphatase more slowly than *p*-nitrophenyl phosphate. V_{max} for the latter is about 100 times greater than that for the thioester, and K_m is about 8 times smaller. The acid-catalysed hydrolysis of the thioester is about 100 times faster than that of the oxygen ester.[331] Other investigations on alkaline phosphatase are described in ref. 332.

The hydrolysis of acyl phosphates catalysed by glyceraldehyde-3-phosphate dehydrogenase has been studied.[333]

Myosin[334] and yeast nucleoside diphosphate kinase[335] have been investigated.

[324] G. C. K. Roberts, D. H. Meadows, and O. Jardetzky, *Biochemistry*, **8**, 2053 (1969).

[325] M. Ruterjans and H. Witzel, *European J. Biochem.*, **9**, 118 (1969).

[326] G. C. K. Roberts, J. Hannah, and O. Jardetzky, *Science*, **165**, 504 (1969).

[327] E. J. del Rosarco and G. G. Hammes, *Biochemistry*, **8**, 1184 (1969); J. Bello and E. Nowoswiat, *ibid.*, p. 628; J. E. Brown and W. A. Klee, *ibid.*, p. 2876; K. S. Chio and A. L. Tappel, *ibid.*, p. 2827; T. Takahashi, M. Irie, and T. Ukita, *J. Biochem.* (Japan), **65**, 55 (1969); M. Irie, *ibid.*, p. 133; F. Sawada and M. Irie, *ibid.*, **66**, 415 (1969); M. C. Lin, W. H. Stein, and S. Moore, *J. Biol. Chem.*, **234**, 6167 (1968); B. K. Joyce and M. Cohen, *ibid.*, **244**, 811 (1969); I. Kato and C. B. Anfinsen, *ibid.*, p. 1004; D. C. Ward, W. Fuller, and E. Reich, *Proc. Nat. Acad. Sci.*, **62**, 581 (1969); A. Deavin, R. Fisher, C. M. Kemp, A. P. Mathias, and B. R. Rabin, *European J. Biochem.*, **7**, 21 (1968).

[328] S. Iida and T. Ooi, *Biochemistry*, **8**, 3897 (1969); V. N. Orekhovich, *Ital. J. Biochem.*, **17**, 241 (1968).

[329] H. N. Fernley and P. G. Walker, *Biochem. J.*, **111**, 187 (1969).

[330] M. Barrett, R. Butler, and I. B. Wilson, *Biochemistry*, **8**, 1042 (1969); see also D. Levine, T. W. Reid, and I. B. Wilson, *ibid.*, p. 2374; T. W. Reid, M. Pavlic, D. J. Sullivan, and I. B. Wilson, *ibid.*, p. 3184.

[331] R. Breslow and I. Katz, *J. Am. Chem. Soc.*, **90**, 7376 (1968).

[332] H. Csopak, *European J. Biochem.*, **7**, 186 (1969); C. Lazdunski and M. Lazdunski, *ibid.*, p. 294; E. Bamann and P. Schwarze, *Arch. Pharm.*, **301**, 839 (1968); R. L. Heinrikson, *J. Biol. Chem.*, **244**, 299 (1969).

[333] D. R. Phillips and T. H. Fife, *Biochemistry*, **8**, 3114 (1969).

[334] B. Finlayson and E. W. Taylor, *Biochemistry*, **8**, 802 (1969).

[335] E. Garces and W. W. Cleland, *Biochemistry*, **8**, 633 (1969).

Sulphur-containing Acids

It has been shown that 4-nitrocatechol cyclic sulphate like 2-hydroxy-5-nitrotoluene-α-sulphonic acid sultone [336] reacts with chymotrypsin to yield a sulphonyl-enzyme. The reaction is competitively inhibited by N-acetyl-L-tryptophan amide and involves nucleophilic attack by the hydroxy group of serine-195 on the sulphur of the cyclic sulphate. Unlike the enzyme sulphonate ester derived from 2-hydroxy-5-nitrotoluene-α-sulphonic acid sulphone, the enzyme-sulphate derived from catechol cyclic sulphate is stable to hydrolysis.[337]

The hydrolyses of 2-hydroxy-5-nitrotoluene-α-sulphonic acid sultone catalysed by carbonic anhydrase [338] and of di-p-nitrophenyl sulphite catalysed by pepsin [339] have also been investigated.

Additional evidence has been presented that the alkaline hydrolysis of o-hydroxytoluene-α-sulphonic acid sultone does not proceed via a carbanion [340] (see also ref. 213, p. 466).

The rate of the acid-catalysed disproportionation of aryl thiolsulphinates is increased in the presence of sulphides. The kinetics of this reaction are very similar to those of the previously studied reaction of aryl thiolsulphinates with aromatic sulphinic acids,[341] except that the rates are much less sensitive to the structure of the sulphide.[342] The sulphinic anhydride $Bu^tS(O)OS(O)Bu^t$ has been synthesized and its hydrolysis shown to be subject to acid catalysis.[343]

The following reactions have also been studied: nucleophilic substitution reactions of sulphonyl halides;[344] hydrolysis of sulphonic anhydrides,[345]

---

[336] See *Org. Reaction Mech.*, **1967**, 367—368.

[337] G. Tomalin, M. Trifunac, and E. T. Kaiser, *J. Am. Chem. Soc.*, **91**, 722 (1969).

[338] E. T. Kaiser and K.-W. Lo, *J. Am. Chem. Soc.*, **91**, 4912 (1969).

[339] S. W. May and E. T. Kaiser, *J. Am. Chem. Soc.*, **91**, 6491 (1969).

[340] P. Müller, D. F. Mayers, O. R. Zaborsky, and E. T. Kaiser, *J. Am. Chem. Soc.*, **91**, 6732 (1969).

[341] See *Org. Reaction Mech.*, **1967**, 91.

[342] J. L. Kice, C. G. Venier, G. B. Large, and L. Heasley, *J. Am. Chem. Soc.*, **91**, 2028 (1969).

[343] J. L. Kice and K. Ikura, *J. Am. Chem. Soc.*, **90**, 7378 (1968).

[344] L. M. Litvinenko, A. F. Popov, and L. I. Sovokina, *Ukr. Khim. Zh.*, **34**, 595 (1968); *Chem. Abs.*, **68**, 95598 (1968); L. M. Litvinenko and V. A. Savelova, *Zh. Obshch. Khim.*, **38**, 747 (1968); *Chem. Abs.*, **69**, 76142 (1968); L. M. Litvinenko and A. F. Popov, *ibid.*, p. 1969; *Chem. Abs.*, **70**, 19407 (1969); V. I. Shishkina and T. I. Proshechkina, *Izv. Vyssh. Ucheb. Saved., Khim. Khim. Tekhnol.*, **12**, 472 (1969); *Chem. Abs.*, **71**, 48925 (1969); A. F. Popov, V. I. Tokarev, L. Litvinenko, and A. I. Toryanik, *Org. Reactivity* (Tartu); *Chem. Abs.*, **90**, 28266 (1969); Y. P. Berkman, G. A. Zemlyakova, and N. P. Lushina, *Ukr. Khim. Zh.*, **34**, 601 (1968); *Chem. Abs.*, **69**, 95597 (1968); A. F. Popov, V. I. Tokanev, and L. M. Litvinenko, *Org. Reactivity* (Tartu), **6**, 690 (1969); J. Preston and R. B. Scott, *J. Org. Chem.*, **33**, 4343 (1968); R. V. Vizgert, I. M. Ozdrovskaya, and E. N. Ozdrovskii, *Zh. Org. Khim.*, **4**, 1812 (1968); *Chem. Abs.*, **70**, 19403 (1969).

[345] N. Christensen, *Dan. At. Energy Comm., Res. Rep.*, 1968; *Chem. Abs.*, **70**, 67297 (1969).

dimethylsulphamoyl chloride,[346] and of cyclic disulphides and their 1,1-dioxides and 1,1,2,2-tetraoxides;[347] esterification of cetyl alcohol with sulphamic acid;[348] and aminolysis of 1,2-oxathiolan-5-one 2-oxide.[349]

Other Acids

Derivatives of nitrous acid,[350] nitric acid,[351] and arsenic acid[352] have been studied.

[346] O. Rogne, *J. Chem. Soc.*(B), **1969**, 663.
[347] L. Field and R. B. Barbee, *J. Org. Chem.*, **34**, 1792 (1969).
[348] E. L. Vulakh, G. M. Pakhomova, and S. M. Loktev, *Kinet. Katal.*, **1968**, 751; *Chem. Abs.*, **70**, 10752 (1969).
[349] Y. H. Chiang, J. S. Luloff, and E. Schipper, *J. Org. Chem.*, **34**, 2397 (1969).
[350] J. Casado and T. Ramos, *An. Quim.*, **64**, 1017, 1027 (1968); J. Tempe, S. Delhaye, H. Heslot, and G. Morel, *Compt. Rend.*, *C*, **266**, 834 (1968); T. Koenig, T. Fithian, M. Tolela, S. Markwell, and D. Rogers, *J. Org. Chem.*, **34**, 952 (1969); A. Aboul-Seoud, *Rocz. Chem.*, **43**, 1513 (1969); T. Fujita, M. Deura, M. Senda, T. Ikegami, and M. Nakajima, *Agric. Biol. Chem.*, **33**, 1192 (1969); A. Abdoul and M. Ahmad, *Bull. Soc. Chim. Belges*, **78**, 5 (1969).
[351] T. M. R. Fraser, U.S. Clearinghouse Fed. Schi. Tech. Inform. AD, 1968, AD-675917; *Chem. Abs.*, **70**, 86755 (1969); O. Mantsch, N. Bodor, and F. Hodosan, *Rev. Roum. Chim.*, **73**, 1435 (1968).
[352] V. E. Bel'skii, M. P. Osipova, G. K. Kamai, and N. A. Chadaeva, *Dokl. Akad. Nauk SSSR*, **185**, 332 (1969).

CHAPTER 13

Photochemistry

R. S. DAVIDSON

Department of Chemistry, The University, Leicester

Woodward and Hoffmann in their much appreciated article entitled "The Conservation of Orbital Symmetry" have discussed the mechanisms of a multitude of photochemical reactions.[1] On the basis of calculations carried out on the cyclization of butadiene to cyclobutene, the premise that the anti-symmetric excited state is wholly responsible for photoinduced reactions has been challenged.[2] It was found that as ring-closure takes place, the potential-energy surface of the antisymmetric excited state (from which reaction starts) crosses that of the forbidden symmetric state (see Figure 1). Because

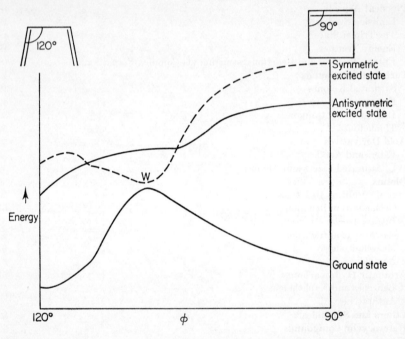

Figure 1. Change in electronic energy of the ground state and two excited states during disrotatory closure.

of this crossing, molecules can pass adiabatically from the antisymmetric to the symmetric state. It is proposed that a molecule in the energy well (W) can then pass to the lower potential surface of the ground state whence cyclization or reversion to diene can occur. The crossing from the anti-symmetric to the symmetric state explains why many dienes and trienes do not fluoresce. Further application of these important findings is awaited with

[1] R. B. Woodward and R. Hoffmann, *Angew. Chem. Internat. Ed. Engl.*, **8**, 781 (1969).
[2] G. Feler, *Theor. Chim. Acta*, **12**, 412 (1968); A. Th. A. M. van der Lugt and L. J. Oosterhoff, *Chem. Comm.*, **1968**, 1235; W. Th. A. M. van der Lugt and L. J. Oosterhoff, *J. Am. Chem. Soc.*, **91**, 6042 (1969).

interest. Zimmerman has discussed the mechanism of a number of reactions investigated by his group in terms of orbital symmetry, and has exemplified the simplicity of predicting the course of concerted reactions by classifying the systems as either Hückel or Möbius systems.[3]

Other exciting developments include the application of nanosecond flash photolysis[4a] and the determination of the lifetimes of excited states which are of the order of a few picoseconds.[4b]

Several good general reviews[5a] and surveys of the literature[5b] have been published.

Physical Aspects

Emission Studies

A review[6a] and books[6b] on luminescence have been published. The use of quinine sulphate as a fluorescence standard has been further evaluated.[7] Pyridine does not phosphoresce because its lowest triplet state is an orbitally forbidden state.[8] Pentan-2-one fluoresces with low efficiency and this is probably due to deactivation by undergoing a Norrish Type II reaction.[9a] Efficient conversion of electronic into thermal energy is suggested as the reason for the low fluorescence yields of a number of pyrazolines.[9b] Deactivation of the excited states of a number of aromatic hydrocarbons has been studied,[10] and in the case of pyrene the rate of intersystem crossing is temperature-dependent.[11] The cooling of α-diketones at low temperatures can "freeze out" non-planar conformations which phosphoresce at different wavelengths to the

[3] H. Zimmerman, *Photochem. Photobiol.*, **7**, 519 (1968); H. Zimmerman, *Angew. Chem. Internat. Ed. Engl.*, **8**, 1 (1969).

[4a] G. Porter and M. R. Topp, *Nature*, **220**, 1228 (1968). Further examples given in text.

[4b] P. M. Rentzepis, *Chem. Phys. Letters*, **3**, 717 (1969).

[5a] J. S. Swenton, *J. Chem. Educ.*, **46**, 7 (1969); E. F. Ullman, *Accounts Chem. Res.*, **1**, 353 (1968); *Adv. Photochem.*, **6** (1968).

[5b] D. Elad, *Current Topics in Organophotochemistry*, Vol. 2, Intra-Science Research Foundation, California, 1968; N. J. Turro, G. S. Hammond, J. N. Pitts, D. Valentine, A. D. Broadbent, W. B. Hammond, and E. Whittle, *Annual Survey of Photochemistry* 1967, Wiley, New York, 1969; A. C. Day and E. J. Forbes, *Ann. Reports*, **65**, *B*, 187 (1968).

[6a] W. R. Ware, *Survey Progr. Chem.*, **4**, 206 (1968).

[6b] R. S. Becker, *Theory and Interpretation of Fluorescence and Phosphorescence*, Wiley-Interscience, New York, 1969; E. C. Lim, *Molecular Luminescence*, Benjamin, New York, 1969.

[7] A. N. Fletcher, *Photochem. Photobiol.*, **9**, 439 (1969); J. E. Gill, *ibid.*, p. 313.

[8] R. J. Hoover and M. Kasha, *J. Am. Chem. Soc.*, **91**, 6508 (1969).

[9a] F. S. Wettack, *J. Phys. Chem.*, **73**, 1167 (1969).

[9b] Z. Raciszewski and J. F. Stephen, *J. Am. Chem. Soc.*, **91**, 4338 (1969).

[10] E. M. Anderson and G. B. Kistiakowsky, *J. Chem. Phys.*, **51**, 182 (1969); G. M. Breuer and E. K. C. Lee, *ibid.*, **51**, 3130 (1969); C. S. Burton and W. A. Noyes, Jr., *ibid.*, **49**, 1705 (1968); J. W. Eastman, *ibid.*, **49**, 4617 (1968); W. P. Helman, *ibid.*, **51**, 354 (1969); G. P. Semeluk, R. D. S. Stevens, and I. Unger, *Can. J. Chem.*, **47**, 597 (1969); R. J. Watts and S. J. Strickler, *J. Chem. Phys.*, **49**, 3867 (1968); J. L. Durham, G. P. Semeluk, and I. Unger, *Can. J. Chem.*, **46**, 3177 (1968); M. E. MacBeath, G. P. Semeluk, and I. Unger, *J. Phys. Chem.*, **73**, 995 (1969).

[11] J. L. Kropp, W. R. Dawson, and M. W. Windsor, *J. Phys. Chem.*, **73**, 1747 (1969).

planar conformation.[12] Emission from transition-metal chelates[13a] and the
part played by spin–orbit coupling in this process[13b] have been discussed.
Emission from the singlet state of azulene is at 700 nm and that from its
triplet at 1000 nm. The lifetime of the singlet state was found to be 8.3×10^{-12}
sec.[4b] Anthracene also has fluorescence bands out in the visible.[14] The participation of upper excited states in chemical reactions has been suggested,[15] and
in some cases emission from these states has been observed.[16] Determinations
of the pK_a values of the singlet and triplet states of a number of molecules have
been made.[17a] In many cases complete equilibration between neutral and
dissociated states is not obtained owing to the rate of the process being less
than that of deactivation.[17b] Fluorescence from the tautomers of indole[18a]
and adenine,[18b] and phosphorescence from the enolate of indan-1-one,[18c] have
been observed. Hydrogen-bonding to the singlet excited state of indole
reduces its fluorescence yield.[19]

The Triplet State

A review on the triplet state has been published.[20] The popular notion that in
the triplet state the uncoupled electrons are further apart has been strongly
criticized.[21] For triplet helium, the electrons are closer to the nucleus and
have a smaller interelectronic separation than in the ground state. Benzophenone phosphoresces in isooctane solution ($\phi_p = 4 \times 10^{-4}$).[22] The lifetime of

[12] M. Almgren, *Photochem. Photobiol.*, **9**, 1 (1969).

[13a] M. K. Dearmond and J. E. Hillis, *J. Chem. Phys.*, **49**, 466 (1968); A. G. Goryushko, Yu. V.
     Naboikin, L. A. Ogurtsova, and A. P. Podgorni, *Optics and Spectroscopy*, **24**, 392 (1968).

[13b] F. E. Lytle and D. M. Hercules, *J. Am. Chem. Soc.*, **91**, 253 (1969).

[14] T. S. Jaseja, V. Parkash, and M. K. Dheer, *J. Appl. Phys.*, **40**, 1882 (1969).

[15] R. C. Dhingra and J. A. Poole, *J. Phys. Chem.*, **72**, 4577 (1968); W. R. Dawson and J. L.
     Kropp, *ibid.*, **73**, 693 (1969); W. R. Dawson and J. L. Kropp, *ibid.*, **73**, 1752 (1969); M. S. S. C.
     Leite and K. R. Naqui, *Chem. Phys. Letters*, **4**, 35 (1969); R. S. Becker, E. Dolan, and D. E.
     Balke, *J. Chem. Phys.*, **50**, 239 (1969).

[16] P. A. Geldof, R. P. H. Rettschnick, and G. J. Hoytink, *Chem. Phys. Letters*, **4**, 59 (1969);
     D. S. Kliger and A. C. Albrecht, *J. Chem. Phys.*, **50**, 4109 (1969); R. A. Keller, *Chem. Phys.
     Letters*, **3**, 27 (1969); A. Yildiz, P. T. Kissinger, and C. N. Reilley, *J. Chem. Phys.*, **49**, 1403
     (1968).

[17a] E. Vander Donckt, J. Nasielski, and P. Thiry, *Chem. Comm.*, **1969**, 1249; L. Avigal, J.
     Feitelson, and M. Ottolenghi, *J. Chem. Phys.*, **50**, 2614 (1969); E. Vander Donckt and
     G. Porter, *Trans. Faraday Soc.*, **64**, 3215, 3218 (1968); A. Grabowska and B. Pakula,
     Photochem. Photobiol., **9**, 377 (1969).

[17b] S. F. Mason and B. E. Smith, *J. Chem. Soc.*(A), **1969**, 325.

[18a] C. M. Chopin and J. H. Wharton, *Chem. Phys. Letters*, **3**, 552 (1969); P.-S. Song and W. E.
     Kurtin, *J. Am. Chem. Soc.*, **91**, 4892 (1969).

[18b] J. W. Eastman, *Ber. Bunsenges. Phys. Chem.*, **73**, 407 (1969).

[18c] Y. Kanda, J. Stanislaus, and E. C. Lim, *J. Am. Chem. Soc.*, **91**, 5085 (1969).

[19] E. Vander Donckt, *Bull. Soc. Chim. Belges*, **78**, 69 (1969); D. L. Horrocks, *J. Chem. Phys.*,
     50, 4151 (1969).

[20] N. J. Turro, *J. Chem. Educ.*, **46**, 2 (1969).

[21] R. P. Messmer and F. W. Birss, *J. Phys. Chem.*, **73**, 2085 (1969).

[22] W. D. K. Clark, A. D. Litt, and C. Steel, *Chem. Comm.*, **1969**, 1087; *J. Am. Chem. Soc.*, **91**,
     5413 (1969).

the triplet in benzene at $0°$ was found to be 6.5×10^{-6} sec, and it was suggested that this short lifetime is due to a specific interaction between the triplet state and benzene. The singlet state of benzene has been observed by nanosecond flash photolysis and found to have a lifetime of 18×10^{-10} sec.[23] The benzene triplet is quite long-lived.[24a] The triplet lifetimes of phenanthrene and carbazole, at low temperature, were found to be dependent upon the molecular packing of the solvent.[24b] The triplet lifetime of naphthalene-2-aldehyde is increased by deuteration of the aldehydic proton.[25] Triplet state energies for a number of compounds have been determined by energy transfer[26] and flash photolysis studies.[27] The concentration of triplets in solution can be determined by ESR,[28] and by using this technique coupled with that of flash photolysis the extinction coefficients of a number of triplet–triplet transitions have been calculated. ESR studies have shown that in the triplet state of naphthalene carbonyl compounds there is greater conjugation than in the naphthalene triplet.[29] The triplet states of benzophenone,[30a] coumarin,[30b] and phenazine[30c] have been studied by ESR.

Energy Transfer

Reviews of the modes of deactivation of excited states[31] and of the use of lasers in studying vibrational energy transfer[32] have been published. Singlet energy transfer from benzene to olefins and other molecules containing π-bonds has been observed[33a,b] and is most efficient when there is good overlap of the emission spectrum of the donor with the absorption of the acceptor.[33a] It has also been suggested that quenching occurs by formation of a complex (exciplex?),[33b] and this explanation has also been invoked to explain the efficient quenching of the fluorescence of 2,3-diazabicyclo[2.2.2]oct-2-ene by

[23] R. Bonneau, J. Joussot-Dubign, and R. Bensasson, *Chem. Phys. Letters*, **3**, 353 (1969).
[24a] R. Astier and Y. H. Meyer, *Chem. Phys. Letters*, **3**, 399 (1969).
[24b] D. J. Morantz and T. E. Martin, *Trans. Faraday Soc.*, **65**, 665 (1969).
[25] N. C. Yang, S. L. Murov, and T.-C. Shieh, *Chem. Phys. Letters*, **3**, 6 (1969).
[26] A. Sykes and T. G. Truscott, *Chem. Comm.*, **1969**, 274; A. Sykes and T. G. Truscott, *ibid.*, p. 929; H. E. A. Kramer, *Z. Phys. Chem.* (Frankfurt), **66**, 73 (1969); H. E. A. Kramer and M. Hafner, *Z. Naturforsch.*, **24b**, 452 (1969); H. E. A. Kramer, M. Hafner, and M. Zügel, *Z. Phys. Chem.* (Frankfurt), **65**, 276 (1969); G. Favaro, F. Masetti, and U. Mazzucato, *ibid.*, **66**, 206 (1969); J. B. Gallivan, *J. Phys. Chem.*, **73**, 3070 (1969).
[27] M. A. Herbert, J. W. Hunt, and H. E. Johns, *Biochem. Biophys. Res. Comm.*, **33**, 643 (1968); W. H. Melhuish, *J. Chem. Phys.*, **50**, 2779 (1969).
[28] J. S. Brinen, *J. Chem. Phys.*, **49**, 586 (1968).
[29] C. H. J. Wells, A. Horsfield, and J. Paxton, *Chem. Comm.*, **1969**, 393.
[30a] M. Sharnoff, *J. Chem. Phys.*, **51**, 451 (1969).
[30b] D. R. Graber, M. W. Grimes, and A. Haug, *J. Chem. Phys.*, **50**, 1623 (1969).
[30c] J. Ph. Grivet and J. M. Lhoste, *Chem. Phys. Letters*, **3**, 445 (1969).
[31] L. M. Stephenson and G. S. Hammond, *Angew. Chem. Internat. Ed. Engl.*, **8**, 261 (1969).
[32] C. B. Moore, *Accounts Chem. Res.*, **2**, 103 (1969).
[33a] E. K. C. Lee, M. W. Schmidt, R. G. Shortridge, Jr., and G. A. Haninger, Jr., *J. Phys. Chem.*, **73**, 1805 (1969).
[33b] A. Morikawa and R. J. Cvetanović, *J. Chem. Phys.*, **49**, 1214 (1968).

conjugated dienes.[34] By examination of the polarization of triplet–triplet absorption spectra, it has been shown that in the transfer of triplet energy from benzophenone to phenanthrene there is not a preferred angular orientation of the benzophenone but it must lie parallel to the phenanthrene.[35] The conclusion[36] that the benzophenone sensitizes the isomerization of hexa-2,4-diene by a chain process has been shown to be incorrect.[37] The sensitized photoisomerization of olefins by triplets with lower energy than that of the olefin[38] has been discussed,[39] and by studying isotope effects[40a] it was concluded that isomerization occurs by way of spectroscopically inaccessible states. However, other kinetic studies support[40b] the diradical mechanism given by Yang.[38] The efficiency of energy transfer from carbonyl compounds with lowest triplet states of the $\pi–\pi^*$ type is concentration-dependent.[41] This phenomenon may be due to the formation, at higher concentrations, of triplet excimers either with a very short lifetime or with a triplet energy lower than the acceptor. Energy transfer from a number of triplets to azulene has been shown to populate the singlet excited state of the hydrocarbon.[42] The participation of the T_2 state of anthracenes in energy transfer processes has been verified,[43a] and their energies have been estimated at between 68 and 74 kcal mole^{-1}.[43b]

Intramolecular energy transfer studies on a number of alkaloids have shown that singlet energy transfer occurs efficiently over distances greater than those allowed by the Förster mechanism.[44] The observation that singlet energy

(1)   (2)

[34] A. C. Day and T. R. Wright, *Tetrahedron Letters*, **1969**, 1067.

[35] K. B. Eisenthal, *J. Chem. Phys.*, **50**, 3120 (1969).

[36] H. L. Hyndman, B. M. Monroe, and G. S. Hammond, *J. Am. Chem. Soc.*, **91**, 2852 (1969).

[37] J. Saltiel, L. Metts, and M. Wrighton, *J. Am. Chem. Soc.*, **91**, 5684 (1969).

[38] N. C. Yang, J. I. Cohen, and A. Shani, *J. Am. Chem. Soc.*, **90**, 3264 (1968).

[39] N. J. Turro, *Photochem. Photobiol.*, **9**, 555 (1969).

[40a] R. A. Caldwell and G. W. Sovocool, *J. Am. Chem. Soc.*, **90**, 7138 (1968).

[40b] J. Saltiel, K. R. Neuberger, and M. Wrighton, *J. Am. Chem. Soc.*, **91**, 3658 (1969).

[41] O. L. Chapman and G. Wampfler, *J. Am. Chem. Soc.*, **91**, 5390 (1969).

[42] A. Martinez, *J. Chim. Phys.*, **65**, 1663 (1968); J. Saltiel and E. D. Megavity, *J. Am. Chem. Soc.*, **91**, 1265 (1969).

[43a] R. S. H. Liu and R. E. Kellog, *J. Am. Chem. Soc.*, **91**, 250 (1969).

[43b] R. S. H. Liu and J. R. Edman, *J. Am. Chem. Soc.*, **91**, 1492 (1969).

[44] R. D. Rauh, T. R. Evans, and P. A. Leermakers, *J. Am. Chem. Soc.*, **90**, 6897 (1968); **91**, 1868 (1969).

transfer does not occur in (**1**)[45] has been shown to be in error.[46] Efficient triplet energy transfer, by an exchange mechanism, occurs in (**2**).[47] In the case of (**3**), triplet energy transfer from the carbonyl group to the double bond

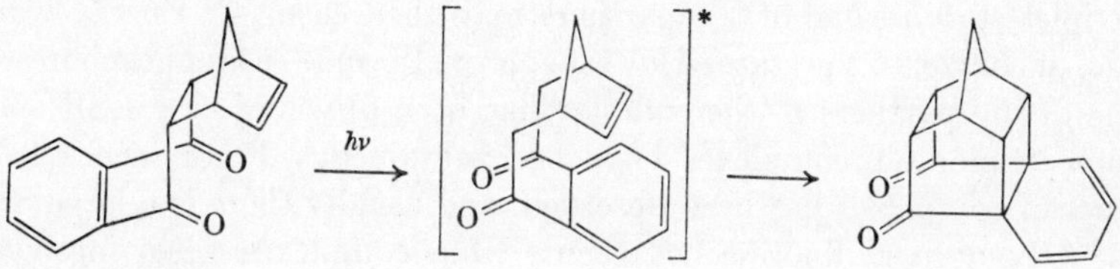

(3)

occurs and excited olefin adds to the aromatic system.[48] Intramolecular triplet energy transfer is only efficient over relatively short distances.[49] Such energy transfer has been shown to occur in polymers,[50a,b] and in the case of poly-(1-vinylnaphthalene) delayed fluorescence produced by triplet–triplet annihilation was observed.[50b] Intramolecular energy transfer in metal chelates has been further investigated.[51]

It is suggested that cyclohexa-1,4-diene quenches triplet carbonyl compounds by vibrational energy transfer.[52] Diborane[53] and iodoacetic acid[54] quench triplets and in doing so are dissociated.

Energy Transfer and Reactions occurring via Complex Formation

It is now recognized that excited states of many molecules form complexes with ground-state molecules. If such a complex is also stable in the ground state it is known as an excited charge-transfer complex (excited CT complex). If the complex is dissociated in its ground state, it is termed an exciplex if the molecules in the complex are different, and an excimer if they are the same.[55] The physical properties and participation of these complexes in energy transfer

[45] *Org. Reaction Mech.*, **1968**, 416.
[46] A. A. Lamola, *J. Am. Chem. Soc.*, **91**, 4786 (1969).
[47] N. Filipescu, J. R. DeMember, and F. L. Minn, *J. Am. Chem. Soc.*, **91**, 4169 (1969).
[48] N. Filipescu and J. M. Menter, *J. Chem. Soc.*(B), **1969**, 616.
[49] R. A. Keller and L. J. Dolby, *J. Am. Chem. Soc.*, **91**, 1293 (1969); M. Zander, *Ber. Bunsenges. Phys. Chem.*, **72**, 1161 (1968).
[50a] W. Klöpffer, *J. Chem. Phys.*, **50**, 2337 (1969).
[50b] R. F. Cozzens and R. B. Fox, *J. Chem. Phys.*, **50**, 1532 (1969).
[51] Y. Matsuda, S. Makishima, and S. Shionoya, *Bull. Chem. Soc. Japan*, **42**, 356 (1969); M. Kleinerman, *J. Chem. Phys.*, **51**, 2370 (1969); M. Kleinerman and S. Choi, *ibid.*, **49**, 3901 (1968).
[52] A. M. Braun, W. B. Hammond, and H. G. Cassidy, *J. Am. Chem. Soc.*, **91**, 6196 (1969).
[53] R. L. Strong, W. M. Howard, and R. L. Tinklepaugh, *Ber. Bunsenges. Phys. Chem.*, **72**, 200 (1968).
[54] H. van Zwet, *Rec. Trav. Chim.*, **87**, 1201 (1968).
[55] J. B. Birks, *Nature*, **214**, 1187 (1967).

and chemical reactions have been the subject of investigations some of which are described hereafter.

Excitation of CT complexes produces the excited singlet state of the complex. Intersystem crossing either to the triplet state of the complex or to the triplet state of one of the partners may then ensue.[56a] Energy transfer occurs, of course, to the state of lowest energy. In some cases phosphorescence of one of the partners in the complex has been observed.[56a,b] ESR spectra of some triplet CT complexes have been obtained.[57] Fluorescence from a number of complexes has been observed, and usually there is a large Stokes shift.[58a] Long-lived fluorescence occurs from complexes when intersystem crossing from the singlet to the triplet state is efficient and when the two states have a small energy gap.[58b] The formation of radical ions by dissociation of excited CT complexes has been observed by flash photolysis.[59]

An excellent review on excimer formation has appeared.[60] The energetics of excimer formation have been investigated,[61a,b] and singlet excimers shown to be more stable than their triplet counterparts.[61a] Fluorescence studies have shown that there is an activation energy for excimer formation.[61c] A number of aromatic hydrocarbons form excimers,[62a] and examples of intramolecular excimer formation have been found.[62b] Excimer formation has been shown to affect transfer of excitation energy from pyrene to perylene.[63]

Singlet exciplexes are more stable than triplet exciplexes.[64] Exciplex formation between amines and hydrocarbons,[65a] and nitro-compounds and

56a G. Briegleb, H. Schuster, and W. Herre, *Chem. Phys. Letters*, **4**, 53 (1969); G. Briegleb, G. Betz, and W. Herre, *Z. Phys. Chem.* (Frankfurt), **64**, 85 (1969).
56b G. Briegleb and H. Schuster, *Angew. Chem. Internat. Ed. Engl.*, **8**, 771 (1969); G. Briegleb, W. Herre, and D. Wolf, *Spectrochim. Acta*, **25**, 39 (1969).
57 H. Hayashi, S. Iwata, and S. Nagakura, *J. Chem. Phys.*, **50**, 993 (1969).
58a T. Himba, J. G. Vegter, and J. Kommandeur, *J. Chem. Phys.*, **49**, 4755 (1968); J. Prochorow and R. Siego Czýnski, *Chem. Phys. Letters*, **3**, 635 (1969); G. Briegleb, J. Trencséni, and W. Herre, *ibid.*, p. 146; N. Mataga and Y. Murata, *J. Am. Chem. Soc.*, **91**, 3144 (1969).
58b G. D. Short, *Chem. Comm.*, **1968**, 1500.
59 R. Potashnik, C. R. Goldschmidt, and M. Ottolenghi, *J. Phys. Chem.*, **73**, 3170 (1969); K. Kiwai, N. Yamamoto, and H. Tsubomura, *Bull. Chem. Soc. Japan*, **42**, 369 (1969); M. Morita and S. Kato, *ibid.*, p. 25.
60 T. Förster, *Angew. Chem. Internat. Ed. Engl.*, **8**, 333 (1969).
61a A. K. Chandra and E. C. Lim, *J. Chem. Phys.*, **49**, 5066 (1968).
61b J. R. Greenleaf, M. D. Lumb, and J. B. Birks, *J. Phys.*(B), **1**, 1157 (1968).
61c R. Speed and B. Selinger, *Austral. J. Chem.*, **22**, 9 (1969).
62a P. Holzman and R. C. Jarnigan, *J. Chem. Phys.*, **51**, 2251 (1969); F. Hirayama and S. Lipsky, *ibid.*, **51**, 1939 (1969); P. K. Ludwig and C. D. Amata, *ibid.*, **49**, 326 (1968); J. B. Birks, J. C. Conte, and G. Walker, *J. Phys.*(B), **1**, 934 (1968); E. Loewenthal, Y. Tomkiewicz, and A. Weinreb, *Spectrochim. Acta*, **25**, *A*, 1501 (1969).
62b F. Schneider and E. Lippert, *Ber. Bunsenges. Phys. Chem.*, **72**, 1155 (1968); J. R. Froines and P. J. Hagerman, *Chem. Phys. Letters*, **4**, 135 (1969); W. Klöpffer, *ibid.*, p. 193.
63 C. R. Goldschmidt, Y. Tomkiewicz, and A. Weinreb, *Spectrochim. Acta*, **25**, *A*, 1471 (1969).
64 D. Greatorex, T. J. Kemp, and J. P. Roberts, *J. Phys. Chem.*, **73**, 1616 (1969).
65a M. G. Kuzmin and L. N. Guseva, *Chem. Phys. Letters*, **3**, 71 (1969); L. N. Guseva, N. A. Sadovskii, and M. G. Kuzmin, *Khim. Vys. Energ.*, **3**, 44 (1969); *Chem. Abs.*, **70**, 77043 (1969).

porphyrins[65b] and chlorophylls,[65c] has been observed. In highly polar solvents, exciplexes can dissociate into radical ions. These may also be formed by electron transfer between the excited molecule and the ground-state molecule. Electron transfer from amines of low ionization potential to benzophenone,[66a] dimethyl isophthalate,[66b] and chlorophyll[66c] has been observed.

The formation of excimers and exciplexes can result in deactivation of excited states by the complexes dissociating to ground-state molecules plus heat or light, or undergoing chemical reaction. In the case of singlet complexes, population of the triplet state of one of the constituent molecules can occur. The occurrence of these processes, in systems in which excimer or exciplex emission was not observed, has been invoked by many authors. Amines quench fluorescence[67a,b] and phosphorescence[67a] of biacetyl. In low-polarity solvents quenching is undoubtedly due to exciplex formation, but in those of high polarity electron transfer may be the process. Sulphides[67b, 68] and tervalent phosphorus compounds[67b] quench the triplet state of carbonyl compounds, and the phosphorus compounds also quench the fluorescence of anthracene.[67b] Amines quench the fluorescence of fluorenone[67b, 69a,b] by exciplex formation. The increased efficiency of quenching in polar solvents may be due to radical ion formation, but this cannot be stated definitely since in polar solvents intersystem crossing to the $n \rightarrow \pi^*$ triplet is much less effective.[69a,b] Calculations have shown that quenching of naphthalene fluorescence by quadricyclene by exciplex formation is energetically feasible.[70] Excimer formation may be the reason for the inefficient dimerization of triplet indene,[71a] and exciplex formation for the isomerization of but-2-ene by benzophenone,[71b] and for the quenching of pyruvonitrile singlets by dienes.[71c] The photoreduction of ketones by amines has been suggested as occurring by excited CT complex (possibly exciplex) formation,[72a] by radical ions,[72b] and in the case of fluorenone

[65b] D. G. Whitten, I. G. Lopp, and P. D. Wildes, *J. Am. Chem. Soc.*, **90**, 7196 (1968); G. R. Seely, *J. Phys. Chem.*, **73**, 125 (1969).

[65c] G. R. Seely, *J. Phys. Chem.*, **73**, 125 (1969).

[66a] R. S. Davidson, P. F. Lambeth, J. F. McKellar, P. H. Turner, and R. Wilson, *Chem. Comm.*, **1969**, 732.

[66b] H. Yamashita, H. Kokubin, and M. Koizumi, *Bull. Chem. Soc. Japan*, **41**, 2312 (1968).

[66c] Y. M. Stolovitskii and V. B. Evstigneev, *Mol. Biol.*, **3**, 176 (1969); *Chem. Abs.*, **71**, 2735 (1969).

[67a] N. J. Turro and R. Engel, *Mol. Photochem.*, **1**, 143 (1969).

[67b] R. S. Davidson and P. F. Lambeth, *Chem. Comm.*, **1969**, 1098.

[68] J. Guttenplan and S. G. Cohen, *Chem. Comm.*, **1969**, 247.

[69a] L. A. Singer, *Tetrahedron Letters*, **1969**, 923; L. A. Singer, G. A. Davis, and V. P. Muralidharan, *J. Am. Chem. Soc.*, **91**, 897 (1969); R. A. Caldwell, *Tetrahedron Letters*, **1969**, 2121.

[69b] J. B. Guttenplan and S. G. Cohen, *Tetrahedron Letters*, **1969**, 2125.

[70] B. S. Solomon, C. Steel, and A. Weller, *Chem. Comm.*, **1969**, 927.

[71a] C. DeBoer, *J. Am. Chem. Soc.*, **91**, 1855 (1969).

[71b] R. A. Caldwell and S. P. James, *J. Am. Chem. Soc.*, **91**, 5184 (1969).

[71c] T. R. Evans and P. A. Leermakers, *J. Am. Chem. Soc.*, **91**, 5898 (1969).

[72a] P. J. Wagner and A. E. Kemppainen, *J. Am. Chem. Soc.*, **91**, 3085 (1969).

[72b] A. Padwa, W. Eisenhardt, R. Gruber, and D. Pashayan, *J. Am. Chem. Soc.*, **91**, 1857 (1969).

by way of its triplet.[69b, 72c] The photoreaction of boron tribromide with alkylbenzenes[73a] and of amines with halogenoaromatics [73b] appears to involve exciplexes as intermediates.

Carbonyl Compounds

Saturated Ketones

The formation of radicals in the photoreduction of carbonyl compounds has been observed by induced dynamic nuclear spin polarization,[74a] ESR,[74b] and flash photolysis.[66a,74c] Ketyl radical ions, formed by dissociation of ketyl radicals, have been observed in the photoreduction of fluorenone and benzophenone by amines in ethanolic and aqueous solution.[75] Kinetics of the combination of ketyl radicals and their anions have been examined, and the role of this process in determining whether pinacol or secondary alcohol formation occurs has been assessed.[76] Ketyl radicals observed in reductions may be formed by photoinduced cleavage of pinacols.[77a] Hydrogen atom transfer from these radicals to ground-state carbonyl molecules has been demonstrated.[77a,b] The rate of deactivation of triplet benzophenone has been determined as 3×10^5 sec^{-1} by two independent methods.[22, 78] There is evidence that triplet benzophenone[79a] and other carbonyl triplets (e.g. phenanthraquinone[79b]) can abstract hydrogen from benzene. The formation of light-absorbing transients[80a] and of unusual ketyl radical coupling products has been further examined.[80b] The bimolecular rate constants for hydrogen abstraction from amines by benzophenone and related compounds are very high (10^7–10^8 M^{-1} sec^{-1}).[81] Rate constants for the reduction of fluorenone by

72c G. A. Davis, P. A. Carapellucci, K. Szoc, and J. D. Gresser, *J. Am. Chem. Soc.*, **91**, 2264 (1969).

73a Y. Ogata, Y. Izawa, H. Tomioka, and T. Ukigai, *Tetrahedron*, **25**, 1817 (1969).

73b T. Latowski, *Z. Naturforsch.*, **23a**, 1127 (1968); T. Tosa, C. Pac, and H. Sakurai, *Tetrahedron Letters*, **1969**, 3635.

74a G. L. Closs and C. E. Closs, *J. Am. Chem. Soc.*, **91**, 4550 (1969).

74b R. Wilson, *J. Chem. Soc.*(B), **1968**, 1581.

74c N. Kanamaru and S. Nagakura, *J. Am. Chem. Soc.*, **90**, 6905 (1968).

75 R. S. Davidson, P. F. Lambeth, F. A. Younis, and R. Wilson, *J. Chem. Soc.*(C), **1969**, 2203.

76 G. O. Schenck, C. Matthias, M. Pape, M. Cziesla, G. von Bünau, E. Roselius, and G. Koltzenburg, *Annalen*, **719**, 80 (1968).

77a R. S. Davidson, F. A. Younis, and R. Wilson, *Chem. Comm.*, **1969**, 826.

77b G. O. Schenck, G. Koltzenburg, and E. Roselius, *Z. Naturforsch.*, **24b**, 222 (1969).

78 P. J. Wagner, *Mol. Photochem.*, **1**, 71 (1969).

79a W. F. Smith, Jr., and B. W. Rossiter, *Tetrahedron*, **25**, 2059 (1969); W. F. Smith, Jr., *ibid.*, p. 2071.

79b M. B. Rubin and Z. Neuwirth-Weiss, *Chem. Comm.*, **1968**, 1607.

80a S. G. Cohen and J. I. Cohen, *Israel J. Chem.*, **6**, 757 (1968); H. L. J. Bäckström and R. J. V. Niklasson, *Acta Chem. Scand.*, **22**, 2589 (1968); N. Filipescu and F. L. Minn, *J. Chem. Soc.*(B), **1969**, 84.

80b B. D. Challand, *Can. J. Chem.*, **47**, 687 (1969).

81 S. G. Cohen and N. Stein, *J. Am. Chem. Soc.*, **91**, 3690 (1969); S. G. Cohen and B. Green, *ibid.*, p. 6824.

tributyltin hydride,[72c] and of β-ketophosphates,[82] have been determined. Benzophenone is reduced by certain phenols[83a] and triphenylsilane.[83b] That 4-aminobenzophenone is reduced by alkanes and not by alcohols has been confirmed, and the unreactivity has been shown to be due to inefficient intersystem crossing to the $n-\pi^*$ triplet state in polar solvents.[84a] This has been confirmed by emission studies.[84b] Further examples of cyclobutanol formation by intramolecular hydrogen abstraction have been reported,[85a]

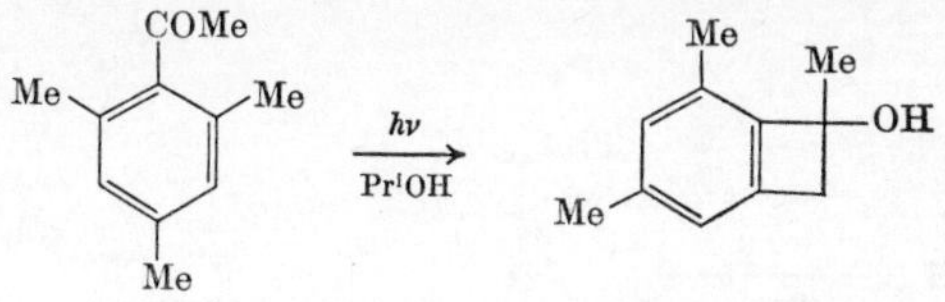

(4)

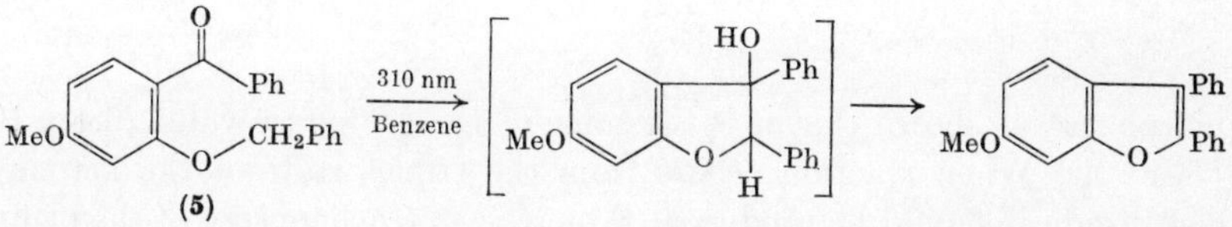

(5)

e.g. (4).[85b] Irradiation of 2-methylbenzophenone in the presence of oxygen gives 2-benzoylbenzoic acid.[86] Intramolecular hydrogen abstraction reactions involving cyclic transition states containing more than six atoms,[87a] e.g. (5),[87b] have been reported. The Type II reaction has again been the subject of a number of investigations. This reaction can lead to the fragmentation of rings,[88a] e.g. (6) and (7).[88b,c] The demethoxylation of α-methoxy-ketones probably occurs via a Type II reaction.[88d] A redetermination of the rate constant for intramolecular hydrogen abstraction in pentan-2-one in the

[82] H. Tomioka, Y. Izawa, and Y. Ogata, *Tetrahedron*, **25**, 1501 (1969).
[83a] H.-D. Becker, *J. Org. Chem.*, **34**, 2472 (1969).
[83b] H.-D. Becker, *J. Org. Chem.*, **34**, 2469 (1969).
[84a] S. G. Cohen, M. D. Saltzman, and J. B. Guttenplan, *Tetrahedron Letters*, **1969**, 4321.
[84b] E. J. O'Connell, *Chem. Comm.*, **1969**, 571.
[85a] B. W. Finucane and J. B. Thomson, *Chem. Comm.*, **1969**, 380; R. Imhof, W. Graf, H. Wehrli, and K. Schaffner, *ibid.*, p. 857; M. Mousseron-Canet and J. P. Chabaud, *Bull. Soc. Chim. France*, **1969**, 239.
[85b] T. Matsuura and Y. Kitaura, *Tetrahedron*, **25**, 4487 (1969).
[86] M. Pfan, E. W. Sarver, and N. D. Heindel, *Compt. Rend., C*, **268**, 1167 (1969).
[87a] R. Breslow and M. A. Winnik, *J. Am. Chem. Soc.*, **91**, 3083 (1969).
[87b] G. R. Lappin and J. S. Zannucci, *Chem. Comm.*, **1969**, 1113.
[88a] A. Padwa and D. Eastman, *J. Am. Chem. Soc.*, **91**, 462 (1969); J. H. Stocker and D. Kern, *Chem. Comm.*, **1969**, 204.
[88b] T.-Y. Chen, *Bull. Chem. Soc. Japan*, **41**, 2540 (1968).
[88c] A. Padwa, E. Alexander, and M. Niemcyzk, *J. Am. Chem. Soc.*, **91**, 456 (1969).
[88d] P. M. Collins and P. Gupta, *Chem. Comm.*, **1969**, 90.

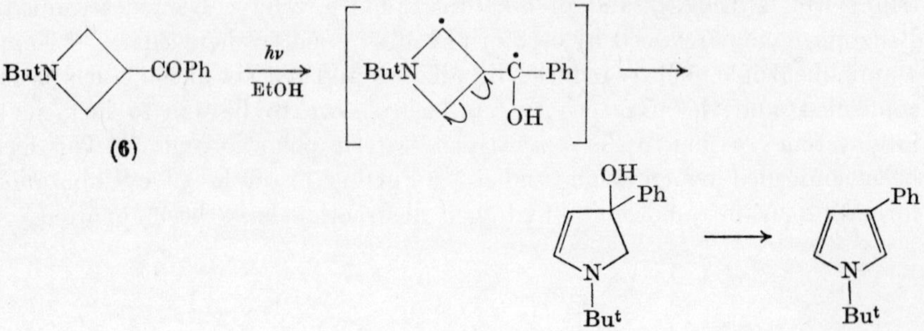

(6)

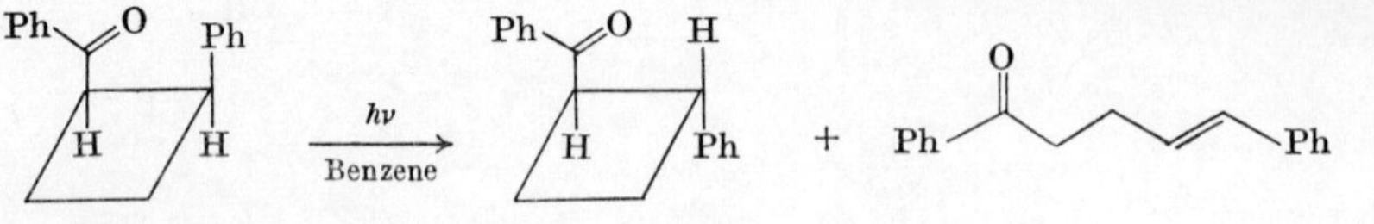

(7)

liquid phase has shown that it is the same as the gas phase value (6.4×10^7 M^{-1} sec^{-1}).[89] When reaction occurs from the triplet state of the ketone a triplet diradical should be produced. From the stereochemistry of the olefins produced in such reactions it has been concluded[90] that the diradical is fairly long-lived and it loses its spin identity.

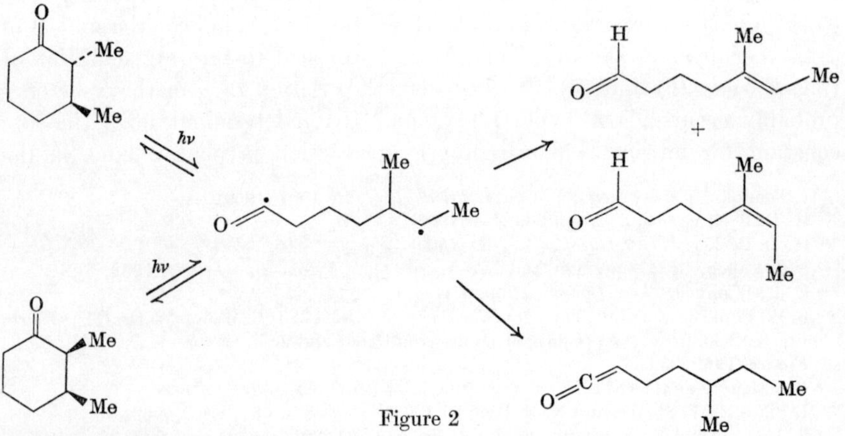

Figure 2

[89] C. H. Bibart, M. G. Rockley, and F. S. Wettack, *J. Am. Chem. Soc.*, **91**, 2802 (1969).
[90] J. E. Gano, *Tetrahedron Letters*, **1969**, 2549; R. A. Caldwell and P. M. Fink, *ibid.*, p. 2987; P. J. Wagner and P. A. Kelso, *ibid.*, p. 4151.

A quantitative examination has been made [91a] of the recyclization and formation of ketenes and alkenals from diradicals produced in the Type I cleavage of triplet cycloalkanones (see Figure 2).[91b] Cleavage of α-substituted cycloalkanones occurs preferentially at the more heavily substituted C–C bond.[91a] β-Substituents retard alkenal formation.[96] Photolysis of camphor gives diradicals by cleavage of both C–C=O bonds,[92a] and carbene formation from the diradical has been detected.[92b] Cleavage of α-hydroxy-ketones to give aldehydes occurs on photolysis.[93] Photolysis of some cyclodecalenones, e.g. (8), has been reported by one group to give oxetans,[94a] and by another

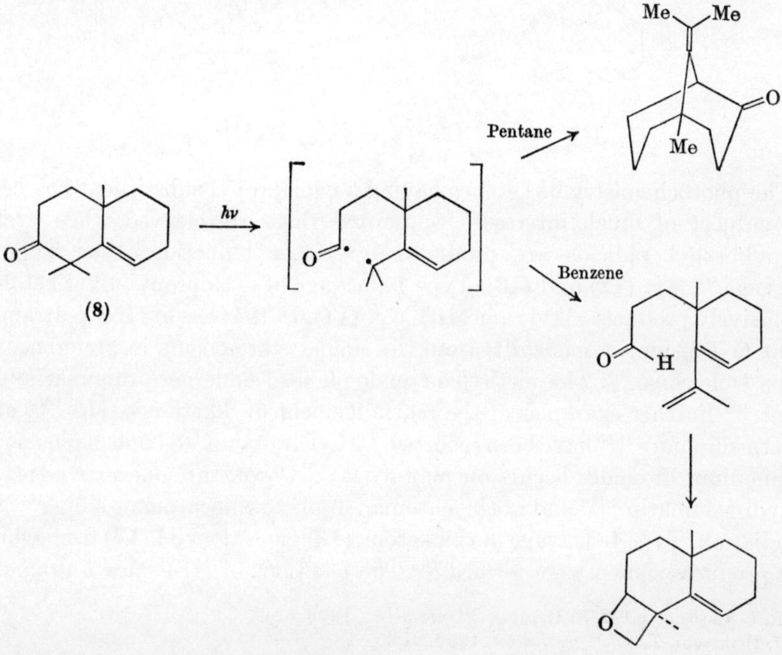

group to give bicyclo[3.3.1]nonanones.[94b] The stereospecific formation of the cyclopropane from (9)[95] is probably a concerted reaction [$_\pi 2 + _\sigma 2$] although it can be visualized as occurring by a diradical mechanism. Many fascinating

[91a] C. C. Badcock, M. J. Perona, G. O. Pritchard, and B. Rickborn, *J. Am. Chem. Soc.*, **91**, 543 (1969); P. J. Wagner and R. W. Spoerke, *ibid.*, p. 4437.

[91b] J. A. Barltrop and J. D. Coyle, *Chem. Comm.*, **1969**, 1081.

[92a] P. Yates and G. Hagens, *Tetrahedron Letters*, **1969**, 3623.

[92b] W. C. Agosta and D. K. Herron, *J. Am. Chem. Soc.*, **90**, 7025 (1968).

[93] E. J. Baum, L. D. Hess, J. R. Wyatt, and J. N. Pitts, *J. Am. Chem. Soc.*, **91**, 2461 (1969).

[94a] K. Kojima, K. Sakai, and K. Tanabe, *Tetrahedron Letters*, **1969**, 3399.

[94b] L. A. Paquette and G. V. Meehan, *J. Org. Chem.*, **34**, 450 (1969).

[95] J. R. Williams and H. Ziffer, *Tetrahedron*, **24**, 6725 (1968).

products, e.g. (10)[96a] and (11),[96b] have been obtained from β,γ-unsaturated ketones by a Type I cleavage reaction.

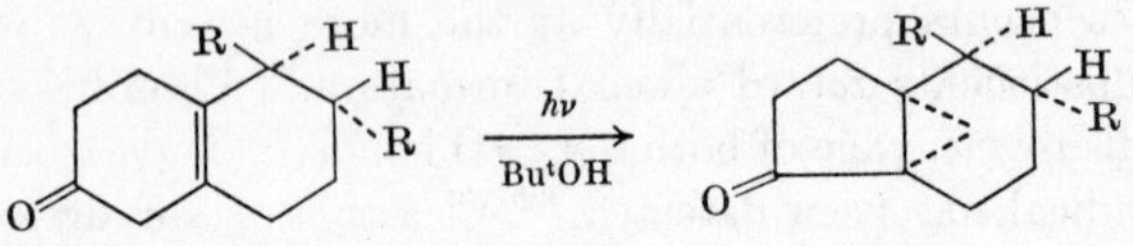

(9)

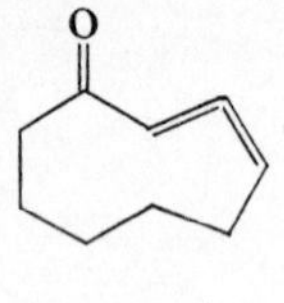

(10) 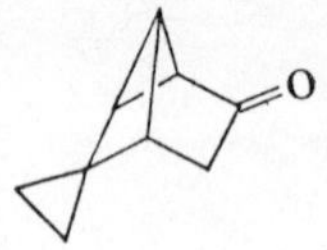   (11)

The photochemistry of ketones having a cyclopropyl substituent has been the subject of much interest. Cyclopropyl rings are cleaved when cyclopropylmethyl radicals are produced by either reduction[97a] or Type I cleavage,[97b] e.g. (12) and (13). Type I cleavage of cyclopropyl alkyl ketones exclusively produces alkyl radicals,[98] e.g. (14). In the case of (15), rearrangement (a 1,3-sigmatropic shift) from the singlet state occurs in preference to Type I cleavage.[99a] The methylene analogue also undergoes this rearrangement.[99b] Further examples of the rearrangement of β-keto-epoxides[100a] and β-keto-sulphides[100b] have been reported. Cleavage of the C–S bond of phenacylsulphonium bromides occurs on photolysis.[101] Chroman-3-one rearranges to dihydrocoumarin,[102a] and isothiochroman-4-one to thiochroman-3-one[102b] on irradiation. Type I cleavage of thioketone (16) gives the ylid (17) from which the products shown were formed.[103] The reactions of (18) show a dramatic

96a R. G. Carlson and D. E. Henton, *Chem. Comm.*, 1969, 674.
96b J. Ipaktschi, *Tetrahedron Letters*, 1969, 2153.
97a W. G. Dauben, L. Schutte, and R. E. Wolf, *J. Org. Chem.*, 34, 1849 (1969); W. G. Dauben and G. W. Shaffer, *ibid.*, p. 2301; W. G. Dauben, L. Schutte, R. E. Wolf, and E. J. Deviny, *ibid.*, p. 2512.
97b R. G. Carlson and E. L. Biersmith, *Chem. Comm.*, 1969, 1049.
98 A. Sonoda, I. Moritani, J. Miki, and T. Tsuji, *Tetrahedron Letters*, 1969, 3187.
99a L. A. Paquette, G. V. Meehan, and R. F. Eizember, *Tetrahedron Letters*, 1969, 995.
99b L. A. Paquette, G. V. Meehan, and R. F. Eizember, *Tetrahedron Letters*, 1969, 999.
100a K. Kojima, K. Sakai, and K. Tanabe, *Tetrahedron Letters*, 1969, 1925; J. P. Pete and M. L. Villaume, *ibid.*, p. 3753.
100b A. Padwa, A. Battisti, and E. Shefter, *J. Am. Chem. Soc.*, 91, 4000 (1969); C. Ganter and J.-F. Moser, *Helv. Chim. Acta*, 52, 725 (1969); C. Ganter and J.-F. Moser, *ibid.*, p. 967.
101 T. Laird and H. Williams, *Chem. Comm.*, 1969, 561.
102a P. K. Grover and N. Anand, *Chem. Comm.*, 1969, 982.
102b W. C. Lumma, Jr., and G. A. Berchtold, *J. Org. Chem.*, 34, 1566 (1969).
103 K. K. Maheshwari and G. A. Berchtold, *Chem. Comm.*, 1969, 13.

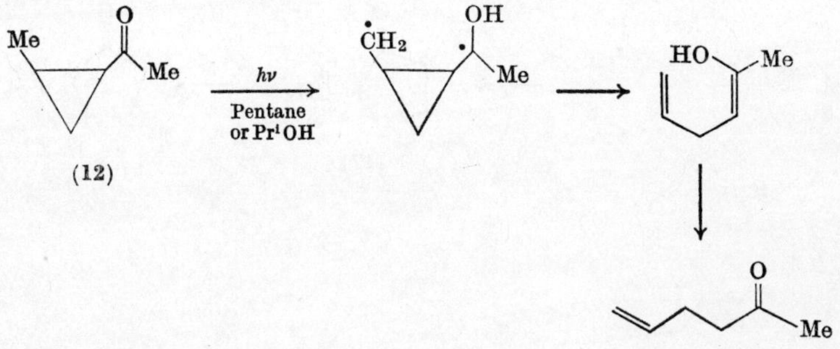

(12)

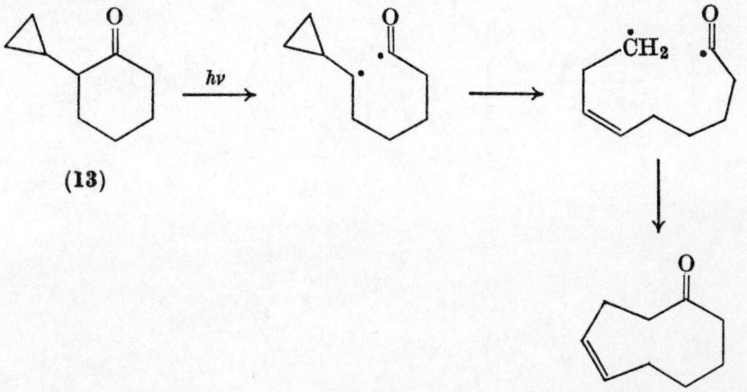

(13)

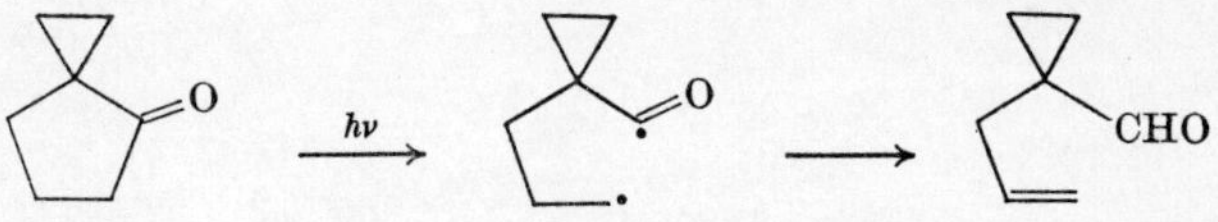

(14)

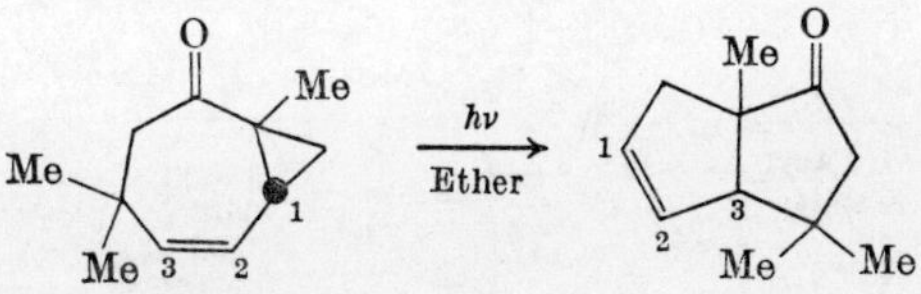

(15)

wavelength dependence.[104] De-t-butylation of 3,5-di-t-butyl-4-hydroxy-benzophenone to give 3-t-butyl-4-hydroxybenzophenone is envisaged as occurring by Type I cleavage of the keto form of the phenol.[105] The decarbonyl-ation of cyclobutane-1,3-dione has been shown to occur by diradicals.[106] The stereospecific isomerization of (19) to (20a) occurs from its singlet state

[104] J. G. Pacifici and C. Diebert, *J. Am. Chem. Soc.*, **91**, 4595 (1969).
[105] T. Matsuura and Y. Kitaura, *Tetrahedron*, **25**, 4487 (1969).
[106] N. J. Turro and T. Cole, *Tetrahedron Letters*, **1969**, 3451.

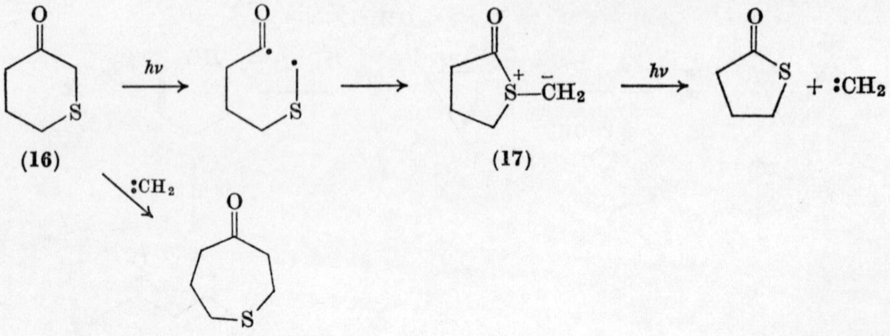

(16)　　　　　　　　　　　　　　　　　　　　　(17)

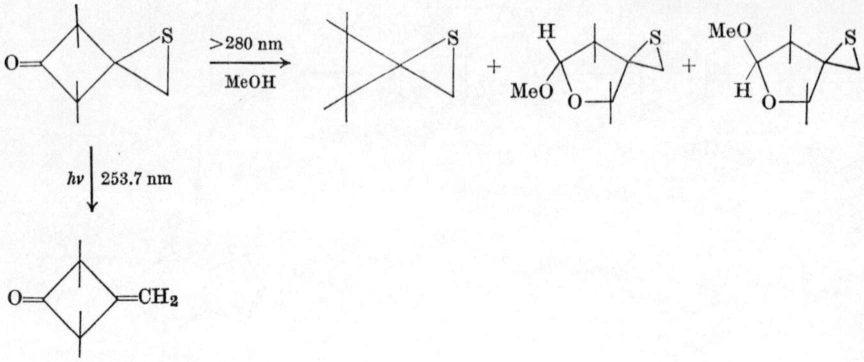

(18)

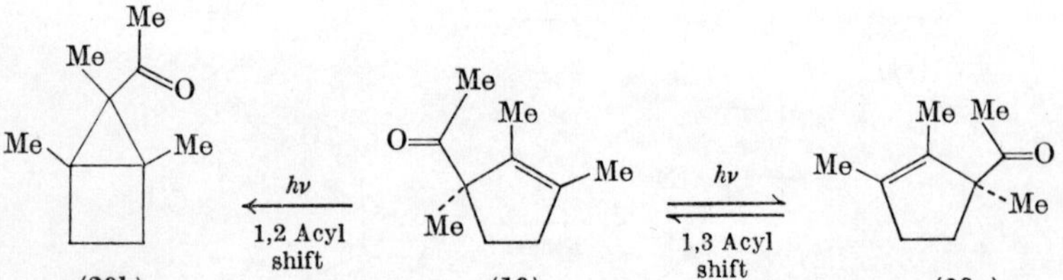

(20b)　　　　　　　　　(19)　　　　　　　　　(20a)

whereas **(20b)** is produced from its triplet state.[107] Both reactions are probably concerted. Several studies of the Type I cleavage of simple carbonyl compounds in the gas phase have been made.[108]

The formation of oxetans by the photocycloaddition of carbonyl compounds to olefins has been reviewed.[109] The singlet state of some carbonyl compounds

107 E. Baggidini, K. Schaffner, and D. Jeger, *Chem. Comm.*, **1969**, 1013.

108 A. T. Blades, *Can. J. Chem.*, **47**, 615 (1969); J. R. Majer, C. Olavesen, and J. C. Robb, *J. Chem. Soc.*(A), **1969**, 893; J. R. Majer and D. Phillips, *J. Chem. Soc.*(B), **1969**, 201; P. Potzinger and G. von Bünau, *Ber. Bunsenges. Phys. Chem.*, **72**, 195 (1968).

109 D. R. Arnold, *Adv. Photochem.*, **6**, 301 (1968).

has occasionally been found to participate.[110] Singlet state addition is stereo-specific. Irradiation of 2-acetylnaphthalene in the presence of methyl cinnamate does not give an oxetan but a product formed by 1,4-addition to the naphthalene nucleus.[111] Product studies on the addition of carbonyl compounds to a variety of olefins have been made,[112a] and a number of intramolecular additions have been reported.[112b]

Enones

Calculations on acraldehyde[113a] and emission studies on steroidal α,β-unsaturated ketones[113b] have shown that the $\pi \rightarrow \pi^*$ triplet is non-planar and the $n \rightarrow \pi^*$ triplet is planar. The $\pi \rightarrow \pi^*$ triplet has the lower energy.[113b] Proton addition to the non planar $\pi \rightarrow \pi^*$ triplet of a steroidal α,β-unsaturated ketone has been invoked to explain its photoisomerization to a β,γ-enone.[114]

$$[O{=}C{-}CH{=}CH{-}CH_2]^* \rightarrow [O{=}C{-}CH_2{-}\overset{+}{C}H{-}CH_2] \rightarrow O{=}C{-}CH_2{-}CH{=}CH{-}$$

An example of a β,γ-unsaturated ketone (21) produced by photoisomerization undergoing a Type I reaction has been reported.[115] The stereospecific rearrangement of (22) to (23) has been reported, and it occurs from the $\pi \rightarrow \pi^*$ triplet state.[116a] Population of the $n \rightarrow \pi^*$ triplet state leads to the formation of the β,γ-isomer. Confirmation that enones of the type (22) do have more than one type of triplet has been obtained by flash photolysis studies.[116b] Use of optically active enones did not produce an optically active triplet in detectable concentrations. Woodward and Hoffmann have shown[1] that the stereochemistry of the rearrangement is in agreement with it being a $[_\sigma 2_a + _\pi 2_a]$ process. The photorearrangement of 4-phenyl-4-(*p*-cyanophenyl)cyclohex-2-enone has been further examined, and the preferential migration of the *p*-cyanophenyl substituent is in accord with the β-carbon atom having odd-electron character.[117a,b] The rearrangement has been found to have an appreciable energy of activation (ca. 10 kcal mole^{-1}).[117c] It has been concluded

[110] N. J. Turro and P. A. Wriede, *J. Org. Chem.*, **34**, 3562 (1969).

[111] D. R. Arnold, L. B. Gillis, and E. B. Whipple, *Chem. Comm.*, **1969**, 918.

[112a] W. G. Bentrude and K. R. Darnall, *Chem. Comm.*, **1969**, 862; S. H. Schroeter, *ibid.*, p. 12; S. H. Schroeter and C. M. Orlando, *J. Org. Chem.*, **34**, 1181 (1969).

[112b] R. R. Sauers, W. Schinski, and M. M. Mason, *Tetrahedron Letters*, **1969**, 79; R. R. Sauers and J. A. Whittle, *J. Org. Chem.*, **34**, 3579 (1969); J. K. Crandall and C. F. Mayer, *ibid.*, p. 2814.

[113a] J. J. McCullough, H. Ohorodnyk, and D. P. Santry, *Chem. Comm.*, **1969**, 570.

[113b] G. Marsh, D. R. Kearns, and K. Schaffner, *Helv. Chim. Acta*, **51**, 1890 (1968); D. R. Kearns, G. Marsh, and K. Schaffner, *J. Chem. Phys.*, **49**, 3316 (1968).

[114] S. Kuwata and K. Schaffner, *Helv. Chim. Acta*, **52**, 173 (1969).

[115] N. Furutachi, Y. Nakadaira, and K. Nakanishi, *J. Am. Chem. Soc.*, **91**, 1028 (1969).

[116a] D. Bellus, D. R. Kearns, and K. Schaffner, *Helv. Chim. Acta*, **52**, 971 (1969).

[116b] G. Rämme, R. L. Strong, and H. H. Richtol, *J. Am. Chem. Soc.*, **91**, 5771 (1969).

[117a] H. E. Zimmerman and N. Levin, *J. Am. Chem. Soc.*, **91**, 879 (1969).

[117b] *Org. Reaction Mech.*, **1967**, 372.

[117c] H. E. Zimmerman and W. R. Elser, *J. Am. Chem. Soc.*, **91**, 887 (1969).

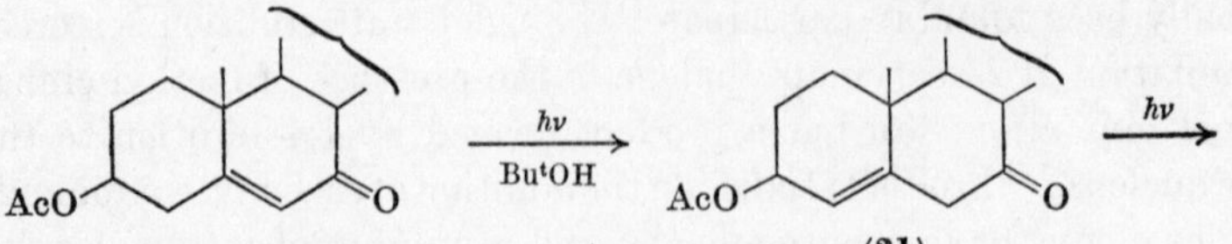

(21)

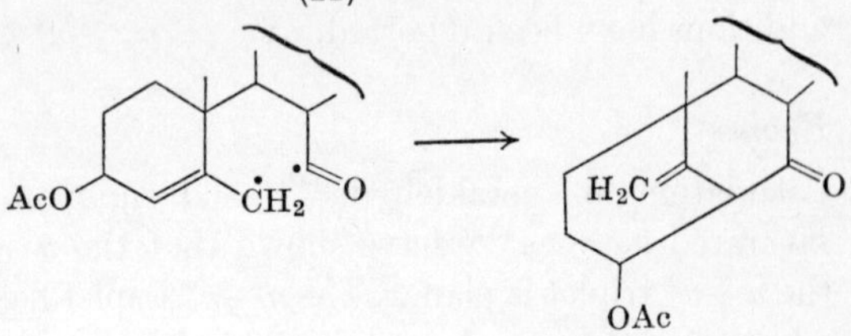

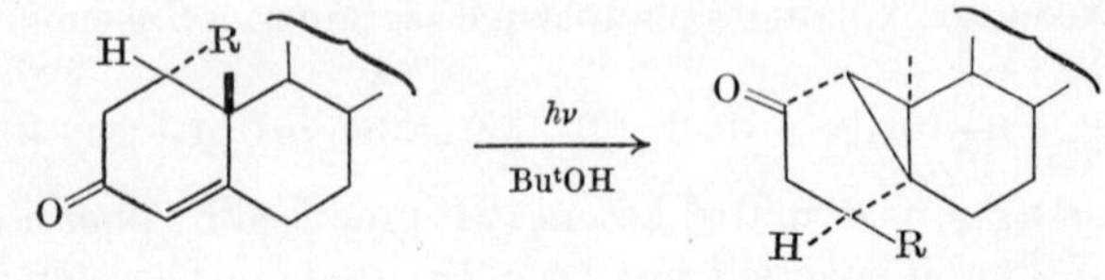

(22)   (23)

that excited-state potential-energy surfaces have thermal activation barriers, similar to those encountered in the ground state. Woodward and Hoffmann consider the rearrangement to be concerted and a $[_\sigma 2_a + _\pi 2_a]$ process,[1] whereas Zimmerman[3] considers it as a series of concerted reactions. Zimmerman's interpretation of the rearrangement of cyclohexa-2,5-dienones to bicyclo-[3.1.0]hexanones, e.g. (24),[3] has been strongly criticized by Woodward and Hoffmann.[1] They state that the physical evidence in favour of this being a multi-step process is not conclusive, and the question of whether it is a single concerted reaction $[_\sigma 2_a + _\pi 2_a]$ or a series of concerted reactions cannot be settled on the available evidence. However, Zimmerman is still providing

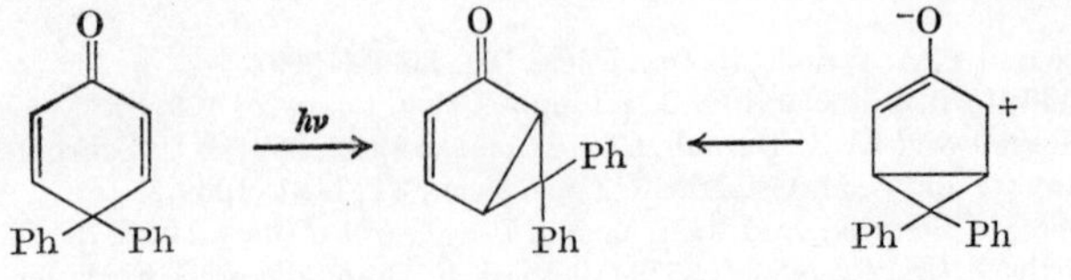

(24)   (25)

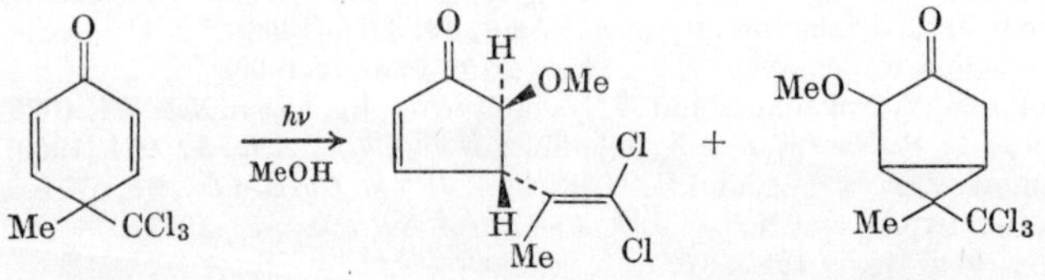

(26)   (27)

persuasive evidence for the intermediacy of zwitterionic species such as (25).[118] Isolation of (27) from the rearrangement of (26) appears indicative of the formation of a zwitterionic intermediate.[119] Enone (28) also ring-contracts

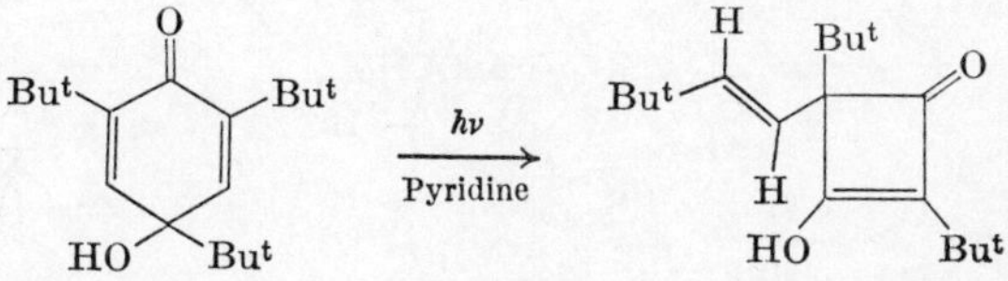

(28)

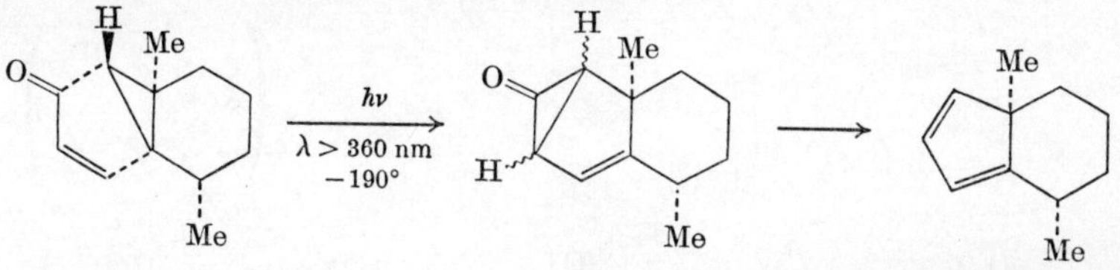

(29)　　　　　(30)

on irradiation.[120] Decarbonylation of (29) was shown, by means of low-temperature techniques, to occur via the cyclopropanone (30). Formation of (30) is envisaged as a $[_\sigma 2_s + _\pi 2_s]$ process.[121] Irradiation of (31)[122a] and (32)[122b] leads to bicyclo[3.1.0]hexanone formation. On irradiation in propan-2-ol,

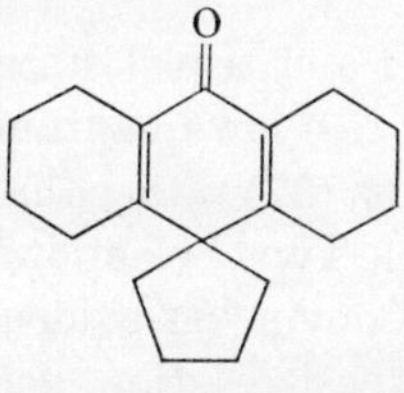

(31) 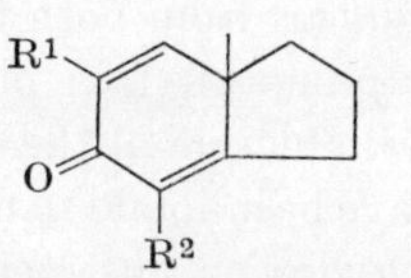　　　　　　　(32) 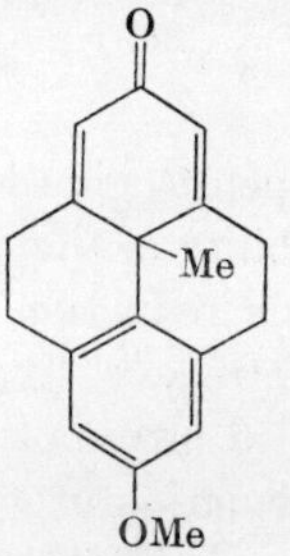　　　　　　　(33)

dienone (33) does not undergo rearrangement and is photoreduced with concomitant loss of the methyl group.[122c] The reduction of steroidal 1,4-dien-3-ones by sodium borohydride in the presence of light leads to phenolic pro-

[118] H. E. Zimmerman, D. S. Crumrine, D. Döpp, and P. S. Huyffer, *J. Am. Chem. Soc.*, **91**, 434 (1969).

[119] D. I. Schuster and V. Y. Abraitys, *Chem. Comm.*, **1969**, 419.

[120] D. A. Plank, J. C. Floyd, and W. H. Stames, *Chem. Comm.*, **1969**, 1003.

[121] L. L. Barber, O. L. Chapman, and J. D. Lassila, *J. Am. Chem. Soc.*, **91**, 3664 (1969).

[122a] H. Hart and D. C. Lankin, *J. Org. Chem.*, **33**, 4398 (1968).

[122b] D. Caine, W. J. Powers, and A. M. Alejandre, *Tetrahedron Letters*, **1968**, 6071.

[122c] H. E. Zimmerman and G. Jones, *J. Am. Chem. Soc.*, **91**, 5678 (1969).

ducts.[123] A number of examples have been reported of the use of low-temperature techniques[124a,b] to detect the formation of ketenes, e.g. (35),[124b] produced by the cleavage of cycloalkenones, e.g. (34). Spirodienones of the type (36)

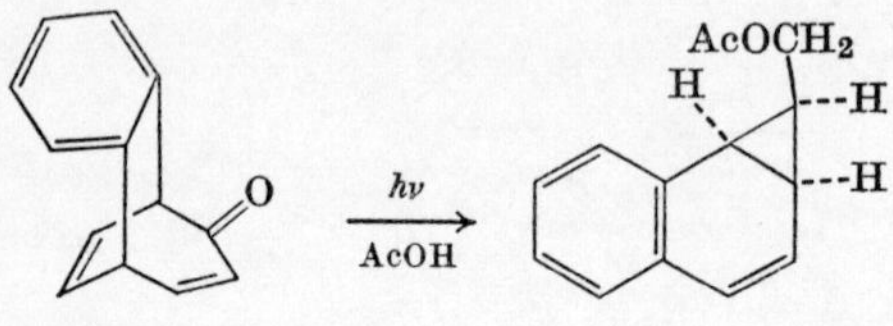

(34) 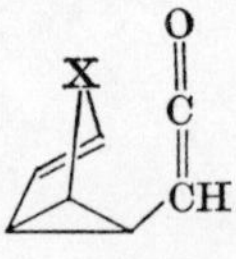   (35)

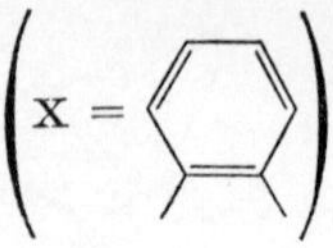

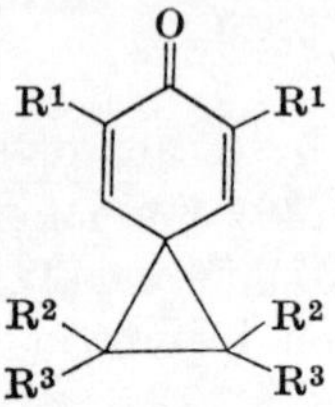

$$\left(X = \right)$$

(36) 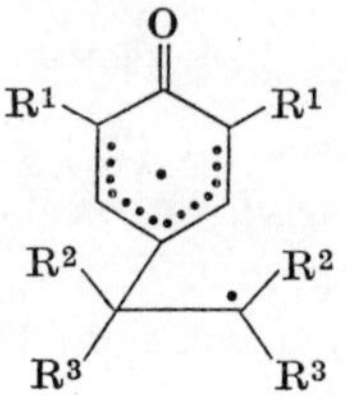   (37) 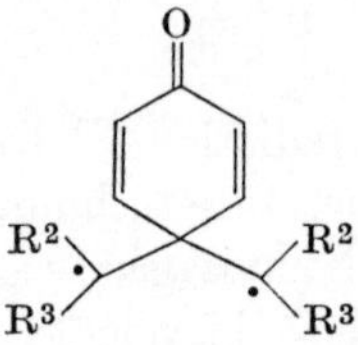   (38)

ring-open to give 1,3-diradicals from both the singlet and triplet state.[125a,b] The ability of the reverse process to take place leads to a low quantum yeild for their reduction.[125a] Formulation of the diradical as (37) rather than (38) is preferred.[125b] Enones have been found to undergo the Type I reaction[126a,b,c] and also intramolecular hydrogen abstraction.[126b] Convincing evidence for the formation of a carbene, as a result of such a Type I reaction, has been presented.[126a] The aldehyde (39) gives the ketone (40) by a 1,5-hydrogen shift.[127] The trienone (41) undergoes a Type I reaction in methanol whereas in

[123] J. Waters and B. Witkop, *J. Org. Chem.*, **34**, 1601 (1969).

[124a] L. L. Barker, O. L. Chapman, J. D. Lassila, *J. Am. Chem. Soc.*, **91**, 531 (1969); M. R. Morris and A. J. Waring, *Chem. Comm.*, **1969**, 526; J. E. Baldwin and S. M. Krueger, *J. Am. Chem. Soc.*, **91**, 2396 (1969).

[124b] O. L. Chapman, M. Kane, J. D. Lassila, R. L. Loeschen, and H. E. Wright, *J. Am. Chem. Soc.*, **91**, 6856 (1969); A. S. Kende, Z. Goldschmidt, and P. T. Izzo, *ibid.*, p. 6858.

[125a] D. I. Schluster and I. S. Krull, *Mol. Photochem.*, **1**, 107 (1969).

[125b] W. H. Pirkle, S. G. Smith, and G. F. Koser, *J. Am. Chem. Soc.*, **91**, 1580 (1969).

[126a] N. J. Turro, E. Lee-Ruff, D. R. Morton, and J. M. Conia, *Tetrahedron Letters*, **1969**, 2991.

[126b] F. A. L. Anet and D. P. Mullis, *Tetrahedron Letters*, **1969**, 737.

[126c] S. Seto, H. Sugiyama, S. Takenaka, and H. Watanabe, *J. Chem. Soc.*(C), **1969**, 1625.

[127] P. W. Schiess, *Chimia*, **22**, 483 (1968).

pentane solution isomerization followed by an intramolecular cycloaddition to give **(42)** occurred.[124a]

Examples of the intermolecular [128a] and intramolecular [128b] cycloaddition

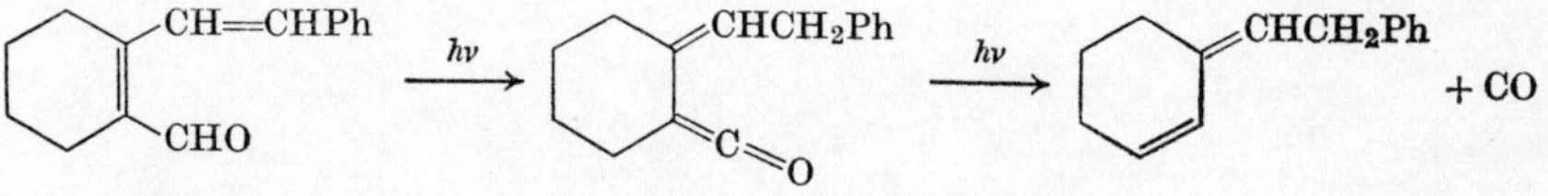

(39) → **(40)** 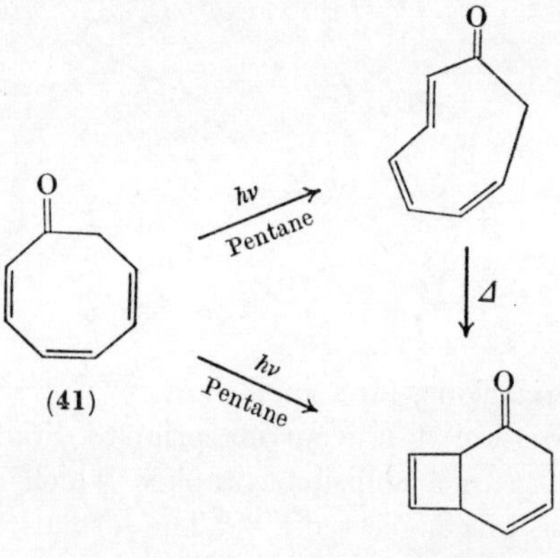 → **(41)** + CO

(42)

of unsaturated ketones to olefins have been reported. Surprisingly, **(43)** undergoes a $[2+4]$-cycloaddition reaction and hydrogen abstraction in preference to a $[2+2]$-cycloaddition reaction.[129] The stereochemistry of the products formed in the normal cycloaddition reactions is indicative of the formation of diradical intermediates. The formation of such species in the addition of cyclohex-2-enone to norbornadiene leads to nortricyclyl derivatives, norbornenyl-substituted cyclohexenones, and the expected cyclobutane.[130] Although enone triplets are efficiently formed, many cycloadditions,

[128a] T. S. Cantrell, W. S. Haller, and J. C. Williams, *J. Org. Chem.*, **34**, 509 (1969); B. D. Challand H. Hikino, G. Kornis, G. Lange, and P. de Mayo, *ibid.*, p. 794; P. J. Nelson, D. Ostrem, J. D. Lassila, and O. L. Chapman, *ibid.*, p. 811; J. W. Hanifin and E. Cohen, *J. Am. Chem. Soc.*, **91**, 4494 (1969); I. W. J. Still, M.-H. Kwan, and G. E. Palmer, *Can. J. Chem.*, **46**, 3731 (1968); G. C. Forward and D. A. Whiting, *J. Chem. Soc.*(C), **1969**, 1868; J. A. Barltrop and D. Giles, *ibid.*, p. 105; L. Duc, R. A. Mateer, L. Brassier, and G. W. Griffin, *Tetrahedron Letters*, **1968**, 6173; T. Matsumoto, H. Shirahama, A. Ichihara, *ibid.*, **1969**, 4103; W. C. Agosta and W. W. Lowrance, Jr., *ibid.*, **1969**, 3053; N. Sugiyama, Y. Sato, M. Yoshiokia, K. Yamada, and H. Kataoka, *Bull. Chem. Soc. Japan*, **42**, 1153 (1969); K. Shima and H. Sakurai, *ibid.*, p. 849.

[128b] C. H. Heathcock and R. A. Badger, *Chem. Comm.*, **1968**, 1510; W. F. Erman and T. W. Gibson, *Tetrahedron*, **25**, 2493 (1969); J. R. Scheffer and B. A. Boire, *Tetrahedron Letters*, **1969**, 4005.

[129] J. Meinwald and J. W. Kobzina, *J. Am. Chem. Soc.*, **91**, 5177 (1969).

[130] J. J. McCullough and P. W. W. Rasmussen, *Chem. Comm.*, **1969**, 387; J. J. McCullough, J. M. Kelly, and P. W. W. Rasmussen, *J. Org. Chem.*, **34**, 2933 (1969).

(43)

including cyclodimerizations, are inefficient.[131a,b,c, 132] Kinetic evidence favours the energy-wasting step occurring prior to diradical formation, and therefore the formation of a collision complex, which can either form the diradical or dissociate, was suggested.[131b,c] The formation of such a complex may well explain the temperature dependence of the quantum yield for the cycloaddition of olefins to cyclopent-2-enone.[132]

Diketones and Quinones

Irradiation of benzil in propan-2-ol gives typical reduction products in addition to some derived from benzoyl radicals.[133] These radicals may be formed from the benzil ketyl radical, $\cdot C(Ph)(OH)COPh$. Excitation of camphorquinone leads to population of the second excited singlet state.[134] The quinone is reduced in alcoholic solution, and its ketyl radical was shown to be produced and is an intermediate.[135a] Some of the products of the reaction are formed by combination of the ketyl radical with a solvent radical.[135b,c] When toluene is the reductant, radical-combination products are formed within a solvent cage. Generation of the two types of radical (quinone ketyl and benzyl radical), at a distance from each other, leads to a lower yield of

131a P. J. Wagner and D. J. Bucheck, *Can. J. Chem.*, **47**, 713 (1969).
131b P. J. Wagner and D. J. Bucheck, *J. Am. Chem. Soc.*, **91**, 5090 (1969).
131c G. Mark, F. Mark, and O. E. Polansky, *Ann. Chem.*, **719**, 151 (1968).
132 R. O. Loutfy, P. de Mayo, and M. F. Tchir, *J. Am. Chem. Soc.*, **91**, 3984 (1969).
133 D. L. Bunbury and T. T. Chuang, *Can. J. Chem.*, **47**, 2045 (1969).
134 L. Tasi and E. Charney, *J. Phys. Chem.*, **73**, 2462 (1969).
135a A. Singh, A. R. Scott, and F. Sopchyshyn, *J. Phys. Chem.*, **73**, 2633 (1969).
135b B. M. Monroe and S. A. Weiner, *J. Am. Chem. Soc.*, **91**, 450 (1969).
135c M. B. Rubin, *Tetrahedron Letters*, **1969**, 3931.

coupling products.[135c] The rate constant for intramolecular hydrogen abstraction in diketones increases as the strength of the C–H bond broken decreases.[136] Irradiation of the diketone (44) produces the cyclized product by intra-

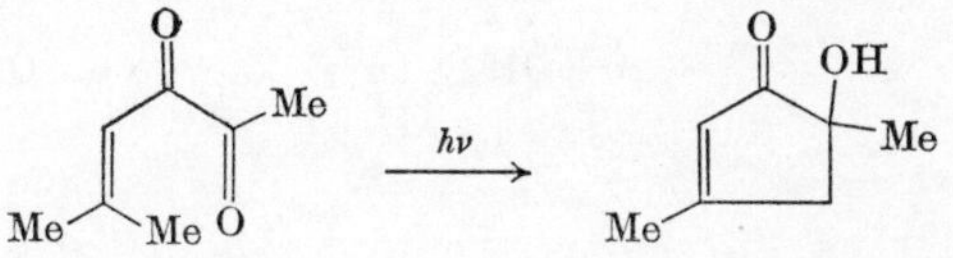

(44) → hv →

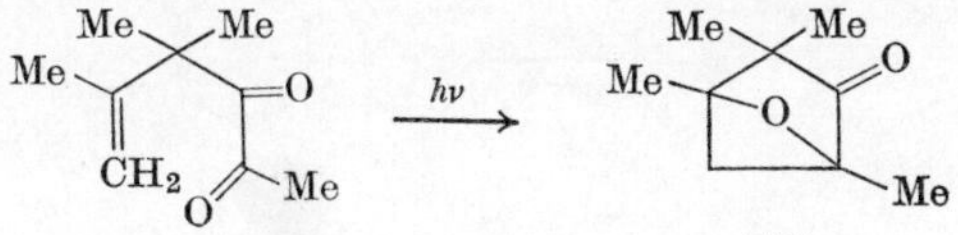

(45)

molecular hydrogen abstraction [137a] whereas the product (45) is formed by an intramolecular cycloaddition reaction.[137b] The 1,4-cycloaddition of enamides to 1,2-diketones gives dioxolens.[138] Anhydrides can be formed by the irradiation of 1,2-diketones in the presence of oxygen.[139]

9,10-Anthraquinone-2-sulphonate produces its radical cation and anion on flashing in alkaline solution.[140] Electron transfer from the anion radical to oxygen produces the superoxide ion which is capable of initiating hydroxylation of the quinone. The reduction of several quinones in hydrogen-donating solvents has been examined,[141a,b,c] and in a number of cases the triplet and ketyl radical of the quinone have been seen by flash photolysis.[141b,c] Phenanthraquinone ketyl radical was observed when the quinone was flashed in benzene.[141c] The products of this reaction include biphenyl,[79b] which suggests that triplet quinone can abstract hydrogen from benzene. Examples of the intramolecular photoreduction of quinones have been reported [142a,b] and the available evidence suggests that abstraction of hydroxylic hydrogens can occur, e.g. (46).[142a]

136 N. J. Turro and T.-J. Lee, *J. Am. Chem. Soc.*, **91**, 5651 (1969).
137a R. Bishop and N. K. Hamer, *Chem. Comm.*, **1969**, 804.
137b R. Bishop and N. K. Hamer, *Chem. Comm.*, **1969**, 804.
138 K. R. Eicken, *Ann. Chem.*, **724**, 66 (1969).
139 A. Cicolella, X. Deglise, M. Bouchy, J.-C. Andre, J. Lemaire, and M. Niclause, *Compt. Rend.*, *C*, **268**, 1929 (1969); C. W. Bird, *Chem. Comm.*, **1968**, 1537; G. E. Gream, J. C. Paice, and C. C. R. Ramsay, *Austral. J. Chem.*, **22**, 1229 (1969).
140 G. O. Phillips, N. W. Worthington, J. F. McKellar, and R. R. Sharpe, *J. Chem. Soc.*(A), **1969**, 767.
141a H. J. Piek, *Tetrahedron Letters*, **1969**, 1169.
141b D. R. Kemp and G. Porter, *Chem. Comm.*, **1969**, 1029; J. F. Brennan and J. Beutel, *J. Phys. Chem.*, **73**, 3245 (1969).
141c P. A. Carapellucci, H. P. Wolf, and K. Weiss, *J. Am. Chem. Soc.*, **91**, 4635 (1969).
142a J. M. Bruce, D. Creed, and K. Dawes, *Chem. Comm.*, **1969**, 594.
142b J. E. Baldwin and J. E. Brown, *Chem. Comm.*, **1969**, 167.

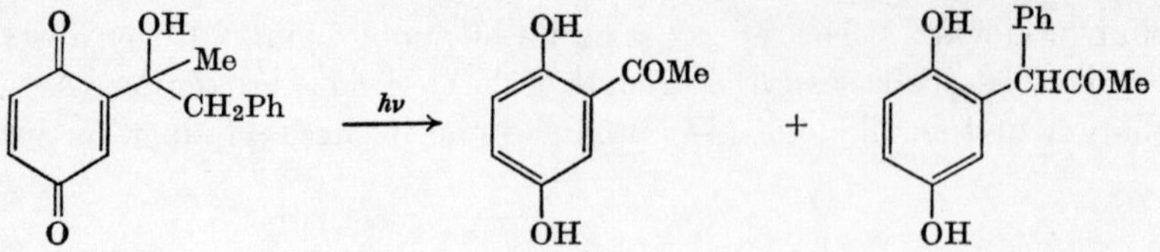

O OH
Me
CH₂Ph
hv →
OH
COMe
OH
+
OH Ph
CHCOMe
OH

(46)

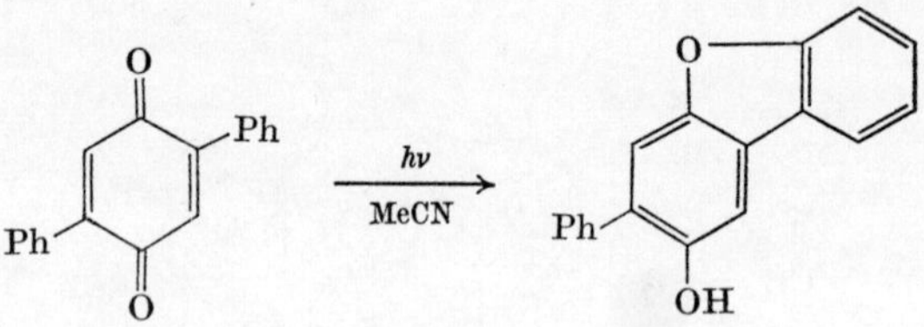

O
Ph
Ph
O
$\xrightarrow{hv}$
MeCN
Ph
O
OH

(47)

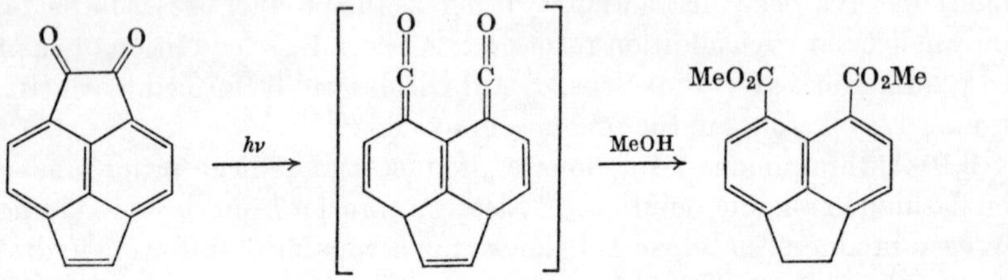

O O
$\xrightarrow{hv}$
$\begin{bmatrix} O & O \\ \parallel & \parallel \\ C & C \end{bmatrix}$
$\xrightarrow{MeOH}$
MeO₂C CO₂Me

(48)

O O
$\xrightarrow{hv}$
+ [C₂O₂] $\longrightarrow$ 2 CO

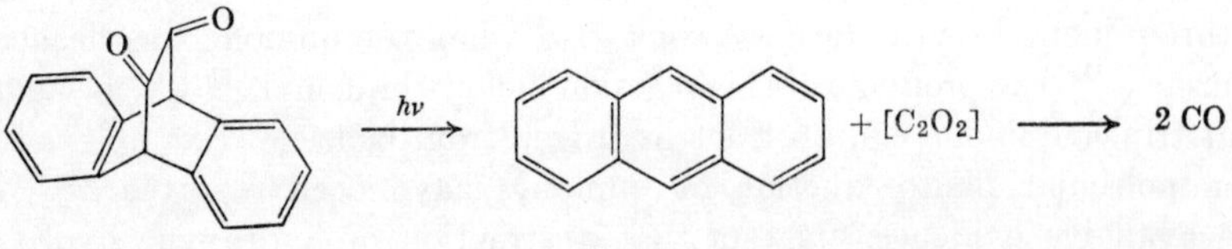

(49)

A number of $[2+2]$-cycloaddition reactions of quinones have been reported.[143] 2,5-Diphenylbenzoquinone (47) cyclodimerizes on irradiation in benzene whereas intramolecular cyclization occurs in acetonitrile.[144] Calculations have shown that $[2+4]$-cycloaddition reactions of quinones could possibly be concerted.[145a] A further example of such a reaction is the addition

143 H. Werbin and E. T. Strom, *J. Am. Chem. Soc.*, **91**, 7296 (1968); S. P. Pappas and N. A. Portnoy, *Chem. Comm.*, **1969**, 597; S. P. Pappas, B. C. Pappas, and N. A. Portnoy, *J. Org. Chem.*, **34**, 520 (1969).

144 H. J. Hageman and W. G. B. Huysmans, *Chem. Comm.*, **1969**, 837.

145a W. C. Herndon and W. B. Giles, *Chem. Comm.*, **1969**, 497.

of phenanthraquinone to isobenzofuran.[145b] Oxetans formed in cycloaddition reactions can undergo further fragmentation.[146] Irradiation of the two quinones (**48**)[147] and (**49**)[148] leads to fragmentation.

Thioketones

Much of the work on the photoaddition of thiobenzophenone to olefins[149a] has been summarized.[149b] Olefins containing electron-donating substituents form thietans and 1,4-dithians by addition of the olefin to the $n \rightarrow \pi^*$ triplet of the thioketone. Electron-deficient olefins give thietans by formation of an excited complex with the $\pi - \pi^*$ singlet of the ketone.

Acid Derivatives

Esters and Amides

Evidence that the photo-Fries rearrangement occurs from the singlet state has been presented.[150a,b] Migration of the acyl group to the *ortho*-position may well be concerted: an example of a 1,3-sigmatropic shift.[150b] Evidence that the rearrangement occurs from the triplet state has also been presented.[150c] Clarification of the multiplicity of the excited state responsible for reaction is awaited with interest. The rearrangement can also be induced by γ-radiolysis.[151] A number of product studies of the photoreaction have been made.[152]

Convincing evidence that the photoinduced rearrangement of anilides to give acylanilines occurs by predissociation of the first excited singlet state has been presented.[153a,b] Rearrangement does not occur in a matrix at low temperature since the radicals formed cannot move far enough apart.[153b] Acyl

[145b] W. Friedrichsen, *Tetrahedron Letters*, **1969**, 1219.

[146] S. Farid and K. H. Scholz, *Chem. Comm.*, **1969**, 572.

[147] F. M. Beringer, R. E. K. Winter, and J. A. Castellano, *Tetrahedron Letters*, **1968**, 6183.

[148] J. Strating, B. Zwanenburg, A. Wagenaar, and A. C. Vading, *Tetrahedron Letters*, **1969**, 125.

[149a] A. Ohno, Y. Ohnishi, and G. Tsuchihashi, *Tetrahedron Letters*, **1969**, 161, 283; A. Ohno, Y. Ohnishi, M. Fukuyama, and G. Tsuchihashi, *J. Am. Chem. Soc.*, **90**, 7038 (1969).

[149b] A. Ohno, Y. Ohnishi, and G. Tsuchihashi, *J. Am. Chem. Soc.*, **91**, 5038 (1969).

[150a] M. R. Sander, E. Hedaya, and D. J. Trecker, *J. Am. Chem. Soc.*, **90**, 7249 (1968).

[150b] H. Shizuka, T. Morita, Y. Mori, and I. Tanaka, *Bull. Chem. Soc. Japan*, **42**, 1831 (1969).

[150c] D. A. Plank, *Tetrahedron Letters*, **1969**, 4365.

[151] D. Belluš, K. Schaffner, and J. Hoigne, *Helv. Chim. Acta*, **51**, 1980 (1968).

[152] D. P. Kelly, J. T. Pinhey, and R. D. G. Rigby, *Austral. J. Chem*, **22**, 977 (1969); H. Obara, H. Takahashi, and H. Hirano, *Bull. Chem. Soc. Japan*, **42**, 560 (1969); P. Sláma, D. Belluš, and P. Hrdlovič, *Coll. Czech. Chem., Comm.* **33**, 3752 (1968); C. Pac, S. Tsutsumi, and H. Sakurai, *J. Chem. Soc. Japan, Ind. Chem. Sect.* **72**, 224 (1969); *Chem. Abs.*, **71**, 38540 (1969).

[153a] H. Shizuka and I. Tanaka, *Bull. Chem. Soc. Japan*, **41**, 2343 (1968); H. Shizuka, *ibid.*, **42**, 52, 57 (1969).

[153b] H. Shizuka and I. Tanaka, *Bull. Chem. Soc. Japan*, **42**, 909 (1969).

radicals, formed by cleavage of N-phenyl cyclic amides (ring size >6), preferentially attack the *ortho*-position of the aromatic ring.[154a,b] If the *ortho*-position is blocked, attack at the *para*-position occurs.[154b,c] α-Lactams,[155a] β-lactams,[155b] and γ-lactams[154b] decarbonylate on irradiation. S,S-Diaryldithiooxalates have been found to decarbonylate to give disulphides.[156] Attempts to prepare the elusive cyclobuta-1,3-dienes by irradiation of substituted cyclobut-3-ene-1,2-dicarboxylic acid anhydrides have been reported.[157] The products indicated that cyclopenta-1,3-dienones and cyclobutadienes were formed. Irradiation of cyclic anhydrides[158a] and cyclic

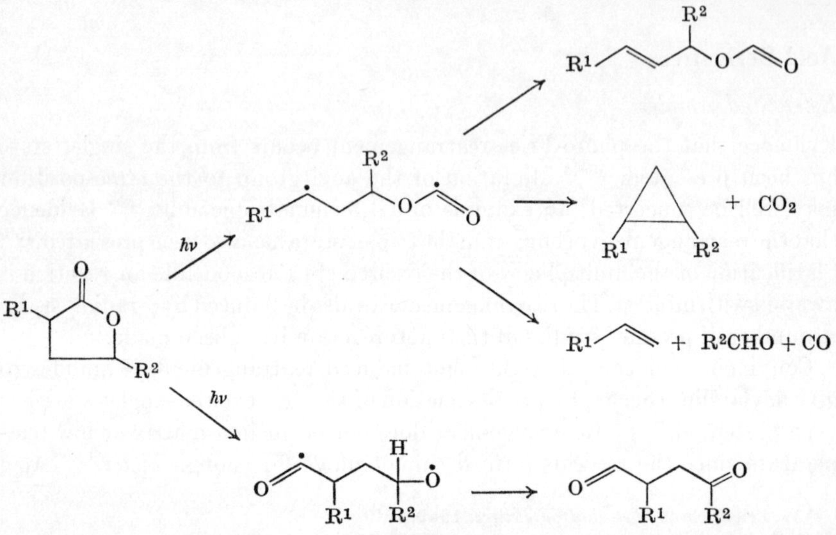

Figure 3

carbonates[158b] was shown to result in decarboxylation. γ-Lactones on irradiation give diradicals which either undergo intramolecular hydrogen abstraction or decarboxylate (see Figure 3).[159] It was suggested that the

[154a] R. W. Hoffmann and K. R. Eicken, *Chem. Ber.*, **102**, 2987 (1969).

[154b] M. Fischer and A. Matthews, *Chem. Ber.*, **102**, 342 (1969).

[154c] M. Fischer, *Tetrahedron Letters*, **1969**, 2281.

[155a] E. R. Talaty, A. E. Dupuy, Jr., and T. H. Golson, *Chem. Comm.*, **1969**, 49; J. C. Sheehan and M. M. Nafissi-V, *J. Am. Chem. Soc.*, **91**, 1176 (1969).

[155b] M. Fischer and F. Wagner, *Chem. Ber.*, **102**, 3486 (1969); M. Fischer, *ibid.*, p. 3495.

[156] H.-G. Heine and W. Metzner, *Ann. Chem.*, **724**, 223 (1969).

[157] G. Maier and U. Mende, *Tetrahedron Letters*, **1969**, 3155.

[158a] I. S. Krull and D. R. Arnold, *Tetrahedron Letters*, **1969**, 4349.

[158b] R. L. Smith, A. Manmade, and G. W. Griffin, *J. Heterocyclic Chem.*, **6**, 443 (1969).

[159] R. Simonaites and J. N. Pitts, *J. Am. Chem. Soc.*, **91**, 108 (1969).

vibrationally excited singlet state is responsible for reaction. The reported stereospecific decarboxylation of (**50**)[160a] and decarboxylation accompanied by rearrangement of (**51**)[160b] are intriguing reactions.

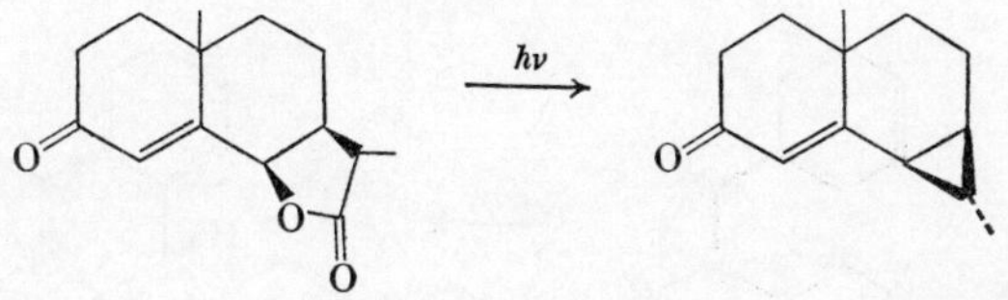

(**50**)

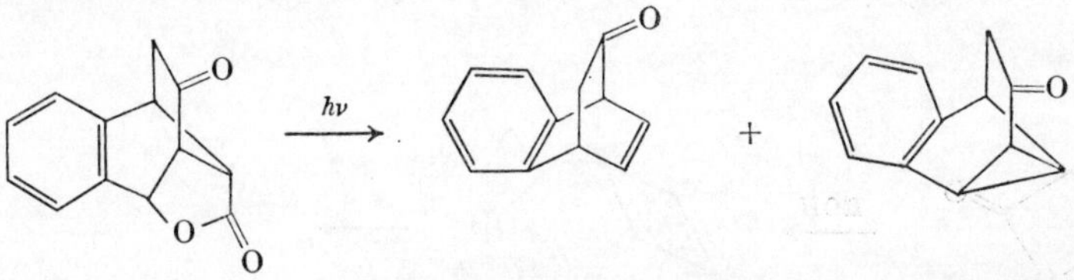

(**51**)

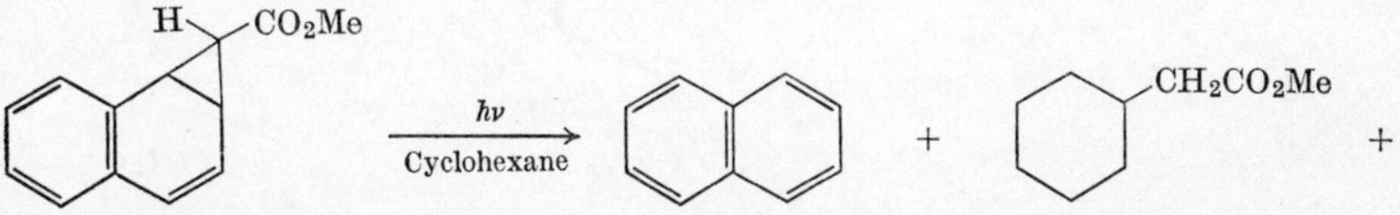

(**52**)

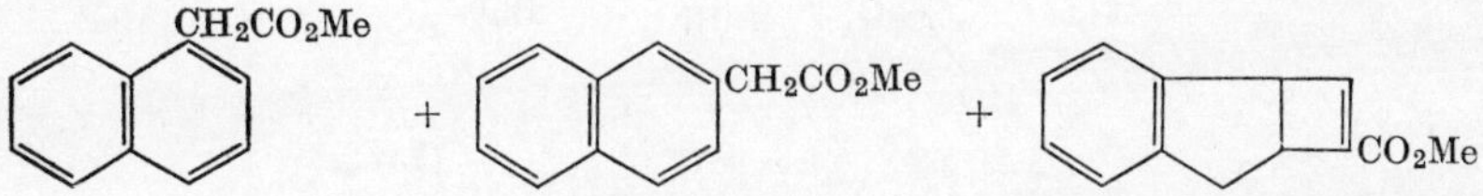

Oxetan formation by the cycloaddition of olefins to aryl esters[161a] and amides[161b] has been reported. Photolysis of the benzonorcaradiene ester (**52**) gives products by fragmentation to the carbene and by rearrangement.[162] 2-Ethoxyethyl phenylacetate photodecomposes by a Type II reaction.[163] The carbonyl of the ester is probably activated by triplet energy transfer from the benzene ring. The corresponding benzoate and phenylpropionate are much less reactive and this unreactivity was ascribed to inefficient energy transfer.

160a G. W. Perold and G. Ourisson, *Tetrahedron Letters*, **1969**, 3871.
160b R. S. Givens and W. F. Oettle, *Chem. Comm.*, **1969**, 1164.
161a Y. Shigemitsu, H. Nakai, and Y. Odaira, *Tetrahedron*, **25**, 3039 (1969).
161b T. Tominaga and S. Tsutsumi, *Tetrahedron Letters*, **1969**, 3175.
162 J. S. Swenton and A. J. Krubsack, *J. Am. Chem. Soc.*, **91**, 786 (1969).
163 H. Morrison, R. Brainard, and D. Richardson, *Chem. Comm.*, **1968**, 1653.

Unsaturated Esters and Amides

On photolysis, enamides[164a] and enol acetates[164b] usually undergo fission at
the CO–N and CO–O bond respectively. Aroyl derivatives of enamines prove
to be the exception and they cyclize to give dihydrocarbostyril derivatives,

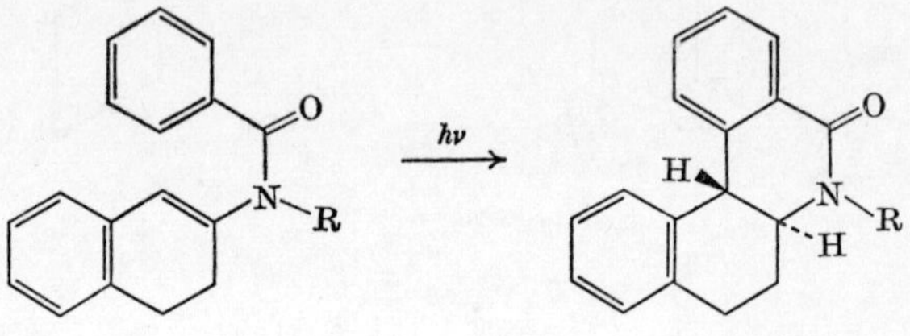

(53)

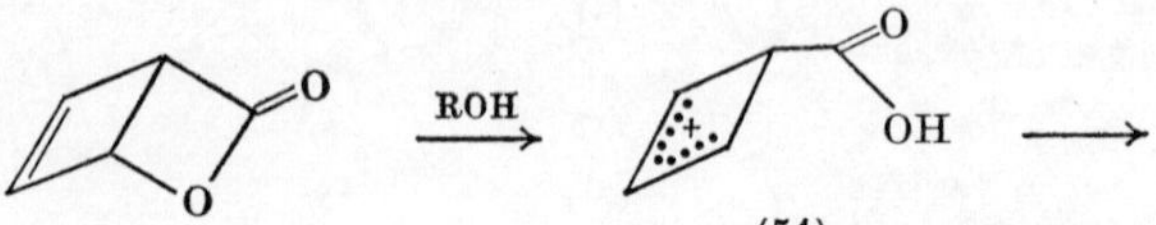

(54)

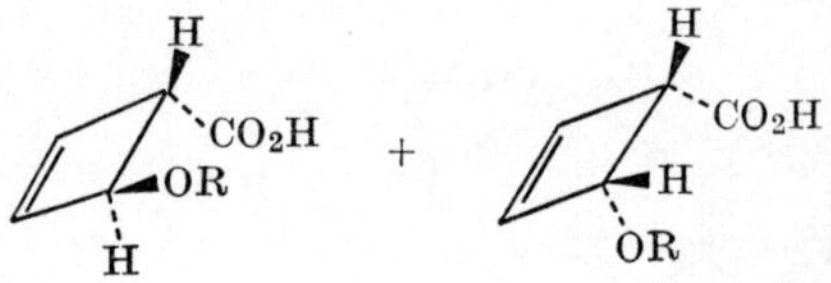

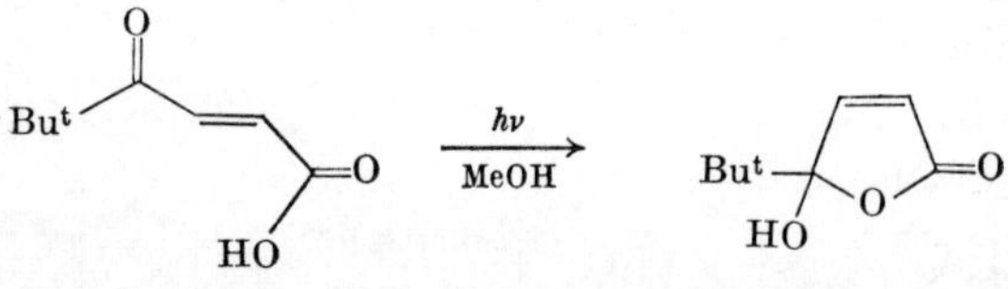

(55)

e.g. (53).[164c] Amides of α,β-unsaturated acids undergo C–N bond fission[165a,b]
from the singlet state. Population of the triplet state results in cyclodimeriza-
tion.[165a] *N*-Phenylmaleimide is photoreduced by *N,N*-dimethylaniline,[166a]
and *N*-acylbenzophenoneimines by propan-2-ol.[166b] Examples of unsaturated
amides and imides undergoing intermolecular[167a] and intramolecu-

164a I. Ninomiya, T. Naito, and T. Mori, *Tetrahedron Letters*, **1969**, 2259.

164b J. Libman, M. Spencer, and Y. Mazur, *J. Am. Chem. Soc.*, **91**, 2062 (1969).

164c I. Ninomiya, T. Naito, and T. Mori, *Tetrahedron Letters*, **1969**, 3643.

165a E. Cavalieri and S. Horoupian, *Can. J. Chem.*, **47**, 2781 (1969).

165b H. Zimmer, D. C. Armbruster, S. P. Kharidia, and D. C. Lankin, *Tetrahedron Letters*, **1969**,
4053.

166a J. M. Fayadh and G. A. Swan, *J. Chem. Soc.*(C), **1969**, 1781.

166b T. Okada, M. Kawanisi, H. Nazaki, N. Toshima, and H. Hirai, *Tetrahedron Letters*, **1969**,
927.

167a D. Bryce-Smith and M. A. Hems, *Tetrahedron*, **25**, 247 (1969).

lar[167b] [2+2]-cycloaddition reactions have been reported. The photochemistry of pyran-2-one has been the subject of further discussion.[168a] Evidence has been presented for the solvolysis of the bicyclo[2.2.0]pyran-2-one formed on irradiation, as occurring via a cyclobutenyl cation (54).[168b] The photoinduced rearrangement of 4-hydroxy-6-methylpyran-2-one to β-methylglutaconic anhydride was suggested to occur via an intermediate bicyclopyranone.[168c] The diverse reactions of unsaturated esters have been the subject of a number of studies. In the examples of the photoinduced addition of nucleophiles to the double bond, the role of the excited state[169a] and of ground-state solvolytic processes is not clear. The formation of β-hydroxycinnamates by irradiation of cinnamates in alkaline solution was suggested as occurring by initial photocyclization to give a β-lactone which subsequently solvolysed.[169b] Irradiation of β-acyloxyacrylic acids gives products, e.g. (55), by intramolecular addition.[169c] β-Alkylcrotonic acids undergo isomerization and intramolecular hydrogen abstraction reactions on irradiation.[170a] The relative efficiency of these reactions is very solvent-dependent. On photolysis, β-cyclopropylcrotonic ester gives 1,3-diradicals by opening of the cyclopropane ring.[170b] A number of cyclic products are obtained by subsequent reaction of the diradical.

Olefins

Ionic Addition Reactions

Much of the earlier work on photosensitized ionic additions to cycloolefins has been reviewed.[171] It is still not certain whether addition occurs to the orthogonally oriented $\pi \rightarrow \pi^*$ triplet of the olefin or to the strained transoid olefin or to both (see Figure 4).[172a,b] Proton transfer to the olefin produces a carbonium ion and the fate of this species has been investigated. Examples of the carbonium ion being deprotonated to produce an isomeric olefin,[172a] of rearrangement and then deprotonation,[172a] and of reaction with the solvent to give an

167b R. T. LaLonde and C. B. Davis, *Can. J. Chem.*, **47**, 3250 (1969); F. C. De Schryver, I. Bhardwaj, and J. Put, *Angew. Chem. Internat. Ed. Engl.*, **8**, 213 (1969).

168a *Org. Reaction Mech.*, **1968**, 435.

168b W. H. Pirkle and L. H. McKendry, *J. Am. Chem. Soc.*, **91**, 1179 (1969).

168c C. T. Bedford and T. Money, *Chem. Comm.*, **1969**, 685.

169a N. Sugiyama, H. Kataoka, C. Kashima, and K. Yamada, *Bull. Chem. Soc., Japan* **42**, 1353 (1969).

169b E. F. Ullman, E. Babad, and M.-E. Sung, *J. Am. Chem. Soc.*, **91**, 5792 (1969).

169c N. Sugiyama, H. Kataoka, C. Kashima, and K. Yamada, *Bull. Chem. Soc. Japan*, **42**, 1098 (1969).

170a M. J. Jorgenson, *J. Am. Chem. Soc.*, **91**, 198 (1969).

170b M. J. Jorgenson, *J. Am. Chem. Soc.*, **91**, 6432 (1969).

171 J. A. Marshall, *Accounts Chem. Res.*, **2**, 33 (1969).

172a J. A. Marshall and A. R. Hochstetler, *J. Am. Chem. Soc.*, **91**, 648 (1969).

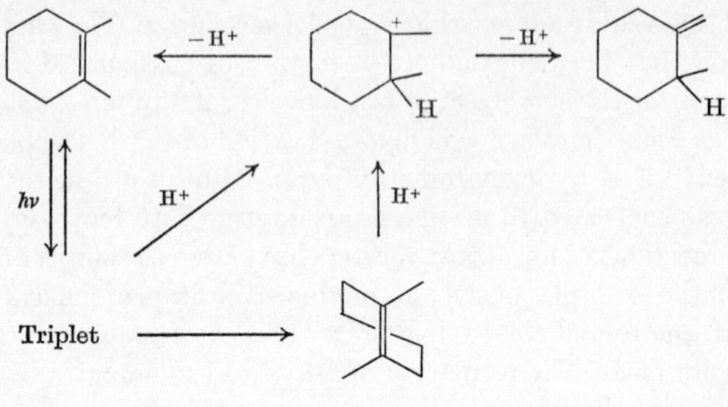

Figure 4

addition product [172b] have all been reported. The flexibility of the alkene ring has an important bearing upon the relative efficiencies of these processes.[172b] There are examples of these reactions occurring only when acid is present, and in its absence, dimerization and radical reactions originating from the triplet state of the olefin occur.[172b] The ionic photoaddition reactions of 1-phenyl-cycloalkenes probably occur by triplet energy transfer from the phenyl ring to the olefin.[172b,c] The reported efficient photoaddition of alcohols to the lactone (56) is particularly interesting.[173] The unsensitized ionic addition of methanol to some steroidal mono-olefins has been reported to occur via a

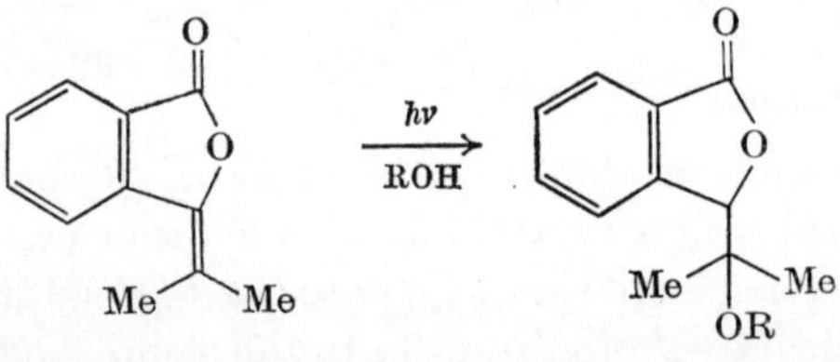

(56)

π–σ^*_{CH} transition.[174] A further example of the photoaddition of methanol to an alicyclic transoid 1,3-diene, via what is probably a bicyclobutane, has been reported.[175a] Addition to acyclic dienes also occurs and some of the products may be derived from a bicyclobutane.[175b] It was suggested that some of the products may arise by protonation of the olefin excited state or its vibrationally excited ground state. Ionic intermediates have been postulated in the photo-rearrangements of allylic alcohols [176a] and *i*-steroids.[176b]

172b P. J. Kropp, *J. Am. Chem. Soc.*, **91**, 5783 (1969).
172c S. Fujita, T. Nômi, and H. Nozaki, *Tetrahedron Letters*, **1969**, 3557.
173 S. F. Nelsen and P. J. Hintz, *J. Am. Chem. Soc.*, **91**, 6190 (1969).
174 H. Compaignon de Marcheville and R. Beugelmans, *Tetrahedron Letters*, **1969**, 1901.
175a J. C. Sircar and G. S. Fisher, *J. Org. Chem.*, **34**, 404 (1969).
175b J. A. Barltrop and H. E. Browning, *Chem. Comm.*, **1968**, 1481.
176a H. Compaignon de Marcheville and R. Beugelmans, *Tetrahedron Letters*, **1968**, 6331.
176b R. Beugelmans and H. Compaignon de Marcheville, *Chem. Comm.*, **1969**, 241.

Cycloaddition Reactions

The stereospecific dimerization of but-2-ene demonstrates the concerted nature of this [2 + 2]-cycloaddition reaction.[177] Examples of [2 + 2]-cycloaddition reactions occurring from the triplet state of the olefin have been reported.[178] *cis,cis*-Cyclodeca-3,8-diene-1,6-dione (**57a**) is photoisomerized to

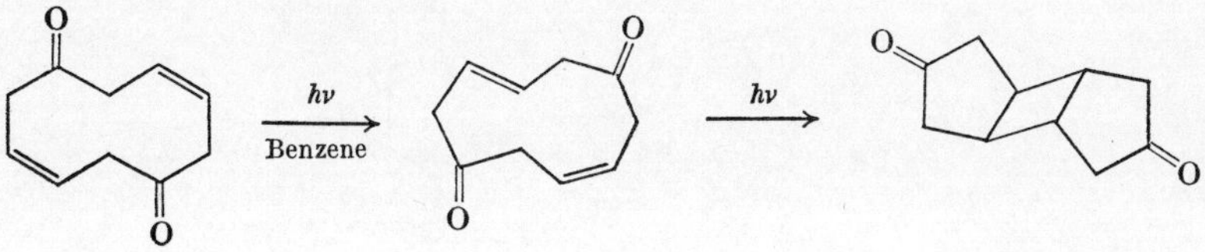

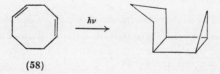

(57a) **(57b)**

the *cis,trans*-dione[179a] which gives the *cis,anti,cis*-tricyclo-compound (**57b**)[179a,b] by an intramolecular [2 + 2]-cycloaddition reaction. The formation of [2 + 2]-cycloaddition products by irradiation of naphthalene[180a] and phenanthrene,[180b] in the presence of diphenylacetylene, occurs from the singlet state of polycyclic hydrocarbon which forms an exciplex with the acetylene. Examples of the type of $[_\pi 2 + _\pi 2]$-cycloaddition shown in Figure 5

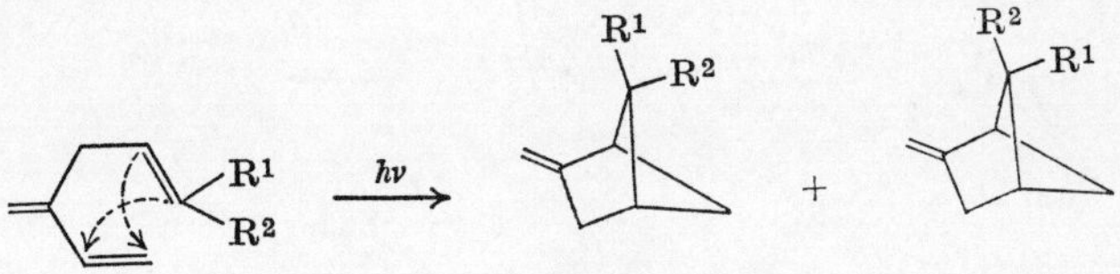

Figure 5

have been reported and shown to occur from the triplet state of the diene.[181a,b] Population of the singlet state of the diene results in cyclobutene formation.[181a]

(58)

[177] H. Yamazaki and R. J. Cvetanovic, *J. Am. Chem. Soc.*, **91**, 520 (1969).

[178] H. Dürr, *Ann. Chem.*, **723**, 102 (1969); C. Giannotti, *Can. J. Chem.*, **46**, 3025 (1968); J. J. McCullough and C. W. Huang, *ibid.*, **47**, 757 (1969); W. Metzner and W. Hartmann, *Chem. Ber.*, **101**, 4099 (1968); H. M. Rosenberg, R. Rondeau, and P. Serve, *J. Org. Chem.*, **34**, 471 (1969).

[179a] J. R. Scheffer and M. L. Lungle, *Tetrahedron Letters*, **1969**, 845.

[179b] J. W. Stankorb and K. Conrow, *Tetrahedron Letters*, **1969**, 2395.

[180a] W. H. F. Sasse, *Austral. J. Chem.*, **22**, 1257 (1969).

[180b] G. Sugowdz, P. J. Collin, and W. H. F. Hasse, *Tetrahedron Letters*, **1969**, 3843.

[181a] J. D. White and D. N. Gupta, *Tetrahedron*, **25**, 3331 (1969).

[181b] S. Kita and K. Fukui, *Bull. Chem. Soc., Japan*, **42**, 66 (1969).

The cyclooctadiene (58) undergoes intramolecular photocyclization by what is probably a $[_\pi 2_s + _\pi 2_s]$-process.[182] The reversible formation of quadricyclenes by photocyclization of norbornadiene has been further investigated, and shown to occur from the singlet state of the diene.[183] Thermal isomerization of (59) has been shown to give a benzene oxide.[184a,b] The singlet state of the oxo-

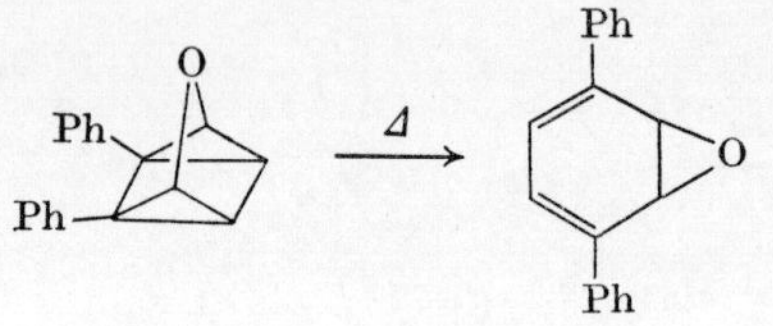

(59)

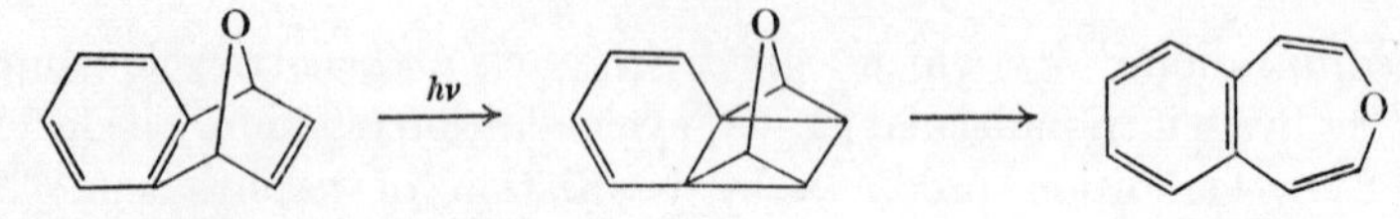

(60)   (61)

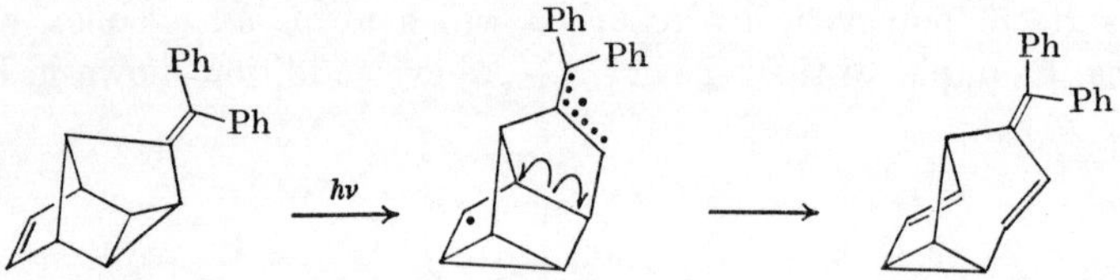

(62)

compound (60) rearranges to the oxepin (61).[184b] Prinzbach has reported[185a,b] a number of interesting $[_\pi 2 + _\sigma 2]$-cycloadditions of which (62)[185b] is an example.

Intramolecular Rearrangements

The rearrangement of divinylmethanes to vinylcyclopropanes (di-π-methane rearrangement) has been shown to be a fundamental process.[186] The rearrangement can be viewed (see Figure 6) as a multi-step process which occurs by an initial π-bonding step (Path A),[186] or a concerted process (Path B),[1] or a two-step process (Path C). Paths B and C occur by an initial sigmatropic

[182] S. Moon and C. R. Ganz, *Tetrahedron Letters*, **1968**, 6275.

[183] G. Kaupp and H. Prinzbach, *Helv. Chim. Acta*, **52**, 956 (1969); R. S. H. Liu, *Tetrahedron Letters*, **1969**, 1409; G. Kaupp and H. Prinzbach, *Ann. Chem.*, **725**, 52 (1969).

[184a] H. Prinzbach and P. Vogel, *Helv. Chim. Acta*, **52**, 396 (1969).

[184b] G. R. Ziegler, *J. Am. Chem. Soc.*, **91**, 446 (1969).

[185a] H. Prinzbach and W. Eberbach, *Chem. Ber.*, **101**, 4083 (1969); H. Prinzbach and M. Klaus, *Angew. Chem. Internat. Ed. Engl.*, **8**, 276 (1969); H. Prinzbach, M. Klaus, and W. Mayer, *ibid.*, p. 883.

[185b] H. Prinzbach and W. Auge, *Angew. Chem. Internat. Ed. Engl.*, **8**, 209 (1969).

[186] H. E. Zimmerman and P. S. Mariano, *J. Am. Chem. Soc.*, **91**, 1718 (1969).

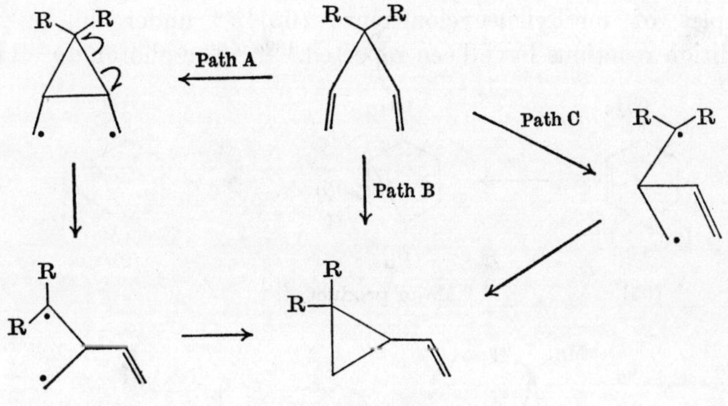

Figure 6

shift. In acyclic and monocyclic systems the rearrangement occurs most readily from the singlet state, and in bicyclic systems from the triplet state.[186]

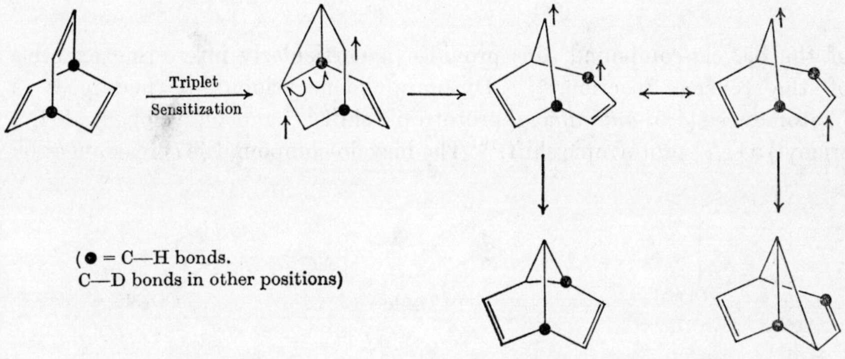

(● = C—H bonds.
C—D bonds in other positions)

(63)   (64)

The rearrangement of barrelene to semibullvalene has been examined by deuterium labelling, and the formation of the two products (63) and (64) is consistent with the formation of triplet diradicals with a finite lifetime.[187] There has been a multitude of examples of rearrangements akin to the barrelene–semibullvalene transformation, e.g. benzobarrelenes,[188a] a variety of dihydrobarrelenes,[188b] and homobarrelenes.[188c]

[187] H. E. Zimmerman, R. W. Binkley, R. S. Givens, G. L. Grunewald, and M. A. Sherwin, *J. Am. Chem. Soc.*, **91**, 3316 (1969).

[188a] R. S. H. Liu and C. G. Krespan, *J. Org. Chem.*, **34**, 1271 (1969); T. D. Walsh, *J. Am Chem. Soc.*, **91**, 515 (1969); N. J. Turro, M. Tobin, L. Friedman, and J. R. Hamilton, *ibid.*, p. 516.

[188b] I. F. Eckhard, H. Heaney, and B. A. Marples, *Tetrahedron Letters*, **1969**, 3273; S. J. Cristol and G. O. Mayo, *J. Org. Chem.*, **34**, 2363 (1969); J. Ipaktschi, *Tetrahedron Letters*, **1969**, 215; H. Hart and R. K. Murray, *ibid.*, p. 379; H. Hart and R. K. Murray, *J. Am. Chem. Soc.*, **91**, 2183 (1969).

[188c] S. J. Cristol, G. O. Mayo, and G. A. Lee, *J. Am. Chem. Soc.*, **91**, 214 (1969); R. C. Hahn and L. J. Rothman, *ibid.*, p. 2408.

17

Examples of methylenecyclohexenes **(65)**[189a] undergoing $[_\sigma 2_a + _\pi 2_a]$-cycloaddition reactions have been reported.[189a,b] The photoinduced opening

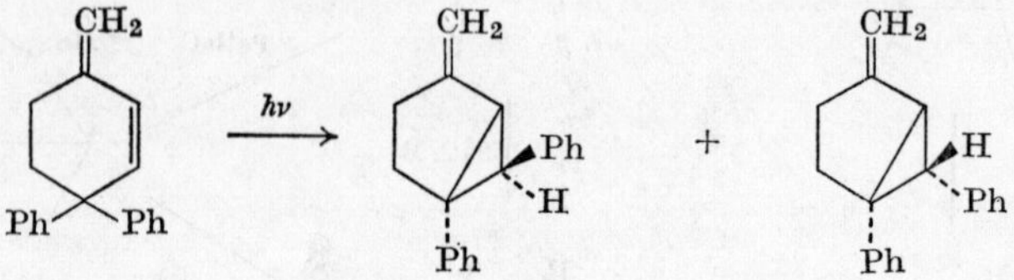

of the bicyclo-compound **(66)** provides a particularly interesting example of the reverse reaction.[189c] Diisopropylidenecyclobutane undergoes a photoinduced [1,5]-*antarafacial* prototropic shift,[190] and bi(cyclohepta-2,4,6-trienyl) a [1,7]-prototropic shift.[191] The bicyclo-compound **(67)** rearranges by

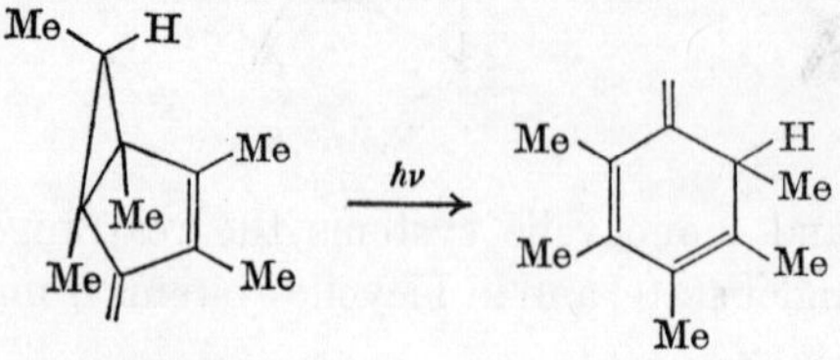

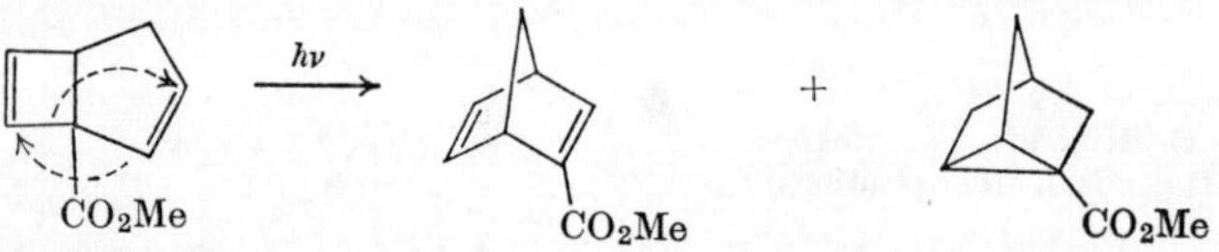

what is probably a [1,3]-sigmatropic shift, to a norbornadiene.[192] 1,1,3,4-Tetramethylcyclohepta-2,4,6-triene undergoes a photoinduced [1,7]-sigmatropic methyl shift.[193] A wide variety of azasemibullvalenes have been

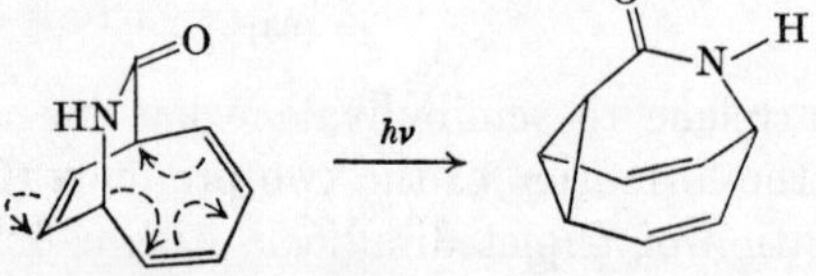

189a H. F. Zimmerman and G. E. Samuelson, *J. Am. Chem. Soc.*, **91**, 5307 (1969).
189b H. Hart, J. D. DeVrieze, R. M. Lange, and A. Sneller, *Chem. Comm.*, **1968**, 1650.
189c H. Hüther and H. A. Brune, *Z. Naturforsch.*, **23b**, 1612 (1968).
190 E. F. Kiefer and J. Y. Fukunaga, *Tetrahedron Letters*, **1969**, 993; E. F. Kiefer and C. H. Tanna, *J. Am. Chem. Soc.*, **91**, 4478 (1969).
191 R. S. Givens, *Tetrahedron Letters*, **1969**, 663.
192 A. A. Gorman and J. B. Sheridan, *Tetrahedron Letters*, **1969**, 2569.
193 L. B. Jones and V. K. Jones, *J. Org. Chem.*, **34**, 1298 (1969).

synthesized from compounds similar to (68) by what is formally a [1,7]-sigmatropic shift.[194] Hepta-1,5,6-triene undergoes a photosensitized Cope rearrangement.[195] The operation of symmetry control in cycloreversion reactions has been clearly demonstrated by the stereospecific ring-opening

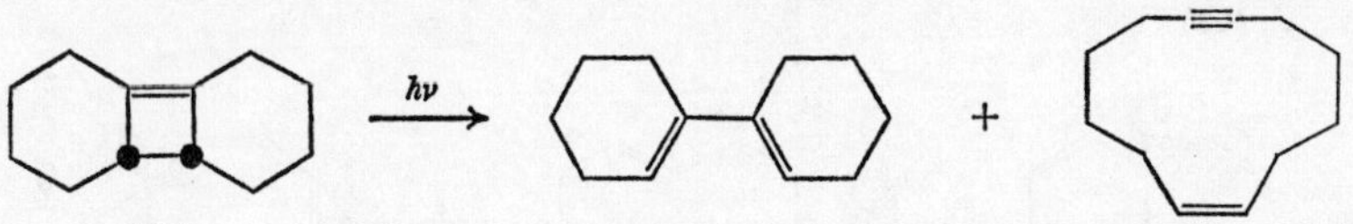

(69)

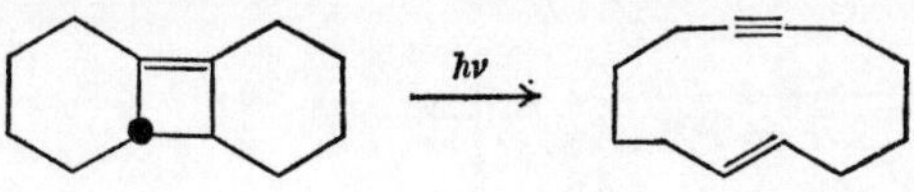

hν

+

(70)

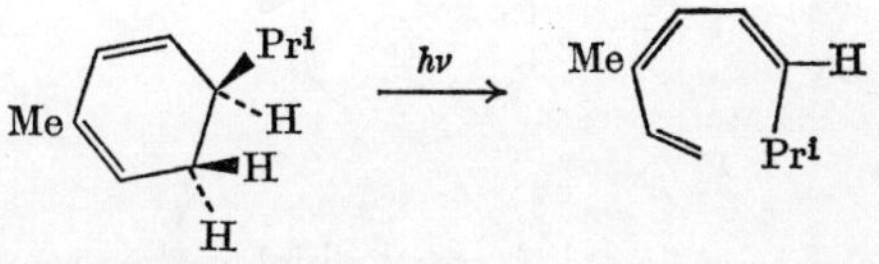

hν

(71)

Me ⟶ Pr^i

hν

Me ⟶ H, Pr^i

(71)

Me, H, H, H ⟶ Pr^i

hν

Me ⟶ Pr^i, H

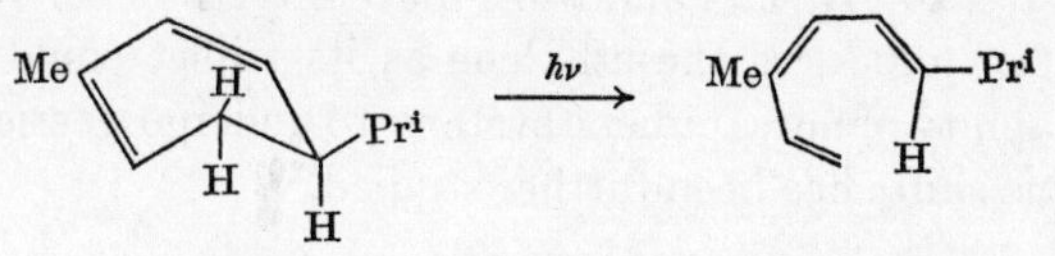

(71)

of the cyclobutenes (69) and (70),[196a] and of the conformers of α-phellandrene (71).[196b] Examples of cyclohexa-1,4-dienes undergoing cycloreversion to hexatrienes have been reported.[197a] The isomerization of the deutero-compound (72) is particularly illuminating.[197b] The cycloreversion of the *cis*-fused dihydronaphthalene (73) gives the shared *trans*-[10]annulene which is photoisomerized to the all-*cis*-[10]annulene.[198] Cyclooctatrienes have been shown to undergo cycloreversion reactions,[199a,b] and in the case of (74)[199a]

[194] L. A. Paquette, J. R. Malpass, G. R. Krow, and T. J. Barton, *J. Am. Chem. Soc.*, **91**, 5296 (1969).
[195] H. R. Ward and E. Karafiate, *J. Am. Chem. Soc.*, **91**, 522 (1969).
[196a] J. Saltiel and L.-S. N. Lim, *J. Am. Chem. Soc.*, **91**, 5404 (1969).
[196b] J. E. Baldwin and S. M. Krueger, *J. Am. Chem. Soc.*, **91**, 6444 (1969).
[197a] M. Mousseron-Canet, and J.-P. Chaubaud, *Bull. Soc. Chim. France*, **1969**, 308; P. Courtot and R. Rumin, *ibid.*, p. 3665.
[197b] R. C. Cookson, S. M. de B. Costa, and J. Hudec, *Chem. Comm.*, **1969**, 1272.
[198] S. Masamune and R. T. Seidner, *Chem. Comm.*, **1969**, 542.
[199a] S. Masamune, P. M. Baker, and K. Hojo, *Chem. Comm.*, **1969**, 1203.
[199b] G. Schroder, W. Martin, and H. Röttell, *Angew. Chem. Internat. Ed. Engl.*, **8**, 69 (1969).

a very strained cyclononatriene is formed which is the precursor of the observed products. 7,8-Diacetoxybicyclo[4.2.0]octa-2,4-diene fragments, on population of its triplet state, to give *cis*- and *trans*-diacetoxyethylene. Whether or not the fragmentation produces the ethylene as its triplet could not be ascertained.[200] The ability of norcaradiene systems to undergo a series of sequential [1,3]-sigmatropic shifts has been further studied.[201]

cis–trans-*Isomerization*

The intramolecularly photosensitized isomerization of substituted butadienes has been studied. In 1,1-dimethyl-1-benzoylpenta-2,4-diene energy transfer from the carbonyl group is so efficient that typical triplet quenchers are unable to compete. The lifetime of the planar triplet was estimated as 5×10^{-10} sec, and of the twisted triplet as 5×10^{-6} sec.[202] The isomerization of stilbenes has been studied by populating the triplet state by excitation through the forbidden singlet–triplet transition. The probability of populating the excited state was increased by carrying out the reaction in the presence of oxygen.[203] The rate constants for isomerization were in excellent agreement with those obtained by classical energy-transfer systems. Quantum yields for the isomerization of *cis*- and *trans*-stilbene have been found to be independent of the

[200] R. A. Caldwell, *J. Org. Chem.*, **34**, 1886 (1969).
[201] T. Toda, M. Nitta, and T. Mukai, *Tetrahedron Letters*, **1969**, 4401; G. W. Gruber and M. Pomerantz, *J. Am. Chem. Soc.*, **91**, 4004 (1969).
[202] P. A. Leermakers, J.-P. Montillier, and R. D. Rauh, *Mol. Photochem.*, **1**, 57 (1969).
[203] A. Bylina and Z. R. Gradowski, *Trans. Faraday Soc.*, **65**, 458 (1969).

stilbene concentration, which demonstrates the absence of self-quenching of *trans*-stilbene.[204] The photoisomerization of styrylnaphthalenes,[205a] styryl-pyridines,[205b] difurylethylenes,[205c] oxindigo,[205d] and some carotenoids[205d,e] has been studied. The isomerization of 1-(1-naphthyl)-2-(4-pyridyl)ethylene by aetioporphyrins has been suggested as occurring by electron transfer.[206] The formation of highly strained *trans*-cycloolefins by isomerization of the *cis*-isomers has been found to occur in a number of systems. Examples include *trans*-cyclooctene,[207a] *cis,trans*-cycloocta-1,3-diene,[207b] *cis,trans*- and *trans,-trans*-cycloocta-1,5-diene,[207b,c] *cis,cis,trans*-octa-1,3,5-triene,[207d] and *cis,-cis,cis,trans*-1,2,4,7-tetraphenylocta-1,3,5,7-tetraene.[207e]

Miscellaneous

The cyclization of stilbenes to form phenanthrenes has been used in the synthesis of polycyclic hydrocarbons,[208a,b] and probably the most outstanding example is that of the helicenes.[208b]

Enolate anions of the type (75) undergo intramolecular acylation reactions on irradiation.[209a] The photoisomerization of the spiro-compound (76) probably occurs by a 1,3 diradical,[209b] and is analogous to the isomerization of the spiro-cyclohexa-2,5-dienones. A most intriguing observation is that aryl-substituted ethylenes (77) can be photoreduced by tertiary amines, and stereospecific *cis*-addition of the amine also occurs.[210] Excitation of the nonatrienide anion increases its basicity to such a degree that it can abstract a proton from hex-1-yne.[211]

[204] H. A. Hammond, D. E. DeMeyer, and J. L. R. Williams, *J. Am. Chem. Soc.*, **91**, 5180 (1969).

[205a] G. S. Hammond, S. G. Shim, and S. P. Van, *Mol. Photochem.*, **1**, 89 (1969).

[205b] D. G. Whitten and M. T. McCall, *J. Am. Chem. Soc.*, **91**, 5097 (1969).

[205c] A. A. Zimmerman, C. M. Orlando, Jr., M. H. Gianni, and K. Weiss, *J. Org. Chem.*, **34**, 73 (1969).

[205d] H. Güsten, *Chem. Comm.*, **1969**, 133.

[205e] M. Mousseron-Canet and J.-L. Olive, *Bull. Soc. Chim. France*, **1969**, 3242.

[206] D. G. Whitten, P. D. Wildes, and I. G. Lopp, *J. Am. Chem. Soc.*, **91**, 3393 (1969).

[207a] J. S. Swenton, *J. Org. Chem.*, **34**, 3217 (1969).

[207b] W. J. Nebe and G. J. Fonken, *J. Am. Chem. Soc.*, **91**, 1249 (1969).

[207c] G. M. Whitesides, G. L. Goe, and A. C. Cope, *J. Am. Chem. Soc.*, **91**, 2608 (1969).

[207d] P. Datta, T. D. Goldfarb, and R. S. Boikess, *J. Am. Chem. Soc.*, **91**, 5429 (1969).

[207e] E. H. White, E. W. Friend, Jr., R. L. Stern, and H. Maskill, *J. Am. Chem. Soc.*, **91**, 523 (1969).

[208a] T. Sato, S. Shimada, and K. Hata, *Bull. Chem. Soc. Japan*, **42**, 766 (1969); G. P. de Gunst, *Rec. Trav. Chim.*, **88**, 801 (1969); F. Dietz and M. Scholz, *Tetrahedron*, **24**, 6845 (1968); J. L. Cooper and H. H. Wasserman, *Chem. Comm.*, **1969**, 200; P. Bortolus, G. Cauzzo, U. Mazzucato, and G. Galiazzo, *Z. Phys. Chem.* (Frankfurt), **63**, 29 (1969); W. M. Horspool, *Chem. Comm.*, **1969**, 467.

[208b] R. H. Martin and M. Deblecker, *Tetrahedron Letters*, **1969**, 3597; R. H. Martin and J. J. Schurter, *ibid.*, p. 3679.

[209a] N. C. Yang, L. C. Lin, A. Shani, and S. S. Yang, *J. Org. Chem.*, **34**, 1845 (1969).

[209b] P. H. Mazzocchi, *Tetrahedron Letters*, **1969**, 989.

[210] R. C. Cookson, S. M. de B. Costa, and J. Hudec, *Chem. Comm.*, **1969**, 753.

[211] J. Schwartz, *Chem. Comm.*, **1969**, 833.

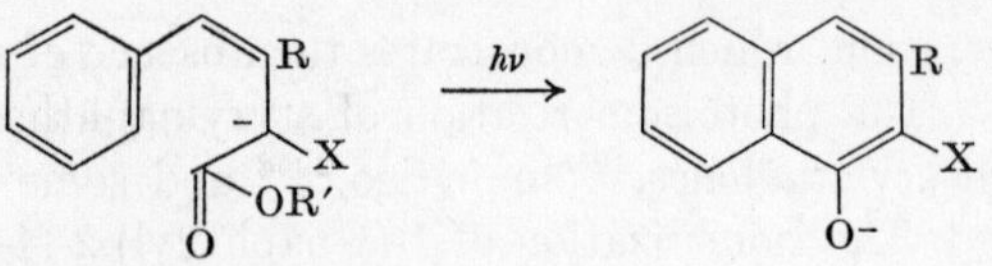

(75)

(X = CO$_2$Et, CN, etc.)

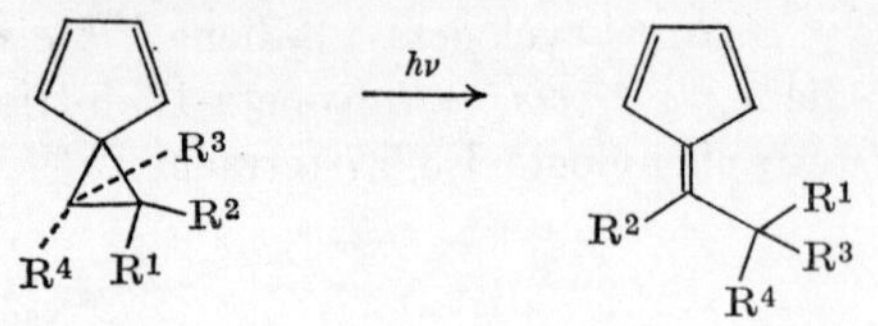

(76)

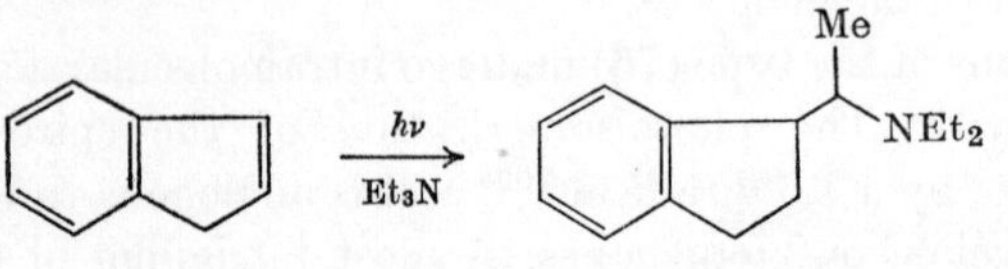

(77)

Acetylenes

Irradiation of conjugated acetylenic esters (e.g. ethyl propiolate) in hydrogen-donating solvents leads to reduction of the triple bond. Reaction in cyclohexane gave products derived by attack of cyclohexyl radicals upon the triple bond.[212] Cd($^3P_1$) photosensitized irradiation of acetylene gives benzene and vinylacetylene.[213] The formation of azulenes **(78)** and **(79)**[214a] is one of the

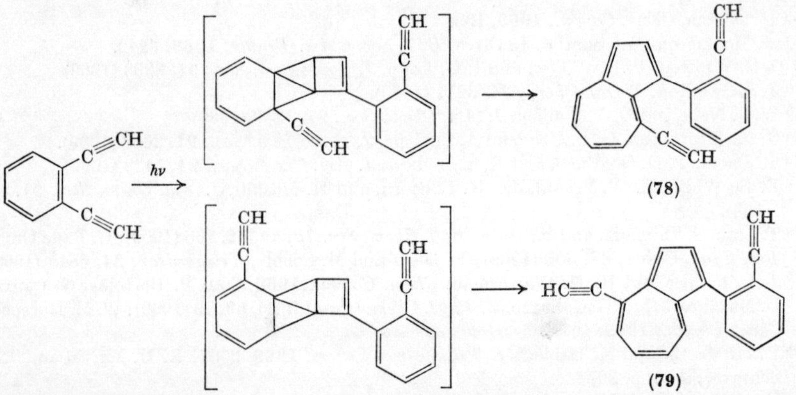

(78)

(79)

[212] G. Büche and S. H. Feairheller, *J. Org. Chem.*, **34**, 609 (1969).
[213] S. Tsunashima and S. Sato, *Bull. Chem. Soc. Japan*, **41**, 2281 (1968).
[214a] E. Müller, M. Sauerbier, and G. Zountsas, *Tetrahedron Letters*, **1969**, 3003.

reported examples of the photocyclization of bisacetylenes.[214b] Photoinduced inter- and intra-molecular nucleophilic additions to the triple bond have been reported, and this appears to be a new class of reaction.[215]

Aromatic Hydrocarbons

Benzenes and Naphthalenes

The physical aspects of the excited electronic states of benzene and naphthalene have been discussed.[216] The efficiency of the isomerization of *trans*-but-2-ene by *o*-xylene falls off at high concentrations of the xylene. The formation of the triplet excimer of *o*-xylene was suggested as being responsible for the short lifetime of the triplet xylene at these concentrations.[217] Tritium-labelling studies have shown[218] that anisole undergoes 1,2-hydrogen shifts (probably via a benzvalene intermediate) on irradiation. The quantum yield for isomerization is very low (0.004). Hexa(trifluoromethyl)benzene photo-isomerizes to its corresponding benzvalene which is further isomerized to a Dewar benzene.[219] This compound further photoisomerizes to hexa(trifluoro-methyl)prismane. The formation of polyolefinic products in the photolysis of degassed solutions of benzene has been studied.[220] Concerted photocyclo-addition reactions of benzene have been discussed in terms of orbital symmetry requirements.[221] The occurrence of 1,2- and 1,4-cycloaddition reactions has been rationalized. The products (see Figure 7) from the photoaddition of *cis*-octene to hexafluorobenzene have been described, and the excited states from which they are derived allocated.[222] Not only does the intramolecularly photosensitized isomerization of 6-phenylhex-2-ene occur on irradiation but also addition of the double bond to the aromatic system.[223] Structures for the adduct formed have been suggested. The photocycloaddition of thiomaleic anhydride to benzene[224a] and of dimethyl acetylenedicarboxylate to naphthalene[224b] has been investigated. Irradiation of the naphthalenide radical ion results in electron ejection.[225]

[214b] B. Bossenbroek, D. C. Sanders, H. M. Curry, and H. Shechter, *J. Am. Chem. Soc.*, **91**, 371 (1969); E. Müller, J. Heiss, and M. Sauerbier, *Ann. Chem.*, **723**, 61 (1969).

[215] T. D. Roberts, L. Ardemagni, and H. Shechter, *J. Am. Chem. Soc.*, **91**, 6185 (1969).

[216] J. B. Birks, L. G. Christophorou, and R. H. Huebner, *Nature*, **217**, 809 (1968).

[217] R. B. Cundall and A. J. R. Voss, *Chem. Comm.*, **1969**, 116.

[218] G. Lodder, P. E. J. du Mee, and E. Havinga, *Tetrahedron Letters*, **1968**, 5949.

[219] M. G. Barlow, R. N. Haszeldine, and R. Hubbard, *Chem. Comm.*, **1969**, 202.

[220] K. H. Grellmann and W. Kühnle, *Tetrahedron Letters*, **1969**, 1537.

[221] D. Bryce-Smith, *Chem. Comm.*, **1969**, 806.

[222] D. Bryce-Smith, A. Gilbert, and B. H. Orger, *Chem. Comm.*, **1969**, 800.

[223] H. Morrison and W. I. Ferree, *Chem. Comm.*, **1969**, 268.

[224a] M. Verbeek, H.-D. Scharf, and F. Korte, *Chem. Ber.*, **102**, 2471 (1969).

[224b] E. Grovenstein, T. C. Campbell, and T. Shibata, *J. Org. Chem.*, **34**, 2418 (1969).

[225] S. Pardhan and L. Fischer, *Arkiv Kemi*, **29**, 577 (1968).

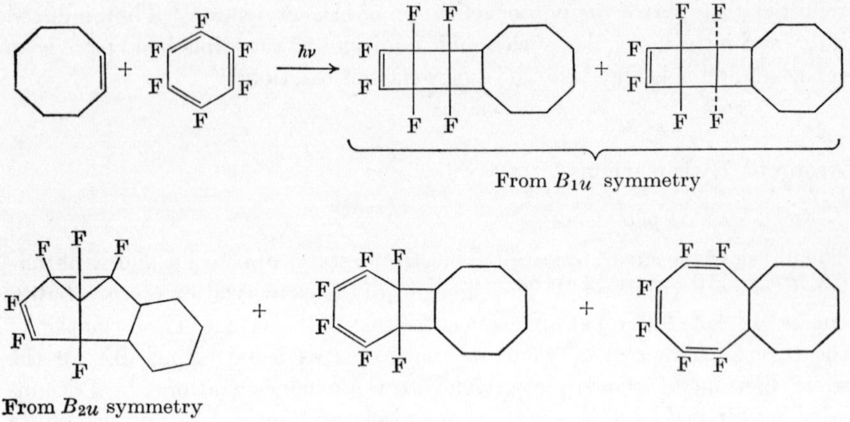

From B_{1u} symmetry

From B_{2u} symmetry

Figure 7

Anthracenes

Further examples of the addition of singlet-state anthracenes to ground-state anthracenes have been studied.[226a,b,c] Anthracenes which do not form photo-dimers, e.g. 9,10-dimethyl- and 9,10-dimethoxy-anthracene, form crossed photoaddition compounds.[226b,c] The yield of these crossed products decreases as the dielectric constant of the solvent increases, and from this observation it was concluded that the addition occurs by an exciplex and not radical ions.[226c]

Anthracene is found to be only slowly photoreduced by tertiary amines in benzene solution whereas the reaction is extremely efficient in acetonitrile solution.[227a,b] In this case it was concluded that exciplex formation leads mainly to deactivation of the singlet state whereas reduction occurs by radical ion formation.[227b] Labelling experiments confirmed the formation of radical ions.[227a] Diels–Alder adducts of anthracene and a number of olefins have been found to be photodecomposed to give anthracene.[228] The fate of the olefin formed in the reaction was not traced.

Ethers and Sulphides

Hydrogen atom ejection has been found to occur in the mercury-photo-sensitized decomposition of methyl ethyl ether,[229a] and on photolysis of

226a M. D. Cohen, Z. Ludmer, J. M. Thomas, and J. O. Williams, *Chem. Comm.*, **1969**, 1172.
226b R. Lapouyade, A. Castellan, and H. Bouas-Laurent, *Compt. Rend., C*, **268**, 217 (1969); R. Lapouyade, A. Castellan, and H. Bouas-Laurent, *Tetrahedron Letters*, **1969**, 3537.
226c H. Bouas-Laurent and R. Lapouyade, *Chem. Comm.*, **1969**, 817.
227a C. Pac and H. Sakurai, *Tetrahedron Letters*, **1969**, 3829.
227b R. S. Davidson, *Chem. Comm.*, **1969**, 1450.
228 H. Nozaki, H. Kato, and R. Noyori, *Tetrahedron*, **25**, 1661 (1969).
229a S. V. Filseth, *J. Phys. Chem.*, **73**, 793 (1969).

aqueous solutions of anisole.[229b] In the latter example the spectra of the hydrated electron and the phenoxymethyl radical were observed. A variety of solvents used for photochemical reactions give, on irradiation in the presence of oxygen, absorbing species.[230] This observation underlines the necessity for

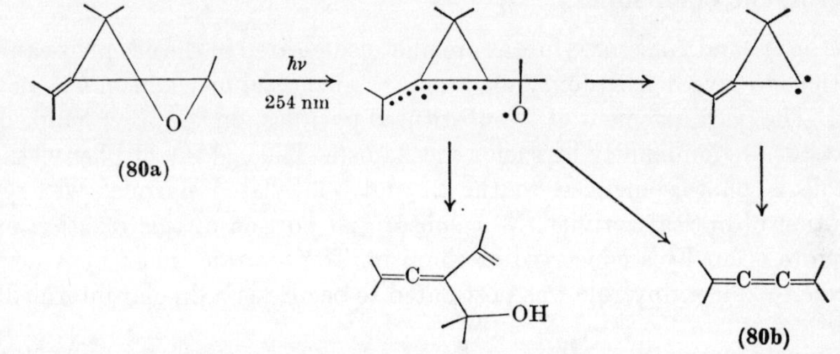

(80a) (80b)

doing solvent blanks in kinetic data determinations. The cumulene (**80b**) is produced on irradiation of the oxide (**80a**), and its formation may be rationalized in terms of an intermediate carbene or diradical species.[231] Decomposition of the ozonide (**81**) gives the dimer of tetramethylbutadiene, which indicates that the singlet state of the diene is formed in the reaction.[232] Intramolecular [2 + 2]-photocycloaddition of the olefinic sulphide (**82**) gives the episulphide (**83**).[233a] The episulphide (**83**)[233a] and a number of transannular sulphides[233b]

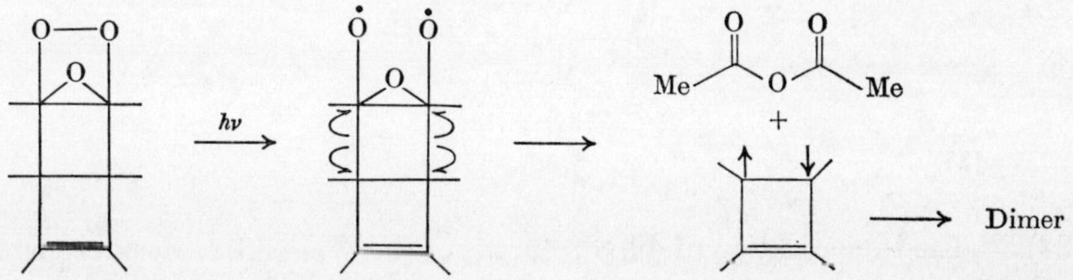

(81)

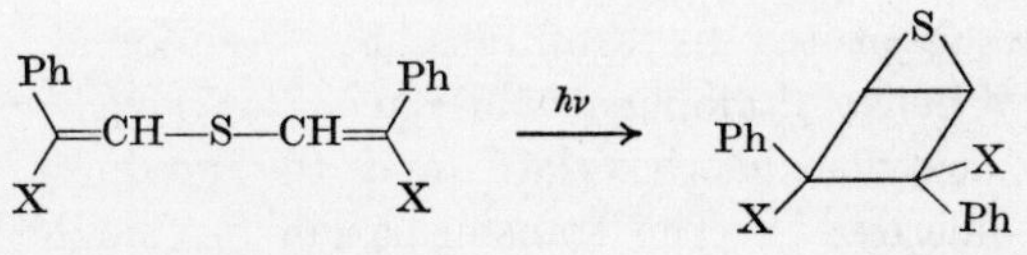

(82) (83)

[229b] H.-I. Joschek and L. I. Grossweiner, *Z. Naturforsch.*, **24b**, 562 (1969).

[230] H. Mauser and V. Bihl, *Z. Naturforsch.*, **24b**, 643 (1969).

[231] J. K. Crandall and D. R. Paulson, *Tetrahedron Letters*, **1969**, 2751.

[232] P. R. Storey, W. H. Morrison, and J. M. Butler, *J. Am. Chem. Soc.*, **91**, 2398 (1969).

[233a] E. Block and E. J. Corey, *J. Org. Chem.*, **34**, 896 (1969).

[233b] E. J. Corey and E. Block, *J. Org. Chem.*, **34**, 1233 (1969).

have been shown to be desulphurized by irradiation in the presence of triphenylphosphine. 2-Methyl-2,3-dihydro-4,5-benzothiophen undergoes C–S bond fission on irradiation.[234]

Heterocyclic Compounds

2-Methoxy- and 2-acetoxy-furans are photoisomerized in the gas phase and in solution, to give dehydrobutyrolactones.[235] A radical mechanism was postulated. The rearrangement of *N*-substituted pyrroles, on the other hand, does not occur predominantly by such a mechanism. Thus, (+)-*N*-(1-phenylethyl)-pyrrole is photoisomerized to the 2- and 3-alkylated pyrroles with 54% retention of optical activity.[236] A substantial portion of this reaction must therefore occur by a concerted mechanism. The rearrangement of *N*-acetyl-pyrrole to 2-acetylpyrrole was postulated to occur via a dipolar intermediate

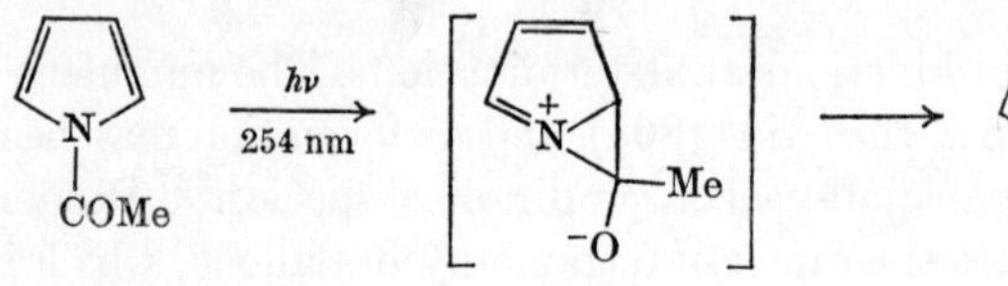

(84)

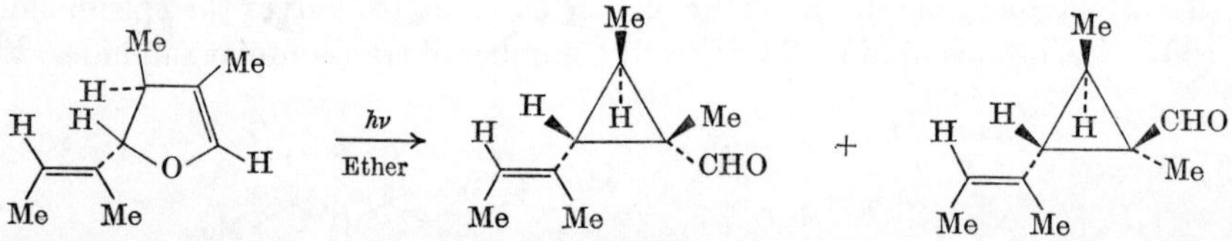

72% yield

(85)   (86)   (86)

(84).[237] The isomerization of dihydrofurans of the type (85) to cyclopropane-carboxaldehydes, e.g. (86), occurs with a degree of stereospecificity, and a concerted mechanism (conrotatory opening of the C-2—O bond) was suggested.[238] Some substituted dihydrofurans have been found to add methanol on irradiation.[239] Benzo[*b*]thiophens undergo a [2 + 2]-cycloaddition reaction with dimethyl acetylenedicarboxylate, and the products, e.g. (87), are further photoisomerized.[240] The transformation of thiophen into pyrroles

[234] D. C. Neckers and J. De Zwaan, *Chem. Comm.*, **1969**, 813.

[235] R. Srinivasan and H. Hiraoka, *Tetrahedron Letters*, **1969**, 2767.

[236] J. M. Patterson and L. T. Burlma, *Tetrahedron Letters*, **1969**, 2215.

[237] H. Shizuka, E. Okutsu, Y. Mori, and I. Tanaka, *Mol. Photochem.*, **1**, 135 (1969).

[238] P. Scribe, D. Hourdin, and J. Wiemann, *Compt. Rend.*, *C*, **268**, 178 (1969).

[239] A. C. Waiss and M. Wiley, *Chem. Comm.*, **1969**, 512.

[240] D. C. Neckers, J. H. Dopper, and H. Wynberg, *Tetrahedron Letters*, **1969**, 2913.

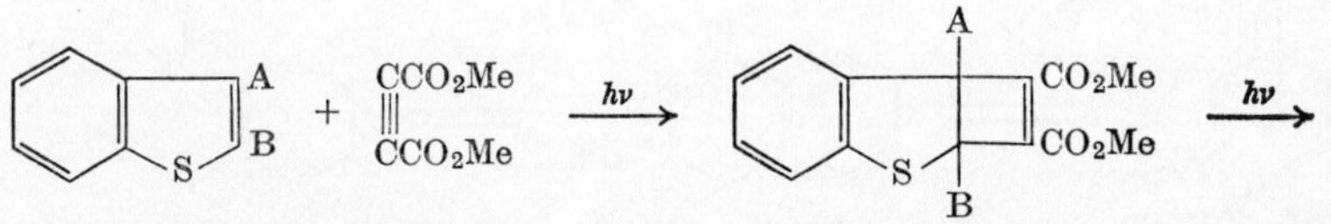

(87)

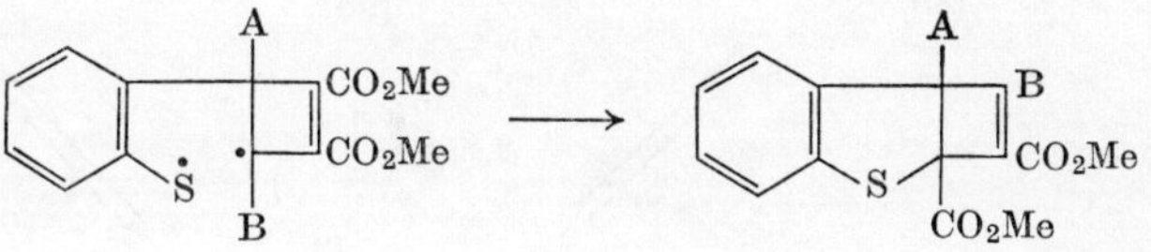

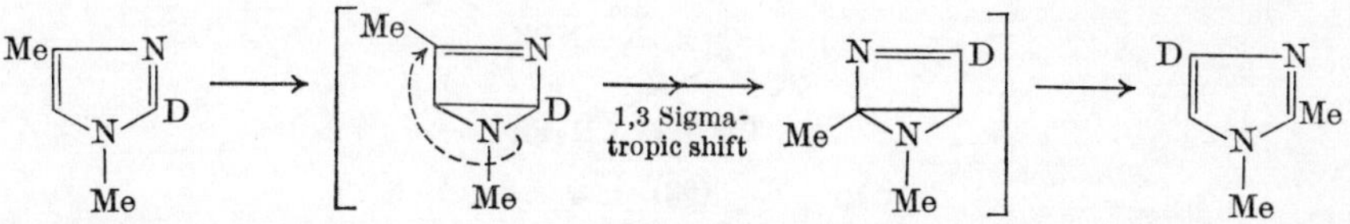

(88)

can be accomplished by irradiating thiophen in the presence of primary amines.[241] A cyclopropenethiocarboxaldehyde was suggested as an intermediate. It was concluded that the rearrangement of the imidazole (88) was concerted and that 1H-azirines were not intermediates.[242] Formation of the stable azomethine imine (92) occurs from the singlet state of phthalazine (90).[243] Isomerization via (89) or (91) as opposed to a direct transformation cannot be ruled out. The anion (93) was shown to undergo stereospecific cyclization to give (94), and this is an excellent example of orbital symmetry control of a carbanion cyclization.[244] Dihydro-1,2-oxazines [245a] and 1,2,5-oxadiazoles [245b] give rearrangement products by initial N–O bond cleavage. The products produced by photorearrangement of 2,5-diphenyloxazole depend upon the solvent used for the reaction.[246] Acylazirines were suggested as intermediates, and one of these was an anti-aromatic 1H-azirine (a 4π system). Azirines were shown to be intermediates in the rearrangement of anthranils to 3,4-azepines.[247] Benzotriazoles photofragment to give nitrenes and nitrogen [248a] whereas *vic*-triazoles give nitriles.[248b] 5-Phenyltetrazole gave

[241] A. Couture and A. Lablache-Combier, *Chem. Comm.*, **1969**, 524.

[242] P. Beak and W. Messer, *Tetrahedron*, **25**, 3287 (1969).

[243] B. Singh, *J. Am. Chem. Soc.*, **91**, 3670 (1969).

[244] D. H. Hunter and S. K. Sim, *J. Am. Chem. Soc.*, **91**, 6202 (1969).

[245a] P. Scheiner, O. L. Chapman, and J. D. Lassila, *J. Org. Chem.*, **34**, 813 (1969).

[245b] T. Mukai, T. Oine, and A. Matsubarn, *Bull. Chem. Soc. Japan*, **42**, 581 (1969); *Org. Reaction Mech.*, **1968**, 449.

[246] M. Kojima and M. Maeda, *Tetrahedron Letters*, **1969**, 2379.

[247] M. Ogata, H. Matsumoto, and H. Kano, *Tetrahedron*, **25**, 5205 (1969); *Org. Reaction Mech.*, **1968**, 447.

[248a] A. J. Hubert, *J. Chem. Soc.*(C), **1969**, 1334.

[248b] J. H. Boyer and R. Selvarajan, *Tetrahedron Letters*, **1969**, 47.

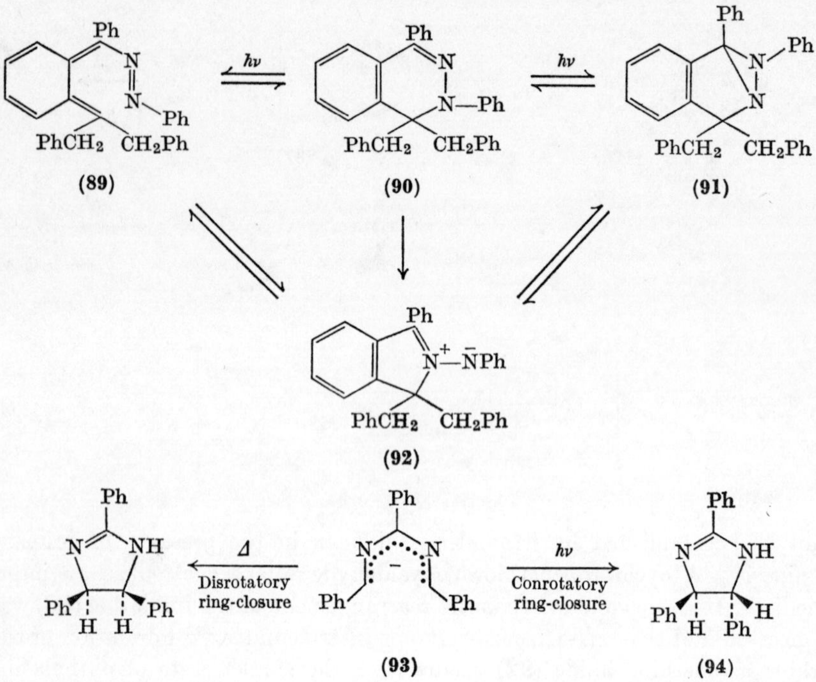

on photolysis a 1,3-dipolar intermediate and phenylcarbene.[249] Further examples of the formation of 1*H*-1,2-diazepines by the photoisomerization of pyridium betaines have been reported.[250] The formation of bicyclo-compounds via the intramolecular cyclization of several azepines has been studied. The choice of which of the two cyclization processes occurs [A or B in formula (95)] is determined by the interaction of any substituents with the nitrogen atom.[251] Disrotatory ring-closure by path A of (95) is hindered by the non-bonded interaction between the methyl group and the nitrogen atom. 2,3-Dihydro-1,2-diazepine ketones undergo similar reactions[252a] whereas (96) undergoes a di-π-methane rearrangement amongst other reactions.[252b] The *N*-ethoxycarbonylaziridine of cyclooctatetraene ring-opens from its triplet state to give an aza-[10]annulene.[253] Several carboxylic acids have been

[249] P. Scheiner, *J. Org. Chem.*, **34**, 199 (1969).

[250] V. Snieckus, *Chem. Comm.*, **1969**, 831; J. Streith and J.-M. Cassal, *Bull. Soc. Chim. France*, **1969**, 2175; T. Sasaki, K. Kanematsu, and A. Kakehi, *Chem. Comm.*, **1969**, 432.

[251] L. A. Paquette and D. E. Kuhla, *J. Org. Chem.*, **34**, 2885 (1969).

[252a] J.-L. Derocque, W. J. Thener, and J. A. Moore, *J. Org. Chem.*, **33**, 4381 (1968).

[252b] M. Ogata, H. Matsumoto, and H. Kano, *Tetrahedron*, **25**, 5217 (1969).

[253] A. G. Anastassiou and J. H. Gebrian, *J. Am. Chem. Soc.*, **91**, 4011 (1969).

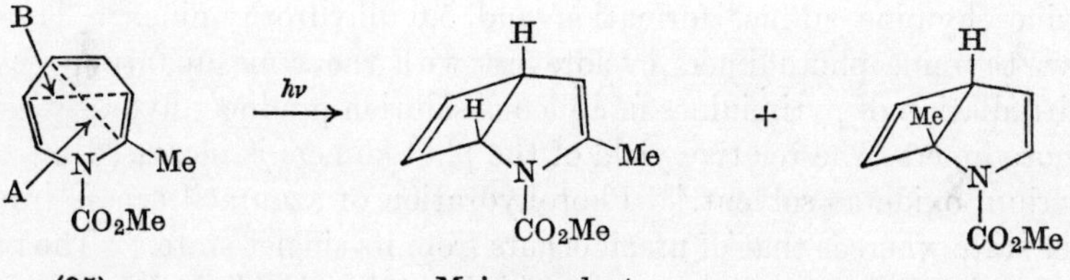

(95)   Major product

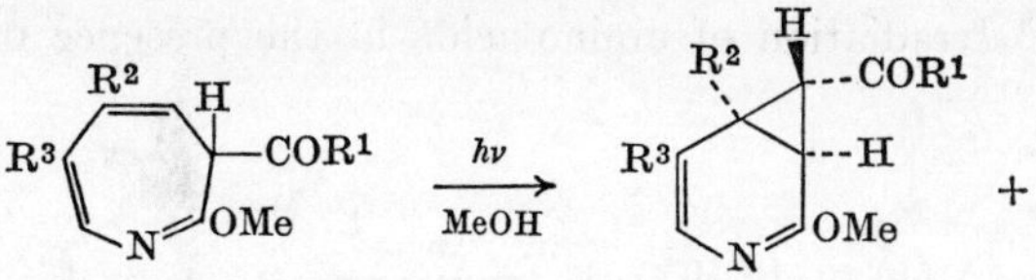

(96)

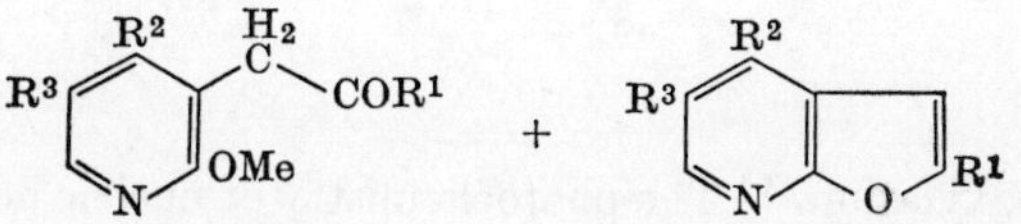

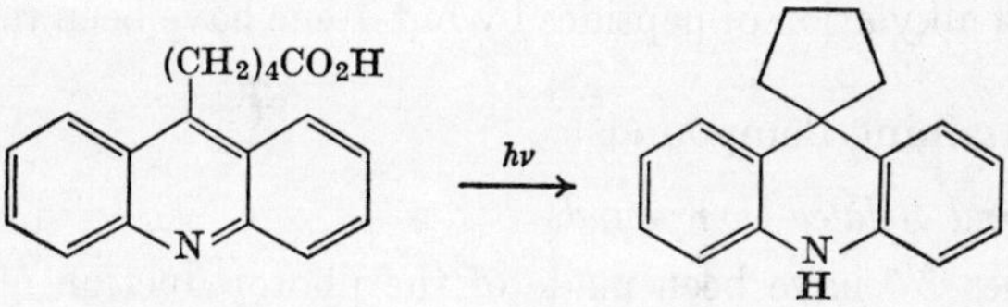

(97)

shown to be decarboxylated by irradiation in the presence of acridine, and this reaction was extended to the synthesis of spiro-compounds, e.g. **(97)**.[254] Reaction via an acridium salt, which decomposed to give an alkyl radical, carbon dioxide, and an acridan radical, was suggested.

Further dimeric compounds have been isolated from irradiated solutions of pyrimidines.[255a] A class of substituted dihydrofurans, the furocoumarins, react with pyrimidines in a [2 + 2]-cycloaddition reaction.[255b] The inefficiency of the photodimerization of 1,3-dimethylthymine in water has been attributed to the pyrimidine forming van der Waals complexes.[256] The dimerization of orotic acid is sensitized by benzophenone.[257] Lamola has shown that triplet sensitization of DNA leads to preferential thymine dimerization as opposed to

[254] R. Noyori, M. Katô, M. Kawanisi, and H. Nozaki, *Tetrahedron*, **25**, 1125 (1969).
[255a] R. O. Rahn and J. L. Hosszu, *Photochem. Photobiol.*, **10**, 131 (1969); M. N. Khattak and S. Y. Wang, *Science*, **163**, 1341 (1969).
[255b] F. Dall'Acqua, S. Marciani, and G. Rodighiero, *Z. Naturforsch.*, **24b**, 307 (1969).
[256] R. Lisewski and K. L. Wierzchowski, *Chem. Comm.*, **1969**, 348.
[257] M. Charlier, C. Helene, and M. Dourlent, *J. Chim. Phys.*, **66**, 700 (1969).

cytosine–thymine adduct formation and 5,6-dihydrothymine.[258] The four known thymine photodimers cyclorevert with the same quantum yield.[259] The irradiation of pyrimidines in aqueous solution produces hydrates as well as photodimers. The relative yield of the photodimers is increased by use of deuterium oxide as solvent.[260] Photohydration of azauracil occurs from its triplet state whereas that of uracil occurs from its singlet state.[261] The reduction of pyrimidines by sodium borohydride is catalysed by light.[262] A number of purines and related compounds are photoreduced by alcohols by a radical mechanism.[263] Irradiation of amino-acids in the presence of caffeine (98)

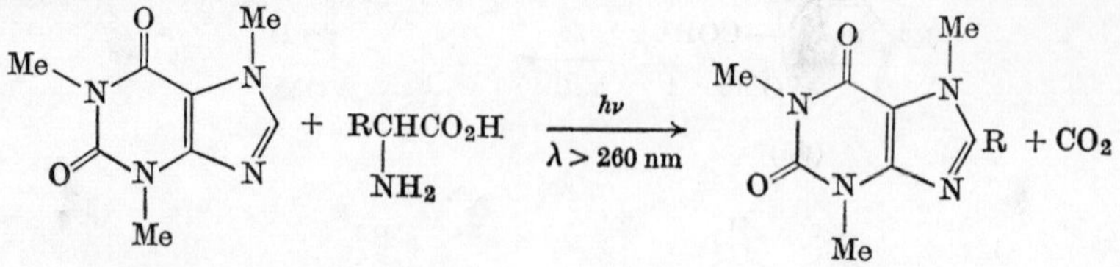

(98)

leads to decarboxylation.[264] The photochemistry of nucleic acids and related compounds has been the subject of a review.[265] Further examples of the photoinitiated alkylation of peptides by but-1-ene have been reported.[266]

Nitrogen-containing Compounds

Azomethines and Related Compounds

Further studies[267a] have been made of the photoreduction of Schiff bases, and the results are in agreement with the mechanism reported last year.[267b] Phenylhydrazones undergo [1,3]-sigmatropic shifts to give azo-compounds, as well as cleavage at the N–N bond (see Figure 8).[268a] In the presence of

258 A. A. Lamola, *Photochem. Photobiol.*, **9**, 291 (1969).

259 M. A. Herbert, J. C. LeBlanc, D. Weinblum, and H. E. Johns, *Photochem. Photobiol.*, **9**, 33 (1969).

260 J. C. Nnadi and S. Y. Wang, *Tetrahedron Letters*, **1969**, 2211.

261 L. Kittler and G. Löber, *Photochem. Photobiol.*, **10**, 35 (1969).

262 Y. Kondo and B. Witkop, *J. Am. Chem. Soc.*, **91**, 5264 (1969).

263 D. Elad, I. Rosenthal, and H. Steinmaus, *Chem. Comm.*, **1969**, 305; H. Steinmaus, I. Rosenthal, and D. Elad, *J. Am. Chem. Soc.*, **91**, 4921 (1969); E. C. Taylor, Y. Maki, and B. E. Evans, *ibid.*, p. 5181.

264 D. Elad and I. Rosenthal, *Chem. Comm.*, **1969**, 905.

265 J. G. Burr, *Adv. Photochem.*, **6**, 193 (1968).

266 D. Elad and J. Sperling, *Chem. Comm.*, **1969**, 234; D. Elad and J. Sperling, *J. Chem. Soc.(C)*, 1579 (1969).

267a E. S. Huyser, R. H. S. Wang, and W. T. Short, *J. Org. Chem.*, **33**, 4323 (1968); N. Toshima H. Hirai, and S. Makishima, *J. Chem. Soc. Japan, Ind. Chem. Sect.*, **72**, 184 (1969); *Chem. Abs.*, **70**, 114378, (1969); G. Balogh and F. C. De Schryver, *Tetrahedron Letters*, **1969**, 1371; A. Pudwa, W. Bergmarhm, and D. Pashayan, *J. Am. Chem. Soc.*, **91**, 2653 (1969).

267b *Org. Reaction Mech.*, **1968**, 452.

268a R. W. Binkley, *Tetrahedron Letters*, **1969**, 1893.

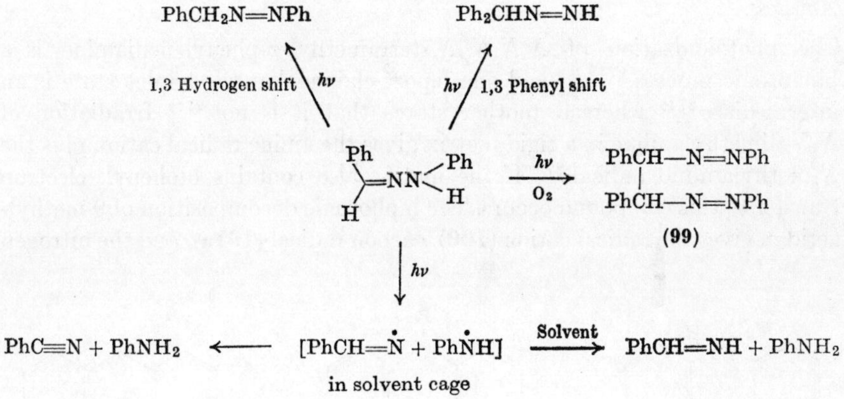

PhCH$_2$N=NPh   Ph$_2$CHN=NH

1,3 Hydrogen shift $\nwarrow$ $h\nu$ $h\nu$ $\nearrow$ 1,3 Phenyl shift

$$\underset{H}{\overset{Ph}{>}}\!=\!NN\!\!<\!\overset{Ph}{\underset{H}{}} \quad \xrightarrow[O_2]{h\nu} \quad \begin{array}{l} \text{PhCH—N=NPh} \\ | \\ \text{PhCH—N=NPh} \end{array}$$

(99)

$\downarrow h\nu$

PhC≡N + PhNH$_2$ ⟵ [PhCH=Ṅ + PhṄH] $\xrightarrow{\text{Solvent}}$ PhCH=NH + PhNH$_2$

in solvent cage

Figure 8

oxygen they undergo photo-oxidative coupling, e.g. to give (99).[268b] Azines photofragment at the N–N bond to give imino radicals which disproportionate.[269a,b] In the presence of oxygen these radicals give nitriles.[269b] The formation of stilbenes by irradiation of azine does not appear to be a simple reaction, and the intermediacy of dimeric compounds which fragment to the olefin has been postulated.[270] Convincing evidence for the formation of oxaziridines from the singlet state of oximes has been reported.[271a,b] Hydrogen abstraction

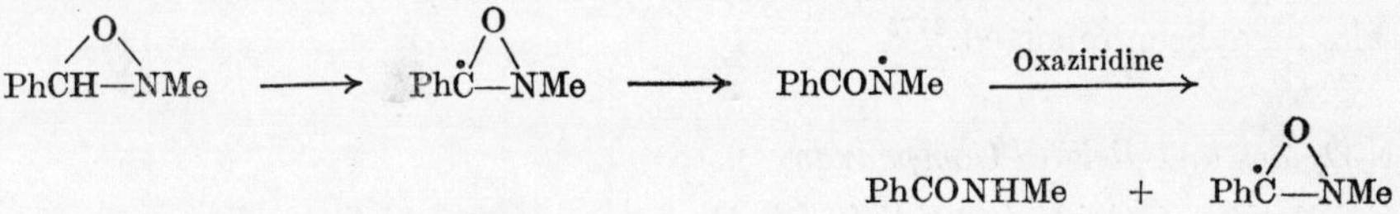

$$\underset{\text{PhCH—NMe}}{\overset{O}{\diagdown}} \longrightarrow \underset{\text{PhĊ—NMe}}{\overset{O}{\diagdown}} \longrightarrow \text{PhCONMe} \xrightarrow{\text{Oxaziridine}}$$

$$\text{PhCONHMe} \quad + \quad \underset{\text{PhĊ—NMe}}{\overset{O}{\diagdown}}$$

from the oxaziridine is the chain-propagating reaction[271a] for its isomerization to an amide.[271a,c] Population of the triplet state of the oxime leads to *syn,anti*-photoisomerization.[271a] Phenylhydrazones and semicarbazones also exhibit this type of photoisomerization.[272] Irradiation of oximes in benzene solution leads to regeneration of the ketone.[273]

[268b] J.-C. Bloch, *Tetrahedron Letters*, **1969**, 4041.

[269a] R. K. Brunton and S. Chang, *Ber. Bunsenges. Phys. Chem.*, **72**, 217 (1968); R. W. Binkley, *J. Org. Chem.*, **34**, 2072 (1969).

[269b] R. W. Binkley, *J. Org. Chem.*, **34**, 3218 (1969).

[270] R. W. Binkley, *J. Org. Chem.*, **34**, 931 (1969).

[271a] H. Izawa, P. de Mayo, and T. Tabata, *Can. J. Chem.*, **47**, 51 (1969).

[271b] T. Oine and T. Mukai, *Tetrahedron Letters*, **1969**, 157.

[272] G. Condorelli and L. L. Costanzo, *Boll. Sedute Accad. Gioenia*, **8**, 753 (1966); *Chem. Abs.*, **70**, 3120 (1969); W. I. Stenberg, P. A. Barks, D. Bays, D. D. Hammargren, and D. V. Rao, *J. Org. Chem.*, **33**, 4402 (1968); B. L. Fox and H. M. Rosenberg, *Chem. Comm.*, **1969**, 1115.

[273] J.-P. Vernes and R. Beugelmans, *Tetrahedron Letters*, **1969**, 2091.

Amines

The photoionization of N,N,N',N'-tetramethyl-p-phenylenediamine is a biphotonic process,[274a,b,c] and one report claims that the triplet state is an intermediate[274b] whereas another states that it is not.[274c] Irradiation of N,N-dimethylaniline in a rigid matrix gives the amine radical cation plus the N-methylanilino radical.[275] If the matrix also contains biphenyl, electron transfer to this compound occurs. The biphotonic decomposition of 9-methyl-acridan gives the radical cation (**100**), carbon radical (**101a**), and the nitrogen

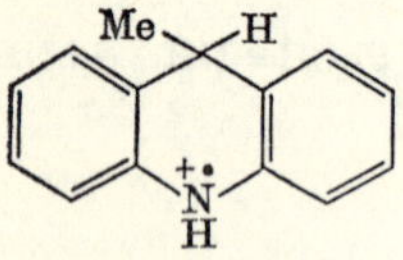

(**100**) 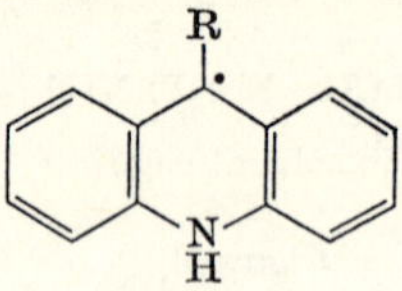   (**101**) 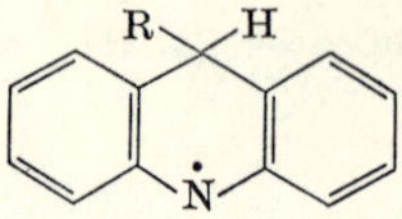   (**102**)

(a, R = Me; b, R = H)

radical (**102a**).[276a,b] The radicals (**101b**) and (**102b**) are also formed on photolysis of biacridan.[276c] Acridan radical cation is also formed by γ-irradiation of the amine.[276d] Triplet energy transfer from acetone to N,N,N-tri-methylanilinium bromide results in fragmentation of the salt to give trimethyl-amine.[277a] Further examples of electron transfer from amines to anilinium salts have been reported.[277b]

N-*Oxides and Related Compounds*

Irradiation of pyridazine N-oxides in the presence of an olefin results in oxygen transfer to give an epoxide and a carbonyl compound.[278] The mechanistic aspects of this intriguing reaction were not probed. Formation of the parent amines by irradiation of N-oxides is facilitated by the presence of oxygen.[279] The types of products derived from N-oxides by initial rearrangement to an

274a H. S. Pilloff and A. C. Albrecht, *J. Chem. Phys.*, **49**, 4891 (1968).
274b K. D. Cadogan and A. C. Albrecht, *J. Chem. Phys.*, **51**, 2710 (1969).
274c R. Potashnik, M. Ottolenghi, and R. Bensasson, *J. Phys. Chem.*, **73**, 1912 (1969).
275 S. Arimitsu, K. Kimura, and H. Tsubomura, *Bull. Chem. Soc. Japan*, **42**, 1858 (1969).
276a V. Zanker and E. Erhardt, *Ber. Bunsenges. Phys. Chem.*, **72**, 267 (1969).
276b V. Zanker and D. Benicke, *Z. Phys. Chem.* (Frankfurt), **240**, 34 (1969).
276c A. Kira and M. Koizumi, *Bull. Chem. Soc. Japan*, **42**, 625 (1969).
276d T. Shida and A. Kira, *Bull. Chem. Soc. Japan*, **42**, 1197 (1969).
277a C. Pac and H. Sakurai, *Chem. Comm.*, **1969**, 20.
277b C. Pac and H. Sakurai, *J. Chem. Soc. Japan, Ind. Chem. Sect.*, **72**, 230 (1969); *Chem. Abs.*, **70**, 114373 (1969).
278 T. Tsuchiya, H. Arai, and H. Igeta, *Tetrahedron Letters*, **1969**, 2747.
279 O. Buchardt, P. L. Kumler, and C. Lohse, *Acta Chem. Scand.*, **23**, 159 (1969).

oxaziridine are dependent upon the solvent used.[279, 280a,b] In one example, the isolation of the intermediate oxaziridine was reported.[280b] Flashing of alcoholic solutions of acridine N-oxides gives a transient (λ_{max} 320 nm; lifetime 0.2 millisec) which may be the intermediate oxaziridine.[281]

(103)

A further example of the formation of a diazo-compound from an N-oxide[282a] is afforded by the azine monoxide (103) which also undergoes a [2 + 2]-cycloaddition reaction.[282b] N-Phenyldibenzoyloxaziridine is produced by irradiation of the corresponding nitrone.[283]

Nitro- and Nitroso-compounds

Pyrochlorophyll photoreduces aromatic nitro-compounds, and the linear relationship between quantum yield and the reduction potential of the nitro-compound was interpreted as being indicative of electron transfer from the chlorophyll to the nitro-compound.[284] Pulse radiolysis studies have shown that radicals of the type $ArNO_2H$ have a p$K \sim 2$.[285] There are conflicting reports as to whether ethers, such as tetrahydrofuran, form CT complexes with aromatic nitro-compounds.[286] Product studies on the photoreduction of nitro-compounds have been made,[287a,b] and in one case a product derived by combination of a solvent radical with the $ArNO_2H$ radical was isolated.[287a] The intramolecular photodecarboxylations of o-nitrophenylacetic acids,[288a] o-nitrophenoxyacetic acid,[288b] and N-(o-nitrophenyl)glycine[288c] have been

280a M. Ishikawa, C. Kaneko, I. Yokoo, and S. Yamada, *Tetrahedron*, **25**, 295 (1969); O. Buchardt, P. L. Kumler, and C. Lohse, *Acta Chem. Scand.*, **23**, 2149 (1969); D. R. Eckroth and R. H. Squire, *Chem. Comm.*, **1969**, 312.

280b G. F. Field and L. H. Sternbach, *J. Org. Chem.*, **33**, 4438 (1968).

281 H. Mantsch, V. Zanker, W. Seiffert, and G. Prell, *Ann. Chem.*, **723**, 95 (1969).

282a *Org. Reaction Mech.*, **1968**, 455.

282b W. R. Dolbier and W. M. Williams, *J. Am. Chem. Soc.*, **91**, 2819 (1969).

283 M. L. Scheinbaum, *Tetrahedron Letters*, **1969**, 4221.

284 G. R. Seely, *J. Phys. Chem.*, **73**, 117 (1969).

285 W. Grünbein and A. Henglein, *Ber. Bunsenges. Phys. Chem.*, **73**, 376 (1969).

286 G. Briegleb and G. Lind, *Z. Naturforsch.*, **23a**, 1747 (1968); G. Briegleb and G. Lind, *ibid.*, p. 1752; D. J. Cowley and L. H. Sutcliffe, *Spectrochim. Acta*, **25A**, 989 (1969).

287a H. Hart and J. W. Link, *J. Org. Chem.*, **34**, 758 (1969).

287b S. Hashimoto, K. Kano, and K. Ueda, *Tetrahedron Letters*, **1969**, 2733.

288a J. D. Margerum and C. T. Petrusis, *J. Am. Chem. Soc.*, **91**, 2467 (1969).

288b P. H. McFarlane and D. W. Russell, *Chem. Comm.*, **1969**, 475.

288c D. J. Neadle and R. J. Pollit, *J. Chem. Soc.*(C), **1969**, 2127.

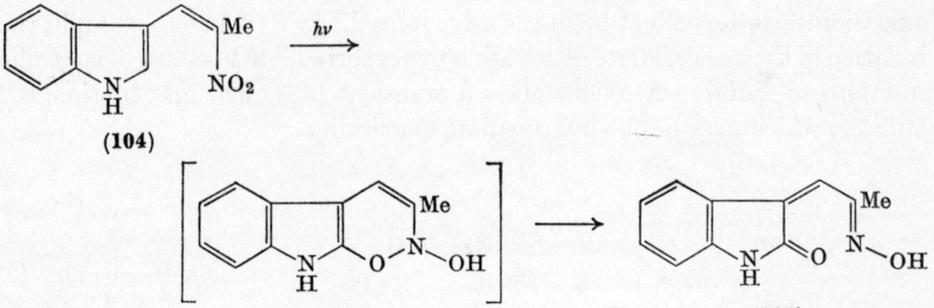

(104)

(105)   **(106)**

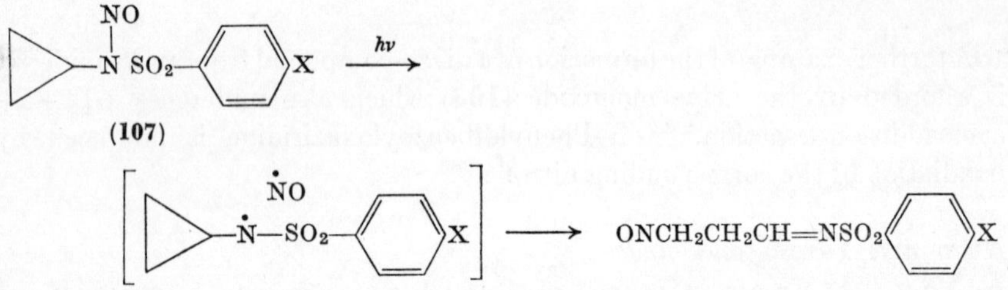

(107)

$ONCH_2CH_2CH=NSO_2C_6H_4X$

reported. The efficiency of these reactions is pH-dependent.[288a,c] The nitro-compound **(104)** undergoes an intramolecular oxygen transfer reaction to give **(106)**, and **(105)** was suggested as an intermediate.[289] Further evidence for the formation of arylnitrenes by the photodeoxygenation of nitro-compounds by triethyl phosphite has been reported.[290] Irradiation of ethanolic solutions of nitrosobenzene with polychromatic light yields a multitude of products some of which are derived by initial cleavage of the nitroso-compounds.[291] The intricate mechanistic details of the photoreactions of 2-chloro-2-nitrosobutane have been probed, and in the oxidation step, reaction of the nitroso group with nitric oxide was postulated.[292] Reported examples[293a,b] of the cleavage of *N*-nitroso-amines to give amino radicals have included the rearrangement of **(107)**.[293b] The formation of nitroso-compounds by the intramolecular rearrangement of *N*-nitroso-amides[294a] and further examples of the Barton reaction[294b] have been reported.

[289] J. S. Cridland and S. T. Reid, *Chem. Comm.*, **1969**, 125.

[290] R. J. Sundberg, B. P. Das, and R. H. Smith, Jr., *J. Am. Chem. Soc.*, **91**, 658 (1969).

[291] R. Tanikaga, *Bull. Chem. Soc. Japan*, **42**, 210 (1969).

[292] L. Creagh and I. Trachtenberg, *J. Org. Chem.*, **34**, 1307 (1969).

[293a] W. B. Watkins and R. N. Seelye, *Can. J. Chem.*, **47**, 497 (1969).

[293b] E. E. J. Dekker, J. B. F. N. Engberts, and T. J. de Boer, *Tetrahedron Letters*, **1969**, 2651.

[294a] Y. L. Chow and J. N. S. Tam, *Chem. Comm.*, **1969**, 747; Y. L. Chow, J. N. S. Tam, and A. C. H. Lee, *Can. J. Chem.*, **47**, 2441 (1969).

[294b] D. H. R. Barton, D. Kumari, P. Welzel, L. J. Danks, and J. F. McGhie, *J. Chem. Soc.*(C), **1969**, 332, D. H. R. Barton, R. P. Budhiraja, and J. F. McGhie, *ibid.*, p. 336; H. Suginome, N. Sato, and T. Masamune, *Bull. Chem. Soc. Japan*, **42**, 215 (1969); J.-M. Surzur, M.-P. Bertrand, and R. Nouguier, *Tetrahedron Letters*, **1969**, 4197.

Azo-compounds and Azides

It has been concluded that acyclic aliphatic azo-compounds dissociate in the vibrationally excited $\pi–\pi^*$ triplet state, and that *cis–trans*-isomerization occurs from a twisted $\pi–\pi^*$ triplet.[295] Population of the triplet state is enhanced by collisional deactivation. However, there is a report[296] that some t-alkyl azo-compounds, on irradiation at low temperature, produce unstable yellow compounds which decompose *thermally* at low temperatures to give nitrogen and alkyl radicals. It was suggested that the unstable intermediate is the *cis*-isomer. Obviously clarification of the nature of this intermediate and of the part played by thermal processes in the decomposition of azo-compounds is required. Decompositions of other azo-compounds have been studied,[297a,b] and in one example a 1,3-prototropic shift was a competing reaction.[297a] It has

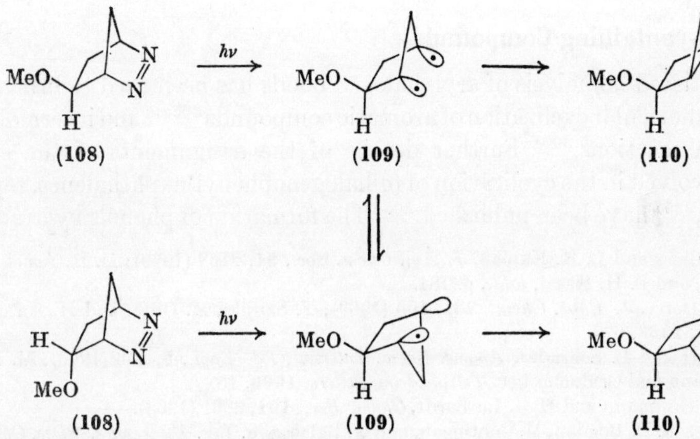

MeO $\xrightarrow{h\nu}$ MeO   MeO

H   H   H

(108)   (109)   (110)

H $\xrightarrow{h\nu}$ MeO   MeO

MeO   H   H

(108)   (109)   (110)

been reported that the ratio of the *exo*- to the *endo*-isomer of (110) is dependent upon whether decomposition of (108) occurs from its singlet or triplet state.[298] The fact that the two isomers of (108) decompose from their singlet states to give a different ratio of isomers of (110) indicates that the ring-closure occurs from the electronically excited diradicals (109). Decomposition from the triplet state gives triplet diradicals which can undergo configurational inter-conversion several times before spin-inversion and ring-closure occur. The ESR

[295] I. I. Abram, G. S. Milne, B. S. Solomon, and C. Steel, *J. Am. Chem. Soc.*, **91**, 1220 (1969).

[296] T. Mill and R. S. Stringham, *Tetrahedron Letters*, **1969**, 1853.

[297a] O. P. Strausz, R. E. Berkley, and H. E. Gunning, *Can. J. Chem.*, **47**, 3470 (1969).

[297b] M. Prochazka, O. Ryba, and D. Lim, *Coll. Czech. Chem. Comm.*, **33**, 3387 (1968); K. Chakravorty, J. M. Pearson, and M. Szwarc, *J. Phys. Chem.*, **73**, 746 (1969); D. G. L. James and R. D. Suart, *Trans. Faraday Soc.*, **65**, 175 (1969).

[298] E. L. Allred and R. L. Smith, *J. Am. Chem. Soc.*, **91**, 6766 (1969).

spectra[299a] and chemistry[299b] of a number of 1,3-diradicals, derived by photolysis of cyclic azo-compounds, have been examined. Further examples of the formation of diazo-compounds as intermediates, in the decomposition of pyrazoles[300a] and of other azo-compounds,[300b] have been reported. The photodecomposition of diazo-compounds has been used extensively as a source of carbenes.[301] 4-Diethylamino-4'-nitroazobenzene is photoreduced by alcohols[302a] and n-butylamine.[302b] On the basis of the products formed, an electron transfer mechanism was postulated for the amine reduction. The photodecomposition of azides has received particular attention.[303a–d] Decomposition can be effected by the use of either singlet[303a,b] or triplet[303c] sensitizers. Singlet sensitization occurs by collisional exchange energy transfer to the bent ground-state of the azide.[303a] The half-lives of a number of triplet nitrenes have been determined by flash photolysis studies of their parent azides.[304]

Halogen-containing Compounds

Photoinitiated homolysis of aryl–halogen bonds has been used to bring about the intramolecular cyclization of aromatic compounds[305a,b] and intermolecular arylation reactions.[305c] Further details of the assignments of the excited states involved in the cyclization of (o-halogenophenyl)naphthalenes, reported last year,[305d] have been published.[305a] The formation of phenols by irradiation

[299a] G. L. Closs and L. R. Kaplan, *J. Am. Chem. Soc.*, **91**, 2168 (1969); D. R. Arnold, A. B. Evnin, and P. H. Kasai, *ibid.*, p. 784.

[299b] F. H. Dorer, *J. Phys. Chem.*, **73**, 3109 (1969); T. Sanjiki, M. Ohta, and H. Kato, *Chem. Comm.*, **1969**, 638.

[300a] H. Dürr and L. Schrader, *Angew. Chem. Internat. Ed. Engl.*, **8**, 446 (1969); M. Franck-Neumann and C. Buchecker, *Tetrahedron Letters*, **1969**, 15.

[300b] R. W. Hoffmann and H. J. Luthardt, *Chem. Ber.*, **101**, 3861 (1968).

[301] T. Oncescu, D. Bogdan, M. Contineau, and G. Balaceanu, *Ber. Bunsenges. Phys. Chem.*, **72**, 274 (1968); J. A. Kaufman and S. J. Weininger, *Chem. Comm.*, **1969**, 593; I. Moritani, T. Hosokawa, and N. Obata, *J. Org. Chem.*, **34**, 670 (1969); T. DoMinh, O. P. Strausz, and H. E. Gunning, *J. Am. Chem. Soc.*, **91**, 1261 (1969); H. Dürr and L. Schrader, *Chem. Ber.*, **102**, 2026 (1969); N. R. Ghosh, C. R. Ghoshal, and S. Shah, *Chem. Comm.*, **1969**, 151.

[302a] G. Irick and J. G. Pacifici, *Tetrahedron Letters*, **1969**, 1303; J. G. Pacifici and G. Irick, *ibid.*, p. 2207.

[302b] J. G. Pacifici, G. Irick, and C. G. Anderson, *J. Am. Chem. Soc.*, **91**, 5654 (1969).

[303a] F. D. Lewis and J. C. Dalton, *J. Am. Chem. Soc.*, **91**, 5260 (1969).

[303b] J. S. Swenton, T. J. Ikeler, and B. H. Williams, *Chem. Comm.*, **1969**, 1263.

[303c] F. D. Lewis and W. H. Saunders, *J. Am. Chem. Soc.*, **91**, 7031, 7033 (1968).

[303d] S.-I. Yamada and S. Terashima, *Chem. Comm.*, **1969**, 511.

[304] A. Reiser, J. W. Willets, G. C. Terry, V. Williams, and R. Marley, *Trans. Faraday Soc.*, **64**, 3265 (1968).

[305a] W. A. Henderson, R. Lopresti, and A. Zweig, *J. Am. Chem. Soc.*, **91**, 6049 (1969).

[305b] W. A. Henderson and A. Zweig, *Tetrahedron Letters*, **1969**, 625; P. J. Grisdale and J. L. R. Williams, *J. Org. Chem.*, **34**, 1675 (1969).

[305c] G. I. Nikishin and M. A. Chel'tsova, *Izv. Akad. Nauk SSSR, Ser. Khim.*, **1968**, 157; *Chem. Abs.*, **69**, 66690 (1968).

[305d] *Org. Reaction Mech.*, **1968**, 459.

of 4-halogenophenols in alcoholic solution is remarkably efficient.[306] Irradiation of allyl chloride at low temperature produces allyl radicals (identified by ESR).[307] Alkyl radicals have been generated by the photolysis of α-halogeno-esters,[308a] amides,[308a] chloromethyl ketones,[308b] and iodoacetylenes.[308c] The very interesting observation has been made that benzene and chloroform form a CT complex at low temperatures which gives a hexatriene on irradiation.[309] Octatetraenes are formed by irradiation of the complex formed between benzene and tetrachloroethylene.

Miscellaneous Compounds

The formation of ethylene by irradiation of ketene probably occurs by attack of the initially formed methylene upon the ketene.[310a] Ketene does not sensitize the phosphorescence of biacetyl, and it was concluded that triplet methylene is derived from a source other than triplet ketene.[310b] N-Vinyl-carbazole gives a cyclobutane dimer when oxygenated solutions are irradiated in the presence of a triplet sensitizer.[311] The necessity for oxygen to be present led to the postulation of the superoxide as an intermediate. The nitration of carbazole occurs when the amine is irradiated in the presence of tetranitro-methane, and the formation of an intermediate excited CT complex was suggested.[312] The quantum yields of the products formed on photodissociation of isopropyl nitrite are wavelength-dependent although the quantum yield for total-product formation is not.[313] The triplet photosensitized isomerization of sulphoxides[314a] and the fragmentation of isocyanates[314b] to give carbenes have been studied. The photoinduced rearrangement of the ylid (111) to the ketone (112) and the fragmentation of other sulphur ylids[315] and phosphorus ylids[316] has been reported. The formation of ethylene and ethane on irradia-

[306] J. T. Pinhey and R. D. G. Rigby, *Tetrahedron Letters*, **1969**, 1267.
[307] R. W. Phillips and D. H. Volman, *J. Am. Chem. Soc.*, **91**, 3418 (1969).
[308a] O. Yonemitsu and S. Naruto, *Tetrahedron Letters*, **1969**, 2387.
[308b] O. Yonemitsu, H. Nakai, Y. Kanaoka, I. L. Karle, and B. Witkop, *J. Am. Chem. Soc.*, **91**, 4591 (1969).
[308c] G. Martelli, P. Spagnolo, and M. Tiecco, *Chem. Comm.*, **1969**, 282.
[309] N. C. Perrins and J. P. Simons, *Trans. Faraday Soc.*, **65**, 390 (1969).
[310a] G. B. Kistiakowsky and T. A. Walter, *J. Phys. Chem.*, **72**, 3952 (1968); H. M. Frey and R. Walsh, *Chem. Comm.*, **1969**, 158.
[310b] M. Grossman, G. P. Semeluk, and I. Unger, *Can. J. Chem.*, **47**, 3079 (1969).
[311] R. A. Carruthers, R. A. Crellin, and A. Ledwith, *Chem. Comm.*, **1969**, 252.
[312] D. H. Iles and A. Ledwith, *Chem. Comm.*, **1969**, 364.
[313] B. E. Ludwig and G. R. McMillan, *J. Am. Chem. Soc.*, **91**, 1085 (1969).
[314a] R. A. Archer and P. V. DeMarco, *J. Am. Chem. Soc.*, **91**, 1530 (1969).
[314b] J. H. Boyer and J. DeJong, *J. Am. Chem. Soc.*, **91**, 5929 (1969).
[315] R. H. Fish, L. C. Chow, and M. C. Caserio, *Tetrahedron Letters*, **1969**, 1259.
[316] Y. Nagao, K. Shima, and H. Sakurai, *J. Pharm. Soc. Japan*, **72**, 236 (1969); *Chem. Abs.*, **70**, 114372 (1969).

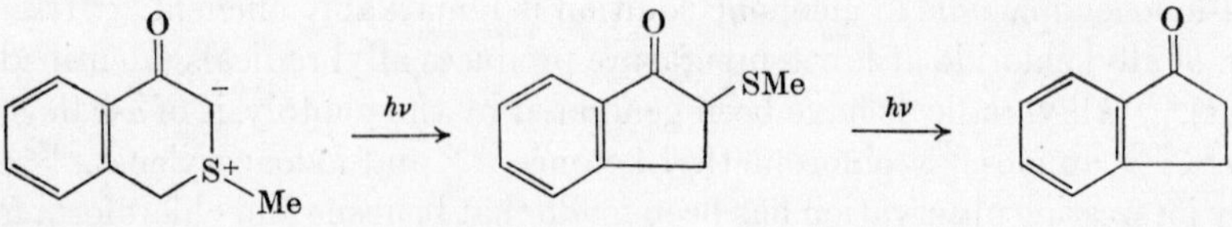

(111)   (112)

tion of ethyl-lithium is an internal process and is suggested to occur through intra-aggregate disproportionations.[317] From labelling studies it has been concluded that some of the biphenyl formed on irradiation of pentaphenyl-antimony in benzene occurs by combination of phenyl radicals derived from the antimony compound.[318a] Such a reaction in benzene solution is most unlikely and the biphenyl may well be produced by an intramolecular process.[318b] Irradiation of lead tetracarboxylates in the region of their CT band gives alkyl radicals (identified by ESR).[319a] A study of the ease of decarboxylation of a number of lead tetracarboxylates has shown that radical formation is the result of a multi-bond cleavage, and does not involve intermediate car-

$$RCO_2Pb(O_2CR)_3 \xrightarrow{h\nu} R\cdot + CO_2 + \cdot Pb(O_2CR)_3$$

boxylate radicals.[319b] ESR studies have indicated that ceric carboxylates decarboxylate by a similar mechanism.[319c]

Other Photoreactions

Photosensitized Oxidation

Two reviews,[320a,b] one being comprehensive,[320a] on photosensitized oxidations have been published. From molecular orbital and state correlation diagrams, selection rules for the reaction of ground state oxygen, singlet oxygen, and excited oxygen with olefins have been derived.[321a,b] Addition of singlet oxygen to mono-olefins is allowed only when the olefin has a low π-ionization potential. The recent finding that singlet oxygen adds to indene to give a

[317] W. H. Glaze and T. L. Brewer, *J. Am. Chem. Soc.*, **91**, 4490 (1969).

[318a] Kei-wen Shen, W. E. McEwen, and A. P. Wolf, *J. Am. Chem. Soc.*, **91**, 1283 (1969).

[318b] P. J. Grisdale, B. E. Babb, J. C. Doty, T. H. Regan, D. P. Maier, and J. L. R. Williams, *J. Organometal. Chem.*, **14**, 63 (1968).

[319a] K. Heusler and H. Loeliger, *Helv. Chim. Acta*, **52**, 1237 (1969).

[319b] J. K. Kochi, R. A. Sheldon, and S. S. Lande, *Tetrahedron*, **25**, 1197 (1969).

[319c] D. Greatorex and T. J. Kemp, *Chem. Comm.*, **1969**, 383.

[320a] F. Gollnick, *Adv. Photochem.*, **6**, 2 (1968).

[320b] C. S. Foote, *Science*, **162**, 963 (1968).

[321a] D. R. Kearns and A. U. Khan, *Photochem. Photobiol.*, **10**, 193 (1969).

[321b] D. R. Kearns, *J. Am. Chem. Soc.*, **91**, 6554 (1969).

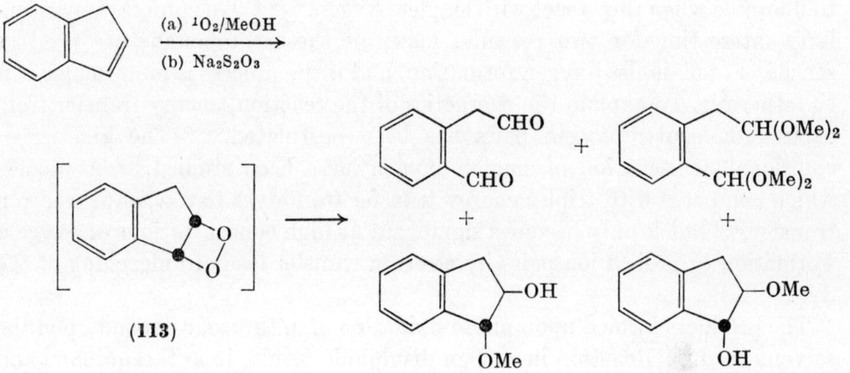

(113)

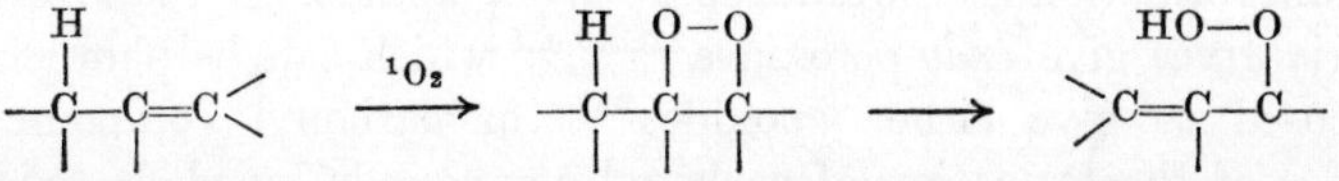

(114)

dioxetan (113)[322a] has led to the suggestion that dioxetans may be the precursors of α-hydroperoxy-olefins, e.g. (114),[321b, 322a] as well as carbonyl compounds.[322a,b] Spectral identification of singlet oxygen in a number of systems (e.g. direct irradiation of oxygen,[323a] energy transfer from triplet molecules[323b]) has been reported. Formation of singlet oxygen by energy transfer from aromatic triplets has been studied,[324a,b] and some compounds are inefficient sensitizers although their triplet states have sufficient energy.[324b] The mechanism of energy transfer from excited singlet-state molecules to oxygen still remains an enigma. A straight energy-transfer mechanism, resulting in the triplet state of the sensitizer being formed, has been ruled out[325] and there is no conclusive evidence for transfer via an excited complex.[326a] Aromatic hydrocarbons form CT complexes with oxygen, and irradiation in the CT band leads to photo-oxidation.[326b] Oxygen also quenches singlet-state excimers.[327] A number of compounds (e.g. Methylene Blue) have been found

[322a] W. Fennical, D. R. Kearns, and P. Radlick, *J. Am. Chem. Soc.*, **91**, 3396 (1969).
[322b] G. Rio and J. Berthelot, *Bull. Soc. Chim. France*, **1969**, 3609.
[323a] D. F. Evans, *Chem. Comm.*, **1969**, 367.
[323b] D. R. Kearns, A. U. Khan, C. K. Duncan, and A. H. Maki, *J. Am. Chem. Soc.*, **91**, 1039 (1969); E. Wasserman, V. J. Kuck, W. M. Delavan, and W. A. Yager, *ibid.*, p. 1040.
[324a] B. Stevens and B. E. Algar, *J. Phys. Chem.*, **73**, 1711 (1969); R. W. Chambers and D. R. Kearns, *Photochem. Photobiol.*, **10**, 215 (1969).
[324b] L. I. Crossweiner, *Photochem. Photobiol.*, **10**, 183 (1969).
[325] C. S. Parmenter and J. D. Rau, *J. Chem. Phys.*, **51**, 2242 (1969).
[326a] N. Kulevsky, C. T. Wang, and V. I. Stenberg, *J. Org. Chem.*, **34**, 1345 (1969).
[326b] K. S. Wei and A. H. Adelman, *Tetrahedron Letters*, **1969**, 3297.
[327] I. B. Berlman, C. R. Goldschmidt, G. Stein, Y. Tomkiewicz, and A. Weinrer, *Chem. Phys. Letters*, **4**, 338 (1969); J. Yguerabide, *J. Chem. Phys.*, **49**, 1018, 1026 (1968).

to fluoresce when they react with singlet oxygen.[328a,b] This process is particularly interesting for two reasons: many of these compounds are used as sensitizers for singlet oxygen formation, and if the process is bimolecular it is endothermic. To explain the energetics of the reaction, energy transfer from excited molecular oxygen pairs has been postulated.[328a] The kinetics of collisional deactivation of singlet oxygen have been studied.[329] A process which competes with triplet energy transfer from dyes to oxygen is electron transfer,[330] but it only becomes significant at high concentrations of oxygen. Formation of radical ion pairs by electron transfer leads to bleaching of the dye.

The products formed upon photo-oxidation of anthracene depend upon the solvent used.[331] Reaction in carbon disulphide results in anthraquinone and bianthrone formation via the anthracene *endo*-peroxide. 1,4-Dialkoxybenzenes and derivatives give *endo*-peroxides [332a,b,c,d] which can be photochemically decomposed to give either epoxides [332a] or carbonyl compounds.[332a,c,d] Examples of the formation of carbonyl compounds by photo-oxidation of dienes [333a] and of hydroperoxides from olefins have been reported.[333b] A number of acyclic conjugated dienes have been found to give cyclic peroxides.[334a,b] Several carotenoid-type compounds fall into this class.[334b] A number of these compounds also yield cumulene hydroperoxides, and it has been remarked that there is a close similarity between the products of photo-oxygenation

328a J. Stauff and H. Fuhr, *Ber. Bunsenges. Phys. Chem.*, **73**, 245 (1969).
328b J. Canva, C. Balny, P. Douzou, and J. Bourdon, *Compt. Rend.*, *C*, **268**, 1027 (1969).
329 T. P. J. Izod and R. P. Wayne, *Proc. Roy. Soc.*, *A*, **308**, 81 (1968); R. P. Steer, R. A. Ackerman and J. N. Pitts, *J. Chem. Phys.*, **51**, 843 (1969); F. D. Findlay, C. J. Fortin, and D. R. Snelling, *Chem. Phys. Letters*, **3**, 204 (1969).
330 M. Koizumi and Y. Usui, *Tetrahedron Letters*, **1968**, 6011; T. Usui, D. Iwanaga, and M. Koizumi, *Bull. Chem. Soc. Japan*, **42**, 1231 (1969).
331 N. Sugiyama, M. Iwata, M. Yoshioka, K. Yamada, and H. Aoyama, *Chem. Comm.*, **1968**, 1563; N. Sugiyama, M. Iwata, M. Yoshioka, K. Yamada, and H. Aoyama, *Bull. Chem. Soc. Japan*, **42**, 1377 (1969).
332a J. Rigaudy, C. Deletang, D. Sparfel, and N. K. Cuong, *Compt. Rend.*, *C*, **267**, 1714 (1968).
332b J. Rigaudy, C. Deletang, and J.-J. Basselier, *Compt. Rend.*, *C*, **268**, 344 (1969).
332c J. Rigaudy, R. Dupont, and N. K. Cuong, *Compt. Rend.*, *C*, **269**, 416 (1969).
332d T. A. Moore and P.-S. Song, *Photochem. Photobiol.*, **10**, 13 (1969).
333a M. Mousseron-Canet and J.-P. Chabaud, *Bull. Soc. Chim. France*, **1969**, 245.
333b S. Itô, H. Takeshita, T. Muroi, M. Itô, and K. Abe, *Tetrahedron Letters*, **1969**, 3091; H. Takeshita, T. Sato, T. Muroi, and S. Itô, *ibid.*, p. 3095; N. Furutachi, Y. Nakadaira, and K. Nakamishi, *Chem. Comm.*, **1968**, 1625; C. D. Snyder and H. Rapoport, *J. Am. Chem. Soc.*, **91**, 731 (1969); E. K. von Gustorf, F. W. Greuels, and G. O. Schenck, *Ann. Chem.*, **719**, 1 (1968); S. Isoe, S. B. Hyeon, H. Ichikawa, S. Katsumura, and T. Sakan, *Tetrahedron Letters*, **1968**, 5561; S. Isoe, S. B. Hyeon, and T. Sakan, *ibid.*, **1969**, 279.
334a G. Bio and J. Berthelot, *Bull. Soc. Chim. France*, **1969**, 1664.
334b M. Mousseron-Canet, J.-P. Dalle, and J.-C. Mani, *Tetrahedron Letters*, **1968**, 6037; J.-P. Dalle, M. Mousseron-Canet, and J.-C. Mani, *Bull. Soc. Chim. France*, **1969**, 232; J.-L. Olive and M. Mousseron-Canet, *ibid.*, p. 3252; C. S. Foote and M. Brenner, *Tetrahedron Letters*, **1968**, 6041.

and those of biosynthetic oxygenation. Photosensitized oxygenation of pyrroles,[335a] purines,[335b] and tropones [335c] has been studied.

The products obtained by photo-oxidation of phenols have been interpreted as occurring by initial phenoxy radical formation.[336] The photo-oxidation of alcohols,[337a] ketones,[337b] amides,[337c] amino-acids,[337d] benzodiazepines,[337e] benzoxazin-4-ones,[337f] and pyrrolines [337g] has been studied, and alkyl radicals have been suggested as intermediates. Azobenzene is produced by the benzophenone-sensitized oxidation of aniline.[338]

Chemiluminescence

A comprehensive review of organic chemiluminescent reactions [339a] and a more selective review on concerted peroxide decompositions [339b] have been published. Singlet oxygen reacts with 10,10'-dimethyl-9,9'-biacridylidene to give N-methylacridone with the emission of light.[340] Formation of an intermediate dioxetan by the addition of singlet oxygen to the 9,9'-double bond was postulated. Thermal decomposition of the dioxetan results in the formation of excited and ground-state N-methylacridone. Decomposition of dioxetans to give an excited and a ground-state carbonyl compound has been predicted from orbital symmetry considerations.[321b, 341] Other dioxetans have been found to decompose with the emission of light.[342] When compounds with triplet energies lower than the excited ketone are present, energy transfer occurs.[343] Thus, decomposition of tetramethyloxetan at 100° in the presence of *trans*-stilbene gives *cis*-stilbene. An even more remarkable example is that of the efficient sensitization of the rearrangement of 4,4'-diphenylcyclohexa-2,5-dienone. Dioxetan formation has been postulated in the chemiluminescent

335a H. H. Wasserman and A. H. Miller, *Chem. Comm.*, **1969**, 199; G. Rio, A. Ranjon, O. Pouchot, and M.-J. Scholle, *Bull. Soc. Chim. France*, **1969**, 1667.

335b T. Matsuura and I. Saito, *Tetrahedron*, **25**, 541, 549, 557 (1969).

335c M. Oda and Y. Kitahara, *Angew. Chem. Internat. Ed. Engl.*, **8**, 673 (1969).

336 T. Matsuura, N. Yoshimura, A. Nishinaga, and I. Saito, *Tetrahedron Letters*, **1969**, 1669; T. Matsuura, A. Nishinaga, N. Yoshimura, T. Arai, K. Omura, H. Matsushima, S. Kato, and I. Saito, *ibid.*, p. 1673.

337a S. M. Eremenko, M. S. Ashinkinazi, and B. Y. Dain, *Ukr. Khim. Zh.*, **34**, 694 (1968); *Chem. Abs.*, **69**, 95694 (1968).

337b A. S. Kallend and J. N. Pitts, *J. Am. Chem. Soc.*, **91**, 1269 (1969).

337c B. Lanska and J. Šebenda, *Coll. Czech. Chem. Comm.*, **34**, 1911 (1969).

337d B. Monties, *Compt. Rend.*, *C*, **269**, 1069 (1969).

337e T. Yonezawa, M. Matsumoto, and H. Kato, *Bull. Chem. Soc. Japan*, **41**, 2543 (1968).

337f M. A. Hems, *Tetrahedron Letters*, **1969**, 375.

337g M. Kawana and S. Emoto, *Bull. Chem. Soc. Japan*, **41**, 2552 (1968).

338 M. Santhanam and V. Ramakrishnan, *Indian J. Chem.*, **6**, 88 (1968).

339a K.-D. Gundermann, *Naturwiss.*, **56**, 62 (1969).

339b M. M. Rauhut, *Accounts Chem. Res.*, **2**, 80 (1969).

340 F. McCapra and R. A. Hann, *Chem. Comm.*, **1969**, 442.

341 F. McCapra, *Chem. Comm.*, **1968**, 155.

342 K. R. Kopecky and C. Mumford, *Can. J. Chem.*, **47**, 709 (1969).

343 E. H. White, J. Wiecko, and D. F. Roswell, *J. Am. Chem. Soc.*, **91**, 5194 (1969).

reaction of Schiff bases with potassium t-butoxide,[344a] and of oxalate amides with hydrogen peroxide.[344b] Ozonization of some compounds (e.g. phenols) is a chemiluminescent reaction and the mechanism is not understood.[345a] The autoxidation of ethylbenzene[345b] and other hydrocarbons[345c] is also accompanied by light emission. Luminol reacts with singlet oxygen to give an *endo*-peroxide which decomposes in chemiluminescent reaction.[346] Further work on the luminol,[347a] luciferin,[347b] and cypridina[347c] chemiluminescent systems has been reported.

Electrochemiluminescence (ECL)

Electron transfer between radical ions to give neutral molecules is often chemiluminescent. Formation of the radical cation and anion in each other's presence can be accomplished electrochemically. In this way the radical cation of N,N,N',N'-tetramethyl-p-phenylenediamine and the radical anion of anthracene were generated and found to combine with the emission of light whose spectrum was identical with that of the fluorescence of anthracene.[348] Electron transfer cannot lead directly to a singlet since this is an endothermic process and therefore the formation of singlets is presumably by triplet–triplet annihilation. The ECL of carbazole has been examined but the emitting species could not be identified.[349] By means of ECL the half-lives of a number of aromatic hydrocarbon radical cations have been determined.[350] ECL reactions occurring via radical ions have been reviewed.[351]

It had been predicted that the process of radical ion combination should

[344a] F. McCapra and R. Wrigglesworth, *Chem. Comm.*, **1969**, 91.

[344b] L. J. Bollyky, B. G. Roberts, R. H. Whitman, and J. E. Lancaster, *J. Org. Chem.*, **34**, 836 (1969).

[345a] R. Iwaki and I. Kamiya, *Bull. Chem. Soc. Japan*, **42**, 855 (1969); I. Kamiya and R. Iwaki, *J. Chem. Soc. Japan, Ind. Chem. Sect.*, **72**, 85 (1969); *Chem. Abs.*, **71**, 49035 (1969).

[345b] M. Höfert, *Photochem. Photobiol.*, **9**, 427 (1969).

[345c] R. E. Kellog, *J. Am. Chem. Soc.*, **91**, 5433 (1969).

[346] K. Kuschnir and T. Kuwana, *Chem. Comm.*, **1969**, 193.

[347a] D. S. Bersis and J. Nikokavouras, *Z. Phys. Chem.* (Frankfurt), **62**, 152 (1968); E. H. White, D. F. Roswell, and O. C. Zafiriou, *J. Org. Chem.*, **34**, 2462 (1969); Y. Omote, H. Yamamoto, S. Tomioka, and N. Sugiyama, *Bull. Chem. Soc. Japan*, **42**, 2090 (1969); K.-D. Gundermann and D. Schedlitzki, *Chem. Ber.*, **102**, 3241 (1969); M. J. Cormier and P. M. Prichard, *J. Biol. Chem.*, **243**, 4706 (1968); T. Förster and K. Rokos, *Z. Phys. Chem.* (Frankfurt), **63**, 208 (1969).

[347b] R. A. Morton, T. A. Hopkins, and H. H. Seliger, *Biochemistry*, **8**, 1598 (1969); G. Mitchell and J. W. Hastings, *J. Biol. Chem.*, **244**, 2572 (1969); M. G. Fracheboud, O. Shimomura, R. K. Hill, and F. H. Johnson, *Tetrahedron Letters*, **1969**, 3951; W. D. McElroy, H. H. Seliger, and E. H. White, *Photochem. Photobiol.*, **10**, 153 (1969).

[347c] T. Goto and H. Fukatsu, *Tetrahedron Letters*, **1969**, 4299.

[348] L. R. Faulkner and A. J. Bard, *J. Am. Chem. Soc.*, **91**, 209 (1969).

[349] K. S. V. Santhanam, R. N. O'Brien, and A. D. Kirk, *Can. J. Chem.*, **47**, 1355 (1969).

[350] S. A. Cruser and A. J. Bard, *J. Am. Chem. Soc.*, **91**, 267 (1969).

[351] A. Zweig, *Adv. Photochem.*, **6**, 425 (1968); D. M. Hercules, *Accounts. Chem. Res.*, **2**, 301 (1969).

be affected by application of a magnetic field,[352] and this has been shown to be the case.[353] Application of this effect has shown that radicals quench triplet states.[354]

The ECL of 10,10'-dimethyl-9,9'-biacridinium nitrate has been examined and 10,10'-dimethyl-9,9'-biacridylidene proposed as an intermediate.[355] Reaction of this compound with the superoxide ion was suggested as being the chemiluminescent reaction. This finding is particularly interesting in the light of the reported reaction of singlet oxygen with this compound.[340]

Solvolysis and Substitution Reactions

The solvolysis of 3,4-dihydrocoumarone has been shown to involve a ketene intermediate[356a] and not a spirocyclobutanone.[356b] It has been reported that cleavage of sugar tosylates to sugars[357a] and of barbitals[357b] by sodium methoxide, and the deboronation of ferroceneboronic acid,[357c] are all accelerated on irradiation. The photohydrolysis of several substituted phenylphosphonates in neutral and weakly basic solution was shown to be by attack of a water molecule on the phosphorus atom.[358] In strongly alkaline solution attack of hydroxyl ions upon the phenyl ring occurs. A number of *m*-halogenophenols undergo photoinduced nucleophilic substitution in alcoholic solution.[359] An important competing reaction is homolysis of the carbon–halogen bond. Photosolvolysis of 4-nitroanisole by hydroxyl[360a] and cyanide ion[360b,c] and methanol[361] has been studied, and in the former reaction participation of the triplet nitro-compound was claimed. Rearrangement of the 4-nitroanisole to 2-nitroso-4-methoxyanisole competes with the substitution reactions.[360b,361] The effect of solvent upon the photosolvolysis of nitro-compounds has been interpreted in terms of the solvent affecting the energy levels of the various excited states.[360c]

[352] B. Brocklehurst, *Nature*, **221**, 921 (1969).
[353] L. R. Faulkner and A. J. Bard, *J. Am. Chem. Soc.*, **91**, 6495 (1969).
[354] L. R. Faulkner and A. J. Bard, *J. Am. Chem. Soc.*, **91**, 6497 (1969).
[355] K. D. Legg and D. M. Hercules, *J. Am. Chem. Soc.*, **91**, 1902 (1969).
[356a] O. L. Chapman and C. L. McIntosh, *J. Am. Chem. Soc.*, **91**, 4309 (1969).
[356b] *Org. Reaction Mech.*, **1968**, 464.
[357a] S. Zen, S. Tashima, and S. Kotō, *Bull. Chem. Soc. Japan*, **41**, 3025 (1968).
[357b] Y. Otsuji, T. Kuroda, and E. Imoto, *Bull. Chem. Soc. Japan*, **41**, 2713 (1968).
[357c] H. C. H. A. van Riel, F. C. Fischer, J. Lugtenburg, and E. Havinga, *Tetrahedron Letters*, **1969**, 3085.
[358] R. O. de Jongh and E. Havinga, *Rec. Trav. Chim.*, **87**, 1318, 1327 (1968).
[359] J. T. Pinhey and R. D. G. Rigby, *Tetrahedron Letters*, **1969**, 1267, 1271.
[360a] R. L. Letsinger and K. E. Steller, *Tetrahedron Letters*, **1969**, 1401.
[360b] R. L. Letsinger and J. H. McCain, *J. Am. Chem. Soc.*, **91**, 6425 (1969).
[360c] R. L. Letsinger and R. R. Hautala, *Tetrahedron Letters*, **1969**, 4205.
[361] L. B. Jones, J. C. Kudrna, and J. P. Foster, *Tetrahedron Letters*, **1969**, 3263.

Photochromism

Some substituted cyclohexadienes undergo photocycloreversion to give coloured hexatrienes which thermally recyclize to the cyclohexadienes.[361] The photocycloreversion of 1,2-dihydroquinolines[363] and the cyclization of 1-aryl-2-nitroalkenes[364] are two very interesting photochromic systems. The coloured products formed on irradiation of bianthrones have been shown to be substituted dihydrophenanthrenes,[365] and those from diarylidene and tetra-arylidenesuccinimides to be substituted dihydronaphthalenes.[366] Formation of radical ions has been shown to be responsible for the photochromic behaviour of phenothiazines[367] and bi-imidazolyls.[368] The photochromism of anils,[369] indolinospirans,[370] and related compounds[371] has been further investigated.

[362] K. R. Huffman, H. Burger, W. A. Henderson, M. Loy, and E. E. Ullman, *J. Org. Chem.*, **34**, 2407 (1969).

[363] J. Kolc and R. S. Becker, *J. Am. Chem. Soc.*, **91**, 6513 (1969).

[364] J. A. Sousa, J. Weinstein, and A. L. Bluhm, *J. Org. Chem.*, **34**, 3320 (1969).

[365] R. Lorenz, V. Wild, and J. R. Huber, *Photochem. Photobiol.*, **10**, 233 (1969); L. J. Dombrowski, C. L. Groncki, R. L. Strong, and H. H. Richtol, *J. Phys. Chem.*, **73**, 3481 (1969); T. Bercovici and E. Fischer, *Israel J. Chem.*, **7**, 127 (1969).

[366] R. J. Hart, H. G. Heller, and K. Salisbury, *Chem. Comm.*, **1968**, 1627.

[367] R. Knoesel, B. Gebus, and J. Parrod, *Compt. Rend.*, *C*, **268**, 727 (1969).

[368] M. A. J. Wilks and M. R. Willis, *J. Chem. Soc.*(B), **1968**, 1526.

[369] W. F. Richey and R. S. Becker, *J. Chem. Phys.*, **49**, 2092 (1968).

[370] T. Bercovici, R. Heiligman-Rim, and E. Fischer, *Mol. Photochem.*, **1**, 23 (1969); I. Shimazu, H. Kokado, and E. Inoue, *Bull. Chem. Soc. Japan*, **42**, 1726, 1730 (1969).

[371] P. H. Vandewyer, J. Hoefnagels, and G. Smets, *Tetrahedron*, **25**, 3251 (1969).

CHAPTER 14

Oxidation and Reduction

M. J. P. HARGER

Department of Chemistry, The University, Leicester

Ozonation and Ozonolysis

There has been considerable support for the proposal by Bailey and his colleagues[1] that conversion of primary ozonide into normal ozonide proceeds by a modified Criegee fragmentation mechanism. Although the *cis:trans* ratio of the stilbene cross-ozonide obtained from each of a series of phenylethylenes ($PhCH=C<$) in CCl_4 at 20° depends on olefin structure and geometry, ozonide labelled exclusively at the ether oxygen is obtained in the presence of $[^{18}O]$benzaldehyde.[2] Even at $-78°$, when reaction of the primary ozonide with aldehyde might be expected to compete more efficiently with its fragmentation, the product contains no ^{18}O in the peroxide bridge,[2] and cannot therefore have been formed according to Story, Murray, and Youssefyeh's primary ozonide aldehyde hypothesis.[3] Alternative modes of reaction between primary ozonide and labelled aldehyde could conceivably lead to normal ozonide labelled at the ether oxygen, as shown in equation (1), but since the *cis:trans* ratio of the stilbene ozonide from each phenylethylene is the same in the presence of added benzaldehyde as it is in its absence, this seems unlikely.[2] Further, the disappearance of primary ozonide (in the absence of added aldehyde) is a first-order process and cannot therefore involve reaction with aldehyde in the rate-determining step.[4] If primary ozonide–aldehyde interactions are unimportant, an alternative theory, possibly similar to that outlined in equation (2), must account for the formation of peroxide-oxygen

[1] *Org. Reaction Mech.*, **1968**, 466—7.
[2] S. Fliszár and J. Carles, *J. Am. Chem. Soc.*, **91**, 2637 (1969).
[3] *Org. Reaction Mech.*, **1966**, 399—401.
[4] F. L. Greenwood and L. J. Durham, *J. Org. Chem.*, **34**, 3363 (1969).

labelled ozonide (**4**) from *trans*-diisopropylethylene and [^{18}O]acetaldehyde. Fliszár and Carles[5] reason that electron release by the isopropyl group could render the zwitterionic C atom in (**1**) less prone to nucleophilic attack, thereby

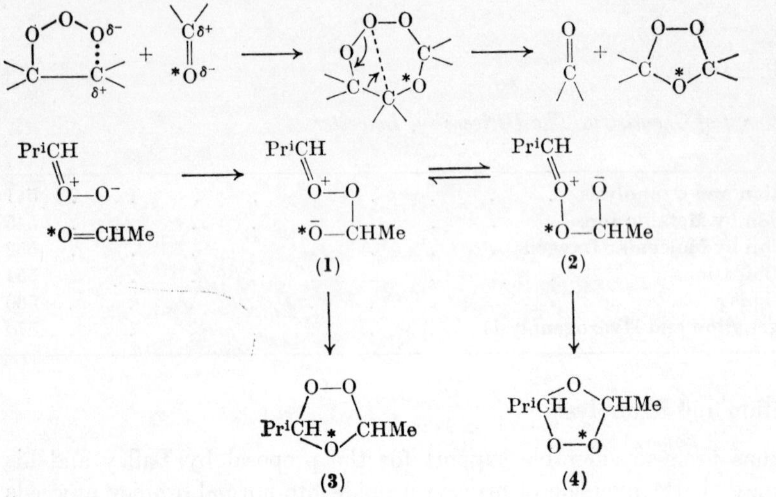

allowing rearrangement of (**1**) to (**2**) to compete with direct formation of ozonide (**3**). These same authors[5] have considered a possible stabilization of the *syn*- and *anti*-configurations of the initial zwitterion, by interaction with unchanged olefin, to account for the dependence of the ozonide *cis*:*trans* ratio (in CCl$_4$) on the initial concentration of olefin, and on olefin geometry even at 25°, when interconversion of zwitterions (*syn*$\rightleftharpoons$*anti*) might otherwise be expected to occur very rapidly.

Experiments with alk-1-enes (H$_2$C=CHR′) have emphasized the important influence of steric factors in both the aldehyde (RCHO) and the zwitterion (or primary ozonide) on the stereochemistry of the resulting ozonide (**5**).[6] They have also shown that a stereospecific concerted reaction of primary ozonide cannot be the cause of the high yield of uncrossed normal ozonide (**6**) obtained from alk-1-enes; *cis*- and *trans*-1-deuterohex-1-enes both give 1:1 mixtures of *cis*- and *trans*-ozonide (**7**).[7] The failure of the ozonide (**8**) from methyl dodec-11-enoate to equilibrate with ^{14}C-labelled methyl 10-formyldecanoate in various solvents at room temperature implies that, in this case at least, ozonide formation is irreversible.[8]

[5] S. Fliszár and J. Carles, *Can. J. Chem.*, **47**, 3921 (1969).
[6] R. W. Murray and G. J. Williams, *J. Org. Chem.*, **34**, 1891 (1969).
[7] R. W. Murray and G. J. Williams, *J. Org. Chem.*, **34**, 1896 (1969).
[8] D. R. Kerur and D. G. M. Diaper, *Can. J. Chem.*, **47**, 43 (1969).

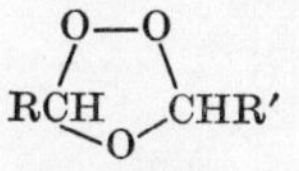

(5) 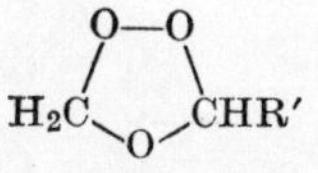   (6) 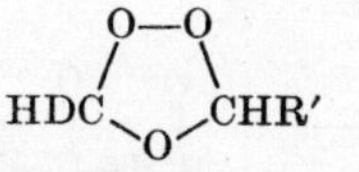   (7) 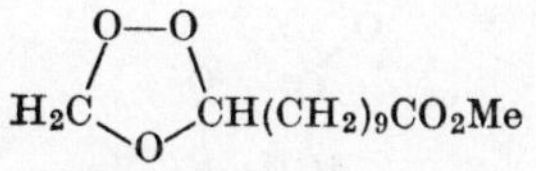   (8)

Much new information is available concerning the direction of fragmentation of an initial ozonide in methanol, where the zwitterion is trapped as the α-methoxy-hydroperoxide. A substituent X in the initial ozonide (**10**) influences the relative extent of cleavage by path a and by path b primarily through the inductive effect of the aryl group in transition states resembling (**9**) and (**11**). Thus, for each of a number of groups R (CO_2Me, CH_2OH, H, Ph, CO_2H, Me, COMe) the proportion of b-cleavage increases as X changes p-$NO_2 \rightarrow m$-$NO_2 \rightarrow m$-Cl$\rightarrow p$-Cl$\rightarrow$H$\rightarrow p$-Me$\rightarrow p$-OMe. Moreover, for any given group R, if α represents the fraction of cleavage by path a, $\log[\alpha/(1-\alpha)]$ is linearly related to the substituent constant σ_x.[9] The results in Table 1 clearly indicate that

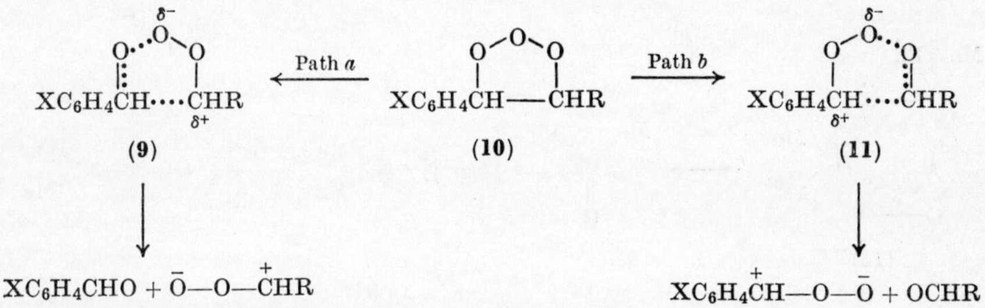

(9)   (10)   (11)

$$XC_6H_4CHO + \bar{O}-O-\overset{+}{C}HR \qquad\qquad XC_6H_4\overset{+}{C}H-O-\bar{O} + OCHR$$

the influence of the group R on the choice of reaction path is not simply inductive. In particular, the remarkable preference for a-cleavage when R = COMe has been attributed to the stabilization of the resulting zwitterion, by resonance contributions from structures (**12**)—(**14**), being reflected in the transition state for path a.[9] Predictably, the direction in which 1,2-dialkylethylene ozonides fragment depends only on the relative inductive effects of the alkyl groups.[10]

Table 1. Cleavage of the initial ozonide from PhCH=CHR in MeOH at 25°

Substituent R	CO_2Me	CH_2OH	H	Ph	CO_2H	Me	COMe
Yield (%) of PhCHO	23	26	41	50	61	82	90

[9] S. Fliszár and M. Granger, *J. Am. Chem. Soc.*, **91**, 3330 (1969); but see also W. P. Keaveney and J. J. Pappas, *Tetrahedron Letters*, **1969**, 841.
[10] S. D. Razumovskii and Y. N. Yur'ev, *Zh. Org. Khim.*, **4**, 1716 (1968); *Chem. Abs.*, **70**, 19488 (1969).

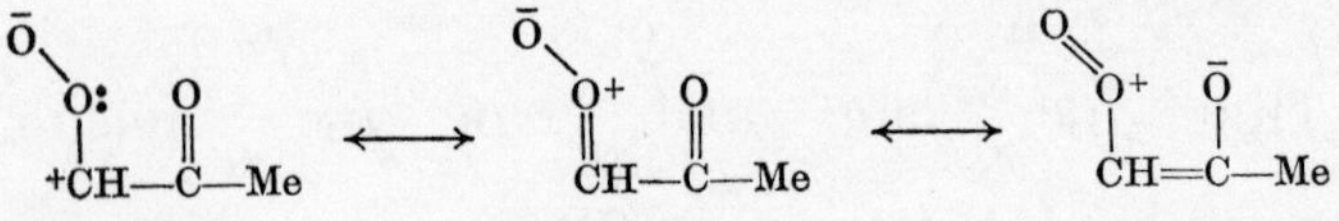

$$
\begin{array}{ccc}
\text{(12)} & \text{(13)} & \text{(14)}
\end{array}
$$

(12) $\overset{\bar{O}}{\underset{}{}}$ $\overset{O:}{\underset{}{}}$ $\overset{O}{\underset{}{}}$ $^{+}CH\!-\!C\!-\!Me$ $\longleftrightarrow$ (13) $\overset{\bar{O}}{\underset{}{}}$ $\overset{O^{+}}{\underset{}{}}$ $\overset{O}{\underset{}{}}$ $CH\!-\!C\!-\!Me$ $\longleftrightarrow$ (14) $\overset{O}{\underset{}{}}$ $\overset{O^{+}}{\underset{}{}}$ $\overset{\bar{O}}{\underset{}{}}$ $CH\!=\!C\!-\!Me$

Ozonolysis of C=N bonds continues to attract attention. Acetophenone dimethylhydrazone gives acetophenone (97%) and *N*-nitrosodimethylamine (65%) in CH_2Cl_2 at $-78°$. While substituents in the aromatic ring have little influence on the rate of reaction, both *O*-methyl-oximes (>C=NOMe) and Schiff bases (>C=NR) are appreciably less reactive than the corresponding dimethylhydrazone. A possible mechanism includes electrophilic attack by ozone on the C=N bond followed by concerted collapse of an intermediate such as (15).[11] Other reports have considered ozonation in the gas phase,[12a]

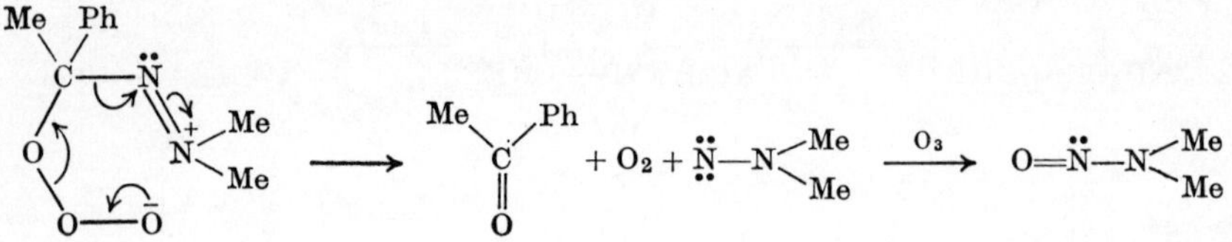

(15)

ozonolysis of tetrahydrochromans in methanol,[12b] and decomposition of normal ozonides by thermal[13] and photochemical[14] means, by electron impact,[15] and by reaction with triphenylphosphine.[16] Experimental results so far available do not reveal whether the singlet oxygen produced by thermal decomposition of a phosphite ester–ozone adduct is $^1\Delta$ or $^1\Sigma$,[17] but theoretical considerations have caused Kearns[18] to conclude that both $^3\Sigma$ and $^1\Sigma$ will be unreactive in concerted cycloaddition to dienes.

[11] R. C. Erickson, P. J. Andrulis, J. C. Collins, M. L. Lungle, and G. D. Mercer, *J. Org. Chem.*, **34**, 2961 (1969).

[12a] W. B. DeMore, *Internat. J. Chem. Kinetics*, **1**, 209 (1969); *Chem. Abs.*, **71**, 49127 (1969).

[12b] I. J. Borowitz and R. D. Rapp, *J. Org. Chem.*, **34**, 1370 (1969).

[13] Y. N. Yur'ev, S. D. Razumovskii, and V. K. Tsyskovskii, *Zh. Org. Khim.*, **4**, 1720 (1968); *Chem. Abs.*, **70**, 19405 (1969).

[14] P. R. Story, W. H. Morrison, and J. M. Butler, *J. Am. Chem. Soc.*, **91**, 2398 (1969).

[15] J. Castonguay, M. Bertrand, J. Carles, S. Fliszár, and Y. Rousseau, *Can. J. Chem.*, **47**, 919 (1969).

[16] J. Carles and S. Fliszár, *Can. J. Chem.*, **47**, 1113 (1969).

[17] R. W. Murray and M. L. Kaplan, *J. Am. Chem. Soc.*, **91**, 5358 (1969).

[18] D. R. Kearns, *J. Am. Chem. Soc.*, **91**, 6554 (1969).

Oxidation by Metallic Ions[19]

Similar rates of oxidation of cyclopentene, cyclohexene, and norbornene by chromic acid in acetic acid suggest little overall relief of strain in the transition states, which probably contain three-membered rings as in (16).[20] With chromyl chloride, however, the rate for norbornene is 300 times that for cyclohexene, and in this case strain may be relieved on formation of a transition state such as (17).[21] The oxidation of styrenes[22] and phenylcyclohexane[23] by chromyl chloride, of olefins by CrO_3–pyridine complexes,[24] and of cyclopropanes,[25] including tricyclenes,[26] by chromic acid, has been investigated.

(16)   (17)   (18)

Wiberg and Schäfer[27] have published details of their spectroscopic investigation of the Cr(VI) oxidation of propan-2-ol in 97% acetic acid. Decomposition of either the monoester ($Me_2CHOCrO_2OH$) or diester ($Me_2CHOCrO_2OCHMe_2$) formed in pre-oxidation equilibria produces acetone and Cr(IV). The latter is oxidized to Cr(V) which reacts with more alcohol. Oxidation of 2-deuteropropan-2-ol by Cr(V) in 97% acetic acid exhibits a larger isotope effect ($k_H/k_D = 9.1$) than that for Cr(VI) oxidation ($k_H/k_D = 6.2$), in contrast to the situation in aqueous solution where the larger isotope effect is associated with Cr(VI).

Aryldiphenylcarbinols ($XC_6H_4CPh_2OH$) are cleaved by Cr(VI) in aqueous acetic acid to mixtures of ketones ($XC_6H_4COPh + PhCOPh$) and phenols ($XC_6H_4OH + PhOH$). The reduced tendency for an aryl group with electron-withdrawing substituents to appear in the phenolic component is consistent with a migration transition state such as (18). The migratory aptitude of XC_6H_4- correlates with σ_x^+ ($\rho = -1.44$), as does the overall rate of oxidation ($\rho = -0.88$).[28]

[19] For convenience, lead tetraacetate and similar reagents will be discussed in this section.
[20] A. K. Awasthy and J. Roček, *J. Am. Chem. Soc.*, **91**, 991 (1969).
[21] F. Freeman and N. J. Yamachika, *Tetrahedron Letters*, **1969**, 3615.
[22] F. Freeman, R. H. DuBois, and N. J. Yamachika, *Tetrahedron*, **25**, 3441 (1969).
[23] I. Schiketanz, I. Necşoiu, M. Renţea, and C. D. Nenitzescu, *Rev. Roum. Chim.*, **13**, 1385 (1968).
[24] W. G. Dauben, M. Lorber, and D. S. Fullerton, *J. Org. Chem.*, **34**, 3587 (1969).
[25] C. Asselineau, H. Montrozier, and J.-C. Promé, *Bull. Soc. Chim. France*, **1969**, 1911.
[26] D. C. Kleinfelter, R. W. Aaron, W. E. Wilde, T. B. Bennett, H. Wei, and J. E. Wiechert, *Tetrahedron Letters*, **1969**, 909.
[27] K. B. Wiberg and H. Schäfer, *J. Am. Chem. Soc.*, **91**, 927, 933 (1969).
[28] R. Stewart and F. Banoo, *Can. J. Chem.*, **47**, 3207 (1969).

18

Lactic acid[29] and mandelic acid[30] are prominent among the other hydroxy-compounds[31] whose oxidation by Cr(vi) has been studied, while aldehydes,[32] esters of both aliphatic and alicyclic alcohols,[33] and oxalic acid[34] have also received attention. Oxidation of lactic acid to give acetaldehyde in a low yield, such as that obtained with Cr(vi), has been claimed to be characteristic of a two-electron oxidant.[29a] On that basis it appears that V(v) is a two-electron oxidant, although Mehrotra[35] dismisses this possibility for the oxidation of citric acid in aqueous H_2SO_4. Mechanistic studies of the V(v) oxidation of the isomeric chlorotoluenes,[36] substituted benzyl alcohols,[37] cyclanols,[38] and D-fructose,[39] and of the oxidative coupling[40] of phenols promoted by $VOCl_3$ or VCl_4 have been reported.

The reactions of benzhydrol and triphenylcarbinol with permanganate are first-order in each reactant at low acidity, but in strongly acidic media ($H_0 <$ -0.5) the concentration of oxidant is kinetically insignificant. Over the region $H_0 = -0.5$ to -2.0, log(first-order rate constant) is linearly related to H_0, the lines having slopes of -1.13 and -0.93 respectively for the two alcohols. Although these reactions are governed by H_0 rather than H_R (which relates to the *equilibrium* between an alcohol and the carbonium ion), Banoo and Stewart[41] propose that the slow step is ionization of the alcohol; the carbonium ion is then rapidly oxidized to benzophenone accompanied, in the case of triphenylcarbinol, by phenol. Other studies with permanganate have focused

[29a] C. L. Jain, R. Shanker, and G. V. Bakore, *Indian J. Chem.*, **7**, 159 (1969); *Chem. Abs.*, **71**, 21445 (1969).
[29b] N. Venkatasubramanian, S. Sundaram, and G. Srinivasan, *Indian J. Chem.*, **6**, 708 (1968); *Chem. Abs.*, **70**, 114354 (1969); N. Venkatasubramanian and S. Sundaram, *J. Inorg. Nucl. Chem.*, **31**, 1761 (1969).
[30] F. B. Beckwith and W. A. Waters, *J. Chem. Soc.*(B), **1969**, 929; S. P. S. Dhakarey and S. Ghosh, *Indian J. Chem.*, **7**, 167 (1969); *Chem. Abs.*, **71**, 21440 (1969).
[31] W. M. Coates and J. R. Corrigan, *Chem. Ind.* (London), **1969**, 1594; J.-C. Richer and N. T. T. Hoa, *Can. J. Chem.*, **47**, 2479 (1969); G. Srinivasan, V. Thiagarajan, and N. Venkatasubramanian, *Current Sci.* (India), **37**, 612 (1968); *Chem. Abs.*, **70**, 28164 (1969); V. V. Egorova, S. N. Ananchenko, A. V. Zakharychev, and I. V. Torgov, *Izv. Akad. Nauk SSSR, Ser. Khim.*, **1969**, 943; *Chem. Abs.*, **71**, 50309 (1969).
[32] S. V. Anantakrishnan and V. Sathyabhama, Proc. 3rd Nucl. Radiat. Chem. Symp., 1967, p. 449; *Chem. Abs.*, **70**, 36738 (1969); K. K. S. Gupta and S. D. Bhattacharya, *Z. Phys. Chem.* (Leipzig), **240**, 279 (1969).
[33] P. S. Radhakrishnamurti and T. C. Behera, *J. Indian Chem. Soc.*, **46**, 92 (1969); *Chem. Abs.*, **71**, 12270 (1969).
[34] S. Chandra, S. N. Shukla, and A. C. Chatterji, *Z. Phys. Chem.* (Leipzig), **237**, 137 (1968).
[35] R. N. Mehrotra, *J. Chem. Soc.*(B), **1968**, 1563.
[36] P. S. Radhakrishnamurti and S. C. Pati, *Z. Phys. Chem.* (Leipzig), **241**, 49 (1969).
[37] G. V. Bakore and R. Shanker, *Indian J. Chem.*, **6**, 699 (1968); *Chem. Abs.*, **70**, 86788 (1969).
[38] P. S. Radhakrishnamurti and S. C. Pati, *Israel J. Chem.*, **7**, 427 (1969).
[39] P. N. Pathak, M. P. Singh, and B. B. L. Saxena, *Z. Phys. Chem.* (Leipzig), **241**, 145 (1969).
[40] W. L. Carrick, G. L. Karapinka, and G. T. Kwiatkowski, *J. Org. Chem.*, **34**, 2388 (1969).
[41] F. Banoo and R. Stewart, *Can. J. Chem.*, **47**, 3199 (1969).

on benzyl alcohol,[42] thymine,[43] 4-thiouracil derivatives,[44] and cycloalkane-nitronate anions.[45] The mechanisms of oxidation of hydroquinone by $Mn(III)$[46] and of benzyl alcohol by MnO_2[47] have been discussed.

The formation constant (K) for the initial 1:1 complex in the oxidation of alcohols by ceric ammonium nitrate in aqueous acetonitrile is markedly insensitive to the electronic effect of substituents in the alcohol, and, more surprisingly, may increase with steric crowding. Thus the value of K is ten times as large for *cis*-4-t-butylcyclohexanol as for its *trans*-isomer. Young and Trahanovsky[48] conclude that the steric requirements of solvation of the alcohol are more demanding than the steric requirements of complex-formation. They emphasize the need for caution in attributing the more rapid metal-ion oxidation of the more hindered of a pair of stereoisomers to relief of strain when the alcohol–oxidant complex decomposes. Ce(IV) oxidations of many other alcohols,[49] oxalic and tartaric acids,[50] hydroquinone,[51] aminoazo-benzenes,[52] and benzaldehyde,[53] where both 1:1 and 1:2 ceric ion–aldehyde complexes may be involved, have been examined.

In ethanol both Ce(IV) and Tl(III) transform monocarboxylic esters of hydroquinones, e.g. (19), into the corresponding quinone and ethyl carboxylate. There are, however, important differences between the reactions: electron-donating groups in Ar increase the yield of ethyl carboxylate with Ce(IV) but retard its formation with Tl(III); an unsubstituted position *ortho* to the ester function in the hydroquinone is essential for acyl transfer to occur with Tl(III) but not with Ce(IV); acyl transfer to methanol occurs more readily than to ethanol with Tl(III), but with Ce(IV) both alcohols display equal reactivity. To account for these differences it is suggested that with Ce(IV) an intermediate such as (20) collapses to an acylium ion, but with Tl(III) nucleophilic attack by the alcohol gives an intermediate (21) which passes directly to product.[54]

[42] K. K. Banerji and P. Nath, *Bull. Chem. Soc. Japan*, **42**, 2038 (1969).

[43] H. Hayatsu and S. Iida, *Tetrahedron Letters*, **1969**, 1031.

[44] H. Hayatsu and M. Yano, *Tetrahedron Letters*, **1969**, 755.

[45] F. Freeman, A. Yeramyan, and F. Young, *J. Org. Chem.*, **34**, 2438 (1969).

[46] G. Davies and K. Kustin, *Trans. Faraday Soc.*, **65**, 1630 (1969).

[47] I. M. Goldman, *J. Org. Chem.*, **34**, 3289 (1969).

[48] L. B. Young and W. S. Trahanovsky, *J. Am. Chem. Soc.*, **91**, 5060 (1969).

[49] W. S. Trahanovsky, L. H. Young, and M. H. Bierman, *J. Org. Chem.*, **34**, 869 (1969); W. S. Trahanovsky, M.G.Young, and P. M. Nave, *Tetrahedron Letters*, **1969**, 2501; W. S. Trahanovsky, P. J. Flash, and L. M. Smith, *J. Am. Chem. Soc.*, **91**, 5068 (1969); D. Greatorex and T. J. Kemp, *Chem. Comm.*, **1969**, 383; M. Rangaswamy and M. Santappa, *Indian J. Chem.*, **7**, 473 (1969); *Chem. Abs.*, **71**, 29820 (1969); N. Venkatasubramanian and T. R. Balasubramanian, *Current Sci.* (India), **37**, 666 (1968); *Chem. Abs.*, **71**, 21434 (1969).

[50] A. A. Luzan and K. B. Yatsimirskii, *Zh. Neorg. Khim.*, **13**, 3216 (1968); *Chem. Abs.*, **70**, 46580 (1969).

[51] C. F. Wells and L. V. Kuritsyn, *J. Chem. Soc.*(A), **1969**, 2575.

[52] M. Matrka and J. Marhold, *Coll. Czech. Chem. Comm.*, **33**, 4273 (1968).

[53] K. B. Wiberg and P. C. Ford, *J. Am. Chem. Soc.*, **91**, 124 (1969).

[54] V. M Clark, M. R. Eraut, and D. W. Hutchinson, *J. Chem. Soc.*(C), **1969**, 79.

(19)   (20)   (21)

Mandelic acid[55] and methyl ketones[56] are oxidized much more quickly by hexacyanoferrate(III) in alkaline solution when a catalytic amount of osmium tetroxide is present. Os(VIII) is the primary oxidant, being reduced to Os(VI) which in turn is rapidly oxidized by hexacyanoferrate(III).[56] Mechanistic studies of the hexacyanoferrate(III) oxidation[57] of formaldehyde,[58] glucose and fructose,[59] thioglycollic acid,[60] phenylhydrazine and its sulphonic acids,[61] ascorbic acid,[62] and aromatic amines[63] have been described, while other investigations have been concerned with the oxidation of various compounds by Cu(II)[64] and Np(VI),[65] oxidative coupling of amines and phenols using silver nitrate[66] or hexachloroiridate(IV),[67] and the Co(III) oxidation of thiourea,[68] 4-methyl-4-phenylpentanoic acid,[69] and several polyhydroxy-compounds.[70]

Whereas the reaction of toluene with Mn(III) acetate[71] in acetic acid resembles the oxidation by Pb(IV) acetate inasmuch as the major product is a mixture of isomeric methylbenzyl acetates, formed by a (simple) radical

[55] N. P. Singh, V. N. Singh, and M. P. Singh, *Austral. J. Chem.*, **21**, 2913 (1968).

[56] V. N. Singh, H. S. Singh, and B. B. L. Saxena, *J. Am. Chem. Soc.*, **91**, 2643 (1969).

[57] M. Spiro and P. W. Griffin, *Chem. Comm.*, **1969**, 262.

[58] V. N. Singh, M. C. Gangwar, B. B. L. Saxena, and M. P. Singh, *Can. J. Chem.*, **47**, 1051 (1969).

[59] K. C. Gupta and M. P. Singh, *Bull. Chem. Soc. Japan*, **42**, 599 (1969).

[60] R. C. Kapoor, O. P. Kachhwaha, and B. P. Sinha, *J. Phys. Chem.*, **73**, 1627 (1969).

[61] A. M. A. Verwey, J. A. Jonker, and S. Balt, *Rec. Trav. Chim.*, **88**, 439 (1969); see also A. M. A. Verwey and S. Balt, *ibid.*, p. 1289.

[62] U. S. Mehrotra, M. C. Agrawal, and S. P. Mushran, *J. Phys. Chem.*, **73**, 1996 (1969).

[63] J. F. Corbett, *J. Chem. Soc.*(B), **1969**, 207.

[64] U. Shanker and M. P. Singh, *Indian J. Chem.*, **6**, 702 (1968); *Chem. Abs.*, **70**, 114353 (1969); S. V. Singh, U. Shanker, and M. P. Singh, *Z. Phys. Chem.* (Leipzig), **240**, 400 (1969).

[65] N. K. Shastri and E. S. Amis, *Inorg. Chem.*, **8**, 2484, 2487 (1969).

[66] F. Effenberger, W. D. Stohrer, and A. Steinbach, *Angew. Chem. Internat. Ed. Engl.*, **8**, 280 (1969).

[67] R. Cecil and J. S. Littler, *J. Chem. Soc.*(B), **1968**, 1420.

[68] A. McAuley and U. D. Gomwalk, *J. Chem. Soc.*(A), **1969**, 977.

[69] P. R. Sharan, P. Smith, and W. A. Waters, *J. Chem. Soc.*(B), **1969**, 857.

[70] A. Meenakshi and M. Santappa, *Current Sci.* (India), **38**, 311 (1969); *Chem. Abs.*, **71**, 69879 (1969).

[71] E. I. Heiba, R. M. Dessau, and W. J. Koehl, *J. Am. Chem. Soc.*, **91**, 138 (1969).

process, oxidation by Co(III) acetate[72] gives predominantly benzyl acetate. Substituted toluenes react with Co(III) acetate at rates which correlate with σ^+ ($\rho = -2.4$), indicating substantial development of positive charge in the transition state; and when the reaction takes place in the presence of a high concentration of chloride ion extensive nuclear halogenation occurs.[72] Electron transfer to the oxidant gives the cation radical (22) which reacts further to yield the observed products as shown in Scheme 1. With easily ionized substrates such as 9,10-dimethylanthracene, the intermediate cation radical can

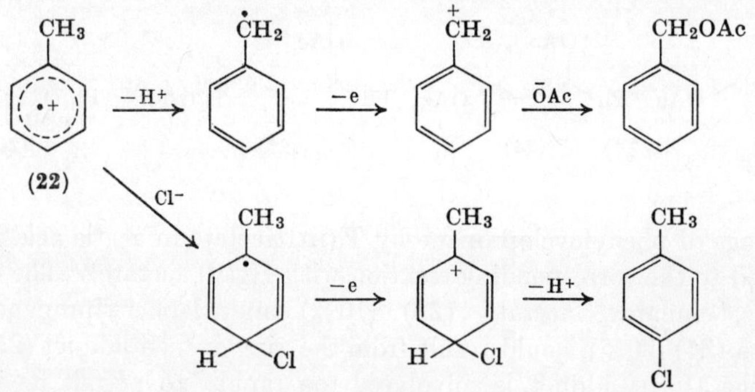

Scheme 1

be observed directly by ESR spectroscopy.[72] *p*-Methoxytoluene gives mainly *p*-methoxybenzyl acetate even with Mn(III) acetate;[71] it seems that substrates having a low ionization potential ($\leqslant 8.0$ ev) can react with Mn(III) acetate by electron transfer mechanisms analogous to those of their reactions with Co(III) acetate.

Since Pd(II) acetate can convert toluene into benzyl acetate in high yield,[73] benzylic oxidation would be expected to compete with the oxidative coupling of toluene using Pd(II) chloride in acetic acid containing sodium acetate.[74] Unger and Fouty[75] now report that side-chain oxidation can be suppressed by performing the coupling in the presence of Hg(II) acetate, when the proportion of 4,4'-bitolyl in the product is also increased. In fact, the bitolyl obtained using a 1:2 ratio of Pd(II) acetate and Hg(II) acetate contains over 70% of the 4,4'-isomer. It appears likely that an organomercury compound is formed initially and then undergoes oxidation by Pd(II). When pre-formed di-(*p*-tolyl)mercury is treated with Pd(II) acetate in acetic acid, metallic palladium is deposited and tolylmercuric acetate and 4,4'-bitolyl are formed with the stoichiometry shown in equation (3). The mechanism probably

[72] E. I. Heiba, R. M. Dessau, and W. J. Koehl, *J. Am. Chem. Soc.*, **91**, 6830 (1969).
[73] C. H. Bushweller, *Tetrahedron Letters*, **1968**, 6123.
[74] D. R. Bryant, J. E. McKeon, and B. C. Ream, *Tetrahedron Letters*, **1968**, 3371.
[75] M. O. Unger and R. A. Fouty, *J. Org. Chem.*, **34**, 18 (1969).

involves electrophilic attack by Pd(II) on the C–Hg bond in ditolylmercury (equation 4) to give tolylpalladium acetate which reacts with a second molecule of ditolylmercury (equation 5).[75]

$$2\,\text{Ar—Hg—Ar} + \text{Pd(OAc)}_2 \longrightarrow 2\,\text{Ar—HgOAc} + \text{Ar—Ar} + \text{Pd} \qquad \dots (3)$$

$$\text{Ar—Hg—Ar} + \text{Pd(OAc)}_2 \longrightarrow \text{Ar—HgOAc} + \text{Ar—Pd(OAc)} \qquad \dots (4)$$

$$\text{Ar—Pd(OAc)} + \text{Ar—Hg—Ar} \longrightarrow \text{Ar—Ar} + \text{Pd} + \text{Ar—Hg(OAc)} \qquad \dots (5)$$

| (23) | (24) | (25) | (26) |

Cleavage of phenylcyclopropane by Tl(III) acetate in acetic acid has been compared to the corresponding reaction with Hg(II) acetate.[76] The observed products, cinnamyl acetate **(23)** (10%) and 1-phenylpropane-1,3-diol diacetate **(24)** (90%), could result from the ring-opened adduct **(25)** which, unlike the Hg(II) adduct, is solvolysed too rapidly to permit its isolation. Although the second-order rate constant for Tl(III) acetate is twelve times as large as for Hg(II) acetate, the reactions of substituted phenylcyclopropanes correlate with σ^+ giving a more negative reaction constant ($\rho = -4.3$ at $50°$) than that for Hg(II) acetate ($\rho = -3.2$ at $50°$). This unusual situation of the faster reaction also being the more discriminating (i.e. the one with greater development of positive charge on the benzylic C atom in the transition state) is discussed by South and Ouellette[76] who conclude that the active electrophile in both cases is the covalent metal acetate. Pb(IV) acetate reacts with phenylcyclopropane much more slowly than does Tl(III) acetate although the same products, **(23)** (29%) and **(24)** (71%), are obtained; but since the rate is still much higher than would be predicted for cleavage by the ion-pair **(26)**, covalent metal acetate is thought to be the reactive entity here also.[77] Pb(IV) acetate is more electrophilic than either Hg(II) or Tl(III) acetates, and its lower reactivity towards phenylcyclopropane may be due to the difficulty of displacing acetate from coordinatively saturated Pb(IV) acetate relative to the ease of displacement of a ligand (acetic acid) from the coordinatively unsaturated mercuric and thallic reagents. The less negative reaction constant ($\rho = -1.75$ at $76°$) for Pb(IV) acetate could arise if displacement of acetate from lead and its attachment to the benzylic C atom were more concerted than with the other oxidants.[77]

[76] A. South and R. J. Ouellette, *J. Am. Chem. Soc.*, **90**, 7064 (1968).
[77] R. J. Ouellette, D. Miller, A. South, and R. D. Robins, *J. Am. Chem. Soc.*, **91**, 971 (1969).

Information concerning the acetoxylation of nitroalkanes[78] and 3-oxo-4,5-epoxy-steroids,[79] the intramolecular cyclization of alcohols,[80] and the Pb(IV) acetate oxidation of cyclopropyl- and cyclobutyl-carbinols,[81] 10-acylphenothiazines,[82] and ketone semicarbazones[83] and phenylhydrazones,[84] has appeared. At room temperature, benzaldehyde phenylhydrazone (**27**) yields benzoylazobenzene (**29**) (ca. 35%) and an azodiacetate (**30**) (⩽5%) accompanied by the azoacetate (**28**) (⩽27%).[85] Azoacetate (**28**) might conceivably

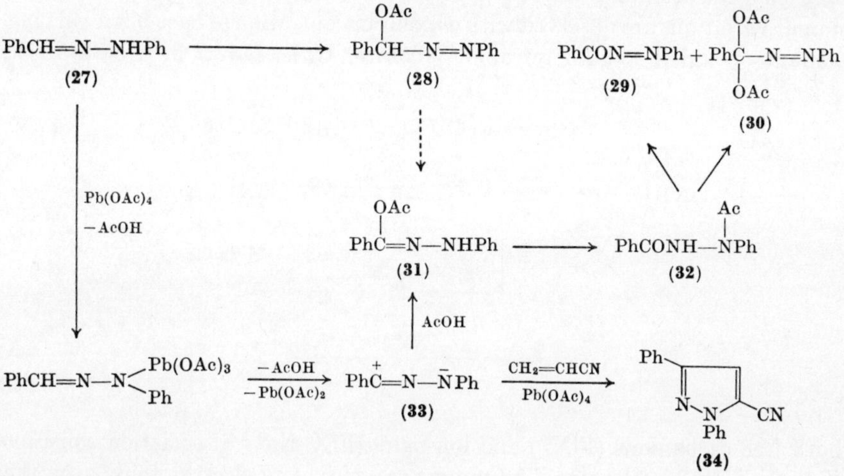

PhCH=N—NHPh
(27)
OAc
PhCH—N=NPh
(28)
PhCON=NPh + PhC—N=NPh
(29)
OAc
OAc
(30)
Pb(OAc)4
−AcOH
OAc
PhC=N—NHPh
(31)
Ac
PhCONH—NPh
(32)
AcOH
PhCH=N—N
Pb(OAc)3
Ph
−AcOH
−Pb(OAc)2
PhC=N—NPh
(33)
CH2=CHCN
Pb(OAc)4
Ph
N
N
CN
Ph
(34)

Scheme 2

be the precursor of the other products, since it is tautomeric with (**31**) and could rearrange to the *N*,*N'*-diacylhydrazine (**32**) which would give (**29**) and (**30**) on reaction with Pb(IV) acetate. Indeed Gladstone[85] reports that (**32**) can be obtained from azoacetate (**28**), but too slowly to be a significant stage in the original reaction. It is proposed instead that nitrileimine (**33**) is the precursor of (**29**) and (**30**), as shown in Scheme 2, and that its genesis is independent of the formation of azoacetate (**28**). Support comes from trapping of the nitrileimine; oxidation of benzaldehyde phenylhydrazone in acrylonitrile gives adduct (**34**) in high yield.[85]

[78] J. J. Riehl and Fr. Lamy, *Chem. Comm.*, **1969**, 406.
[79] M. L. Mihailović, J. Foršek, L. Lorenc, Z. Maksimović, H. Fuhrer, and J. Kalvoda, *Helv. Chim. Acta*, **52**, 459 (1969).
[80] M. L. Mihailović, S. Konstantinović, A. Milovanović, J. Janković, Ž. Čeković, and D. Jeremić, *Chem. Comm.*, **1969**, 236.
[81] M. L. Mihailović and Ž. Čeković, *Helv. Chim. Acta*, **52**, 1146 (1969).
[82] B. D. Podolešov, *Croat. Chem. Acta*, **40**, 201 (1968); *Chem. Abs.*, **70**, 11655 (1969).
[83] A. M. Cameron, P. R. West, and J. Warkentin, *J. Org. Chem.*, **34**, 3230 (1969).
[84] G. B. Gubelt and J. Warkentin, *Can. J. Chem.*, **47**, 3983 (1969).
[85] W. A. F. Gladstone, *Chem. Comm.*, **1969**, 179.

Oxidation by Molecular Oxygen

Involvement of ion-pairs in the formation of fluorenone from fluorene and molecular oxygen in t-butanol solutions of t-butoxide, suggested last year,[86] has been confirmed using 9-substituted fluorenes (FlHX; $X = CN$, SO_2Ph).[87] In dilute solutions of sodium t-butoxide each substrate exists entirely as the anion, and the rate of formation of fluorenone can be expressed by equation (6). However, the rate varies with the concentration of t-butoxide, increasing with base concentration for $X = CN$ but decreasing for $X = SO_2Ph$. Moreover, similar variations are observed with a fixed concentration of base but a varying amount of added sodium tetraphenylborate. If, as shown in equation (7),

$$\text{Rate} = k_{obs}[\text{FlHX}] = k[O_2][\text{FlHX}] \qquad \ldots (6)$$

$$\text{FlXH} \longrightarrow \text{FlX}^- \text{Na}^+ \underset{}{\overset{K_d}{\rightleftharpoons}} \text{FlX}^- + \text{Na}^+ \qquad \ldots (7)$$

$$O_2 \Big\downarrow \qquad \qquad \Big\downarrow O_2$$

$$\underset{k_{\pm}}{\longrightarrow} \text{Fl}{=}O \overset{k_-}{\longleftarrow}$$

$$k_{obs} = \frac{(k_{\pm}[\text{Na}^+] + k_- K_d)}{([\text{Na}^+] + K_d)} \qquad \ldots (8)$$

both free carbanions (FlX$^-$) and ion-pairs (FlX$^-$Na$^+$; dissociation constant K_d) participate in the rate-determining transfer of an electron to oxygen, the observed rate constant must be given by the expression in equation (8). This is indeed the case, $k_{obs}([\text{Na}^+] + K_d)$ being linearly related to [Na$^+$] for either substrate. Analysis of the data reveals that when $X = CN$ the ion-pair is rather more reactive than the free carbanion, but when $X = SO_2Ph$ the free anion is the more reactive by a factor of 20.[87] Other hydrocarbons whose reactions with molecular oxygen have been investigated include ethylene,[88] cyclooctene,[89] anthracene,[90] benzene,[91a] toluene,[91] and ethylbenzene.[92] Both steps in the sequence *p*-xylene→*p*-toluic acid→terephthalic acid are catalysed equally efficiently by bromine and cobalt naphthenate individually, but mixed catalysts accelerate the second step much more than the first.[93]

[86] *Org. Reaction Mech.*, **1968**, 474—5.
[87] D. Bethell, R. J. E. Talbot, and R. G. Wilkinson, *Chem. Comm.*, **1968**, 1528.
[88] P. François and Y. Trambouze, *Bull. Soc. Chim. France*, **1969**, 51.
[89] J. C. Giacomoni, A. Cambon, and R. Guedj, *Bull. Soc. Chim. France*, **1969**, 4077.
[90] N. A. Vasil'eva, V. A. Proskuryakov, A. N. Chistyakov, and T. P. Soboleva, *Zh. Prikl. Khim.*, **41**, 2802 (1968); *Chem. Abs.*, **70**, 95959 (1969).
[91a] H. Hotta, N. Suzuki, and T. Komori, *Bull. Chem. Soc. Japan*, **42**, 2041 (1969).
[91b] Y. Ogata and T. Morimoto, *J. Chem. Soc.*(B), **1969**, 74.
[92] J. Blum, J. Y. Becker, H. Rosenman, and E. D. Bergmann, *J. Chem. Soc.*(B), **1969**, 1000.
[93] J. Rouchaud and L. Sondengam, *Bull. Soc. Chim. Belges*, **78**, 11 (1969).

Attention has also been paid to the autoxidation of disulphides,[94] esters of thiodipropionic acid,[95] hydrazines,[96] enamines,[97] tetra(diethylamino)ethylene,[98] acetaldehyde,[99] and butyraldehyde[100] where the intermediate $PrCO_2OCH(OH)Pr$ can be isolated at $-40°$.

Xanthene instantly forms the 9-xanthyl cation by reaction with dissolved oxygen in 80% H_2SO_4, but in 45% acid the reaction is very slow. Each mole of oxygen gives rise to two equivalents of cation, and the reaction is first-order in substrate concentration and H^+ activity. Since H_2SO_4 is not the oxidant, it appears that hydride abstraction by oxygen is acid-catalysed, perhaps because of the presence of HO_2^+ in low, but kinetically significant, concentrations.[101]

In accord with the observed kinetics, the (simplified) mechanisms of Scheme 3 have been proposed for the autoxidation of ascorbic acid (H_2A) catalysed by vanadyl and uranyl ions in acid solution.[102] Transfer of a single electron within the catalyst–ascorbate–oxygen complex, **(36)** or **(38)**, formed in a series of equilibria, is the rate-limiting step. Since the uranyl–ascorbate complex **(37)** is no less stable than the vanadyl–ascorbate complex **(35)**, it is perhaps surprising that oxidation catalysed by VO^{2+} is at least 10^3 times as

$$VO^{2+} + H_2A + H^+ \rightleftharpoons VOH_3A^{3+} \overset{O_2}{\rightleftharpoons} VOH_3A(O_2)^{3+}$$
$$\qquad\qquad\qquad\qquad\quad \textbf{(35)} \qquad\qquad\qquad \textbf{(36)}$$

$$VOH_3A(O_2)^{3+} \overset{slow}{\longrightarrow} VOH_3A\cdot(O_2\cdot)^{3+} \overset{fast}{\longrightarrow} VO^{2+} + H_2O_2 + H^+ + A$$

$$UO_2^{2+} + H_2A + H^+ \rightleftharpoons UO_2H_3A^{3+} \overset{O_2}{\rightleftharpoons} UO_2H_3A(O_2)^{3+}$$
$$\qquad\qquad\qquad\qquad\quad \textbf{(37)} \qquad\qquad\qquad \textbf{(38)}$$

$$UO_2H_3A(O_2)^{3+} \overset{slow}{\longrightarrow} UO_2H_3A\cdot(O_2\cdot)^{3+} \overset{fast}{\longrightarrow} UO_2^{2+} + H_2O_2 + H^+ + A$$

Scheme 3

fast as with UO_2^{2+}, and has a much greater entropy of activation (VO^{2+}, 4.1 e.u.; UO_2^{2+}, -18 e.u.). Khan and Martell[102] suggest that the vanadyl complex **(36)** may be formed by attack of oxygen on the back side of the metal

[94] F. E. Murray and H. B. Rayner, *Pulp. Pap. Mag. Can.*, **69**, 64 (1968); *Chem. Abs.*, **69**, 51334 (1968).
[95] G. Scott, *Chem. Comm.*, **1968**, 1572.
[96] H. Dorn, H. Dilcher, and K. Walter, *Ann. Chem.*, **720**, 111 (1968).
[97] R. A. Jerussi, *J. Org. Chem.*, **34**, 3648 (1969).
[98] A. N. Fletcher, *J. Phys. Chem.*, **73**, 3686 (1969).
[99] G. Gut and M. Wirth, *Chimia*, **22**, 425 (1968).
[100] O. P. Yablonskii, M. G. Vinogradov, R. V. Kereselidze, and G. I. Nikishin, *Izv. Akad. Nauk SSSR, Ser. Khim.*. **1969** 318; *Chem. Abs.*, **70**, 114419 (1969).
[101] N. C. Deno, E. L. Booker, K. E. Kramer, and G. Saines, *J. Am. Chem. Soc.*, **91**, 5237 (1969).
[102] M. M. T. Khan and A. E. Martell, *J. Am. Chem. Soc.*, **91**, 4668 (1969).

and that such attack might be hindered by the two oxo oxygens of the uranyl ion; the difference in $\Delta S^{\ddagger}$ indicates that vanadyl ion is better able to bring together the oxygen and the substrate. Several other papers relating to heterogeneous[103] and non-enzymic homogeneous[104] catalysis of oxidation and dehydrogenation[105] have appeared. Oxygen must be present for the oxidation of 4-thiouridine by sodium sulphite to proceed.[106]

For each mole of hydroxylated product obtained with γ-butyrobetaine hydroxylase, one mole of α-ketoglutaric acid (the cofactor) is decarboxylated to succinic acid.[107] In accord with a mechanism in which one atom of the oxygen molecule is incorporated into the succinate and one into the hydroxylated product, $^{18}O_2$ yields succinate containing one ^{18}O atom per molecule.[107] In addition to those already recognized, an important step in the enzymic oxidation of glucose is thought to be the substrate-catalysed decomposition of the glucose–enzyme complex to gluconolactone and reduced enzyme.[108] A review[109] and many important communications concerning the mechanisms of oxygenases[110] and dehydrogenases[111] have been published.

Other Oxidations

A conceivable mechanism (equation 9) for the conversion of Schiff bases (**39**) into oxazirans (**40**) with m-chloroperoxybenzoic acid in t-butanol comprises nucleophilic attack by the oxidant followed by slow intramolecular nucleophilic attack of nitrogen on oxygen. Madan and Clapp[112] claim, however, that such a mechanism is inconsistent with the experimental rate law if $k_1 > k_{-1}$, and incompatible with the influence of substituents (see below) if $k_1 < k_{-1}$. They postulate a concerted mechanism (equation 10) in which the less hindered hydroxylic O atom of the solvated peroxyacid suffers nucleophilic attack by the π-electrons of the C=N bond. This is consistent with the observed reaction

103 E. F. Lutz and P. H. Williams, *J. Org. Chem.*, **34**, 3656 (1969); W. Hanke, *Z. Chem.*, **9**, 1 (1969).

104 Y. Ogata and Y. Kosugi, *Tetrahedron*, **25**, 1055 (1969); N. Ohta, *Ann. N.Y. Acad. Sci.*, **158**, 560 (1969); several papers of interest appear in *Discuss. Faraday Soc.*, **46** (1968).

105 M. Hájek and K. Kochloefl, *Coll. Czech. Chem. Comm.*, **34**, 2739 (1969).

106 H. Hayatsu, *J. Am. Chem. Soc.*, **91**, 5693 (1969).

107 B. Lindblad, G. Lindstedt, M. Tofft, and S. Lindstedt, *J. Am. Chem. Soc.*, **91**, 4604 (1969).

108 F. R. Duke, M. Weibel, D. S. Page, V. G. Bulgrin, and J. Luthy, *J. Am. Chem. Soc.*, **91**, 3904 (1969).

109 O. Hayaishi and M. Nozaki, *Science*, **164**, 389 (1969).

110 J. C. Allen, *Chem. Comm.*, **1969**, 906; B. G. Malmström, A. F. Agró, and E. Antonini, *European J. Biochem.*, **9**, 383 (1969); D. J. T. Porter and H. J. Bright, *Biochem. Biophys. Res. Comm.*, **36**, 209 (1969).

111 W. S. Allison, M. J. Connors, and D. J. Parker, *Biochem. Biophys. Res. Comm.*, **34**, 503 (1969); A. S. Mildvan and H. Weiner, *Biochemistry*, **8**, 552 (1969); H. Weiner, *ibid.*, p. 526 (1969).

112 V. Madan and L. B. Clapp, *J. Am. Chem. Soc.*, **91**, 6078 (1969).

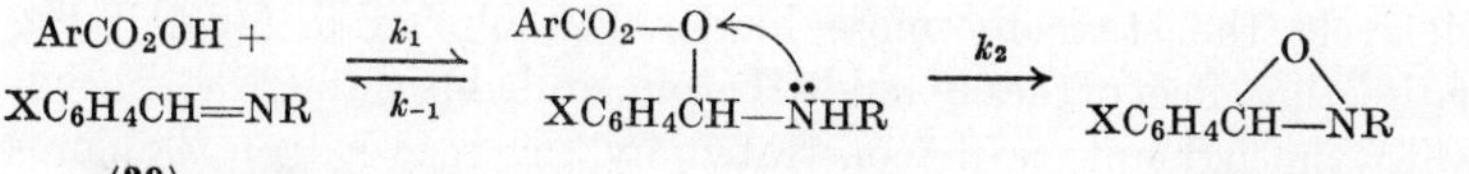

$$ \cdots (9) $$

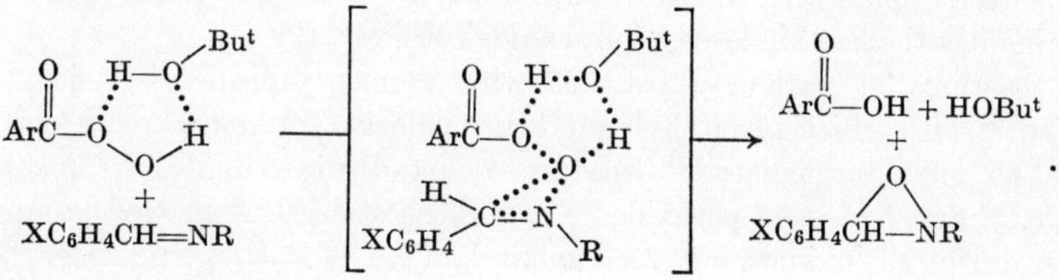

$$ \cdots (10) $$

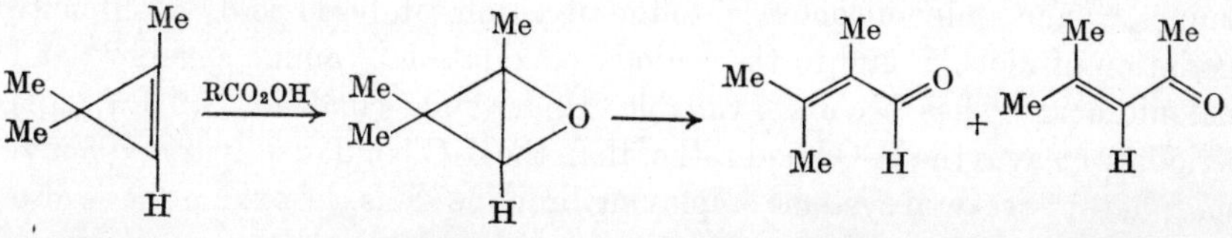

| (41) | (42) | (43) | (44) |

constants ($\rho = -1.75$ for various X, $R = Bu^t$; $\rho = -0.98$ for various Y in $R = YC_6H_4CH_2$, $X = p$-NO_2), the small kinetic solvent isotope effect [$k(Bu^tOH)/k(Bu^tOD) = 1.10$ at $25°$], and the values of the activation parameters ($E_{act} = 5.0$ kcal mole^{-1}, $\Delta S^{\ddagger} = -47$ e.u., for $X = p$-NO_2, $R = Bu^t$).[112] Oxabicyclobutane (**42**) appears to be an intermediate in the formation of α,β-unsaturated carbonyl compounds, (**43**) and (**44**), from cyclopropene (**41**) and peroxyacids.[113]

Several papers have been concerned with the oxidation of ketones,[114] sulphides,[115] and sulphoxides [116] by peroxyacid. Phenoxymethylpenicillin

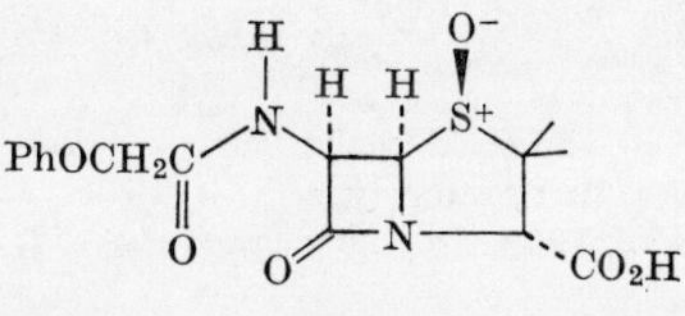

| (45) | (46) |

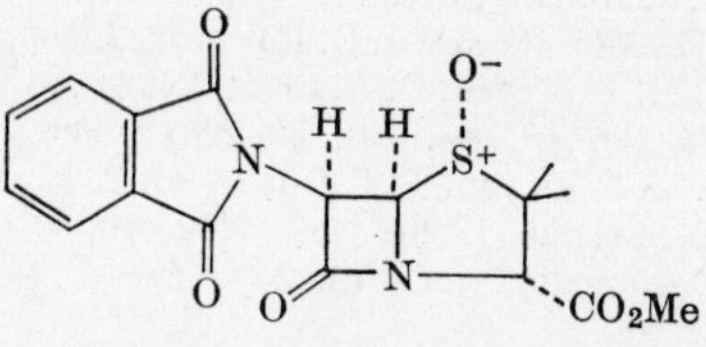

[113] J. Ciabattoni and P. J. Kocienski, *J. Am. Chem. Soc.*, **91**, 6534 (1969); see also L. E. Friedrich and R. A. Cormier, Abstracts, 158th National Meeting of Amer. Chem. Soc., New York, N.Y., Sept. 7—12, 1969.

[114] J. S. E. Holker, W. R. Jones, and P. J. Ramm, *J. Chem. Soc.*(C), **1969**, 357.

[115] N. Kucharczyk and V. Horák, *Coll. Czech. Chem. Comm.*, **34**, 2417 (1969); K. Undheim and V. Nordal, *Acta Chem. Scand.*, **23**, 1966 (1969).

[116] S. A. Khan, M. Ashraf, A. B. Chughtai, and I. Ahmad, *Pakistan J. Sci. Ind. Res.*, **10**, 151 (1967); *Chem. Abs.*, **69**, 58672 (1968).

gives exclusively the sterically more hindered β-sulphoxide (45);[117] the important directing effect of the 6β-amido H atom, probably through hydrogen-bonding with the oxidant, is demonstrated by the reaction of m-chloro-peroxybenzoic acid with methyl phthalimidopenicillinate which yields predominantly the less hindered α-sulphoxide (46).[118]

The reactions of various substrates with peroxy radicals,[119] benzoyl peroxide,[120] with which tetraethylammonium chloride forms benzoyl hypochlorite as an intermediate,[121] diisopropyl peroxydicarbonate,[122] nickel peroxide,[123] and hydrogen peroxide[124] have been studied. A methyl group migrates during the oxidation of 2,4-dimethylaniline by H_2O_2–peroxidase.[125] Further information relating to the peroxydisulphate oxidation of alcohols,[126] amines,[127] and amino-alcohols,[128] to the silver ion catalysed peroxydisulphate oxidation of diols,[129] and to the periodate oxidation of amino-sugars[130] and malonic acid[131] has become available. Under mild conditions (pH 7, 25°) periodate cleaves the P–O bond rather than the C–O bond in p-hydroxyphenyl phosphate;[132] several systems employing bromine[133] as the oxidant have also been considered as models for oxidative phosphorylation.[134]

[117] R. D. G. Cooper, P. V. DeMarco, J. C. Cheng, and N. D. Jones, J. Am. Chem. Soc., 91, 1408 (1969).

[118] R. D. G. Cooper, P. V. DeMarco, and D. O. Spry, J. Am. Chem. Soc., 91, 1528 (1969); D. H. R. Barton, F. Comer, and P. G. Sammes, ibid., p. 1529.

[119] D. F. Bowman, B. S. Middleton, and K. U. Ingold, J. Org. Chem., 34, 3456 (1969).

[120] C. G. Reid and P. Kovacic, J. Org. Chem., 34, 3308 (1969).

[121] N. J. Bunce and D. D. Tanner, J. Am. Chem. Soc., 91, 6096 (1969).

[122] P. Kovacic, C. G. Reid, and M. E. Kurz, J. Org. Chem., 34, 3302 (1969).

[123] R. Konaka, S. Terabe, and K. Kuruma, J. Org. Chem., 34, 1334 (1969).

[124] G. V. Bakore and V. R. Dwivedi, Indian J. Chem., 6, 651 (1968); Chem. Abs., 70, 36814 (1969); J. Pasek and J. Jarkovsky, Khim. Prom., 44, 261 (1968); Chem. Abs., 69, 35134 (1968); R. Granger, J. Boussinesq, J. P. Girard, and J. C. Rossi, Bull. Soc. Chim. France, 1969, 2801; R. Granger, J. Boussinesq, J. P. Girard, J. C. Rossi, and J. P. Vidal, ibid., p. 2806.

[125] V. R. Holland, B. M. Roberts, and B. C. Saunders, Tetrahedron, 25, 2291 (1969).

[126] J. E. McIsaac and J. O. Edwards, J. Org. Chem., 34, 2565 (1969).

[127] N. M. Beileryan, O. A. Chaltykyan, and G. A. Esayan, Dokl. Akad. Nauk Arm. SSR, 44, 174 (1967); Chem. Abs., 69, 2267 (1968).

[128] N. M. Beileryan, T. T. Gukasyan, R. M. Akopyan, and O. A. Chaltykyan, Armyan. Khim. Zh., 21, 225 (1968); Chem. Abs., 70, 3050 (1969); A. L. Samvelyan, N. M. Beileryan, and O. A. Chaltykyan, Uch. Zap., Erevan. Gos. Univ., 1968, No. 1, 61; Chem. Abs., 71, 38069 (1969).

[129] G. D. Menghani and G. V. Bakore, Bull. Chem. Soc. Japan, 41, 2574 (1968); G. D. Menghani and G. V. Bakore, Current Sci. (India), 37, 641 (1968); Chem. Abs., 70, 41151 (1969); G. D. Menghani and G. V. Bakore, Indian J. Chem., 7, 786 (1969); G. D. Menghani and G. V. Bakore, Z. Phys. Chem. (Leipzig), 241, 153 (1969); M. M. Khan and S. P. Srivastava, J. Indian Chem. Soc., 46, 574 (1969); Chem. Abs., 71, 90507 (1969).

[130] C. B. Barlow and R. D. Guthrie, Carbohydrate Res., 11, 53 (1969).

[131] B. P. Yadava and H. Krishna, Indian J. Chem., 6, 514 (1968); Chem. Abs., 70, 31954 (1969).

[132] C. A. Bunton and J. Hellyer, Tetrahedron Letters, 1969, 187.

[133] E. Bäuerlein and T. Wieland, Chem. Ber., 102, 1299 (1969); T. Wieland and H. Aquila, ibid., p. 2285; D. O. Lambeth and H. A. Lardy, Biochemistry, 8, 3395 (1969).

[134] W. N. Aldridge and M. S. Rose, FEBS Letters, 4, 61 (1969).

Deno and Potter's suggestion[135] that oxidation of alcohols by aqueous bromine (at pH $\leqslant$ 6) involves removal of an α-hydrogen as a proton was criticized last year,[136] and now another aspect of their theory—that the increasing rate in the pH range 6—8 is due to increasing participation by HOBr—has come under attack.[137] It is claimed that the inhibitory effect of added bromide ion at pH 6.7, where Br_2 and HOBr have approximately equal concentrations, results from removal of Br_2, the active oxidant, as kinetically inactive tribromide; and that the initially slow but gradually accelerating oxidation by HOBr in unbuffered solution is due to the liberation of bromide ion which in turn releases Br_2:

$$RCH_2OH + HOBr \rightarrow RCHO + H_3O^+ + Br^-$$

$$Br^- + HOBr \rightleftharpoons HO^- + Br_2$$

The dependence of rate on pH is attributed to the reactivity of bromine towards alkoxide ions being greater than towards alcohol molecules, and a correlation is claimed between the pH at which the rate begins to increase and the pK_a of the alcohol.[137] Other workers have examined the reactions of norbornan-2-ols with bromine,[138] o-(methylthio)benzoic acid with iodine,[139] bornane-2,3-diols with iodine–mercuric oxide,[140] and cyclobutanone with hypochlorous acid.[141] Chemical and spectroscopic evidence has been obtained for a tetravalent-sulphur intermediate in the oxidation of a sulphide to its sulphoxide with t-butyl hypochlorite.[142] Treatment of solutions of thian ($\bigcirc$S) in CH_2Cl_2 at $-78°$ with the reagents, *in order*, shown in equations (11) and (12) gives the t-butoxy- and the ethoxy-sulphonium trichloromercurate respectively as the major product. When an ethanolic solution of thian is treated with t-butyl hypochlorite and the reaction mixture then stirred with [1-^{14}C]ethanol at room temperature, the ethoxysulphonium salt obtained on addition of mercuric chloride after 30 minutes contains no radiolabel. If, however, the experiment is performed in the presence of a catalytic amount of chloride ion, the ethoxy group is found to be statistically scrambled. These results can be rationalized as in Scheme 4, where the chlorosulphonium salt (47) reacts quickly, even at $-78°$, to give the tetravalent-sulphur intermediate (48) (*not* an ion pair) which is long-lived, even at room temperature, in the absence of chloride ion.[142]

[135] *Org. Reaction Mech.*, **1967**, 426.
[136] *Org. Reaction Mech.*, **1968**, 480, ref. 127; but see also *Tetrahedron Letters*, **1969**, Issue No. 1 Errata.
[137] B. Perlmutter-Hayman and Y. Weissmann, *J. Am. Chem. Soc.*, **91**, 668 (1969).
[138] V. Thiagarajan and N. Venkatasubramanian, *Proc. Indian Acad. Sci.*, **67**, 37 (1968); *Chem. Abs.*, **70**, 56929 (1969).
[139] W. Tagaki, M. Ochiai, and S. Oae, *Tetrahedron Letters*, **1968**, 6131.
[140] A. Goosen and H. A. H. Laue, *J. Chem. Soc.*(B), **1969**, 995.
[141] J. A. Horton, M. A. Laura, S. M. Kalbag, and R. C. Petterson, *J. Org. Chem.*, **34**, 3366 (1969).
[142] C. R. Johnson and J. J. Rigau, *J. Am. Chem. Soc.*, **91**, 5398 (1969).

$$\text{>S} + Bu^tOCl + EtOH + HgCl_2 \longrightarrow \text{>}\overset{+}{S}\text{—}OBu^t \; HgCl_3^- \quad \ldots (11)$$

$$\text{>S} + EtOH + Bu^tOCl + HgCl_2 \longrightarrow \text{>}\overset{+}{S}\text{—}OEt \; HgCl_3^- \quad \ldots (12)$$

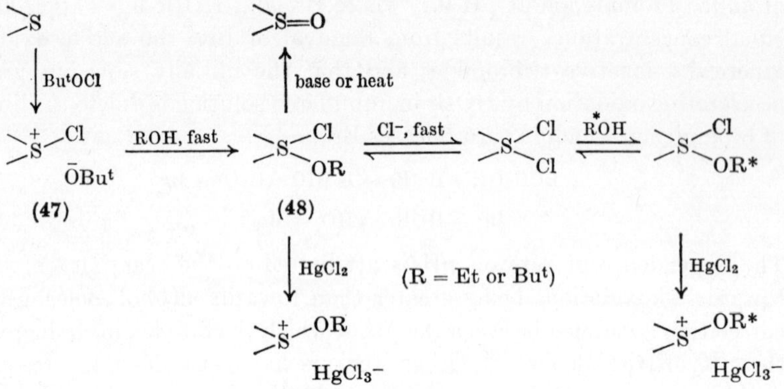

$$\text{>S} \qquad\qquad\qquad \text{>S=O}$$

$$\Big\downarrow Bu^tOCl \qquad\qquad \Big\downarrow \text{base or heat}$$

$$\text{>}\overset{+}{\underset{\bar{O}Bu^t}{S}}\text{—Cl} \xrightarrow{ROH,\,fast} \text{>S}\overset{Cl}{\underset{OR}{\diagdown}} \xrightleftharpoons{Cl^-,\,fast} \text{>S}\overset{Cl}{\underset{Cl}{\diagdown}} \xrightleftharpoons{\overset{*}{R}OH} \text{>S}\overset{Cl}{\underset{OR^*}{\diagdown}}$$

$$(47) \qquad\qquad\qquad (48)$$

$$\Big\downarrow HgCl_2 \qquad (R = Et\ or\ Bu^t) \qquad\qquad \Big\downarrow HgCl_2$$

$$\text{>}\overset{+}{S}\text{—OR} \qquad\qquad\qquad\qquad \text{>}\overset{+}{S}\text{—OR}^*$$

$$HgCl_3^- \qquad\qquad\qquad\qquad HgCl_3^-$$

Scheme 4

Further work has revealed mechanistic similarities between the degradation of aliphatic amines by reaction with chlorine dioxide and by anodic oxidation; in both instances removal of an electron from nitrogen gives initially a cation radical.[143] The kinetics of the oxidation of alcohols by N-bromosuccinimide have been studied[144] in the presence of mercuric acetate when, it is claimed,[144a] oxidation by molecular bromine is completely suppressed. Several reports of the reactions of N-chlorinated oxidants,[145] including 1-chlorobenzotriazole[146] and chloramine T,[147] and of 2,3-dichloro-5,6-dicyanobenzoquinone[148] have appeared.

A cyclic intermediate containing two molecules of oxidant might account for the reaction of thioanisole with N_2O_4 in CCl_4 being second-order in oxidant.[149] The kinetics of the nitric acid oxidation of oxalic acid catalysed by

[143] L. A. Hull, G. T. Davis, D. H. Rosenblatt, and C. K. Mann, *J. Phys. Chem.*, **73**, 2142 (1969); see also L. A. Hull, W. P. Giordano, D. H. Rosenblatt, G. T. Davis, C. K. Mann, and S. B. Milliken, *ibid.*, p. 2147; L. A. Hull, G. T. Davis, and D. H. Rosenblatt, *J. Am. Chem. Soc.*, **91**, 6247 (1969).

[144a] N. Venkatasubramanian and V. Thiagarajan, *Can. J. Chem.*, **47**, 694 (1969).

[144b] N. S. Srinivasan, V. Thiagarajan, and N. Venkatasubramanian, *Current Sci.* (India), **38**, 138 (1969); *Chem. Abs.*, **70**, 106102 (1969).

[145] T. Higuchi, A. Hussain, and I. H. Pitman, *J. Chem. Soc.*(B), **1969**, 626; I. H. Pitman, H. Dawn, T. Higuchi, and A. A. Hussain, *ibid.*, p. 1230.

[146] C. W. Rees and R. C. Storr, *J. Chem. Soc.*(C), **1969**, 1474.

[147] K. Weber and F. Valić, *Z. Phys. Chem.* (Leipzig), **238**, 353 (1968); F. Valić, *ibid.*, **239**, 24 (1968).

[148] H.-D. Becker, *J. Org. Chem.*, **34**, 1198, 1203, 1211 (1969).

[149] H. G. Hauthal, H. Onderka, and W. Pritzkow, *J. Prakt. Chem.*, **311**, 82 (1969); *Chem. Abs.*, **70**, 86796 (1969).

Mn^{2+},[150] of deoxybenzoin,[151] of diphenyl sulphide,[152] and of diphenyl-methane[153] have been discussed. Any mechanism (e.g. Scheme 5) for the latter reaction must be compatible with the following observations: the induction period is eliminated by addition of $NaNO_2$; the rate constant for the early stages of the reaction increases with the concentration of $NaNO_2$ at

$$HNO_3 + HNO_2 \rightleftharpoons 2\,NO_2\cdot + H_2O$$

$$NO_2\cdot + H^+ \rightleftharpoons HNO_2\cdot{}^+$$

$$Ph_2CH_2 + HNO_2\cdot{}^+ \quad (or\ NO_2\cdot) \xrightarrow{slow} Ph_2\overset{\bullet}{C}H + H_2NO_2{}^+ \quad (or\ HNO_2)$$

$$Ph_2\overset{\bullet}{C}H + NO_2\cdot \longrightarrow Ph_2CH(ONO) \xrightarrow{H_2O} Ph_2CHOH + HNO_2$$

$$Ph_2CHOH + HNO_2\cdot{}^+ \quad (or\ NO_2\cdot) \longrightarrow Ph_2\overset{\bullet}{C}OH + H_2NO_2{}^+ \quad (or\ HNO_2)$$

$$Ph_2\overset{\bullet}{C}OH + NO_2\cdot \longrightarrow Ph_2C(OH)ONO \longrightarrow Ph_2CO + HNO_2$$

Scheme 5

low concentrations; the rate is proportional to h_0; the rates of substituted derivatives $XC_6H_4CH_2Ph$ correlate with $\sigma_x{}^+$ ($\rho = -1.7$ at $90°$).[153]

The stereochemistry of the oxidative ring-opening of aziridine derivatives by dimethyl sulphoxide,[154] and the mechanisms of oxidation of alcohols,[155] where a ketenimine ($Ph_2C=C=NAr$) may replace dicyclohexylcarbodiimide in the Moffatt reaction,[156] have been discussed. Reactions in which pyridine N-oxide acts as an oxidant towards ketenes and acid chlorides have been described,[157] and a mechanism has been proposed to account for the formation of methanesulphonic acid and formaldehyde in its acid-catalysed reaction with dimethyl sulphoxide.[158] Other reports have considered the oxidation of alkyl aryl ethers,[159] and the reactions of thiols with flavins,[160] cinnamaldehyde with xenon trioxide,[161] and 4-benzeneazo-1-aminonaphthalene with aniline hydrochloride.[162]

[150] V. S. Koltunov, *Kinet. Katal.*, **9**, 1034 (1968); *Chem. Abs.*, **70**, 36821 (1969).
[151] Y. Ogata and H. Tezuka, *Tetrahedron*, **25**, 4797 (1969).
[152] N. Marziano, E. Maccarone, U. Romano, P. Fiandaca, and R. Passerini, *Ann. Chim.* (Italy), **59**, 565, 573 (1969).
[153] Y. Ogata, H. Tezuka, and T. Kamei, *J. Org. Chem.*, **34**, 845 (1969); Y. Ogata, H. Tezuka, and T. Kamei, *Tetrahedron*, **25**, 4919 (1969).
[154] S. Fujita, T. Hiyama, and H. Nozaki, *Tetrahedron Letters*, **1969**, 1677.
[155] W. H. Clement, T. J. Dangieri, and R. W. Tuman, *Chem. Ind.* (London), **1969**, 755.
[156] R. E. Harmon, C. V. Zenarosa, and S. K. Gupta, *Chem. Comm.*, **1969**, 327; R. E. Harmon, C. V. Zenarosa, and S. K. Gupta, *Tetrahedron Letters*, **1969**, 3781.
[157] T. Koenig and T. Barklow, *Tetrahedron*, **25**, 4875 (1969).
[158] M. E. C. Biffin, J. Miller, and D. B. Paul, *Tetrahedron Letters*, **1969**, 1015; see also V. J. Traynelis and K. Yamauchi, *ibid.*, p. 3619.
[159] O. C. Musgrave, *Chem. Rev.*, **69**, 499 (1969).
[160] M. J. Gibian and D. V. Winkelman, *Tetrahedron Letters*, **1969**, 3901.
[161] H. J. Rhodes, R. P. Shiau, and M. I. Blake, *J. Pharm. Sci.*, **57**, 1706 (1968).
[162] G. Schroeder and W. Lüttke, *Ann. Chem.*, **723**, 83 (1969).

Electrochemical oxidation of hexamethylbenzene $(ArCH_3)$[163] in MeCN–AcOH (99:1) affords a mixture of pentamethylbenzylacetamide $(ArCH_2NHAc)$ and pentamethylbenzyl acetate $(ArCH_2OAc)$ via the pentamethylbenzyl cation.[164] With Bu_4NBF_4 as supporting electrolyte, the ester:amide ratio (81:19) is much higher than when Bu_4NClO_4 is used (22:78), presumably because AcOH preferentially solvates the fluoroborate anion and is held close to the anode surface.[164] Aliphatic hydrocarbons,[165] benzpinacol,[166] hydroquinone,[167] ascorbic acid,[168] phenols,[169] and amines[170] are among the other compounds studied. Electrolysis of cyclopropanecarboxylate (49) in methanol at $-40°$ gives *cis*-olefin (51), while the stereoisomeric carboxylate (52) gives *trans*-olefin.[171] Since neither cyclopropyl radicals nor cations would be

$$
(49) \xrightarrow{-2e} (50) \xrightarrow[\text{MeOH}]{-CO_2} (51)
$$

expected to undergo stereospecific ring-opening, it seems likely that an acyloxonium ion, e.g. (50), is first formed and then suffers concerted decomposition.[171]

Reductions[172]

Second-order rate constants (k_2) for the reduction of 4-t-butylcyclohexanone (55), *cis*-3,5-dimethylcyclohexanone (56), 4-t-butyl-2,2-dimethylcyclohexanone (57), and 3,3,5-trimethylcyclohexanone (58) by lithium tri-t-butoxyaluminium hydride have been measured at $30°$; these are shown in Table 2 together with the rate constants for axial attack $(k_a$, giving equatorial OH) and equatorial attack $(k_e$, giving axial OH) deduced from the values of k_2 and

163 L. Eberson and B. Olofsson, *Acta Chem. Scand.*, **23**, 2355 (1969).
164 K. Nyberg, *Chem. Comm.*, **1969**, 774.
165 M. Fleischmann and D. Pletcher, *Tetrahedron Letters*, **1968**, 6255.
166 W. Kemula, Z. R. Grabowski, and M. K. Kalinowski, *J. Am. Chem. Soc.*, **91**, 6863 (1969); R. F. Michielli and P. J. Elving, *ibid.*, p. 6864.
167 B. R. Eggins and J. Q. Chambers, *Chem. Comm.*, **1969**, 232; V. D. Parker, *ibid.*, p. 716; G. Wegner, T. F. Keyes, N. Nakabayashi, and H. G. Cassidy, *J. Org. Chem.*, **34**, 2822 (1969).
168 K. Kretzschmar and W. Jaenicke, *Tetrahedron Letters*, **1969**, 3811.
169 V. D. Parker, *J. Am. Chem. Soc.*, **91**, 5380 (1969).
170 P. J. Smith and C. K. Mann, *J. Org. Chem.*, **34**, 1821 (1969); N. A. Hampson, J. B. Lee, J. R. Morley, and B. Scanlon, *Can. J. Chem.*, **47**, 3729 (1969).
171 T. Shono, I. Nishiguchi, S. Yamane, and R. Oda, *Tetrahedron Letters*, **1969**, 1965; see also L. Eberson, *J. Am. Chem. Soc.*, **91**, 2402 (1969).
172 R. L. Augustine (Ed.), *Reductions; Techniques and Applications in Organic Synthesis*, Marcel Dekker, New York, 1968.

the composition of the cyclohexanol products.[173] Chérest and Felkin[174] suggested last year that the stereochemical outcome of hydride reduction of a cyclohexanone depended on the relative magnitude of the torsional strain between the C-1–H and C-2–H_a bonds in transition state (53) for equatorial attack, and the steric strain between the axial substituent R on C-3 (or C-5) and the attacking hydride H in transition state (54) for axial attack. The new kinetic measurements are in complete accord with their theory. Ketones (56)

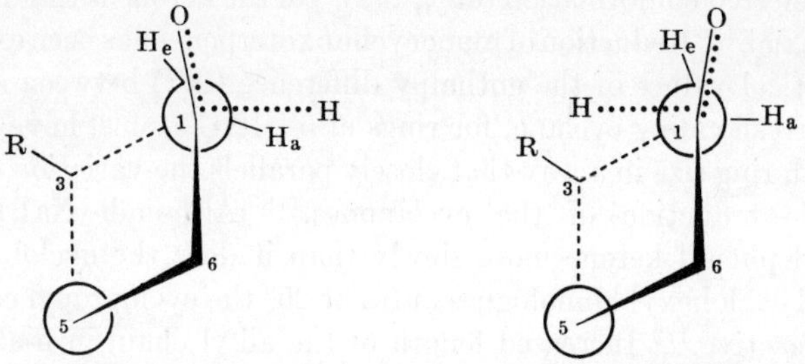

(53) Equatorial attack   (54) Axial attack

Table 2. Rate constants for reduction of substituted cyclohexanones by $LiAlH(OBu^t)_3$ at 30°

Substituted cyclohexanone	R	k_2 ($M^{-1}\,sec^{-1}$)	k_a ($M^{-1}\,sec^{-1}$)	k_e ($M^{-1}\,sec^{-1}$)
(55), 4-t-Butyl	H	1.40	1.27	0.13
(56), *cis*-3,5-Dimethyl	H	0.99	0.88	0.11
(57), 4-t-Butyl-2,2-dimethyl	H	0.91	0.84	0.07
(58), 3,3,5-Trimethyl	Me	0.084	0.009	0.075

and (57) each carry two Me substituents, but in common with (55) the axial positions on C-3 and C-5 are occupied by hydrogen in the preferred conformation. The three ketones have similar values of k_2, and in each case $k_a > k_e$ in agreement with the view[174] that when R = H eclipsing interactions in transition state (53) outweigh steric interactions in (54). Moving to ketone (58) with an axial Me group on C-3, the greatly diminished overall rate of reduction is due almost entirely to a 100-fold decrease in k_a; clearly steric interaction in transition state (54) is now of major importance. Other workers[175] have emphasized that factors in addition to the size of axial substituents on C-3 and C-5 can influence the stereochemical course of reduction of cyclohexanones by Grignard reagents. Thus the proportion of *cis*-isomer in the cyclohexanol

[173] J. Klein, E. Dunkelblum, E. L. Eliel, and Y. Senda, *Tetrahedron Letters*, **1968**, 6127.
[174] *Org. Reaction Mech.*, **1968**, 483—4.
[175] G. Chauvière, Z. Welvart, D. Eugène, and J.-C. Richer, *Can. J. Chem.*, **47**, 3285 (1969).

obtained from 2-methyl- or 2-t-butyl-cyclohexanone increases substantially with alkyl substitution at C-2 of the reducing agent, and with each reducing agent is greater for the t-butyl ketone.[175]

Hydride reduction of 4-chlorocyclohexanone yields a substantially greater proportion of *cis*-alcohol than does 4-methylcyclohexanone. However, the various theories advanced to explain the supposed facility of equatorial attack on 4-chlorocyclohexanone have become largely redundant with the discovery that the preferred conformation (69% at 27°) of the ketone is that in which the halogen is axial.[176] Reduction of monocyclic oxoterpenes has been examined,[177] and theoretical values of the enthalpy difference (ΔH) between a cyclanone and the corresponding cyclane, for rings of 5—10 C atoms, have been found to vary with ring size in a way that closely parallels the variation of log k_2 for borohydride reduction of the cyclanones.[178] Although $NaBH_4$ reduces cyclopropyl phenyl ketone more slowly than it does the cyclobutyl, cyclopentyl, and cyclohexyl homologues at 0°, at 25° the cyclopropyl compound is the most reactive.[179] Increased length of the alkyl chain in n-alkyl methyl ketones produces a significant decrease in the rate of reduction on going from Et to Pr, but further changes have little effect; interaction between the carbonyl O atom and the H atoms on C-γ is thought to reduce the reactivity of the C=O group.[180] Although borohydride reductions of *o*-hydroxyaryl ketones give the expected secondary alcohols, the esters of these ketones are reduced to *o*-alkylphenols.[181]

The Grignard reagent from (R)-$(-)$-1-chloro-3-phenylbutane (**59-R**) reduces t-butyl phenyl ketone to α-t-butylbenzyl alcohol which contains an excess of the (R)-enantiomer. Morrison, Black, and Ridgway[182] suggested that asymmetric induction resulted from unequal rates of transfer of the H atoms (H_a and H_b) on C-2 of the Grignard reagent to the carbonyl C of the ketone, in cyclic transition states. A predominance of (R)-alcohol would require that H_a in (**59-R**) [and, by analogy, H_b in its enantiomer (**59-S**)] were preferentially transferred. To test their theory, Morrison and Ridgway[183] have now measured the ratio of H to D (k_H/k_D) incorporated at C-1 of the alcohol obtained from t-butyl phenyl ketone with each of the *racemic* Grignard reagents (**60**) and (**61**). Preferential transfer of D from either enantiomer of (**60**) would be predicted, while for (**61**) the reverse would be true. The experimental results, $k_H/k_D = 0.4$ using (**60**) and $k_H/k_D = 8$ using (**61**), are in full

176 D. N. Kirk, *Tetrahedron Letters*, **1969**, 1727.
177 H. Rothbächer and F. Suteu, *Pharmazie*, **24**, 222 (1969); *Chem. Abs.*, **71**, 50239 (1969).
178 N. L. Allinger, J. A. Hirsch, M. A. Miller, and I. J. Tyminski, *J. Am. Chem. Soc.*, **91**, 337 (1969).
179 S. F. Sun and P. R. Neidig, *J. Org. Chem.*, **34**, 1854 (1969).
180 P. Geneste, G. Lamaty, and B. Vidal, *Bull. Soc. Chim. France*, **1969**, 2027.
181 B. J. McLoughlin, *Chem. Comm.*, **1969**, 540.
182 *Org. Reaction Mech.*, **1968**, 484—5.
183 J. D. Morrison and R. W. Ridgway, *J. Am. Chem. Soc.*, **91**, 4601 (1969).

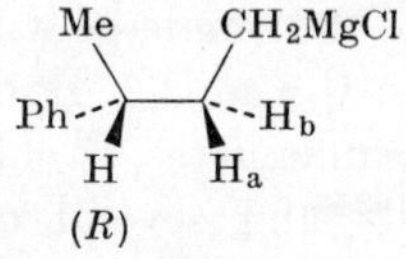

Me CH₂MgCl
Ph- -Hb
H Ha
(R)

(59-R)

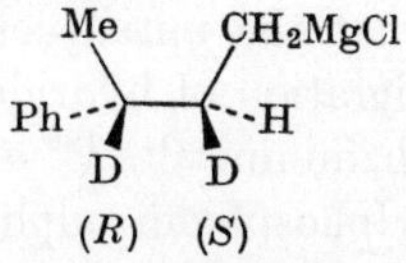

Me CH₂MgCl
Ph- -H
D D
(R) (S)

(60-R)

Me CH₂MgCl
Ph- -D
D H
(R) (R)

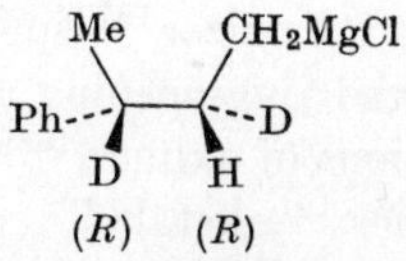

(61-R)

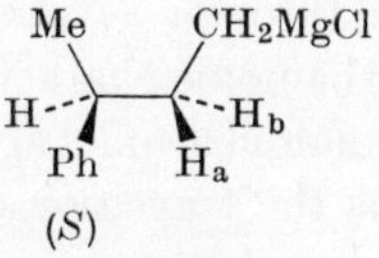

Me CH₂MgCl
H- -Hb
Ph Ha
(S)

(59-S)

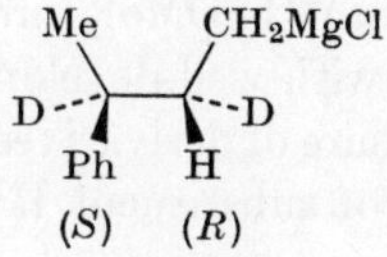

Me CH₂MgCl
D- -D
Ph H
(S) (R)

(60-S)

Me CH₂MgCl
D- -H
Ph D
(S) (S)

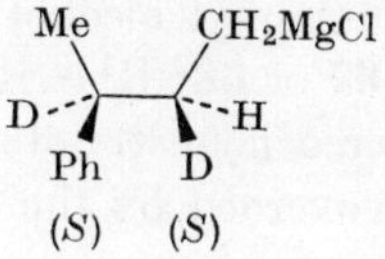

(61-S)

accord with prediction. Reduction of the same ketone by the Grignard reagent from $(\pm)$-*threo*-1-chloro-3-phenyl-2,3-dideuteropropane, in which C-3 is asymmetric only by virtue of D-labelling, shows the "normal" value of the isotope effect to be $k_H/k_D = 2.3$. Clearly the observed isotope effects with (60) and (61) arise because asymmetric induction enforces the "normal" isotope effect in the case of (61), but opposes, and actually reverses, it with (60).[183] Asymmetric induction in the reaction of β-amino-ketones with LiAlH₄ has also been discussed.[184]

Reduction of epoxides by borohydride[185] and by mixed hydride reagents[186] continues to attract attention. Both *endo*-oxide (62) and its *exo*-isomer (65) react with LiAlD₄–AlCl₃ (ratio 1.8:2.5; effective reagent AlDClX where X = D or Cl) to give a mixture of alchols (63) (15—16%) and (64) (84—85%).

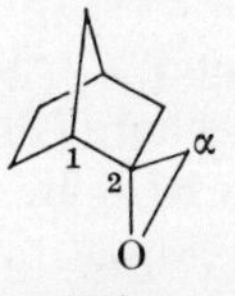

(62) 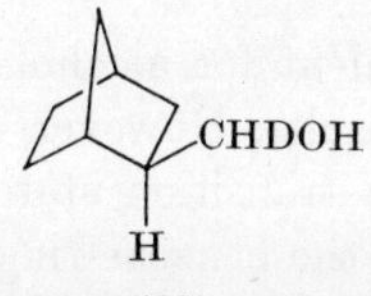 (63) 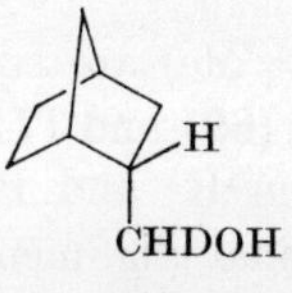 (64) 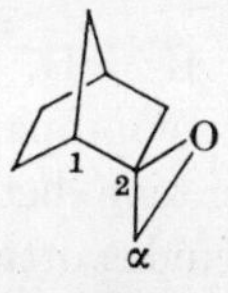 (65)

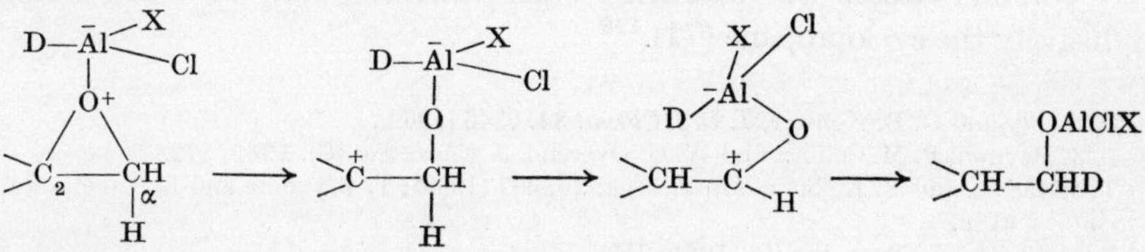

(66)

[184] M.-J. Brienne, C. Fouquey, and J. Jacques, *Bull. Soc. Chim. France*, **1969**, 2395; M. J. Lyapova and B. J. Kurtev, *Chem. Ber.*, **102**, 3739 (1969).
[185] A. Rashid and G. Read, *J. Chem. Soc.*(C), **1969**, 2053.
[186] E. Laurent and P. Villa, *Bull. Soc. Chim. France*, **1969**, 249.

Bly and Konizer[187] suggest that the initial complex (66) undergoes non-concerted ring-opening and migration of hydride from C-α to C-2. Hydride reductions of oximes,[188] quinolizinium salts,[189] o-nitrocinnamates,[190] chlorobenzenes,[191] ketals,[192] alkynylphosphine sulphides,[193] and several other compounds[194] have also been studied this year.

Kinetic measurements suggest that the formation of phenylcyclopropane from cinnamyl alcohol with $LiAlH_2(OMe)_2$ proceeds through a transition state (67 or 68; $R^1 = R^2 = H$) with well developed carbanionic character.[195] The stereochemistry of ring-closure of derivatives of cinnamyl alcohol appears to be governed by the nature of substituent R^1. Thus the transition states

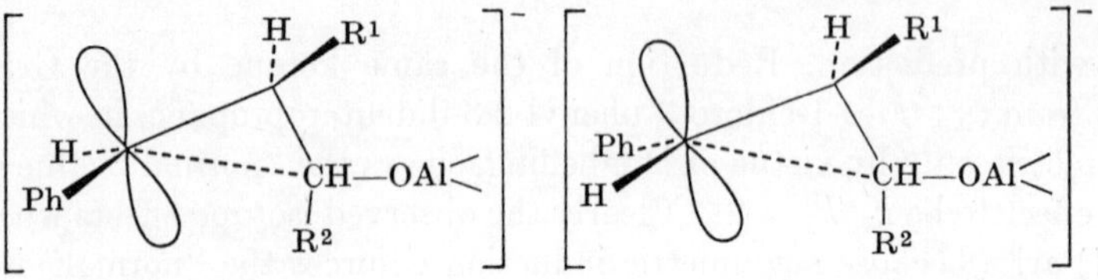

(67)   (68)

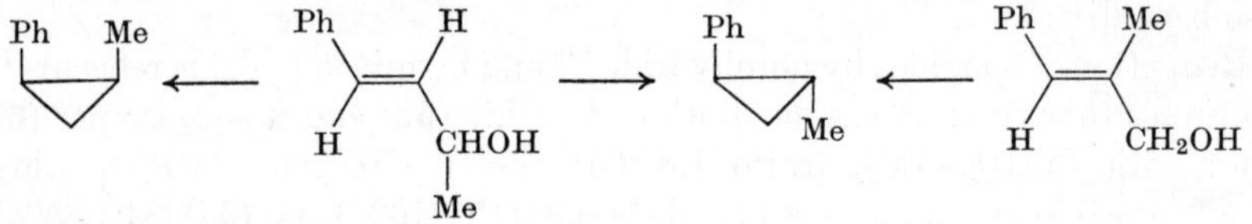

(69)   (70)   (71)   (72)

(67 and 68; $R^1 = H$, $R^2 = Me$) are equivalent for alcohol (70) and a 1:1 mixture of cyclopropanes (69) and (71) results. However, for alcohol (72) eclipse interaction between R^1 and Ph in transition state (67; $R^1 = Me$, $R^2 = H$) precludes orthogonal alignment of the benzene ring with respect to the carbanion; charge delocalization is greatly impeded with the consequence that reaction occurs via transition state (68; $R^1 = Me$, $R^2 = H$) giving exclusively the cyclopropane (71).[195]

[187] R. S. Bly and G. B. Konizer, *J. Org. Chem.*, **34**, 2346 (1969).
[188] P. J. Beynon, P. M. Collins, and W. G. Overend, *J. Chem. Soc.*(C), **1969**, 272.
[189] T. Miyadera and Y. Kishida, *Tetrahedron*, **25**, 397 (1969); T. Miyadera and R. Tachikawa, *ibid.*, p. 5189.
[190] R. T. Coutts, *J. Chem. Soc.*(C), **1969**, 713.
[191] P. Olavi, I. Virtanen, and P. Jaakkola, *Tetrahedron Letters*, **1969**, 1223.
[192] P. C. Loewen, W. W. Zajac, and R. K. Brown, *Can. J. Chem.*, **47**, 4059 (1969).
[193] A. M. Aguiar, J. R. S. Irelan, and N. S. Bhacca, *J. Org. Chem.*, **34**, 3349 (1969).
[194] A. P. Gray and D. E. Heitmeier, *J. Org. Chem.*, **34**, 3253 (1969); R. H. Starkey and W. H. Reusch, *ibid.*, p. 3522; K. Wiesner and T. Inaba, *J. Am. Chem. Soc.*, **91**, 1036 (1969).
[195] M. J. Jorgenson and A. F. Thacher, *Chem. Comm.*, **1969**, 1290.

High stereoselectivity in the reduction of a number of organomercurials RHgX (X = OH, OAc, Cl etc.) by NaBH$_4$ in aqueous solution appeared to demand a mechanism such as equation (13) (X = OH) in which the intermediate alkylmercuric hydride decomposes by a concerted intramolecular process. It has now been found, however, that although hydroxymercuration of *cis*- and *trans*-but-2-enes is entirely stereospecific, NaBD$_4$ reduces either the *threo*- or *erythro*-MeCH(OH)CH(HgOAc)Me so obtained to a 1:1 mixture of *erythro*- and *threo*-3-deuterobutan-2-ols.[196] Moreover, borohydride reduction of nor-bornenylmercuric chloride (**73**; Y = HgCl) and nortricyclylmercuric chloride (**75**; Y = HgCl) gives the same mixture of (**73**; Y = H) (6%), (**74**) (34%), and (**75**; Y = H) (60%).[197] To account for non-stereospecific reduction and

$$\underset{\text{R---Hg---X}}{\overset{\text{H}\underset{\cdot}{\overset{\cdot}{\leftharpoondown}}\bar{\text{B}}\text{H}_3}{}} \longrightarrow \underset{\text{R---Hg}}{\overset{\text{H}}{}} + \underset{\text{X}}{\overset{\bar{\text{B}}\text{H}_3}{|}} \longrightarrow \text{RH} + \text{Hg} + \bar{\text{B}}\text{H}_3\text{X} \quad \dots (13)$$

(73)   (74)   (75)

$$\text{RHgX} \xrightarrow{\text{BH}_4^-} \text{RHgH} \longrightarrow \text{R}\cdot + \cdot\text{HgH} \longrightarrow \text{RH} + \text{Hg} \quad \dots (14)$$

formation of rearranged products, it is suggested that the intermediate alkyl-mercuric hydride dissociates in a non-concerted radical process (equation 14).[196] Deoxygenation of 2-ferrocenylnorbornan-2-ols with LiAlH$_4$–AlCl$_3$[198] and reduction of alkyl(tri-n-butylphosphine)copper(I) compounds to the corresponding alkanes with hydrido(tri-n-butylphosphine)copper(I)[199] both proceed with retention of configuration at the reaction centre.

Although homolytic mechanisms are well established for many dehalo-genation reactions using organotin hydrides, both tin and germanium hydrides[200] can reduce carbonium ions by hydride transfer and are, in fact, more reactive than the corresponding silicon[201] compounds. For example, olefin formation accompanying reduction of the 4-t-butylcyclohexyl cation with Ph$_3$SnH or Ph$_3$GeH (12%) is much less extensive than it is with Ph$_3$SiH (72%).[200] Diborane in nitromethane dehalogenates (some) arylhalogeno-methanes,[202] and optically active LiN(Ph)CH(Ph)Me can induce asymmetry

[196] D. J. Pasto and J. A. Gontarz, *J. Am. Chem. Soc.*, **91**, 719 (1969); but see ref. **197**, footnote 6.
[197] G. A. Gray and W. R. Jackson, *J. Am. Chem. Soc.*, **91**, 6205 (1969).
[198] M. J. A. Habib and W. E. Watts, *J. Chem. Soc.*(C), **1969**, 1469.
[199] G. M. Whitesides, J. S. Filippo, E. R. Stredronsky, and C. P. Casey, *J. Am. Chem. Soc.*, **91**, 6542 (1969).
[200] F. A. Carey and H. S. Tremper, *Tetrahedron Letters*, **1969**, 1645.
[201] F. A. Carey and H. S. Tremper, *J. Org. Chem.*, **34**, 4 (1969).
[202] S. Matsumura and N. Tokura, *Tetrahedron Letters*, **1969**, 363.

in the product from reduction of a ketone, by way of hydride transfer in a six-membered cyclic transition state.[203]

Reduction of ketones by aluminium isopropoxide[204] still attracts attention. The effects of substituents on the reduction of acetophenone fail to correlate with either σ or σ^{+}.[205] The stereochemistry of the product from the reduction of 2-ethylchroman-4-one is highly dependent on whether or not acetone is continuously removed; its presence assists isomerization of the first-formed *cis*-chromanol.[206]

Controlled reduction of a polycyclic aromatic hydrocarbon with lithium in liquid ammonia generally gives, in high yield, only the dihydro-compound resulting from addition of hydrogen at the positions predicted by MO calculations to have greatest charge-density in the intermediate di-anion.[207] 9,10-Diethylanthracene gives the 9,10-dihydro-derivative consisting entirely of the *trans*-isomer (**80**; R = Et, axial,equatorial). Harvey and coworkers[207] suggest that stabilization of di-anion (**76**; R = Et) by overlap of orbitals on C-9 and C-10 with the aromatic rings requires that the ethyl groups be equatorial, and that initial protonation at C-9 from the axial direction gives

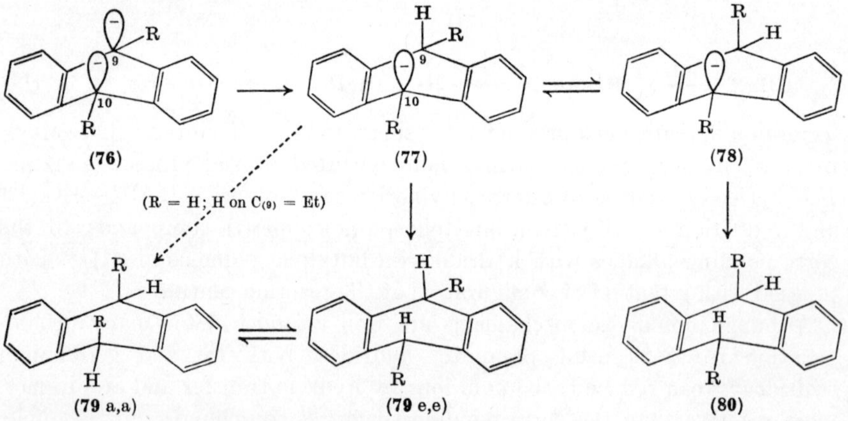

(R = H; H on C$_{(9)}$ = Et)

(**76**)   (**77**)   (**78**)

(**79** a,a)   (**79** e,e)   (**80**)

mono-anion (**77**; R = Et) which can invert to (**78**; R = Et). Because steric interaction between the equatorial ethyl group on C-9 and the *peri* hydrogens destabilizes conformation (**77**) relative to (**78**), protonation of the preferred conformation of the mono-anion will lead to *trans*-dihydro-product. To account

[203] G. Wittig and U. Thiele, *Ann. Chem.*, **726**, 1 (1969).

[204] V. J. Shiner and D. Whittaker, *J. Am. Chem. Soc.*, **91**, 394 (1969).

[205] Z. Csuros, V. Kalman, A. Lengyel-Meszaros, and J. Petro, *Period. Polytech., Chem. Eng.* (Budapest), **12**, 161 (1968); *Chem. Abs.*, **70**, 19347 (1969).

[206] K. Hanaya and K. Furuse, *J. Chem. Soc. Japan*, **89**, 1002 (1968); *Chem. Abs.*, **70**, 56922 (1969).

[207] R. G. Harvey, L. Arzadon, J. Grant, and K. Urberg, *J. Am. Chem. Soc.*, **91**, 4535 (1969).

for the product from 9,10-dimethylanthracene consisting of equal amounts of the *cis*-(**79**; R = Me, diequatorial or diaxial) and *trans*-(**80**; R = Me) isomers, it is argued that steric interactions in the mono-anion are no longer significant and that reaction occurs equally via conformations (**77**; R = Me) and (**78**; R = Me). Alkylation of the di-anion (**76**; R = H) from anthracene with ethyl bromide[208] gives predominantly *cis*-9,10-diethyl-9,10-dihydroanthracene and is compatible with the view that conformation (**77**; R = H, H on C-9 = Et) is preferred, but it seems surprising that alkylation with methyl bromide[208] should exhibit comparable stereoselectivity.

Anisole,[209] *N*-phenylindole,[210] and [2.2]metacyclophane[211] are among the aromatic compounds whose reduction by sodium in liquid ammonia has been discussed. Whereas 1,4-dimethylnaphthalene gives the 5,8-dihydro-derivative, 1,4-bistrimethylsilylnaphthalene is reduced in the substituted ring, in accord with calculations suggesting that the π-spin population is greatest at positions 1 and 4 because of interaction of unoccupied silicon *d*-orbitals with the π-electron system.[212] The alkali-metal reductions of tetraalkylammonium halides,[213] alkyl halides,[214] α,β-unsaturated ketones,[215] and cholesta-3,5-

(81) (82) (83) (84)

dieno[3,4-*b*]oxathian[216] have been examined, and Taylor[217] has disputed the view, discussed last year,[218] that reduction of a simple cyclohexanone proceeds mainly through an anion radical rather than a di-anion intermediate. The products of sodium reduction of the non-enolizable 1,3-diketone (**81**) have been rationalized by supposing that the anion radical (**82**) can be further reduced to the di-anion diradical (**83**) which can rearrange to di-anion (**84**).[219] The reactions of hydrated electrons have been discussed.[220]

[208] R. G. Harvey and L. Arzadon, *Tetrahedron*, **25**, 4887 (1969); R. G. Harvey and C. C. Davis, *J. Org. Chem.*, **34**, 3607 (1969).
[209] D. R. Burnham, *Tetrahedron*, **25**, 897 (1969).
[210] B. Heath-Brown, *Chem. Ind.* (London), **1969**, 1595.
[211] J. Reiner and W. Jenny, *Helv. Chim. Acta*, **52**, 1624 (1969).
[212] H. Alt, E. R. Franke, and H. Bock, *Angew. Chem. Internat. Ed. Engl.*, **8**, 525 (1969).
[213] R. R. Dewald and K. W. Browall, *Chem. Comm.*, **1968**, 1511.
[214] J. Jacobus and J. F. Eastham, *Chem. Comm.*, **1969**, 138.
[215] W. Herz and J. J. Schmid, *J. Org. Chem.*, **34**, 3473 (1969).
[216] A. Ishida, Y. Hiyoshi, T. Koga, and M. Tomoeda, *Chem. Pharm. Bull.* (Japan), **17**, 355 (1969); *Chem. Abs.*, **70**, 106752 (1969).
[217] D. A. H. Taylor, *Chem. Comm.*, **1969**, 476.
[218] *Org. Reaction Mech.*, **1968**, 487.
[219] R. Le Goaller, M. Rougier, C. Zimero, and P. Arnaud, *Tetrahedron Letters*, **1969**, 4193.
[220] M. H. Studier and E. J. Hart, *J. Am. Chem. Soc.*, **91**, 4068 (1969); E. J. Hart, *Accounts Chem. Res.*, **2**, 161 (1969); M. Anbar, *Adv. Phys. Org. Chem.*, **7**, 115 (1969).

Reduction of phenyl(trichloromethyl)carbinol with zinc, zinc amalgam, or zinc–copper in neutral or acidic media gives β,β-dichlorostyrene (**86**) together with a small amount of (**85**) or (**87**).[221] In the mechanism outlined in Scheme 6 reduction is envisaged as a heterogeneous reaction on the metal surface, with

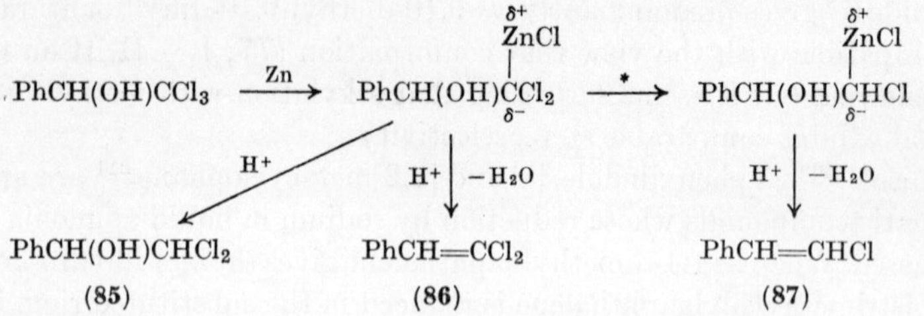

$$\text{PhCH(OH)CCl}_3 \xrightarrow{\text{Zn}} \overset{\overset{\delta+}{\underset{|}{\text{ZnCl}}}}{\text{PhCH(OH)}\underset{\delta-}{\text{C}}\text{Cl}_2} \xrightarrow{\quad * \quad} \overset{\overset{\delta+}{\underset{|}{\text{ZnCl}}}}{\text{PhCH(OH)}\underset{\delta-}{\text{C}}\text{HCl}}$$

$$\overset{\text{H}^+}{\swarrow} \qquad \text{H}^+ \Big| -\text{H}_2\text{O} \qquad\qquad \text{H}^+ \Big| -\text{H}_2\text{O}$$

$$\text{PhCH(OH)CHCl}_2 \qquad\qquad \text{PhCH}{=}\text{CCl}_2 \qquad\qquad \text{PhCH}{=}\text{CHCl}$$
$$(\textbf{85}) \qquad\qquad\qquad (\textbf{86}) \qquad\qquad\qquad (\textbf{87})$$

* Elimination of Cl$^-$ followed by H$^-$ transfer from the metal surface to the electron-deficient C atom.

Scheme 6

the extent to which each reaction path is followed depending on the experimental conditions.[221] Clemmensen reduction of diketones has been reviewed.[222] Investigations of the reduction of 1-tetralones[223] and arylsulphonyl halides[224] by zinc, of 3-carboxamidoquinolinium salts by HCO_2D and DCO_2H in triethylamine,[225] of bipyridylium salts by dithionite,[226] and of S,S-dimethyl-sulphiminium salts by iodide ion[227] have been described. Equations (15)—(18)

$$\text{Me}_2\text{SNH}_2{}^+ + \text{H}^+ \rightleftharpoons \text{Me}_2\text{SNH}_3{}^{2+} \qquad \ldots (15)$$

$$\text{Me}_2\text{SNH}_3{}^{2+} + \text{I}^- \xrightarrow{\text{slow}} \text{Me}_2\text{SI}^+ + \text{NH}_3 \qquad \ldots (16)$$

$$\text{Me}_2\text{SI}^+ + 2\,\text{I}^- \longrightarrow \text{Me}_2\text{S} + \text{I}_3{}^- \qquad \ldots (17)$$

$$\text{NH}_3 + \text{H}^+ \rightleftharpoons \text{NH}_4{}^+ \qquad \ldots (18)$$

are in accord with the kinetics of the latter reaction; the unusually pronounced solvent isotope effect $[k(\text{H}_2\text{O})/k(\text{D}_2\text{O}) = 0.27]$ for a mechanism involving specific proton transfer prior to the rate-determining step probably arises from ND_3, rather than NH_3, being the leaving group in the slow step in D_2O.[227]

[221] E. Kiehlmann, R. J. Bianchi, and W. Reeve, *Can. J. Chem.*, **47**, 1521 (1969).
[222] J. G. St. C. Buchanan and P. D. Woodgate, *Quart. Rev.*, **23**, 522 (1969).
[223] J. Gardent, G. Hazebroucq, and G. Cormier, *Bull. Soc. Chim. France*, **1969**, 4001.
[224] N. Kunieda, K. Sakai, and S. Oae, *Bull. Chem. Soc. Japan*, **41**, 3015 (1968).
[225] L. R. Gizzi and M. M. Joullié, *Tetrahedron Letters*, **1969**, 3117.
[226] J. G. Carey, J. F. Cairns, and J. E. Colchester, *Chem. Comm.*, **1969**, 1280.
[227] J. H. Krueger, *J. Am. Chem. Soc.*, **91**, 4974 (1969).

Other reports have considered the mechanisms of reductions with hydrazine,[228a] and the reactions of sulphoxides with optically active phosphonothioic acids,[228b] benzenesulphonyl chloride with cuprous chloride,[229] and pyracylene with di-iron nonacarbonyl.[230]

Stereochemical similarities between electrochemical and Zn–AcOH reduction of *gem*-dihalogenocyclopropanes have become apparent.[231] For example, both procedures exclusively remove bromine from each isomer of 7-bromo-7-chlorobicyclo[4.1.0]heptane, and give predominantly that chlorocyclopropane in which the original configuration is retained.[231] Electrochemical dehalogenation has also been studied with compounds containing epoxide,[232] ether,[233] ketone,[234] and carboxylate[235] functions in addition to one or more halogen atoms. Papers concerning the cathodic reduction of aromatic aldehydes[236a] and ketones,[236] allenic ketones,[237] α,β-unsaturated aldehydes,[238] 1,3-diketones,[239] and thioketones[240] have been presented. Although C-3—C-5 bond formation may be significant in the excited ($n \rightarrow \pi^*$) state of 4,4-diphenylcyclohexa-2,5-dienone, cross-bond formation does not occur when the 7 π-electron system is generated electrochemically.[241] The first polarographic wave of 2-t-butyl-3-phenyloxaziridine in aqueous alcohol occurs at very positive potentials and corresponds to the formation of PhCH(OH)NHBu^t. At pH 1—4 competitive elimination of t-butylamine and water produces benzaldehyde and *N*-t-butylbenzaldimine which are further reduced to benzyl alcohol and *N*-t-butylbenzylamine.[242] Among the many other electrochemical reductions investigated are those of olefinic[243] and acetylenic[243a]

[228a] G. K. Koch, *J. Labelled Compounds*, **5**, 99 (1969); *Chem. Abs.*, **71**, 48961 (1969).

[228b] M. Mikolajczyk and M. Para, *Chem. Comm.*, **1969**, 1192.

[229] A. Orochov, M. Asscher, and D. Vofsi, *J. Chem. Soc.*(B), **1969**, 255.

[230] B. M. Trost and G. M. Bright, *J. Am. Chem. Soc.*, **91**, 3689 (1969).

[231] R. E. Erickson, R. Annino, M. D. Scanlon, and G. Zon, *J. Am. Chem. Soc.*, **91**, 1767 (1969); see also A. J. Fry and R. H. Moore, *J. Org. Chem.*, **33**, 1283 (1968).

[232] A. Cisak, *Rocz. Chem.*, **42**, 907 (1968); *Chem. Abs.*, **69**, 64112 (1968).

[233] K. P. Butin, N. A. Belokoneva, V. N. Eldikov, N. S. Zefirov, I. P. Beletskaya, and O. A. Reutov, *Izv. Akad. Nauk SSSR, Ser. Khim.*, **1969**, 254; *Chem. Abs.*, **70**, 114463 (1969).

[234] N. Limosin and E. Laviron, *Bull. Soc. Chim. France*, **1969**, 4189.

[235] B. Czochralska, *Rocz. Chem.*, **42**, 895 (1968); *Chem. Abs.*, **69**, 64111 (1968).

[236a] J. H. Stocker, R. M. Jenevein, and D. H. Kern, *J. Org. Chem.*, **34**, 2810 (1969).

[236b] J. H. Stocker and R. M. Jenevein, *J. Org. Chem.*, **34**, 2807 (1969); M. K. Kalinowski and A. Lasia, *Rocz. Chem.*, **43**, 1265 (1969).

[237] P. Martinet, J. Simonet, and M. Morenas, *Compt. Rend.*, *C*, **268**, 253 (1969).

[238] D. Barnes and P. Zuman, *Trans. Faraday Soc.*, **65**, 1668, 1681 (1969).

[239] T. J. Curphey, C. W. Amelotti, T. P. Layloff, R. L. McCartney, and J. H. Williams, *J. Am. Chem. Soc.*, **91**, 2817 (1969).

[240] R. M. Elofson, F. F. Gadallah, and L. A. Gadallah, *Can. J. Chem.*, **47**, 3979 (1969); W. Kemula, H. Kryszczynska, and M. K. Kalinowski, *Rocz. Chem.*, **43**, 1071 (1969); *Chem. Abs.*, **71**, 87134 (1969).

[241] A. Mazzenga, D. Lomnitz, J. Villegas, and C. J. Polowczyk, *Tetrahedron Letters*, **1969**, 1665.

[242] H. Lund, *Acta Chem. Scand.*, **23**, 563 (1969).

[243a] L. Horner and H. Röder, *Ann. Chem.*, **723**, 11 (1969).

compounds, phenylazonaphthols,[244] pyridazines,[245] imines and semicarbazones,[246] dinitrobenzenes,[247] sulphones,[248] benzenesulphonamides and carboxamides,[249] diazonium salts,[250] naphthoic acids,[251] and aryl nitriles.[252]

Hydrogenation and Hydrogenolysis

The isomerization of damsin catalysed by $RhCl(PPh_3)_3$ suggested that a normal homogeneous hydrogenation proceeds by *reversible* formation of the olefin–$RhH_2Cl(PPh_3)_2$ complex, followed by *non-synchronous* transfer of two H atoms from the metal to the substrate with formation of an alkylrhodium intermediate.[253] Results now available fully support that view. Addition of D_2 to 1,4-dimethylcyclohexene in benzene yields equal amounts of *cis*- and *trans*-1,4-dimethylcyclohexane but the deuterium distribution in the *cis*-isomer (3.9% D_1, 88% D_2, 7.0% D_3) differs from that in the *trans*-isomer (3.4% D_1, 93.7% D_2, 1.1% D_3). Moreover, a small amount of the recovered olefin consists of D_1-species. Hussey and Takeuchi[254] suggest that the *cis*- and *trans*-isomers of (88; X = D, Y = H) are reversibly formed at similar rates, and rearrange reversibly to the isomeric alkylrhodium intermediates (89;

(88)　　　　　　(89)

X = D, Y = H). These then collapse to *cis*- and *trans*-1,4-dimethyldideuterocyclohexane. Because it is more strained, *trans*-(89) is formed rather more slowly than *cis*-(89), but because *trans*-1,4-dimethylcyclohexane is more stable than the *cis*-compound, *trans*-(89) collapses to dideutero-product more quickly than does *cis*-(89). For *cis*-(89), therefore, alternative reactions such as formation of *cis*-(88; X = H, Y = D) followed by HD–D_2 exchange to

243b J. F. Archer and J. Grimshaw, *J. Chem. Soc.*(B), **1969**, 266.

244 S. Millefiori and G. Campo, *Ann. Chim.* (Italy), **59**, 128 (1969); S. Millefiori, *ibid.*, p. 138.

245 S. Millefiori, *Ann. Chim.* (Italy), **59**, 15 (1969).

246 A. J. Fry and R. G. Reed, *J. Am. Chem. Soc.*, **91**, 6448 (1969); D. Fleury, *Bull. Soc. Chim. France*, **1969**, 3763.

247 A. Tallec, *Ann. Chim.* (France), **4**, 67 (1969); *Chem. Abs.*, **71**, 55929 (1969).

248 O. Manoušek, O. Exner, and P. Zuman, *Coll. Czech. Chem. Comm.*, **33**, 3988 (1968).

249 L. Horner and R.-J. Singer, *Ann. Chem.*, **723**, 1 (1969); O. Manoušek, O. Exner, and P. Zuman, *Coll. Czech. Chem. Comm.*, **33**, 4000 (1968).

250 F. F. Gadallah and R. M. Elofson, *J. Org. Chem.*, **34**, 3335 (1969).

251 N. M. Przhiyalgovskaya, N. S. Yares'ko, and V. N. Belov, *Zh. Org. Khim.*, **4**, 896 (1968); *Chem. Abs.*, **69**, 18383 (1968).

252 O. Manoušek, P. Zuman, and O. Exner, *Coll. Czech. Chem. Comm.*, **33**, 3979 (1968).

253 *Org. Reaction Mech.*, **1968**, 494.

254 A. S. Hussey and Y. Takeuchi, *J. Am. Chem. Soc.*, **91**, 672 (1969).

give *cis*-(**88**; X = Y = D) and ultimately trideutero-product, can compete relatively effectively.[254] Complications are not encountered in the addition of D_2 to straight-chain olefins.[255]

Heathcock and Poulter[256] invoke alkylrhodium intermediates to rationalize (Scheme 7; L = PPh_3) the formation of ethylcyclopropane (85%) and n-pentane (14%) in the $RhCl(PPh_3)_3$-catalysed hydrogenation of vinylcyclopropane. Initial transfer of a single hydrogen to the C atoms α and β to the cyclopropane ring (paths a and b) would give two alkylrhodium intermediates, leading to

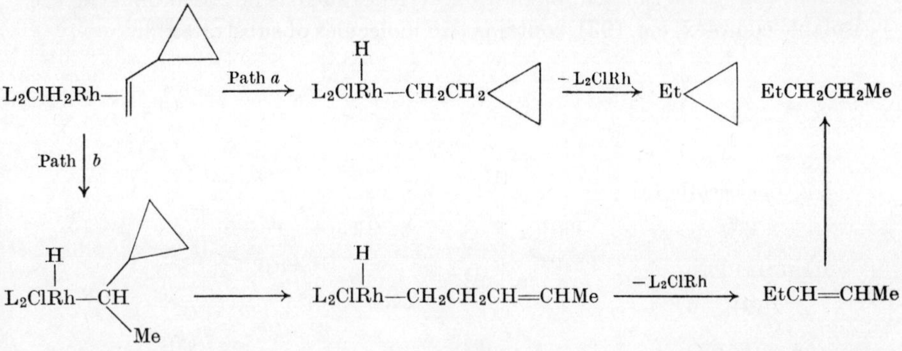

$$L_2ClH_2Rh- \quad \xrightarrow{\text{Path } a} \quad L_2ClRh-\overset{H}{C}H_2CH_2\triangleleft \quad \xrightarrow{-L_2ClRh} \quad Et\triangleleft \quad EtCH_2CH_2Me$$

$$\text{Path} \downarrow b$$

$$L_2ClRh-\overset{H}{C}H\underset{Me}{\triangleleft} \quad \longrightarrow \quad L_2ClRh-CH_2CH_2CH=CHMe \quad \xrightarrow{-L_2ClRh} \quad EtCH=CHMe$$

Scheme 7

the products of hydrogenation and hydrogenolysis respectively. Assuming a secondary alkylrhodium bond to be of lower energy than the corresponding tertiary bond, the smaller proportion of hydrogenolysis (path b, 1.5%) accompanying hydrogenation of 2-cyclopropylpropene can be readily understood.[256]

Although oxidation of ligand phosphine could conceivably account for the observed acceleration of $RhCl(PPh_3)_3$-catalysed hydrogenations by small amounts of oxygen,[257] it seems more likely that removal of traces of unco-ordinated phosphine—a known poison—is responsible.[258] Horner and co-workers[259] have explained how α-ethylstyrene and α-methoxystyrene can complex to $RhClH_2[(S)\text{-}MePhPr^nP]_2$ so that (S)-2-phenylbutane and (R)-1-methoxy-1-phenylethane respectively are preferentially formed. Reports of homogeneous catalysis by rhodium complexes with aminophosphines,[260] amino-acids,[261] and sulphides[262] as ligands have appeared.

[255] J. R. Morandi and H. B. Jensen, *J. Org. Chem.*, **34**, 1889 (1969).

[256] C. H. Heathcock and S. R. Poulter, *Tetrahedron Letters*, **1969**, 2755.

[257] H. van Bekkum, F. van Rantwijk, and T. van de Putte, *Tetrahedron Letters*, **1969**, 1.

[258] C. O'Connor and G. Wilkinson, *Tetrahedron Letters*, **1969**, 1375.

[259] L. Horner, H. Siegel, and H. Büthe, *Angew. Chem. Internat. Ed. Engl.*, **7**, 942 (1968).

[260] Y. Chevallier, R. Stern, and L. Sajus, *Tetrahedron Letters*, **1969**, 1197.

[261] O. N. Efimov, M. L. Khidekel, V. A. Avilov, P. S. Chekrii, O. N. Eremenko, and A. G. Ovcharenko, *Zh. Obshch. Khim.*, **38**, 2668 (1968); *Chem. Abs.*, **70**, 78344 (1969).

[262] B. R. James, F. T. T. Ng, and G. L. Rempel, *Inorg. Nucl. Chem. Letters*, **4**, 197 (1968).

The disproportionation that accompanies heterogeneous hydrogenation of 1,4-dihydroaromatic compounds can be avoided by using RhCl(PPh₃)₃ as catalyst,[263] although a number of iridium complexes catalyse the disproportionation of cyclohexa-1,4-diene under mild conditions.[264] Variation of either phosphine or halogen ligands in iridium catalyst (**90**) influences its reactivity; of the ligands examined, iodine and triphenylphosphine cause most rapid formation of the hydride complex.[265] Reaction of fumaronitrile or some other activated olefin with iridium complex (**91**) allows a product such as (**92**) to be isolated, but with acetylenes such as hexafluorobutyne the isolable complex, e.g. (**93**), contains two molecules of substrate.[266]

Ir(CO)halogen(R₃P)₂

(**90**)

IrH(CO)(PPh₃)₃

(**91**)

(**92**)

(**93**)

Homogeneous catalytic hydrogenation has been reviewed,[267] and investigations of hydrogenation and other reactions employing complexes of chromium,[268] molybdenum and tungsten,[269] cobalt,[270] nickel,[271] titanium,[272] rhodium,[273] and ruthenium[274] have been described. Protonation of the bridged carboxylate Rh₂(OAc)₄ by HBF₄ gives the binuclear cationic species

[263] J. J. Sims, V. K. Honwad, and L. H. Selman, *Tetrahedron Letters*, **1969**, 87.

[264] J. E. Lyons, *Chem. Comm.*, **1969**, 564.

[265] W. Strohmeier and T. Onoda, *Z. Naturforsch.*, **24b**, 461, 515 (1969).

[266] W. H. Baddley and M. S. Fraser, *J. Am. Chem. Soc.*, **91**, 3661 (1969).

[267] G. Wilkinson, *Bull. Soc. Chim. France*, **1968**, 5055; M. E. Vol'pin and I. S. Kolomnikov, *Uspekhi Khim.*, **38**, 561 (1969); *Chem. Abs.*, **71**, 21337 (1969). Several papers of interest are contained in *Discuss. Faraday Soc.*, **46** (1968).

[268] A. I. Yakubchik, B. I. Tikhomirov, and I. A. Klopotova, Katal. Reakts. Zhidk. Faze, 2nd Tr. Vses. Konf., Alma-Ata, Kaz. SSR, 1966, p. 436; *Chem. Abs.*, **69**, 2259 (1968).

[269] A. Miyake and H. Kondo, *Angew. Chem. Internat. Ed. Engl.*, **7**, 880 (1968).

[270] W. Strohmeier and N. Iglauer, *Z. Phys. Chem.* (Frankfurt), **61**, 29 (1968); D. Seyferth and R. J. Spohn, *J. Am. Chem. Soc.*, **91**, 6192 (1968).

[271] E. W. Duck, J. M. Locke, and C. J. Mallinson, *Ann. Chem.*, **719**, 69 (1968).

[272] R. Stern, G. Hillion, and L. Sajus, *Tetrahedron Letters*, **1969**, 1561.

[273] C. K. Brown and G. Wilkinson, *Tetrahedron Letters*, **1969**, 1725; R. E. Harmon, J. L. Parsons, D. W. Cooke, S. K. Gupta, and J. Schoolenberg, *J. Org. Chem.*, **34**, 3684 (1969); W. Strohmeier and W. Rehder-Stirnweiss, *J. Organometal. Chem.*, **18**, P28 (1969).

[274] D. Rose, J. D. Gilbert, R. P. Richardson, and G. Wilkinson, *J. Chem. Soc.*(A), **1969**, 2610; B. Hui and B. R. James, *Chem. Comm.*, **1969**, 198; A. C. Skapski and F. A. Stephens, *ibid.*, p. 1008.

Rh_2^{4+} which in the presence of stabilizing ligands forms a catalyst for hydrogenation.[275]

While being an efficient catalyst for homogeneous hydrogenation of olefins,[276] the complex $[py_2(HCONMe_2)RhCl_2(BH_4)]$ obtained from trichlorotripyridylrhodium and $NaBH_4$ in dimethylformamide displays, under certain conditions, characteristics usually associated with heterogeneous catalysts.[277] Thus, for each of a series of cyclic alkenes hydrogenation proceeds at a rate proportional to olefin concentration at low olefin-to-catalyst ratios, but at higher ratios (>ca. 100:1) is zero-order in alkene; hydrogen transfer is now rate-limiting, as it is in heterogeneous (Pd–C) hydrogenation of cycloalkenes. Moreover, the relative rate for each olefin under these conditions (norbornene > cyclohexene > cycloheptene > cyclopentene > cyclooctene) is that found in heterogeneous hydrogenation, and correlates with the heat of hydrogenation. Presumably the transition state is alkane-like, with the energy of activation closely related to the energy of the $sp^2 \rightarrow sp^3$ transformation.[277] Unlike many homogeneous catalysts, this complex is effective in the hydrogenation of 3-oxo-$\Delta^{4,5}$-steroids,[278] and gives a ratio of products (5α-H:5β-H) similar to that obtained over Rh–C.[279]

Hydrogenation of 4-t-butylmethylenecyclohexane over Pd–C yields quantities of *cis*- and *trans*-4-t-butylmethylcyclohexanes, as well as isomerized olefin, similar to those obtained homogeneously with $RhCl(PPh_3)_3$. Augustine and Van Peppen[280] have considered this and other results in their mechanistic comparison of heterogeneous and homogeneous catalysis. Many other authors have considered heterogeneous hydrogenation of acyclic olefins,[281] including the styrene derivatives p-$Me_3MC_6H_4CH=CH_2$ where reactivity over Raney nickel decreases as M changes $Si \rightarrow C \rightarrow Ge \rightarrow Sn$,[282] cyclic olefins,[283] allenes,[284]

[275] P. Legzdins, G. L. Rempel, and G. Wilkinson, *Chem. Comm.*, **1969**, 825.
[276] I. Jardine and F. J. McQuillin, *Chem. Comm.*, **1969**, 477; P. Abley and F. J. McQuillin, *ibid.*, p. 477.
[277] I. Jardine and F. J. McQuillin, *Chem. Comm.*, **1969**, 502.
[278] I. Jardine and F. J. McQuillin, *Chem. Comm.*, **1969**, 503.
[279] I. Jardine, R. W. Howsam, and F. J. McQuillin, *J. Chem. Soc.*(C), **1969**, 260.
[280] R. L. Augustine and J. Van Peppen, *Ann. N.Y. Acad. Sci.*, **158**, 482 (1969).
[281] G. C. Bond and J. M. Winterbottom, *Trans. Faraday Soc.*, **65**, 2779 (1969); J. Horiuti and K. Miyahara, *Z. Phys. Chem.* (Frankfurt), **64**, 36 (1969); C. A. Brown, *J. Am. Chem. Soc.*, **91**, 5901 (1969); F. Bozon-Verduraz and S. J. Teichner, *J. Catal.*, **11**, 7 (1968).
[282] E. A. Chernyshev, I. F. Zhukova, T. L. Krasnova, and L. K. Freidlin, *Zh. Obshch. Khim.*, **38**, 504 (1968); *Chem. Abs.*, **69**, 51331 (1968).
[283] W. C. Baird, B. Franzus, and J. H. Surridge, *J. Org. Chem.*, **34**, 2944 (1969); A. S. Hussey and G. P. Nowack, *ibid.*, p. 439; C. A. Brown, *Chem. Comm.*, **1969**, 952; H. van Bekkum, F. van Rantwijk, G. van Minnen-Pathuis, J. D. Remijnse, and A. van Veen, *Rec. Trav. Chim.*, **88**, 911 (1969); G. V. Smith and M. C. Menon, *Ann. N.Y. Acad. Sci.*, **158**, 501 (1969); E. S. Balenkova, V. I. Alekseeva, G. I. Khromova, and S. I. Khromov, *Neftekhim.*, **9**, 184 (1969); *Chem. Abs.*, **71**, 21426 (1969).
[284] L. Crombie and P. A. Jenkins, *Chem. Comm.*, **1969**, 394.

conjugated dienes,[285] simple acetylenes[286] and ethynylcarbinols,[287] α,β-unsaturated carbonyl compounds,[279, 288] and aromatic hydrocarbons.[289] Hydrogenolysis of the cyclopropane ring in tricyclo[4.4.1.0]undecane over PtO$_2$ yields *cis-* and *trans*-9-methyldecalins in equal amounts; since the products do not isomerize under the reaction conditions, hydrogenolysis is clearly non-stereospecific.[290]

While important mechanistic information still comes from stereochemical studies of the catalytic hydrogenolysis of derivatives of benzyl alcohol,[291] attention this year has been focused primarily on benzylamines,[292] including

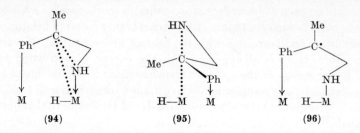

(94)   (95)   (96)

phenylaziridines.[293] Optically active 2-methyl-2-phenylaziridine invariably gives a high yield of optically active 2-phenylpropylamine, but predominant inversion of configuration over Pd contrasts with retention over Pt. Doubtless the affinity between N and the catalyst favours a transition state like (94)

[285] J. J. Phillipson, P. B. Wells, and G. R. Wilson, *J. Chem. Soc.*(A), **1969**, 1351; P. B. Wells and A. J. Bates, *ibid.*, **1968**, 3064.

[286] G. C. Bond, G. Webb, and P. B. Wells, *J. Catal.*, **12**, 157 (1968); R. S. Mann and K. C. Khulbe, *ibid.*, **10**, 401 (1968); R. S. Mann and S. C. Naik, *Indian J. Technol.*, **6**, 57 (1968); *Chem. Abs.*, **69**, 51330 (1968); R. S. Mann and K. C. Khulbe, *J. Phys. Chem.*, **73**, 2104 (1969).

[287] G. V. Movsisyan, G. A. Chukhadzhyan, and A. A. Aleksanyan, *Armyan. Khim. Zh.*, **21**, 474 (1968); *Chem. Abs.*, **70**, 10993 (1969).

[288] R. L. Augustine, D. C. Migliorini, R. E. Foscante, C. S. Sodano, and M. J. Sisbarro, *J. Org. Chem.*, **34**, 1075 (1969); F. J. McQuillin, R. W. Howsam, and I. Jardine, *Ann. N.Y. Acad. Sci.*, **158**, 492 (1969).

[289] H. van Bekkum, H. M. A. Buurmans, G. van Minnen-Pathuis, and B. M. Wepster, *Rec. Trav. Chim.*, **88**, 779 (1969); O. N. Efimov, O. N. Eremenko, A. G. Ovcharenko, M. L. Khidekel, and P. S. Chekrii, *Izv. Akad. Nauk SSSR, Ser. Khim.*, **1969**, 855; *Chem. Abs.*, **71**, 29824 (1969); N. E. Zlotina and S. L. Kiperman, *Kinet. Katal.*, **8**, 1335 (1967); *Chem. Abs.*, **69**, 35081 (1968); Y. Bahurel, G. Descotes, and J. Sabadie, *Bull. Soc. Chim. France*, **1969**, 3232; see also N. Belorizky and D. Gagnaire, *Compt. Rend.*, *C*, **268**, 688 (1969); A. B. Vol-Epshtein, S. G. Gagarin, and I. N. Dobrushkina, *Kinet. Katal.*, **10**, 581 (1969); *Chem. Abs.*, **71**, 48957 (1969).

[290] Z. Majerski and P. von R. Schleyer, *Tetrahedron Letters*, **1968**, 6195.

[291] S. Mitsui, Y. Kudo, and M. Kobayashi, *Tetrahedron*, **25**, 1921 (1969); S. Mitsui, M. Fujimoto, Y. Nagahisa, and T. Sukegawa, *Chem. Ind.* (London), **1969**, 241.

[292] C. Ó. Murchú, *Tetrahedron Letters*, **1969**, 3231; H. Dahn and J. Garbarino, *Bull. Soc. Vaud. Sci. Nat.*, **70**, 27 (1968); *Chem. Abs.*, **69**, 76410 (1968).

[293] S. Mitsui and Y. Sugi, *Tetrahedron Letters*, **1969**, 1287, 1291.

with Pt, in contrast to one like (95) with Pd.[293] It might seem surprising that Ni and Co, which have especially strong affinities for N, give 2-phenylpropylamine with a relatively small excess of retention over inversion. Mitsui and Sugi[293] attribute this lower stereospecificity to significant participation by the benzylic radical (96), resulting from complete rupture of the benzylic C–N bond on the catalyst.

Heterogeneous hydrogenation of aldehydes[294] and ketones[295] has been further studied. Although bulky *o*-substituents greatly retard reaction, the rate of hydrogenation of acetophenone over Pd increases with the size of alkyl groups in the *m*- and *p*-positions.[296] The influence of substituents on the rate of hydrogenation of nitrobenzene[297] and 2-nitrofuran[298] has been examined, and reviews of the mechanisms of desulphuration by Raney nickel,[299] and the use of D_2 as a tracer in catalytic reactions,[300] have appeared.

[294] V. V. Abalyaeva, A. S. Astakhova, E. N. Bakhanova, and M. L. Khidekel, *Izv. Akad. Nauk SSSR, Ser. Khim.*, **1969**, 89; *Chem. Abs.*, **70**, 114740 (1969).

[295] Y. Watanabe, Y. Mizuhara, and M. Shiota, *Can. J. Chem.*, **47**, 1495 (1969); see also K. Harada and K. Matsumoto, *J. Org. Chem.*, **33**, 4467, 4526 (1968).

[296] H. van Bekkum, A. P. G. Kieboom, and K. J. G. van de Putte, *Rec. Trav. Chim.*, **88**, 52 (1969).

[297] V. Růžička and H. Šantrochová, *Coll. Czech. Chem. Comm.*, **34**, 2999 (1969); see also A. M. Sokol'skaya, D. V. Sokol'skii, and S. M. Reshetnikov, *Kinet. Katal.*, **8**, 1331 (1967); *Chem. Abs.*, **69**, 35127 (1969).

[298] A. V. Finkel'shtein and G. A. Reutov, *Reakts. Sposob. Org. Soed.*, **5**, 909 (1968); *Chem. Abs.*, **71**, 48938 (1969).

[299] W. A. Bonner and R. A. Grimm, *Org. Sulfur Compounds*, **2**, 35 (1966).

[300] R. L. Burwell, *Accounts Chem. Res.*, **2**, 289 (1969).

Author Index 1969

19

20

Cumulative Subject Index, 1965—1969

Anhydrolactams, rearrangement, **66**, 239

Aniline, effect of coordination to cobalt on rate of bromination, **67**, 196

Anils, conversion into benzimidazoles and quinoxalines, **68**, 251

Anion radicals, *see* Radical anions

Anisole,
electrophilic substitution in, **65**, 163–164; **66**, 197; **68**, 215
in the presence of cyclohexa-amylose, **69**, 250
photolysis, **69**, 520–521
protonation, **66**, 200

Annulenes, electrophilic substitution in, **67**, 194

[10]Annulenes, as intermediates in the photo-isomerization of dihydro-naphthalenes, **69**, 515–516

[16]Annulene,
di-anion, **69**, 127
photorearrangement, **67**, 396

Anthracene,
electrophilic substitution, **66**, 206; **67**, 192; **68**, 206, 248
fluorescence, **69**, 484, 486
intersystem crossing of, **67**, 406
oxidation, **67**, 265; **68**, 482; **69**, 552
photochemistry, **69**, 520
photodimerization, **67**, 406; **68**, 444; **69**, 520
photo-oxidation, **69**, 536
photoreduction, **69**, 520
proton adduct of, **67**, 198
radical anion, **65**, 219; **69**, 538

Anthracene-9-carboxaldehyde, photolysis, **68**, 444

Anthranils,
electrophilic substitution, **66**, 208
photoisomerization, **68**, 447–448

Antimony, nucleophilic displacement at, **69**, 109

Aromatic hydrocarbons, protonation, **65**, 159; **66**, 200; **68**, 218

Arsenic, nucleophilic displacement at, **65**, 72; **68**, 96; **69**, 109

1,2-Aryl shifts,
in carbenes, **67**, 283
in nitrenes, **68**, 340

1,2-Aryl shifts—*continued*
ionic, **65**, 4, 31–36; **66**, 19–23; **67**, 21–27; **68**, 28, 30; **69**, 23–25, 31
photochemical, **66**, 371–372
radical, **65**, 184–186; **66**, 229, 232; **67**, 226, 267; **68**, 293; **69**, 339

1,4-Aryl shifts, **65**, 98

Arylcarbinols, brominative cleavage, **65**, 167

1-Arylethyl acetates, correlation of rates of pyrolysis with aromatic reactivity, **69**, 236

t-Aryl phosphines, electrophilic substitution in, **68**, 217; **69**, 251

Aspirins, hydrolysis, **65**, 271; **67**, 337–340; **68**, 383–384; **69**, 447–448

Association-prefaced catalysis, *see* Catalysis, association-prefaced

Autoxidation, **65**, 201, 202, 304, 307; **66**, 411–412; **67**, 266, 430–432; **68**, 129–130, 263, 315–317, 469, 473, 474; **69**, 297, 301, 358–362, 532, 538, 553
inhibition, **67**, 266; **68**, 315; **69**, 361

Autoxidation, of
acetyldehyde, **67**, 432; **69**, 553
acetylene, **67**, 432
alcohols, **66**, 411; **67**, 432
aldehydes, **65**, 307; **66**, 411; **69**, 360
alkenes, **65**, 304; **66**, 412; **67**, 432; **68**, 315, 475; **69**, 358, 359
N-alkylamides, **66**, 412; **67**, 431; **69**, 359
alkylchlorobenzenes, **67**, 431
1-alkylpyrroles, **67**, 431
2-arylindanones, **65**, 359
ascorbic acid, **67**, 431; **68**, 475; **69**, 553
benzaldehyde, **67**, 206
benzene, **68**, 475
boron compounds, **68**, 315; **69**, 362
butan-2-one, **68**, 475
N-butyldihydroisoindole, **68**, 317
N-butylquinol, **68**, 316
butyraldehyde, **69**, 553
chloroacetaldehyde, **67**, 266
chloroacetyl chloride, **67**, 266
chloroform, **67**, 266
cumene, **66**, 412; **67**, 431, 432; **68**, 315, 473; **69**, 359
cyclohexane, **69**, 359

Catalysis—*continued*

general acid, in—*continued*

hydrolysis of enol ethers, **65**, 253; **66**, 330; **67**, 322–323; **68**, 162, 351, 362; **69**, 418, 420

hydrolysis of orthoesters, **65**, 276

hydrolysis of oxazolines, **68**, 359

ketonization of enols, **69**, 420

nitrone formation, **66**, 317

nucleophilic aromatic substitution, **65**, 135; **66**, 163; **69**, 216

oxime formation, **65**, 317

phenylhydrazone formation, **66**, 318

reaction of formaldehyde with urea, **69**, 413

reduction of azobenzene-*p*-sulphonate, **67**, 435

Schiff base formation, **65**, 242

semicarbazone formation, **65**, 242

thiosemicarbazone formation, **69**, 413

general base in,

addition to double bonds, **68**, 171

amide hydrolysis, **69**, 435–436

dissociation of aldehyde–hydrogen peroxide adducts, **68**, 357

enolization, **65**, 246; **66**, 323

ester aminolysis, **67**, 333; **69**, 440–442

ester hydrolysis, **66**, 339; **67**, 330, 334; **68**, 381

hemithioacetal formation, **66**, 315

hydration of ketones, **68**, 357

hydrazinolysis of phenyl acetate, **65**, 262; **66**, 341

hydrolysis of benzoyl cyanide, **66**, 340; **67**, 331

hydrolysis of 1,3-diphenyl-2-imidazolinium chloride, **68**, 372

hydrolysis of ethyl trifluorothiolacetate, **65**, 260; **66**, 340; **67**, 330–331

nucleophilic aromatic substitution, **65**, 133–135; **66**, 160–162; **67**, 166–170; **68**, 188; **69**, 215–216

reaction of formaldehyde with urea, **69**, 413

reaction of formamide with hydroxylamine, **65**, 260–261

Catalysis—*continued*

general base in—*continued*

ring closure of 4-chlorobutanol, **65**, 56

solvolysis of methyl perchlorate, **67**, 99

transamination, **65**, 244

heterogeneous, in hydrogenation, **66**, 417–419; **67**, 439–441; **68**, 491–494; **69**, 573–575

homogeneous, in hydrogenation, **66**, 417; **67**, 438; **68**, 494–495; **69**, 570–573

intromolecular, in

acetal hydrolysis, **65**, 238; **67**, 307; **69**, 400

additions, **65**, 111; **69**, 193

amide hydrolysis, **65**, 263; **67**, 340, 342; **69**, 448–450, 455

azide decomposition, by iron carbonyl, **68**, 346

electrophilic aromatic substitution, **66**, 196

enolization, **65**, 248; **66**, 322–323; **67**, 319; **69**, 421–422

ester aminolysis, **67**, 344; **69**, 445–446

ester hydrazinolysis, **65**, 262

ester hydrolysis, **65**, 264–270; **66**, 342–350; **67**, 337–344; **68**, 382–389; **69**, 445–446, 451–454, 455

glycoside hydrolysis, **66**, 312; **68**, 353; **69**, 400

imide hydrolysis, **69**, 445

methanolysis of 2-silylpyridines, **68**, 220

mutarotation reactions of sugars, **66**, 315

nitrile hydrolysis, **69**, 453

oxetan-ring opening, **66**, 61

phosphonate ester hydrolysis, **69**, 473, 474

reaction of phthalaldehydic acid and indole, **65**, 258

Schiff-base hydrolysis, **65**, 243; **66**, 319

Schiff-base isomerization, **69**, 416

sulphate hydrolysis, **67**, 365–366; **68**, 409

thioester hydrolysis, **69**, 451, 453

23

Errata for Organic Reaction Mechanisms 1966

P. 399: The *cis:trans* ratios shown in the diagram are correct; those in the accompanying text are incorrect.

P. 406: The left-hand structure in Scheme 1 should have a positive charge on the S atom.

Errata for Organic Reaction Mechanisms 1967

P. 21, lines 6–7: *For syn-* and *anti*-homocub-1,3-ylene di(toluene-*p*-sulphonates) *read syn-* and *anti*-1,3-bishomocubyl toluene-*p*-sulphonates.

P. 159, formula (**59**): *For* p *read* P.

Errata for Organic Reaction Mechanisms 1968

P. 47, formula (**144**): An extra bond should be drawn at the bottom to give the same ring system as in (**147**).

P. 91, formula (**57**): The left-hand six-membered ring should be aromatic.

P. 95, line 10: *For* yields (**67**) not (**68**) *read* yields (**67**) not (**69**).

P. 107, ref. 254: The names of two authors, M. G. Cattania and F. Guella, have been omitted.

P. 133: Formula (**36**) should be

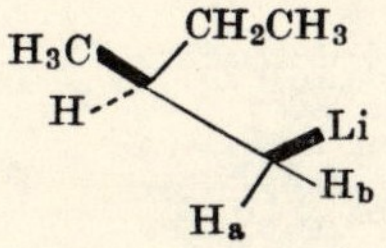

$$
\begin{array}{c}
H_3C \quad\quad CH_2CH_3 \\
H \cdots\!\!-\!\!-\!\!\!\diagdown\!\!\!\diagup\!\!-\!\!-\text{Li} \\
\diagup\!\!\diagdown \\
H_a \quad H_b
\end{array}
$$

P. 266, Table 2: The first two product ratios refer to (**4**), the next two to (**5**), and the last two to the remaining formulae respectively.

P. 373, equation (2), left-hand formula: *For* PH *read* Ph.

P. 389, ref. 78: *This should read* See also pp. 111 and 189.

P. 517, Author index: The Ham . . . and Hal . . . entries should be interchanged.

Errata for Organic Reaction Mechanisms 1966

P. 399: The *cis:trans* ratios shown in the diagram are correct; those in the accompanying text are incorrect.

P. 406: The left-hand structure in Scheme 1 should have a positive charge on the S atom.

Errata for Organic Reaction Mechanisms 1967

P. 21, lines 6–7: *For syn-* and *anti*-homocub-1,3-ylene di(toluene-*p*-sulphonates) *read syn-* and *anti*-1,3-bishomocubyl toluene-*p*-sulphonates.

P. 159, formula (**59**): *For* p *read* P.

Errata for Organic Reaction Mechanisms 1968

P. 47, formula (**144**): An extra bond should be drawn at the bottom to give the same ring system as in (**147**).

P. 91, formula (**57**): The left-hand six-membered ring should be aromatic.

P. 95, line 10: *For* yields (**67**) not (**68**) *read* yields (**67**) not (**69**).

P. 107, ref. 254: The names of two authors, M. G. Cattania and F. Guella, have been omitted.

P. 133: Formula (**36**) should be

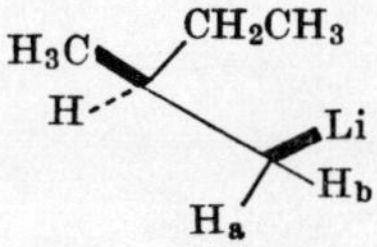

P. 266, Table 2: The first two product ratios refer to (**4**), the next two to (**5**), and the last two to the remaining formulae respectively.

P. 373, equation (2), left-hand formula: *For* PH *read* Ph.

P. 389, ref. 78: *This should read* See also pp. 111 and 189.

P. 517, Author index: The Ham . . . and Hal . . . entries should be interchanged.